# Grundlehren der mathematischen Wissenschaften 273

*A Series of Comprehensive Studies in Mathematics*

N. K. Nikol'skiĭ

# Treatise on the Shift Operator

*Spectral Function Theory*

With an Appendix
by S. V. Hruščev and V. V. Peller

Translated from the Russian
by Jaak Peetre

Springer-Verlag
Berlin Heidelberg New York Tokyo

Nikolaĭ K. Nikol'skiĭ
Leningrad Branch of the Steklov Mathematical Institute
Fontanka 27, 191011 Leningrad, USSR

Jaak Peetre
Lund Institute of Technology, Department of Mathematics
Box 118, S-22100 Lund, Sweden

Title of the Russian original edition: *Lektsii ob operatore sdviga*
Publisher Nauka, Moscow 1980

This volume is part of the *Springer Series in Soviet Mathematics*
Advisers: L. D. Faddeev (Leningrad), R. V. Gamkrelidze (Moscow)

Mathematics Subject Classification (1980): 47A45, 47A20, 47A15, 47A60, 47A70, 47B35, 47B40, 47C05, 46J15, 30B30, 30B50, 30D50, 30D55, 30E05, 46J15, 41A20, 41A50, 60G25.

ISBN 978-3-642-70153-5 ISBN 978-3-642-70151-1 (eBook)
DOI 10.1007/978-3-642-70151-1

Library of Congress Cataloging in Publication Data
Nikol'skiĭ, N. K. (Nikolaĭ Kapitonovich)
Treatise on the shift operator. (Grundlehren der mathematischen Wissenschaften ; 273)
Translation of: Lektsii ob operatore sdviga.
Bibliography: p. Includes indexes.
1. Shift operators (Operator theory) 2. Invariant subspaces. 3. Spectral theory (Mathematics) I. Title. II. Series: Grundlehren der mathematischen Wissenschaften in Einzeldarstellungen mit besonderer Berücksichtigung der Anwendungsgebiete ; Bd. 273.
QA329.2.N5513 1985 515.7'24 84-26869



Typesetting: Asco Trade Typesetting Ltd., Hong Kong

2141/3140-543210

# Preface

This book is an elementary introduction to non-classical spectral theory. After the basic definitions and a reduction to the study of the functional model the discussion will be centered around the simplest variant of such a model which, formally speaking, comprises only the class of contraction operators with a one-dimensional rank of non-unitarity ($\text{rank}(I - T^*T) = \text{rank}(I - TT^*) = 1$). The main emphasis is on the technical side of the subject, the book being mostly devoted to a development of the analytical machinery of spectral theory rather than to this discipline itself. The functional model of Sz.-Nagy and Foiaş reduces the study of general operators to an investigation of the compression

$$T = PS|K$$

of the shift operator $S$, $Sf = zf$, onto coinvariant subspaces (i.e. subspaces invariant with respect to the adjoint shift $S^*$). In the main body of the book (the "Lectures" in the proper meaning of the word) this operator acts on the Hardy space $H^2$ and is itself a part of the operator of multiplication by the independent variable in the space $L^2$ (in the case at hand $L^2$ means $L^2(\mathbb{T})$, $\mathbb{T}$ being the unit circle), this operator again being fundamental for classical spectral theory. The language of functional models is the language of function theory – or more specifically – Nevanlinna multiplicative function theory, which allows one to apply in connection with problems of spectral theory, not only the rather limited set of tools which were used in the classical approach (the theorems of Liouville and Phragmén-Lindelöf, various compactness principles) but literally the entire theory of boundary behavior of analytic functions as developed in the past 60 years. In the last few years an opposite trend has also made itself noticeable. With the aid of the concept of functional model there are formulated and solved problems which traditionally belonged to the theory of functions of one complex variable. By way of illustration we can mention the theorem on interpolation with unbounded multiplicities, as set forth in Lecture IX, and many further classical results on boundary behavior, which in these Lectures are derived by purely operator-theoretic considerations (uniqueness theorems, the theorem of the Riesz brothers on analytical measures etc.). Quite generally one may say that the clarification of the connections between the fundamental problems of function theory and spectral theory is also the main subject of these Lectures.

The bulk of the "Lectures" (Lecture IV, VI–X) is devoted to a clarification of the rôle of the celebrated corona theorem by Carleson. Around this central theme, in spectral language amounting to a study of unconditionally convergent spectral expansions, we have grouped some other themes – interpolation in various classes of analytic functions, the description of invariant subspaces in terms of multiplicative representations of the "characteristic function" of the operator, approximation theoretic consequences etc. In particular these last consequences will be paid special attention; we shall notably study the question of the completeness of various systems of functions connected with the shift operator: completeness of polynomials with respect to weights, remainders of the Taylor series ($S^{*n}f$, $n \geq 0$), powers of suitable convolution operators, systems of rational functions. These approximation problems are connected with one of three digressions from the "scalar" variant of the functional model referred to above. We consider here problems on the multiplicity of the spectrum and on generating sets for the multiple shift (the operator $S_E$ in the space of vector valued $H^2$-functions), which are essential both for the theory itself as well as for its ramifications (for instance, in the theory of stationary stochastic processes). The other two digressions concern the theory of "the class $C_0$", defined by the condition that for its members an analogue of the Hamilton-Cayley theorem (the existence of non-trivial functions $f$ annihilating the operator $T: f(T) = \mathbb{O}$) holds, and theorems on unconditionally convergent spectral expansions for operators with a simple point spectrum. The first of these questions is included in a natural way in the framework of the "scalar" theory, because it leads essentially to a study of the maximal functional calculus $f \mapsto f(T)$ for the class of all bounded analytic functions $f$ in the disk. Several Lectures will be devoted to the particulars of this calculus. The general problem of bases of eigenspaces is included for two reasons. It admits a very transparent solution in the language of the characteristic functions and it demonstrates the still unsurmountable difficulties of the theory of operator (matrix) analytic functions with non-commutative values, the principal one being how to adequately translate "additive" problems into the "multiplicative" language of characteristic functions.

The book is divided into 11 Lectures and 5 Appendices. Most of the Lectures are divided vertically: a main canonical part dedicated to the one dimensional model plus a part entitled "Supplements and Bibliographical Notes" developing the first part in various directions. The main part is thereby included in the framework of the Hardy space $H^2$ and has the character of a special "spectral function theory". The second part usually goes far beyond this framework. The first part is written in great detail and, as far as possible, in an elementary fashion, while the second one is far more condensed and sketchy. Each Lecture concludes with a brief review of the literature and hints to unsolved questions in the portion of the theory in question. What the Appendices 1–5 are about is clear from a glance at the table of contents.

The embryo of this book was the text of lectures I delivered between 1967 and 1972 to an audience of mathematics students at the University of Leningrad. The whole framework has now been modernized, including also the latest achievements in the topics touched upon, some results even being published for the first time. To be able to read the book formally one needs only the standard material in mathematical analysis taught in undergraduate courses (measure theory, Fourier series in $L^2$, analytic and harmonic functions, the Fourier transform, linear algebra, Banach spaces, the Stone-Weierstrass theorem, the spectral theorem). A number of more special theorems, which are constantly used in the book (and as a rule concern the theory of the Hardy classes $H^p$), are proved in Appendix 2. The notation employed is usually explained only at the moment of its first appearance[1]; the sign □ stands for the end of a proof or an argument; $\operatorname{clos}_X A$ denotes the closure of the set $A$ in the space $X$; $\bigvee(\ldots)$ the closed linear hull of the set $(\ldots)$; Lat $T$ the set of all subspaces invariant with respect to the transformation $T$.

While working on the book I profited from the support and advice of a great number of persons and to all of them – from the very first listeners to the most severe critics among the specialists – I hereby express my gratitude. With regard to mathematical ideas which have influenced various parts of the book, I owe much to V. I. Vasyunin, S. A. Vinogradov and S. V. Hruščev. I have also received invaluable advice in the final polishing of the text by V. I. Vasyunin, many of whose methodological ideas have been used in the definitive version. Many essential remarks and a lot of friendly assistance came from V. P. Havin, who also read the entire manuscript. I received much further help from V. I. Vasyunin, L. N. Dovbyš, S. V. Kislyakov, V. V. Peller, L. G. Hanin, S. V. Hruščev in composing the necessary technical apparatus connected with the manuscript (bibliography, indices etc.) and likewise I must mention I. V. Bykova, E. B. Eryhova and L. P. Vinogradova for their selfdenying work in preparing the typescript. To all of them I extend my sincere thanks. And finally, but perhaps really in the first place, I am grateful to my family for the encouragement and the enthusiasm with which they endure all the hardships of a mathematical life.

---

[1] Cf. also the List of Symbols.

# Preface to the English Edition

I am very pleased that Springer-Verlag has decided to translate my book thereby making it more readily available to Western readers. I would like to express here my gratitude to the Editors for their goodwill and readiness in taking into account all my requests.

This edition is very significantly enlarged, as compared to the Russian one. Besides smaller local corrections and improvements let me mention here the following additions and changes.

Section 4 of Lecture II has been revised in the spirit of the paper Vasyunin-Nikol'skiĭ (Izv. Akad. Nauk SSSR Ser. Mat. **47**, 942–960 (1983).

Lecture VIII (Section 4) contains now a new approach to the problem of free interpolation in $H^\infty$ (the interpolation operator of Jones-Vinogradov).

There are also four entirely new appendices (Appendix 2–5).

Appendix 2 contains proofs of all assertions about Hardy classes used throughout the book. This makes the book accessible to everyone with just rudimentary background in Analysis.

Appendix 3 contains the modern proof of the Carleson Corona Theorem and its operator theoretic generalization (Tom Wolff's proof as modified by Tolokonnikov et al., with the best estimates for the "solving functions" known at present (1983)).

Appendix 4 gives a rather extensive survey of the theory of Toeplitz and Hankel operators connected with the general orientation of the book – the spectral function theory. I have tried to collect here practically all known results of the theory pertaining to the above orientation. Many of these results are given new proofs and some are published for the first time.

S. V. Hruščev and V. V. Peller have had the kindness to write Appendix 5 specially for this edition. It contains results on singular numbers of Hankel operators and their application to the theory of random processes. It further gives all preliminaries on random processes which are necessary for the proper understanding of the subject.

I would like to thank some of my colleagues for helping me with useful discussions and other support during my work with the appendices, more precisely V. A. Tolokonnikov (Appendix 3), V. I. Vasyunin (Appendix 3 and 4), V. V. Peller, A. L. Volberg, L. N. Dovbyš, S. V. Hruščev (Appendix 4), and finally V. P. Havin who helped me in a variety of ways. L. N. Dovbyš, D. V. Yakubovič, S. V. Kislyakov, N. G. Makarov, V. V. Peller and B. M. Solomyak have assisted me with the proofreading.

Lastly, I am also deeply grateful to the translator Prof. Jaak Peetre for his thorough work.

# Contents

# Introductory Lecture. What This Book is About

Spectral theory deals with linear transformations or families of linear transformations in a vector space.

Every object on which such a transformation acts is subject to spectral analysis: the evolution induced by this action makes it possible to distinguish certain simplest "harmonic" constituents and then to reconstruct from these the whole object as well as all details of its variation, of interest for the theory. This program ceases to be academic only if the "simplest constituents" are really simple and the contribution to the general evolution of each harmonic is taken into account with sufficient accuracy. The most important example, forming the cornerstone of classical spectral theory, is the Spectral Theorem.

**Spectral Theorem.** *Let $H$ be a Hilbert space and $T$ a normal operator with simple spectrum. Then there exists a compactly supported Borel measure $\mu$ in the plane $\mathbb{C}$ such that $T$ is unitarily equivalent to the operator of multiplication by the independent variable*

$$f \mapsto zf \tag{1}$$

*in the space $L^2(\mu)$.*

Here, as well as elsewhere in this book, the word *operator* stands for *continuous linear transformation,* "normal" means that $T$ commutes with its adjoint $T^*$ ($TT^* = T^*T$) and "simple spectrum" amounts to the existence of a generating (cyclic) vector $x$ ($x \in H$), i.e. we have $\bigvee(T^n x : n \geq 0) = H$. The most important requirement in the Spectral Theorem is that of normality, whereas passage to the general case of a non-trivial spectral multiplicity leads only to a minor complication – one has to replace $L^2(\mu)$ by a direct integral

$$\int \oplus H(\zeta)\, d\mu(\zeta)$$

of Hilbert spaces $H(\zeta)$. The latter are now playing the rôle of the "simplest harmonics" in the above general scheme for spectral analysis, the action of $T$ in $H(\zeta)$ being multiplication by (the number) $\zeta$.

All, or almost all, information on normal operators (and, in particular, on self-adjoint and unitary ones) can be obtained from a study of the *model* (1), which renders extraordinarily transparent many computations with the operator $T$. If we drop the assumption of normality (i.e. the commutativity of the $*$-algebra generated by $T$) we get major complications already in the finite dimensional case ($\dim H < \infty$). For a general finite rank operator spectral analysis is accomplished by its *Jordan model* (decomposition into Jordan blocks), which in its informativity, its capacity to answer all sorts of questions of interest for the theory however by no means can be compared with the Spectral Theorem. Nevertheless the Jordan model is so useful in the finite dimensional case and some other instances that the attempts to construct analogues for general classes of operators have given rise to an independent branch of spectral theory. The difficulty here consists not so much in isolating the elementary "Jordan blocks" but in the circumstance that the initial operator only in a very weak sense can be reconstructed from the blocks so that in order to derive real consequences concerning the nature of its action (as in the case $\dim H < \infty$) this block structure often turns out to be useless.

The basic idea of the new non-classical spectral theory is to abstain from looking at the linear operator as a sum of simple transformations (of the type Jordan blocks) and instead consider it as a "part" of a complicated universal mapping, allotted (in compensation) with many auxiliary structures. This theory has as its objective to obtain analogues of the spectral resolution for unitary (normal) operators or – bypassing this analogy – a direct deduction of the basic consequences of the latter for an as large as possible collection of transformations: the description of functions of the operator and estimates for their norms, formulae for the computation of the functionals trace, det, ind, mult etc. Usually this last problem will be solved with the aid of certain "canonical representations" or "models" making the computations mentioned obvious. While the finite dimensional theory deals with diagonal, triangular or Jordan representations (models) of matrices, for general transformations from the early 60's on one of the basic (and, as we will see, universal) models has been the operator of multiplication with the independent variable in spaces of analytic functions (= the *shift operator*, acting in these spaces by formula (1)) and its compressions onto coinvariant subspaces. And so the present course is precisely devoted to the analysis of these operators.

**1. Basic Objects.** We denote by $H^2$ the class of all functions $f$ analytic in the disk $\mathbb{D} = \{\zeta : \zeta \in \mathbb{C}, |\zeta| < 1\}$ of the complex plane $\mathbb{C}$ such that

$$f(\zeta) = \sum_{n \geq 0} \hat{f}(n)\zeta^n, \quad \zeta \in \mathbb{D}; \qquad \|f\|^2 \overset{\text{def}}{=} \sum_{n \geq 0} |\hat{f}(n)|^2 < \infty. \tag{2}$$

It is clear that this is a Hilbert space and usually it is convenient to imbed it in the natural way into the space $L^2 = L^2(\mathbb{T}, m)$ where $\mathbb{T} = \{\zeta : \zeta \in \mathbb{C}, |\zeta| = 1\}$ is the unit circle and $m$ the normalized Lebesgue measure on $\mathbb{T}$ ($m\mathbb{T} = 1$). The symbol $z$ will everywhere in this book designate the identity map on a portion of the plane (basically either the disk $\mathbb{D}$ or the circle $\mathbb{T}$), $z(\zeta) \equiv \zeta$, so that the family of functions $\{z^n : n \in \mathbb{Z}\}$ is the standard trigonometric basis of the space $L^2$. In order to be able to keep apart the two function $z : \mathbb{D} \to \mathbb{D}$ and $z : \mathbb{T} \to \mathbb{T}$ in the rare cases when they could be confused we are going to write $z|\mathbb{D}$ and $z|\mathbb{T}$ respectively; in general $f|A$ denotes the restriction of the map (function) $f$ to the set $A$. Thus we imbed the space $H^2$ into $L^2$ observing that the map

$$j : z^n|\mathbb{D} \mapsto z^n|\mathbb{T}, \quad n \geq 0,$$

extends to an isometric (linear) operator from $H^2$ into $L^2$. The image $jH^2$ consists of precisely those (classes of) functions $F$, $F \in L^2$, for which $\hat{F}(n) = 0$, $n < 0$; moreover if $F = jf$ then $\hat{f}(n) = \hat{F}(n)$, $n \geq 0$ (here, in agreement with (2), the $\hat{f}(n)$, $n \geq 0$, are the Taylor coefficients of $f$, $f \in H^2$, and the $\hat{F}(n)$, $n \in \mathbb{Z}$, the Fourier coefficients of $F$, $F \in L^2$, i.e. $\hat{F}(n) \stackrel{\text{def}}{=} (F, z^n)$ where

$$(\varphi, \psi) = \int_{\mathbb{T}} \varphi \bar{\psi} \, dm$$

is the scalar product in $L^2$). The function $f$ can be reconstructed from its "boundary values" $F$ with the aid of the Poisson intergral:

$$f(\zeta) = \int_{\mathbb{T}} F(t) P_{|\zeta|}(\bar{t}\zeta/|\zeta|) \, dm(t), \quad \zeta \in \mathbb{D},$$

where $P_r(t) = (1 - r^2)(1 - 2r \operatorname{Re} t + r^2)^{-1}$ is the Poisson kernel. This last fact, along with the previous ones, is completely elementary (for polynomials $f$ it is evident if one notes that $\hat{P}_r(n) = r^{|n|}$, $n \in \mathbb{Z}$, and the rest follows from the Banach-Steinhaus theorem). In what follows, unless this will lead to ambiguity, we will *identify* $f$ and $jf$ so that $H^2$ will consist of those elements of $L^2$ which can be analytically continued into the disk $\mathbb{D}$. Later on we will need considerably deeper facts on the class $H^2$ as well as the "analytical" subspaces $H^p$ of the spaces $L^p = L^p(\mathbb{T}, m)$; these facts are summarized in App. 2 and the reader can refer to it, whenever he needs it. The $L^p$ norm will be denote by the symbol $\|\cdot\|_p$, $1 \leq p \leq \infty$.

The two operators fundamental for this course are defined by the following formulae:

the *bilateral shift*

$$\mathscr{S} f = zf,$$

where $f$ is a function on the circumference $\mathbb{T}$, $z = z|\mathbb{T}$;

the *unilateral shift*

$$Sf = zf,$$

where $f$ is a function regular in the disk $\mathbb{D}$, $z = z|\mathbb{D}$. As a rule $\mathscr{S}$ is considered as a (unitary) operator in $L^2$ and $S$ as an (isometric) operator in $H^2$. The term "shift" may be associated with the action of $\mathscr{S}$ and $S$ on the Fourier (Taylor) coefficients: $(\mathscr{S}f)^\wedge(n) = \hat{f}(n-1)$, $n \in \mathbb{Z}$; $(Sf)^\wedge(n) = \hat{f}(n-1)$, $n \geq 1$, and $(Sf)^\wedge(0) = 0$; or in the form of the equations $\mathscr{S} z^n = z^{n+1}$, $n \in \mathbb{Z}$. This implies in particular that $\mathscr{S} H^2 \subset H^2$ and $S = \mathscr{S}|H^2$.

In an analogous fashion we define the classes $L^2(E)$ and $H^2(E)$ of vector functions, with values in an auxiliary Hilbert space $E$. For example, $H^2(E)$ consists of all Taylor series $f$, $f(\zeta) = \sum_{n \geq 0} \hat{f}(n)\zeta^n$, $\zeta \in \mathbb{D}$, where $\hat{f}(n) \in E$, $n \geq 0$, $\sum_{n \geq 0} \|\hat{f}(n)\|_E^2 < \infty$. The corresponding shift operators will be denoted by the symbols $\mathscr{S}_E$ and $S_E$. It is clear that $H^2(E) = \sum_{i=1}^d \oplus H^2$, $S_E = \sum_{i=1}^d \oplus S$ where $d = \dim E$ and analogously for $\mathscr{S}_E$ (for the proof it suffices to take any orthonormal basis in $E$). It is clear that $\mathscr{S}_E^* f = \bar{z}f$ and that

$$S_E^* f = S_E^*\left(\sum_{n\geq 0}\hat{f}(n)z^n\right) \equiv \sum_{n=1}^{\infty}\hat{f}(n)z^{n-1} = \frac{f - f(0)}{z}.$$

The spectral theory developed in this book is based on a detailed study of the *invariant subspaces* of the operators $T$ under consideration, i.e. subspaces $E$ such that $x \in E \Rightarrow Tx \in E$. (Unless we say the contrary we have always *closed subspaces* in mind.) It is appropriate to recall here that already the classical theorem of Schur on the representation of matrices in triangular form is equivalent to the existence of an increasing chain of invariant subspaces. Analogous statements for other transformations are as useful as Schur's theorem – for the calculation of spectra, traces and determinants – and we associate such triangular representations with natural chains of invariant subspaces of the shift operator and its "projections" (compressions); a special case of such a representation is the classical Fourier representation (or transform). Besides the already mentioned fundamental purpose – the construction of a non-classical spectral theory for the shift operator – the study of invariant subspaces is stimulated in particular by their simple but most fruitful connections with problems of approximation. E.g. the vector $x$, $x \neq \mathbb{0}$, is cyclic with respect to the operator $T$ defined in the space $X$ iff it does not belong to any non-trivial $T$-invariant subspace, so that the knowledge of all these subspaces usually also gives criteria for the completeness of "generalized polynomials" $\sum_{k=0}^{n} a_k T^k x$. Cyclicity criteria constitute a part of the so-called *individual theorems*, which describe the invariant subspace $\bigvee(T^n x : n \geq 0)$, $x \in X$, in terms of the generating vector $x$; such theorems too obtain due attention in this book. Let us remark that even the well-known theorem of Weierstrass on approximation by polynomials can be interpreted as an individual theorem: it says that for the multiplication operator $f \mapsto zf$ in the space of continuous functions on a compact interval of the real axis $\mathbb{R}$ all functions $f$ for which $f(t) \neq 0$ for all $t$ are cyclic.

**2. The Functional Model** for a contraction operator $T$ on a Hilbert space is constructed with the aid of the multiple shift operator $S_E$. This makes it possible to lend a precise meaning to its being universal. We limit ourselves presently to proving the following simple proposition.

**Model Theorem.** *Let $T$ be a linear contraction of the Hilbert space $H$* ($|T| \leq 1$) *such that*

$$\lim_n T^n x = \mathbb{0}, \quad \forall x \, (x \in H).$$

*Then there exists a Hilbert space $E$, a subspace $K$, $K \subset H^2(E)$, invariant with respect to the adjoint shift $S_E^*$ ($S_E^* K \subset K$), and a unitary transformation $V$ from $H$ onto $K$ such that*

$$T = V^{-1}(S_E^* | K)V.$$

*Proof*. Let $D = (1 - T^*T)^{1/2}$ (the positive branch of the square root),

$E = \operatorname{clos} DH$ and

$$Vx \stackrel{\text{def}}{=} \sum_{n \geq 0} z^n DT^n, \quad x \in H.$$

Then

$$\| Vx \|^2_{H^2(E)} = \sum_{n \geq 0} \| DT^n x \|^2_H = \sum_{n \geq 0} ((I - T^*T)T^n x, T^n x)$$
$$= \sum_{n \geq 0} (\| T^n x \|^2_H - \| T^{n+1} x \|^2_H) = \| x \|^2_H,$$

because $\sum_{n=0}^m (\| T^n x \|^2_H - \| T^{n+1} x \|^2_H) = \| x \|^2_H - \| T^{m+1} \|^2_H$ and $\lim_m \| T^m x \|_H = 0$. Thus $V$ gives a unitary imbedding of $H$ onto $K$, $K \stackrel{\text{def}}{=} VH$. Moreover

$$x \in H \Rightarrow S_E^* Vx = S_E^* \sum_{n \geq 0} z^n DT^n x = \sum_{n \geq 1} z^{n-1} DT^n x = VTx.$$

It follows that $S_E^* K \subset K$ and $(S_E^* | K)V = VT$. □

The operator $S_E^* | K$ will therefore be called the *functional model* for $T$; notice that $\dim E \leq \dim(I - T^*T)H$. In equal measure this term will be connected with the adjoint operator $P_K S_E | K$ serving as a model for $T^*$ (here $P_K$ is the orthogonal projection of $H^2(E)$ onto $K$). There exists an extensive theory of functional models which establishes deep connections between the spectral properties of the operator under consideration and the structure of analytic vector functions along with the Riesz-Nevanlinna factorization theorems. A central object in the theory is the so-called "characteristic function" of the operator $T$ which as its main part contains the resolvent $(\lambda I - T)^{-1}$ and provides a possibility to translate spectral theory problems into the language of function theory. The most important and most characteristic feature of this approach to the study of operators is perhaps the complete overlap of interests of two directions of mathematical analysis intersecting at this juncture: the problems of spectral theory expressed in the language of characteristic functions contain all questions traditionally of interest for function theory – on uniqueness sets, on free interpolation by analytic functions, on the multiplicative nature of these functions, on approximation by polynomials and by rational functions, on analytic continuation etc.

The functional model has most intensively been studied in the case of a finite rank deviation of the operator $T$ from unitary operators ($d \stackrel{\text{def}}{=} \operatorname{rank}(I - T^*T) < \infty$). The main and most complete part of these Lectures is devoted to the case $d = 1$, that is, the operators $S$ and $S^*$ in the Hardy space $H^2$ and their restrictions $(S^* | K)$ and compressions $(P_K S | K)$ to invariant subspaces. The completeness of the theory in this simplest case compensates the absence of generality and provides the key to the search of theorems which are unknown or little understood in the case of higher dimensional defect operators $I - T^*T$ and $I - TT^*$. In fact, the functional model has even for $d = 1$ realistic consequences for the equations of mathematical physics (cf. Pavlov [3]) and the case $d < \infty$ (and, more generally, the case with nuclear defects $I - T^*T$ and $I - TT^*$) leads, for instance, to a complete spectral classification of many physically relevant scattering problems (cf. Lax-Phillips [1], [2]). Here we cannot discuss these topics in greater detail referring instead the reader to the books mentioned where examples of this type are studied. Finally, we wish to emphasize that the operators considered in these "Lectures" as a rule operate on subspaces of $L^2$ (or its vector valued analogues) so that the theory of these operators constitutes a peculiar "spectral function theory", most of its results possessing an independent value even without direct reference to these operators.

**3. The Details of the Plan**, which are enumerated in this Section are destined to help the reader to get better oriented in the "Lectures" and, if possible, to determine in advance those parts which are for him most (un)interesting and useful (useless). Let us however point out that these "details" poorly cover the (significant; cf. the Preface) portion of the contents which falls under the heading "Supplements and bibliographical remarks", in an essential way extending the main body of the text.

The main part of the contents of the "Lectures" is concerned with the spectral analysis of the *functional model*, i.e. a *coinvariant* subspace $K = H^2 \ominus \Theta H^2$ for the shift operator $S$ and the compression $PS|K$ of $S$ onto $K$. This theme covers a wide circle of ideas – the decomposition into invariant subspaces induces an analysis of multiplicative representations of analytical functions (the characteristic function $\Theta$, functions in the subspace $K$ etc.), the study of the convergence of series in the spectral components leads to problems of interpolation by analytic functions and to delicate questions of the $H^\infty$-functional calculus (including the Carleson corona theorem), the last topic again to the known imbedding theorems for analytic functions and to the classification of subsets of the plane (spectra) according to the degree of thinness.

More general operators than the ones comprised in the functional model just described will be studied in two ways – either we construct for them a sufficiently rich functional calculus (which in itself serves as a spectral theorem) or we study a dilation of such operators $T$. If $T$ is a contraction on a Hilbert space $H$ then by a *unitary dilation* of $T$ we mean any unitary operator $U$ acting on a space $\mathscr{H}$, $\mathscr{H} \supset H$, such that

$$T^n = P_H U^n | H, \quad n \geq 0,$$

$P_H$ being the orthogonal projection onto the subspace $H$.

**Dilation Theorem.** *Let $H$ be a Hilbert space and $T$ a linear contraction operator on $H$ ($\|T\| \leq 1$). Then there exists a Hilbert space $\mathscr{H}$ and a unitary operator $U$ in $\mathscr{H}$ such that $U$ is a dilation of $T$ and in addition $\mathscr{H} = \bigvee (U^n H : n \in \mathbb{Z})$.*

A dilation having this last property will be termed *minimal*. The minimal dilation is uniquely determined by $T$ in the following sense: if $U_1$ and $U_2$ are two such dilations (acting in spaces $\mathscr{H}_1$ and $\mathscr{H}_2$ respectively) then there exists a unitary operator $W$ from $\mathscr{H}_1$ onto $\mathscr{H}_2$ such that $WU_1 = U_2 W$ and $W|H = I_H$. To prove the uniqueness it suffices to put $W(\sum_{n \in \mathbb{Z}} U_1^n x_n) = \sum_{n \in \mathbb{Z}} U_2^n x_n$ (where $x_n \in H$ and $x_n = \mathbb{0}$ outside a finite set of indices $n$) and to check the unitarity of this $W$.

*The proof of the theorem* can easily be reduced to the special case when $T$ is an isometric operator. In fact, consider the Hilbert space $l^2(H)$ of all $H$-valued sequences $x = \{x_n\}_{n\geq 0}$ with finite norm $\|x\| = (\sum_{n\geq 0}\|x_n\|_H^2)^{1/2}$ and define an isometric operator $V$ by the formula

$$V(x_0, x_1, \dots) = (Tx_0, Dx_0, x_1, \dots), \quad x \in l^2(H),$$

where $D = (I - T^*T)^{1/2}$. We consider $H$ as imbedded in $l^2(H)$ via the map $x \mapsto (x, \mathbb{O}, \mathbb{O}, \dots)$, $x \in H$. It is clear then that $P_H V^n x = T^n x$, $n \geq 0$, that is, $V$ is an isometric dilation of $T$.

It remains to construct a unitary dilation of the isometric operator $V$. This can readily be achieved by taking the following unitary operator $\mathscr{U}$ in the space $l^2(H) \oplus l^2(H)$: $\mathscr{U}|(l^2(H) \oplus \mathbb{O}) = V \oplus \mathbb{O}$, $\mathscr{U}|(\mathbb{O} \oplus l^2(H)) =$ any isometric map of this space onto $(l^2(H) \ominus Vl^2(H)) \oplus l^2(H)$. It is clear that $\mathscr{U}^n|l^2(H) = V^n$, $n \geq 0$.

In order to get a minimal dilation once one dilation (say $\mathscr{U}$) has been found it suffices to set $U = \mathscr{U}|\bigvee(\mathscr{U}^n H : n \in \mathbb{Z})$. □

It is clear that in studying the dilation of contraction operators it could be convenient first to subtract the "unitary" part of the operator from the proper contractive part. The following lemma shows that this is always possible.

**Langer's Lemma.** *Let $T$ be a linear contraction of the space $H$. Then $H = H_u \oplus H_0$, $TH_u \subset H_u$, $TH_0 \subset H_0$, where $T|H_u$ is unitary while as $H_0$ does not contain any (non-trivial) $T$ invariant subspace on which $T$ is unitary.* (Such an operator $T|H_0$ will be called *completely non-unitary.*)

*Proof.* Put

$$H_u = \{x : x \in H;\ \|x\| = \|T^n x\| = \|T^{*n} x\|, n \geq 0\}.$$

This is a linear subspace, because $H_u = \bigcap_{n\in\mathbb{Z}} \operatorname{Ker}(I - T_n^* T_n)$ where $T_n = T^n$ for $n \geq 0$ and $T_n = T^{*(-n)}$ for $n \leq 0$. That $H_u$ is invariant with respect to $T$ and $T^*$ is seen as follows: Let $x \in H_u$. Then

$$\|T^n Tx\| = \|T^{n+1}x\| = \|x\| = \|Tx\|, \quad n \geq 0,$$

and, since $\|Tx\| = \|x\|$ entails $T^*Tx = x$, also

$$\|T^{*n}Tx\| = \|T^{*(n-1)}T^*Tx\| = \|T^{*(n-1)}x\| = \|x\| = \|Tx\|,$$

that is, $Tx \in H_u$. In the same way one checks that $T^*x \in H_u$. It is now clear that the operator $T|H_u$ is unitary and that the subspace $H_0$ does not contain any non-zero vector $x$ such that all its $T_n$-images have the same norm as $x$. □

One can say a lot about an operator $T$ once one knows its minimal unitary dilation. In the first place we will study the spectral properties of such a dilation (absolute continuity of the spectrum, invariant subspaces etc.). Part of this analysis carries over also to the operator $T$ (the compression of $U$): description of the invariant subspaces using multiplicative properties of its characteristic function, description of cyclic (generating) subspaces etc. This last thing is also of interest from other points of view – function theory (of course) and stationary stochastic processes (because many results on the spectral type of the stochastic process have precisely such an approximation theoretic character). The consequences of such a (sufficiently general) analysis comprise many classical theorems in function theory – on analytic measures (Riesz brothers), on weighted polynomial approximation (Szegö), on pointwise interpolation (Nevanlinna, Carathéodory, Pick), on the support of a convolution (Titchmarsh), on exponential bases (Paley, Wiener etc.).

A major rôle in the construction of the spectral theory of model operators is occupied by the Carleson corona theorem (for its formulation see Lecture III): On it we will base the spectral mapping theorem for the $H^\infty$-calculus (Lecture III), with the aid of it we construct the family of spectral projectors (Lecture IV, IX) etc. Another method which we employ in the study of the model operator is the introduction of special Hankel and Toeplitz operators, which leads to the proof of important theorems on the "lifting of the commutant" (the theorems of Sarason and of Sz.-Nagy and Foiaş, cf. Lecture VIII), description of exponential bases in the space $L^2(0, a)$, proof of a whole series of theorems on interpolation (the theorems of Carleson, Carathéodory, Nevanlinna-Pick). Finally, we remark that in each of the directions of our theory just referred to the "abstract" methods (operator methods or even methods involving the geometry of Hilbert space, as in the problem of unconditional bases and the interpolation problem, cf. Lecture VI–VIII) are closely intervowen with very concrete constructions in the spirit of classical function theory.

Let us express the hope that this general view on known purely "analytical" facts as well as new ones, connecting spectral theory with the traditional theory of analytic functions, will turn out to be useful for both disciplines.

**4. Concluding Remarks.** The connections between the shift operator and the rest of mathematics are manifold and besides the disciplines touched upon in this Lecture this operator plays an essential rôle also in many others – from scattering theory to stationary stochastic processes. Such connections and applications of the general theory developed in these "Lectures" could have enriched this course but unfortunately because of limited space it has not been possible to study them in any greater detail here. Therefore we refer the reader to the following monographs: Lax-Phillips [1], Rozanov [2], Dym-MacKean [1], Helson [1], [2], Sz.-Nagy and Foiaş [1]. There one also gets supplementary information on the general functional model of Sz.-Nagy and Foiaş. We remark that the foundations of the theory developed in this course were laid by Livšic in 1945 in a different – but as it appears now – equivalent form; for more details see Sz.-Nagy and Foiaş [1]. A fundamental rôle played also the famous paper by Beurling (1949) on invariant subspaces which we are going to set forth in Lecture I. However, in general spectral theory the shift operator probably first made its appearance in Plessner's note [1] where it is shown in particular that $S_E$ provides the general form of a completely non-unitary isometric operator, a fact that has been rediscovered by many mathematicians.

The Dilation Theorem is due to Sz.-Nagy who derived it (1954) from an earlier (and equivalent) theorem of Naĭmark's [2] on dilations of non-negative operator measures. Our presentation follows Douglas [1]. The Model Theorem (in fact for general contractions $T$) was deduced by Sz.-Nagy and Foiaş from the dilation theorem; concerning the relevant papers cf. the bibliographical comments in the book Sz.-Nagy and Foiaş [1]. A proof close to the one of the text can be found in Halmos's problem book [2]. Regarding Langer's Lemma, cf. Langer [1], Sz.-Nagy and Foiaş [1].

# Lecture I. Invariant Subspaces

The study of invariant subspaces of the operators $S$ and $\mathscr{S}$ has a fundamental significance for our subject. Such a study was initiated essentially already by Szegö and Wiener but the really basic contribution came with the paper by Beurling from the year 1949 which we have already referred to. Later Beurling's methods were perfected by many authors and led to the most transparent proofs which we set forth in this Lecture.

**1. The Fundamental Theorem.** Let us agree to denote the *multiplication operator* corresponding to the bounded function $\varphi$ by the boldface letter of the same denomination: $\boldsymbol{\varphi}: f \mapsto \varphi f$.

**Beurling-Helson Theorem.**[1] *If $E$ is a subspace of $L^2$ invariant with respect to the operator $\mathscr{S}$ then there are only two possibilities:*

1. *$\mathscr{S}E = E$ in which case there exists an m-measurable subset $e$, $e \subset \mathbb{T}$, such that $E = \boldsymbol{\chi}_e L^2$, where $\chi_e$ is the characteristic function of $e$.*

2. *$\mathscr{S}E \neq E$ in which case there exists a measurable function $\Theta$ on $\mathbb{T}$, with $|\Theta| = 1$ almost everywhere (a.e.), such that $E = \boldsymbol{\Theta} H^2$.*

*Proof.* 1. Denote by $P_E$ the orthogonal projection (in $L^2$) onto $E$ and let us put $\varphi = P_E \mathbb{1}$. Then $\mathbb{1} - \varphi$ is orthogonal to $\mathscr{S}^n \varphi$ for every $n$ ($n = 0, \pm 1, \ldots$), i.e.

$$\int_{\mathbb{T}} (\varphi - |\varphi|^2) z^n \, dm = 0, \quad n \in \mathbb{Z}.$$

Consequently $\varphi = |\varphi|^2$ and there exists a set $e$, $e \subset \mathbb{T}$, with $\varphi = \chi_e$. It is clear that $\bigvee(\mathscr{S}^n \chi_e : e \in Z) = \boldsymbol{\chi}_e L^2 \subset E$. If $G \in E \ominus \boldsymbol{\chi}_e L^2$ then, on the one hand, $\mathbb{1} - \chi_e$ is orthogonal to $\mathscr{S}^n G$, $n \in \mathbb{Z}$, so as before $\bar{G}(\mathbb{1} - \chi_e) = \mathbb{0}$, on the other hand, $G$ is orthogonal to $\mathscr{S}^n \chi_e$, $n \in \mathbb{Z}$, so $\bar{G} \chi_e = \mathbb{0}$. It follows that $G = \mathbb{0}$ and therefore $E = \boldsymbol{\chi}_e L^2$.

2. Let $\mathscr{S}E \neq E$. Then $\mathscr{S}E$ is a proper subspace of $E$; pick up $\Theta \in E \ominus \mathscr{S}E$, $\|\Theta\|_2 = 1$. As

$$(\mathscr{S}^n \Theta, \Theta) = \int_{\mathbb{T}} z^n |\Theta|^2 \, dm = 0$$

for all $n$, $n \geq 1$, and as further $|\Theta|^2 \, dm$ is a real measure, we must have $|\Theta| = \text{const}$ and consequently $|\Theta| = 1$ a.e. on $\mathbb{T}$. Moreover the operator $\boldsymbol{\Theta}$ is isometric in $L^2$ and therefore $\bigvee(\mathscr{S}^n \Theta : n \geq 0) = \boldsymbol{\Theta} \bigvee(\mathscr{S}^n \mathbb{1} : n \geq 0) = \boldsymbol{\Theta} H^2$. Thus $\boldsymbol{\Theta} H^2 \subset E$. Let $G \in E \ominus \boldsymbol{\Theta} H^2$. Then, on the one hand, $(G, \mathscr{S}^n \Theta) = 0$ for $n \geq 0$ and, on the

[1] In references Theorem BH.

other hand, $\mathscr{S}^n G \in \mathscr{S} E$ for $n \geq 1$ so that $(\Theta, \mathscr{S}^n G) = 0$, $n \geq 1$. Together this gives $\int_{\mathbb{T}} \Theta \bar{G} z^n \, dm = 0$ for all $n$, $n \in \mathbb{Z}$, that is, $\Theta \bar{G} = \mathbb{0}$, $G = \mathbb{0}$. Consequently $E = \boldsymbol{\Theta} H^2$. □

*Remark.* It is clear that the converse statements too are true (and trivial): every subspace of the form $\chi_e L^2$ is invariant with respect to $\mathscr{S}$ and $\mathscr{S}^{-1}$ and every subspace of the form $\boldsymbol{\Theta} H^2$, $|\Theta| = 1$ a.e., is invariant with respect to $\mathscr{S}$ but not with respect to $\mathscr{S}^{-1}$. Moreover a subspace $E$ cannot be written in more than one way in the form $\chi_e L^2$ or $\boldsymbol{\Theta} H^2$, $|\Theta| = 1$ a.e. (in the last case up to a multiplicative constant): for $\chi_e L^2$ this is evident; if $\boldsymbol{\Theta} H^2 = \boldsymbol{\Theta}' H^2$ then $\Theta/\Theta' \in H^2$ and $\overline{\Theta/\Theta'} = \Theta'/\Theta \in H^2$ so that $\Theta/\Theta' = \text{const}$.

We mention one important special case of Theorem *BH*.

**Corollary 1.** *If $E \subset H^2$, $E \neq \{\mathbb{0}\}$ and $SE \subset E$, then $E = \boldsymbol{\Theta} H^2$, where $\Theta \in H^2$, $|\Theta| = 1$ a.e. on $\mathbb{T}$.*

For the proof it suffices to remark that $\mathscr{S} E \subset E$ but $\mathscr{S}^{-1} E \not\subset E$ (if $f \in E$ and $\hat{f}(n) \neq 0$ for some $n$, $n \geq 0$, then $\mathscr{S}^{-n-1} f \notin H^2$); by theorem *BH* then $E = \boldsymbol{\Theta} H^2$ and as $E \subset H^2$ we must have $\Theta \in H^2$. □

Functions in $H^2$ of modulus one on $\mathbb{T}$ will, following Beurling, be called *inner*. These functions play an outstanding rôle in many questions of spectral analysis and function theory, as we will convince ourselves very soon.

The most unexpected consequence of theorem *BH* is perhaps the "geometric" derivation of the well-known uniqueness theorem for the class $H^2$ (theorem of the Riesz brothers).

**Corollary 2.** *If $f \in H^2$ and $m\,\{\zeta : \zeta \in \mathbb{T}, f(\zeta) = 0\} > 0$ then $f = \mathbb{0}$.*

It is clear that the set $\{\zeta : \zeta \in \mathbb{T}, f(\zeta) = 0\}$, $f$ an element of $L^2$ (and, in particular, of $H^2$), is uniquely defined up to a null set and that all functions in the $\mathscr{S}$-invariant subspace $E_f$,

$$E_f \overset{\text{def}}{=} \bigvee (\mathscr{S}^n f : n \geq 0),$$

vanish a.e. on this set. If now $f \neq \mathbb{0}$ then $E_f \neq \{\mathbb{0}\}$ and so by Corollary 1 $E_f = \boldsymbol{\Theta} H^2$, $\Theta \in E_f$ and $|\Theta| = 1$ a.e. Contradiction. □

## 2. The Inner-Outer Factorization.

**Definition.** $f$, $f \in H^2$, is called *outer* if $E_f = H^2$.

**Factorization Theorem.** *If $f \in H^2$, $f \neq \mathbb{0}$, then $f$ can be uniquely (up to multiplicative constants) be written as product $f = f^i f^e$ of an inner factor $f^i$ and an outer one $f^e$; moreover $E_f = \mathbf{f}^i H^2$.*

*Proof.* By Corollary 1 $E_f = \boldsymbol{\Theta} H^2$ where $\Theta$ is an inner function. Put $f^i = \Theta$, $f^e = f\bar{\Theta} (= f/\Theta)$. Then $f^e \in H^2$. As $\mathbf{f}^i$ is an isometric operator in $L^2$ (and then also in $H^2$) it follows that

$$E_f = \bigvee (z^n f^i f^e : n \geq 0) = \mathbf{f}^i \bigvee (z^n f^e : n \geq 0) = \mathbf{f}^i E_{f^e}.$$

Hence $\mathbf{f}^i H^2 = \mathbf{f}^i E_{f^e}$ and $E_{f^e} = H^2$ so that $f^e$ is an outer function. The uniqueness of the inner-outer factorization follows from the formula $E_f = \mathbf{f}^i H^2$ and the remark following Theorem *BH*. □

For outer functions we have an integral analogue of the classical maximum principle (estimating the function inside the domain in terms of its boundary values).

**Corollary 3.** *Let $f \in H^2$. The following statements are equivalent.*
1. *$f$ is an outer function.*
2. *$g \in H^2$, $g/f \in L^2 \Rightarrow g/f \in H^2$.*

The implication $2 \Rightarrow 1$ is trivial: if $g = f^e$ then $g/f = 1/f^i = \overline{f^i} \in L^2$ so $\overline{f^i} \in H^2$, $f^i \in H^2$ and $f^i = \text{const}$, that is, $f$ is outer. Let us prove that $1 \Rightarrow 2$. Let $g \in H^2$, $g/f \in L^2$, and let $\{p_n\}_{n \geq 1}$ be a sequence of polynomials in $z$ such that $\lim_n \| p_n f - \mathbb{1} \|_2 = 0$. Then $\int_{\mathbb{T}} | p_n g - g/f | \, dm = \int_{\mathbb{T}} | p_n f - \mathbb{1} | | g/f | \, dm \leq \| p_n f - \mathbb{1} \|_2 \| g/f \|_2$. Hence $(L^1)$-$\lim_n p_n g = g/f$ and therefore $(g/f)^\wedge(k) = 0$ for $k < 0$, that is, $g/f \in H^2$. □

From this follow also the following properties: if $f$ and $g$ are outer functions and $|f| = |g|$ a.e. then $f = \lambda g$ with $\lambda \in \mathbb{T}$ ($k$ and $1/k = \bar{k}$ belong to $H^2$ if $k = f/g$); if $f \in L^2$ then there exists an outer function $g$ with the same modulus $|g| = |f|$ iff $\mathscr{S}^{-1} E_f \not\subset E_f$, i.e. $E_f$ is a non-reducing (definition on p. 14) $\mathscr{S}$-invariant subspace (apply Theorem *BH* and the Factorization Theorem).

Further progress is connected with the utilization of the Riesz-Nevanlinna factorization theory (cf. App. 2), more exactly with the theorem stating that a non-negative function $h$, $h \in L^2$, is the modulus of an outer function iff

$$\int_{\mathbb{T}} \log h \, dm > -\infty.$$

**Corollary 4.** *Let $f \in L^2$. Then*
1. *$\mathscr{S} E_f = E_f \Leftrightarrow \log |f| \in L^1$.*
2. *$E_f = L^2 \Leftrightarrow \log |f| \notin L^1$ but $f \neq 0$ a.e.*

The first statement has essentially already been proved (Theorem *BH* plus the preceding remark), the second follows from the first one: if $E_f = L^2$ then $\mathscr{S} E_f = E_f$ and therefore $\log |f| \notin L^1$; the inequality $f \neq 0$ a.e. follows from the general fact that

$$E_f \subset \chi_e L^2, \quad e \stackrel{\text{def}}{=} \{\zeta : \zeta \in \mathbb{T}, f(\zeta) \neq 0\}.$$

Conversely, the right hand side of Statement 2 implies that $\mathscr{S} E_f = E_f$ but then $E_f = \chi_e L^2$ (Theorem *BH*) and $me = 1$, that is, $E_f = L^2$. □

It is useful to think of the approximation theoretic meaning of Statement 2, which gives a description of the cyclic elements for the operator $\mathscr{S}$ (it is really question about the completeness of *the set of all polynomials* $\mathscr{P}_A$ in $z$ in the weighted space $L^2(|f|^2 \, dm)$).

**3. The Arithmetic of Inner Functions.** Let us begin with the divisibility properties. We say that the inner function $\Theta_1$ *divides* the inner function $\Theta_2$ if $\Theta_2/\Theta_1 \in H^2$ (so that $\Theta_2/\Theta_1$ too is an inner function).

**Corollary 5.** $\Theta_1 H^2 \supset \Theta_2 H^2 \Leftrightarrow \Theta_1$ *divides* $\Theta_2$.

Indeed, if $\Theta_2 = \Theta_1 \mathscr{I}$, $\mathscr{I}$ an inner function, then $\Theta_2 H^2 = \Theta_1 \mathscr{I} H^2 \subset \Theta_1 H^2$. Conversely, from the inclusion it follows that $\Theta_2 \in \Theta_1 H^2$ so that $\Theta_2/\Theta_1 \in H^2$. □

**Corollary 6.** $\Theta_1 H^2 \cap \Theta_2 H^2 = \Theta H^2$ *where* $\Theta$ *is the least common multiple* (LCM) *of the functions* $\Theta_1$ *and* $\Theta_2$ *(in the sense that both* $\Theta_1$ *and* $\Theta_2$ *divide* $\Theta$ *and if they divide an inner function* $\Theta'$ *then* $\Theta$ *in turn divides* $\Theta'$).

Indeed, the subspace $\Theta_1 H^2 \cap \Theta_2 H^2$ being invariant with respect to the operator $S$ contains $\Theta_1\Theta_2$ ($\neq \mathbb{0}$) and therefore must have the form $\Theta H^2$. By Corollary 5 $\Theta$ is divisible by $\Theta_1$ and $\Theta_2$. If $\Theta'$ is divisible by $\Theta_1$ and $\Theta_2$ then by the same corollary $\Theta_k H^2 \supset \Theta' H^2$, $k = 1, 2$, so that $\Theta H^2 \supset \Theta' H^2$ implying that $\Theta$ divides $\Theta'$. □

In a similar way one can also prove the existence of the LCM of an arbitrary family of inner functions. The existence of the *greatest common divisor* (GCD) of the functions $\Theta_1$ and $\Theta_2$ is connected with a corresponding study of the invariant subspace $\bigvee(\Theta_1 H^2, \Theta_2 H^2)$ but we will not dwell on this question here. The greatest common divisor of $\Theta_1$ and $\Theta_2$ will sometimes be denoted by $\Theta_1 \wedge \Theta_2$ and the least common multiple by $\Theta_1 \vee \Theta_2$.

**4. The Adjoint Operator $S^*$.**

**Corollary 7.** 1. *Let* $E$ *be a (closed) subspace of* $H^2$ *such that* $E \neq H^2$ *and* $S^*E \subset E$. *Then* $E = H^2 \ominus \Theta H^2$ *where* $\Theta$ *is an inner function. Conversely, every subspace of the form* $H^2 \ominus \Theta H^2$ *is invariant with respect to* $S^*$.

2. *If* $E$ *is a subspace of* $H^2$ *such that* $SE \subset E$, $S^*E \subset E$, *then either* $E = \{\mathbb{0}\}$ *or* $E = H^2$.

The first assertion of Corollary 7 follows at once from the general relation between the invariant subspaces of an (any) operator $A$ and its adjoint $A^*$:

$$A^*E \subset E \Leftrightarrow AE^\perp \subset E^\perp$$

(here $E^\perp$ denotes the orthogonal complement of the subspace $E$ in the initial space), which again follows at once from the definition of the adjoint operator and the orthogonal complement. In order to prove the second assertion we notice that if $\{\mathbb{0}\} \neq E \neq H^2$ then

$$E = \Theta H^2, \quad E = H^2 \ominus \Theta' H^2,$$

where $\Theta$ and $\Theta'$ are inner functions. These relations contradict each other because they imply $(\Theta\Theta', \Theta'\Theta) = \mathbb{0}$. □

**Corollary 8.** *Let* $\{\Theta\}$ *be a family of inner functions and set* $K_\Theta \overset{\text{def}}{=} H^2 \ominus \Theta H^2$. *Then*

$$K_{\Theta_1} \subset K_{\Theta_2} \Leftrightarrow \Theta_1 \text{ divides } \Theta_2;$$
$$\bigvee(K_\Theta : \Theta \in \{\Theta\}) = K_{\Theta'}, \quad \Theta' = \mathrm{LCM}\{\Theta\}.$$
$$\bigcap_{\Theta \in \{\Theta\}} K_\Theta = K_{\Theta'}, \quad \Theta' = \mathrm{GCD}\{\Theta\}.$$

(In the limiting cases $\bigvee(K_\Theta : \Theta \in \{\Theta\}) = H^2$ and $\bigcap_\Theta K_\Theta = \{\mathbb{0}\}$ we put *by definition* $\mathrm{LCM}\{\Theta\} = \mathbb{0}$ and $\mathrm{GCD}\{\Theta\} = \mathbb{1}$ respectively.)

This follows from Corollaries 5–7. □

Concluding this Lecture let us remark that among other things we have also obtained so-called *individual theorems* for the operators $S$ and $\mathscr{S}$, describing the subspace $E_f$ either in terms of the zeros of $f$ ($E_f = \chi_e L^2$, $e = \{\zeta : f(\zeta) \neq 0\}$, provided $\log|f| \notin L^1$) or with the aid of the outer function $F$ with modulus $|F| = |f|$ $\left(E_f = \frac{\mathbf{f}}{\mathbf{F}} H^2\text{, provided } \log|f| \in L^1\right)$. In particular, every $S$- (or $\mathscr{S}$-) invariant subspace is cyclic, that is, of the form $E_f$ for some $f$. What concerns the operators $S^*$ the problem of the effective description of the cyclic subspaces $E_f^* \overset{\text{def}}{=} \bigvee(S^{*n}f : n \geq 0)$, $f \in H^2$, requires an additional study and will be solved in the following Lecture.

## Supplements and Bibliographical Notes

Now we turn to a considerably more abstract situation, viz. general unitary operators. From rather simple geometric facts concerning their invariant subspaces we derive some classical results in function theory (for instance, the theorem of the Riesz brothers on analytic measures). We describe further the lattice $\mathrm{Lat}\, S_E$, discuss briefly various generalizations of the theory developed in this Lecture as well as the connections with other subjects, and finally conclude with the history of the subject.

**5. Invariant Subspaces.** Let $H$ be a Hilbert space and $T$ a (continuous linear) operator in $H$. The subspaces in $\mathrm{Lat}\, T \cap \mathrm{Lat}\, T^*$ will be called *reducing subspaces* (for the operator $T$).

**Decomposition Lemma.** *If $E \in \mathrm{Lat}\, T$ then $E = E_1 \oplus E_2$ where $E_1$ is a reducing and $E_2$ a completely non-reducing invariant subspace for $T$ (i.e. $E' \subset E_2$, $E' \in \mathrm{Lat}\, T \cap \mathrm{Lat}\, T^* \Rightarrow E' = \{\mathbb{0}\}$).*

*Proof.* Take $E_1$ to be the closed linear hull of all $T$-reducing subspaces contained in $E$ and set $E_2 = E \ominus E_1 = E \cap E_1^\perp$. □

**Wold-Kolmogorov Lemma.** *Let $U$ be an isometric operator and let $E \in \mathrm{Lat}\, U$. Then*

$$E = \left(\bigcap_{n \geq 0} U^n E\right) \oplus \left(\sum_{n \geq 0} \oplus\, U^n D\right)$$

*where* $D = E \ominus UE$. *Here the first term in this sum is a reducing subspace on which the operator* $U$ *is unitary while the part* $U|\sum_{n\geq 0} \oplus U^n D$ *of* $U$ *on the subspace* $\sum_{n\geq 0} \oplus U^n D$ *is completely non-unitary.*

*Remark.* If $U$ is unitary then this lemma gives the decomposition of a $U$-invariant subspace $E$ into a sum of a reducing subspace and a completely non-reducing one.

*Proof of the Lemma.* If $x \in D$ and $y \in E$ arbitrary then $x \perp U^n y$ for $n \in \mathbb{N}$. Therefore in particular $U^n D \perp U^m D$ for $n \neq m$ $(n, m \geq 0)$. The difference $E \ominus \sum_{n\geq 0} \oplus U^n D$ is contained in $\bigcap_{n\geq 0} U^n E \stackrel{\text{def}}{=} E_\infty$, because from $y \in E$, $y \perp D$, it follows that $y \in UE$, similarly from $y = Uy_1$, $y \perp U^2 D$ that $y_1 \in UE$, that is, $y \in U^2 E$, and so forth. Conversely, if $x = U^{n+1}e$, $e \in E$, and $d \in D$ then $(x, U^n d) = (U^{n+1}e, U^n d) = (Ue, d) = 0$, i.e. $x \perp U^n D$, $n \geq 0$, so that $E_\infty \subset E \ominus \sum_{n\geq 0} \oplus U^n D$. It is likewise clear that $UE_\infty = E_\infty$ and that $E_\infty$ is a reducing subspace ($U^*U = I$, consequently $U^* E_\infty \subset E_\infty$) while the operator $U|\sum_{n\geq 0} \oplus U^n D$ must be completely non-unitary, since from $E' \subset \sum_{n\geq 0} \oplus U^n D$ follows necessarily $UE' \neq E'$. □

**Corollary.** *If* $U$ *is a completely non-unitary isometric operator in* $H$ *then*

$$H = \sum_{n\geq 0} \oplus U^n D, \quad D = H \ominus UH,$$

*so that* $U$ *is unitarily equivalent to the shift operator* $S_D$ *in the space* $H^2(D)$. □

Let us connect the above lemmata with the Spectral Theorem.

**Theorem on Invariant Subspaces.** *Let* $H$ *be a separable Hilbert space and* $U$ *a unitary operator in* $H$ *with the spectral resolution*

$$H \cong \int_{\mathbb{T}} \oplus H(\zeta)\, d\mu(\zeta), \quad U \cong \mathbf{z}.$$

*Finally, let* $E \in \operatorname{Lat} U$ *and let* $E = E_1 \oplus E_2$ *be the decomposition of* $E$ *into a reducing and a completely non-reducing part. Then there exist two* ($\mu$-*measurable*) *operator functions* $P$ *and* $\Theta$ *and a Hilbert space* $D$ *such that*

a) $P(\zeta)$ *is an orthogonal projection in* $H(\zeta)$ *for* $\mu$-*a.e.* $\zeta$, $\zeta \in \mathbb{T}$;

b) $E_1 = PH \stackrel{\text{def}}{=} \{g \in H : g(\zeta) \in P(\zeta)H(\zeta)\, \mu\text{-a.e.}\}$;

c) $\Theta(\zeta)$ *is an isometric operator from* $D$ *into* $H(\zeta)$ *for* $\mu$-*a.e.* $\zeta$, $\zeta \in \mathbb{T}$, *and* $\Theta(\zeta) = \mathbb{0}$ *for* $\mu_s$-*a.e.* $\zeta$, $\zeta \in \mathbb{T}$; *here* $\mu_s$ *is the singular constituent of the measure* $\mu$;

d) *Either* $E_2 = \{\mathbb{0}\}$ *or* $E_2 = \boldsymbol{\varphi}\boldsymbol{\Theta} H^2(D)$, $\boldsymbol{\Theta} H^2(D) = \{g \in H : g(\zeta) = \Theta(\zeta) f(\zeta)$, $\zeta \in \mathbf{T}$; $f \in H^2(D)\}$, *where* $\varphi = (d\mu_a/dm)^{-1/2}$, $\mu_a$ *being the absolutely continuous constituent of* $\mu$; *the two measures* $\mu_a$ *and* $m$ *are then equivalent.*

*The proof* falls naturally into two parts corresponding to the Decomposition Lemma: $E = E_1 \oplus E_2$ and we can thus consider separately $E_1$ and $E_2$.

*Description of* $E_1$. Let $\tau$ be a countable dense subset of $E_1$. The elements of $\tau$ can be considered as functions on $\mathbb{T}$ choosing for each of them a representative from the corresponding equivalence class (an element of the space $\int_{\mathbb{T}} \oplus H(\zeta)\, d\mu(\zeta)$). Then the relation $g \in (E_1)^{\perp}$ is equivalent to the identities $(z^n f, g) = 0$ $(f \in \tau, n \in \mathbb{Z})$, i.e. to

$$\int_{\mathbb{T}} \zeta^n (f(\zeta), g(\zeta))\, d\mu(\zeta) = 0, \quad f \in \tau, \quad n \in \mathbb{Z},$$

i.e. to the identities $(f(\zeta), g(\zeta)) = 0$ $\mu$-a.e. $(f \in \tau)$. From this it follows that the projections $P(\zeta)$ onto the subspaces $E(\zeta) \overset{\text{def}}{=} \bigvee(f(\zeta) : f \in \tau)$, $\zeta \in \mathbb{T}$, indeed supply us with the desired function $P$.

*Description of* $E_2$. The Wold-Kolmogorov expansion

$$E_2 = \sum_{n \geq 0} \oplus \mathbf{z}^n D, \quad D = E_2 \ominus \mathbf{z} E_2,$$

at once gives the representation of $E_2$ as an $H^2$-space. There arise however some difficulties connected with the computations of the values of "functions" in $D$. In order to surmount these difficulties pick an (at most countably infinite) orthonormal basis $\tau$ for $D$ and for every element in $\tau$ a representative (a function defined everywhere on $\mathbb{T}$). For any $f \in D$ we have $(f, z^n f) = 0$ for $n \geq 1$, by the definition of $D$. Then also $(f, z^n f) = 0$ for $n \leq -1$ and there exists therefore a constant $c = c(f)$ such that

$$\| f(\lambda) \|^2\, d\mu(\lambda) = c\, dm \tag{1}$$

(in the sense that the measure $\sigma \mapsto \int_\sigma \| f(\lambda) \|^2\, d\mu(\lambda)$, $\sigma \subset \mathbb{T}$, is a multiple of Lebesgue measure $m$). By integration we find

$$\| f \|^2 = \int_{\mathbb{T}} \| f(\lambda) \|^2\, d\mu(\lambda) = c \int_{\mathbb{T}} dm = c.$$

From this we can draw several consequences. First $f = 0$ $\mu_s$-a.e., so unless $E_2 = \{\mathbb{O}\}$, which case we henceforth exclude from our discussion, the measures $\mu_a$ and $m$ must be equivalent. Secondly $\| f(\lambda) \|_{H(\lambda)} = \| f \| \varphi(\lambda)$ ($m$-)a.e. and similarly, by polarization or by a direct argument, imitating the deduction of (1), $(f(\lambda), g(\lambda))_{H(\lambda)} = (f, g) \varphi^2(\lambda)$ a.e. In particular it follows that outside a suitable set, $e \subset \mathbb{T}$, $me = 0$, these relations hold identically and simultaneously for all $f$ and $g$ belonging to our basis $\tau$; remember that $\tau$ was at most countably infinite. Therefore the map $\Theta(\lambda)$ defined on the elements of $\tau$ by the formula $\Theta(\lambda) f = (\varphi(\lambda))^{-1} f(\lambda)$ defines an isometric imbedding of $D$ into $H(\lambda)$ $(\lambda \notin e)$; we can assume that $\varphi \neq 0$ on $e$. Next let $g \in E_2$, $g = \sum_{n \geq 0} \mathbf{z}^n g_n$, $g_n \in D$, with all coefficients $g_n$ vanishing outside a finite set of indices. Consider $f = \sum_{n \geq 0} \mathbf{z}^n g_n$, $f \in H^2(D)$; for all practical purposes $f$ and $g$ can be identified. Then

$$(\boldsymbol{\Theta} f)(\lambda) = \Theta(\lambda) f(\lambda) = \Theta(\lambda) \sum_{n \geq 0} \lambda^n g_n = \sum_{n \geq 0} \lambda^n \Theta(\lambda) g_n$$
$$= \sum_{n \geq 0} \lambda^n (\varphi(\lambda))^{-1} g_n(\lambda) = (\varphi(\lambda))^{-1} g(\lambda)$$

for a.e. $\lambda$, $\lambda \in \mathbb{T}$. It is clear that if we extend $\boldsymbol{\Theta}$ by continuity to the whole of $H^2(D)$ we get the desired representation $E_2 = \boldsymbol{\varphi} \boldsymbol{\Theta} H^2(D)$. (Alternatively, we could have assumed from the outset that $\mu_a = m$, $\varphi = 1$, which would have simplified somewhat the computations; the measure $\mu$ in the Spectral Theorem is determined only up to equivalence.) □

**Corollary 1.**

$$\operatorname{Lat} U = \operatorname{Lat} U \cap \operatorname{Lat} U^* \Leftrightarrow \exists \sigma, \quad \sigma \subset \mathbb{T} : m\sigma > 0, \quad \mu\sigma = 0.$$

For in the said condition $\mu_a$ and $m$ cannot be equivalent and $E_2 = \{\mathbb{0}\}$. □

Let us also remark that the equality $\mu\sigma = 0$ is equivalent to the condition that the spectral measure $\mathscr{E}_U$ of the operator $U$ vanishes on $\sigma$.

**Corollary 2.** *If $UE \subset E$ then $E = E_s \oplus E_a$ where $E_s$ is a $U$-invariant subspace lying in the singular subspace $\int \oplus H(\zeta)\, d\mu_s(\zeta)$ and $E_a$ one in the absolute continuous subspace $\int \oplus H(\zeta)\, d\mu_a(\zeta)$.* □

**Corollary 3.** $\dim D \leq m\text{-ess}\inf\{\dim H(\zeta) : \zeta \in \mathbb{T}\}$. □

**Corollary 4.** *Let $\mu$ be a measure on $\mathbb{T}$ and let $E$ be a $\mathbf{z}$-invariant subspace of $L^2(\mu)$. Then $E = E_s \oplus E_a$ and we have either $E_a = \chi_\sigma \cdot L^2(\mu_a)$ (the case of a reducing subspace $E$) or $E_a = \boldsymbol{\psi} H^2$, $|\psi|^2\, d\mu_a = dm$ (the opposite case).*

For the proof we need only to produce the expression for a non-reducing subspace $E_a$. But in this case $E_a = \boldsymbol{\varphi\Theta} H^2(D)$, $\dim D = 1$ (with $\varphi$, $\Theta$, $D$ as in the Theorem on Invariant Subspaces); identifying $H^2$ and $H^2(D)$, and regarding $\Theta$ as a scalar function, we get $E_a = \boldsymbol{\varphi\Theta} H^2 = \boldsymbol{\psi} H^2$, with $\psi = \varphi\Theta$; since $|\Theta| = 1$ a.e. clearly $|\psi|^2\, d\mu_a = dm$. □

**Corollary 5 (Theorem of the Riesz Brothers on Analytic Measures).** *Let $\mu$ be a Borel mass distribution (signed measure) on the circle $\mathbb{T}$ and assume that $\int_{\mathbb{T}} z^n\, d\mu = 0$, $n \geq 1$. Then $d\mu = h\, dm$, $h \in H^1$.*

In fact, let $|\mu|$ denote the variation of the mass distribution $\mu$ (a (positive) measure on $\mathbb{T}$) so that $d\mu = \varepsilon |d\mu|$ with $|\varepsilon| = 1$ $|\mu|$-a.e. on $\mathbb{T}$. Consider $E = \bigvee_{L^2(|\mu|)} (z^n : n \geq 1)$. As $z \neq 0$ $|\mu|$-a.e. on $\mathbb{T}$ (and thus also $|\mu_s|$-a.e.) we have $E = L^2(|\mu_s|) \oplus E_a$. But as $\bar{\varepsilon} \perp E$ and $\varepsilon \neq 0$ $|\mu_s|$-a.e. then $L^2(|\mu_s|) = \{\mathbb{0}\}$ and $\mu_s = \mathbb{0}$. Since $E \neq L^2(|\mu|)$ and $z \neq 0$ $|\mu|$-a.e. we must then have $E = \boldsymbol{\psi} H^2$, $|\psi|^2\, d|\mu| = dm$. Thus $d\mu = h\, dm$, $h \in L^1(\mathbb{T})$, and since

$$\int_{\mathbb{T}} z^n h\, dm = \int_{\mathbb{T}} z^n\, d\mu = 0 \quad \text{for} \quad n \geq 1$$

we get $h \in H^1$. □

**6. The Shift of Arbitrary Multiplicity.** Let us use the Theorem on Invariant Subspaces to derive some results on the shift operator $S_D$ in the Hardy space $H^2(D)$, $D$ any Hilbert space.

**Corollary 6 (Lax-Halmos Theorem).** *The subspace $E$ belongs to* Lat $S_D$ *iff*

$$E = \boldsymbol{\Theta} H^2(D'), \tag{2}$$

*where $D'$ is a subspace of $D$ and $\Theta$ an ($m$-measurable) operator valued function on the* *circle $\mathbb{T}$ whose values $\Theta(\zeta)$ are isometric operators from $D'$ into $D$ (for a.e. $\zeta, \zeta \in \mathbb{T}$) and which admits an analytic continuation into the disk $\mathbb{D}$ (in the sense of $H^p$ theory); such functions will be called "left inner functions". The representation (2) is unique in the sense that*

$$\Theta H^2(D') = \Theta' H^2('D) \Leftrightarrow \Theta = \Theta' V$$

*where $V$ is a unitary operator from $D'$ onto $'D$.*

Indeed, the representation (2) follows at once if we notice that $E$ is a completely non-reducing invariant subspace for the operator $\mathscr{S}_D$ in $L^2(D)$ (also $\mu_a = m$, $\varphi = \mathbb{1}$). That $\Theta$ is analytic in the sense required follows from the fact that $\Theta(d \cdot \mathbb{1}) \in H^2(D)$ for all $d, d \in D'$. (Analytic vector functions are discussed in detail in e.g. Sz.-Nagy and Foiaş [1], Rudin [4]. The class of all $X$-valued function $f$ bounded on $\mathbb{T}$ and analytic in $\mathbb{D}$ (as well as the class of its boundary values) will be denoted by $H^\infty(X)$; in particular thus $\Theta \in H^\infty(L(D', D))$, where $L(D', D)$ is the space of all linear operators from $D'$ into $D$; we write $L(D) = L(D, D)$.)

In order to establish the uniqueness let us define a unitary map $U$ from $H^2(D')$ onto $H^2('D)$ by the identity

$$Uf = g \quad \text{if} \quad \Theta f = \Theta' g.$$

We consider the spaces $D'$ and $'D$ as imbedded in the corresponding $H^2$ spaces (as the subspaces of constants: $d \mapsto d \cdot \mathbb{1}$). Let us check that $UD' = {'D}$. If we set $V = U|D'$ the desired result ($\Theta d = \Theta' V d, d \in D'$) will then follow by the very definition of $V$. It is clear that the operator $U$ commutes with multiplication by $z$: $U\mathbf{z} = \mathbf{z}U$, which in particular gives $U(\mathbf{z}H^2(D')) = \mathbf{z}(UH^2('D)) = \mathbf{z}H^2('D)$. The operator $U$ being unitary we therefore get

$$UD' = U\{(\mathbf{z}H^2(D'))^\perp\} = \{U(\mathbf{z}H^2(D'))\}^\perp = \{\mathbf{z}H^2('D))\}^\mathbb{T} = {'D}. \square$$

**Corollary 7 (Model Theorem, II).** *Let $T$ be a contraction in the Hilbert space $H$ and assume that* (s)-$\lim_n T^{*n} = \mathbb{0}$ *(that is,* $\lim_n T^{*n}x = \mathbb{0}$, $\forall x \in H$*). Then $T$ is unitarily equivalent to its model, the operator*

$$f \mapsto P_K zf, \quad f \in K,$$

*where $P_K$ is the orthogonal projection onto the subspace $K$,*

$$K = H^2(D) \ominus \Theta H^2(D'), \tag{3}$$

*the function $\Theta = \Theta_T$ having the same properties as in the preceding corollary and* $D = \operatorname{clos}(I - TT^*)H$.

For the proof it suffices to apply to the operator $T^*$ the Model Theorem of the Introductory Lecture and to the subspace $E = H^2(D) \ominus K$ the Lax-Halmos Theorem (where $K$ is the subspace in the Model Theorem). $\square$

The operator valued function $\Theta = \Theta_T$ in formula (3) will be called the *characteristic function* of $T$. Later on we will only be interested in the case when $\Theta_T$ takes on unitary values a.e. in $\mathbb{T}$. In Lecture II we will see that this latter requirement distinguishes precisely the "class $C_{00}$" of all contractions $T$ such that (s)-$\lim_n T^n = \mathbb{0}$, (s)-$\lim_n T^{*n} = \mathbb{0}$. By analogy with scalar analytic functions an element $\Theta$ in $H^\infty(L(D', D))$ will be called *inner* if $\Theta(\zeta)$ is a unitary operator from $D'$ into $D$ for a.e. $\zeta$, $\zeta \in \mathbb{T}$. Obviously the set of these operators is non-empty only if $\dim D = \dim D'$. If this condition is fulfilled, then with the aid of an auxiliary unitary correspondence $U$ we can identify $D'$ and $D$, replace $\Theta$ by $\Theta U$ and always employ the functional model $f \mapsto P_K zf$ in the space $K = H^2(D) \ominus \Theta H^2(D)$, which we likewise also will do. The inner operator functions characterize $S_D$-invariant subspaces for which the analogues of the "scalar" results of this Lecture turn out to be particularly simple.

**Corollary 8.** *Let $\Theta$ and $\Theta'$ be inner functions belonging to $H^\infty(L(D))$. Then:*

1. $\boldsymbol{\Theta} H^2(D) \subset \boldsymbol{\Theta}' H^2(D) \Leftrightarrow \Theta'$ *divides $\Theta$ in the sense that there exists an inner function $\Omega$ with $\Theta = \Theta'\Omega$.*

2. $\boldsymbol{\Theta} H^2(D) = \boldsymbol{\Theta}' H^2(D) \Leftrightarrow \Theta(\zeta) \equiv \Theta'(\zeta)\Omega$ *where $\Omega$ is a (constant) unitary operator in $D$.*

3. *If $D'$ is a Hilbert space and $\vartheta$ a left inner function with values in $L(D', D)$ then $\boldsymbol{\Theta} H^2(D) \subset \boldsymbol{\vartheta} H^2(D')$ entails that $\vartheta$ is an inner function.*

4. *Set $K_{\boldsymbol{\Theta}} \stackrel{\text{def}}{=} H^2(D) \ominus \boldsymbol{\Theta} H^2(D)$. Every invariant subspace of the operator $S_D^* | K$ has the form $K_\vartheta$ where $\vartheta$ is a left inner divisor of $\Theta : \Theta = \vartheta \cdot \vartheta'$, $\vartheta'$ being an inner function.*

5. *Let $\{\vartheta\}$ be any family of left inner divisors of $\Theta$. Then there exist inner functions $\vartheta'$ and $\vartheta''$ such that*

$$\bigcap_{\vartheta \in \{\vartheta\}} K_\vartheta = K_{\vartheta'}, \quad \bigvee (K_\vartheta : \vartheta \in \{\vartheta\}) = K_{\vartheta''}.$$

(Recall the convention made at the end of Corollary8 of Theorem *BH*.)

Let us verify Statement 1: regarding $\boldsymbol{\Theta}$ and $\boldsymbol{\Theta}'$ as multiplication operators in the space $L^2(D)$ we have $(\boldsymbol{\Theta}')^* \cdot \boldsymbol{\Theta} H^2(D) \subset H^2(D)$ and $(\boldsymbol{\Theta}')^* f = (\Theta')^* f$, $\forall f$, so that $(\Theta')^*\Theta \stackrel{\text{def}}{=} \Omega \in H^\infty(L(D))$ and $\Theta = \Theta'\Omega$. The converse is trivial.

Let us verify Statement 2: from Statement 1 it follows that $\Omega \stackrel{\text{def}}{=} (\Theta')^*\Theta \in H^\infty$ and $\Omega^* \in H^\infty$ so that we must have $\Omega \equiv$ const.

Let us verify Statement 3: for each function $f$ in $H^2(D)$ there exists a unique function $g$, $g \in H^2(D')$, such that $\boldsymbol{\Theta} f = \boldsymbol{\vartheta} g$. As $\vartheta(\zeta)$ is an isometric operator for a.e. $\zeta$, $\zeta \in \mathbb{T}$, we must thus have $g(\zeta) = \vartheta(\zeta)^*\Theta(\zeta)f(\zeta)$, $\zeta \in \mathbb{T}$. This gives $\| g(\zeta) \|_{D'}^2 \le \| f(\zeta) \|_D^2$, $\zeta \in \mathbb{T}$, so that, since $\int_{\mathbb{T}} \| g(\zeta) \|_{D'}^2 \, dm = \int_{\mathbb{T}} \| f(\zeta) \|_D^2 \, dm$, we get $\| g(\zeta) \|_{D'}^2 = \| f(\zeta) \|_D^2$ for a.e. $\zeta$, $\zeta \in \mathbb{T}$. Using the separability of $D$ we can find a countable dense subset $\mathscr{D}$ in $D$ and a subset $\sigma$ of $\mathbb{T}$ with $m\sigma = 0$ such that $\| \vartheta(\zeta)^*\boldsymbol{\Theta}(\zeta)d \|_{D'} = \| d \|_D$ for all $\zeta$, $\zeta \in \mathbb{T} \setminus \sigma$, and *all* $d$, $d \in \mathscr{D}$. But this shows that $\vartheta^*(\zeta)$ is an isometric operator for a.e. $\zeta$, $\zeta \in \mathbb{T}$, so that $\vartheta$ must indeed be an inner function.

Statement 4 follows from Statement 3 and the Lax-Halmos Theorem. Finally Statement 5 follows from Statement 4. □

The function $\vartheta'$ in Statement 5 will be called the *greatest common divisor* of the family $\{\vartheta\}$ and $\vartheta''$ the *least common multiple*. They may be described in the language of "left divisibility" in a similar way as was done in Sec. 4 for scalar functions.

What is the "spectral multiplicity" of the multiple shift $S_D$? In order to be able to formulate the answer we first define the spectral multiplicity for any linear operator $T$ acting in a Banach space $X$ as the number (or the symbol $\infty$) $\mu_T$ given by the formula

$$\mu_T = \inf \dim L$$

where $L \subset X$, the inf being taken over all *generating* (or *cyclic*) subspaces $L$, that is, subspaces such that $E_L \overset{\text{def}}{=} \bigvee(T^n L : n \geq 0) = X$. In the case $T = S_D$ it is easy to see that each such subspace $L$ must have a dimension at least $\dim D$: if $\dim \bigvee(l(0) : l \in L) < \dim D$ then every constant $e \cdot \mathbb{1}$ will be orthogonal to $E_L$, provided $e \perp \bigvee(l(0) : l \in L)$. This argument has a general character and shows that the dimension of a generating subspace for the operator $T$ always must exceed $\dim \operatorname{Ker} T^*$.

What concerns the shift operator then, besides the inequality $\mu_{S_D} \geq \dim D$, we trivially also get the opposite inequality $\mu_{S_D} \leq \dim D$ (the subspace $L = D\mathbb{1}$ of constant functions is $S_D$ cyclic) so that indeed $\mu_{S_D} = \dim D$. The question of determining all cyclic subspaces is however not that simple. Let us define *outer* operator functions as follows. Let $E$ be any Hilbert space and $F$ a function with values in $L(E, D)$ such that $F(\cdot)e \in H^2(D)$ for each $e$, $e \in E$ (to put it differently, $F$ is a *strong $H^2$-function*). If $p$ is a polynomial with values in $E$, i.e. $p(\zeta) = \sum_{k=0}^{n} \hat{p}(k)\zeta^k$, $\hat{p}(k) \in E$ (all such polynomials constitute (definition) the set $\mathscr{P}_A(E)$) then the function $\mathbf{F}p = \sum_{k=0}^{n} z^n \mathbf{F}\hat{p}(k)$ belongs to the space $H^2(D)$. The strong $H^2$-function $F$ will be called *outer* if

$$\operatorname{clos} \mathbf{F} \cdot \mathscr{P}_A(E) = H^2(D).$$

Using this notion we can prove an analogue of the scalar Factorization Theorem (cf. Sect. 2 above).

**Corollary 9 (Inner-Outer Factorization).** *Every strong $H^2$-function $F$ can be expressed in the form*

$$F = \Theta F^e$$

*where $F^e$ is an outer function with values in $L(E, D')$ and $\Theta$ a left inner function with values in $L(D, D')$, $D'$ being an auxiliary Hilbert space.*

For the proof it suffices to represent the $S_D$-invariant subspace $\operatorname{clos} \mathbf{F} \cdot \mathscr{P}_A(E)$ in the canonical form $\mathbf{\Theta} H^2(D')$ setting $F^e = \mathbf{\Theta}^* F$. Since the images $F(\zeta)E$ are contained in the images $\Theta(\zeta)D'$ we have $F = \mathbf{\Theta} F^e$. The operator $\mathbf{\Theta}$ being isometric it follows that $F^e$ must be outer. □

The strong outer functions determine cyclic subspaces for the operator $S_D$, exactly as the scalar outer functions do it in the case $\dim D = 1$: if $E \subset H^2(D)$ then we have $\bigvee(S_D^n E : n \geq 0) = H^2(D)$ iff the "evaluation" function $F$,

$$F(\zeta)e \overset{\text{def}}{=} e(\zeta), \quad e \in E, \quad \zeta \in \mathbb{D},$$

is outer.

Thus for a complete description of the cyclic subspaces for the operator $S_D$

there remains only to investigate the structure of outer operator functions. In this direction we prove only one result, where in order to get a uniform formulation the scalar function $\mathbb{0}$, identically equal to 0, will be considered to be inner (and consequently $\mathbb{0}^i = \mathbb{0}$), denoting by $\pi$ an arbitrary orthogonal projection in the space $E$.

**Theorem on Outer Functions.** *Let $D$ and $E$ be Hilbert spaces,* $\dim D \le \dim E$, $\dim D < \infty$, *and let $F$ be a strong $H^2$-functions with values in $L(E, D)$. The following statements are equivalent:*

1. *$F$ is an outer function.*
2. $\mathrm{GCD}((\det\{F|\pi E\})^i : \dim \pi E = \dim D\} = 1$.

In order to clarify the meaning of Statement 2 let us remark that the operators $F(\zeta)|\pi E$ run through all "minors" of order $\dim D$ of the operator $F(\zeta)$ but that it suffices to consider only minors of the *matrix* $\{(F(\zeta)e_i, d_j) : 1 \le i \le \dim E, 1 \le j \le \dim D\}$ with respect to two fixed bases in $E$ and $D$ respectively. The theorem is non-trivial also in the one dimensional case $\dim D = 1$. However it is then properly speaking contained (implicitly) in the results of Sect. 3 of this Lecture. It takes the form: the set $E$, $\{\mathbb{0}\} \ne E \subset H^2$ generates a trivial $S$-invariant subspace iff $\mathrm{GCD}\{f^i : f \in E\} = \mathbb{1}$.

*Proof of the Theorem.* $1 \Rightarrow 2$. If $F$ is an outer function there exist polynomials $p_{n,j}$, $p_{n,j} \in \mathscr{P}_A(E)$, $n \ge 1$, $1 \le j \le \dim D \overset{\text{def}}{=} m$ such that $\lim_n F p_{n,j} = d_j \mathbb{1}$, $1 \le j \le m$, and $\{d_j\}_{1 \le j \le m}$ is a suitable orthonormal basis in $D$. Now we use some simple facts on the Hardy spaces $H^p$ for $0 < p < 1$ set forth in Appendix 2. Let us consider the determinant $\Delta_n(\zeta) \overset{\text{def}}{=} \det\{F(\zeta)p_{n,j}(\zeta)\}_{1 \le j \le m}$ of the family of vectors $\{F p_{n,j} : 1 \le j \le m\}$. Then $\Delta_n$ is the sum of functions each of which is the product of $m$ functions of the form $(F p_{n,j}, d_k)$, $1 \le k, j \le m$ and these functions belong to $H^2$ (since $F p_{n,j} \in H^2(D)$) and converge as $n \to \infty$ in the norm of that space to the functions $\delta_{jk}\mathbb{1}$ respectively. Therefore $\Delta_n \in H^{2/m}$ and

$$(H^{2/m})\text{-}\lim_n \Delta_n = \mathbb{1}.$$

One the other hand one can express the determinant in terms of minors of dimension $m = \dim D$ of the operator $F(\zeta)$. If $\dim E < \infty$ the formula required is well-known from linear algebra (the Binet-Cauchy formula):

$$\Delta_n(\zeta) = \sum_{\kappa} \det(F(\zeta)|\pi_\kappa E) \det\{\pi_\kappa p_{n,j}(\zeta) : 1 \le j \le m\},$$

where $\kappa$ runs through all subsets of the set $\{1, \dots, \dim E\}$ with $\operatorname{card} \kappa = m$ and $\pi_\kappa$ stands for the orthogonal projection onto the subspace $\bigvee(e_i : i \in \kappa)$ ($\{e_i\}_{i \ge 1}$ being some orthogonal basis in $E$). If however $\dim E = \infty$ it suffices to use the representation

$$\Delta_n = (H^{2/m})\text{-}\lim_k \det\{F Q_k p_{n,j}\}_{1 \le j \le m},$$

where $Q_k \overset{\text{def}}{=} \pi_{\{1,\dots,k\}}$, each determinant $\det\{(F Q_k)(Q_k p_{n,j})\}_{1 \le j \le m}$ being expressed in terms of the Binet-Cauchy formula.

In any case, setting $\mathrm{GCD}\{\det(F|\pi_\kappa E))^i : \mathrm{card}\,\kappa = \dim D\} \stackrel{\mathrm{def}}{=} \vartheta$ we get $\Delta_n \in \vartheta H^{2/m}$ for all $n$, $n \geq 1$. Hence $\mathbb{1} \in \vartheta H^{2/m}$ and consequently $\vartheta = \mathbb{1}$ (cf. Appendix 2).

$2 \Rightarrow 1$. Assume that $\mathrm{clos}(\mathbf{F} \cdot \mathscr{P}_A(E)) \neq H^2(D)$. Then there exists a left inner function $\Theta$, $\Theta \in H^\infty(L(D', D))$ such that $\Theta H^2(D') = \mathrm{clos}\,\mathbf{F}\mathscr{P}_A(E)$, $\dim D' \leq \dim D$. If $\{e_j : 1 \leq j \leq m\}$ is an orthonormal set in $E$ then

$$\det\{Fe_j : 1 \leq j \leq m\} = \det\{\Theta f_j : 1 \leq j \leq m\}, \tag{4}$$

where $f_j \in H^2(D')$, $1 \leq j \leq m$. If $\dim D' < \dim D$ then for $m > \dim\ D'$ we have $\det\{\Theta f_j(\zeta) : 1 \leq j \leq m\} \equiv 0\,(\zeta \in \mathbb{D})$ which is impossible. Therefore $\dim D' = \dim D$ and $\Theta$ is an inner function. Then $\det \Theta$ too must be inner and the right hand side of the identity (4) equals $\det \Theta \cdot \det\{f_j : 1 \leq j \leq m\}$. From Statement 2 follows now that $\det \Theta \equiv \mathrm{const}$. Therefore also $\Theta^* = \Theta^{-1} = \Omega/\det \Theta$, where $\Omega(\zeta)$ is the operator whose matrix elements are the algebraic complements of the matrix $\Theta(\zeta)$, $\zeta \in \mathbb{T}$, must be inner. Thus $\Theta \equiv \mathrm{const}$ and $F$ is outer. □

The preceding theorem takes on a particularly simple form if

$$\dim E = \dim D.$$

**Corollary (Helson-Lowdenslager Theorem).** *If* $\dim D < \infty$ *and* $F$ *is a strong* $H^2$*-function with values in* $L(D)$ *then* $F$ *is outer iff* $\det F$ *is outer.* □

**7. Concluding Remarks.** As already has been noted, the main contents of Lecture I are classical and have been known since 1949. From this time on begins also the systematic interaction between spectral theory and the Nevanlinna theory of boundary values, which is also essentially the basic object of study of the present book. In particular, the operator interpretation makes it possible to supply the transparent and rather simple proofs of many important results in the theory of $H^p$-classes which were presented in Sects. 1–5 of Lecture I (the boundary uniqueness theorem, the inner-outer factorization, the description of analytic measures).

The first part of the Beurling-Helson Theorem was obtained already by Wiener in connection with his study of the cyclic vectors of the *group* of shifts in the space $L^1(\mathbb{R})$ (Tauberian theorems) but this as well as the corresponding part of a more general theorem on invariant subspaces can also be derived from von Neumann's general results proven in the 30's and the 40's (the Spectral Theorem and the theorem on the commutant of a normal operator). The elegant proof of Theorem *BH* presented in the text is due to Srinivasan; cf. Helson [2]. The idea of projecting the function $\mathbb{1}$ for the description of $\mathbf{z}$-invariant subspaces was used by Beurling [1], Halmos (cf. Hoffman [1]) and others but goes of course back to the Wold-Kolmogorov decomposition. The original version of Beurling's theorem comprised strictly speaking only Corollary 1 of Theorem *BH* and the formula $E_f = \mathbf{f}^i H^2$.

Every "individual theorem", including notably Corollary 4, Sect. 2, admits an approximation theoretic interpretation (cf. Introductory Lecture). It is not hard to see that the relation $\mathscr{S} E_f = E_f$ of Statement 1 in Corollary 4, Sect. 2, is equivalent to the possibility of approximating the function $\mathbb{1}$:

$$\inf_{p\in\mathscr{P}_A^o}\int_{\mathbb{T}}|\mathbb{1}-p|^2|f|^2\,dm=0,$$

where $\mathscr{P}_A^o$ is the set of all polynomials in $z$ vanishing at the origin. The qualitative result on approximation, thus settled by Corollary 4, Sect. 2, is a limiting case of an important relation discovered already in the 30's by several authors:

**Szegö-Kolmogorov Theorem.**

$$\inf_{p\in\mathscr{P}_A^o}\int_{\mathbb{T}}|\mathbb{1}-p|^2\,d\mu=\exp\left(\int_{\mathbb{T}}\log h\,dm\right),\tag{5}$$

*where $\mu$ is a Borel measure on $\mathbb{T}$ and $h=d\mu_a/dm$.*

*Proof*. It is easy to derive formula (5) from the Theorem on Invariant Subspaces (cf. Douglas [2], p. 175). Indeed, let $E=\operatorname{clos}_{L^2(\mu)}\mathscr{P}_A^o$. By Corollary 2 of the theorem just mentioned we have $E=L^2(\mu_s)\oplus E_a$ so that $\operatorname{dist}(\mathbb{1},E)=\operatorname{dist}(\mathbb{1}_a,E_a)$, $\mathbb{1}_a\overset{\text{def}}{=}P_{L^2(\mu_a)}\mathbb{1}$. If $\operatorname{dist}(\mathbb{1}_a,E_a)>0$ then $\mathbf{z}E_a\neq E_a$ so by Corollary 4 of the same theorem $E_a=\varphi H^2$, $|\varphi|^2\,d\mu_a=dm$. Therefore $1/\varphi\overset{\text{def}}{=}\Phi\in L^2(\mathbb{T})$ and as $\varphi\in E_a\ominus\mathbf{z}E_a$ (cf. the proof of Corollary 4, Sect. 3) it follows that $0=\int\varphi\bar{z}^n\,d\mu_a=\int\bar{\Phi}\bar{z}^n\,dm$, $n\geq 2$, that is, $\Phi\bar{z}\in H^2$. In the formula $d\mu_a=|\Phi|^2\,dm$ it is only the modulus $|\Phi|$ that matters; taking $\Phi$ outer we therefore get

$$\begin{aligned}\operatorname{dist}(\mathbb{1},E)^2&=\inf_{p\in\mathscr{P}_A^o}\int_{\mathbb{T}}|\mathbb{1}-p|^2|\Phi|^2\,dm=\inf_{p\in\mathscr{P}_A^o}\int_{\mathbb{T}}|\Phi-p\Phi|^2\,dm\\&=\|\Phi-P_{zH^2}\Phi\|^2=|\Phi(0)|^2.\end{aligned}$$

Finally, by a certain property of outer function (see Appendix 2)

$$|\Phi(0)|^2=\exp\left(\int_{\mathbb{T}}\log|\Phi|^2\,dm\right)=\exp\left(\int_{\mathbb{T}}\log(d\mu_a/dm)\,dm\right).$$

If however $\operatorname{dist}(\mathbb{1}_a,E_a)=0$ then $h\in\operatorname{clos}_{L^2}\mathbf{h}\mathscr{P}_A^o$ and therefore $\operatorname{clos}_{L^2}\mathbf{h}\mathscr{P}_A^o$ is a reducing subspace for the operator $\mathscr{S}$. By Corollary 4 of Theorem *BH* it follows that $\log h\notin L^1$. □

**Corollary.** *Let $H^2(\mu)=\operatorname{clos}_{L^2(\mu)}\mathscr{P}_A$. Then*

1. $H^2(\mu)\doteq H^2(\mu_a)\oplus L^2(\mu_s)$.
2. $H^2(\mu)=L^2(\mu)\Leftrightarrow\log\dfrac{d\mu_a}{dm}\notin L^1(\mathbb{T})$.
3. *If* $\log\dfrac{d\mu_a}{dm}\in L^1(\mathbb{T})$, *then* $\bigcap_{n\geq 0}z^nH^2(\mu)=L^2(\mu_s)$.

Indeed, statement 1 follows from Corollary 2, Sect. 5. By the Theorem the subspace $H^2(\mu)$ reduces the operator $\mathbf{z}$ (in $L^2(\mu)$) iff $\log\dfrac{d\mu_a}{dm}\notin L^1(\mathbb{T})$. Thus Statements 2, 3 follow from Corollary 4, Sect. 5. □

On the distance of $\mathbb{1}$ from $\mathscr{P}_A^o$ in other metrices ($L^p$, Orlicz etc) cf. Ahiezer [1] Urbanik [1], Gamelin [1].

The derivation of the uniqueness theorem (Corollary 2) from Theorem *BH* is due to Helson [2] and belongs in its character to so-called "abstract function theory". This theory sets out with a uniform algebra (a closed subalgebra of $C(X)$, $X$ a compact Hausdorff space) and a measure $m$ on $X$, representing a multiplicative functional $\varphi : f \mapsto \int_X f\,dm$, $f \in A$. Under suitable assumptions on $A$ (for instance, if $A$ is a *Dirichlet algebra,* i.e. if $\operatorname{clos}_{C(X)} \operatorname{Re} A = C_{\mathbb{R}}(X)$, $\operatorname{Re} A \stackrel{\text{def}}{=} \{\operatorname{Re} f : f \in A\}$) one can develop an elaborate theory of $A$-invariant subspaces of $L^p(m)$ analogous to the one of this Lecture (where $\varphi(f) = f(0)$, $f \in \mathscr{P}_A$ and $m$ is Lebesgue measure on the circle $\mathbb{T}$). If however $A$ is not an algebra with just one single generator then the $A$ invariant subspaces cannot any longer be defined by inner functions; sometimes one succeeds in describing them in terms of special families of unimodular functions (cocycles; Helson-Lowdenslager). Much information on "abstract function theory" is contained in the books by Gamelin [1] and by Barbey and König [1]; for its applications or development in the case of concrete spaces we refer to the papers by Sarason [1], Neville [1], Nakazi [1] where one likewise can find ample bibliographical references.

The inner-outer factorization belongs to the theory of boundary values founded by Nevanlinna [1] and by F. and M. Riesz [1]; regarding this subject see Nevanlinna [1], Privalov [1], Hoffman [1] (cf. Appendix 2). This theory was subsequently developed within the framework of "abstract function theory" as well as for operator valued analytic functions, about which a few words below. In the classical approach outer functions are defined as not having any inner factor in their canonical factorizations (cf. Appendix 2). The treatment in this Lecture of this circle of ideas is close to the one of Douglas's book [2]. The words "inner" and "outer" were introduced by Beurling [1] (in the scalar case).

Already Smirnov [1] was aware of the importance of outer functions in connection with problems of approximation theory. In particular he proved the universal maximum principle, a special case of which is our Corollary 3 of Theorem *BH* and which also contains all theorems expressing the "Phragmén-Lindelöf principle". From Corollary 3 follows, for instance, the following result of the latter type.

**Phragmén-Lindelöf Theorem.** *Let $F$ and $G$ be analytic in $\mathbb{D}$, $G$ being an outer function of "Nevanlinna class" (i.e. the quotient of two outer $H^2$ functions). If $|F| \leq |G|$ in $\mathbb{D}$ and $F|\mathbb{T} \in L^2$ ($F|\mathbb{T} \in L^\infty$) then follows $F \in H^2$ ($F \in H^\infty$ respectively).*

*Proof*. Let $G = G_1/G_2$ with $G_i$ outer in $H^2$ ($i = 1, 2$). By the integral representation of outer functions (see Appendix 2) we can write $G_i$ as the quotient $G_i = G_i''/G_i'$ of two outer functions in modulus not exceeding one. (Take $G_i' \stackrel{\text{def}}{=}$ the outer function with modulus $\min(\mathbb{1}, |G_i|)$, $i = 1, 2$.) Therefore $G = G'/G''$ where $G'$ and $G''$ are outer functions in $H^\infty$.

Since $|G''F| \leq |G'|$ in $\mathbb{D}$ and the function $g \stackrel{\text{def}}{=} G''F$ belongs to $H^2$ (by the hypothesis) we can apply Corollary 3 with this $g$ and with $f \stackrel{\text{def}}{=} G''$. We obtain $F \in H^2$. The rest follows from the relation $H^\infty = L^\infty \cap H^2$ explained in Appendix 2. □

The range of applicability of theorems of this kind can be enlarged with the aid of conformal mapping. If $\Omega$ is a domain conformally equivalent to the disk $\mathbb{D}$ we will call a function $f$ defined in $\Omega$ outer if its "pullback" $f \circ \omega$ has this property ($\omega : \mathbb{D} \to \Omega$ is the conformal equivalence). We can subsequently apply the result already established. Here is a simple example of a result obtained along this route.

**Corollary.** *Let $f$ be regular in $\mathbb{D}$. Assume that*

$$|f(\zeta)| \le c|e^{\frac{c}{1-\zeta}}|, \quad \zeta \in \mathbb{D},$$

*with a suitable $c$, $c > 0$, and that $f$ stays bounded near the point $\zeta = 1$ if we approach it along the circumference or along the radius:*

$$\overline{\lim_{\zeta \to 1, \zeta \in \mathbb{T}}} |f(\zeta)| < \infty, \qquad \overline{\lim_{\zeta \to 1, \operatorname{Im}\zeta = 0}} |f(\zeta)| < \infty.$$

*Then $f \in H^\infty$.*

Indeed, it suffices to verify that the restrictions

$$e^{\frac{c}{1-z}}|\mathbb{D}_\pm, \quad \mathbb{D}_\pm = \mathbb{D} \cap \mathbb{C}_\pm,$$

where $\mathbb{C}_\pm = \{\zeta : \zeta \in \mathbb{C}, \operatorname{Im}\zeta \gtrless 0\}$, are outer functions. Returning to the disk (with the aid of conformal mapping) we get the function

$$G \overset{\text{def}}{=} \exp\left\{-c'\left(i\frac{1+z}{1-z}\right)^{1/2}\right\}, \quad c' > 0.$$

This function is outer in $\mathbb{D}$, because $G \in H^\infty$; $G(\zeta) \neq 0$, $\zeta \in \mathbb{D}$, and if there were a singular part then its representing measure must be concentrated to the set $\{1\}$ (this is rather simple; it will also follow trivially from the description of the "spectrum of an inner function" to be given in Lecture III) but there we cannot have any mass:

$$|G(r)| = \exp\left\{-\frac{c'}{\sqrt{2}}\left(\frac{1+r}{1-r}\right)^{1/2}\right\}, \quad 0 \le r < 1. \square$$

The theorems on the inner-outer factorization for operator valued functions analytic in the disk $\mathbb{D}$ and on the representation of positive semi-definite functions $W$ in the form $W(\zeta) = F(\zeta)^* F(\zeta)$ with $F$ outer are of great importance not only in operator theory (cf. e.g. Sz.-Nagy and Foiaş [1]) but also in the applications to stationary stochastic processes, where the formula for the best prediction of the process at a given time in terms of its past in an essential manner depends on this factorization. Such theorems were therefore proved by many authors starting with Wiener, Masani, Rozanov, Lowdenslager; concerning this question see Rozanov [2], Helson [1, 2], Sz.-Nagy and Foiaş [1] and for a more general formulation and also the historical side of the question Gellar-Page [1], Page [1], Suciu-Valusescu [1, 2]. The last paper contains in particular Corollary 9 (Sect. 6) on the Inner-Outer Factorization and it likewise introduces the notion of strong $H^2$-functions. The definition of outer operator functions utilized in Sect. 6 is due to Helson [2], in whose book one can also find the Corollary of the

Theorem on Outer Functions, as well as a detailed bibliography (until 1964). In this theorem itself one can apparently replace the hypothesis $\dim D < \infty$ by that of the existence of a "scalar multiple" (cf. Lecture III) for the function $F$ (making corresponding modifications in Statement 2). The rôle of outer functions in the question of generating subspaces for the bilateral shift $\mathscr{S}_E$ is similar to their rôle in the analogous problem for scalar functions. Everything reduces to finding the conditions under which a positive operator function $A$ can be written as a square of an analytic (outer) function $F$ ($A(\zeta) = F(\zeta)^* F(\zeta)$). In the finite dimensional case ($\dim E < \infty$) and a.e. invertible functions $A(\cdot)$ the answer is again (as in the Theorem on Outer Functions) expressible in terms of the determinant: $\int_{\mathbb{T}} \log \det A(\zeta)\, dm(\zeta) > -\infty$; the non-invertible case is slightly more complicated, cf. Helson [2]. Operator analogues of the Szegö-Kolmogorov Theorem are also of great significance in this circle of ideas; one can look for the latter in the books and articles listed in this paragraph (cf. further Moore [1]). Concluding our discussion of theorems on inner-outer factorization, let us give an "abstract" result of this type which can be stated purely in Hilbert space terms (Page [1]): let $U$ be a unitary operator in a Hilbert space $H$, $K \in \operatorname{Lat} U$ and $A$ an invertible operator commuting with $U$; then $A = WB$ where the operators $W$ and $B$ too commute with $U$, $W$ being unitary and $BK = K$. At the price of an unessential complication of the formulation one can also relax the assumption of $A$ being invertible replacing it by the requirement of the existence of an operator $D$ such that $DAK \subset K$, $DU = UD$. If we want to deal with (operator) functions we have to put $U = \mathscr{S}_E$, $K = H^2(E)$. Then the condition $DAH^2(E) \subset H^2(E)$ turns out to be a crude analogue of the conditions $\log |a| \in L^1(\mathbb{T})$ (if $A$ is the operator of multiplication with $a$). Cf. also Cambern [1].

The Beurling-Helson Theorem and its generalizations (in particular the above Theorem on Invariant Subspaces) translates the study of geometric objects (for instance, invariant subspaces) into the language of the multiplicative structure of analytic functions. Their joint study allows us for instance to give a description of the arithmetic of inner functions (cf. Sect. 4). The construction of the GCD and the LCM of a family $\{\vartheta\}$ of inner functions can be written down explicitly. If $\vartheta$ is any inner functions put $\mu^\vartheta = \mu^B + \mu^s$ where $\mu^s$ is a singular measure on $\mathbb{T}$, the *representative* of the singular part of $\vartheta$ (cf. Appendix 2) and further

$$\mu^B \sigma \overset{\text{def}}{=} \frac{1}{2} \sum_{\lambda \in \sigma \cap \vartheta^{-1}(0)} k_\vartheta(\lambda)(1 - |\lambda|^2), \quad \sigma \subset \mathbb{D},$$

$k_\vartheta(\lambda)$ being the multiplicity of the zero of $\vartheta$ at the point $\lambda$, $\lambda \in \mathbb{D}$. Then the functions $\Theta^\vee = \mathrm{LCM}\{\vartheta\}$ and $\Theta^\wedge = \mathrm{GCD}\{\vartheta\}$ are given by the measures

$$\mu^{\Theta^\vee} = \bigvee \{\mu^\vartheta : \vartheta \in \{\vartheta\}\}, \quad \mu^{\Theta^\wedge} = \bigwedge \{\mu^\vartheta : \vartheta \in \{\vartheta\}\}$$

respectively, where the symbols $\bigvee$ and $\bigwedge$ stand for the sup and the inf respectively in the lattice of (non-negative) Borel measures on the closed disk $\operatorname{clos} \mathbb{D}$. In particular $k_{\Theta^\vee}(\lambda) = \max_\vartheta k_\vartheta(\lambda)$ (provided $\Theta^\vee \neq 0$, cf. the note at the end Corollary

8 (Sect. 4)) and $k_{\Theta^\wedge}(\lambda) = \min_\vartheta k_\vartheta(\lambda)$, $\lambda \in \mathbb{D}$. Also the singular part of the measure $\mu^{\Theta^\wedge}$ takes on the set $\sigma$, $\sigma \subset \mathbb{T}$, the value $\inf \sum_\vartheta \mu^\vartheta(\sigma_\vartheta)$ where the inf goes over all decompositions $\{\sigma_\vartheta\}$ of $\sigma$. For details see Hoffman [1].

The Decomposition Lemma plays a major rôle in the theory of dilations and has been known for times immemorial. Concerning the Wold-Kolmogorov Lemma see Wold [1]. In connection with $\mathscr{S}_E$ invariant subspaces it was apparently first used by Lax [1] and Halmos [1].

The theorem on Invariant Subspaces is proved in the paper Nikol'skiĭ [1] but important special cases were of course known earlier ($H(\zeta) = E$, $\forall\zeta$, Lax [1], Halmos [1]; $\dim H(\zeta) \leq 1$, Hoffman [1]). The treatment of Sect. 5 is close to the one in Douglas [2].

The idea of deriving the Theorem of the Riesz Brothers on Analytic Measures (Corollary 5 of the Theorem on Invariant Subspaces) from the description of $\mathbf{z}$-invariant subspaces in $L^2(\mu)$ goes back to Lowdenslager and is used in the papers by Sarason [2] and Douglas [2]. Corollary 1 of the same theorem was first obtained by Wermer [1] (in a different way) in connection with the problem of spectral synthesis for normal operators (for particulars see Lecture IV).

The results of this lecture admit various generalizations and applications in function theory as well as in operator theory; some idea of the possible directions of research can be obtained from the papers Nikol'skiĭ [2, 4] (the last one contains an extensive bibliography). One such direction is to seek analogues of Theorem *BH* in spaces other than $L^2(\mathbb{T})$. In the sequel we will only need the generalizations of this theorem contained in the following proposition. Let $X$ be a *Banach lattice* (or *Köthe space*)[2] of measurable functions on the circumference $\mathbb{T}$, i.e. a Banach space $X \subset L^1(\mathbb{T})$ with the property $f \in X, g \in L^1(\mathbb{T}), |g| \leq |f|$ a.e. in $\mathbb{T}$ then $g \in X$ and $\|g\|_1 \leq \|g\|_X \leq \|f\|_X$. Set $X_A = \{f \in X : \hat{f}(n) = 0, n < 0\}$.

**Theorem on Invariant Subspaces of a Banach Lattice.** *Let $X$ be a Banach lattice. Assume that $C(\mathbb{T}) \subset X$ (continuous imbedding), $\operatorname{clos}_X C(\mathbb{T}) = X$, $\operatorname{clos}_X C_A = X_A$. Let moreover $E$ be any subspace of $X$ with $\mathbf{z}E \subset E$. Then*

1) *either $\mathbf{z}E = E$ and there exists a subset $e$, $e \subset \mathbb{T}$, such that $E = \chi_e E$; or*

2) *$\mathbf{z}E \neq E$ and there exists a measurable function $\Theta$, $|\Theta| = 1$ a.e., such that $E = \Theta X_A$.*

*The same statement holds also for $\mathbf{z}$-invariant subspaces in the dual space $X^*$, provided they are assumed to be closed in the weak topology $\sigma(X^*, X)$.*

*Our proof* will be short but complete (in the sense that we put out all essential details). We divide it into several independent propositions.

A. *Every continuous linear functional $F$ in $X$ has the form $F = F_y$; $F_y(x) \overset{\text{def}}{=} \int_{\mathbb{T}} xy\,dm$, where $y \in L^1(\mathbb{T})$, $\int_{\mathbb{T}} |xy|\,dm < \infty$, $\forall x \in X$.*

---

[2] *Translator's note.* The established Russian word for "Banach lattice" is "Banach ideal"; "ideal" refers to the property $f \in X, |g| \leq |f|$ a.e. implies $g \in X$. The corresponding English terminology seems to be rather shaky. Thus in the sequel we will usually say "ideal property" (not lattice).

Let us set $\mu e = F(\chi_e)$, $e \subset \mathbb{T}$. If $me = 0$ then $\mu e = 0$ and $\mu$ as a set function is countably additive, because $\lim_n \|\chi_{e_n}\|_X = 0$ if $e_1 \supset e_2 \cdots$, $\lim_n me_n = 0$ (if this were not the case, so that $\lim_k \sup_{i>k} \|\chi_{e_k} - \chi_{e_i}\|_X > 0$, there would exist pairwise disjoint sets $\sigma_j$, $j \geq 1$, such that $\inf_{j \geq 1} \|\chi_{\sigma_j}\|_X > 0$ (take $\sigma_j \stackrel{\text{def}}{=} e_{k_j} \backslash e_{k_{j+1}}$ with $k_j$ increasing); the presence of such a family of sets $\sigma_j$ contradicts the separability of $X$:

$$\left\| \sum_{j \geq 1} c_j \chi_{\sigma_j} \right\|_X \geq \sup_{j \geq 1} |c_j| \, \|\chi_{\sigma_j}\|_X \geq \text{const} \sup_{j \geq 1} |c_j| \quad \text{if} \quad (c_j \in \mathbb{C}, j \geq 1).$$

$\mu$ is thus absolutely continuous with respect to the measure $m$ so that $d\mu = y\,dm$, $y \in L^1(\mathbb{T})$. Since simple functions are dense in $X$ we get $F(x) = \int_{\mathbb{T}} xy\,dm$, $x \in X$. □

From this proof it follows that if $x_n, x \in X$, $|x_n(\zeta)| \leq |x(\zeta)|$ for a.e. $\zeta$, $\zeta \in \mathbb{T}$, and $\lim_n x_n(\zeta) = x(\zeta)$ for a.e. $\zeta$, $\zeta \in \mathbb{T}$, then $x_n \xrightarrow{\text{weak}} x$.

B. *If* $\mathbf{z}E = E$ *then* $F_y | E \equiv 0 \Leftrightarrow xy \equiv 0$, $\forall x \in E$.

The non-trivial one of these two implications ($\Rightarrow$) is obtained by the following reasoning: if $x \in E$ then $z^n x \in E$ for $n \in \mathbb{Z}$ and therefore $0 = F_y(z^n x) = \int z^n xy\,dm$, $n \in \mathbb{Z}$. As $xy \in L^1(\mathbb{T})$ it follows that $xy = \mathbb{0}$. □

Statement 1 of the theorem is now clear.

C. $\text{clos}_X(E \cap L^\infty) = E$.

Let $x \in E$ and put $h_n = \min(\mathbb{1}, n/|x|)$ $(n = 1, 2, \ldots)$. Then $0 \leq h_n \leq 1$, $\log h_n \in L^1(\mathbb{T})$, so that there exists an outer function $\varphi_n$ with $|\varphi_n| = h_n$ $(n = 1, 2, \ldots)$. Now each $\varphi_n$ can be realized as the bounded pointwise limit of a suitable sequence of polynomials $\{\varphi_n^k\}_{k \geq 1}$ (take the Fejér means of the Taylor series). Then in particular $\varphi_n^k x \xrightarrow{\text{weak}} \varphi_n x$ so that from $\varphi_n^k x \in E$ $(n, k = 1, 2, \ldots)$ it follows that $\varphi_n x \in E$, $n = 1, 2, \ldots$. By the construction $h_n(\zeta) \leq h_{n+1}(\zeta)$, $\lim_n h_n(\zeta) = 1$ $(\zeta \in \mathbb{T})$ so that $\lim_n \|\varphi_n\|_{H^2} = 1$, $\lim_n \varphi_n(\lambda) = 1$, $\forall \lambda \in \mathbb{D}$ (the last assertion follows from the integral representation for $\varphi_n$; Appendix 2). Consequently $\lim_n \|\varphi_n - \mathbb{1}\|_{H^2}^2 = \lim_n (\|\varphi_n\|_{H^2}^2 + 1 - 2\,\text{Re}\,\varphi_n(0)) = 0$. Therefore a suitable subsequence $\{\varphi_{n_k}\}_{k \geq 1}$ tends to the function $\mathbb{1}$ a.e. on the circumference $\mathbb{T}$ and finally $\varphi_{n_k} x \xrightarrow{\text{weak}} x$. □

D. *If* $L^2(\mathbb{T}) \subset X$, $\mathbf{z}E \neq E$ *then* $E = \mathbf{\Theta} X_A$.

In fact, in the case at hand $E = \text{clos}_X \text{clos}_{L^2}(E \cap L^\infty)$. If $\text{clos}_{L^2}(E \cap L^\infty)$ were both $\mathbf{z}$ and $\bar{\mathbf{z}}$-invariant so would $E$. Therefore by Theorem *BH* we conclude that $\text{clos}_{L^2}(E \cap L^\infty) = \mathbf{\Theta} H^2$ where $|\Theta| = 1$ a.e. on $\mathbb{T}$. Taking account of the fact that $\mathbf{\Theta}$ acts as an isomorphism on $X$ we get $E = \text{clos}_X \mathbf{\Theta} H^2 = \mathbf{\Theta}\,\text{clos}_X H^2 = \mathbf{\Theta} X_A$. □

E. *If the Statements* 1 *and* 2 *hold in the Banach lattice* $X$ *then they likewise hold for any subspace* $E$ *of the dual* $X^*$ *that is closed in the (weak) topology* $\sigma(X^*, X)$ *(with respect to the natural duality of Statement* 1).

With this duality ($\langle x, y \rangle = \int xy\,dm$) the operator conjugate to multiplication by $z$ is again multiplication by $z$ and the annihilator $E^\perp = \{y : y \in X^*; \langle x, y \rangle = 0, \forall x \in E\}$ of a subspace $E$ of the form $\chi_e X$ or $\mathbf{\Theta} X_A$ is $\chi_{\mathbb{T} \backslash e} X^*$ or $\bar{\mathbf{\Theta}} \mathbf{z} (X^*)_A$ respectively. □

F. *Completion of the proof*. By virtue of $A$ and $C$

$$E = \text{clos}_X \text{clos}_{\sigma^*}(E \cap L^\infty)$$

where $\sigma^*$ denotes the weak topology $\sigma(L^\infty, L^1)$ of the space $L^\infty(\mathbb{T})$. On the other hand $\text{clos}_{\sigma^*}(E \cap L^\infty)$ is a weakly $\sigma(L^\infty, L^1)$ closed subspace of $L^\infty(\mathbb{T})$ invariant with respect to $\mathbf{z}$ but not with respect to $\bar{\mathbf{z}}$. From Statements D and E it therefore follows that this subspace has the form $\Theta H^\infty$ ($H^\infty = L^\infty_A$). Finally

$$E = \text{clos}_X \boldsymbol{\Theta} H^\infty = \boldsymbol{\Theta}\, \text{clos}_X H^\infty = \boldsymbol{\Theta} X_A. \ \square$$

**Special Cases.** $X = L^p(\mathbb{T})$, $1 \le p < \infty$; $L^\infty(\mathbb{T})$ in the weak topology $\sigma(L^\infty, L^1)$. For these spaces the theorem was established by Srinivasan, see Helson [2].

The multiplication operator $\mathbf{z}$ plays a leading rôle in many questions of analysis; it is connected with many interesting and difficult problems (and gives itself rise to new problems of this kind). Let us mention e.g. the problem of the existence of non-trivial invariant subspaces for so-called *subnormal operators* (parts of normal ones). This problem is equivalent to the following one: let there be given a Borel measure $\mu$ in the plane $\mathbb{C}$ with compact support; does the operator $\mathbf{z}; f \mapsto zf$ in the subspace $H^2(\mu) \stackrel{\text{def}}{=} \text{clos}_{L^2(\mu)} \mathscr{P}_A$ have at least one non-trivial invariant subspace? This question has a long history; there are known many sufficient conditions on $\mu$ guaranteeing a positive solution (cf. the survey article by Nikol'skiĭ [4]). All these conditions yield indeed more than required, namely the presence of the following alternative: either $H^2(\mu) = L^2(\mu)$ or there exists a point $\zeta$, $\zeta \in \mathbb{C}$, such that the functional $p \mapsto p(\zeta)$, $p \in \mathscr{P}_A$, is continuous in the metric $H^2(\mu)$. Not long ago the American mathematician Brown [1] actually proved the existence of invariant subspaces for *any* subnormal operator but this solution bypasses the alternative just mentioned.

# Lecture II. Individual Theorems for the Operator $S^*$

Only a small portion of our knowledge on $S^*$-invariant subspaces can be obtained directly from the canonical representation $K = H^2 \ominus \Theta H^2$, $\Theta$ an inner function. For instance, one can establish that the hull $\bigvee(K_i : i = 1, 2,)$ of two such subspaces $K_1$ and $K_2$ is again a non-trivial $S^*$-invariant subspace in its own right, because $\bigvee(K_i : i = 1, 2)^{\perp} = \Theta_1 H^2 \cap \Theta_2 H^2 = \Theta H^2$ ($\neq \{\mathbb{0}\}$ where $\Theta$ is the LCM of $\Theta_1$ and $\Theta_2$. However many other theorems require a deeper analysis. This Lecture conventionally falls into two parts: the first one is devoted to a detailed analysis of $S^*$-cyclic vectors (that is, elements $f$ in $H^2$ such that $E_f^* \overset{\text{def}}{=} \bigvee(S^{*n}f : n \geq 0) = H^2$) and the second one to an approximation theoretic characterization of $S^*$-invariant subspaces in terms of the root vectors of that operator.

### 1. Pseudocontinuation of $H^2$-Functions and $S^*$-Cyclicity.

**Definition.** Let $f \in H^2$. We say that the meromorphic function $\tilde{f}$ defined in $\hat{\mathbb{C}} \setminus \operatorname{clos} \mathbb{D}$ is a *pseudocontinuation* of $f$ if $\tilde{f}$ is a function of bounded type (in the sense of Nevanlinna, cf. Appendix 2) and if $\tilde{f}(\zeta) = f(\zeta)$ for a.e. $\zeta$, $\zeta \in \mathbb{T}$.

The notion of pseudocontinuation turns out to be most useful in the description of cyclic vectors for the operator $S^*$; let us denote by $C$ the set of all such vectors and let us further set $N \overset{\text{def}}{=} H^2 \setminus C$.

**Theorem on Cyclic Vectors.** *Let $f \in H^2$. The following statements are equivalent.*

1. *$f \in N$.*
2. *$f$ has a pseudocontinuation.*

Half of this theorem at once follows from the following elementary description of $S^*$-invariant subspaces.

**Lemma on** Lat $S^*$. *If $\Theta$ is an inner function then*

$$H^2 \ominus \Theta H^2 = H^2 \cap \bar{z}\overline{\Theta H^2}$$

*where the bar stands for complex conjugation.*

*Proof of the lemma.* Let $f \in H^2$. Then

$$f \in H^2 \ominus \Theta H^2 \Leftrightarrow f \perp \Theta H^2 \Leftrightarrow \bar{\Theta} f \perp H^2 \Leftrightarrow z\bar{\Theta} f \perp zH^2$$

$$\Leftrightarrow z\bar{\Theta} f \in \overline{H^2} \Leftrightarrow f \in \mathbf{z}\overline{\Theta H^2}. \square$$

*Proof of the theorem.* $1 \Rightarrow 2$. As $E_f^* \overset{\text{def}}{=} \bigvee(S^{*n}f : n \geq 0) \neq H^2$ then by Corollary 7 or Theorem $BH$ (Lecture I) $H^2 \ominus E_f^* = \Theta H^2$, that is, we must have $f \in H^2 \ominus \Theta H^2$ where $\Theta$ is a suitable inner function. By the Lemma on Lat $S^*$ $f =$

$\bar{z}\Theta\bar{g}$, $g \in H^2$. Now the inner function $\Theta$ has a standard pseudocontinuation $\tilde{\Theta}$ obtained by the reflection principle

$$\tilde{\Theta}(\zeta) = 1/\overline{\Theta(1/\bar{\zeta})}, \quad |\zeta| > 1. \tag{1}$$

Therefore $\tilde{f} \stackrel{\text{def}}{=} \bar{z}\Theta\bar{g}$ is a pseudocontinuation of $f$.

$2 \Rightarrow 1$. Let $\tilde{f}$ be a pseudocontinuation of $f$ and write $\tilde{f} = H/G$ where $H$ and $G$ are bounded analytic functions in $\hat{\mathbb{C}}\backslash\operatorname{clos}\mathbb{D}$, $G \neq \mathbb{0}$. Then $fG = H \in L^1(\mathbb{T})$ and $\hat{H}(n) = 0$ for $n > 0$; from this it follows that

$$0 = \int_{\mathbb{T}} \bar{z}^n f G \, dm = (S^{*n}f, \bar{G}) = (S^{*(n-1)}f, z\bar{G}), \quad n > 0,$$

that is, $z\bar{G} \in H^2 \ominus E_f^*$, $z\bar{G} \neq \mathbb{0}$, and therefore $f \in N$. □

Let us now mention a few properties of the operation of pseudocontinuation, trivial or at least trivial consequences of wellknown facts on the Nevanlinna class (cf. Appendix 2):

a) *uniqueness*: if $\tilde{f}$ exists and $\tilde{f} = \mathbb{0}$ then $f = \mathbb{0}$;

b) *linearity*: if $\tilde{f}$ and $\tilde{g}$ exist then $(\lambda f + \mu g)^{\sim} = \lambda\tilde{f} + \lambda\tilde{g}$;

c) *permanence*: if $\tilde{f}$ exists and if the function $f$ admits analytic continuation (in the usual sense) through the point $\zeta$, $|\zeta| = 1$, then this continuation coincides with $\tilde{f}$ (and thus defines a univalent meromorphic function $\hat{\mathbb{C}}\backslash\operatorname{clos}\mathbb{D}$);

d) *stability for multiplication*: if $\tilde{f}$ and $\tilde{g}$ exist and $fg \in H^2$ then $(fg)^{\sim} = \tilde{f}\tilde{g}$.

**Corollary 1.** *$N$ is a linear subspace; $N + C \subset C$; $\mathscr{P}_A \subset N$.*

Indeed, the two first statements follow from Property b) of pseudocontinuation and the last one is trivial. □

**Corollary 2.** $C \neq \varnothing$; $\operatorname{clos} C = H^2$; $S^*C \subset C$.

As examples of cyclic elements for the operator $S^*$ we can take such functions as $\log(\mathbb{1} - z)$, $(\mathbb{1} - z)^{1/2}$ etc. The set $C$ is dense in $H^2$ since (for instance) $p + \lambda\log(\mathbb{1} - z) \in C$ for each polynomial $p$ and each complex number $\lambda$, $\lambda \neq 0$. The last statement of the corollary can easily be deduced from the identity $(S^{*n}S^*f, g) = (S^{*n}f, Sg)$, $n \geq 0$. □

**Corollary 3.** *Assume that $f \in H^2$ and let $f = f^e f^i$ be its canonical factorization. Then*

$$f \in N \Leftrightarrow f^e \in N.$$

Every inner function has the standard pseudocontinuation (1); the rest follows from d). □

**Corollary 4.** *Let $f \in H^2$ with $\overline{\lim}_n |\hat{f}(n)|^{1/n} < 1$. Then $f \in N \Leftrightarrow f$ is a rational function.*

The implication $\Leftarrow$ follows at once from the theorem; on the other hand, if $f \in N$ then Property c) of pseudocontinuation shows that $f$ must be a rational function. □

**Example.** $\exp\dfrac{1}{z-2}\in C$. □

Below we will denote by $P_+$ the orthogonal projection of $L^2$ onto the subspace $H^2$ (*the Riesz projection*): $P_+(\sum_{n\in\mathbb{Z}}\hat{f}(n)z^n)=\sum_{n\geq 0}\hat{f}(n)z^n$.

**Corollary 5.** *If* $u\in L^2$, $u\neq\mathbb{O}$ *and* $\log|u|\notin L^1$, *then* $P_+u\in C$; *if* $f\in H^2, f\neq\text{const}$ *and* $\log|\operatorname{Re} f|\notin L^1$ *then* $f\in C$.

Indeed, let the function $f=P_+u$ have the pseudocontinuation $\tilde{f}$. Then the sum $Q(\zeta)=\tilde{f}(\zeta)+\sum_{n\geq 1}\hat{u}(-n)\zeta^{-n}$, $|\zeta|>1$, belongs to the Nevanlinna class in $\hat{\mathbb{C}}\backslash\operatorname{clos}\mathbb{D}$ and $Q=u$ a.e. on $\mathbb{T}$. As $\int_{\mathbb{T}}\log|Q|\,dm=-\infty$ then $Q=\mathbb{O}$, which is impossible. Consequently $P_+u\in C$. The second assertion of the corollary is verified in an analogous way. □

Here are a few more properties of (non) cyclic vectors: 1) $C+C=H^2$; 2) $f\in N\Rightarrow\sum_{n\geq 0}z^n\operatorname{Re}\tilde{f}(n)\in N$; 3) $f\in N\Leftrightarrow\operatorname{sign}f^2=\varphi/\psi$ where $\varphi$ and $\psi$ are inner functions; $\operatorname{sign}f^2\overset{\text{def}}{=}f^2/|f^2|$; 4) if $\inf_{k\geq 1}n_{k+1}/n_k>1$, $\sum_{k\geq 1}|a_k|^2<\infty$ and $a_k\neq 0$ for an infinite sequence of indices $k$ then $\sum_{k\geq 1}a_kz^{n_k}\in C$; 5) in Corollary 4 one cannot replace the exponential decrease of the Taylor coefficients by something weaker; more precisely, if $\varepsilon_k>0$, $\varepsilon_k\downarrow 0$, and $\sum_{k\geq 1}c_kk^2\leq 1$, $\sum_{k\geq(2\varepsilon_n)^{-1/2}}c_k\leq\exp(-n\varepsilon_n)$ for $n\geq 1$, then the function

$$f=\sum_{k\geq 1}\frac{c_k}{z-\left(1+\dfrac{1}{k^2}\right)}$$

belongs to $N$ and admits the estimate $|\hat{f}(n)|<\exp(-n\varepsilon_n)$, $n\geq 1$.

The first three of these assertions follow easily from the theorem on cyclic vectors and its corollaries.

**2. Approximation by Rootspaces.** Among the non-cyclic vectors $f$ of the operator $S^*$ the simplest and most important ones are those which generate a finite dimensional invariant subspace $E_f^*=\bigvee(S^{*n}f:n\geq 0)$. Below we will verify that such vectors $f$ are rational functions ("Kronecker's theorem") and also that with the aid of the corresponding finite dimensional subspaces one can approximate any element in $\operatorname{Lat}S^*$. Moreover the elements of $N$ (the non-cyclic vectors) are distinguished by the fact that they admit approximation by rational functions the "capacity" of which (in a sense made precise below) is uniformly bounded. The corresponding theorem is intimately connected with the so-called spectral synthesis problem for $S^*$-invariant subspaces, otherwise put, with the construction of such subspaces with the aid of finite dimensional ones. Part of Lecture IV is devoted to a discussion of the spectral synthesis problem (and to a discussion from that point of view of the results set forth in the present Section).

Let us begin by describing the root spaces of the operator $S^*$. Recall that an *eigenvalue* of the operator $T$ is a number $\lambda$, $\lambda\in\mathbb{C}$, with the property that $\operatorname{Ker}(T-\lambda I)\neq\{\mathbb{O}\}$ and an *eigenvector* (or eigenfunction) any nonzero element of $\operatorname{Ker}(T-\lambda I)$. The *point spectrum* $\sigma_p(T)$ is the set of all eigenvalues.

**Beurling's Lemma.** *The point spectrum of the operator $S^*$ coincides with the disk $\mathbb{D}$. If $|\lambda| < 1$ then a normalized eigenfunction $x_\lambda$ ($\|x_\lambda\| = 1$) corresponding to the eigenvalue $\bar{\lambda}$ is provided by $x_\lambda \overset{\text{def}}{=} (1 - |\lambda|^2)^{1/2}(\mathbb{1} - \bar{\lambda}z)^{-1}$ and $\operatorname{Ker}(S^* - \bar{\lambda}I)^n$ consists of all rational functions with the only pole at the point $1/\bar{\lambda}$, of order not exceeding $n$.*

*Proof*. Identifying the Taylor coefficients of $S^*f$ and $\bar{\lambda}f$ we get $\hat{f}(n) = \bar{\lambda}^n\hat{f}(0)$, $n \geq 0$; this proves the first two assertions of the lemma. The rest follows from the following series of evident identities:

$$\operatorname{Ker}(S^* - \bar{\lambda}I)^n = H^2 \ominus (S - \lambda I)^n H^2,$$

$$(S - \lambda I)^n H^2 = \{f : f \in H^2, f(\lambda) = \cdots = f^{(n-1)}(\lambda) = 0\};$$

$$f(\lambda) = (f, (\mathbb{1} - \bar{\lambda}z)^{-1}),$$

$$f^{(j)}(\lambda) = (f, j!\, z^j(\mathbb{1} - \bar{\lambda}z)^{-j-1}), \quad 0 \leq j \leq n-1. \tag{2}$$

Thus $\operatorname{Ker}(S^* - \bar{\lambda}I)^n = \bigvee((\mathbb{1} - \bar{\lambda}z)^{-k}z^{k-1} : 1 \leq k \leq n)$. □

**Corollary.** $\operatorname{Ker}(S^* - \bar{\lambda}I)^n = (\mathbf{b}_\lambda^n H^2)^\perp$ *where*

$$b_\lambda(\zeta) = \frac{\lambda - \zeta}{1 - \bar{\lambda}\zeta}\frac{|\lambda|}{\lambda}, \quad \zeta \in \mathbb{D},$$

*for* $\lambda \in \mathbb{D}\backslash\{0\}$ *and* $b_0 = z$. □

It is now appropriate to recall (for details see Appendix 2) that every inner functions $\varphi$ has the canonical decomposition $\varphi = BV$ where $B$ is a *Blaschke product*, $B = \prod_{k(\lambda)>0} b_\lambda^{k(\lambda)}$, $\sum_{\lambda \in \mathbb{D}} k(\lambda)(1 - |\lambda|) < \infty$, and $V$ an inner function having no zeros in $\mathbb{D}$ (*singular function*). The Corollary of Beurling's Lemma shows that to the part $\operatorname{Ker}(S^* - \bar{\lambda}I)^n$ of the linear manifold of root vector belonging to the eigenvalue $\bar{\lambda}$ there corresponds in the canonical representation the simplest Blaschke product $b_\lambda^n$. Here is a generalization of this fact.

**Lemma on Finite Dimensional Subspaces.** *Let $\Theta$ be an inner function. Then* $\dim(\Theta H^2)^\perp < \infty \Leftrightarrow \Theta$ *is a finite Blaschke product.*

*Proof*. $\Leftarrow$. If $\Theta = \prod_\lambda b_\lambda^{k(\lambda)}$ is a finite Blaschke product then $\Theta H^2 = \bigcap_\lambda \mathbf{b}_\lambda^{k(\lambda)} H^2$ (Corollary 6 of Theorem *BH*, Lecture I) and

$$\dim(\Theta H^2)^\perp = \dim \bigvee((\mathbf{b}_\lambda^{k(\lambda)} H^2)^\perp : \lambda \in \mathbb{D}) < \infty.$$

$\Rightarrow$. By a fundamental theorem in linear algebra $(\Theta H^2)^\perp$ is generated by root vectors of the operators $S^*$; consequently $(\Theta H^2)^\perp = \bigvee(\operatorname{Ker}(S^* - \bar{\lambda}I)^{k(\lambda)} : \lambda)$ for some finite collection of points $\lambda$, $\lambda \in \mathbb{D}$, and suitable numbers $k(\lambda)$, $k(\lambda) > 1$, so by the Corollary of Beurling's Lemma and by Corollary 6 of Theorem *BH*, Lecture I $\Theta = \prod_\lambda b_\lambda^{k(\lambda)}$. □

Let us show two lemmata which will be useful not only in this Lecture but also in several of the following ones. Let us recall that by $P_E$ we denote the orthogonal projection onto the subspace $E$ and that $P_\Theta \overset{\text{def}}{=} P_{(\Theta H^2)^\perp}$. ($P_+$ is as before the Riesz projection.)

**Projection Lemma.** *Let $\Theta$ be an inner function. Then $P_\Theta = I - \Theta P_+ \bar{\Theta}$.*

*Proof*. Put $P'_\Theta = I - P_\Theta$ and let $f \in H^2$. By definition of orthogonal projection $P'_\Theta f$ is a function in $\Theta H^2$ such that $(P'_\Theta f, h) = (f, h)$ for all $h, h \in \Theta H^2$. Let us write $P'_\Theta f = \Theta g$, $h = \Theta k$ with $g, k \in H^2$. Then $(g, k) = (\Theta g, \Theta k) = (f, \Theta k) = (\bar{\Theta} f, k)$ for every $k, k \in H^2$. Consequently $g = P_+ \bar{\Theta} f$ and $P'_\Theta f = \Theta P_+ \bar{\Theta} f$. □

In an analogous manner we find that the orthogonal projection onto the subspace $H^2 \ominus \Theta H^2$ *in the space* $L^2$ is given by $P_+ - \Theta P_+ \bar{\Theta}$.

**Convergence Lemma.** *Let $\varphi_n \in H^\infty$, $n \geq 1$, $\sup_n \|\varphi_n\|_\infty < \infty$.*

1. *If $\varphi \in H^\infty$ and $\varphi(\lambda) = \lim_n \varphi_n(\lambda)$ $(\forall \lambda \in \mathbb{D})$ then* (s)-$\lim_n P_+ \bar{\varphi}_n = P_+ \bar{\varphi}$ *(strong (pointwise) convergence of operators in $H^2$).*
2. *If $\varphi \in H^\infty$ and $\lim_n \|\varphi - \varphi_n\|_2 = 0$ then* (s)-$\lim_n \varphi_n P_+ \bar{\varphi}_n = \varphi P_+ \bar{\varphi}$.

*Proof*. 1. It is clear that $\sup_n |P_+ \bar{\varphi}_n| \leq \sup_n \|\varphi_n\|_\infty < \infty$. By the Banach-Steinhaus theorem it suffices therefore to consider the behavior of $P_+ \bar{\varphi}_n f$ on a suitable total subset of $H^2$. It is convenient to take $f = (1 - \bar{\lambda} z)^{-1}$, $|\lambda| < 1$, because then $P_+ \bar{\varphi}_n f = \overline{\varphi_n(\lambda)} f$ (this follows readily if we first notice that $P_+ \bar{z}^n f = \bar{\lambda}^n f$, $n \geq 0$) and the desired convergence ($\lim_n \|P_+ \bar{\varphi}_n f - P_+ \bar{\varphi} f\|_2 = 0$) is clear.

2. If $f \in H^2$ then

$$\|\varphi_n P_+ \bar{\varphi}_n f - \varphi P_+ \bar{\varphi} f\|_2 < \|\varphi_n P_+ (\bar{\varphi}_n - \bar{\varphi}) f\|_2$$
$$+ \|(\varphi_n - \varphi) P_+ \bar{\varphi} f\|_2 \leq (\sup_{n \geq 1} \|\varphi_n\|_\infty) \cdot \left( \int_{\mathbb{T}} |\varphi_n - \varphi|^2 |f|^2 \, dm \right)^{1/2}$$
$$+ \left( \int_{\mathbb{T}} |\varphi_n - \varphi|^2 |P_+ \bar{\varphi} f|^2 \, dm \right)^{1/2}.$$

Both integrals tend to zero (as $n \to \infty$), because the sequence $\{\varphi_n\}_{n \geq 1}$ is uniformly bounded and tends to the function $\varphi$ in measure. □

Our new (approximation theoretic) description of the $S^*$-invariant subspaces uses the following

**Definition.** Let $X$ be a normed space and $\{E_n\}_{n \geq 1}$ a sequence of subspaces. The *lower limit* of this sequence is the set $\underline{\lim}_n E_n$,

$$\underline{\lim}_n E_n \stackrel{\text{def}}{=} \left\{ x : x \in X,\ \lim_n \operatorname{dist}(x, E_n) = 0 \right\}$$
$$= \left\{ x : x \in X,\ \exists x_n (x_n \in E_n) : \lim_n x_n = x \right\}.$$

It is clear that $\underline{\lim}_n E_n$ is a (closed) subspace of $X$ and if all the $E_n$ are invariant for a given operator $T$ then so will be the lower limit $\underline{\lim}_n E_n$. Let us describe the lower limits of $S^*$-invariant subspaces.

**Theorem on Lower Limits.** *Let $\Theta_n$ be inner functions $(n = 1, 2, \ldots)$, $(\Theta_n H^2)^\perp$ the* *corresponding subspaces and* $\{\Theta_n\}'$ *the set of all limit points of the sequence* $\{\Theta_n\}_{n\geq 1}$ *in the weak topology of the space* $H^2$. *Then*

$$\underline{\lim}_n\,(\Theta_n H^2)^\perp = (\Theta H^2)^\perp$$

*where* $\Theta = \mathrm{GCD}(\varphi^i : \varphi\in\{\Theta_n\}')$; *here* $\varphi^i$ *denotes the inner factor of* $\varphi$, $\mathbb{0}^i = \mathbb{0}$ *and* $\mathrm{GCD}(\mathbb{0},\varphi^i) = \varphi^i$, $\forall\varphi^i$.

*Proof*. Let us first check that

$$\underline{\lim}_n\,(\Theta_n H^2)^\perp = \bigcap \underline{\lim}_k\,(\Theta_{n_k} H^2)^\perp,$$

where the intersection is taken over all weakly convergent subsequences $\{\Theta_{n_k}\}_{k\geq 1}$ of the sequence $\{\Theta_n\}_{n\geq 1}$. From the definition it is clear that left hand side of this identity is contained in the right hand side. Let us prove the reverse inclusion.

Let $f\in\bigcap\underline{\lim}_k\,(\Theta_{n_k}H^2)^\perp$. If now $f\notin\underline{\lim}_n\,(\Theta_n H^2)^\perp$ then there exists a $\delta$, $\delta>0$, and a sequence of natural numbers $\{n_k\}_{k\geq 1}$, $n_k<n_{k+1}$, such that $\mathrm{dist}(f,(\Theta_{n_k}H^2)^\perp)\geq\delta$, $k\geq 1$. Pick a weakly convergent subsequence $\{\Theta_{n_{k_i}}\}_{i\geq 1}$. Then $\mathrm{dist}(f,(\Theta_{n_{k_i}}H^2)^\perp)\geq\delta$, $i\geq 1$, so $f\notin\underline{\lim}_i\,(\Theta_{n_{k_i}}H^2)^\perp$, in contradiction with the initial hypothesis.

It remains to show that

$$\underline{\lim}_n\,(\varphi_n H^2)^\perp = (\varphi^i H^2)^\perp$$

if $\{\varphi_n\}_{n\geq 1}$ is a sequence of inner functions weakly convergent to the function $\varphi$. Then $\varphi(\lambda) = \lim_n\varphi_n(\lambda)$, $\lambda\in\mathbb{D}$, so that $\varphi\in H^\infty$, $\|\varphi\|_\infty\leq 1$. Moreover

$$\mathrm{dist}(f,(\varphi_n H^2)^\perp) = \|f - P_{\varphi_n}f\|_2 = \|P_+\bar\varphi_n f\|_2$$

for every function $f, f\in H^2$. By the Convergence Lemma $\underline{\lim}_n\,P_+\bar\varphi_n f = P_+\bar\varphi f$ and we have to verify that

$$P_+\bar\varphi f = \mathbb{0} \Leftrightarrow f\in(\varphi^i H^2)^\perp.$$

By the definition of the Riesz projection

$$(\varphi^i H^2)^\perp = \mathrm{Ker}\,P_+\,\bar\varphi^i \subset \mathrm{Ker}\,P_+\,\bar\varphi^i\bar\varphi^e = \mathrm{Ker}\,P_+\,\bar\varphi.$$

Conversely if $f\in H^2$, $P_+\bar\varphi f = \mathbb{0}$, then $f\perp\varphi\mathscr{P}_A$ and therefore $f\perp\mathrm{clos}\,\varphi\mathscr{P}_A = \varphi^i H^2$, that is, we have $f\in(\varphi^i H^2)^\perp$. □

*Remark*. The inclusion $P_+\bar\varphi(\Theta H^2)^\perp\subset(\Theta H^2)^\perp$ with $\varphi$ and $\Theta$ in $H^\infty$ is essentially contained in the preceding reasoning.

**Corollary.** *If* $\Theta_n$ *are inner functions* ($n = 1, 2, \ldots$) *then*

$$\underline{\lim}_n\,(\Theta_n H^2)^\perp = H^2 \Leftrightarrow \lim_n \Theta_n(\lambda) = 0\ (\forall\lambda\in\mathbb{D}).$$

Indeed, it suffices only to recall once more a fact which we have already used:

$$\Theta_n \xrightarrow{\text{weak}} \Theta \Leftrightarrow \lim_n \Theta_n(\lambda) = \Theta(\lambda)(\forall\lambda\in\mathbb{D}). \ \square$$

This theorem and its corollary allow us to approximate any $S^*$-invariant subspace by finite dimensional ones. At the same time we get a new description of $S^*$-cyclic vectors. We need however also the following result due to Schur.

**Schur's Theorem.** *If $\varphi \in H^\infty$, $\|\varphi\|_\infty \le 1$, then there exists a sequence $\{B_n\}_{n\ge 1}$ of finite Blaschke products which converges uniformly to $\varphi$ on compact subsets of $D$.*

*Proof*. If $f$ is any analytic function let us agree to write $f^t(z) \overset{\text{def}}{=} \overline{f(\bar{z})}$. Let $n \in \mathbb{N}$ be given and pick a partial sum (of sufficiently high order) $p$, say, of the Taylor series of the function $\varphi(rz)$ where $r$ is sufficiently close to 1, $0 \le r < 1$. Then

$$|\varphi(\zeta) - p(\zeta)| < \frac{1}{n+1} \quad \text{for} \quad |\zeta| \le 1 - \frac{1}{n}$$

and $\|p\|_\infty \le 1$. Let us introduce a number $N$ so large that the product $z^N p^t(1/z)$ will be a polynomial in $z$ and so that

$$|\zeta^N| < \frac{1}{n+1}, \quad |\zeta^N p^t(1/\zeta)| < \frac{1}{n+1} \quad \text{for} \quad |\zeta| \le 1 - \frac{1}{n}.$$

Set

$$B_n = (p + z^N)(1 + z^N p^t(1/z))^{-1}.$$

This is a rational function regular in $\mathbb{D}$ (because $|\zeta^N p^t(1/\zeta)| < \|z^N p^t(1/z)\|_\infty \le \|p\|_\infty \le 1$ by the maximum principle) and on $\mathbb{T}$ we have

$$|B_n| = |p + z^N| \cdot |1 + z^N \overline{p(z)}|^{-1} = |\bar{z}^N p + 1| \cdot |1 + z^N \bar{p}|^{-1} = 1$$

outside the zeros of the denominator (and the latter are only at most finite in number). Consequently $B_n$ is a finite Blaschke product. Finally if $|\zeta| \le 1 - 1/n$ then

$$\begin{aligned} |\varphi(\zeta) - B_n(\zeta)| &< \frac{1}{n+1} + |p(\zeta) - B_n(\zeta)| \\ &= \frac{1}{n+1} + |\zeta^N p^t(1/\zeta) p(\zeta) - \zeta^N| \cdot |1 + \zeta^N p^t(1/\zeta)|^{-1} \\ &< \frac{1}{n+1} + \frac{\dfrac{1}{n+1} + \dfrac{1}{n+1}}{1 - \dfrac{1}{n+1}} < \frac{3}{n}. \ \square \end{aligned}$$

**Definition.** Let $f$ be a rational function all of whose poles belong to the set $\hat{\mathbb{C}} \backslash \operatorname{clos} \mathbb{D}$. By the *capacity* of $f$ we mean the number

$$\operatorname{cap}(f) \overset{\text{def}}{=} \sum k_f(\lambda)(1 - 1/|\lambda|)$$

where the sum is taken over all poles $\lambda$ of $f$, $k_f(\lambda)$ denoting the multiplicity of the pole at the point $\lambda$.

**Theorem on Finite Dimensional Approximation.** 1. *If* $E \in \operatorname{Lat} S^*$ *then there exist finite Blaschke products* $B_n$, $n \geq 1$, *such that*

$$E = \underline{\lim_n} (\mathbf{B}_n H^2)^\perp.$$

2. *If* $B_n$, $n \geq 1$, *are finite Blaschke products then*

$$\underline{\lim_n} (\mathbf{B}_n H^2)^\perp \neq H^2 \Leftrightarrow \underline{\lim_n} \operatorname{cap}(B_n) < \infty$$

3. $f \in N \Leftrightarrow f \in H^2$ *and there exists a sequence* $\{f_n\}_{n \geq 1}$ *of rational functions with the poles contained in* $\hat{\mathbb{C}} \backslash \operatorname{clos} \mathbb{D}$ *such that* $\underline{\lim}_n \| f - f_n \|_2 = 0$, $\sup_{n \geq 1} \operatorname{cap}(f_n) < \infty$.

*Proof*. The Statements 1 and 2 follow from the description of Lat $S^*$, Schur's Theorem, the Theorem on Lower Limits and the latter's corollary; it suffices just to remark that for the Blaschke products $B_n$ holds

$$\lim_n B_n(\lambda) = 0, \quad \forall \lambda \in \mathbb{D} \Leftrightarrow \lim_n \left| \frac{B_n}{b_{\lambda_1} \cdots b_{\lambda_N}}(0) \right| = 0, \quad \forall N \geq 1$$

$$\Leftrightarrow \lim_n \operatorname{cap}(B_n) = \infty,$$

where $\lambda_1, \ldots, \lambda_N$ stand for the first $N$ zeros of $B_n$. Statement 3 follows from the Statements 1 (take $E = E_f^*$) and 2, from the Lemma on Finite Dimensional Subspaces and from the identity $\operatorname{cap}(f_n) = \operatorname{cap}(B_n)$, where $B_n$ is the Blaschke product representing the subspace $E_{f_n}^*$ ($E_{f_n}^* = H^2 \ominus \mathbf{B}_n H^2$). □

**Corollary.** $N$ *is an* $F_\sigma$ *set of the first category* (*that is, a countable union of nowhere dense subsets*) *in* $H^2$.

In fact, the theorem gives the identity $N = \bigcup_{n \geq 1} \operatorname{clos}_{H^2} R_n$ where $R_n$ is the set of all rational functions $f$ with poles in $\hat{\mathbb{C}} \backslash \operatorname{clos} \mathbb{D}$ and such that $\operatorname{cap}(f) \leq n$. Each of the sets clos $R_n$ is closed and nowhere dense, because they all do not intersect the everywhere dense subset $C$. □

Concluding this Lecture let us remark that *every $S^*$-invariant subspace $E$ admits a generating vector*: $E = E_f^*$ for some $f, f \in H^2$. We get such a vector by setting, for instance, $f = S^*\Theta$ if $E = (\Theta H^2)^\perp$ is the canonical representation of $E$. Indeed, $(S^*\Theta, \Theta h) = (\mathbb{1}, zh) = 0$ for every $h$, $h \in H^2$, and therefore $f \in E$; on the other hand, the relation $0 = (S^{*n}\Theta, h) = (\bar{z}^n, \bar{\Theta} h)$, $n \geq 1$, implies that $\bar{\Theta} h \in H^2$ so that if such an $h$ belongs to $E$ it must necessarily vanish. Thus $E = \bigvee (S^{*n} f : n \geq 0)$. □

## Supplements and Bibliographical Notes

First we take up a general analysis of the new description of $S^*$-invariant subspaces obtained in this Lecture. Then we give multidimensional analogues (for the operator $S_E^*$, $\dim E > 1$) of some of the results of Sects. 1 and 2.

**3. More General Capacities** appear in the study of the shift operator in spaces of analytic functions other than $H^2$. Let us consider a Banach space $X$ and an operator $T$ acting in $X$ and possessing a total (in $X$) collection of *root vectors* $x$ (that is, elements $x$ of the set $\bigcup_{\lambda, n \geq 1} \operatorname{Ker}(T - \lambda I)^n$):

$$\bigvee(\operatorname{Ker}(T - \lambda I)^n : \lambda \in \sigma_p(T), n \geq 1) = X$$

(such operators will be called *complete*). In this section we always assume that $\dim \operatorname{Ker}(T - \lambda I) = 1$, $\lambda \in \sigma_p(T)$. If $x' \in X^*$ and $\lambda \in \sigma_p(T)$ we set

$$k_{x'}(\lambda) \stackrel{\text{def}}{=} \sup\{n : \operatorname{Ker}(T - \lambda I)^n \subset \operatorname{Ker} x' : n \geq 0\}.$$

Then for any integer valued function (divisor) $k$ defined on $\sigma_p(T)$, choosing some element $x$, $x \in X$, not contained in $\bigvee(\operatorname{Ker}(T - \lambda I)^{k(\lambda)} : \lambda \in \sigma_p(T))$, we can define the *T-capacity* of $k$:

$$\operatorname{cap}_x k = \inf\{\|x'\| : k_{x'} \geq k, |\langle x, x' \rangle| \geq 1\};$$

here and in the sequel the form $\langle .\,, . \rangle$ realizes the duality between the space $X$ and its dual $X^*$. The following example is a basic but not the only illustration of the situation just described: $X$ is a space of functions analytic in $\mathbb{D}$ containing the rational fractions $(\mathbb{1} - \lambda z)^{-1}$, $\lambda \in \mathbb{D}$, as a total subset; it is assumed that the dependence of this fraction on the parameter $(\lambda \mapsto (\mathbb{1} - \lambda z)^{-1})$ is analytic (in the metric $X$). Let further the left shift $S^* : f \mapsto \dfrac{f - f(0)}{z}$ act continuously in $X$. In that case the dual space $X^*$ can in a natural way be realized as a space of analytic functions in $\mathbb{D}$: if $\varphi \in X^*$ we set

$$\Phi(\lambda) = \langle (\mathbb{1} - \lambda z)^{-1}, \varphi \rangle, \quad \lambda \in \mathbb{D}.$$

Then $\operatorname{Ker}(S^* - \lambda I)^n = \bigvee((\mathbb{1} - \lambda z)^{-k} z^{k-1} : 1 \leq k \leq n)$ and that this subspace is orthogonal to the functional $\Phi$ means that $\Phi^{(k)}(\lambda) = 0$, $0 \leq k < n$, so that $k_\Phi$ is a divisor associated with $\Phi$, counting the multiplicity of its zeros. As "supporting" element (in the definition of the $T$-capacity) we can take one of the eigenfunctions $x = (\mathbb{1} - \lambda z)^{-1} (\lambda \in \mathbb{D})$ but, better, any cyclic vector (or family such vectors) for the operator $S^*$. In order to get a better possibility to compare with the previous results it is convenient to set

$$\operatorname{cap} k = \inf\{n + \|\Phi\|_{X^*} : k_\Phi(\lambda) \geq k(\lambda), \lambda \in \mathbb{D}; |\hat{\Phi}(n)| \geq 1, n \geq 1\}. \tag{3}$$

Similarly also in the case of a general operator $T$: if $\{x_n : n \geq 1\}$ is a total family in $X$ (or an arbitrary $T$-cyclic family, i.e. a family such that $\bigvee(T^k x_n : k \geq 0, n \geq 1) = X$) then by definition

$$\operatorname{cap} k = \inf\{n + \|x'\| : k_{x'}(\lambda) \geq k(\lambda), \lambda \in \sigma_p(T), |\langle x_n, x' \rangle| \geq 1\}, \tag{4}$$

and if $E \in \operatorname{Lat} T$ then

$$\operatorname{cap} E \stackrel{\text{def}}{=} \operatorname{cap} k_E,$$

$$k_E(\lambda) \stackrel{\text{def}}{=} \sup \dim(\operatorname{Ker}(T - \lambda I)^n \cap E) \; (\lambda \in \sigma_p(T))$$

and finally $\operatorname{cap} x = \operatorname{cap} E_x^T$, $E_x^T = \bigvee(T^n x : n \geq 0)$, $x \in X$. The following abstract version of the Theorem on Finite Dimensional Approximation is proved in the paper by Gribov and Nikol'skiĭ [1].

**Theorem on Finite Dimensional Approximation, II.** *Let $T$ be a complete operator in a Banach space $X$.*

1. *If $E_m \in \operatorname{Lat} T$, $\dim E_m < \infty$ $(m = 1, 2, \ldots)$ then*

$$\underline{\lim}\, E_m \neq X \Leftrightarrow \underline{\lim}_m \operatorname{cap} E_m < \infty.$$

2. *If $y_m$, $m \geq 1$, are linear combinations of root vectors of $T$, $y = \lim_m y$ and $\sup_m \operatorname{cap} y_m < \infty$ then $E_m \neq X$, i.e. $y$ is a non-cyclic vector for $T$.*

Sometimes it can be convenient to replace this characteristic of the massivity of $T$-invariant subspaces by a "cruder" one, putting

$$\operatorname{Cap} k = \inf\{n + |\varphi(T)| : k_\varphi \geq k,\ \|\varphi(T)x_n\| \geq 1\},$$

$$\operatorname{Cap} E = \operatorname{Cap} k_E, \quad \operatorname{Cap} f = \operatorname{Cap} E_f^T;$$

in this definition the inf is taken over all $n$, $n = 1, 2, \ldots$, and all functions $\varphi$ such that $\varphi(T)$ "makes sense". The last words can be given various meaning ($\varphi$ can run through the rational functions with poles outside the spectrum $\sigma(T)$, the closure of these functions in some operator topology, the uniform or the weak etc.). If $E \in \operatorname{Lat} T$, $\dim E < \infty$, then from $k_\varphi \geq k_E$ it follows that $\varphi(T)E = \{\mathbb{0}\}$ and therefore

$$\operatorname{dist}(x_n, E)|\varphi(T)| \geq \operatorname{dist}(\varphi(T)x_n, \varphi(T)E) = \|\varphi(T)x_n\| \geq 1$$

so that $\operatorname{cap} E \leq \operatorname{Cap} E$. Keeping the notation of the preceding theorem we get

**Corollary 1.** *If $\underline{\lim}_m \operatorname{Cap} E_m < \infty$ then $\underline{\lim}_m E_m \neq X$.* □

The capacity Cap gives the possibility to employ here *functional calculi* for the operator $T$. By a calculus we mean a continuous homomorphism $\varphi \mapsto \varphi(T)$ of some topological function algebra $A$ containing the polynomials (usually as a dense subset) such that $\mathbb{1} \mapsto I$, $z \mapsto T$. For example, any operator $T$ admits a calculus of class $l_A^1(w_n) = \{\varphi : \varphi = \sum_{n\geq 0} \hat{\varphi}(n)z^n, \sum_{n \geq 0} |\hat{\varphi}(n)| w_n \stackrel{\text{def}}{=} \|\varphi\|_{l_A^1(w_n)} < \infty\}$, $w_n = |T^n|$, $n \geq 0$, with the natural definition of the action:

$$\varphi(T) = \sum_{n \geq 0} \hat{\varphi}(n) T^n, \quad \varphi \in l_A^1(w_n).$$

From the Dilation Theorem it follows readily that general contractions in Hilbert space always admit a $C_A$-calculus (where $C_A$ is the algebra of functions continuous in clos $\mathbb{D}$ and analytic in the disk $\mathbb{D}$) while completely non-unitary ones (i.e. contractions without a unitary part) admit even an $H^\infty$-calculus (regarding these calculi see Lecture III). If $\mathfrak{A}$ is one of these algebras then $|\varphi(T)| \leq \|\varphi\|_{\mathfrak{A}}$, $\varphi \in \mathfrak{A}$, and, for operators $T$ such that $\bigvee(\operatorname{Ker}(T - \lambda I) : \lambda_n \in \sigma_p(T)) = X$, we can replace in the statement of the above corollary the capacity Cap by $\inf\{n + \|\varphi\|_{\mathfrak{A}} : k_\varphi \geq k$, $|\varphi(\lambda_n)| \geq 1\}$. Finally, the last step in this series of simplifications is connected with the uniqueness theorems for analytic functions: if $\sigma_p(T)$ is a uniqueness set for the algebra $\mathfrak{A}$ ($\varphi \in \mathfrak{A}$, $\varphi | \sigma_p(T) = \mathbb{O} \Rightarrow \varphi = \mathbb{O}$) then one can do with an even cruder capacity $\mathrm{Cap}_{\mathfrak{A}}$, having the advantage of not depending on the operator:

$$\mathrm{Cap}_{\mathfrak{A}} k \stackrel{\text{def}}{=} \inf\{n + \|\varphi\|_{\mathfrak{A}} : k_\varphi \geq k, |\hat{\varphi}(n)| \geq 1\}. \tag{5}$$

In the following corollary we assume for simplicity that $\mathfrak{A}$ is either $H^\infty$ or $l_A^1(w_n)$ ($w_n = |T^n|$, $n \geq 0$) although its proof is based on only a small number of simple properties of $\mathfrak{A}$, which allows us to apply it to many other similar algebras.

**Corollary 2.** *Assume that the operator $T$ has a complete set of eigenspaces in $X$, that $\sigma_p(T) \subset \mathbb{D}$ and is a uniqueness set for $\mathfrak{A}$, and finally that $T$ admits an $\mathfrak{A}$-calculus. Then*

$$\varliminf_m \mathrm{Cap}_{\mathfrak{A}} E_m < \infty \Rightarrow \varliminf_m E_m \neq X.$$

The definitions (3), (4) and (5) indicate[1] that the capacities in question may serve as a quantitative measure for the nearness of the set (divisor) to being a set (divisor) of uniqueness in the spaces $\mathfrak{A}$ and $X^*$ respectively and that the "right" estimates for these capacities should be equivalent to uniqueness theorems in the corresponding classes of analytic functions. In fact this is the case and one can show, for instance (cf. Gribov-Nikol'skiĭ [1]), that

$$\mathrm{Cap}_{H^\infty} k \approx \sum_{\lambda \in \mathbb{D}} k(\lambda)(1 - |\lambda|),$$

$$\mathrm{Cap}_{l_A^1(w_n)} k \approx -\int_{\mathbb{T}} \log\left(\min_{k(\lambda)>0} |\zeta - \lambda|\right) dm(\zeta) + \sum_{\lambda \in \mathbb{D}} k(\lambda)(1 - |\lambda|),$$

where $cn \leq w_n \leq cn^\alpha$ ($n \geq 1$) for some $c$ and $\alpha > 0$, and the sign $\approx$ means that two quantities connected with it either both tend to infinity, as $k$ varies, or not. The first of these relations readily follows from classical facts on the spaces $H^p$ (see Appendix 2) while the second one can be deduced from comparatively new results on uniqueness of analytic functions that are smooth up to the boundary.

**4. The Operator $S_E^*$.** In this Section we will study cyclic subspaces of the operator $S_E^*$ and derive multidimensional analogues of some of the auxiliary results of Sects. 1 and 2. First of all let us remark that what concerns cyclicity $S_E^*$ differs in an essential way from its adjoint $S_E$ and this is seen already in the scalar case $\dim E = 1$ (which we have investigated): whereas for the operator $S_E$ in order that a function $f = f^i \cdot f^e$, $f \in H^2$, chosen "at random", should be cyclic we must impose the supplementary property $f^i = 1$, for the operator $S_E^*$ on the contrary such a function must belong to a very special class (functions admitting pseudo-continuation) in order to be *non*-cyclic; there are in general rather "few" functions of the latter type in $H^2$, as is seen from the Corollary of the Theorem on Finite Dimensional Approximation. A similar distinction prevails also for $\dim E > 1$;

[1] Let us recall that (3) refers to the most interesting case, viz. the one of the shift operator $T = S^*$.

for example, $S_E^*$ is by necessity *cyclic*, i.e. has a cyclic vector (of course in the hypothesis of the separability of the space $E$; notice that we always assume that the spaces encountered are separable, unless such an assumption leads to a contradiction), although an analogous conclusion for the restrictions $S_E^*|K$ to its invariant subspaces (as was shown in Sect. 2 for $\dim E = 1$) in general is not true. Later on (Lecture IV and Appendix 1) we will subject this question to a much deeper analysis but presently we show only the following simple result.

**Cyclicity Theorem for $S_E^*$.** *Let $E$ be a separable Hilbert space.*

1. *The cyclic vectors of $S_E^*$ form a dense subset of $H^2(E)$.*
2. *Let $K \in \operatorname{Lat} S_E^*$. The operator $S_E^*|K$ has a generating subspace of dimension not greater than and, in general, also not less than* $\dim E$.

*Proof* 1. Let $\{e_k\}_{k\geq 1}$ be a sequence of vectors which is dense in the unit ball of $E$. Let

$$f = \sum_{k\geq 0} c_k z^{n(k)} e_k$$

where we assume that $n(k+1) - n(k) > 0$ $(k \geq 0)$, $\lim_k (n(k+1) - n(k)) = \infty$,

$$\lim_k \frac{1}{|c_k|^2} \sum_{i\geq k+1} |c_i|^2 = 0.$$

Then $E_f^* \overset{\text{def}}{=} \bigvee(S_E^{*n} f : n \geq 0) = H^2(E)$. In fact, it suffices to show that $z^m e \in E_f^*$ for all $m$, $m \geq 0$, and $e$, $e \in E$; but it is clear that for $n(k) - n(k-1) > m$ and $n = n(k) - m$

$$\left\| z^m e - \frac{1}{c_k} S^{*n} f \right\|_{H^2(E)}^2 = \|e - e_k\|_E^2 + \frac{1}{|c_k|^2} \sum_{i\geq k+1} |c_i|^2 \|e_i\|_E^2,$$

which gives the desired result.

2. This statement will be divided into two parts in the first of which we prove slightly more than written in the formulation. By the Lax-Halmos Theorem

$$K = H^2(E) \ominus \Theta H^2(E')$$

where $E' \subset E$, $\Theta \in H^\infty(L(E', E))$, $\Theta$ a left inner function (i.e. the values $\Theta(\zeta)$ are isometric for a.e. $\zeta$, $\zeta \in \mathbb{T}$).

**Lemma 1.** *If* $\dim E' = \dim E < \infty$ *then* $\mu_{S_E^*|K} \leq \dim E$ *and as a generating subset we can take the subspace* $L \overset{\text{def}}{=} S_E^*(\Theta E')$.

2. *If* $\dim E' < \dim E$ *then* $\mu_{S_E^*|K} \leq \dim E' + 1$.

*Proof of the Lemma.* 1. In this case $\Theta$ is an inner function and for convenience we can, as always, assume that $E' = E$ and that this subspace is imbedded in $H^2(E)$ in the standard way $(e \mapsto e \cdot \mathbb{1})$. If $e \in E$ then

$$(S_E^* \Theta e, \Theta f) = (\Theta e, z\Theta f) = (e\mathbb{1}, zf) = 0 \tag{6}$$

for all $f$, $f \in H^2(E)$, so that $L \subset K$. If now $f \in K$ and $f \perp S_E^{*n} L$, $n \geq 0$, then for every $e$, $e \in E$,

$$0 = (\bar{z}^{n+1}\Theta e, f) = (\bar{z}^{n+1}e, \Theta^* f), \quad n \geq 0,$$

i.e. $\Theta^* f \in H^2(E)$ and $f \in \Theta H^2(E)$. This means that indeed $\bigvee(S_E^{*n}L : n \geq 0) = K$.

2. Let $L$ be the same subspace as in the formulation of Statement 1. From (6) it follows again that $L \subset K$, whereas the orthogonality $f \perp S_E^{*n}L$, $n \geq 0$, entails that $(\bar{z}^{n+1}\Theta e, f) = 0, \forall n \geq 0, \forall e \in E'$, i.e. $\Theta H_-^2(E') \perp f$; here $H_-^2(E) \overset{\text{def}}{=} L^2(E) \ominus H^2(E)$. If in addition $f \in K$ then $f \perp \Theta H^2(E')$ and therefore $f \perp \Theta L^2(E')$. It is clear that also conversely $f \in H^2(E)$, $f \perp \Theta L^2(E')$, implies $f \in K$, $f \perp S^{*n}L$, $n \geq 0$. Thus if $M \overset{\text{def}}{=} \bigvee(S_E^{*n}L : n \geq 0)$ then

$$K \ominus M = H^2(E) \cap (L^2(E) \ominus \Theta L^2(E')).$$

This last subspace (let us denote it by $M'$) is trivially invariant for the shift $S_E$ and as $M' \subset H^2(E)$ then by the Corollary of the Wold-Kolmogorov Lemma (Lecture I) the restriction $S_E|M'$ must be unitarily equivalent to the operator $S_D$, $D = M' \ominus S_E M'$. The part of the theorem already established guarantees that $(S_E|M')^* = P_{M'}S_E^*|M'$ is cyclic. To complete the proof of the lemma it suffices to apply the following general auxiliary proposition, putting in it $X = K$, $T = S_E^*|K$ and noticing that then $X/M = K \ominus M$ and $T/M = P_{M'}S_E^*|M'$ (more precisely, the corollary of the Sublemma which we employ here amounts to the following: if $K$ is a Hilbert space, $T \in L(K)$ and $M \in \operatorname{Lat} T$ then the spectral multiplicity of the operator $P_{M^\perp}T|M$ does not exceed $\mu_T$).

**Sublemma.** *Let $X$ be a Banach space and $T \in L(X)$. Let $M \in \operatorname{Lat} T$ and consider the corresponding factor operator $T/M$ acting in the factor space $X/M$. Then*

$$\mu_{T/M} \leq \mu_T \leq \mu_{T|M} + \mu_{T/M}. \tag{7}$$

*Moreover the canonical projection of any $T$-cyclic set is $T/M$-cyclic in $X/M$.*

*The proof* of the sublemma will be based on the Hahn-Banach theorem, telling us that there is plenty of linear functionals. For the sake of brevity let us set $\dot{T} = T/M$ and $\dot{X} = X/M$. Consider $C \subset M$ with $\dim C = \mu_{T|M}$, $\bigvee(T^n C : n \geq 0) = M$ and $C' \subset X$ with $\dim C' = \dim \dot{C}' = \mu_{\dot{T}}$, $\bigvee(\dot{T}^n \dot{C}' : n \geq 0) = \dot{X}$. If $f$ is a continuous linear functional in $X$ and $T^n C \subset \operatorname{Ker} f$, $n \geq 0$, then $M \subset \operatorname{Ker} f$ so that $f$ defines a continuous linear functional $\dot{f}$ on $\dot{X}$. But since $\langle \dot{T}^n \dot{C}', \dot{f} \rangle = \langle T^n C', f \rangle$, $n \geq 0$, then from $T^n C' \subset \operatorname{Ker} f$, $n \geq 0$, and by the choice of $C'$ we obtain $\dot{f} = \mathbb{0}$ and therefore $f = \mathbb{0}$. This gives

$$X = \bigvee(T^n(C + C') : n \geq 0), \quad \mu_T \leq \dim C + \dim C'$$
$$= \mu_{T|M} + \mu_{\dot{T}}.$$

What concerns the first inequality of (7) we have essentially already proved it: if $C \subset X$, $\mu_T = \dim C$ and $\bigvee(T^n C : n \geq 0) = X$, then it is clear that $(T^n C)^{\cdot} = \dot{T}^n \dot{C}$ and $\bigvee(\dot{T}^n \dot{C} : n \geq 0) = \dot{X}$. Thus $\mu_{T/M} \leq \dim \dot{C} \leq \dim C = \mu_T$. □□

Continuing the proof of the second statement of the Cyclicity Theorem for $S_E^*$ we have to display an example which confirms the exactness of the estimate

$\mu_{S_E^*|K} \leq \dim E$ already established. To this end it suffices to set $K = H^2(E) \ominus \mathbf{z}H^2(E)$, because then $S_E^* K = \{\mathbb{0}\}$ and $\dim K = \dim E$. □

Other more "full-fledged" examples of the computation of the spectral multiplicity $\mu_{S_E^*|K}$ for the restrictions of the operator $S_E^*$ will be given in Appendix 1. Presently we turn to the characterization of generating subspaces of $S_E^*$ itself. First however we need the formula for the projection onto an $S_E^*$ invariant subspace.

**Projection Lemma, II.** *Let E and E' be Hilbert spaces and* $\Theta$ *a left inner function with values in* $L(E', E)$. *Then the orthogonal projection*

$$P_\Theta \overset{\text{def}}{=} P_{K_\Theta}, \quad K_\Theta \overset{\text{def}}{=} H^2(E) \ominus \Theta H^2(E'),$$

*is given by the formula*

$$P_\Theta = I - \Theta P_+ \Theta^*.$$

*Proof*. As in the case $\dim E = 1$ the operator $P' \overset{\text{def}}{=} \Theta P_+ \Theta^*$ has the following obvious properties which together imply that $I - P'$ is the orthogonal projection (in $H^2(E)$) onto $K$: *a)* $P'f \in \Theta H^2(E')$, $\forall f$; b) $P'f = f$, $\forall f \in \Theta H^2(E')$; c) $P' \geq \mathbb{0}$. □

**Corollary.** $S_E^*|K \in C_{00} \Leftrightarrow \Theta$ *is an inner function. Therefore a linear contraction T in Hilbert space is unitarily equivalent to the model operator* $S_E^*|K$ *with an inner characteristic function* $\Theta = \Theta_T$ *iff* $T \in C_{00}$.

In proving this corollary we first recall that $T \in C_{00} \Leftrightarrow ((s)\text{-}lim_n\, T^n = \mathbb{0}$, $(s)\text{-}\lim_n T^{*n} = \mathbb{0})$ (by definition) and that if $(s)\text{-}\lim_n T^n = \mathbb{0}$ (and only in this case) the operator $T$ will be equivalent to a model operator $S_E^*|K$ with a left inner function $\Theta$ (Model Theorem, II, Lecture I). Therefore it remains to be shown that

$$(s)\text{-}\lim_n (S_E^*|K_\Theta)^{*n} = \mathbb{0} \Leftrightarrow (\Theta \text{ is an inner function}).$$

But $(S_E^*|K_\Theta)^{*n} = P_\Theta S_E^n|K_\Theta$, $n \geq 0$ (trivial computation, which besides will be given in detail in the main body of Lecture III) and

$$\lim_n P_\Theta S_E^n f = \mathbb{0}, \quad \forall f \in K_\Theta \Leftrightarrow \lim_n P_\Theta S_E^n f = \mathbb{0}, \quad \forall f \in H^2(E).$$

Further

$$\begin{aligned} \|P_\Theta S_E^n f\|_{H^2(E)}^2 &= \|(I - \Theta P_+ \Theta^*) S^n f\|_{H^2(E)}^2 = \|f\|_{H^2(E)}^2 - \|P_+ \Theta^* S_E^n f\|_{H^2(E')}^2 \\ &= \|f\|_{H^2(E)}^2 - \|\Theta^* S_E^n f\|_{L^2(E')}^2 + \|(I - P_+)\Theta^* S^n f\|_{L^2(E')}^2, \end{aligned}$$

where the last term tends to zero (as $n \to \infty$) for any $\Theta$ and $f$, $f \in H^2(E)$. Hence

$$\lim_n \|P_\Theta S^n f\|_{H^2(E)}^2 = \|f\|_{H^2(E)}^2 - \|\Theta^* f\|_{L^2(E')}^2,$$

The right hand side of this identity is zero for all $f$ in $H^2(E)$ iff $\Theta^*$ is an isometric operator from $H^2(E)$ into $L^2(E')$ but this again is equivalent to the values $\Theta(\zeta)^*$ being isometric for a.e. $\zeta$, $\zeta \in \mathbb{T}$, that is, to $\Theta$ being an inner function. □

Returning to the problem of cyclic subspaces for the operator $S_E^*$ let us first of all mention the existence of a certain auxiliary structure ("degree of cyclicity") on the set of all non-cyclic subspaces (or vectors). It is clear that, for example, the vector $f = \binom{f_1}{f_2}$ with the condition $f_i \in N$ ($i = 1, 2$) and the one with the condition $f_1 \in N$, $f_2 \in C$ are non-cyclic "of different degree" (in the space $H^2(\mathbb{C}^2)$). This degree of cyclicity will for us be on the one hand the main object of study, and on the other hand a tool for reducing the problem which interests us to the known case $\dim E = 1$ (i.e. to the question of deciding when a family of functions admits pseudocontinuation, or not). The desired classification of subspaces $F$, $F \subset H^2(E)$, with respect to their "degree of cyclicity" arises if we consider locally (in $\mathbb{D}$) the rank of the orthogonal complement $(E_F^*)^\perp$ of the space $E_F^*$,

$$E_F^* \stackrel{\text{def}}{=} E_F^{S^*} = \bigvee(S_E^{*n} F : n \geq 0).$$

**Definition.** Let $F \subset H^2(E)$ and assume that $d \stackrel{\text{def}}{=} \dim E < \infty$. By the *degree of cyclicity* of $F$ we mean the number

$$\operatorname{dc}(F) = \operatorname{coDim}(H^2(E) \ominus E_F^*) \stackrel{\text{def}}{=} d - \max_{\zeta \in \mathbb{D}} \dim\{g(\zeta) : g \in H^2(E) \ominus E_F^*\}.$$

If $s \geq 0$, $s$ an integer, we set

$$N_s = N_s(E) \stackrel{\text{def}}{=} \{F : F \subset H^2(E), \operatorname{dc}(F) \leq s\}.$$

Directly from the definition we see that a) $N_s \subset N_{s+1}$, $s \geq 0$; b) $N_s = N_d$ for $s \geq d$; c) $N_{d-1}$ is the set of all non-cyclic subspaces $F$ (i.e. subspaces such that $E_F^* \neq H^2(E)$), $N_d \backslash N_{d-1}$ being thus the set of all cyclic ones, to be denoted by the symbol $C^E$.

First we connect the characteristic dc with the dimension of the auxiliary space $E'$ in the Lax-Halmos Theorem (p. 17–18), according to which $E_F^* = H^2(E) \ominus \Theta H^2(E')$ for some left inner function $\Theta$ and some $E'$, $0 \leq \dim E' \leq d = \dim E$.

**Lemma on the Local Rank.** *Let $E$ and $D$ be Hilbert spaces,* $\dim E < \infty$. *Let $\Phi$ be a strong $H^2$-function with values in $L(D, E)$, with the inner-outer factorization $\Phi = \Theta\Phi^e$ (cf. Corollary 9, p. 20). Denote* $\operatorname{Rank}\Phi \stackrel{\text{def}}{=} \max_{\zeta \in \mathbb{D}} \operatorname{rank}\Phi(\zeta)$, *where* $\operatorname{rank}\Phi(\zeta) = \dim \Phi(\zeta)D$, *and let us suppose that* $\operatorname{Rank}\Phi < \infty$. *Then*

1. $\operatorname{clos}\Phi\mathscr{P}_A(D) = \bigvee(S_E^n \Phi D : n \geq 0) = \Theta H^2(E')$ *and* $\dim E' = \operatorname{Rank}\Phi = \operatorname{Rank}\Theta$.

2. $\{f : f \in H^2(D), \Phi(\zeta)f(\zeta) \equiv \mathbb{O}\} = \vartheta H^2(D')$ *and* $\dim D = \dim D' + \operatorname{Rank}\Phi$, $\vartheta$ *being a left inner function.*

*Proof.* 1. The identity

$$\operatorname{clos}\Phi\mathscr{P}_A(D) = \Theta H^2(E')$$

is contained in Corollary 9, p. 20, just referred to. Thus we have just to check that $\dim E' = \operatorname{Rank}\Theta = \operatorname{Rank}\Phi$. For the analytic function $\Phi$ clearly holds $\operatorname{rank}\Phi(\zeta) = \text{const}$ outside the zero set of a certain $H^\infty$ function. As $\Theta$ takes on isometric boundary values, we must have $\operatorname{Rank}\Theta = \dim E'$ and $\operatorname{Rank}\Phi = \operatorname{Rank}\Phi^e \leq \dim E'$. But the strict inequality $\operatorname{Rank}\Phi^e < \dim E'$ is in contradiction with the Theorem on Outer Functions (p. 21).

2. The subspace $\operatorname{Ker}\mathbf{\Phi}(\subset H^2(D))$ is invariant for $S_D$ and thus has the form $\vartheta H^2(D')$ with a suitable left inner function $\vartheta$. If $\dim D = \infty$ it is clear that $\dim[\operatorname{Ker}\Phi](\zeta) = \infty$ for all $\zeta \in \mathbb{D}$, i.e. $\dim D' = \infty$. Hence we can assume further that $\dim D < \infty$. It is clear that for $\zeta \in \mathbb{D}$

$$\dim \vartheta(\zeta)D' = \dim[\operatorname{Ker}\Phi](\zeta) \leq \dim \operatorname{Ker}[\Phi(\zeta)] = \dim D - \operatorname{rank}\Phi(\zeta).$$

From the analyticity of $\vartheta$ and $\Phi$ it follows that there exists a point $\zeta_1, \zeta_1 \in \mathbb{D}$, with $\dim D' = \dim \vartheta(\zeta_1)D'$, $\operatorname{rank}\Phi(\zeta_1) = \operatorname{Rank}\Phi$. Hence $\dim D' \leq \dim D - \operatorname{Rank}\Phi$.

For the proof of the opposite inequality let us consider a principal minor $\Phi_1(\zeta_1)$ of the matrix of the operator $\Phi(\zeta_1)$ (with respect to two arbitrary fixed bases in $D$ and $E$ respectively). Then $\det\Phi_1 \in \mathrm{H}^\infty$, $\det\Phi_1 \not\equiv 0$. Let $D = D_1 \oplus D_2$, $E = E_1 \oplus E_2$ ($\dim D_1 = \dim E_1 = \operatorname{rank}\Phi(\zeta_1)$) be the decompositions of the spaces $D$ and $E$ corresponding to this minor and let

$$\Phi(\zeta) = \begin{pmatrix} \Phi_1(\zeta) & \Phi_2(\zeta) \\ \varphi_1(\zeta) & \varphi_2(\zeta) \end{pmatrix}, \quad \zeta \in \mathbb{D},$$

be the matrix representation of $\Phi(\zeta)$ with respect to this decomposition. For the proof of the inequality $\dim[\operatorname{Ker}\mathbf{\Phi}](\zeta_1) \geq \dim \operatorname{Ker}\Phi(\zeta_1)$ we use the relation

$$\varphi_2 \det\Phi_1 = \varphi_1 \Phi_1^{co}\Phi_2 (\Phi_1^{co} \overset{\text{def}}{=} \det\Phi_1 \cdot \Phi_1^{-1}).$$

From this relation we get the inclusion

$$\Omega H^2(D_2) \subset \operatorname{Ker}\mathbf{\Phi},$$

where $\Omega \in H^\infty(L(D_2, D))$,

$$\Omega = \begin{pmatrix} \Phi_1^{co}\Phi_2 \\ -\det\Phi_1 \end{pmatrix},$$

and $\operatorname{Rank}\Omega = \dim D_2 = \dim D - \operatorname{Rank}\Phi = \dim\operatorname{Ker}\Phi(\zeta_1)$, i.e. $\dim[\operatorname{Ker}\mathbf{\Phi}](\zeta_1) \geq \dim\operatorname{Ker}[\Phi(\zeta_1)]$. □

*Remark.* From the above proof it is clear that the space of bounded solutions of the equation $\mathbf{\Phi}(\zeta)f(\zeta) \equiv \mathbb{O}$ ($f \in H^\infty(D)$) has the same local dimension as $\operatorname{Ker}\mathbf{\Phi}$. If $\varphi_1, \ldots, \varphi_k$ ($k = \dim D$) are the columns of the matrix $\Phi$ then this dimension equals the local dimension of the space of $H^\infty$-solutions of the equation $\sum \varphi_i(\zeta)\psi_i(\zeta) \equiv 0$, $\psi_i \in H^\infty$.

Our (inductive) study of the classes $N_s$ will be based on the following lemma on the change of the cyclicity degree dc under the action of so-called "Toeplitz operators" (definition in Appendix 4).

**dc-Lemma.** *Let $F \subset H^2(E)$, let $E_F^* = H^2(E) \ominus \Theta H^2(E')$ be the canonical representation of the corresponding $S_E^*$-invariant subspace and let $\Omega \in H^\infty(L(D, E))$, $D$ and $E$ being finite dimensional Hilbert spaces. Then*

$$\operatorname{dc}(P_+\Omega^*F) = \operatorname{dc}(F) - [\dim E - \operatorname{Rank}(\Omega, \Theta)],$$

*where $(\Omega, \Theta)$ denotes an $L(D \times E', E)$-valued operator function defined in the obvious way.*

*Proof*. Set

$$d'(F) = \dim E - \mathrm{dc}(F)\ (= \dim E')$$

and let us find $d'(P_+\Omega^*F)$, i.e. the quantity

$$\begin{aligned}&\mathrm{Dim}\{\varphi : \varphi \in H^2(D),\ P_+\Omega^*F \perp \varphi\mathscr{P}_A\}\\ &\overset{\text{def}}{=} \max_{\zeta\in\mathbb{D}} \dim\{\varphi(\zeta) : \varphi \in H^2(D),\ P_+\Omega^*F \perp \varphi\mathscr{P}_A\}.\end{aligned}$$

It is clear that

$$\begin{aligned}P_+\Omega^*F \perp \varphi\mathscr{P}_A &\Leftrightarrow \Omega^*F \perp \varphi\mathscr{P}_A \Leftrightarrow F \perp \Omega\varphi\mathscr{P}_A \Leftrightarrow \Omega\varphi\mathscr{P}_A \subset \Theta H^2(E')\\ &\Leftrightarrow \Omega\varphi \in \Theta H^2(E').\end{aligned}$$

This shows that the set of functions $\{\varphi\}$ which interests us is in one-to-one correspondence with the set of pairs $\binom{\varphi}{\psi}$, $\psi \in H^2(E')$, such that $(\Omega, -\Theta)\binom{\varphi}{\psi} = \Omega\varphi - \Theta\psi = \mathbb{O}$ (in fact, if to some vector function $\varphi$ there correspond two functions $\psi_1$ and $\psi_2$ then $\Theta(\psi_1 - \psi_2) = \mathbb{O}$ which gives $\psi_1 = \psi_2$).

Thus

$$\begin{aligned}d'(P_+\Omega^*F) &= \mathrm{Dim}\,\mathrm{Ker}(\Omega, \Theta) = \mathrm{co\,Rank}(\Omega, \Theta)\\ &= \dim D + \dim E' - \mathrm{Rank}(\Omega, \Theta)\\ &= \dim D + \dim E - \mathrm{dc}(F) - \mathrm{Rank}(\Omega, \Theta),\end{aligned}$$

that is,

$$\begin{aligned}\mathrm{dc}(P_+\Omega^*F) &= \dim D - d'(P_+\Omega^*F)\\ &= \mathrm{dc}(F) - [\dim E - \mathrm{Rank}(\Omega, \Theta)].\ \square\end{aligned}$$

**Corollary 1.**

$$\mathrm{dc}(F) - [\dim E - \mathrm{Rank}\,\Omega] \le \mathrm{dc}(P_+\Omega^*F) \le \mathrm{dc}(F).$$

Indeed, we always have $\mathrm{Rank}(\Omega, \Theta) \ge \mathrm{Rank}\,\Omega$. □

Now we return to the study of the classes $N_s$. As an analogue of the scalar pseudocontinuable functions we have the following concept: the function $f$ in $H^2(E)$ is said to admit pseudocontinuation if there exist functions $h$, $h \in H^2_-(E)$, and $H$, $H \in H^\infty_-(L(E))$, where $H(\zeta)$ is an invertible operator in $E$ for all points $\zeta$, $\zeta \in \hat{\mathbb{C}}\backslash\mathrm{clos}\,\mathbb{D}$, off a suitable discrete set, such that $f(\zeta) = H(\zeta)^{-1}h(\zeta)$ a.e. $\zeta$, $\zeta \in \mathbb{T}$.

Restricting ourselves to the case $\dim E < \infty$ we remark that in this situation pseudocontinuation is then equivalent to the existence of a scalar function $G$, $G \in H^\infty_-$, and a function $g$ in $H^2_-(E)$ such that $f(\zeta) = g(\zeta)/G(\zeta)$ for a.e. $\zeta$, $\zeta \in \mathbb{T}$ (as $G(\zeta)$ it suffices to take $\det H(\zeta)$, then $g = GH^{-1}h$). In the finite dimensional case ($\dim E < \infty$) it also makes sense to speak of "coordinatewise" pseudocontinuation of the function $f$, $f \in H^2(E)$, if we fix a suitable basis $\{e_i\}$ in $E$ (so that it is question about the pseudocontinuation of the scalar coordinate functions $f_i$ of $f$, $f = \sum_{i=1}^n f_i e_i$) or, more generally, if we consider one dimensional sections $\pi f$

of $f$ ($\pi$, any rank one projection in $E$). The following result shows that this continuation coincides with the previous "vector" pseudocontinuation.

**Corollary 2** (Description of $N_0$). *Let* $F \in H^2(E)$, $\dim E < \infty$, $\dim F < \infty$, *and let* $E_F^* = H^2(E) \ominus \Theta H^2(E')$ *be the canonical representation of* $E_F^*$. *The following statements are equivalent.*

1. $F \in N_0$.
2. $\Theta$ *is an inner function.*
3. *All function in* $E_F^*$ *admit pseudocontinuation.*
4. *The functions in F admit pseudocontinuation.*
5. *The functions in F are coordinatewise pseudocontinuable.*

In fact, $1 \Rightarrow 2$ by the definition of the class $N_0$ and by the Lemma on Local Rank.

$2 \Rightarrow 3$ from the fact that $f \in K_\Theta \Rightarrow \Theta^* f \in H_-^2(E)$ and $f = \Theta(\Theta^* f)$, observing furthermore that as in the scalar case the inner function $\Theta$ has the standard pseudocontinuation

$$\zeta \mapsto (\Theta(1/\bar{\zeta})^*)^{-1}, \quad |\zeta| > 1.$$

$3 \Rightarrow 4$ is trivial.

$4 \Rightarrow 5$ from the above mentioned possibility of taking the determinant.

$5 \Rightarrow 1$. As $\dim F < \infty$ it is clear that there exists a scalar inner function $\vartheta$ such that $\bar{\vartheta} f \in H_-^2(E)$, $\forall f \in F$. Put now $\Omega = \vartheta I_d$, $d = \dim E$, in the dc-Lemma. Then $P_+ \Omega^* F = \mathbb{O}$ and by that lemma

$$\mathrm{dc}(F) = \dim E - \mathrm{Rank}(\vartheta I_d, \Theta) = 0. \ \square$$

The dc-Lemma allows us to establish some simple but useful transformations which do not alter the "cyclicity degree" of a family of functions. Indeed, the following holds true.

**Corollary 3.** *In the hypotheses of the* dc-*Lemma*

1. *if* $\mathrm{Rank}\,\Omega = \dim E$ *then* $\mathrm{dc}(P_+ \Omega^* F) = \mathrm{dc}(F)$;

2. *if* $\Omega' \in H^\infty(L(E))$, $\mathrm{Rank}\,\Omega' = \dim E$, *and the functions* $\Omega'(\cdot)e$ *admit pseudo-continuation*[2] ($\forall e \in E$) *then* $\mathrm{dc}(\Omega' F) = \mathrm{dc}(F)$;

3. *if* $\pi$ *is any (coordinate) projection in E, all functions in* $(I - \pi)F$ *admit pseudocontinuation and further* $\dim(I - \pi)F < \infty$ *then* $\mathrm{dc}(F) = \mathrm{dc}(\pi F)$.

In fact, the first statement follows from Corollary 1.

The second statement follows from the first one, because by the definition of pseudocontinuation there exists a scalar inner function $\vartheta$ such that $\bar{\vartheta} \cdot \Omega'(\cdot)e \in H_-^2(E)$, $\forall e \in E$ (which gives $\vartheta(\Omega')^* \in H^\infty(L(E))$) so that

$$\mathrm{dc}(F) = \mathrm{dc}(P_+(\bar{\vartheta}\Omega' F)) = \mathrm{dc}(P_+(\bar{\vartheta} I_E)\Omega' F) = \mathrm{dc}(\Omega' F).$$

Statement 3 follows at once from Statement 1 if we set $\Omega = \pi + \vartheta(I - \pi)$, $\vartheta$

---

[2] To phrase it otherwise: the matrix elements $\Omega'(\cdot)$ with respect to (arbitrary) bases in $D$ and $E$ admit pseudocontinuation (cf. Corollary 2).

being an inner function with the property $\bar{\vartheta}(I-\pi)F \subset H^2_-(E)$:

$$\mathrm{dc}(F) = \mathrm{dc}(P_+\Omega^*F) = \mathrm{dc}(\pi F). \ \square$$

In particular, the last statement of Corollary 3 shows that the degree of cyclicity $\mathrm{dc}(F)$, $F \subset H^2(E)$, does not change if we imbed $E$ in a larger space $\tilde{E}$, thereby regarding $F$ as a part of $H^2(\tilde{E})$.

It is further of interest to note that Corollary 3 allows us to strengthen somewhat the dc-Lemma itself.

**Corollary 4.** *Let the assumptions of the* dc-*Lemma be fulfilled and let* $\Omega'$ *be as in Corollary* 3, *Statement* 2. *Then*

$$\mathrm{dc}(P_+\Omega^*\Omega'F) = \mathrm{dc}(P_+\Omega^*F) = \mathrm{dc}(F) - [\dim E - \mathrm{Rank}(\Omega, \Theta)].$$

In fact, $\mathrm{dc}(P_+\Omega^*\Omega'F) =$ (dc-Lemma) $\mathrm{dc}(\Omega'F) - [\dim E - \mathrm{Rank}(\Omega, \Theta)] =$ (Corollary 3) $\mathrm{dc}(F) - [\dim E - \mathrm{Rank}(\Omega, \Theta)]$. $\square$

The connection between the characteristic dc and the inner-outer factorization is not so simple as in the case of scalar functions (for which $f \in C \Leftrightarrow f^e \in C$). In this direction we limit ourselves to the following result.

**Corollary 5.** *Let* $F \in H^2(E)$ *and let* $F = F^iF^e$ *be the inner-outer factorization of* $F$ *(as strong* $H^2$*-function). Then*

$$\mathrm{dc}(F^e) \le \mathrm{dc}(F) \le \mathrm{dc}(F^e) + [\dim E - \mathrm{Rank}\, F^i].$$

*In particular*

$$\mathrm{Rank}\, F^i = \dim E \Rightarrow \mathrm{dc}(F) = \mathrm{dc}(F^e)$$

*and if* $\mathrm{dc}(F^e) = 0$ *(i.e.* $F^e$ *has a pseudocontinuation) then* $F \in N_{d-1}$ *(i.e.* $F \notin C^E$*).*

In fact, by the dc-Lemma

$$\mathrm{dc}(F^e) = dc(P_+F^{i*}F) = \mathrm{dc}(F) - [\dim E - \mathrm{Rank}(F^i, \Theta_F)]$$

and this formula makes evident all statements of the corollary. $\square$

Let us also remark that the implication $\mathrm{Rank}\, F^i = \dim E \Rightarrow \mathrm{dc}(F) = \mathrm{dc}(F^e)$ likewise is contained in Corollary 3, Statement 2.

One of the forms of induction which permits one to pass from classes $N_s$ with small values of $s$ to ones with large degrees of cyclicity consists of adding (algebraically or set theoretically) subspaces (or just vectors) $F_i$, $i = 1, 2, \ldots$, from the classes $N_s$ in question. Let us begin with the following simple proposition.

**Corollary 6.** *Let* $F$ *and* $G$ *be subspaces of* $H^2(E)$. *Then*:
1. *If* $F \subset G$ *then* $\mathrm{dc}(F) \le \mathrm{dc}(G)$.
2. $\mathrm{dc}(F + G) \le \mathrm{dc}(F) + \mathrm{dc}(G)$.
3. *If* $E_F^* = H^2(E) \ominus \Theta_F H^2(D)$ *is the canonical representation of* $E_F^*$ *then*

$$\mathrm{dc}(F + G) = \mathrm{dc}(F) + \mathrm{dc}(P_+\Theta_F^*G). \tag{8}$$

In fact, Statement 1 is self-evident and Statement 2 follows at once from

Statement 3 if we use the estimate in Corollary 1: $\mathrm{dc}(P_+\Theta_F^*G) \leq \mathrm{dc}(G)$. There remains the verification of Statement 3.

To this end let us set $\Omega = \Theta_F$ in the dc-Lemma. Since $E_{F+G}^* \stackrel{\mathrm{def}}{=} H^2(E) \ominus \Theta H^2(E') \supset E_F^* = H^2(E) \ominus \Theta_F H^2(D)$ we have $\Theta H^2(E') \subset \Theta_F H^2(D)$ and therefore $\Theta = \Theta_F \Phi$ with $\Phi \in H^\infty(L(E', D))$. It is also clear (cf. the Lemma on the Local Rank) that $\mathrm{Rank}(\Theta_F, \Theta) = \mathrm{Rank}(\Theta_F, \Theta_F\Phi) = \mathrm{Rank}\,\Theta_F = \dim D = \dim E - \mathrm{dc}(F)$. From the dc-Lemma we now get

$$\begin{aligned}\mathrm{dc}(F+G) &= \mathrm{dc}(P_+\Theta_F^*(F+G)) + [\dim E - \mathrm{Rank}(\Theta_F, \Theta)]\\ &= \mathrm{dc}(P_+\Theta_F^*G) + \mathrm{dc}(F). \ \square\end{aligned}$$

The identity (8) shows in particular that the degree of cyclicity of the subspace $F$ does not change when we add $G$ iff $G$ is "inessential modulo $F$", i.e. when $G = G' + \Theta_F G''$ where $G' \subset E_F^*$ (i.e. $G'$ can be arbitrarily well approximated by combinations of shifts $S^{*n}F, n \geq 0$) and $G''$ is a subspace admitting pseudocontinuation ($G'' = P_+\Theta_F^*G$; cf. Projection Lemma, II).

In the language of the classes $N_s$ inequality 2 of Corollary 6 can be written as an inclusion $N_{s_1} + N_{s_2} \subset N_{s_1+s_2}$, $s_i \geq 0$, $i = 1, 2$. This inclusion can also be considered as a kind of analogon of the linearity of the class $N$ (the scalar case). The converse inclusion $N_{s_1+s_2} \subset N_{s_1} + N_{s_2}$ would mean that every subspace $F$ of class $N_{s_1+s_2}$ is contained in the sum of two subspaces $F_i$ ($i = 1, 2$) of class $N_{s_i}$ ($i = 1$, 2) respectively: $F \subset F_1 + F_2$. In the following theorem we will not only establish this converse inclusion (which together with the previous one gives the symbolic identity $N_{s_1+s_2} = N_{s_1} + N_{s_2}$) but we will also display a linear operator which decomposes $F$ into a "sum" of subspaces $F_i$, $i = 1, 2$. The second statement of the theorem gives an alternative description of the classes $N_s$, based on induction over the dimension $\dim E$ of the coefficient space.

**Theorem on Noncyclic Vectors.** *Let $E$ be a Hilbert space, $\dim E < \infty$, and let $F \subset H^2(E)$, $\dim F < \infty$.*

1. *For any $s_1, s_2 \geq 0$ holds $N_{s_1} + N_{s_2} = N_{s_1+s_2}$ in the sense that the algebraic sum of subspaces in $N_{s_1}$ and $N_{s_2}$ belongs to $N_{s_1+s_2}$ and for any subspace $F$ of the last type there exists a linear operator $A$ such that $AF \subset N_{s_1}$, $(I - A)F \in N_{s_2}$.*
2. *$F \in N_s(E) \Leftrightarrow (\pi F \notin C^{\pi E}$ for every orthogonal projection $\pi$ in $E$ with $\dim \pi E = s + 1)$.*

For the proof of the first statement of the theorem we need an auxiliary result on the possibility of "complementing" an operator function to a function admitting a scalar multiple, that is, roughly speaking to an $H^\infty$-function whose values on the circumference $\mathbb{T}$ constitute a uniform family of isomorphisms (more precisely, a function $\Phi$ in $H^\infty(L(E))$ is said to admit a scalar multiple if there exist functions $\Omega, \Omega \in H^\infty(L(E))$, and $\varphi, \varphi \in H^\infty$, such that $\Phi\Omega \equiv \Omega\Phi = \varphi \cdot I_E$). In Appendix 3 we will prove and discuss also other more sophisticated results of this type.

**Complementing Lemma.** *Let $E$ and $E'$ be Hilbert spaces, $E' \subset E$, $\dim E < \infty$, and let $\Phi' \in H^\infty(L(E', E))$. The following statements are equivalent.*

1. $\operatorname{ess\,inf}_{\zeta \in \mathbb{T}} \min\{\|\Phi'(\zeta)x\| : \|x\| = 1\} > 0$.

2. *There exist functions* $\Phi$ *and* $\Omega$ *in* $H^\infty(L(E))$ *with* $\Phi|E' = \Phi'$, $\Phi|(E \ominus E')$ *being a left inner function, such that* $\Omega\Phi = mI_E$, *where* $m$ *is a scalar inner function.*

*Remark.* This lemma allows us in particular to complement an arbitrary left inner function $\Phi'$ to a "uniform isomorphism" with the aid of a function $\Phi|(E \ominus E')$ which is left inner (see the formulation). However, the function $\Phi$ arising in this way need not be inner. For instance, it is easy to prove that $\Phi$ can be taken inner iff $\Phi'e$ admits pseudocontinuation, $\forall e \in E'$.

Let us further point out that complementing of a left inner function $\Phi'$ to an a.e. (on $\mathbb{T}$) nondegenerate function $\Phi$ (without a uniform estimate from above for the norm $\|\Phi(\zeta)^{-1}\|$, $\zeta \in \mathbb{T}$) can be achieved using a constant component $\Phi|(E \ominus E')$ (i.e. one which does not depend on $\zeta$); namely, from Corollary 2 of the Theorem on Noncyclic Vectors and from the dc-Lemma it follows that there exist an orthogonal projection in $E$, $\operatorname{rank} \pi = \dim E - \dim E'$, such that $\operatorname{Rank}(\pi, \Phi') = \dim E$.

We use now the Complementing Lemma for the proof of the Theorem on Noncyclic Vectors.

*Proof of the Theorem.* 1. The inclusion $N_{s_1} + N_{s_2} \subset N_{s_1+s_2}$ has already been established (Corollary 6 of the dc-Lemma). Let us therefore prove the opposite inclusion.

Without loss of generality we may assume that $s \overset{\text{def}}{=} s_1 + s_2 \leq \dim E$. Let $F \subset H^2(E)$ with $\operatorname{dc}(F) = s$. Define the left inner function $\Theta$ by the identity $E_F^* = H^2(E) \ominus \Theta H^2(E')$. Let us put $\Phi' = \Theta$ and apply the Complementing Lemma, denoting by $\Phi$, $\Omega$ and $m$ the functions the existence of which is guaranteed by that lemma. As $\dim E = \dim E' + \operatorname{dc}(F)$ (Lemma on the Local Rank) there exist subspaces $E_i$, $E_i \subset E$ $(i = 1, 2)$ with $E = E_1 \oplus E_2 \oplus E'$ and $\dim E_i = s_i$ $(i = 1, 2)$. Denote by $\pi_i$ $(i = 1, 2)$ and $\pi'$ the orthogonal projections in $E$ onto the subspaces $E_i$ $(i = 1, 2)$ and $E'$ respectively and let

$$\Omega_2 \overset{\text{def}}{=} \Phi\pi_2\Omega, \quad \Omega_1 \overset{\text{def}}{=} \Phi(I_E - \pi_2)\Omega.$$

By the definition of $\Phi$ and $\Omega$ we have $\Omega_1 + \Omega_2 = \Phi\Omega = mI_E$ so that setting $Af = P_+ m\Omega_1^* f$, $f \in H^2(E)$, we get $(I - A)f = P_+(I - m\Omega_1^*)f = P_+ m\Omega_2^* f$.

The operator $A$ is indeed the one we are looking for (use Corollary 4 of the dc-Lemma):

$$\begin{aligned}
\operatorname{dc}(AF) &= \operatorname{dc}(F) - [\dim E - \operatorname{Rank}(\Omega_1, \Theta)] \\
&= s - [\dim E - \operatorname{Rank}(\Phi\pi_1\Omega + \Theta\pi'\Omega, \Theta)] \\
&= \big(\operatorname{Rank}(A + BC, B) = \operatorname{Rank}(A, B)\big) \quad s - [\dim E - \operatorname{Rank}(\Phi\pi_1\Omega, \Phi\pi')] \\
&= \big(\Phi\pi' = \Theta\pi' = \Theta \text{ and } \Phi(\zeta) \text{ is an isomorphism for all } \zeta, |\zeta| < 1, \\
&\qquad \text{off an (at most) countable set of values}\big) \quad s - [\dim E - \operatorname{Rank}\Phi\pi_1\Omega - \operatorname{Rank}\Theta] \\
&= s - [s - \operatorname{Rank}\Phi\pi_1\Omega] = s_1, \\
\operatorname{dc}((I - A)F) &= \operatorname{dc}(F) - [\dim E - \operatorname{Rank}(\Omega_2, \Theta)] = s - [\dim E - \operatorname{Rank}(\Phi\pi_2\Omega, \Phi\pi')] \\
&= s - [s - \operatorname{Rank}\Phi\pi_2\Omega] = s_2.
\end{aligned}$$

2. One half of this statement (the implication $\Rightarrow$) is a trivial special case of Corollary 1 of the dc-Lemma: if $\Omega = \pi$ then

$$\mathrm{dc}(\pi F) = \mathrm{dc}(P_+\Omega^* F) \le \mathrm{dc}(F) \le s < \dim \pi E.$$

In order to prove the opposite inclusion ($\Leftarrow$) we pick an arbitrary orthonormal basis $\{e_j\}_{j=1}^d$ ($d = \dim E$) in $E$ and denote for any subset $\kappa$, $\kappa \subset \{1, \dots, d\}$, $\operatorname{card}\kappa = d - s - 1$, by $\pi^\kappa$ the orthogonal projection onto the subspace $\bigvee(e_j : j \notin \kappa)$. As $\dim \pi^\kappa E = s + 1$ then by assumption $\pi^\kappa F \notin C^{\pi^\kappa E}$, $\forall \kappa$. Therefore there exist non-zero functions $\psi_\kappa$ such that $\psi_\kappa \in H^2(E)$, $\pi^\kappa \psi_\kappa = \psi_\kappa$ and $\pi^\kappa F \perp \psi_\kappa H^2$ (whence $F \perp \psi_\kappa H^2$). Let $\Psi = \{\psi_\kappa\}_\kappa$ be the corresponding strong $H^2$-function (the matrix function whose columns are the vectors $\psi_\kappa$, $\operatorname{card}\kappa = d - s - 1$). By the Lemma on the Local Rank it suffices to show that

$$\operatorname{Rank}\Psi \ge d - s \tag{9}$$

because then $\mathrm{dc}(F) \le \dim E - \operatorname{Rank}\Psi \le s$.

Inequality (9) is an immediate consequence of the Lemma on the Local Rank and the following simple result on the rank of matrices having many zeros on every column.

**Lemma.** *Consider subsets $\kappa \subset \{1, \dots, d\}$, $\operatorname{card}\kappa = m$, and assume that for every such $\kappa$ there corresponds a vector $\psi_\kappa$, $\psi_\kappa \in \mathbb{C}^d$, with the property that $\psi_\kappa \ne \mathbb{0}$ and $(\psi_\kappa)_j = 0$ for $j \in \kappa$. Then the matrix $\Psi = \{\psi_\kappa\}$ having all the $\psi_\kappa$ as columns has* $\operatorname{rank}\Psi \ge m + 1$.

*Proof of the Lemma.* The assertion is true if $m = 1$: if we would have $\operatorname{rank}\ \Psi = 1$ then $\forall i, j \in \{1, \dots, d\} \exists \alpha, \beta$ ($|\alpha| + |\beta| \ne 0$): $\alpha\psi_i + \beta\psi_j = \mathbb{0}$. As $\psi_i \ne \mathbb{0}$ $\forall i$ then $\alpha\beta \ne 0$ and this gives $(\psi_i)_j = 0$ $\forall i, j$. Contradiction.

Induction: Let us consider the matrix $\Phi^j \overset{\text{def}}{=} \{\psi_\kappa\}_{\kappa \ni j}$; omitting in this matrix the row with number $j$ and using the induction hypothesis we get $\operatorname{rank}\Psi^j \ge (m - 1) + 1 = m$. If for some $j$ ($1 \le j \le d$) and $\sigma$, $\operatorname{card}\sigma = m$, $\operatorname{rank}(\Psi^j, \psi_\sigma) \ge m + 1$ then a fortiori $\operatorname{rank}\Psi \ge m + 1$. In the opposite case $\psi_\sigma$ must be a linear combination of vectors $\psi_\kappa, j \in \kappa$, which gives $(\psi_\sigma)_j = 0$, that is, ($j$ being arbitrary) $\psi_\sigma = \mathbb{0}$. Thus always $\operatorname{rank}\Psi \ge m + 1$. □□

*Remark.* From Statement 2 of the theorem (and its proof) it follows that

$$\mathrm{dc}(F) = \max\{s : \exists \pi \text{ (an orthogonal projection), } \operatorname{rank}\pi = s, \pi F \in C^{\pi E}\}$$

$$= \max\{s : \exists \kappa, \kappa \subset \{1, \dots, d\}, \operatorname{card}\kappa = s, \pi_\kappa F \in C^{\pi_\kappa E}\},$$

where $\pi_\kappa$ is the orthogonal projection onto the coordinate space $\bigvee(e_i : i \in \kappa)$, $\{e_i\}_{i=1}^d$ being an (arbitrary) orthonormal basis in $E$.

*Proof of the Complementing Lemma.* It is clear that $2 \Rightarrow 1$. Let us prove the opposite inclusion.

Pick an orthonormal basis $\{e_i\}_{i=1}^d$, $d = \dim E$, in $E$ such that $E' = \bigvee(e_i : 1 \le i \le d')$, $d' = \dim E'$. Let $\pi_\kappa$ be orthogonal projections in $E$ onto the subspace $\bigvee(e_i : i \in \kappa)$, where $\kappa \subset \{1, \dots, d\}$, $\operatorname{card}\kappa = d'$ ($= \operatorname{Rank}\Phi'$) and set $\Phi_\kappa = \pi_\kappa \Phi'$.

Statement 1 is equivalent to the requirement that $\Phi'(\zeta)^*\Phi'(\zeta) \geq \delta I_E$, for some $\delta$, $\delta > 0$, and for a.e. $\zeta$ ($\zeta \in \mathbb{T}$), or

$$\det \Phi'^*\Phi' \geq \delta \tag{10}$$

a.e. on $\mathbb{T}$ (with another $\delta$). On the other hand by the Binet-Cauchy formula (cf. Gantmaher [1])

$$\det \Phi'^*\Phi' = \sum_{\substack{\kappa \\ \operatorname{card}\kappa = d'}} |\det \Phi_\kappa|^2, \tag{11}$$

where all determinants $\det \Phi_\kappa$ are $H^\infty$-functions.

Set $E'' = E \ominus E'$. The desired function $\Phi$ will be constructed according to the following plan: first we find a function $G$, $G \in H^\infty(L(E, E'))$, such that $G\Phi' = mI_E$, where $m$ is a scalar inner function, then we show that there exists a left inner function $\Phi''$ with $\operatorname{Ker} \mathbf{G} = \mathbf{\Phi}'' H^2(E'')$, we set $\Phi | E' = \Phi'$, $\Phi | E'' = \Phi''$ and finally we verify the existence of a function $\Omega$ satisfying the desiderata of Statement 2.

Construction of $G$. It suffices to find functions $g_\kappa$ in $H^\infty$ such that

$$\sum_{\substack{\kappa \\ \operatorname{card}\kappa = d'}} g_\kappa \det \Phi_\kappa \equiv m, \tag{12}$$

with a suitable inner function $m$. In fact, knowing such functions, we construct the matrix $\{G_{ij}\}_{1 \leq i \leq d', 1 \leq j \leq d}$ of the operator function $G$ (in the bases $\{e_j\}_{1 \leq j \leq d}$, $\{e_i\}_{1 \leq i \leq d'}$) by the formula

$$G_{ij} = \sum_{\substack{\kappa \\ \kappa \ni j}} g_\kappa \Phi_\kappa^{ij}$$

where $\Phi_\kappa^{ij}$ is the algebraic complement of the element $\Phi'_{ij}$ with indices $i, j$ in the matrix of the operator function $\Phi_\kappa$. It is clear that then $G \in H^\infty(L(E, E'))$ and that for all $i, l$ ($1 \leq i \leq d'$, $1 \leq l \leq d'$) holds

$$\begin{aligned} \sum_{j=1}^{d} G_{ij}\Phi'_{jl} &= \sum_{\substack{\kappa \\ \operatorname{card}\kappa = d'}} g_\kappa \sum_{\substack{j \\ j \in \kappa}} \Phi_\kappa^{ij}\Phi'_{jl} \\ &= \sum_{\substack{\kappa \\ \operatorname{card}\kappa = d'}} g_\kappa \delta_{jl} \det \Phi_\kappa = \delta_{il} \cdot m, \end{aligned}$$

that is, $G\Phi' = mI_{E'}$.

The existence of functions $g_\kappa$ solving equation (12) follows from condition (10), formula (11) and the following lemma which has a purely function theoretic character and which therefore has been relegated to Appendix 2.

**Lemma on the Algebra $L^\infty \cap \mathrm{Nev}$.** *Denote by $L^\infty \cap \mathrm{Nev}$ the algebra of all bounded functions on the circumference $\mathbb{T}$ which can be written as the ratio $\varphi/\psi$ of two functions in $H^\infty$. Let $f_i \in L^\infty \cap \mathrm{Nev}$, $1 \leq i \leq k$. The following statements are equivalent*:

1. $\operatorname{ess}_{\mathbb{T}} \inf \sum_{i=1}^{k} |f_i|^2 > 0$.
2. $\exists g_i : g_i \in L^\infty \cap \mathrm{Nev}$ ($1 \leq i \leq k$), $\sum_{i=1}^{k} g_i f_i \equiv \mathbb{1}$.

3. $\exists g_i: g_i \in H^\infty$ $(1 \le i \le k)$, $\sum_{i=1}^k g_i f_i$ *is an inner function.*
4. $\exists g_i: g_i \in H^\infty$ $(1 \le i \le k)$, $\text{ess}_{\mathbb{T}} \inf |\sum_{i=1}^k g_i f_i| > 0$.

Thus we have constructed a function $G$ in $H^\infty(L(E, E'))$ with the property $G\Phi' = mI_{E'}$, $m$ an inner function.

Construction of $\Phi$. This construction is contained in the following Sublemma which we shall use once more (in Appendix 3); that is why this Sublemma is stated in somewhat more general hypotheses than is needed in the proof of the Complementing Lemma.

**Sublemma.** *Let $\Phi' \in H^\infty(L(E', E))$, $E' \subset E$, $\dim E' < \infty$ and let $m$ be a scalar inner function. The following statements are equivalent:*

a. *There exists a function $G$, $G \in H^\infty(L((E, E'))$ such that $G\Phi' = mI_{E'}$.*

b. *There exist functions $\Phi$ and $\Omega$ in $H^\infty(L(E))$ such that $\Phi | E' = \Phi'$ and $\Omega\Phi = \Phi\Omega = mI_E$.*

*Moreover, if* a. *holds then the function $\Phi$ in* b. *may be chosen in such a way that the restriction $\Phi | (E \ominus E')$ is a left inner function.*

*Proof of Sublemma.* It is clear that b $\Rightarrow$ a (put $G = P_{E'}\Phi^{-1}$).

To prove the opposite inclusion (a $\Rightarrow$ b) let us mention that the subspace $\operatorname{Ker} G = \{f \in H^2(E): G(\zeta)f(\zeta) \equiv \mathbb{O}\}$ is invariant with respect to the shift $S_E$ and so by the Lax-Halmos Theorem (Corollary 6 on p. 17–18) must have the form $\Theta H^2(D)$ where $\Theta$ is a left inner function with values in $L(D, E)$ and $\dim D \le \dim E$. By the Lemma on the Local Rank $\dim D = \dim E - \operatorname{Rank} G =$ (in view of $G\Phi' = mI_{E'}$, $m \not\equiv 0$) $\dim E - \dim E' = \dim E''$. Let $U$ be a unitary operator acting from $E''$ into $D$ and set $\Phi'' \overset{\text{def}}{=} \Theta U$. Then $\operatorname{Ker} \mathbf{G} = \mathbf{\Phi}'' H^2(E'')$ and we get our $\Phi$ in $H^\infty(L(E))$ setting

$$\Phi(\zeta)|E' = \Phi'(\zeta), \quad \Phi(\zeta)|E'' = \Phi''(\zeta), \quad \zeta \in \mathbb{D}.$$

Construction of $\Omega$. Let us first remark that $(mI_E - \mathbf{\Phi}'\mathbf{G})H^2(E) \subset \mathbf{\Phi}''H^2(E'') = \operatorname{Ker} \mathbf{G}$, because $G(mI_E - \Phi'G)f = (Gm - G\Phi'G)f = (Gm - mG)f = \mathbb{O}$, $\forall f \in H^2(E)$. Consequently $(\Phi'')^*(mI_E - \Phi'G) \in H^\infty(L(E, E''))$ and we can set

$$\Omega = G \oplus (\Phi'')^*(mI_E - \Phi'G).$$

It is clear that $\Omega \in H^\infty(L(E))$ and that

$$\Phi\Omega = \Phi'G + \Phi''(\Phi'')^*(mI_E - \Phi'G) = mI_E,$$

$$\begin{aligned}\Omega\Phi &= G\Phi'P_{E'} + (\Phi'')^*(mI_E - \Phi'G)(\Phi'P_{E'} + \Phi''P_{E''}) \\ &= mP_{E'} + (\Phi'')^*(m\Phi'P_{E'} - \Phi'G\Phi'P_{E'} + m\Phi''P_{E''}) \\ &= mP_{E'} + (\Phi'')^*m\Phi''P_{E''} = mI_E. \ \square\end{aligned}$$

Let us now indicate some concrete methods of constructing functions in the various classes $N_s$ which arise either from the Theorem on Noncyclic Vectors or from the dc-Lemma or directly from the basic definitions. First of all we can use induction on the index of the class $s$ and the dimension of the space $E$, since the case $s = 0$ and $\dim E = 1$ can be considered as settled. One way of carrying out the induction over the dimension $\dim E$ is to use the coordinate representation $f = \begin{pmatrix} f_1 \\ \vdots \\ f_d \end{pmatrix}$, $d = \dim E$, of the function $f$ in $H^2(E)$ (below we assume that a basis is fixed in $E$ once for all, the coordinate functions $f_i$ being scalar $H^2$-functions).

**Corollary 1.** *Let* $f, x \in H^2(E)$ *and assume that there exists an index* $j$ $(1 \le j \le d)$ *such that* $f_j - x_j \ne \mathbb{0}$. *If in addition the ratios*

$$\frac{f_i - x_i}{f_j - x_j} \text{ admit pseudocontinuation } (1 \le i \le d) \tag{13}$$

*then* $\mathrm{dc}(f) \le \mathrm{dc}(x) + 1$. *In particular* $f \in N_1$ *if the quotients* $f_i/f_j$ *are pseudocontinuable* $(f_j \not\equiv 0;\ 1 \le i \le d)$.

Indeed, with no loss of generality we may assume that $j = 1$ and that

$$\frac{f_i - x_i}{f_1 - x_1} = \frac{\bar{a}_i}{\bar{b}_i} \quad (i = 2, \ldots, d). \tag{14}$$

where $a_i, b_i \in H^\infty$, $b_i \not\equiv 0$, $2 \le i \le d$. Setting

$$\Omega = \begin{pmatrix} 0 & -a_2 & \cdots & -a_d \\ 0 & b_2 & & \mathbb{0} \\ \vdots & & \ddots & \\ 0 & \mathbb{0} & & b_d \end{pmatrix}$$

and using Corollary 1 of the dc-Lemma and the identity $\Omega^* f = \Omega^* x$, resulting from (14), we get

$$\begin{aligned} \mathrm{dc}(f) - 1 &= \mathrm{dc}(f) - [\dim E - \operatorname{Rank} \Omega] \le \mathrm{dc}(P_+ \Omega^* f) \\ &= \mathrm{dc}(P_+ \Omega^* x) \le \mathrm{dc}(x). \ \square \end{aligned}$$

*Remark.* Yet another way of getting functions of class $N_s$, close to Corollary 1 but perhaps simpler, amounts to the following. Let $h \in C$ (a scalar $S^*$-cyclic function) and $\varphi_i \in H^\infty$, $1 \le i \le d$, with $\varphi_i \ne \mathbb{0}$ for at least one $i$. Then for the vector $f = \{f_i\}_{i=1}^d$, $f_i \overset{\text{def}}{=} P_+ \bar{\varphi}_i h$, holds $\mathrm{dc}(f) = 1$.

For the proof one can again use the dc-Lemma or its Corollary 3:

$$\Omega = (\varphi_1, \ldots, \varphi_d), \quad P_+ \Omega^* h = f, \quad \operatorname{Rank} \Omega = 1 \Rightarrow \mathrm{dc}(f) = \mathrm{dc}(h) = 1. \ \square$$

As the Theorem on Noncyclic Vectors shows, one can construct examples of functions (subspaces) of class $N_s$ starting with known $N_1$ functions (because $N_s = N_1 + \cdots + N_1$ ($s$ times)); for example, one may take the $N_1$ functions of the preceding corollary. Now the condition of pseudocontinuability of the ratios (13) for $x_i \in N_0$ means that the coordinates $f_i$ are linearly independent over the ring of pseudocontinuable functions, so one may hope that for the vector function $f$ the presence of a group of $s$ "strongly independent over $N_0$" coordinates will allow us to give the estimate $\mathrm{dc}(f) \ge s$. We will substantiate this in one simple situation

(Corollary 2, below), for which formulation we first must make the notion of "singularity of an analytic function" somewhat more clear.

Let $f \in H^2(E)$. Consider the set $G$ of all $E$-valued function $g$ meromorphic (that is, with meromorphic coordinates) in corresponding domains $\Omega_g \subset \mathbb{C}$ and such that: each $\Omega_g$ has a piecewise smooth boundary; $\partial\Omega_g \cap \mathbb{T}$ contains a (non-empty) arc of $\mathbb{T}$; $g$ has on this arc angular boundary values (see Appendix 2) and the latter coincide a.e. with the values of $f$. The *set of (essential) singularities* of $f$ is by definition the set $\operatorname{sing}(f) = \mathbb{C}\backslash\bigcup_{g\in G}\Omega_g$. For example, if $|a| \geq 1$ then $\operatorname{sing}(\log(z-a)) = \{a\}$, $\operatorname{sing}\left(\exp\dfrac{b}{z-a}\right) = \{a\}$ (with a corresponding $b$, $b \neq 0$); here we can think of both scalar functions and vector valued ones, putting $a \overset{\text{def}}{=} \{a_1, \ldots, a_d\} \subset \mathbb{C}^d$, $\log(z-a) \overset{\text{def}}{=} \{\log(z-a_i)\}_{i=1}^d$, $\exp\dfrac{b}{z-a} \overset{\text{def}}{=} \left\{\exp\dfrac{b_i}{z-a_i}\right\}_{i=1}^d$. We have always $\operatorname{sing}(f) \supset \bigcup_{i=1}^d \operatorname{sing}(f_i)$.

**Corollary 2.** *Let* $f = \{f_i\}_{i=1}^d = \sum_{i=1}^d f_i e_i$, $f \in H^2(E)$ (*i.e. the* $f_i$ *are the coordinate functions of* $f$ *with respect to the basis* $\{e_i\}$).

1. *If* $\operatorname{sing}(f_1)\backslash\mathbb{T} \not\subset \operatorname{sing}\{f_i\}_{i=2}^d$ *then* $H^2 \cdot e_1 \subset E_f^*$.
2. *If* $\operatorname{sing}(f_i)\backslash\mathbb{T} \not\subset \operatorname{sing}\{f_j\}_{j=1, i\neq j}^d$, $1 \leq i \leq d$, *then* $f \in C^E$.
3. *If* $\operatorname{sing}(f_i)\backslash\mathbb{T} \not\subset \operatorname{sing}\{f_j\}_{j\in\kappa\backslash i}$, $\forall i \in \kappa$, $\kappa \subset \{1, \ldots, d\}$, *then* $\operatorname{dc}(f) \geq \operatorname{card}\kappa$.

In fact, if $\varphi = \{\varphi_i\}_{i=1}^d \perp S_E^{*n} f$, $n \geq 0$, then $\sum_{i=1}^d f_i\bar{\varphi}_i = \bar{z}\bar{\psi}$, $\psi \in H^2$; if $\varphi_1 \neq \mathbb{0}$ then $f_1 = (\bar{z}\bar{\psi} - \sum_{i=2}^d f_i\bar{\varphi}_i)/\bar{\varphi}_1$ has no more singularities in $\mathbb{C}\backslash\operatorname{clos}\mathbb{D}$ than the vector function $\{f_i\}_{i=2}^d$. Hence $\varphi_1 = \mathbb{0}$, which proves Statement 1.

It is clear that $1 \Rightarrow 2$ and that by Statement 2 $\pi_\kappa f \in C^{\pi_\kappa E}$ ($f$ being the function in Statement 3), $\pi_\kappa E = \bigvee(e_i : i \in \kappa)$. By Statement 2 of the Theorem on Noncyclic Vectors then $\operatorname{dc}(f) \geq \operatorname{card}\kappa$. □

*Examples.* Let $|a_i| > 1$ and $a_i \neq a_j$ for $i \neq j$, $1 \leq i \leq d$, $1 \leq j \leq d$. Then $f = \{f_i\}_{i=1}^d$ is cyclic in $H^2(\mathbb{C}^d)$ if

$$f_i = \exp\frac{b_i}{z-a_i} \quad (b_i \neq 0, \forall i);$$

$$f_i = \log(z-a_i);$$

$$f_i = (z-a_i)^{\alpha_i}, \quad \alpha_i \in \mathbb{C}\backslash(\mathbb{Z} + i\mathbb{Z}).$$

*Remark.* In all three cases one can by an analogous reasoning establish the cyclicity for $|a_i| = 1$, $a_i \neq a_j$ ($i \neq j$) and $b_i\bar{a}_i > 0$, $\operatorname{Re}\alpha_i \in (-\frac{1}{2}, \infty)\backslash\mathbb{Z}$, too.

The following result on cyclic vector functions can be interpreted in the same spirit as the foregoing Corollary 2: the cyclicity of $f = \{f_i\}_{i=1}^d$ signifies a certain "independence" of the singularities of the coordinates $f_i$.

**Corollary 3.** *Let* $\psi = \{\psi_i\}_{i=1}^d \in H^2(\mathbb{C}^d)$. *The following statements are equivalent.*

1. $\psi$ *is a cyclic vector in* $H^2(\mathbb{C}^d)$.
2. *For any cyclic vector* $\varphi$ *in* $H^2(\mathbb{C}^{d-1})$ *there exists an index* $i$, $1 \leq i \leq d$, *such that* $\binom{\varphi}{\psi_i}$ *is cyclic in* $H^2(\mathbb{C}^d)$.

In fact, let us first establish $1 \Rightarrow 2$. Let $\psi$ and $\varphi$ be cyclic vector functions in $H^2(\mathbb{C}^d)$ and $H^2(\mathbb{C}^{d-1})$ respectively and assume that for each $i$, $1 \le i \le d$, there exists a non-zero vector function $\binom{\alpha_i}{\beta_i}$, with $\alpha_i \in H^2(\mathbb{C}^{d-1})$ and $\beta_i \in H^2$, which is orthogonal to $S^{*n}_{\mathbb{C}^d}\binom{\varphi}{\psi_i}$, $n \ge 0$. In view of the cyclicity of $\varphi$ we must have $\beta_i \ne \mathbb{0}$ for all $i$, $1 \le i \le d$. Consider the matrix whose columns are the $\alpha_i$, $1 \le i \le d$. As $\operatorname{Rank} \alpha \le d - 1$ then (cf. the Lemma on the Local Rank and the remark following it) there exist functions $\gamma_i$, $\gamma_i \in H^\infty$ $(1 \le i \le d)$, not all of them zero, such that $\sum_{i=1}^d \gamma_i \alpha_i \equiv \mathbb{0}$. From this it follows that

$$\alpha_i^* \varphi + \bar{\beta}_i \psi_i \in H^2_-, \quad 1 \le i \le d,$$

$$\sum_{i=1}^d \bar{\gamma}_i \bar{\beta}_i \psi_i \in H^2_-,$$

but this last inclusion contradicts the cyclicity of $\psi$.

Before proving the opposite inclusion $(2 \Rightarrow 1)$ let us mention a special case of the statement already proved $(1 \Rightarrow 2)$, which will be used in the second half of the proof.

*Special Case.* For any function $\varphi$, $\varphi \in C^{E_1}$, there exists a function $\psi$, $\psi \in H^2(E_2)$, such that $\varphi + \psi \in C^{E_1 \oplus E_2}$.

In order to see this it suffices to consider first the special case $\dim E_2 = 1$ and then to proceed by induction. If now $\dim E_2 = 1$ (and $E_1 = \mathbb{C}^d$, $E_2 = \mathbb{C}$) then we can take any cyclic vector in $H^2(\mathbb{C}^{d+1})$ $\left(\text{for example, } \psi = \left\{\exp \dfrac{1}{z - k - 1}\right\}_{k=1}^{d+1}\right)$ and increase the "dimension" of $\varphi$ (without destroying the cyclicity) with the aid of the implication $1 \Rightarrow 2$ already established. □

In fact, it is not hard to see that for any function $\varphi$, $\varphi \in H^2(\mathbb{C}^d)$, and for all numbers $a$, $|a| > 1$, except for only $\operatorname{dc}(\varphi)$ exceptional values, we have $\operatorname{dc}\begin{pmatrix} \varphi \\ \exp(z - a)^{-1} \end{pmatrix} = \operatorname{dc}(\varphi) + 1$ (apply the remark following the Theorem on Noncyclic Vectors and the part of Corollary 3 already proved).

Let us now turn to the implication $2 \Rightarrow 1$. Assuming that $\psi$ is non-cyclic in $H^2(\mathbb{C}^d)$ we will construct a $(d - 1)$-dimensional vector $\varphi$ which contradicts Statement 2. Let us first consider the following

*Special Case.* If $\psi_i$, $1 \le i \le d - 1$, are arbitrary functions in $H^2$ (and $\psi_d \equiv 0$, for example) then there exists a $(d - 1)$-dimensional cyclic vector $\varphi$ such that all $d$-dimensional vectors $\binom{\varphi}{\psi_i}$ are non-cyclic.

In order to prove this special case of the general result we are going to use the remark after the Theorem on Noncyclic Vectors and consider a $k$ dimensional coordinate projection $\pi$, $\pi : \mathbb{C}^{d-1} \to \mathbb{C}^{d-1}$, $\operatorname{rank} \pi = k \stackrel{\text{def}}{=} \operatorname{dc}(\psi)$, with the property that $\pi\psi \in C^{\pi\mathbb{C}^{d-1}}$. Let us complete this vector function $\pi\psi$ with new coordinate functions to a vector function $\varphi$ which is cyclic in $H^2(\mathbb{C}^{d-1})$. This $\varphi$ is indeed the desired one: if the coordinate $\psi_i$ is contained in $\pi\psi$ then the non-cyclicity of $\binom{\varphi}{\psi_i}$ is trivial, if not then, on the one hand, $k < d - 1$ and, on the other hand, (by virtue of the remark to the theorem) $(\pi + \pi_i)\psi \notin C^{(\pi+\pi_i)\mathbb{C}^{d-1}}$; that is, also in this case $\binom{\varphi}{\psi_i}$ cannot be cyclic. □

Continuing the proof of the implication $2 \Rightarrow 1$ pick a non-zero function $\vartheta$, $\vartheta \in H^2(\mathbb{C}^d)$, which is orthogonal to $E^*_\psi$, i.e.

$$\sum_{i=1}^{d-1} \bar{\vartheta}_i \psi_i + \bar{\vartheta}_d \psi_d \in H^2_-. \tag{15}$$

Without loss of generality we can assume that $\vartheta_d \not\equiv 0$. Using the special case just established we find a $(d-1)$-dimensional cyclic vector $\varphi$ such that the $\binom{\varphi}{\psi_i}$ are non-cyclic in $H^2(\mathbb{C}^d)$ for all $i$, $1 \le i \le d-1$. There exist then vectors $\binom{\beta_i}{\alpha_i}$, orthogonal to the corresponding subspaces, i.e.

$$\bar{\alpha}_i \psi_i + \beta_i^* \varphi \in H^2_-, \quad 1 \le i \le d-1, \tag{16}$$

with $\alpha_i \not\equiv 0$ (cyclicity of $\varphi$). Let us set $\alpha = \prod_{i=1}^{d-1} \alpha_i$ and $A_i = \alpha/\alpha_i$, let us multiply (16) by the function $\bar{A}_i \bar{\vartheta}_i$ and take the sum over $i$ and let us multiply (15) by $\bar{\alpha}$ and subtract the left hand sides of the resulting inclusions from each other:

$$\left( \sum_{i=1}^{d-1} \bar{A}_i \bar{\vartheta}_i \beta_i^* \right) \varphi - \bar{\alpha} \bar{\vartheta}_d \psi_d \in H^2_-.$$

As $\alpha\vartheta_d \not\equiv 0$, $\binom{\varphi}{\psi_d}$ cannot be cyclic. This contradiction proves that $2 \Rightarrow 1$. □

Concluding our discussion of the classes $N_s$ let us make a few more observations connected with the dc-Lemma and its corollaries. The attempt to use this lemma for an inductive (with respect to $s$) description of the classes $N_s$ is based on the possibility of lowering the degree of cyclicity $\mathrm{dc}(F)$ when passing to the subspace $P_+\Omega^* F$. Such a lowering is possible only if $\operatorname{Rank}\Omega < \dim E$ but an explicit computation of the difference $\mathrm{dc}(F) - \mathrm{dc}(P_+\Omega^* F) = \dim E - \operatorname{Rank}(\Omega, \Theta_F)$ is obstructed by the non-effective definition of $\Theta_F$. Therefore it is of interest to have exact bounds for the drop of the dc-characteristic when one varies the various parameters in the dc-formula. The results are the following (but we are not going to prove them here):

1) Let us fix the subspace $F$, $F \subset H^2(E)$, and let $0 \le \alpha \le \mathrm{dc}(F)$. Which is the largest $\operatorname{Rank}\Omega$ for which $P_+\Omega^* F \in N_\alpha$? Answer: $\dim E + \alpha - \mathrm{dc}(F)$; more exactly

$$\mathrm{dc}(F) = \dim E + \alpha - \max\{\operatorname{Rank}\Omega : P_+\Omega^* F \in N_\alpha\}.$$

The case $\alpha = 0$ is interesting.

2) Let us fix $F$, $F \subset H^2(E)$, and a number $\gamma$, $\dim E - \mathrm{dc}(F) \le \gamma \le \dim E$. How much can we lower the value of $\mathrm{dc}(F)$, using up all possible subspaces $P_+\Omega^* F$ with $\operatorname{Rank}\Omega = \gamma$? Answer: at most to the value $\dim E - \gamma$ (the answer is independent of $F$); more exactly

$$\mathrm{dc}(F) = \dim E - \gamma + \min\{\mathrm{dc}(P_+\Omega^* F) : \operatorname{Rank}\Omega = \gamma\}.$$

The case which can be used in an inductive description of the classes $N_s : \gamma = \dim E - 1$, $F \notin N_0$.

3) Let us fix $\Omega, \Omega \in H^\infty(L(D,E))$. How much can we lower $\mathrm{dc}(F)$ when passing to $P_+\Omega^*F$, employing all possible subspaces $F$? Answer: at most to the value $\dim E - \operatorname{Rank}\Omega$ and this value is attained for some $F$, $F \in N_\alpha$, provided $\alpha \leq \operatorname{Rank}\Omega$; more exactly

$$\alpha \leq \operatorname{Rank}\Omega \Rightarrow \operatorname{Rank}\Omega = \dim E + \alpha - \max\{\mathrm{dc}(F) : P_+\Omega^*F \in N_\alpha\}.$$

The case $\alpha = 0$ is interesting (then the formula holds for all functions $\Omega$)

And finally, concluding this long Section, let us mention the (trivial) generalizations of two "scalar" results from Sects. 1 and 2.

**Lemma on** Lat $S_E^*$, **II.** *Let $\Theta$ be a left inner function. Then*

$$f \in H^2(E) \ominus \Theta H^2(E') \Leftrightarrow \bar{z}\Theta^t(z)f(\bar{z}) \in H^2(E'), \quad f \in H^2(E)$$

*where* $\Theta^t(z) \overset{\text{def}}{=} (\Theta(\bar{z}))^*$. □

**Beurling's Lemma, II.** $\sigma_p(S_E^*) = \mathbb{D}$. *If $\lambda \in \mathbb{D}$ then*

$$\operatorname{Ker}(S_E^* - \bar{\lambda}I)^n = \sum_{k=1}^{n} (\mathbb{1} - \bar{\lambda}z)^{-k} z^{k-1} E;$$

*the right hand side of this identity consists (by the definition of the algebraic sum of subspaces) of all functions of the form $\sum_{k=1}^{n} (\mathbb{1} - \bar{\lambda}z)^{-k} z^{k-1} e_k$, $e_k \in E$, $1 \leq k \leq n$, that is, of all E-valued rational functions with the only pole $1/\bar{\lambda}$ having an order not exceeding n.*

**5. Concluding Remarks.** The connection between various kinds of "pseudo-analytic" continuations and the property for the function of being orthogonal to a z-invariant subspace was noticed rather early. Let us here mention the papers by Bochner and Bohnenblust [1], Carleman [1], Nyman [1], Tumarkin [1]. A general analysis of this relationship (to tell the truth from a quite different point of view) was given by Domar [1] and Shapiro [1], along with a manifold of applications. In particular, in the last paper one can clearly see the rôle of approximation by rational functions with given poles in questions of pseudocontinuation (so-called overconvergence).

The Theorem on Cyclic Vectors was proved by Douglas, Shapiro and Shields [1]. Related ideas can essentially be found in earlier papers on prediction of stochastic processes (Masani-Wiener [1, 2], Rozanov [1], Matveev [1]) but the problem solved there concerning the regularity of a process is immediately connected only with the description of those sets $F$, $F \subset L^2(E)$, for which the invariant subspace $\bigvee(\mathscr{S}_E^{-n}F : n \geq 0)$ does not contain any reducing subspace for the operator $\mathscr{S}_E$. Here is one characterization of such sets $F$ (to be found e.g. in Rozanov's book [2]): there exists a principal minor $M(\zeta) = \det\{f_{ik}(\zeta)\}_{i\in I, k\in K} \neq 0$ a.e. with $\int_{\mathbb{T}} \log M(\zeta)\, dm > -\infty$ such that the ratios $M_{sk}/M$ admit pseudocontinuation, where the minor $M_{sk}$ is obtained from $M$ upon substituting the $s$-th row ($s \notin K$) for the $k$-th row ($k \in K$). Solev observed (unpublished) that from this result one can derive the (scalar) Theorem on Cyclic Vectors proved in Sect. 1. For $\dim E = 1$ the Rozanov-Wiener-Masani result may be supplemented by the following fact, which readily follows from the Beurling-Helson Theorem: a finite set of functions $F$ generates a non-reducing $S$-invariant subspace $\bigvee(\mathscr{S}^n F : n \geq 0)$ iff $\int_{\mathbb{T}} \log|f| > -\infty$, $\forall f \in F$, and the ratios $f'/f$, $f' \in F$, are meromorphic functions of Nevanlinna class in $\mathbb{D}$.

The corollaries of the Theorem on Cyclic Vectors set forth in Sect. 1 are taken over from the paper Douglas-Shapiro-Shields [1].

For Beurling's Lemma see Beurling [1] and for the description of $S^*$-invariant subspaces and the projection $P_\Theta$ see Sz.-Nagy and Foiaş [1]. In the last book (as well as in Helson's book [2]) one can also read about analytic continuation of functions in $(\Theta H^2)^\perp$ over those points on the circle where $\Theta$ is analytic. Indeed, Sz.-Nagy and Foiaş consider the case of the most general functional model for contraction operators (the scalar variant of the corresponding statement is the Corollary of the Livšic-Moeller Theorem, Lecture III).

The Convergence Lemma and the Theorem on Lower Limits are from the paper Gribov-Nikol'skiĭ [1]; in the case when the $\Theta_n$ are Blaschke products the last theorem was in a somewhat different form proved by Tumarkin [2].

The Theorem on Finite Dimensional Approximation is close to one of the main results of the paper Douglas-Shapiro-Shields [1].

The most important Schur's Theorem is in this Lecture developed following an idea of Rudin's [1] but there are other approaches too to approximation of this type (cf. Lecture VIII, Sect. 6; still another proof is in Carathéodory's book [1], p. 13). E.g. when the functions to be approximated are inner, one can use the following theorem due to Frostman [1].

**Frostman's Theorem.** *If $\Theta$ is an inner function then for a.e. $\lambda$, $\lambda \in \mathbb{D}$* (and even for quasi-all points, i.e. for all points outside a set of logarithmic capacity zero) *the function* $\dfrac{\Theta - \lambda \mathbb{1}}{\mathbb{1} - \bar{\lambda}\Theta}$ *will be a Blaschke product.*

*Outline of the proof* (cf. Collingwood-Lohwater [1], Hoffman [1]). The function $u \overset{\text{def}}{=} \int_0^\delta \log|b_t(z)|^{-1}\,dt$ is bounded in clos $\mathbb{D}$ and continuous in (clos $\mathbb{D}$)$\setminus$ $[0,\delta]$, $0 < \delta < 1$, and the family of non-negative functions $g_r$, $g_r(t) = \int_{\mathbb{T}} \log|b_t \circ \Theta_r|^{-1}\,dm$, $t \in (0,\delta)$, depends monotonically on $r$ ($\Theta_r \overset{\text{def}}{=} \Theta(rz)$). Since also $\lim_{r\uparrow 1} u(\Theta_r(\zeta)) = 0$ for a.e. $\zeta$, $\zeta \in \mathbb{T}$, we get $\int_0^\delta (\lim_{r\uparrow 1} g_r(t))\,dt = \lim_{r\uparrow 1} \int_0^\delta g_r(t)\,dt = \lim_{r\uparrow 1} \int_{\mathbb{T}} (u \circ \Theta_r)\,dm = 0$ so that $\lim_{r\uparrow 1} g_r(t) = 0$ for a.e. $t$, $t \in (0,\delta)$. Now from the canonical factorization of the inner function $b_t \circ \Theta$ as a product of a Blaschke product and a singular factor it is clear that $g_r(t) \geq \mu_t(t)$ where $\mu_t$ is the singular measure corresponding to $b_t \circ \Theta$. Therefore $b_t \circ \Theta$ must be a Blaschke product for a.e. $t$. The same reasoning works for any other radius $[0, \delta\lambda]$, $|\lambda| = 1$. □

As $\lim_{\lambda\to 0} \left\| \Theta - \dfrac{\Theta - \lambda\mathbb{1}}{\mathbb{1} - \bar{\lambda}\Theta} \right\|_\infty = 0$ thus the corollary of Schur's Theorem which interests us follows from this after first approximating the Blaschke products $\dfrac{\Theta - \lambda\mathbb{1}}{\mathbb{1} - \bar{\lambda}\Theta}$ with their partial products. This approach, thus based on an application of Frostman's Theorem, has been substantially extended in Ginzburg's paper [3] where one can find not only operator analogues of this theorem but also interesting applications to the determination of the "typical" spectral structure of contraction operators $T$ with finite rank defects ($\operatorname{rank}(I - T^*T)) = \operatorname{rank}(I - TT^*) < \infty$). In connection with the convergence of sequences of inner functions let us mention a very general theorem due to Tumarkin [3], [4] which can be applied e.g. to the computation of the GCD in the Theorem on Lower Limits. Here is the formulation of his result: let $f_n$ ($n = 1, 2, \ldots$) be analytic functions in $\mathbb{D}$ with

$$\sup\left\{\int_{\mathbb{T}} \log^+|f_n(r\zeta)|\,dm(\zeta) : 0 \le r < 1, n \ge 1\right\} < \infty;$$

in order that the sequence $\{f_n\}_{n\ge1}$ should converge uniformly on compact subsets of $\mathbb{D}$ it is necessary and sufficient that the sequence of measures $d\mu_n = \log|f_n|\,dm + d\mu_n^s + d\mu_n^B$ in the canonical representation of the functions $f_n$ ($d\mu_n^s$, a singular measure on $\mathbb{T}$, $d\mu_n^B$, a discrete one with the masses $\frac{1}{2}k_{f_n}(\zeta)(1 - |\zeta|^2)$ at the points $\zeta$, $\zeta \in \mathbb{D}$) converges weakly to a limit $\log|f|\,dm + d\mu^s + d\mu^B$ (from which the limit function $\lim_n f_n$ then can be recovered). This result can be deduced from an analogous proposition due to Fihtengol'c [1] on the convergence of sequences of harmonic functions.

For Sect. 3 of this Lecture see Gribov-Nikol'skiĭ [1]. The Theorem on Finite Dimensional Approximation, II covers the main result of Hilden-Wallen [1]; we have just to remark (and this is quite easy) that in the hypotheses of the theorem of that paper one by necessity has $\sum_{\lambda\in\sigma_p(T)}(1 - |\lambda|) = \infty$, i.e., the point spectrum $\sigma_p(T)$ is a set of uniqueness for the class $H^\infty$. On the whole one can say that the capacity cap, which is constructed using the dual space $X^*$ and which adequately corresponds to the uniqueness theorem there, opens up more vistas than the capacity Cap connected with the calculus. A simple example is contained in Hilden-Wallen [1]: if $X$ is the space of regular functions in $\mathbb{D}$ admitting a finite Dirichlet integral $\iint_{\mathbb{D}}|f'|^2\,dx\,dy$ and $T = S^*$ is the inverse shift in $X$, then it is easy to see that $\mathfrak{A}(S^*) = H^\infty$ (the maximal calculus can be constructed on the basis of the algebra $H^\infty$) and the capacity $\mathrm{Cap}_{\mathfrak{A}} = \mathrm{Cap}_{H^\infty}$ is insufficient for the description of all $S^*$-invariant subspaces. The corresponding counter-example in that paper is indeed based on the existence of a set of non-uniqueness $\sigma$ for the space $X^*$ (with respect to the standard Cauchy duality $\langle f, g\rangle = \sum_{n\ge0}\hat{f}(n)\hat{g}(n)$ this is the Bergman space $\{g : g$ regular in $\mathbb{D}$, $\iint_{\mathbb{D}}|g|^2\,dx\,dy < \infty\}$) with the property $\sum_{\lambda\in\sigma}(1 - |\lambda|) = \infty$. It is clear that a use of the capacity cap allows one to bypass this obstruction; moreover one has reasons to believe (cf. Gribov-Nikol'skiĭ [1], Nikol'skiĭ [5]) that with the aid of cap one can prove the general analogue of the main statement (Statement 1) of the Theorem on Finite Dimensional Approximation. Vector variants of this theorem and of the Theorem on Lower Limits seem not to be known.

The Cyclicity Theorem for $S_E^*$ has an illustrative character and can be found e.g. in Halmos's problem book [2] but both of its statements admit essential strengthenings. Concerning Statement 2 we said already a few words after the proof of the theorem and an extension of it in a certain direction is contained in the lemma forming a part of the proof. The proof of Statement 1 likewise contains a slightly stronger fact which may be formulated as follows: we have really constructed not only a cyclic but indeed a "hypercyclic" vector $f$, i.e. an element such that

$$\operatorname{clos}\{\lambda S^{*n}f : n \geq 0, \lambda \in \mathbb{C}\} = H^2(E).$$

It is also know (Wogen [1]) that there exist functions $f$ (defined by requirements of the same character as in the proof of the Cyclicity Theorem for $S_E^*$) which are cyclic respect to all *Toeplitz operators* $g \mapsto P_+\bar{\varphi}g$ ($g \in H^2$) with anti-analytic symbols $\bar{\varphi}$, $\varphi \in H^\infty$, differing from a constant ($\varphi \not\equiv$ const). There likewise a more general fact is established: the operator $T$ in the space $l^2(E)$ of vector valued sequences has a cyclic vector if its matrix $\{T_{ij} : 1 \leq i, j < \infty\}$ (with respect to the standard decomposition as an orthogonal sum) is such that $T_{ij} = \mathbb{O}$ for $j < i + n$ and $\operatorname{clos} T_{i,i+n}E = E$, $i \geq 1$ (with a suitable $n$, $n \geq 0$).

The Theorem on Noncyclic Vectors for the operator $S_E^*$ was proved by Vasyunin and Nikol'skiĭ and is here published for the first time. The Sublemma used in the proof is suggested by a result of Tolokonnikov (see App. 3, Sect. 10) and the trick with the functions $\Phi_\kappa$ is borrowed from the paper Fuhrmann [1]. In fact, the Complementing Lemma holds under the weaker hypothesis $\dim E' < \infty$ (see Vasyunin-Nikol'skiĭ [1]).

The Corollary of the Projection Lemma, II (description of the class $C_{00}$ in terms of characteristic functions) is due to Sz.-Nagy and Foiaş; cf. their book [1].

# Lecture III. Compressions of the Shift and the Spectra of Inner Functions

Following the program outlined in the Introductory Lecture let us proceed to the main object of this course, viz. the operator $S^*|K$, where $K$ is a subspace of $H^2$-invariant for the operator $S^*$. Such restrictions will be called *parts* of the original operator. Together with these it will also be convenient to consider the adjoint operators, the so-called *compressions of the shift operator*. (We will also refer to the operator $S^*|K$, as well as to its adjoint, as the "truncated shift".) This terminology is motivated by the following

**Lemma on the Compression of the Shift.** *Let $K \in \operatorname{Lat} S^*$; $P = P_K$. Then*

$$(S^*|K)^* = PS|K \text{ and } (PS|K)^n = PS^n|K, \quad n \geq 0.$$

*The proof* follows at once from the definition of orthogonal projection and the identity $(S^{*n}x, y) = (x, S^n y)$ valid for any $x$, $y$ in $K$ and all $n$, $n \geq 0$. □

The aim of this lecture is to determine the spectra of the operators $PS|K$ and $S^*|K$, to build up a $H^\infty$-functional calculus and to prove for it a spectral mapping theorem.

**1. The Spectrum of an Operator and the Spectrum of a Function.**

**Definition.** Let $\Theta$ be an inner function. By the *spectrum* $\sigma(\Theta)$ of $\Theta$ we mean the complement (with respect to the whole closed disk clos $\mathbb{D}$) of the set of all points $\lambda$, $\lambda \in \operatorname{clos} \mathbb{D}$, such that the function $1/\Theta$ can be continued analytically into a (full) neighborhood of $\lambda$.

This definition is in good agreement with the usual conception in the theory of analytic functions of the spectrum as being the set of all "singularities" of the function in question. In what follows we will denote by $\sigma_p(T)$ the point spectrum of the linear operator $T$ (i.e. $\sigma_p(T) = \{\lambda : \lambda \in \mathbb{C}, \operatorname{Ker}(T - \lambda I) \neq \{0\}\}$) and by $\sigma(T)$ its spectrum, $R(\lambda, T) = (\lambda I - T)^{-1}$ being the resolvent. Furthermore $\Theta^{-1}(0) = \{\lambda : \lambda \in \mathbb{D}, \Theta(\lambda) = 0\}$ is the set of all zeros of $\Theta$ and $\operatorname{supp} \mu$ denotes the support of the measure (any measure) $\mu$.

**Livšic-Moeller Theorem**[1]. *Let $\Theta$ be an inner function and set $K = (\Theta H^2)^\perp$, $P = P_K$, $T = PS|K$. Then*

$$\sigma(T) = \sigma(\Theta), \quad \sigma_p(T) = \sigma(\Theta) \cap \mathbb{D}.$$

It is clear that this theorem gains the greater significance the better we can describe the spectrum $\sigma(\Theta)$ in terms of suitable natural parameters for the function $\Theta$. The proof of Theorem LM will likewise be based on such a descrip-

[1] In references Theorem LM.

tion, using the canonical factorization of $\Theta$ as the product of a Blaschke product and a singular function (see Appendix 2).

**Theorem on the Spectrum of a Function.** *Let $\Theta$ be an inner function and let $\Theta = B\cdot\exp\left(-\int_{\mathbb{T}}\frac{\zeta+z}{\zeta-z}d\mu^s(\zeta)\right)$ be its canonical factorization. Then*

$$\begin{aligned}\sigma(\Theta) &= \operatorname{clos}\Theta^{-1}(0)\cup\operatorname{supp}\mu^s\\ &= \left\{\lambda:\lambda\in\operatorname{clos}\mathbb{D},\ \underline{\lim}_{\zeta\to\lambda,\zeta\in\mathbb{D}}|\Theta(\zeta)|=0\right\}.\end{aligned}$$

*Proof.* Let us first verify that

$$\operatorname{clos}\Theta^{-1}(0)\cup\operatorname{supp}\mu^s\supset\sigma(\Theta), \tag{1}$$

i.e. that the function $1/\Theta$ can be continued analytically to the complement of the set standing in the left hand side of this inclusion. This is trivial for the factor $\exp\left(\int_{\mathbb{T}}\frac{\zeta+z}{\zeta-z}d\mu^s\right)$ of the function $1/\Theta$, while the analytic continuation of $B$ (and thereby of $1/B$ too) across points of the set $\mathbb{T}\backslash\operatorname{clos}\Theta^{-1}(0)$ follows from the estimate

$$\begin{aligned}|1-b_\lambda(\zeta)| &= \left|\frac{1}{|\lambda|}-\frac{1}{|\lambda|}\frac{|\lambda|^2-\bar\lambda\zeta}{1-\bar\lambda\zeta}+1-\frac{1}{|\lambda|}\right|\\ &\le\frac{1-|\lambda|}{|\lambda|}+\frac{1-|\lambda|^2}{|\lambda|}\frac{1}{|1-\bar\lambda\zeta|}\\ &\le\operatorname{const}(1-|\lambda|)\frac{1}{\operatorname{dist}(\zeta,\{\bar\lambda^{-1}:\Theta(\lambda)=0\})},\end{aligned}$$

if we recall that $B=\prod_\lambda b_\lambda^{k(\lambda)}$, $\sum_\lambda k(\lambda)(1-|\lambda|)<\infty$. This establishes (1).

As clearly

$$\sigma(\Theta)\supset\left\{\lambda:\lambda\in\operatorname{clos}\mathbb{D},\ \underline{\lim}_{\zeta\to\lambda,\zeta\in\mathbb{D}}|\Theta(\zeta)|=0\right\}$$

it remains to be verified that this last set contains $\operatorname{clos}\Theta^{-1}(0)\cup\operatorname{supp}\mu^s$. This is a triviality what concerns the first term (i.e. $\operatorname{clos}\Theta^{-1}(0)$). Let us therefore check that

$$\lambda\in\mathbb{T},\ \underline{\lim}_{\zeta\to\lambda,\zeta\in\mathbb{D}}|\Theta(\zeta)|>0\Rightarrow\lambda\notin\operatorname{supp}\mu^s.$$

In fact, if $\Delta$ is an open neighborhood of the point $\lambda$ such that the function $|\Theta|$ is bounded away from 0 in $\Delta$ (or better, in $\Delta\cap\mathbb{D}$) and if $\delta$ is a compact arc of the circumference $\mathbb{T}$ contained in $\Delta$ then the singular function

$$\Theta_\delta=\exp\left(-\int_\delta\frac{\zeta+z}{\zeta-z}d\mu^s(\zeta)\right)$$

too is bounded away from 0 near the circle $\mathbb{T}$ (as $|\Theta_\delta|\ge|\Theta|$ in $\mathbb{D}$). Therefore $1/\Theta_\delta\in H^2$ and $\Theta_\delta\equiv\operatorname{const}$, that is, $\mu^s\delta=0$. □

**2. Functional Calculus and Derivation of Theorem LM.** Before proceeding to Theorem LM let us establish a lemma which also has an independent interest.

**Calculus Lemma.** *In the notation of Theorem LM let us set*

$$\varphi(T) = P_K \boldsymbol{\varphi}|K, \quad \varphi(T^*) = P_+ \bar{\boldsymbol{\varphi}}^t|K,$$

*where* $\varphi \in H^\infty$, *with* $\varphi^t \stackrel{\text{def}}{=} \overline{\varphi(\bar{z})}$. *Then*

a) $|\varphi(T)| \le \|\varphi\|_\infty$;

b) *the mapping* $\varphi \mapsto \varphi(T)$ *is linear, multiplicative, and coincides on polynomials with the natural one (i.e.* $p(T) = \sum_{k\ge 0} \hat{p}(k) T^k$ *if* $p \in \mathscr{P}_A$*); moreover* $\varphi(T^*) = \varphi^t(T)^*$;

c) $\operatorname{Ker} \varphi(T) = \frac{\boldsymbol{\Theta}}{\boldsymbol{\psi}} H^2 \ominus \boldsymbol{\Theta} H^2$, $\quad \operatorname{Ker} \psi^t(T^*) = H^2 \ominus \boldsymbol{\psi} H^2 \quad$ *where* $\psi = \operatorname{GCD}(\Theta, \varphi^i)$; *in particular* $\varphi(T) = \mathbb{O} \Leftrightarrow \varphi^t(T^*) = \mathbb{O} \Leftrightarrow \varphi/\Theta \in H^\infty$.

d) $(s)\text{-}\lim_n T^n = (s)\text{-}\lim_n T^{*n} = \mathbb{O}$.

*Proof.* a) Trivial.

b) The identity $z^n(T) = T^n$, $n \ge 0$, is contained in the Lemma on the Compression of the Shift. The multiplicative property follows from the following identity (with $f \in H^2$, $g = g_f \in H^2$):

$$P_K \varphi P_K \psi f = P_K \varphi(\psi f + \Theta g) = P_K \varphi \psi f.$$

c) $\operatorname{Ker} \varphi(T) = \{f \in K : \varphi f \in \boldsymbol{\Theta} H^2\}$. But $\varphi f \in \boldsymbol{\Theta} H^2 \Leftrightarrow$ the function $\Theta$ divides the product $\varphi^i f^i$. As $\Theta/\psi$ and $\varphi^i/\psi$ are relatively prime this last condition is equivalent to the assertion that $f^i$ is divided by $\Theta/\psi$. Consequently $\operatorname{Ker} \varphi(T) = \left\{f \in K : f \in \frac{\boldsymbol{\Theta}}{\boldsymbol{\psi}} H^2\right\} = \frac{\boldsymbol{\Theta}}{\boldsymbol{\psi}} H^2 \ominus \boldsymbol{\Theta} H^2$.

In an analogous way we prove the second identity: We have

$$\operatorname{Ker} \varphi^t(T^*) = \operatorname{Ker} \varphi(T)^* = \{f \in K : P_K \bar{\varphi} f = \mathbb{O}\}$$

but as $P_+ \bar{\varphi} f \in K$ (cf. the remark following the Theorem on Lower Limits in Lecture II) we obtain

$$\begin{aligned} \operatorname{Ker} \varphi^t(T^*) &= \{f \in K : P_+ \bar{\varphi} f = 0\} \\ &= \{f \in K : f \perp \boldsymbol{\varphi} H^2\} \\ &= (H^2 \ominus \boldsymbol{\Theta} H^2) \cap (H^2 \ominus \boldsymbol{\varphi}^i H^2) \\ &= H^2 \ominus \boldsymbol{\psi} H^2. \end{aligned}$$

d) The identity $(s)\text{-}\lim T^{*n} = \mathbb{O}$ is trivial, because $\|T^{*n} f\|_2^2 = \|S^{*n} f\|_2^2 = \sum_{k\ge n} |\hat{f}(k)|^2$ for any $n$, $n \ge 0$, and any $f$ in $K$. The corresponding statement for the operator $T$ follows from the identity

$$\begin{aligned} \|P_K z^n f\|_2 &= \|z^n f - \Theta P_+ \bar{\Theta} f z^n\|_2 \\ &= \|z^n \bar{\Theta} f - P_+ \bar{\Theta} f z^n\|_2 \\ &= \|(I - P_+) \bar{\Theta} f z^n\|_2. \ \square \end{aligned}$$

*Remark.* We have already encountered the last statement of the lemma (see "Supplements and Bibliographical Notes" to Lecture II). From c) it follows notably that

$$\operatorname{Ker}\varphi(T) = \{\mathbb{0}\} \Leftrightarrow \operatorname{Ker}\varphi(T)^* = \{\mathbb{0}\}$$

so that

$$\operatorname{Ker}\varphi(T) = \{\mathbb{0}\} \Leftrightarrow \operatorname{clos}\varphi(T)K = K.$$

*Proof of Theorem LM.* Let us first verify that $\sigma(T) \subset \sigma(\Theta)$. As $|T| \leq 1$ we have at least $\sigma(T) \subset \operatorname{clos}\mathbb{D}$. Let $\lambda \in \operatorname{clos}\mathbb{D}\backslash\sigma(\Theta)$. Then $\Theta$ is regular in a neighborhood of $\lambda$ and $\Theta(\lambda) \neq 0$. Therefore the function $u$,

$$u \stackrel{\text{def}}{=} (\Theta(\lambda)\mathbb{1} - \Theta)(\lambda\mathbb{1} - z)^{-1}$$

belongs to $H^\infty$ and by the Calculus Lemma

$$(\lambda I - T)u(T) = u(T)(\lambda I - T) = \Theta(\lambda)I - \Theta(T) = \Theta(\lambda)I.$$

Thus $\lambda \notin \sigma(T)$ and $R(\lambda, T) = u(T)/\Theta(\lambda)$.

Let us prove that $\sigma_p(T) = \sigma(\Theta) \cap \mathbb{D}$. From the Calculus Lemma, part d), it follows that $\sigma_p(T) \subset \mathbb{D}$. Therefore it suffices to show that $\lambda \in \mathbb{D}$, $\Theta(\lambda) = 0$ entails $\lambda \in \sigma_p(T)$. But, in the previous notation, $\mathbb{0} = \Theta(\lambda)I = (T - \lambda I)u(T)$ so if we assume that $T - \lambda I$ is injective we must have $u(T) = \mathbb{0}$ and it follows again from the Calculus Lemma, Statement c), that $u/\Theta \in H^\infty$. This contradiction proves that $\sigma_p(T) = \sigma(\Theta) \cap \mathbb{D}$.

It remains to prove that $\sigma(\Theta) \cap \mathbb{T} \subset \sigma(T)$. Let $\lambda \in \mathbb{T}\backslash\sigma(T)$. Let us verify that in this case every function in $K = H^2 \ominus \Theta H^2$ admits an analytic continuation across the point $\lambda$; we can then apply this to the function

$$f \stackrel{\text{def}}{=} P_\Theta(\mathbb{1} - \bar{\zeta}z)^{-1} = (\mathbb{1} - \bar{\zeta}z)^{-1}(\mathbb{1} - \Theta\bar{\Theta}(\zeta))$$

where $\zeta$ is some point in the disk with $\Theta(\zeta) \neq 0$, which gives $\lambda \notin \sigma(\Theta)$ concluding the proof. Thus let $f \in K$, $\bar{\lambda} \in \mathbb{T}\backslash\sigma(T^*)$. Then the vector function

$$t \mapsto (I - tS^*)^{-1}f, \ |t| < 1$$

can be analytically continued to a neighborhood of the point $\lambda$ and therefore a fortiori also the function

$$((I - tS^*)^{-1}f, \mathbb{1}) = \sum_{n\geq 0} t^n(S^{*n}f, \mathbb{1}) = \sum_{n\geq 0} t^n\hat{f}(n) = f(t). \ \square$$

**Corollary.** *The inner function $\Theta$ admits an analytic continuation across a point of the circle $\mathbb{T}$ iff every function in the subspace $H^2 \ominus \Theta H^2$ too has this property.*

*The proof* of this statement is contained in the last but one paragraph. □

**3. The Spectrum of the Operator $\varphi(T)$.** The operator calculus constructed in Sect. 2 will play a significant rôle in the following analysis of the operators $PS|K$ and $S^*|K$. Let us establish a spectral mapping theorem for this calculus. A peculiar feature of this theorem is that in the formula we have in mind, viz.

$\sigma(\varphi(T)) = \varphi(\sigma(T))$, we must pay special attention to the meaning of the right hand side, because for functions in $H^\infty$ the values at points of the circumference $\mathbb{T}$ are not defined everywhere. Our idea to overcome this difficulty is to consider instead of the "value" $\varphi(\lambda)$, $\lambda \in \mathbb{T}$, the set of all limiting values of the function $\varphi$ at the point $\lambda$, i.e. the set $\{\zeta : \underline{\lim}_{w \to \lambda, w \in \mathbb{D}} |\varphi(w) - \zeta| = 0\}$.

**Spectral Mapping Theorem.** *Let $\Theta$ be an inner function and set $T = P_\Theta S|K$, $K = H^2 \ominus \Theta H^2$. Let $\varphi \in H^\infty$. Then*

$$\sigma(\varphi(T)) = \left\{\zeta : \zeta \in \mathbb{C},\ \inf_{\mathbb{D}}(|\Theta| + |\varphi - \zeta \mathbb{1}|) = 0\right\}. \tag{2}$$

In particular, if $\varphi$ has a continuous extension to clos $\mathbb{D}$ then $\sigma(\varphi(T)) = \varphi(\sigma(\Theta)) = \varphi(\sigma(T))$. The Spectral Mapping Theorem is intimately connected with another important result for the algebra $H^\infty$, namely the Corona Theorem by Carleson, which will play an active rôle in our analysis of the model operators. We give here its formulation with a somewhat simplified estimate; for a sharper form and for the proof see Appendix 3.

**Corona Theorem.** *Let $f_i \in H^\infty$, $i = 1, \ldots, n$, and assume that*

$$\inf_{\mathbb{D}} \sum_{i=1}^{n} |f_i|^2 \overset{\text{def}}{=} \delta^2 > 0.$$

*Then there exist functions $g_i$, $i = 1, \ldots, n$, in $H^\infty$ such that $\sum_{i=1}^n f_i g_i = 1$. Moreover $\sup_{\mathbb{D}} (\sum_{i=1}^n |g_i|^2)^{1/2} \leq 11\delta^{-3}$, provided $\sum_{i=1}^n |f_i|^2 \leq 1$.*

*Proof of the Spectral Mapping Theorem.* The identity (2) consists of two inclusions $\supset$ and $\subset$, the proof of the first of which does not call for the introduction of any additional tools. In fact, if $\psi \in H^\infty$ then $\psi(T)^* = P_+\bar{\psi}|K$ and therefore

$$P_+\bar{\psi}\frac{\mathbb{1}}{1 - \bar{\lambda}z} = \frac{\overline{\psi(\lambda)}}{\mathbb{1} - \bar{\lambda}z}, \quad \lambda \in \mathbb{D},$$

$$\begin{aligned}\psi(T)^* P_\Theta \frac{\mathbb{1}}{\mathbb{1} - \bar{\lambda}z} &= P_+\bar{\psi}\left\{\frac{\mathbb{1}}{\mathbb{1} - \bar{\lambda}z} - \Theta P_+ \frac{\bar{\Theta}}{\mathbb{1} - \bar{\lambda}z}\right\} \\ &= \frac{\overline{\psi(\lambda)}}{\mathbb{1} - \bar{\lambda}z} - \overline{\Theta(\lambda)} \cdot P_+\bar{\psi}\left(\frac{\Theta}{\mathbb{1} - \bar{\lambda}z}\right),\end{aligned}$$

$$\left\|\psi(T)^* P_\Theta \frac{\mathbb{1}}{\mathbb{1} - \bar{\lambda}z}\right\|_2 \leq (|\psi(\lambda)| + |\Theta(\lambda)| \cdot \|\psi\|_\infty)(1 - |\lambda|^2)^{-1/2},$$

while

$$\begin{aligned}\left\|P_\Theta \frac{\mathbb{1}}{\mathbb{1} - \bar{\lambda}z}\right\|_2^2 &= \left\|\frac{\mathbb{1}}{\mathbb{1} - \bar{\lambda}z}\right\|_2^2 - \left\|\Theta P_+ \frac{\bar{\Theta}}{\mathbb{1} - \bar{\lambda}z}\right\|_2^2 \\ &= \frac{1 - |\Theta(\lambda)|^2}{1 - |\lambda|^2}.\end{aligned}$$

These computations show that

$$\inf\{\|(\varphi(T)^* - \bar{\zeta}I)f\|_2 : \|f\|_2 = 1\} = 0$$

if $\inf_{\mathbb{D}}(|\Theta| + |\varphi - \zeta\mathbb{1}|) = 0$ (take $\psi = \varphi - \zeta\mathbb{1}$, $f = P_\Theta(1 - \bar{\lambda}z)^{-1}$. $\|P_\Theta(\mathbb{1} - \bar{\lambda}z)^{-1}\|_2^{-1}$, $\lambda \in \mathbb{D}$), that is, in this hypothesis holds $\zeta \in \sigma(\varphi(T))$.

The reverse inclusion ($\subset$) is obtained from the Corona Theorem in conjunction with the Calculus Lemma. Indeed, if $\inf_{\mathbb{D}}(|\Theta| + |\varphi - \zeta\mathbb{1}|) > 0$ then there exist functions $u$ and $v$ in the algebra $H^\infty$ such that

$$u\Theta + v(\varphi - \zeta\mathbb{1}) = \mathbb{1}.$$

But then also

$$\begin{aligned} u(T)\Theta(T) + v(T)(\varphi(T) - \zeta I) &= v(T)(\varphi(T) - \zeta I) \\ &= (\varphi(T) - \zeta I)v(T) = I, \end{aligned}$$

that is, $\zeta \notin \sigma(\varphi(T))$. □

Returning to Theorem LM let us point out that it of course perfectly agrees with formula (2). Indeed, it can be reformulated as a special case of the Spectral Mapping Theorem:

$$\sigma(T) = \left\{\zeta : \zeta \in \mathbb{C},\ \inf_{\mathbb{D}}(|\Theta| + |z - \zeta\mathbb{1}|) = 0\right\}.$$

## Supplements and Bibliographical Notes

**4. The Cyclic Vectors for the Operators** $T = PS|K$ **and** $T^*$ can be described in the spirit of Lecture II. In fact, from the proof of the Theorem on Cyclic Vectors it follows that the function $f$ belonging to $K = H^2 \ominus \Theta H^2$ is a cyclic element for the operator $T^* = S^*|K$ iff its pseudocontinuation $\tilde{f}$ has at least as "much singularities" in $\hat{\mathbb{C}}\backslash\operatorname{clos}\mathbb{D}$ as the function $\tilde{\Theta} : \zeta \mapsto \frac{1}{\zeta}\overline{\Theta\left(\frac{1}{\bar{\zeta}}\right)}^{-1}$, $|\zeta| > 1$ (more exactly, when from $\vartheta\tilde{f} \in H^2_-$ follows that the function $\tilde{\Theta}^{-1}$, which is inner in $\hat{\mathbb{C}}\backslash\operatorname{clos}\mathbb{D}$, divides the function $\vartheta$, inner in the same domain). This description is not always very effective but if $H^2 \ominus \Theta H^2$ contains a dense set of rational functions (that is, when $\Theta$ is a Blaschke product $\prod_{\lambda \in \sigma_p(T)} b_\lambda^{k(\lambda)}$) then this criterion for cyclicity turns out to be sufficiently transparent: $f$ cyclic means that $\tilde{f}$ has at the point $1/\bar{\lambda}$ a pole of order at least $k(\lambda)$, $\lambda \in \mathbb{D}$. If we take as $f$ the series $\sum_{\lambda \in \sigma_p(T)} \sum_{k=1}^{k(\lambda)} \frac{c_{k,\lambda}}{(\mathbb{1} - \bar{\lambda}z)^k}$ summed in a special way, this says that $c_{k(\lambda),\lambda} \neq 0\ \forall \lambda \in \sigma_p(T)$ (regarding this matter, cf. Lecture VIII).

There is yet a quite different method of deciding whether an element $f$ is cyclic for the operator $T^* = S^*|K$ or not, and that method does not only apply to $T^*$ and the adjoint operator $T = P_K S|K$ but works also in much more general circumstances, namely in the framework of "$C_0$ operators" (see below Sect. 6), for which the problem of cyclicity (of the operator), as well as the one of (determining all) cyclic vectors will be considered in Lecture IV and in Appendix 1.

**Cyclicity Theorem.** *Put* $T = P_K S|K$, $K = H^2 \ominus \Theta H^2$ *and let* $C^T$ *be the set of all cyclic vectors for* $T$.

1. $f \in C^T \Leftrightarrow (\varphi \in H^\infty, \varphi(T)f = \mathbb{O} \Rightarrow \varphi/\Theta \in H^\infty)$.
2. $f \in C^{T^*} \Leftrightarrow (\varphi \in H^\infty, \varphi(T^*)f = \mathbb{O} \Rightarrow \varphi^t/\Theta \in H^\infty)$.

*Proof.* 1. If $f \in C^T$ and $\varphi(T)f = \mathbb{O}$ then $T^n\varphi(T)f = \varphi(T)T^nf = \mathbb{O}$, $n \geq 0$, so $\varphi(T) = \mathbb{O}$. By the Calculus Lemma (part c)) $\varphi/\Theta \in H^\infty$. The converse follows from the same lemma: the orthogonal complement of the $T$-invariant subspace $E_f^T = \bigvee(T^nf : n \geq 0)$ belongs to the lattice $\operatorname{Lat}(S^*|K)$ and therefore has the form $H^2 \ominus \vartheta H^2$, $\Theta/\vartheta \in H^\infty$. Hence $E_f^T = \vartheta H^2 \ominus \Theta H^2$ and, by the lemma quoted, $(\Theta/\vartheta)(T)f = \mathbb{O}$. The hypothesis therefore gives $\vartheta = \text{const}$ so that indeed $f \in C^T$.
2. Analogous proof. □

The description of $C^T$ can be made even more transparent: $f \in C^T$ iff the inner part $f^i$ of $f$ is relative prime to $\Theta$ (for details see Corollary 1 of the Theorem on Lat $T_\Theta$, Lecture IV).

Let us also remark that the preceding criteria for cyclicity are direct generalizations of the corresponding results in the theory of elementary divisors in linear algebra (see Gantmaher [1]) and coincide with the latter if $\dim K_\Theta < \infty$ (i.e. when $\Theta$ is a finite Blaschke product). However the real depth of this analogy will be apparent only in the framework of the theory of the class $C_0$ (cf. Sect. 6 below, Lecture IV and Appendix 1). As we have already noticed, for this class, the problem of defining the multiplicity $\mu_T$ of the spectrum (in terms of the characteristic function $\Theta_T$) is not at all simple. Not less (if not more) interesting is the problem of a detailed study of the functions $\varphi(T)$ of the operator $T$ ($T = PS|K$, $K = H^2 \ominus \Theta H^2$) introduced in the Calculus Lemma: for which $\varphi$ ($\varphi \in H^\infty$) are the operator $\varphi(T)$ and $\varphi(T^*)$ cyclic? Which are their cyclic vectors? More generally, what can be said about Lat $\varphi(T)$ and Let $\varphi(T^*)$? Let us remark that similar questions are of interest (and turn out to be difficult) already for the operators $\varphi$ and $P_+\bar{\varphi}$ in the space $H^2$ (i.e. without any projection on the subspace $K$); regarding this question see Lecture IV, Sect. 7 and also Sarason [3], Nikol'skiĭ [6].

**5. A Calculus for Completely Non-Unitary Contractions** can be based on the Dilation Theorem (Intr. Lecture) but first it has to be shown that the minimal unitary dilation has an absolutely continuous spectrum. We begin with the following result which generalizes the Lemma on the Compression of the Shift.

**Sarason's Projection Lemma.** *Let* $H$ *be a Hilbert space,* $A$ *a linear operator in* $H$ *and* $K$ *a subspace of* $H$. *Set*

$$B = P_K A|K.$$

*The operator $A$ is a dilation of the operator $B$* (i.e. $B^n = P_K A^n | K$, $n \geq 0$) *iff* $K = H_1 \ominus H_2$ *where* $H_i \in \operatorname{Lat} A$, $i = 1, 2$, $H_2 \subset H_1$.

*Proof.* The sufficiency of this condition follows at once if we notice that with respect to the decomposition

$$H = H_1^{\perp} \oplus K \oplus H_2$$

the operator $A$ has the lower triangular matrix

$$A \cong \begin{pmatrix} * & \mathbb{0} & \mathbb{0} \\ * & B & \mathbb{0} \\ * & * & * \end{pmatrix},$$

so that the crucial dilation identity $B^n = P_K A^n | K$, $n \geq 0$, follows immediately from the rules of matrix multiplication.

For the proof of the necessity set $H_1 = \bigvee (A^n K : n \geq 0)$, $H_2 = H_1 \ominus K$. Let us verify that $AH_2 \subset H_2$. But if $x \in K$ then $P_K A A^n x = B B^n x = B P_K A^n x$, $n \geq 0$, so that $P_K A | H_1 = B P_K | H_1$. From this it follows that $P_K A H_2 = \{\mathbb{0}\}$ and $AH_2 \subset H_2$. □

**Theorem of Sz.-Nagy and Foiaş.** *Let $T$ be a contraction in the Hilbert space $H$ and $U$ its minimal unitary dilation. If $T$ is completely non-unitary then the spectrum of $U$ is absolutely continuous.*

*Proof.* Let $\mathscr{H}$ be the space where the dilation $U$ is defined and let $\mathscr{H} = \mathscr{H}_s \oplus \mathscr{H}_a$ be its decomposition as a sum of a singular and an absolutely continuous subspace with respect to the spectral measure of $U$ (regarding this matter cf. the "Supplements and Bibliographical Notes" to Lecture I). By Sarason's Projection Lemma $H = \mathscr{H}^1 \ominus \mathscr{H}^2$ where $\mathscr{H}^i \in \operatorname{Lat} U$, $i = 1, 2$, $\mathscr{H}^2 \subset \mathscr{H}^1$. By Corollary 2 of the Theorem on Invariant Subspaces (Lecture I, Sect. 5) $\mathscr{H}^i = \mathscr{H}_s^i \oplus \mathscr{H}_a^i$ where $\mathscr{H}_s^i$, $\mathscr{H}_a^i \in \operatorname{Lat} U$ $(i = 1, 2)$; $\mathscr{H}_s^i \subset \mathscr{H}_s$, $\mathscr{H}_a^i \subset \mathscr{H}_a$ $(i = 1, 2)$, the subspaces $\mathscr{H}_s^i$ $(i = 1, 2)$ being reducing (consequence of the same theorem). Thus

$$H = (\mathscr{H}_s^1 \ominus \mathscr{H}_s^2) \oplus (\mathscr{H}_a^1 \ominus \mathscr{H}_a^2)$$

where the term $\mathscr{H}_s^1 \ominus \mathscr{H}_s^2$, being the orthogonal difference of two imbedded reducing subspaces, too reduces the operator $U$. Therefore

$$T | (\mathscr{H}_s^1 \ominus \mathscr{H}_s^2) = U | (\mathscr{H}_s^1 \ominus \mathscr{H}_s^2)$$

and, as $T$ is a completely non-unitary contraction, we must have $\mathscr{H}_s^1 \ominus \mathscr{H}_s^2 = \{\mathbb{0}\}$ and $H \subset \mathscr{H}_a$. But the minimality of $U$ means that

$$\mathscr{H} = \bigvee (U^n H : n \in \mathbb{Z}),$$

while as the preceding inclusion gives $U^n H \subset \mathscr{H}_a$, $n \in \mathbb{Z}$. Therefore $\mathscr{H} \subset \mathscr{H}_a$ and $\mathscr{H}_s = \{\mathbb{0}\}$. □

It is now easy to construct the *calculus of Sz.-Nagy and Foiaş* for the operator $T$, for we know that its dilation $U$ has a spectral resolution

$$\mathscr{H} = \int_{\mathbb{T}} H(\zeta)\,d\mu(\zeta)$$

with an absolutely continuous measure $\mu$. Indeed, in this case $\varphi(U)$ is defined for every function $\varphi$ belonging to the algebra $H^\infty$ (and as a matter of fact even to $L^\infty$, but non-analytic functions presently won't interest us): $\varphi(U) = \boldsymbol{\varphi}$, i.e. $(\varphi(U)f)(\zeta) = \varphi(\zeta)f(\zeta)$, $\zeta \in \mathbb{T}$, $f \in \mathscr{H}$. Set

$$\varphi(T) = P_H \varphi(U)|H, \quad \varphi \in H^\infty.$$

Exactly as in the Calculus Lemma we verify that the calculus of Sz.-Nagy and Foiaş is linear, multiplicative, does not increase the norm ($|\varphi(T)| \leq \|\varphi\|_\infty$) and extends to the algebra $H^\infty$ the standard polynomial calculus. Let us also remark that it is *continuous in the following sense*:

(*): if $\varphi_n \in H^\infty$, $\sup_n \|\varphi_n\|_\infty < \infty$, $\lim_n \varphi_n(\zeta) = 0$ for a.e. $\zeta$, $\zeta \in \mathbb{T}$, then for any $f$, $f \in H$, holds

$$\|\varphi_n(T)f\|^2 \leq \|\varphi_n(U)f\|^2 = \int_{\mathbb{T}} |\varphi_n(\zeta)|^2 \|f(\zeta)\|^2\,d\mu(\zeta),$$

so that $\lim_n \|\varphi_n(T)f\| = 0$.

If $(s)$-$\lim_n T^n = \mathbb{0}$ we can say more: already from $\sup_n \|\varphi_n\|_\infty < \infty$, $\lim_n \varphi_n(\lambda) = 0$ ($\lambda \in \mathbb{D}$), it follows that $\lim_n \|\varphi_n(T)f\| = 0$, $f \in H$. In fact, the operator $T$ is unitarily equivalent to the model operator $S_E^*|K$, $K \subset H^2(E)$, and $\varphi_n(S_E^*|K) = \varphi_n(S_E^*)|K$: therefore it suffices to verify the relation in question for $\varphi_n(S_E^*)$ instead of $\varphi_n(T)$ and this again is most conveniently done by checking it on the dense (in $H^2(E)$) set of eigenfunctions $f = (\mathbb{1} - \lambda z)^{-1}e$:

$$\varphi_n(S_E^*)(\mathbb{1} - \lambda z)^{-1}e = \varphi_n(\lambda)(\mathbb{1} - \lambda z)^{-1}e \quad (\lambda \in D, e \in E),$$

for which the desired result ($\lim_n \|\varphi_n(S_E^*)f\| = 0$) is trivial. □

From (*) we deduce the *permanence of the calculus constructed* compared with the Riesz-Dunford calculus (the latter will be treated in detail in the following Lecture):

$$\varphi \mapsto \frac{1}{2\pi i}\int_\gamma \varphi(\zeta)R(\zeta, T)\,d\zeta,$$

where $\varphi$ is any function regular in the interior of a smooth Jordan curve $\gamma$ enclosing the spectrum $\sigma(T)$ of $\mathbb{T}$. In fact, if $\varphi \in H^\infty$ and if there exists a neighborhood of $\sigma(T)$ into which $\varphi$ has a single valued analytic continuation, then of course we can take this neighborhood $\mathcal{O}$ to be starshaped ($\zeta \in \mathcal{O} \Rightarrow r\zeta \in \mathcal{O}$, $0 \leq r \leq 1$) so that for a curve $\gamma$ contained in $\mathcal{O}$ holds

$$\lim_{r\to 1}\max_{\zeta\in\gamma} |\varphi(\zeta) - \varphi_r(\zeta)| = 0 \quad (\varphi_r(\zeta) \overset{\text{def}}{=} \varphi(r\zeta), 0 \leq r < 1)$$

$$\varphi_r(T) = \frac{1}{2\pi i}\int_\gamma \varphi_r(\zeta)R(\zeta, T)\,d\zeta \tag{3}$$

(one can deform the contour $\gamma$ in such a way that it lies entirely inside the circle of convergence of the series $\sum_{n\geq 0} \hat{\varphi}(n) r^n z^n$). It remains then only to pass to the limit in this formula (3) for $\varphi_r$; to the left we use the (*)-continuity of the calculus of Sz.-Nagy and Foiaş, to the right the uniform convergence of $\varphi_r$ to $\varphi$ on the curve $\gamma$.

**6. The Class $C_0$.** The calculus of Sz. Nagy and Foiaş allows us to construct a remarkable extension of the theory of elementary divisors, which is based on the multiplicative structure of $H^\infty$-functions and comprises operators which are of interest both for the general theory as well as for the applications. The basic idea is to consider operators for which there holds an analogue of the Cayley-Hamilton theorem ($\varphi(T) = \mathbb{O}$ for suitable analytic functions $\varphi$), and to give a subsequent analysis of their properties in relation to the dependence on the multiplicative nature of the annihilating functions.

**Definition.** The *class* $C_0$ is the collection of all completely non-unitary contractions $T$ such that there exists a function $\varphi$, $\varphi \in H^\infty$, $\varphi \not\equiv 0$, annihilating $T$ ($\varphi(T) = \mathbb{O}$).

**Lemma on the Minimal Annihilator.** *Let $T \in C_0$. There exists an inner function $m_T$ such that $m_T(T) = \mathbb{O}$ and*

$$\varphi \in H^\infty, \quad \varphi(T) = \mathbb{O} \Rightarrow \varphi/m_T \in H^\infty.$$

*Proof.* Put $E = \{\varphi : \varphi \in H^\infty, \varphi(T) = \mathbb{O}\}$ and let us show that $E$ is $S$-invariant and in addition weak-star closed in $H^\infty$. The first statement is trivial so let us prove the second one. Let $\psi \in H^\infty \backslash E$. Then there exist elements $x$ and $y$ in $H$ (the space on which $T$ acts) such that $(\psi(T)x, y) \neq 0$. But then in the notation of the previous Section

$$\begin{aligned} 0 \neq (\psi(T)x, y) &= (P_H \psi(U)x, y) = (\psi(U)x, y) \\ &= \int_{\mathbb{T}} \psi(\zeta)(x(\zeta), y(\zeta))\, d\mu(\zeta) \\ &= \int_{\mathbb{T}} \psi(\zeta)(x(\zeta), y(\zeta)) \left(\frac{d\mu}{dm}\right)(\zeta)\, dm, \end{aligned}$$

where $h \overset{\text{def}}{=} (x(\cdot), y(\cdot)) \dfrac{d\mu}{dm}$ is summable on $\mathbb{T}$. As

$$\int_{\mathbb{T}} \varphi h\, dm = (\varphi(T)x, y) = 0$$

for any $\varphi$, $\varphi \in E$, we see that $\psi \notin \operatorname{clos}_{\sigma(H^\infty, L^1)} E$.

Thus $E$ is a weakly closed shift invariant subspace of $H^\infty$ and therefore (Lecture I, Sect. 7) $E = \mathbf{m} \cdot H^\infty$ where $m$ is an inner function. Thus we can take $m_T = m$. □

The function $m_T$ will be called the *minimal annihilator of the operator T*; it is obviously unique up to a factor of modulus 1. $m_T$ plays the same rôle in the theory of the class $C_0$ as does the minimal polynomial in the spectral theory of matrices.

We will encounter several instances illustrating the validity of this thesis, the first being the following theorem.

**Theorem on the Spectrum of a $C_0$-Operator.** *If $T \in C_0$ then*

$$\sigma(T) = \sigma(m_T).$$

*Also $\sigma(m_T) \cap \mathbb{D} = \sigma_p(T)$.*

*Proof.* The inclusion $\sigma(T) \subset \sigma(m_T)$ is established exactly as the corresponding part of Theorem LM. The relation $\sigma_p(T) = \sigma(m_T) \cap \mathbb{D}$ is also proved in an analogous way, with the differences that instead of Statement *d*) of the Calculus Lemma we use the fact that $T$ is completely non-unitary (this gives the inclusion $\sigma_p(T) \subset \mathbb{D}$) and instead of Statement c) of the same lemma the definition of the minimal annihilator.

Only the inclusion $\sigma(m_T) \cap \mathbb{T} \subset \sigma(T)$ requires some new ideas. Let $\lambda \in \mathbb{T} \backslash \sigma(T)$. If we would have $\lambda \in \sigma(m_T)$ then $m_T$ would admit a decomposition into inner factors $m_T = m_1 m_2$ such that $m_1 \not\equiv \text{const}$ and $\sigma(m_1) \cap \sigma(T) = \varnothing$ (to achieve this put

$$m_1 = \exp\left(-\int_{\mathbb{T} \cap \mathcal{O}} \frac{\zeta + z}{\zeta - z} d\mu(\zeta)\right)$$

where $\mathcal{O}$ is a sufficiently small neighborhood of the point $\lambda$, $\mu$ being the representing measure for the singular part of $m_T$). It is clear that $m_1$ can be continued analytically to a neighborhood of $\sigma(T)$ and that this continuation never vanishes. By the remarks made in Sect. 5 we can likewise interpret $m_1(T)$ in the Riesz-Dunford calculus and, since the function $m_1^{-1}$ is analytic in a neighborhood of $\sigma(T)$, the operator $m_1(T)$ is invertible. Thus $m_2(T) = (m_1(T))^{-1} m_T(T) = \mathbb{O}$ which contradicts the definition of the minimal annihilator. □

In the sequel the operators of class $C_0$ will be subject to a more detailed analysis but presently we restrict ourselves to some observations connected with the Spectral Mapping Theorem. It is not hard to show that the operators of class $C_0$ have the property

$$(s)\text{-}\lim_n T^n = \mathbb{O}, \quad (s)\text{-}\lim_n T^{*n} = \mathbb{O}.$$

It is clear that $T \in C_0 \Leftrightarrow T^* \in C_0$ and also that $m_{T^*} = m_T^\dagger$ so that it is sufficient to check only one of the formulae; to this end one first verifies that $(s)\text{-}\lim_n U^{-n}T^n$ exists and defines an operator $L$ from $H$ into $\mathscr{H}$ (the space where the dilation $U$ acts) such that $L\varphi(T) = \varphi(U)L$, $\forall \varphi \in H^\infty$. This again gives $0 = \|Lm_T(T)x\| = \|m_T(U)Lx\| = \|Lx\| = \lim_n \|U^{-n}T^n x\| = \lim_n \|T^n x\|$. □

Therefore the functional model for the operator $T$ of class $C_0$ has the form

$$T \cong P_K S | K, \; K = H^2(E) \ominus \Theta H^2(E),$$

where $\Theta = \Theta_T$ is an inner function, $\dim E = \dim(I - T^*T)H$. This gives the following

**Theorem on the Scalar Multiple.** *Let $\varphi \in H^\infty$ and assume $T \in C_{00}$. The following statements are equivalent.*

1. $\varphi(T) = \mathbb{0}$.
2. *There exists a bounded analytic function $\Omega$, $\Omega(\zeta) \in L(E)$ $(\zeta \in \mathbb{D})$, such that*

$$\Theta\Omega = \Omega\Theta = \varphi I.$$

*Proof.* As $\varphi(T) = P_K \boldsymbol{\varphi}|K$ it is clear that the relation $\varphi(T) = \mathbb{0}$ is equivalent to the inclusion $\boldsymbol{\varphi}H^2(E) \subset \boldsymbol{\Theta}H^2(E)$ or $\boldsymbol{\Theta}^*\boldsymbol{\varphi}H^2(E) \subset H^2(E)$. We then get $1 \Rightarrow 2$ (take $\Omega = \Theta^*\varphi$) as well as $2 \Rightarrow 1$ ($\boldsymbol{\varphi}H^2(E) = \boldsymbol{\Theta\Omega}H^2(E) \subset \boldsymbol{\Theta}H^2(E)$). □

**Corollary 1.** *If* $\dim E < \infty$ *then the determinant $d$, $d(\zeta) = \det \Theta(\zeta)$, $\zeta \in \mathbb{D}$, annihilates $T$.*

Indeed, $d(\zeta)I = \Theta(\zeta)\Omega(\zeta) = \Omega(\zeta)\Theta(\zeta)$ where $\Omega$ is the matrix whose elements are the minors of order $\dim E - 1$ of the matrix $\Theta(\zeta)$. □

**Corollary 2.** $d/m_T \in H^\infty$. □

We can now carry over part of the proof of the Spectral Mapping Theorem to operators of class $C_0$, but not the entire proof. Let $\psi \in H^\infty$. If

$$\inf_{\mathbb{D}}(|\psi| + |m_T|) > 0,$$

then we can as before apply the Corona Theorem of Carleson and, having found solutions in the class $H^\infty$ of the equation $u\psi + vm_T = \mathbb{1}$, invert the operator $\psi(T)$:

$$u(T)\psi(T) = u(T)\psi(T) + v(T)m_T(T) = I.$$

The converse conclusion does not cause any difficulties provided the greatest lower bound $\inf_{\mathbb{D}}(|\psi| + |m_T|) = 0$ is attained in $\mathbb{D}$: by what we have just shown the corresponding point must be an eigenvalue for $T$ and its $\psi$-image one for $\psi(T)$. Also in the general case we can make the same computations as in the previous proof, only we get a different result:

$$\left\| \psi(T)^* \mathrm{P}_\Theta \frac{e}{\mathbb{1} - \bar{\lambda}z} \right\|_2 \leq (|\psi(\lambda)| + \|\Theta(\lambda)^* e\| \cdot \|\psi\|_\infty) \cdot (1 - |\lambda|^2)^{-1/2}, \tag{4}$$

$$\left\| P_\Theta \frac{e}{\mathbb{1} - \bar{\lambda}z} \right\|_2^2 = \frac{1 - \|\Theta(\lambda)^* e\|^2}{1 - |\lambda|^2} \tag{5}$$

where $e$ is an arbitrary vector in the coefficient space $E$, of norm 1. Consequently $0 \in \sigma(\psi(T))$ if

$$\inf_{\lambda \in \mathbb{D}} \inf_{\|e\|=1} \{|\psi(\lambda)| + \|\Theta(\lambda)^* e\|\} = 0 \tag{6}$$

It is only in the case of a finite dimensional defect space $E$ however that the preceding operator condition can easily be replaced by a scalar one: if $\inf_{\mathbb{D}}(|\psi| + |m_T|) = 0$ then by Corollary 2 of the Theorem on the Scalar Multiple we have also $\inf_{\mathbb{D}}(|\psi| + |\det \Theta|) = 0$ but the determinant $\det \Theta(\zeta)$, $\zeta \in \mathbb{D}$, is the product of the $\dim E$ eigenvalues (counted with multiplicity) so that simultaneously with

$\det\Theta(\zeta)$ at least one of the latter will tend to 0, which establishes condition (6). We have proved

**Corollary 3.** *If* $\varphi\in H^\infty$, $T\in C_0$, *then*

$$\left\{\zeta\in\mathbb{C}:\inf_{\lambda\in\mathbb{D}}\inf_{\|e\|=1}(|\varphi(\lambda)-\zeta|+\|\Theta(\lambda)^*e\|)=0\right\}$$

$$\subset\sigma(\varphi(T))\subset\left\{\zeta\in\mathbb{C}:\inf_{\mathbb{D}}(|\varphi-\zeta\mathbb{1}|+|m_T|)=0\right\},$$

*where in the case* $\dim E<\infty$ *the two extremal sets coincide.* □

Let us check that in condition (6) for $\dim E=\infty$ *one cannot replace the term* $\|\Theta(\lambda)^*e\|$ neither by $|m_T(\lambda)|$ nor by $|\det\Theta(\lambda)|$. (In this example (which can be omitted at the first reading) we use some results from Lecture VII.) Let $B=\prod_{n\geq 1}b_{\lambda_n}$ be a Blaschke product with the zeros on the ray $(0,1)$ and such that $\lim_{x\to 1-0}|B(x)|=0$. Then $\lim_{\zeta\to 1,\zeta\in\Gamma}|B(\zeta)|=0$ where $\Gamma$ is any Stolz angle, $\Gamma=\{\zeta:\arg(1-\zeta)\leq\alpha\}$, $0<\alpha<\frac{\pi}{2}$. (The analogous property is valid for a general function $B$ in $H^\infty$. In fact, the family $\{B_\varepsilon\}_{\varepsilon>0}$, $B_\varepsilon\overset{\text{def}}{=}B(1+\varepsilon(z-1))$ is normal in $\mathbb{D}$ (in Montel's sense) and each of its limit functions (for $\varepsilon\to 0$) is zero because of $\lim_{\varepsilon\to 0}B_\varepsilon(x)=\lim_{y\to 1-0}B(y)=0$, $-1<x<1$. Hence $\lim_{\varepsilon\to 0}B_\varepsilon(\zeta)=0$ uniformly on compact subsets of $\mathbb{D}$. This means that the angular limit $\lim_{\zeta\to 1,\zeta\in\Gamma}B(\zeta)$ is also zero. □) As $\psi$ we can likewise choose a Blaschke product but one with the zeros $\{\mu_n\}_{n\geq 1}$ on the sides $\partial\Gamma$ of the angle $\Gamma$, accumulating at the point 1 and defining a "Carleson sequence" (for this notion see Lecture VII); for instance, $\mu_n=1-q^ne^{i\alpha}$, $n\geq 1; 0<q<1$, will do. Finally we take $T$ to be the normal operator with simple eigenvalues $\{\lambda_n:n\geq 1\}$. Then $T\in C_0$, $m_T=B$, and $\psi(T)$ is normal with the eigenvalues $\psi(\lambda_n):n\geq 1$. As the set $\{\mu_k\}_{k\geq 1}\cup\{\lambda_n\}$, $n=1,2\ldots$ is contained in the angle $\Gamma$ and satisfies uniformly (in $n$) a "sparseness condition" (cf. Lecture VII) we have $\inf_{n\geq 1}|\psi(\lambda_n)|>0$. Consequently $\psi(T)$ is an invertible operator. On the other hand

$$\inf_{\mathbb{D}}(|\psi|+|m_T|)\leq\inf_{n\geq 1}(|\psi(\mu_n)|+|B(\mu_n)|)=0.\ \square$$

Incidentally, it is clear that in the above example $\Theta_T(\zeta)$ is a normal operator with simple point spectrum $\{b_{\lambda_n}(\zeta):n\geq 1\}$, $\zeta\in\mathbb{D}$, but condition (6) is not satisfied.

All this indicates that for the description of the spectrum $\sigma(\varphi(T))$ in the general case the "scalar" Corona Theorem is *not sufficient* and we have to invoke its operator analogues. Thus it would be sufficient to be able to find criteria for the solvability of the equation

$$I=\psi U+\Theta V \tag{7}$$

for the unknown ($L(E)$-valued) functions $U$, $V$ analytic and bounded in the disk $\mathbb{D}$, when similar functions $\psi$, $\Theta$ are given in advance. In the case just discussed $\Theta$ was an inner function and $\psi$ a scalar one. The trivial necessary condition

$$\|e\| \le \|U(\zeta)^*\psi(\zeta)^*e\| + \|V(\zeta)^*\Theta(\zeta)^*e\|$$
$$\le \mathrm{const}(\|\psi(\zeta)^*e\| + \|\Theta(\zeta)^*e\|); \quad e \in E, \zeta \in \mathbb{D},$$

is also sufficient for operators with a finite dimensional defect space $E$ ($\dim E < \infty$, $\Theta = \Theta_T$ an inner function); if $\dim E = 1$ this follows from the Corona Theorem of Carleson and if $\dim E > 1$ from its generalization by Fuhrmann [1]. The difficulties in the general case ($\dim E = \infty$) are discussed in Sz.-Nagy [1] and also in Appendix 3. Concluding this Section let us make the following auxiliary remarks: a) Fuhrmann's theorem describes the spectrum not only for functions of $P_\Theta S|K_\Theta$ but for all operators commuting with this operator (regarding such operators see Lecture VIII); b) if we seek those solutions of equation (7) for which $U$, as well as $\psi$, is scalar then by the Theorem of the Scalar Multiple we anew arrive at the previous criterion: $\inf_{\mathbb{D}}(|\psi| + |m_T|) > 0$; c) other operator interpretations of the Corona Theorem are proposed by Arveson [1], Sz.-Nagy and Foiaş [10], Schubert [1]; for the formulation and proof see Appendix 4, Sect. 169; d) a few years before Fuhrmann's paper Moeller [1] proved a spectral mapping theorem in the special case of a function with only one singular point $\left(\text{the operator } \varphi(P_\Theta S|K_\Theta),\ \varphi = \exp\dfrac{z+1}{z-1}\right)$; e) compare also Foiaş-Mlak [1] and Appendix 3.

**7. The Characteristic Function and the Spectrum** of a general model operator of class $C_{00}$ are connected roughly as in the one dimensional case ($\dim E = 1$), i.e. as in Theorem LM.

**Theorem on the Spectrum.** *Let $E$ be a Hilbert space and $\Theta$ an inner function with values in $L(E)$. Set $K = K_\Theta \overset{\text{def}}{=} H^2(E) \ominus \Theta H^2(E)$ and $T = \mathbf{T}_\Theta \overset{\text{def}}{=} P_K S_E|K$. Then*:

1. $\sigma(T) = \sigma(\Theta)$ *where the set $\sigma(\Theta)$, called the spectrum of the function $\Theta$, is defined by the relation* $(\operatorname{clos}\mathbb{D})\backslash\sigma(\Theta) = \{\lambda : \lambda \in \operatorname{clos}\mathbb{D},\ \Theta(\zeta)^{-1}$ *exists in a certain relative neighborhood of $\lambda$ and admits analytic continuation to a (full) neighborhood of $\lambda$*$\}$.
2. $\sigma_p(T) = \{\lambda : \lambda \in \mathbb{D}, \operatorname{Ker}\Theta(\lambda) \ne \{\mathbb{0}\}\}$,
   $\sigma_p(T^*) = \{\bar{\lambda} : \lambda \in \mathbb{D}, \operatorname{Ker}\Theta(\lambda)^* \ne \{\mathbb{0}\}\}$.

*The proof* of this theorem will be based on two lemmas which will also be used in the following lectures.

**Lemma on the Function $\Theta^t$.** *Let the hypotheses of the Theorem on the Spectrum be fulfilled. Then $T^* = S^*|K_\Theta$ is unitarily equivalent to the operator $\mathbf{T}_{\Theta^t}$, $\Theta^t = \Theta(\bar{z})^*$.*

*Proof.* Put $Vf = \bar{z}\Theta^t f(\bar{z})$, $f \in K_\Theta$. Let us verify that the operator $V$ establishes an isomorphism between the spaces $K$ and $K_{\Theta^t} \overset{\text{def}}{=} H^2(E) \ominus \Theta^t H^2(E)$ mapping the operator $\mathbf{T}_\Theta^*$ onto $\mathbf{T}_{\Theta^t}$.

The first of these assertions is contained in the Lemma on Lat $S_E^*$ (Lecture II,

Sect. 4):

$$f \in K_\Theta \Leftrightarrow f \in H^2(E), \quad \bar{z}\Theta^t f(\bar{z}) \in H^2(E).$$

Further, using this lemma once more with the function $g = \bar{z}\Theta^t(z)f(\bar{z})$ and taking into account that $f = \bar{z}\Theta(z)g(\bar{z}) = \bar{z}(\Theta^t)^t g(\bar{z})$ we get

$$f \in K_\Theta \Leftrightarrow \bar{z}\Theta^t(z)f(\bar{z}) \in K_{\Theta^t}.$$

That $V$ is unitary follows from the fact that the values $\Theta(\zeta)$, $\zeta \in \mathbb{T}$, are unitary.

Using the Projection Lemma, II (Lecture II, Sect. 4) we find the operator $V^{-1}\mathbf{T}_{\Theta^t}V$:

$$\begin{aligned} V^{-1}\mathbf{T}_{\Theta^t}Vf &= V^{-1}P_{\Theta^t}z(\bar{z}\Theta^t f(\bar{z})) = V^{-1}\Theta^t(I - P_+)(\Theta^{t*}\Theta^t f(\bar{z})) \\ &= V^{-1}\Theta^t(I - P_+)f(\bar{z}) = V^{-1}\Theta^t(f(\bar{z}) - f(0)\mathbb{1}) \\ &= \bar{z}(f(z) - \mathrm{f}(0)\mathbb{1}) = S_E^* f. \ \square \end{aligned}$$

**Lemma on Eigenfunctions.**

$$\operatorname{Ker}(\mathbf{T}_\Theta^* - \bar{\lambda}I) = \{(\mathbb{1} - \bar{\lambda}z)^{-1}e : e \in \operatorname{Ker}\Theta(\lambda)^*\}.$$

$$\operatorname{Ker}(\mathbf{T}_\Theta - \lambda I) = \{(z - \lambda\mathbb{1})^{-1}\Theta e : e \in \operatorname{Ker}\Theta(\lambda)\}.$$

*Proof.* It is sufficient to find $\operatorname{Ker}(T_\Theta^* - \bar{\lambda}I)$ and then apply the Lemma on the Function $\Theta^t$. But $\mathbf{T}_\Theta^* = S_E^*|K$ and every eigenfunction for $S_E^*$ is of the form $(\mathbb{1} - \bar{\lambda}z)^{-1}e$, $e \in E$ (Beurling's Lemma, II; Lecture II, Sect. 4). It remains to determine those vectors $e$, $e \in E$, for which $(\mathbb{1} - \bar{\lambda}z)^{-1}e \in K_\Theta$:

$$\begin{aligned} (\mathbb{1} - \bar{\lambda}z)^{-1}e \in K_\Theta &\Leftrightarrow (\mathbb{1} - \bar{\lambda}z)^{-1}e \perp \Theta H^2(E) \\ &\Leftrightarrow (\Theta(\lambda)h(\lambda), e) = 0, \quad \forall h \in H^2(E) \\ &\Leftrightarrow \Theta(\lambda)^* e = \mathbb{0}. \ \square \end{aligned}$$

Thus Statement 2 of the Theorem on the Spectrum is already proved (and we have indeed also computed the corresponding eigenspace). Turning to the proof of Statement 1, we remark that for points $\lambda$ in the disk $\mathbb{D}$ the analyticity of $\zeta \mapsto \Theta(\zeta)^{-1}$ in a neighborhood of $\lambda$ is a consequence of the invertibility of $\Theta(\lambda)$ and that for points $\lambda$ on the circumference ($\lambda \in \mathbb{T}$) by this analyticity we mean (as in the scalar case) the possibility of analytic continuation into a (full) neighborhood of the point under consideration.

Let us first verify that

$$\sigma(T) \cap \mathbb{D} = \sigma(\Theta) \cap \mathbb{D}. \tag{8}$$

If $\Theta(\lambda)$ is invertible ($\lambda \in \operatorname{clos}\mathbb{D}$) and $\Theta$ has an analytic continuation to a neighborhood of $\lambda$ let us set

$$u = \frac{\Theta - \Theta(\lambda)}{z - \lambda\mathbb{1}}\Theta(\lambda)^{-1}, \quad U = P_K\mathbf{u}|K. \tag{9}$$

Since $(\lambda\mathbb{1} - z)u = I - \Theta\Theta(\lambda)^{-1}$ we have

$$P_K(\lambda \mathbb{1} - z)uf = f, \quad f \in K,$$

$$P_K u(\lambda \mathbb{1} - z)f = f, \quad f \in K,$$

and it follows that $\lambda I - \mathbf{T}_\Theta$ is invertible, provided we can also show that

$$P_K(\lambda \mathbb{1} - z)(I - P_K)uf = \mathbb{0},$$

$$P_K u(I - P_K)(\lambda \mathbb{1} - z)f = \mathbb{0},$$

for any $f$, $f \in K$. The first of these relations is trivial because the subspace $(I - P_K)H^2(E) = \mathbf{\Theta} H^2(E)$ is invariant for multiplication by $z$. The second one is a consequence of the invariance of the subspace $\mathbf{\Theta} H^2$ with respect to the operator $\mathbf{u}$; this last property ($\mathbf{u}\mathbf{\Theta} H^2(E) \subset \mathbf{\Theta} H^2(E)$) is equivalent to the inclusion $\mathbf{\Theta}^* \mathbf{u} \mathbf{\Theta} H^2(E) \subset H^2(E)$, that is, equivalent to the analyticity of $\Theta^* u \Theta$: $\Theta^* u \Theta \in H^\infty(L(E))$. But it is readily seen that $\Theta^* u \Theta = \dfrac{\Theta(\lambda)^{-1}\Theta - I}{z - \lambda \mathbb{1}}$ so that the required analyticity is trivial.

Thus we have shown that $\sigma(T) \subset \sigma(\Theta)$ and that the resolvent $R(\lambda, T)$ at the point $\lambda$, $\lambda \in \operatorname{clos} \mathbb{D} \backslash \sigma(\Theta)$, coincides with the operator $U$ in (9).

The converse statement will first be established only for the portion of the spectrum which is contained in the disk $\mathbb{D} : \sigma(\Theta) \cap \mathbb{D} \subset \sigma(T) \cap \mathbb{D}$. With the aid of Statement 2, which we have already verified, this inclusion is at once reduced to the study of only those points $\lambda$, $\lambda \in \sigma(\Theta) \cap \mathbb{D}$, for which $\operatorname{Ker} \Theta(\lambda) = \{\mathbb{0}\}$, $\operatorname{Ker} \Theta(\lambda)^* = \{\mathbb{0}\}$. If now $\lambda$ is such a point then the operator $\Theta(\lambda)^*$ is not invertible and moreover $\inf\{\|\Theta(\lambda)^* e\| : \|e\|_E = 1\} = 0$. The formulae (4) and (5) show, as in the scalar case, that $\lambda \in \sigma(T)$. This finishes the proof of (8).

It remains to be shown that $\sigma(\Theta) \cap \mathbb{T} \subset \sigma(T)$. This is done about as in the scalar case (see the last paragraph of the proof of the Livšic-Moeller Theorem, Sect. 3). One has just to remark in advance that at this juncture one can without loss of generality assume that $\mathbb{D} \not\subset \sigma(T)$ and that as $f$ one can, instead of $P_\Theta(\mathbb{1} - \bar{\zeta} z)^{-1}$, take the function $f = P_\Theta(\mathbb{1} - \bar{\zeta} z)^{-1} e = (\mathbb{1} - \bar{\zeta} z)^{-1}(e \cdot \mathbb{1} - \Theta\Theta(\zeta)^* e)$, $e \in E$, $\zeta \in \mathbb{D} \backslash \sigma(T)$. □

In conclusion let us remark that for operators $T$ of class $C_0$ the computation of the spectrum $\sigma(\Theta_T)$, as was seen in Sect. 6, is reduced to the study of the spectrum of the scalar function $m_T$ (or $d_T$ if $\det \Theta_T$ exists). Also it is not hard to verify that the inner function $\Theta$ has an analytic continuation to a neighborhood of the point $\lambda$, $\lambda \in \mathbb{T}$, iff $\overline{\lim}_{\zeta \to \lambda, \zeta \in \mathbb{D}} |\Theta(\zeta)^{-1}| < \infty$.

**8. Concluding Remarks.** Sarason's Projection Lemma can be found in Sarason [8]. This lemma shows in particular that the projections $P_K S | K$, $K_\Theta = H^2 \ominus \mathbf{\Theta} H^2$, of the scalar shift operator are precisely those completely non-unitary contractions whose minimal dilation is the simplest operator with an absolutely continuous spectrum, viz. the operator $\mathscr{S}$ in the space $L^2$. The lemma further shows that the space $\mathscr{H}$, in which the dilation $U$ of the operator $T$, $T \in L(H)$, acts, admits the representation $\mathscr{H} = G^* \oplus H \oplus G$ where $G \in \operatorname{Lat} U$,

$G^* \in \operatorname{Lat} U^*$. This representation also constitutes the basis upon which one constructs the functional model for the contraction $T$; regarding the physical interpretation of $G^*$ and $G$ ("incoming" and "outgoing" spaces) see Lax-Phillips [1].

The Livšic-Moeller Theorem is proved in Livšic [1], Moeller [2] and its analogue for general completely non-unitary contractions $T$ (i.e. the Theorem on the Spectrum, Sect. 7) in Sz.-Nagy and Foiaş [4], Helson [2]. The variant of this theorem for contractions of class $C_0$ (as well as other facts on $C_0$-operators) is taken over from Sz.-Nagy and Foiaş [3]. This class was independently (and simultaneously with Sz.-Nagy and Foiaş) in a somewhat different form studied by Ginzburg [1, 4]. Regarding this circle of ideas cf. further: Sz.-Nagy and Foiaş [1], Lax-Phillips [1], Douglas [1].

The Corollary of the Livšic-Moeller Theorem can be stated in the following way: all functions in the space $K_\Theta = H^2 \ominus \Theta H^2$ can be continued analytically across the point $\zeta$, $\zeta \in \mathbb{T}$, iff $\mu^\Theta O_\zeta = 0$ for some neighborhood $O_\zeta$ (in clos $\mathbb{D}$) of $\zeta$, $\mu^\Theta = \mu^s + \mu^B$ being the representing measure of the inner function $\Theta$. This formulation adds plausibility to the following general principle: if the measure $\mu^\Theta$ is "small" near $\zeta$ then the functions in $K_\Theta$ must be "smooth" near that point. This is confirmed by the following theorem by Ahern-Clark [2] according to which the following conditions are equivalent:

1. For every function $f$ in $K_\Theta$ the derivatives $f^{(i)}$, $0 \le i \le N$, have finite radial (non-tangential) limits at the point $\zeta$, $\zeta \in \mathbb{T}$.
2. $\mathbf{T}_\Theta^{*N}(P_\Theta \mathbb{1}) \in (I - \bar{\zeta}\mathbf{T}_\Theta)^{N+1} K_\Theta$.
3. $\displaystyle\int_{\operatorname{clos}\mathbb{D}} \frac{d\mu^\Theta(\lambda)}{|\mathbb{1} - \bar{\zeta}\lambda|^{2N+2}} < \infty.$

It is interesting to compare Condition 3 (or rather the part corresponding to the term $\mu^B$ in the measure $\mu^\Theta$) with Frostman's condition considered in Lecture VII and further to remark that from the Ahern-Clark theorem follows the existence of an infinite dimensional subspace $K_\Theta$ such that all functions in $K_\Theta$ and all their derivatives have finite radial (non-tangential) limits at all points of the circumference $\mathbb{T}$; however $K_\Theta \not\subset H^\infty$.

The spectra of inner functions were studied rather long ago (roughly in the same way as was done in this Lecture); cf., for instance, Hoffman [1], Helson [2]; our exposition is close to Helson's book [2]. The known theorems on factorization of bounded operator functions, similar to the ones of the Nevanlinna-Riesz theory, concern in general functions $\Theta_T$ which admit a scalar multiple ($\Theta_T \Omega = m_T \cdot I$) but even in this case the formulae, expressing $\sigma(\Theta_T)$ in terms of the "parameters" of the multiplicative representations in question, are not used very often because of the non-commutativity of the corresponding (operator) products and integrals, as this non-commutativity obstructs a translation of the "parameters" into the "additive" language of the space $K_\Theta$ and the operator $\mathbf{T}_\Theta$; a few more details concerning all this can be found in Lecture IV, Sect. 5. Regarding the multiplicative representations themselves see Brodskiĭ [6], Brodskiĭ-Livšic [1] as well as the bibliographical survey by Nikol'skiĭ [4].

The $H^\infty$-functional calculus for completely non-unitary contractions was constructed by Sz.-Nagy and Foiaş after they had first established the absolute continuity of the spectrum of the minimal unitary dilation; cf. [1]. In the present Lecture all this is done somewhat differently. There are however some subtleties of the Sz.-Nagy and Foiaş calculus for which we have to refer to the book [1]. Let us also notice that already Plesner [1] used the $H^\infty$-calculus in the case of partially isometric operators. Bibliographical comments on the Spectral Mapping Theorem were given in the preceding discussion (Sects. 5–6).

The Corona Theorem was proved by Carleson [2] with the estimate $\delta^{-A}$ for the norms $\|g_i\|_\infty$ ($A$ being a constant depending on $n$ only). The estimate $11\delta^{-3}$ is a consequence of the sharper form of the theorem due to Wolff, Tolokonnikov and Uchiyama; the detailed statements, proofs and comments can be found in Appendix 3.

The spectrum $\sigma(T)$ of the operator $T$ is not a continuous "function" of the argument $T$, not even in the weakest interpretation of these words: for instance, the sequence of operators $T_n$,

$$T_n f = zf - (1 - n^{-1})\hat{f}(-1)\mathbb{1},$$

in the space $L^2$ converges in norm to an operator $T$, unitarily equivalent to the sum $S^* \oplus S$, while as $\sigma(T_n) \subset \mathbb{T}$, $\forall n$; $\sigma(T) = \operatorname{clos}\mathbb{D}$; cf. Halmos [2]. Nevertheless one can get theorems of the spectral mapping type by a "limiting procedure" from the trivial case of the polynomial calculus or from the Riesz-Dunford calculus. Here is one variant based on the theory of Banach algebras.

**Rough Spectral Mapping Theorem.** *Let $T$ be a linear operator in a Banach space $X$ and let $\mathscr{A}(T)$ be the closure in $L(X)$ of the algebra $R$ of all rational functions $f(T)$ in $T$. Then:*

1. *There exists a unique continuous homomorphism $A \mapsto a$ from $\mathscr{A}(T)$ into $C(\sigma(T))$ which on $R$ coincides with the mapping $f(T) \mapsto f|\sigma(T)$.*
2. *$\sigma(A) = a(\sigma(T))$.*

*Proof.* If $f$ is a rational function with poles in $\mathbb{C}\backslash\sigma(T)$ then $\sigma(f(T)) = f(\sigma(T))$ and therefore $\max_{\sigma(T)}|f| \le |f(T)|$. The homomorphism in Statement 1 is the continuous extension of the map $f(T) \mapsto f|\sigma(T)$, $f(T) \in R$.

If an operator $A$ is invertible in the algebra $\mathscr{A}(T)$ then it is of course invertible in $L(X)$ too. Converesely, if $A$ is invertible in $X$ and $\lim_n f_n(T) = A$, $f_n(T) \in R$, then from some index $n$ on also all the operators $f_n(T)$ are invertible and $\lim_n f_n(T)^{-1} = A^{-1}$ (the continuity of the operation of passing to the inverse in the open set of invertible elements). As $f_n(T)^{-1} = f_n^{-1}(T)$ it follows that $A^{-1} \in \mathscr{A}(T)$.

It therefore only remains to show that every complex homomorphism $\varphi$ of the algebra $\mathscr{A}(T)$ has the form

$$\varphi(A) = a(\zeta), \quad A \in \mathscr{A}(T), \tag{10}$$

for some $\zeta$, $\zeta \in \sigma(T)$. Set $\zeta = \varphi(T)$. Then $\varphi(p(T)) = p(\zeta)$ for any polynomial $p$, $p \in \mathscr{P}_A$, and so also for all rational functions: if $f = (z - \lambda\mathbb{1})^{-1}$, $\lambda \in \mathbb{C}\backslash\sigma(T)$, then

$$1 = \varphi(f(T)(T - \lambda I)) = \varphi(f(T))(\zeta - \lambda),\ \varphi(f(T)) = f(\zeta).$$

This establishes the inclusion $\zeta \in \sigma(T)$ and therefore the formula (10). □

In particular we get a spectral mapping theorem for the $l_A^1(|T^n|)$-calculus mentioned in Lecture II and for the $C_A$-calculus for contractions in Hilbert space. However, to go beyond the limits of the algebra $\mathscr{A}(T)$ along this route succeeds only rarely (what we have in mind is the passage to the limit from rational functions using a weaker convergence than the one in the above theorem); regarding this matter cf. Naimark [1], p. 242. On the spectral mapping theorem for more subtle calculi; cf. Dyn'kin [1], Solomyak [1].

The Lemma on Eigenfunctions (Sect. 7) is contained in the papers Lax-Phillips [2] (for a halfplane) and Nikol'skiĭ-Pavlov [2] (for the disk).

# Lecture IV. Decomposition into Invariant Subspaces

Beginning with this Lecture we will use the invariant subspaces of the shift operator $S$ for the analysis of the spectral properties of the compressions

$$\mathbf{T}_\Theta \stackrel{\text{def}}{=} P_\Theta S|K, \quad P_\Theta = P_{K_\Theta}, \quad K_\Theta = \mathrm{H}^2 \ominus \Theta H^2,$$

of this operator onto a coinvariant subspace $K_\Theta$. In this Lecture we only manage to take the first preliminary step in this direction. We try to decompose the operator $\mathbf{T}_\Theta$ into a sum of simpler parts, connecting such a decomposition with a refinement of the spectrum $\sigma(\mathbf{T}_\Theta)$. When a further refinement of the spectrum is not any longer possible, it will be necessary to single out a chain of invariant subspaces, which leads to integral representations for the operator which are analogous to the triangular form for matrices in linear algebra. It turns however out to be most difficult to reconstruct the properties of the operator subject to our investigation from the corresponding properties of its restrictions to the terms of the decomposition. Several of the following lectures will be devoted to this reconstruction. Here we just study the simplest problems connected with spectral subspaces and with finite dimensional subspaces.

**1. Spectral Synthesis.** The finite dimensional $T$-invariant subspaces should look simpler than others, and among them the simplest are the *rootspaces*, that is, spaces of the form $\operatorname{Ker}(T - \lambda I)^n$, $\lambda \in \mathbb{C}$, $n \geq 0$. An explicit formula for the latter can be obtained from a general description of Lat $T$.

**Theorem on** Lat $\mathbf{T}_\Theta$. *Let $\Theta$ be an inner function and set*

$$\mathrm{K}_\Theta = H^2 \ominus \Theta H^2, \quad \mathbf{T}_\Theta = P_\Theta S|K,$$

*where $P_\Theta = P_{K_\Theta}$. Then the lattice* Lat $\mathbf{T}_\Theta$ *consists of the subspaces $E$,*

$$E = \mathrm{K}_\Theta \cap \vartheta H^2 = \vartheta H^2 \ominus \Theta H^2,$$

*where $\vartheta$ is an (inner) divisor of $\Theta$. Indeed $\vartheta$ is the greatest common divisor* (GCD) *of the inner parts of the functions in $E$.*

*Proof.* Let $T = \mathbf{T}_\Theta$, $E \in \operatorname{Lat} T$ and $E' = K \ominus E$. Then $T^*E' \subset E'$, i.e. $S^*E' \subset E'$, and there exists an inner function $\vartheta$ such that $E' = H^2 \ominus \vartheta H^2$. As $E' \subset K$, $\vartheta$ is a divisor of $\Theta$ (Corollary 8 of Lecture I, Sect. 4). Finally

$$\begin{aligned} E = K \ominus E' &= \{f : f \in K, f \perp (H^2 \ominus \vartheta H^2)\} \\ &= \{f : f \in K, f \in \vartheta H^2\} = K \cap \vartheta H^2 \\ &= \{f : f \in \vartheta H^2, f \perp \Theta H^2\} = \vartheta H^2 \ominus \Theta H^2. \end{aligned}$$

The last assertion of the theorem follows from the definition of the GCD and the identity

$$V \overset{\text{def}}{=} \bigvee (S^n(\vartheta H^2 \ominus \Theta H^2) : n \geq 0) = \vartheta H^2.$$

In order to verify this identity we make use of the fact that $\vartheta$ is an isometric operator:

$$\bigvee (S^n(\vartheta H^2 \ominus \Theta H^2) : n \geq 0) = \vartheta \bigvee \left( S^n \left( H^2 \ominus \frac{\Theta}{\vartheta} H^2 \right) : n \geq 0 \right) \overset{\text{def}}{=} \vartheta V_0$$

and further notice that $V_0$ is an invariant subspace for $S$ (by the construction) as well as for $S^*$:

$$S^*(\Theta\vartheta^{-1}H^2)^\perp \subset (\Theta\vartheta^{-1}H^2)^\perp \subset V_0,$$

$$S^*S^n(\Theta\vartheta^{-1}H^2)^\perp = S^{n-1}(\Theta\vartheta^{-1}H^2)^\perp \subset V_0 \quad \text{for } n \geq 1.$$

By Corollary 7 of Lecture I, Sect. 4, therefore $V_0 = H^2$ and $V = \vartheta H^2$. □

**Corollary 1.** *The function $f$, $f \in H^2 \ominus \Theta H^2$ is T-cyclic ($T = \mathbf{T}_\Theta$) iff $\Theta$ and $f^i$ are relatively prime*: $\mathrm{GCD}(\Theta, f^i) = \mathbb{1}$. *More generally*

$$E_f^T = \varphi H^2 \ominus \Theta H^2, \text{ where } \varphi = \mathrm{GCD}(\Theta, f^i). \ \square$$

**Corollary 2.** *Let $E \in \operatorname{Lat} \mathbf{T}_\Theta$. Then*

$$\dim E < \infty \Leftrightarrow E = \Theta\vartheta^{-1}(H^2 \ominus \vartheta H^2),$$

*where $\vartheta$ is a divisor of $\Theta$ which is at the same time a finite Blaschke product.* □

**Corollary 3.** *If $T = \mathbf{T}_\Theta$, $\lambda \in \sigma_p(T)$, then*

$$\bigcup_{n \geq 1} \operatorname{Ker}(T - \lambda I)^n = \operatorname{Ker}(T - \lambda I)^k = \Theta \mathbf{b}_\lambda^{-k}(H^2 \ominus \mathbf{b}_\lambda^k H^2), \tag{1}$$

*$k = k(\lambda)$ being the multiplicity of the zero of the function $\Theta$ at the point $\lambda$. If $\bar{\lambda} \in \sigma_p(T^*)$ then*

$$\bigcup_{n \geq 1} \operatorname{Ker}(T^* - \bar{\lambda} I)^n = \operatorname{Ker}(T^* - \bar{\lambda} I)^k = H^2 \ominus \mathbf{b}_\lambda^k H^2. \tag{2}$$

*Also $\lambda \in \sigma_p(T) \Leftrightarrow \bar{\lambda} \in \sigma_p(T^*)$.*

This corollary follows readily from the Theorem on $\operatorname{Lat} \mathbf{T}_\Theta$, Corollary 2 and Beurling's Lemma (Lecture II) but can also be obtained from the Calculus Lemma (Lecture III). Indeed, according to Statement c) of that lemma

$$\begin{aligned} \operatorname{Ker}(T - \lambda I)^n &= K_\Theta \cap \{f : f \in H^2, b_\lambda^n f \in \Theta H^2\}, \\ \operatorname{Ker}(T^* - \bar{\lambda} I)^n &= H^2 \ominus \mathrm{GCD}(\Theta, b_\lambda^n) H^2, \end{aligned} \tag{3}$$

from which we deduce (2) and the first equality in (1); putting $n = k$ in (3) we also get the second one. □

**Corollary 4.** *If $E \in \operatorname{Lat} \mathbf{T}_\Theta$ and $\vartheta$ denotes the inner divisor of the function $\Theta$ corresponding to this subspace according to the Theorem on* $\operatorname{Lat} \mathbf{T}_\Theta$ *(i.e. $E = \vartheta H^2 \ominus \Theta H^2$), then the part $\mathbf{T}_\Theta | E$ is unitarily equivalent to the operator $\mathbf{T}_{\Theta/\vartheta}$.*

Indeed, the multiplication operator $\bar{\vartheta}$, which is unitary in $L^2(\mathbb{T})$, maps $E$ onto the subspace $H^2 \ominus \bar{\vartheta}\Theta H^2$ and further $\mathbf{T}_\Theta|E = \vartheta T_{\Theta/\vartheta}\bar{\vartheta}|E$: if $f \in E$ then $P_\Theta zf = \Theta(I - P_+)\bar{\Theta}zf = \vartheta((\bar{\vartheta}\Theta)(I - P_+)(\vartheta\bar{\Theta})\bar{\vartheta}zf$. □

**Corollary 5.** *Let $\Theta = B \cdot \Theta^s$ be the canonical factorization of the function $\Theta$, $B$ a Blaschke product and $\Theta^s$ a singular function. Then the subspaces generated by (all) the root vectors of the operators $T$ and $T^*$ are given respectively by the formulae:*

$$\bigvee(\operatorname{Ker}(T - \lambda I)^{k(\lambda)} : \lambda \in \sigma_p(T)) = \mathbf{\Theta}^s H^2 \ominus \mathbf{\Theta} H^2,$$

$$\bigvee(\operatorname{Ker}(T^* - \bar{\lambda} I)^{k(\lambda)} : \bar{\lambda} \in \sigma_p(T^*)) = H^2 \ominus \mathbf{B} H^2.$$

In fact by the Theorem on Lat $\mathbf{T}_\Theta$ the first of these hulls equals $\vartheta H^2 \ominus \mathbf{\Theta} H^2$ where $\vartheta$ is the greatest common divisor of the inner parts of the functions belonging to $\operatorname{Ker}(T - \lambda I)^{k(\lambda)}$, $\lambda \in \sigma_p(T)$. From formula (1) it follows that this divisor coincides with $\Theta^s$. The hull of the root vectors of $T^*$ is obtained from formula (2) and Corollary 8 of Lecture I, Sect. 4, if we notice that $\operatorname{LCM}\{b_k^{k(\lambda)}\} = B$. □

One of the basic results in linear algebra concerns the completeness of the family of eigen and root vectors for a finite dimensional operator. A general linear map $T$, having a similar property of completeness, is in some sense "approximately" close to the finite dimensional operators.

**Definition.** An operator $T$ defined in a space $X$ is called *complete* if the family of root vectors of $T$ is total, i.e. if

$$\bigvee(\operatorname{Ker}(T - \lambda I)^n : \lambda \in \sigma_p(T), n \geq 1) = X.$$

Completeness is an important property of a linear operator; the presence of this property at once transfers its study to the framework of "discrete" spectral analysis.

**Completeness Theorem.** *Let $\Theta$ be an inner function, $K = H^2 \ominus \mathbf{\Theta} H^2$, $T = P_K S|K$. The following statements are equivalent:*

1. *$T$ is complete.*
2. *$T^*$ is complete.*
3. *$\Theta$ is a Blaschke product.*

*The proof* follows immediately from Corollary 5. □

However completeness alone does not guarantee that an operator is sufficiently "close" to the finite dimensional case. We have to further clarify to what extent the properties of the initial operator can be reconstructed from its finite dimensional restrictions. One of the fundamental objects which one would like to recover is the family of its invariant subspaces Lat $T$.

**Definition.** Let $T$ be a complete operator. We say that the operator $T$ *admits spectral synthesis* if every $T$-invariant subspace is generated by root vectors belonging to the subspace in question.

Thus the structure of the invariant subspaces of an operator admitting spectral synthesis is as simple as in the finite dimensional case: they are precisely those subspaces which are spanned by an (ascending) family of finite dimensional invariant subspaces.

**Theorem on Spectral Synthesis.** *Every complete truncated shift admits spectral synthesis.*

*Proof.* By the Theorem on Lat $T_\Theta$ such an operator is given by $T = P_K S|K$ or $T^* = S^*|K$, with $K = H^2 \ominus \mathbf{B}H^2$, $B$ a Blaschke product. In view of Corollary 4 of the Theorem on Lat $\mathbf{T}_\Theta$ the restriction of $T$ to an invariant subspace is unitarily equivalent to an operator $T_{B'}$ where $B'$ is an inner divisor of $B$. As $B'$ again is a Blaschke product, upon applying once more the Theorem on Lat $\mathbf{T}_\Theta$, we get the desired result. □

Another important object which one would like to recover from the finite dimensional restrictions of an operator $T$ is its spectrum $\sigma(T)$. Complete truncated shifts have this property: $\sigma(T)(\sigma(T^*))$ is the closure of the set of eigenvalues of the operator $T(T^*)$. This assertion is an immediate consequence of the Livšič-Moeller Theorem and the Completeness Theorem.

**2. Spectral Subspaces.** The decomposition of an operator into summands with "small" spectra is one of the basic devices of spectral analysis. The simplest instance of this method is the projection onto an isolated part of the spectrum, which can easily be accomplished in the framework of the classical holomorphic Riesz-Dunford calculus. Let us begin by constructing the latter. Let $\sigma$ be a compact subset of the plane $\mathbb{C}$ and let $A(\sigma)$ denote the algebra of all functions analytic near the points of $\sigma$ (every function $f$ is defined and analytic in a corresponding neighborhood $\mathcal{O}_f$ of the set $\sigma$; strictly speaking, $A(\sigma)$ is thus an algebra of germs of functions). Let $X$ be a Banach space and $T$ a continuous linear operator in $X$. Consider $f \in A(\sigma(T))$ and let $\gamma_f$ be the oriented boundary of the compact set $\gamma_f \cup \operatorname{int}\gamma_f$ consisting of a finite number of piecewise smooth curves, each homeomorphic to the circle $\mathbb{T}$, assuming moreover that $\gamma_f \cup \operatorname{int}\gamma_f \subset \mathcal{O}_f$ and that $\sigma(T)$ lies in the interior $\operatorname{int}\gamma_f$ of this compact set. Define

$$f(T) = \frac{1}{2\pi i}\int_{\gamma_f} R(\lambda, T) f(\lambda)\, d\lambda,$$

where $R(\lambda, T) = (\lambda I - T)^{-1}$ is the resolvent of $T$.

**Lemma on the Riesz-Dunford Calculus.** 1. *The map*

$$f \mapsto f(T), \quad f \in A(\sigma(T)),$$

*is linear and multiplicative, and it coincides on the algebra $\mathcal{P}_A$ with the natural polynomial calculus.*

2. $\sigma(f(T)) = f(\sigma(T))$.

*Proof.* 1. The linearity is obvious. If $f \in \mathcal{P}_A$ then the integration along $\gamma_f$ can be replaced by integration along a large circle $\mathbb{T}_r = \{\lambda : |\lambda| = r\}$ with (say) $r = |T| + 1$, because the function $\lambda \mapsto f(\lambda)R(\lambda, T)$ is regular in $\mathbb{C}\backslash\sigma(T)$; therefore

$$f(T) = \frac{1}{2\pi i}\int_{\mathbb{T}_r} R(\lambda, T)f(\lambda)\,d\lambda$$

$$= \sum_{n\geq 0} T^n \frac{1}{2\pi i}\int_{\mathbb{T}_r} f(\lambda)\lambda^{-n-1}\,d\lambda = \sum_{n\geq 0}\hat{f}(n)T^n.$$

The multiplicative property for the functions $f$ and $g$ in $A(\sigma(T))$ is most conveniently obtained assuming that $\gamma_g \subset \operatorname{int}\gamma_f$, $\gamma_f \cup \operatorname{int}\gamma_f \subset \mathcal{O}_f \cap \mathcal{O}_g$, which of course does not diminish the generality:

$$f(T)g(T) = \frac{-1}{4\pi^2}\int_{\gamma_f} f(\lambda)R(\lambda, T)\,d\lambda \int_{\gamma_g} g(\zeta)R(\zeta, T)\,d\zeta$$

$$= \frac{-1}{4\pi^2}\int_{\gamma_f}\int_{\gamma_g} f(\lambda)g(\zeta)\frac{R(\lambda, T) - R(\zeta, T)}{\zeta - \lambda}\,d\zeta\,d\lambda$$

$$= \frac{-1}{4\pi^2}\int_{\gamma_g} f(\lambda)R(\lambda, T)\int_{\gamma_f}\frac{g(\zeta)}{\zeta - \lambda}\,d\zeta\,d\lambda$$

$$+ \frac{1}{4\pi^2}\int_{\gamma_f} g(\zeta)R(\zeta, T)\int_{\gamma_g}\frac{f(\lambda)}{\zeta - \lambda}\,d\lambda\,d\zeta$$

$$= \frac{1}{2\pi i}\int_{\gamma_g} f(\zeta)g(\zeta)R(\zeta, T)\,d\zeta = (fg)(T).$$

2. If $\lambda \in \sigma(T)$, $f \in A(\sigma(T))$ then $f - f(\lambda) = (z - \lambda\mathbb{1})g$ where $g \in A(\sigma(T))$ so that $f(T) - f(\lambda)I = (T - \lambda I)g(T)$. Therefore $f(T) - f(\lambda)I$ is not invertible and $f(\sigma(T)) \subset \sigma(f(T))$. Conversely, if $\mu \notin f(\sigma(T))$ then the function $f - \mu\mathbb{1}$ is invertible in $A(\sigma(T))$ so that from $(f - \mu\mathbb{1})g = \mathbb{1}$, $g \in A(\sigma(T))$, it follows that $(f(T) - \mu I)g(T) = I$ or $\mu \notin \sigma(f(T))$. Consequently $\sigma(f(T)) \subset f(\sigma(T))$. □

Consider now an *isolated portion* $\sigma$ of the spectrum $\sigma(T)$, i.e. $\sigma(T) = \sigma \cup \sigma'$, where both $\sigma$ and $\sigma'$ are closed subsets, $\sigma \cap \sigma' = \varnothing$. Then there exists a function $f_\sigma$, $f_\sigma \in A(\sigma(T))$, equal to 1 in a neighborhood of $\sigma$ and to 0 in a neighborhood of $\sigma'$. The operator $\mathcal{P}_\sigma$ (the *Riesz projection*),

$$\mathcal{P}_\sigma \stackrel{\text{def}}{=} f_\sigma(T),$$

has the properties:
(a) $\mathcal{P}_\sigma^2 = \mathcal{P}_\sigma$,
(b) $I - \mathcal{P}_\sigma = \mathcal{P}_{\sigma'}$,
(c) $\mathcal{P}_\sigma X = \operatorname{Ker}\mathcal{P}_{\sigma'}$ and $\mathcal{P}_{\sigma'}X = \operatorname{Ker}\mathcal{P}_\sigma$ are closed subspaces of $X$,
(d) $\mathcal{P}_\sigma X + \mathcal{P}_{\sigma'}X = X$,
(e) $T\mathcal{P}_\sigma X \subset \mathcal{P}_\sigma X$,
(f) $\sigma(T|\mathcal{P}_\sigma X) = \sigma$
(g) $E \in \operatorname{Lat} T$, $\sigma(T|E) \subset \sigma \Rightarrow E \subset \mathcal{P}_\sigma X$.

The properties (a)–(d) are trivial while (e) follows from (c) and $T\mathscr{P}_\sigma = \mathscr{P}_\sigma T$ and (f) from Statement 2 of the previous lemma. In order to prove (g) we need the following

**Lemma on the Spectrum of a Restriction.** *Let $T$ be a linear operator, $\lambda \notin \sigma(T)$, $E \in \operatorname{Lat} T$. Then*

$$\lambda \notin \sigma(T|E) \Leftrightarrow R(\lambda, T)E \subset E,$$

*and if one of these condition is fulfilled then $R(\lambda, T|E) = R(\lambda, T)|E$.*

*Proof.* If $\lambda \notin \sigma(T|E)$ then clearly $R(\lambda, T|E)x = R(\lambda, T)x$, $\forall x \in E$, from which follows that $R(\lambda, T)E \subset E$.

The other implication ($\Leftarrow$) follows from the following identity (where $x \in E$)

$$\begin{aligned}(\lambda I - T|E)R(\lambda, T)x &= (\lambda I - T)R(\lambda, T)x = x \\ &= R(\lambda, T)(\lambda I - T)x = R(\lambda, T)(\lambda I - T|E)x. \ \square\end{aligned}$$

From this lemma follows indeed property (g): if $x \in E$, $E \in \operatorname{Lat} T$, then

$$x = \frac{1}{2\pi i}\int_{\gamma_{f_\sigma}} R(\lambda, T|E)x\,d\lambda = \frac{1}{2\pi i}\int_{\gamma_{f_\sigma}} R(\lambda, T)xf_\sigma(\lambda)\,d\lambda = \mathscr{P}_\sigma x,$$

so that $E \subset \mathscr{P}_\sigma X$. $\square$

Thus if the spectrum $\sigma(T)$ is a sum of pairwise disjoint summands $\sigma_1, \ldots, \sigma_n$ then $T$ is the direct sum of its *spectral components* $T_i = T|\mathscr{P}_{\sigma_i}X$ and $\sigma(T_i) = \sigma_i$, $i = 1, \ldots, n$.

For the decomposition of connected parts of the spectrum we need the notion of (maximal) spectral subspace, thus postulating the properties e)–g) of the subspaces $\mathscr{P}_\sigma X$.

**Definition.** The subspace $E$ is said to be a *spectral subspace* of the operator $T$ if $E \in \operatorname{Lat} T$ and

$$E' \in \operatorname{Lat} T, \quad \sigma(T|E') \subset \sigma(T|E) \Rightarrow E' \subset E.$$

**Definition.** The operator $T$ is called *decomposable* if to each finite covering of $\sigma(T)$ by open subsets $\mathcal{O}_\alpha$, $\alpha \in A$ ($A$ a finite set) there corresponds a family of spectral subspaces $\{E_\alpha : \alpha \in A\}$ such that

$$\sigma(T|E_\alpha) \subset \mathcal{O}_\alpha, \quad \alpha \in A; \quad X = \sum_{\alpha \in A} E_\alpha.$$

To every spectral subspace there corresponds a well-defined portion of the spectrum $\sigma(T)$. Indeed, (as is easily seen) $R(\lambda, T)E \in \operatorname{Lat} T$ for $\lambda \notin \sigma(T)$ and moreover $\sigma(T|R(\lambda, T)E) = \sigma(T|E)$ so that (if $E$ is a spectral subspace) $R(\lambda, T)E \subset E$ and, by the Lemma on the Spectrum of a Restriction, $\sigma(T|E) \subset \sigma(T)$. That $T$ is decomposable means therefore that we have a property similar to the decomposition into spectral components: to some (arbitrary fine but possibly non-disjoint) decompositions of $\sigma(T)$ there corresponds a decomposition of the space into spectral subspaces. This is one of the weakest (and also most general) analogues of the Spectral Theorem, retaining at least some of its features.

Let us now return to our model operators $\mathbf{T}_\Theta$ and $\mathbf{T}_\Theta^*$ and show that they are indeed decomposable. For every subset $\sigma$ of the closed disk $\operatorname{clos}\mathbb{D}$ we let $\Theta_\sigma$ be the following divisor of $\Theta$:

$$\Theta_\sigma = \prod_{\lambda\in\sigma} b_\lambda^{k_\Theta(\lambda)} \exp\left(-\int_{\sigma\cap\mathbb{T}} \frac{\zeta+z}{\zeta-z}\, d\mu^s(\zeta)\right). \tag{4}$$

**Theorem on Spectral Subspaces.** *Let $\Theta$ be an inner function and consider the operator $T = \mathbf{T}_\Theta$.*

1. *The subspace $E$ is a spectral subspace for $T$ iff*

$$E = \mathbf{\Theta}_{\mathcal{O}} H^2 \ominus \mathbf{\Theta} H^2 \quad \text{and} \quad \operatorname{clos}\mathbb{D}\backslash\mathcal{O} = \sigma(T|E).$$

2. *The subspace $E$ is a spectral subspace for $T^*$ iff*

$$E = H^2 \ominus \mathbf{\Theta}_F H^2 \quad \text{and} \quad F^* = \sigma(T^*|E), \quad F^* \overset{\text{def}}{=} \{\lambda : \bar{\lambda}\in F\}.$$

3. *The operators $T$ and $T^*$ are decomposable.*

*Proof.* The proofs of Statements 1 and 2 are similar to each other so we write down the spectral subspaces only for the operator $T^* = S^*|K$. If $E\in\operatorname{Lat} T^*$ then $E = H^2 \ominus \vartheta H^2$, where $\vartheta$ is a suitable divisor of $\Theta$, and $\sigma(T^*|E) = \sigma(\vartheta)^*$. It is however clear that among all these subspaces with a fixed spectrum $F^* = \sigma(T^*|E)$ the largest one is $H^2 \ominus \mathbf{\Theta}_F H^2$ (cf. Corollary 5 of Lecture I, Sect. 3 and the Theorem on the Spectrum of a Function, Lecture III).

Passing to the decomposability let us again consider just one of the operators, this time the operator $T$. Let $\{\mathcal{O}_\alpha\}_{\alpha\in A}$ be a finite open cover of $\operatorname{clos}\mathbb{D}$ and let $\mathcal{O}'_\alpha$ be open subsets with the property that $\bigcup \mathcal{O}'_\alpha \supset \operatorname{clos}\mathbb{D}$, $F_\alpha \overset{\text{def}}{=} \operatorname{clos}\mathcal{O}'_\alpha \subset \mathcal{O}_\alpha$, $\alpha\in A$. As $\bigcap_\alpha(\operatorname{clos} D\backslash\mathcal{O}'_\alpha) = \varnothing$ we have

$$\min_{\zeta\in\operatorname{clos}\mathbb{D}} \max_\alpha \operatorname{dist}(\zeta, (\operatorname{clos}\mathbb{D})\backslash\mathcal{O}'_\alpha) \overset{\text{def}}{=} \delta > 0.$$

Let us observe that if $\varphi$ is an inner function then

$$\inf\{|\varphi(\zeta)| : \zeta\in\mathbb{D},\quad \operatorname{dist}(\zeta,\sigma(\varphi)) \ge \delta\} \overset{\text{def}}{=} \varepsilon_\varphi(\delta) > 0,$$

if $\delta > 0$; this follows from the Theorem on the Spectrum of a Function (Lecture III). Set $\Theta_\alpha = \Theta_{\operatorname{clos} D\backslash\mathcal{O}'_\alpha}$, $\varepsilon = \min_\alpha \varepsilon_{\Theta_\alpha}$. Then $\sum_\alpha |\Theta_\alpha(\zeta)|^2 \ge \max|\Theta_\alpha(\zeta)|^2 \ge \varepsilon(\delta)^2$ for all $\zeta$, $\zeta\in\mathbb{D}$. By Carleson's Corona Theorem (Lecture III) there exist then functions $g_\alpha$, $g_\alpha\in H^\infty$, such that $\sum_\alpha g_\alpha\Theta_\alpha = \mathbb{1}$. Consequently $\sum_\alpha \Theta_\alpha(T)g_\alpha(T) = I$ and

$$\begin{aligned}\Theta_\alpha(T)K \subset \operatorname{Ker}(\Theta/\Theta_\alpha)(T) &= \operatorname{Ker}\Theta_{\mathcal{O}'_\alpha}(T)\\ &= \mathbf{\Theta}_\alpha H^2 \ominus \mathbf{\Theta} H^2 \subset \mathbf{\Theta}_{(\operatorname{clos}\mathbb{D})\backslash F_\alpha} H^2 \ominus \mathbf{\Theta} H^2\\ &= \operatorname{Ker}\Theta_{F_\alpha}(T) \overset{\text{def}}{=} E_\alpha.\end{aligned} \tag{5}$$

Thus each element $x$ in $K$ is a sum of terms $x_\alpha$, $x_\alpha = \Theta_\alpha(T)g_\alpha(T)x$, and each $x_\alpha$ belongs to the spectral subspace $E_\alpha$. Finally $\sigma(T|E_\alpha) \subset F_\alpha \subset \mathcal{O}_\alpha$. □

**3. Unicellular Operators.** There is a natural limit for the decomposition of the operator $\mathbf{T}_\Theta$ into spectral subspaces. This is possible only as long as one does not encounter subspaces $E$ in $\operatorname{Lat}\mathbf{T}_\Theta$ with a one point spectrum, more exactly, subspaces such that $\sigma(\mathbf{T}_\Theta|E)$ consists of one single point. Such subspaces (if there exist any) are not subject to further spectral decomposition. Therefore we have to try to associate with them a "Jordan structure", in a similar way as one does this in linear algebra with matrices with a one point spectrum. As a part of a model operator is unitarily equivalent to another model operator (Corollary 4 of the Theorem on Lat $\mathbf{T}_\Theta$) we can limit ourselves to such linear operators only.

**Lemma on Jordan blocks.** *Let $\Theta$ be an inner function and consider $K_\Theta = H^2 \ominus \mathbf{\Theta} H^2$ and the corresponding operator $\mathbf{T}_\Theta$. Assume that $\sigma(T_\Theta) = \sigma(\Theta) = \{\lambda\}$ where $\lambda \in \operatorname{clos} \mathbb{D}$. Then*

$$\dim K_\Theta < \infty \Leftrightarrow \lambda \in \mathbb{D}$$

*and if one of these (equivalent) conditions is fulfilled then $\mathbf{T}_\Theta$ (as well as $\mathbf{T}_\Theta^*$) is a Jordan block (i.e. the matrix of $\mathbf{T}_\Theta$ in a suitable basis is a Jordan block).*

*Proof.* By the Lemma on Finite Dimensional Subspaces (Lecture II): $\dim K_\Theta < \infty \Leftrightarrow$ ($\Theta$ is a finite Blaschke product). The spectrum of such a product is the (finite) set of its zeros, which is contained in $\mathbb{D}$. This establishes the implication "$\Rightarrow$" but "$\Leftarrow$" too follows from the same Lemma on Finite Dimensional Subspaces combined with the Theorem on the Spectrum of a Function (Lecture III), according to which the condition $\sigma(\Theta) = \{\lambda\}$, $\lambda \in \mathbb{D}$, entails $\Theta = b_\lambda^n$ (for some $n$, $n \geq 1$).

In the proof of Beurling's Lemma (Lecture II) we have essentially already given bases for the space $H^2 \ominus \mathbf{b}_\lambda^n H^2$ in which $\mathbf{T}_\Theta$ and $\mathbf{T}_\Theta^*$ are given by a single Jordan block. As

$$(z - \lambda \mathbb{1})^{n-k}(\mathbb{1} - \bar{\lambda} z)^{-n} \perp \mathbf{b}_\lambda^n H^2, \quad 1 \leq k \leq n,$$

this family is indeed the desired basis for the operator $\mathbf{T}_\Theta$; for $\mathbf{T}_\Theta^* = S^* | K_\Theta$ it follows from formula (2) (Lecture II, p. 33) that

$$(S^* - \bar{\lambda} I)\frac{z^k}{(\mathbb{1} - \bar{\lambda} z)^{k+1}} = \frac{z^{k-1}}{(\mathbb{1} - \bar{\lambda} z)^k}, \quad 1 \leq k \leq n-1,$$

i.e. the basis $\{z^k(\mathbb{1} - \bar{\lambda} z)^{-k-1}\}_{k=0}^{n-1}$ transforms the operator $\mathbf{T}_\Theta^*$ into Jordan form. □

As we already remarked in the Introductory Lecture each linear operator $T$ acting in a finite dimensional space is a sum of Jordan blocks, i.e. $X = \sum_{k=1}^m X_k$ (direct sum), $TX_k \subset X_k$, and $T|X_k = \lambda_k I + J_k$ ($k = 1, \ldots, m$) where $\{\lambda_1, \ldots, \lambda_m\} = \sigma(T)$ and $J_k$ is an operator in $X_k$ which in a suitable basis for that space acts by the formula $J_k e_0 = \mathbb{0}$, $J_k e_i = e_{i-1}$, $1 \leq i \leq \dim X_k - 1$. In the basis $\{e_i\}$ the matrix of $J_k$ is the elementary Jordan block

$$\begin{pmatrix} 0 & 1 & & & & \\ & 0 & & & \mathbb{0} & \\ & & \cdot & & & \\ & & & \cdot & & \\ & & & & \cdot & \\ & \mathbb{0} & & & 0 & 1 \\ & & & & & 0 \end{pmatrix}.$$

From this decomposition theorem it follows easily that the operator $T$ is given by one single Jordan block (i.e. $m = 1$ in the above notation) iff the lattice Lat $T$ is linearly ordered with respect to set theoretic inclusion:

$$E, E' \in \operatorname{Lat} T; \quad E' \not\subset E \Rightarrow E \subset E'.$$

This geometrical description of elementary Jordan blocks turns out to be suitable for developing an analogous notion in a considerably more general situation.

**Definition.** Let $X$ be a Banach space and $T$ a continuous linear operator in $X$. $T$ is said to be *unicellular* if the lattice Lat $T$ of its invariant subspaces is linearly ordered by set theoretic inclusion.

It is clear that, as in the finite dimensional case, the spectrum of an unicellular operator at least must be connected (for to disjoint parts of the spectrum there correspond noncomparable (spectral) invariant subspaces). Pursuing this analogy we may hope that an arbitrary operator in some sense can be expressed as a sum of unicellular operators. But despite the fact that for some important classes of operators in Hilbert space these words may be given a precise meaning, in general the analogy does not extend very far. Nevertheless already in this Section we will see how fruitful the study of unicellular operators can be in infinite dimensional spaces.

**Unicellularity Theorem.** *Let $\Theta$ be an inner function and set as usual*

$$K_\Theta = H^2 \ominus \Theta H^2, \quad P = P_{K_\Theta} \quad \textit{and} \quad T = \mathbf{T}_\Theta = PS|K_\Theta.$$

*Then $T$ is unicellular iff $\sigma(T)$ is a singleton, i.e. $\sigma(T) = \{\lambda\}$, and then either $\Theta = b_\lambda^n$ (the case $|\lambda| < 1$) or $\Theta = \exp\left(-a\dfrac{\lambda + z}{\lambda - z}\right)$ (the case $|\lambda| = 1$) for suitable $n$ ($n \geq 1$) and $a$ ($a > 0$).*

*Proof I.* By the Theorem on Lat $\mathbf{T}_\Theta$ the $T$-invariant subspaces are in one-to-one correspondence with the inner divisors of $\Theta$. Also (Corollary 5 of Theorem BH, Lecture I) the relation of inclusion between subspaces is translated into divisibility of the corresponding inner functions. Therefore $T$ is unicellular iff the set of all divisors of $\Theta$ is linearly ordered: if $\vartheta_k$, $k = 1, 2$, divide $\Theta$ then either $\vartheta_1$ divides $\vartheta_2$ or $\vartheta_2$ divides $\vartheta_1$. Let us check that the condition appearing in the theorem is necessary for $T$ to be unicellular. Indeed, if $\sigma(T) = \sigma(\Theta)$ contains more than one point then either one of the sets $\sigma(\Theta) \cap \mathbb{T}$ and $\sigma(\Theta) \cap \mathbb{D}$ has the same property or $\sigma(\Theta) \cap \mathbb{D} \neq \varnothing$ and $\sigma(\Theta) \cap \mathbb{T} \neq \varnothing$. In the first case we have either the non-comparable divisors $b_{\lambda_1}$ and $b_{\lambda_2}$, $\lambda_1 \neq \lambda_2$, $\lambda_k \in \Theta^{-1}(0)$, $k = 1, 2$, or the divisors

$$\exp\left(-\int_{\Delta_k} \frac{\zeta + z}{\zeta - z} d\mu^s(\zeta)\right), \quad k = 1, 2,$$

where $\Delta_k \subset \mathbb{T}$, $\mu^s \Delta_k > 0$, $k = 1, 2$, $\Delta_1 \cap \Delta_2 = \varnothing$, $\mu^s$ being the measure on $\mathbb{T}$ corresponding to the singular factor $\Theta^s$ of the function $\Theta$; in the second case we take $\Theta^s$ and the Blaschke product $B$ ($\Theta = B\Theta^s$).

The sufficiency of our condition is trivial: if $\sigma(\Theta) = \{\lambda\}$ then either $|\lambda| < 1$ in which case $\Theta = b_\lambda^n$ or $|\lambda| = 1$ in which case $\Theta = \exp\left(-\frac{\lambda + z}{\lambda - z}\mu^s\{\lambda\}\right)$; in both cases the divisor are ordered in the required way. □

*Proof II.* If $\sigma(\Theta) = \{\lambda\}$, $\lambda \in \operatorname{clos} \mathbb{D}$, we check that $T_\Theta$ is unicellular exactly as before. If $\sigma(\Theta)$ is not a singleton then there exists an open covering $\mathcal{O}_1 \cup \mathcal{O}_2$ of the closed disk $\operatorname{clos} \mathbb{D}$ such that $\sigma(\mathbf{T}_\Theta) \not\subset \mathcal{O}_i$, $i = 1, 2$. The spectral subspaces of $\mathbf{T}_\Theta$ corresponding to this covering are non-comparable (by the Theorem on Spectral Subspaces and by the definition of decomposability). □

This result has interesting applications to the study of the operator of indefinite integration and to Titchmarsh's well-known convolution theorem. Let us define in the space $L^2(0,1)$ of (classes of) square integrable functions on the interval $(0,1)$ the operator $J$ by the formula

$$Jf(x) = \int_0^x f(t)\,dt, \quad 0 < x < 1.$$

It is clear that the subspaces

$$L_a^2 \stackrel{\text{def}}{=} \{f : f \in L^2(0,1), f = 0 \text{ a.e. on } (0,a)\}, \quad 0 \le a \le 1,$$

are invariant for $J$. The converse is also true but far less trivial. Let us prove this (that is, the fact that $J$ is a unicellular operator: $\operatorname{Lat} J = \{L_a^2 : 0 \le a \le 1\}$) invoking the associated truncated shift. Let us first notice that from this again follows the following *cyclicity criterion* for the space $L^2(0,1)$: the sequence of primitive functions $\{J^n f : n \ge 0\}$ is total in $L^2(0,1)$ iff $0 \in \operatorname{supp} f$ (where $\operatorname{supp} f$ denotes the support of the measure $f\,dt$), that is, when there does not exist any number $a$, $a > 0$, with the property that $f = 0$ a.e. in $(0,a)$. This result answers a question raised by Gel'fand already in 1937.

We will base our study of the operator $J$ on the following result due to Sarason: *Set* $\Theta = \exp\frac{z + \mathbb{1}}{z - \mathbb{1}}$, $K_\Theta = H^2 \ominus \Theta H^2$, $P = P_{K_\Theta}$ *and* $T = \mathbf{T}_\Theta = PS|K_\Theta$. *Then the operators*[1] $\frac{1}{2}(I + T)$ *and* $(I + J)^{-1}$ *are unitarily equivalent, that is,* $\frac{1}{2}V^{-1}(I + T)V = (I + J)^{-1}$ *where* $V$ *is a suitable isometric map of the space* $L^2(0,1)$ *onto* $K_\Theta$.

This proposition will be obtained in Lecture $V$ as a simple consequence of the Theorem on the Triangular Form of the Truncated Shift. As by the Unicellularity Theorem $T$ has a linearly ordered lattice of invariant subspaces, the same will be true for the operator $\frac{1}{2}(I + T)$ as well as for $(I + J)^{-1}$. The remaining argument constitutes the content of the

**Lemma on Components.** *Let $A$ be a linear operator in a Banach space $X$ and $E$ a subspace of $X$. Then the set*

[1] That $-1 \notin \sigma(J)$ will be seen below.

$$\{\lambda : \lambda \in \hat{\mathbb{C}}, R(\lambda, A)E \subset E\}$$

*contains with every point also the connected component in the complement* $\hat{\mathbb{C}} \setminus \sigma(A)$ *to which this point belongs; here we by definition say that* $R(\infty, A)E \subset E \Leftrightarrow AE \subset E$.

*Proof.* The set defined in the lemma is open and closed in $\hat{\mathbb{C}} \setminus \sigma(A)$: if $R(\lambda_n, A)E \subset E$ and $\lambda = \lim_n \lambda_n$, $\lambda \in \hat{\mathbb{C}} \setminus \sigma(A)$ then $R(\lambda, A)E \subset E$, since $E$ is closed (for $\lambda = \infty : A = \lim_n \lambda_n(\lambda_n R(\lambda_n, A) - I)$); if $R(\lambda, A)E \subset E$ then all points $\mu$ close to $\lambda$ have the same property, as is seen by expanding $R(\mu, A)$ in a Neumann series in powers of $R(\lambda, A)$. □

Thus $J$ is a unicellular operator. As a consequences of this we can obtain the following result of importance in analysis and in the theory of differential equations.

**Titchmarsh's Convolution Theorem.** *Let us define the convolution of the summable functions* $f$ *and* $g$ *on the half-axis* $\mathbb{R}_+ = [0, \infty)(f, g \in L^1(\mathbb{R}_+))$ *by the formula*

$$(f * g)(x) = \int_0^x f(t)g(x - t)\,dt, \; x > 0.$$

*Assume that* $f * g = 0$ *a.e. on* $(0, \omega)$, $\omega > 0$. *Then there exist non-negative numbers* $a$ *and* $b$ *with* $a + b = \omega$ *such that* $f = 0$ *a.e. on* $(0, a)$ *and* $g = 0$ *a.e. on* $(0, b)$. *In particular, if* $0 \in \operatorname{supp} f$ *then* $g = 0$ *a.e. on* $(0, \omega)$. *Finally, if* $f * g = \mathbb{O}$ *then either* $f = \mathbb{O}$ *or* $g = \mathbb{O}$.

Proof. Without loss of generality we can take $\omega = 1$. Let us first notice that $f_1 \stackrel{\text{def}}{=} Jf \in L^2(0, 1)$, $g_1 \stackrel{\text{def}}{=} Jg \in L^2(0, 1)$ and furthermore that $Jf = (f * \chi)|(0, 1)$ where $\chi(t) = \chi_{[0,1]}(t)$, $t > 0$. From the hypothesis of the theorem and from the commutativity of convolution (which can easily be verified on the basis of Fubini's theorem) it follows that $\chi * \chi * (f * g) = f_1 * g_1 = 0$ a.e. on $(0, 1)$ and likewise that $((\underbrace{\chi * \cdots * \chi}_{n \text{ times}}) * f_1) * g_1 = 0$ a.e. on $(0, 1)$. In other words $J^n f_1 * g_1 = 0$ a.e. on $(0, 1)$, that is,

$$\int_0^x (J^n f_1)(t) g_1(x - t)\,dt = 0 \quad \text{for a.e. } x, \; x \in (0, 1).$$

But this intégral defines a continuous function in $x$. Therefore in particular $(J^n f_1, \overline{\tilde{g}_1})_{L^2(0,1)} = 0$ for $n \geq 0$ where $\tilde{g}_1(t) = \overline{g_1(1 - t)}$, $t \in (0, 1)$. As $J$ is a unicellular operator, we must have $E^J_{f_1} = L^2_a$ for some $a$, $0 \leq a \leq 1$. Consequently $g_1 = 0$ a.e. on the interval $(a, 1)$ and $g_1 = 0$ a.e. on the interval $(0, 1 - a)$. Thus

$$\int_0^x f(t)\,dt = 0 \quad \text{for a.e. } x, \; x \in (0, a),$$

$$\int_0^x g(t)\,dt = 0 \quad \text{for a.e. } x, \; x \in (0, 1 - a),$$

from which we deduce the first statement of the theorem. The remaining statements are trivial consequences of the first one. □

*Remark.* From this proof we see that $J^n f = \chi_n * f$, with $\chi_n = \chi * \cdots * \chi|(0,1) = J^{n-1}\chi$, that is, $\chi_n(t) = t^{n-1}/(n-1)!$, $0 < t < 1$. Hence $|J^n| \le 1/(n-1)!$ so that $\sigma(\mathrm{J}) = \{0\}$.

Finally let us notice that Titchmarsh's Convolution Theorem also can be expressed as an identity of the form

$$\inf \operatorname{supp} f * g = \inf \operatorname{supp} f + \inf \operatorname{supp} g;$$

the inequality $\ge$ here is trivial while $\le$ constitutes precisely the contents of the theorem in question.

In the following Lecture we investigate in some detail special "natural" chains of invariant subspaces of the model operator $\mathbf{T}_\Theta$ and in connection with each such chain the integral operator analogous to $J$.

## Supplements and Bibliographical Notes

**4. On Invariant Subspaces.** Part of the results of Sect. 1 can with appropriate modifications be carried over to the general model operators $\mathbf{T}_\Theta = P_K S|K$, $K = H^2(D) \ominus \boldsymbol{\Theta} H^2(D)$ with the inner characteristic function $\boldsymbol{\Theta}$. This is for instance the case with the Theorem on Lat $\mathbf{T}_\Theta$ and some of its corollaries.

**Theorem on** Lat $\mathbf{T}_\Theta$, II. *If $\Theta$ is an inner $L(D)$-valued function and*

$$\mathbf{T}_\Theta = P_K S|K, \quad K = H^2(D) \ominus \boldsymbol{\Theta} H^2(D),$$

*then the lattice* Lat $\mathbf{T}_\Theta$ *consists of those subspaces $E$ of $K$ which have the form*

$$E = \boldsymbol{\vartheta} H^2(D) \ominus \boldsymbol{\Theta} H^2(D),$$

*where $\vartheta$ is an inner left divisor of $\Theta$.*

*The proof* is essentially a repetition of the argument in the corresponding "scalar" case: the subspace $K \ominus E$ is invariant with respect to the adjoint operator $S^*$ and therefore of the form $K \ominus E = H^2(D) \ominus \boldsymbol{\vartheta} H^2(D)$ (Lecture I, Sect. 6, Corollary 8) and, since it is contained in $K$ and $\boldsymbol{\Theta}$ is an inner function, it follows that $\vartheta$ must be an inner left divisor of that function. □

*Remark.* One can also get vector analogues of Corollary 1 (on cyclic vectors; here it is more proper to speak about cyclic subspaces) and of the part of the Theorem on Lat $\mathbf{T}_\Theta$ which concerns the description of the function $\vartheta$ in terms of the "inner parts" of the functions in the subspace $E$. However, in the present higher dimensional case ($\dim E > 1$) it turns out that these analogues are not very effective. It may happen for example that the functions $\boldsymbol{\Theta}$ and $\vartheta$ coincide on a portion of $D$. Therefore this second part of the Theorem on Lat $\mathbf{T}_\Theta$ takes the form: Set $D' = \{d : d \in D;\ \vartheta^*(\zeta)\Theta(\zeta)d = \text{const}, \zeta \in \mathbb{T}\}$ and let $U$ be a unitary operator in $D$ such that $UD' = \vartheta^*(\zeta)\Theta(\zeta)D'$, $\zeta \in \mathbb{T}$. Then $E = \boldsymbol{\vartheta} H^2(D) \ominus \boldsymbol{\Theta}_0 H^2(D) = \boldsymbol{\vartheta} H^2(D) \ominus \boldsymbol{\Theta} U^{-1} H^2(D) = \boldsymbol{\vartheta}_0 H^2(D_0) \ominus \boldsymbol{\Theta}_0 H^2(D_0)$, with $D_0 = D \ominus UD'$, $\vartheta_0 =$

$\vartheta|D_0$, $\Theta_0 = \Theta U^{-1}|D_0$, the left inner function $\vartheta_0$ being the "greatest common divisor" of the functions in $E$ in the sense that $\bigvee(S_D^n E : n \geq 0) = \vartheta_0 H^2(D_0)$.

With similar modifications we have also an analogue of Corollary 4 of the Theorem on Lat $\mathbf{T}_\Theta$.

**Corollary 1.** *Let* $\mathbf{T}_\Theta$, $E$, $\vartheta$ *have the same meaning as in the Theorem on* Lat $\mathbf{T}_\Theta$, II. *Then the restriction* $\mathbf{T}_\Theta|E$ *is unitarily equivalent to* $\mathbf{T}_{\vartheta^*\Theta}$.

The formal proof follows the "scalar" pattern (except that in place of $\bar{\vartheta}$ we must write $\vartheta^*$) but one has to add to this proof what was said in the preceding remark on the Theorem on Lat $\mathbf{T}_\Theta$, II itself: it may occur that the function $\vartheta^*\Theta$ degenerates to a constant on some portion of $D$ and the canonical form of the operator $\mathbf{T}_\Theta|E$ is then provided by the operator $\mathbf{T}_{\vartheta_0^*\Theta_0}$ on the space $H^2(D_0) \ominus \vartheta_0^*\Theta_0 H^2(D_0)$. □

Let us recall that the invariant subspaces of the adjoint operator $T^*$ (i.e. the operator $S_D^*|K_\Theta$) were written down already in Lecture I, Sect. 6 (Corollary 8 of the Theorem on Invariant Subspaces). They all have the form

$$H^2(D) \ominus \vartheta H^2(D), \tag{6}$$

$\vartheta$ being a left inner divisor of $\Theta$.

**Corollary 2.** *To the eigenspace* $E = \operatorname{Ker}(\mathbf{T}_\Theta - \lambda I)$ *there corresponds the left inner divisor* ${}_\lambda\Theta$ *given by*

$$\Theta(\zeta) = {}_\lambda\Theta(\zeta)\{b_\lambda(\zeta)\Delta_\lambda + (I - \Delta_\lambda)\},$$

*where* $\Delta_\lambda$ *is the orthogonal projection in* $D$ *onto the subspace* $\operatorname{Ker}\Theta(\lambda)$, *and to the eigenspace* $E^* = \operatorname{Ker}(\mathbf{T}_\Theta^* - \bar{\lambda} I)$ *the left inner divisor* ${}_\lambda\Delta b_\lambda + (I - {}_\lambda\Delta)$,

$$\Theta(\zeta) = \{b_\lambda(\zeta){}_\lambda\Delta + (I - {}_\lambda\Delta)\}\Theta_\lambda(\zeta),$$

*where* ${}_\lambda\Delta$ *is the orthogonal projection onto the subspace* $\operatorname{Ker}\Theta(\lambda)^*$.

Let us check for instance that the function $\vartheta = b_\lambda\, {}_\lambda\Delta + (I - {}_\lambda\Delta)$ indeed is a left inner divisor of $\Theta$ representing the invariant subspace $\operatorname{Ker}(\mathbf{T}_\Theta^* - \bar{\lambda} I)$ in the sense of formula (6). In fact, since $\Theta(\lambda)^*{}_\lambda\Delta = \mathbb{0}$ we have ${}_\lambda\Delta\Theta(\lambda) = \mathbb{0}$ so ${}_\lambda\Delta\Theta/b_\lambda$ is analytic (and bounded) in the disk $\mathbb{D}$. It follows that the quotient $\Theta_\lambda \stackrel{\text{def}}{=} \{{}_\lambda\Delta\bar{b}_\lambda + (I - {}_\lambda\Delta)\}\Theta$ is an inner function. The formula

$$\operatorname{Ker}(\mathbf{T}_\Theta^* - \bar{\lambda} I) = H^2(D) \ominus \vartheta H^2(D)$$

holds true in view of the Lemma on Eigenfunctions (Lecture III).

The first statement of Corollary 2 can be proved in an analogous way. One can also derive it from the second one invoking the Lemma on the Function $\Theta^t$ (Lecture III) stating the equivalence of the operators $\mathbf{T}_\Theta^*$ and $\mathbf{T}_{\Theta^t}$. □

It is not that simple to get results similar to the Completeness Theorem of Sect. 1. We will say a few words about this after we have analyzed the case of operators whose characteristic functions have a *scalar multiple* (operators of class $C_0$).

**5. Synthesis for $C_0$-Operators.** Let $T$ be an operator of class $C_0$ in the Hilbert space $H$. First we write down a few properties of the family of subspaces $\operatorname{Ker}\varphi(T)$, $\varphi \in H^\infty$.

(a) *If* $\mathrm{GCD}(\varphi^i, m_T) = \mathbb{1}$ *then* $\operatorname{Ker}\varphi(T) = \{\mathbb{0}\}$.

Indeed, the minimal annihilator of the restriction $T|\operatorname{Ker}\varphi(T)$ divides the functions $m_T$ and $\varphi$ (and thus $\varphi^i$ too); therefore it must coincide with the function $\mathbb{1}$ whereby $\operatorname{Ker}\varphi(T) = \{\mathbb{0}\}$. □

From (a) we get the following corollary: *if $T$ is a completely non-unitary contraction in the Hilbert space $\mathscr{H}$ and $\varphi$ an outer function belonging to $H^\infty$ then* $\operatorname{Ker}\varphi(T) = \{\mathbb{0}\}$. In fact, if $H = \operatorname{Ker}\varphi(T)$ then $TH \subset H$ and $\varphi(T|H) = \mathbb{0}$ so that $T|H \in C_0$. But from (a) we get $\operatorname{Ker}\varphi(T|H) = \{\mathbb{0}\}$, i.e. $H = \{\mathbb{0}\}$. □

(b) $\operatorname{Ker}\varphi(T) = \operatorname{Ker} m'(T)$ *where* $m' = \mathrm{GCD}(\varphi^i, m_T)$.

In fact, $\varphi = \varphi^e\varphi' \cdot m'$ where $\varphi'$ is an inner function which is relatively prime to $m_T$. Therefore $(\varphi^e\varphi')(T)$ is injective and $\operatorname{Ker}\varphi(T) = \operatorname{Ker}(\varphi^e\varphi')(T)\cdot m'(T) = \operatorname{Ker} m'(T)$. □

(c) *If* $E = \operatorname{Ker}\varphi(T)$ *then* $m_{T|E} = m' = \mathrm{GCD}(\varphi^i, m_T)$.

Indeed, using (b) we see that $m_{T|E}$ divides $m'$; but the function $m_{T|E}\cdot(m_T/m')$ is an annihilator of the operator $T$ (since $(m_T/m')(T)H \subset \operatorname{Ker} m'(T) = E$) so that if we would have $m'/m_{T|E} \not\equiv \text{const}$ it would follow that $m_T$ is not the minimal annihilator. □

(d) *The spectral subspace of $T$ corresponding to a closed part $\sigma$ of the spectrum $\sigma(T)$ is* $\operatorname{Ker} m_\sigma(T)$ *where* $m_\sigma = (m_T)_\sigma$ *in the sense of formula* (4).

It is clear that $\operatorname{Ker} m_\sigma(T) \stackrel{\text{def}}{=} E$ is an invariant subspace and that $\sigma(T|E) = \sigma(m_{T|E}) = \sigma(m_\sigma) = \sigma$; on the other hand, if $E'$ is any $T$-invariant subspace then $m_{T|E'}$ divides $m_T$ and further if $\sigma(T|E') \subset \sigma$ then $\sigma(m_{T|E'}) \subset \sigma$ so $m_{T|E'}$ must divide $m_\sigma$, i.e. $E' \subset \operatorname{Ker} m_\sigma(T)$. □

(e) *Let $M$ be any family of inner divisors of $m_T$ and set* $m = \mathrm{LCM}\, M$. *Then*

$$\bigvee(\operatorname{Ker} m'(T) : m' \in M) = \operatorname{Ker} m(T).$$

In fact, the inclusion $\subset$ is trivial. In order to prove the converse let us denote by $E$ the orthogonal difference of the subspaces in question and let us set $A = P_E T|E$. By Sarason's Projection Lemma (see Lecture III) $T$ is a dilation of $A$ and $\varphi(A) = P_E\varphi(T)|E$ $(\forall\varphi \in H^\infty)$ so that $A \in C_0$. Let us show that $m_A = \mathbb{1}$, which will then imply the desired result, viz. $E = \{\mathbb{0}\}$. For this again it suffices to check that $m_A$ divides all the functions $m/m'$, $m' \in M$. But if $x \in E$ then $(m/m')(A)x = P_E(m/m')(T)x \in P_E \operatorname{Ker} m'(T)$ (since $x \in \operatorname{Ker} m(T)$) and the subspace $\operatorname{Ker} m'(T)$ is orthogonal to $E$ by the definition of the latter so that $(m/m')(A)x = \mathbb{0}$. Thus $(m/m')(A) = \mathbb{0}$ and $m_A$ divides $m/m'$, $m' \in M$. □

(f) *In the hypothesis of the preceding Statement*

$$\bigcap_{m' \in M} \operatorname{Ker} m'(T) = \operatorname{Ker}\mu(T), \quad \mu = \mathrm{GCD}\, M.$$

Analogous proof. □

(g) *For* $\lambda \in \mathbb{D}$ *let* $E_\lambda = \bigcup_{n\geq 1} \operatorname{Ker}(\lambda I - T)^n$ *be the root manifold corresponding to the point* $\lambda$ *and denote by* $k = k(\lambda)$ *the multiplicity with which the elementary factor* $b_\lambda$ *enters into the minimal annihilator* $m_T$. *Then*

$$E_\lambda = \operatorname{Ker} b_\lambda^k(T).$$

In fact, by (b) we have $\operatorname{Ker}(\lambda I - T)^n = \operatorname{Ker} b_\lambda^n(T) = \operatorname{Ker} b_\lambda^k(T)$ if $n \geq k$, so that $E_\lambda \subset \operatorname{Ker} b_\lambda^k(T)$. The converse is trivial. □

These properties of the operators of class $C_0$ are entirely sufficient for deducing a criterion for the completeness of the root manifolds and for the possibility of spectral synthesis. But before passing to the proof let us formulate an important result, which supplements the Statements (a)–(g), characterizing those operators of class $C_0$ for which the lattice of invariant subspaces has an especially simple form – it consists of subspaces of the type $\operatorname{Ker} \varphi(T)$ only.

**Theorem on the Simplicity of the Spectrum.** *Let* $T \in C_0$. *The following statements are equivalent*:

1. $C^T \neq \varnothing$ (*i.e.* $T$ *is cyclic*).
2. $\operatorname{Lat} T = \{\operatorname{Ker} \varphi(T) : \varphi \in H^\infty\}$.

*If* $T = P_K S|K, K = H^2(E) \ominus \Theta H^2(E)$, *with* $\dim E < \infty$, *then the Statements* 1 *and* 2 *are equivalent to another one*:

3. $m_T = \det \Theta$.

The proof of this theorem will be deferred to Appendix 1 where we also list some other properties of the operator $T$ equivalent to the simplicity of the spectrum (e.g. the relation $\mu_T = 1$ signifying the cyclicity of $T$). In Appendix 1 we will see that this theorem contains the criterion which we promised in Lecture III, Sect. 4, for the cyclicity of a vector $x$ with respect to a $C_0$-operator ($x \in C^T \Leftrightarrow (\varphi(T)x = \mathbb{0} \Rightarrow \varphi \in m_T H^\infty)$). There we will further study various connections between the Theorem on the Simplicity of the Spectrum and other properties of the class $C_0$. This theorem will also be used in Sect. 7 below for the study of unicellular operators of class $C_0$. But now let us turn to the problem of spectral synthesis.

**Completeness Criterion, I.** *Let* $T \in C_0$. *The following Statements are equivalent.*

1. $T$ *is complete.*
2. $T^*$ *is complete.*
3. $m_T$ *is a Blaschke product.*

*Proof.* If $m_T = B \cdot V$ (the canonical factorization as a product of a Blaschke product and a singular function) then by (g) and (e)

$$\bigvee(\operatorname{Ker}(\lambda I - T)^n : n \geq 1, \lambda \in \sigma_p(T)) = \operatorname{Ker} B(T).$$

Hence by the minimality of $m_T$ we infer that $1 \Leftrightarrow 3$. The rest is evident because $m_{T^*} = m_T^t$. □

**Corollary.** *Every complete operator in* $C_0$ *admits synthesis.*

In fact, the completeness of the restriction $T|E$ of $T$ onto any invariant subspace $E$ follows from the preceding Completeness Criterion, because the divisor $m_{T|E}$ of the Blaschke product $m_T$ is again a Blaschke product. □

This corollary (from which incidentally follows anew the Theorem on Spectral Synthesis for a "scalar" operator $\mathbf{T}_\Theta$) can be proved in yet another – very simple, but still noteworthy – way. Indeed, set $B_r = \prod_{|\lambda|>r} b_\lambda^{k(\lambda)}$, $0 \le r < 1$; then

$$B_r(T)H \subset \operatorname{Ker} \prod_{|\lambda|\le r} b_\lambda^{k(\lambda)}(T) = \bigvee(\operatorname{Ker}(\lambda I - T)^{k(\lambda)} : |\lambda| \le r).$$

Moreover $(s)\text{-}\lim_{r\to 1} B_r(T) = I$ (this follows from the $(*)$-continuity of the functional calculus (Lecture III) or by an application of the Banach-Steinhaus theorem after first having checked it on root vectors). Thus if $E \in \operatorname{Lat} T$ then every element $x$ belonging to $E$ is the limit

$$x = \lim_{r\to 1-0} B_r(T)x \tag{7}$$

of combinations $B_r(T)x$ of root vectors contained in $E$. □

This second proof has some preferences compared with the preceding one, since it not only establishes the fact that the vector $x$ can be synthesized by its "harmonics" (i.e. the root vectors belonging to $E_x^T = \bigvee(T^n x : n \ge 0)$) but gives also a linear procedure (7) for accomplishing this. It is not hard to see that (7) is a linear summation method for the Fourier series

$$x \sim \sum_{\lambda \in \sigma_p(T)} \mathscr{E}_\lambda x, \quad x \in H,$$

where $\mathscr{E}_\lambda$ denotes the spectral projection, corresponding to the point $\lambda$ (i.e. $\mathscr{E}_\lambda | E_\mu = \delta_{\lambda\mu} I$; $\lambda, \mu \in \sigma_p(T)$). We shall return to this matter in Lecture VIII.

To what has been said we have to add just a little bit more to obtain: a) the synthesis of an arbitrary contraction $T$ such that $\mathbb{D} \not\subset \sigma_p(T)$, with a nuclear (trace class one) defect operator $I - T^*T$ (i.e. a completely continuous operator with the series of its eigenvalues, each taken with its multiplicity, being convergent); b) the synthesis of a complete dissipative operator $A$ (i.e. $A_J \ge 0$, $A_J \stackrel{\text{def}}{=} 1/2i(A - A^*)$) with a nuclear imaginary part $A_J$. Assertion b), constituting the contents of a theorem of Brodskiĭ's [4], follows from a), if we apply the Cayley transform

$$A \mapsto T \stackrel{\text{def}}{=} (A - iI)(A + iI)^{-1}$$

and remark that $1 \notin \sigma(T)$ (so that $\sigma_p(T) \not\supset \mathbb{D}$) and $\operatorname{Lat} T = \operatorname{Lat} A$ (by virtue of the identities $T = I + 2iR(-i, A)$ and $A = 2iR(1, T) - iI$, noting that the resolvent sets $\mathbb{C}\backslash\sigma(A)$ and $\mathbb{C}\backslash\sigma(T)$ are connected; cf. the Lemma on Components (Sect. 3)).

An interesting example (Leggett [1]): the operator

$$(Af)(x) = 2i \int_{[0,x]} f\, d\mu, \quad 0 \le x \le 1,$$

in the space $L^2(\mu)$, $\mu$ a finite point measure (atomic) on the segment $[0, 1]$; in this case $A_J \ge 0$ with $A_J$ nuclear so $A$ admits spectral synthesis.

All the operators discussed in the preceding paragraph have the property that for the corresponding characteristic function $\Theta$ the determinant $\det\Theta$ makes sense. The above Completeness Criterion can be stated in terms of this determinant, as we will also do in the following theorem (for the special case of an operator $T$ with finite dimensional defect spaces $\dim(I - T^*T)K = \dim(I - TT^*)K < \infty$).

**Completeness Criterion, II.** *Let $T = P_K S|K$ be an invertible model operator acting in the space $K = H^2(E) \ominus \Theta H^2(E)$, with $\dim E < \infty$, $\Theta$ an inner function. The following statements are equivalent:*

1. *$T$ is complete.*
2. *$|\det\Theta(0)| = \prod_{\lambda\in\sigma_p(T)} |\lambda|^{k(\lambda)}$, $k(\lambda)$ denoting the multiplicity of the corresponding root space.*
3. *$\det(TT^*) = \prod_{\lambda\in\sigma_p(T)} |\lambda|^{2k(\lambda)}$.*

Before proving this result let us remark that the existence of $\det(TT^*)$ will be obtained during the process of the proof of the theorem (as a matter of fact $\dim(I - T^*T)K = \dim E$) and that the last statement does not depend on the knowledge of the functional model of the operator but uses only the fact that it belongs to the class $C_{00}$ (or satisfies some other condition of the same type) and that the defect spaces are finite dimensional. The hypothesis that $T$ is invertible can be eliminated, if we introduce instead a condition at some (any) other point $\mu$ of the disk $\mathbb{D}$ ($\mu \neq 0$) with $\det\Theta(\mu) \neq 0$.

*Proof.* The determinant $d_T \overset{\text{def}}{=} \det\Theta$ is an inner function which is divisible by the minimal annihilator $m_T$ (Lecture III, Sect. 6) and divides $m_T^n$, $n = \dim E$ (as $m_T I \equiv \Omega\cdot\Theta$ as have $m_T^n = \det\Omega\cdot\det\Theta$). Let us write $d_T$ as a product

$$d_T = d_T^s\cdot B,\ B = \prod_{\lambda\in\sigma_p(T)} b_\lambda^{k(\lambda)},$$

of a singular function $d_T^s$ and a Blaschke product $B$ (it follows from the Theorem on the Spectrum of a $C_0$-Operator, Property (g) of such operators and the connection just established between $m_T$ and $d_T$ that the multiplicities $k(\lambda)$ of the zeros coincide with the dimensions $\dim E_\lambda$ of the root subspaces $E_\lambda$). Since $d_T^s$ is inner we have $d_T^s \equiv \text{const}$ iff $|d_T^s(0)| = 1$, that is, Statement 2 holds true. Using the Completeness Criterion I we conclude that $1 \Leftrightarrow 2$.

It remains to show that

$$|\det\Theta(0)|^2 = \det(T^*T), \tag{8}$$

To this end let us replace the first of these determinants by $\det\Theta(0)\overline{\det\Theta(0)} = \det\Theta(0)\Theta(0)^*$ and let us compute the operator $TT^*$: if $f\in K$ then

$$TT^*f = P_K SS^*f = P_K(f - f(0)) = f - P_K f(0).$$

Without loss of generality we may assume that already in the very formulation of the theorem the operator $T$ is given in canonical form (i.e. the function $\Theta(\cdot)e$ is not constant for any $e$, $e\in E$; equivalently $\|\Theta(0)e\| < \|e\|$, $\forall e \neq 0$; cf. Lecture I, Sect. 6). Therefore the operator $V$,

$$Ve \overset{\text{def}}{=} P_K(e\mathbb{1}) = e - \Theta\Theta(0)^*e,\quad e\in E,$$

is an isomorphism of $E$ onto its image. From this it follows also that $VE$ is the image of the map $I - TT^*$ and that on the orthogonal complement $K \ominus VE$ the operator $TT^*$ coincides with $I$. This last fact allows us to conclude that $\det TT^* = \det(TT^*|VE)$ (or even *define* $\det TT^*$ by this identity) and then to compute it using the isomorphism $V$: if $e \in E$ then

$$\begin{aligned} V^{-1}TT^*Ve &= V^{-1}(Ve - P_K(Ve(0))) \\ &= V^{-1}(\Theta(0)\Theta(0)^*e - \Theta\Theta(0)^*(\Theta(0)\Theta(0)^*e)) \\ &= \Theta(0)\Theta(0)^*e. \end{aligned}$$

Thus $\det TT^* = \det \Theta(0)\Theta(0)^*$. □

Using the general notion of determinant for operators differing from the identity operator by a nuclear term it is easy to transfer the preceding proof to the contractions $T$ mentioned under the heading a) on p. 96; cf. also Gohberg-Kreĭn [1].

It is an interesting and important problem to describe contractions $T$ admitting spectral synthesis in the language of the characteristic function $\Theta_T$ beyond the bounds of the class $C_0$. Apparently such a description has to be preceded by a criterion of completeness for the family of root spaces (be it under the assumption that the resolvent is meromorphic in the disk $\mathbb{D}$), i.e. by the clarification of the question which operator functions should be regarded as "Blaschke products"; cf. Nikol'skiĭ [7, 6] (in the last paper please do not read the erroneous line $95^8$). If $T$ is a model operator, $T = S^*|K$, where $K = H^2(E) \ominus \boldsymbol{\Theta} H^2(E')$ and $\boldsymbol{\Theta}$ is a left inner function, then the completeness of this operator can be described in terms of the multiplicative structure of the operator function $\boldsymbol{\Theta}$. In fact, if $\vartheta_\lambda$, $\vartheta_\lambda \in H^\infty(L(E_\lambda, E))$ are the left divisors of $\boldsymbol{\Theta}$ representing the $T$-invariant subspaces $\operatorname{clos} K_\lambda$, $K_\lambda \overset{\text{def}}{=} \bigcup_{n \geq 1} \operatorname{Ker}(T - \lambda I)^n$, $\lambda \in \sigma_p(T)$, then

$$\bigvee(K_\lambda : \lambda \in \sigma_p(T)) = K \Leftrightarrow \boldsymbol{\Theta} H^2(E') = \bigcap_{\lambda \in \sigma_p(T)} \boldsymbol{\vartheta}_\lambda H^2(E_\lambda).$$

This formula is a kind of multiplicative expansion (cf. Corollary 8 in Lecture I, Sect. 6). If the linear manifolds $K_\lambda$ are closed then they must have finite length: there exists a natural number $k$ such that $K_\lambda = \operatorname{Ker}(T - \lambda I)^k$ (the least such number will be denoted by $k(\lambda)$). In this case $K_\lambda \cap K_{\lambda'} = \{\mathbb{0}\}$, $\lambda \neq \lambda'$ (i.e. the functions $\vartheta_\lambda$ and $\vartheta_{\lambda'}$ are relatively prime), $T|K_\lambda \in C_0$ and $m_{T|K_\lambda} = b_\lambda^{k(\lambda)}$, $\lambda \in \sigma_p(T)$. If we further assume that $\boldsymbol{\Theta}$ is an inner function (i.e. $T \in C_{00}$) then we can take the $\vartheta_\lambda$ to be inner too (cf. Lecture I, Sect. 6) and the relation

$$\boldsymbol{\Theta} H^2(E) = \bigcap_{\lambda \in \sigma_p(T)} \boldsymbol{\vartheta}_\lambda H^2(E), \tag{9}$$

is equivalent to completeness and corresponds exactly to the above completeness criterion for $C_0$-operators ($m_T = \prod_{\lambda \in \sigma_p(T)} b_\lambda^{k(\lambda)}$). In many cases (for instance, under the additional assumption that $I - T^*T$ is nuclear) the "additive" geometric condition (9) can be transformed into a multiplicative one, viz. the possibility of representing $\boldsymbol{\Theta}$ as a Blaschke-Potapov product (converging in a suitable sense)

$$\Theta = \prod_{n\geq 1} (b_{\lambda_n} P_n + (I - P_n)),$$

where $\{\lambda_n\}_{n\geq 1}$ is some enumeration of the point spectrum $\sigma_p(T)$ (counted with multiplicity) and the $P_n$ are corresponding orthogonal projections in $E$. Regarding such products consult Potapov [1], [2], Ginzburg [4], Brodskiĭ [6]; we will say a few more words on the factorization of operator functions in Sect. 8 of the present Lecture. The projections $P_n$ depend in a complicated way on the enumeration of the spectrum and in praxis it can be rather difficult to check that a given inner function $\Theta$ is a Blaschke-Potapov product, unless we do not restrict ourselves to functions admitting a scalar multiple (i.e. operators of class $C_0$). In that case, as we have seen, everything is reduced to scalar inner functions, either the minimal annihilator $m_T$ or the determinant $d_T$ (if it exists). For scalar inner functions $\varphi$ it is easy to establish (cf. the proof of Frostman's Theorem, p. 59) the following criterion for the absence of the singular factor $\varphi^s$ (so that $\varphi$ is indeed a Blaschke product):

$$\lim_{r\uparrow 1} \int_{\mathbb{T}} \log|\varphi(r\zeta)|\, dm(\zeta) = 0.$$

Concluding this Section we would like to say a few words regarding the spectral synthesis for operators which do not fit well into the functional model of Sz.-Nagy and Foiaş. The problem of the possibility of synthesis is non-trivial already for normal operators in Hilbert space.

**Wermer-Sarason Theorem.** *Let $T$ be a normal operator with a total set of eigenvectors. The following statements are equivalent:*

1. *$T$ admits spectral synthesis.*
2. $E \in \operatorname{Lat} T$, $E \neq \{\mathbb{O}\} \Rightarrow \sigma_p(T|E) \neq \varnothing$.
3. $\operatorname{Lat} T \subset \operatorname{Lat} T^*$.
4. *If $\mathscr{E}_T$ is the spectral measure of $T$ and $\mu$ any finite mass distribution (signed measure) in the plane $\mathbb{C}$, absolutely continuous with respect to $\mathscr{E}_T$, then $\int p\, d\mu = 0$, $\forall p \in \mathscr{P}_A$, implies $\mu = \mathbb{O}$.*
5. *$T^* \in \mathfrak{R}(T)$ where $\mathfrak{R}(T)$ is the weakly closed algebra generated by $T$ and $I$.*
6. *There exists no bounded domain $G$, $G \subset \mathbb{C}$, with the property that $\sup_G |f| = \sup_{G\cap\sigma_p(T)} |f|$ for every function $f$ bounded and regular in $G$.*

The Statements 3, 4 and 5 are equivalent also for general normal operators; the most important feature is the connection between Statement 6 and all the rest (Sarason [4]). This theorem gives a rather clear picture of the synthesizable normal operators mainly because of Statements 4 and 6, which allow one to display many concrete examples of normal operators not admitting (or, on the contrary, admitting) spectral synthesis. Here is one such example (Piranian's theorem): if the spectrum $\sigma(T)$ is contained in the disk $\mathbb{D}$ and has points of accumulation only at the boundary $\mathbb{T} = \partial\mathbb{D}$ then $T$ is non-syntesizable iff a.e. point of $\mathbb{T}$ is an angular limit point of $\sigma_p(T)$ (in the sense of the theory of boundary values, Appendix 2). From the Wermer-Sarason Theorem one can see that the result a), p. 96, on synthesis of contractions is exact (the ideal of nuclear operators cannot be replaced by any larger symmetric normed ideal). For details and for the history of the question we refer to the surveys Nikol'skiĭ [4, 7].

For completely continuous operators $T$ and more generally for operators with a "sparse" spectrum the problem of spectral synthesis can be transformed into the following purely geometric question involving the family of eigenspaces $\{E_\lambda : \lambda \in \sigma_p(T)\}$. (For simplicity, take $E_\lambda = \operatorname{Ker}(\lambda I - T)$, $\dim E_\lambda = 1$ $(\lambda \in \sigma_p(T))$.) Let $\mathscr{E}_\lambda$ be the spectral projection onto $E_\lambda$ (i.e. the operator defined by the formulae $\mathscr{E}_\lambda | E_\mu = \delta_{\lambda\mu} I$, $\mu \in \sigma_p(T)$) and set

$$X_\sigma \overset{\text{def}}{=} \bigvee (E_\lambda : \lambda \in \sigma),\ X^\sigma \overset{\text{def}}{=} \bigcap_{\lambda \in \sigma_p(T) \backslash \sigma} \operatorname{Ker} \mathscr{E}_\lambda .$$

Is it true that $X_\sigma = X^\sigma$ for every set $\sigma$? If the answer to this question is positive the family $\{E_\lambda\}$ will be called *hereditarily total* (or complete). If the complement $\mathbb{C} \backslash \sigma(T)$ is connected ("sparseness" of the spectrum) the hereditary completeness of the family $\{E_\lambda\}$ is equivalent to the possibility of spectral synthesis for the operator $T$. The hereditary completeness is of course a simple consequence of the existence of a linear method of summation of the Fourier series

$$x \sim \sum_{\lambda \in \sigma_p(T)} \mathscr{E}_\lambda x,$$

even in the weakest sense of the word (for example, as in (7) we may require that there exist sets $\sigma_n$, $\sigma_n \subset \sigma_{n+1} \subset \sigma_p(T)$, $\bigcup \sigma_n = \sigma_p(T)$, and operators $V_n$ such that $V_n \mathscr{E}_\lambda X \subset \mathscr{E}_\lambda X$, $\lambda \in \sigma_p(T)$, $V_n \mathscr{E}_\lambda = \mathbb{O}$, $\lambda \in \sigma_p(T) \backslash \sigma_n$ and $\lim_n V_n x = x$ for each $x$, $x \in X$). The converse is not true (Figiel-Johnson [1]). For all details we refer to the papers Markus [1], Nikol'skiĭ [2, 4, 7], Dovbyš-Nikol'skiĭ-Sudakov [1]. Let us just remark that already for a compact operator spectral synthesis is in general not only impossible but as a part of such an operator we can take any compact operator (in particular, one not having any eigenvalues at all); cf. Nikol'skiĭ [2] where other classes of operators are considered too.

The problem of spectral synthesis is indirectly related to the topics of Lecture II, Sect. 2–3, where we studied invariant subspaces $E$ of the operator $T$ as lower limits $E = \underline{\lim}_n E_n$ of finite dimensional $T$-invariant subspaces $E_n$. The possibility of choosing the subspaces $E_n$ in $E$ itself (i.e. $E_n \subset E$, $n \geq 1$) is equivalent to the synthesizability of this subspace (i.e. the completeness of $T|E$), so that in the general case the formula $E = \underline{\lim}\, E_n$ can be interpreted as a kind of "approximative synthesizability" of $E$. References connected with this point of view can be found in Lecture II.

**6. On Spectral Subspaces.** It is not always possible to write down explicit formulae for the spectral subspaces to some extent reminding of those obtained in Sect. 2. But for operators of class $C_0$ this is indeed possible, although an explicit computation of the corresponding "spectral" divisors $\Theta_\sigma$ of the characteristic function $\Theta_T$ in terms of some multiplicative representation, apparently, is exceedingly difficult; the function $\Theta_\sigma$ is defined by the formula

$$\Theta_\sigma H^2(E) \ominus \Theta_T H^2(E) = \operatorname{Ker} m_\sigma(T)$$

where $m_\sigma$ is the spectral divisor of the minimal annihilator $m_T$ (in Property $d$) in Sect. 5) and $\Theta_T$, $E$ etc. the usual data for the functional model of $T$. However, the proof that every $C_0$-operator is decomposable does not depend on these explicit formulae, because the reasoning in question can be copied verbatim from the proof of the decomposability of the "one dimensional" model operator (Statement 3 of the Theorem on Spectral Subspaces), once we have remarked that this proof can be expressed exclusively in the language of the functional calculus.

**Foiaş's Theorem.** *Every operator of class $C_0$ is decomposable.*

*Proof.* Let $T$ be a $C_0$-operator acting in the space $K$. Set $m = m_T$. If we replace everywhere in the proof of Statement 3 of the Theorem on Spectral Subspaces the letter $\Theta$ by the letter $m$ and omit the line (5), we get the desired result. □

Let us now turn to the general discussion of the decomposition of an operator into its spectral components begun in Sect. 2. If the operator $T$, $T \in L(X)$, is decomposable then the corresponding spectral subspaces $X_\sigma^T$ can be described in terms of the "local spectrum":

$$X_\sigma^T = \{x \in X\colon \text{there exists an } X\text{-valued analytic function } f \text{ on the set } \mathbb{C}\backslash\sigma \text{ such that } (\lambda I - T)f(\lambda) \equiv x,\ \lambda \in \mathbb{C}\backslash\sigma\}. \tag{10}$$

The very possibility to distinguish non-trivial spectral subspaces is usually achieved either on the basis of a sufficiently rich functional calculus (the functions which one can apply the operator to must dissect the spectrum and the spectral mapping theorem must hold for them) or thanks to the property of the resolvent being an "analytic majorant" (if $\mathcal{O}$ is a neighborhood of the spectrum $\sigma(T)$ and $A(\mathcal{O})$ stands for the space of all functions analytic in $\mathcal{O}$ then the set $\{f : f \in A(\mathcal{O}), |f(\zeta)| < |R(\zeta, T)|, \zeta \in \mathcal{O}\backslash\sigma(T)\}$ has to be relatively compact in $A(\mathcal{O})$; this property allows us to keep the local spectrum within given bounds when passing to the limit). A discussion of this (as well as many other related) questions can be found in Colojoara-Foiaş [1], Dunford-Schwartz [3], Nikol'skiĭ [4]. Let us finally remark that decomposability is only one of several possible extensions of the Spectral Theorem; another one – spectrality in the sense of Dunford and its relation to decomposability – will be the subject of Lecture VIII-X (in connection with the model operator). In these Lectures we will in particular exhibit that "decomposition of the spectrum" sometimes (under an assumption of thinness of the spectrum) can be successful even if it is realized by analytic functions.

**7. Concerning Unicellular Operators** without any additional assumptions we know essentially as little as on arbitrary operators in general. However the three properties established in the following lemma deserve special attention and will also be used in the following considerations.

**Lemma on Unicellular Operators.** *Let $H$ be a (separable) Hilbert space and $T$ a unicellular operator. Then*

1. *The set* $C^T$ *of cyclic vectors of* $T$ *is non-void. Also its complement* $H \backslash C^T$ *is an at most countably infinite sum of proper (and therefore nowhere dense)* $T$-*invariant subspaces.*

2. *If* $E_n \in \operatorname{Lat} T$, $n \geq 1$, *and the vectors* $x_n$ *are such that* $\bigvee(T^k x_n : k \geq 0) = E_n$, *then there exist scalars* $c_n$ *with* $\sum_{n\geq 1} |c_n| \, \|x_n\| < \infty$ *such that*

$$\bigvee(T^k x : k \geq 0) = \bigvee(E_n : n \geq 0) \quad \textit{for } x = \sum_{n \geq 1} c_n x_n.$$

3. *If* $E \in \operatorname{Lat} T$ *and* $AT = TA$ $(A \in L(H))$ *then* $AE \subset E$.

*Proof.* 1. Let $\{e_n\}_{n\geq 1}$ be an orthonormal basis in $H$ and set

$$\alpha(E) \stackrel{\text{def}}{=} \sum_{n \geq 1} 2^{-n} \| P_E e_n \|^2, \quad E \in \operatorname{Lat} T.$$

Clearly $\alpha(E') \leq \alpha(E)$ if $E' \subset E$ and if $\alpha(E') = \alpha(E)$ then, in view of

$$0 = \alpha(E) - \alpha(E') = \sum_{n \geq 1} 2^{-n}((P_E - P_{E'})e_n, e_n),$$

we must have $P_{E \ominus E'} e_n = \mathbb{O}$ for all $n$, $n \geq 1$, so that $E = E'$. $T$ being unicellular we have likewise the reverse inclusion

$$E, E' \in \operatorname{Lat} T, \quad \alpha(E') \leq \alpha(E) \Rightarrow E' \subset E.$$

Let $\{E_n\}_{n \geq 1}$ be an ascending sequence (possibly finite) such that

$$E_n \in \operatorname{Lat} T, \quad E_n \neq H,$$

$$\lim_n \alpha(E_n) = \sup\{\alpha(E) : E \in \operatorname{Lat} T, \, E \neq H\}.$$

Then

$$E \in \operatorname{Lat} T, \quad E \neq H \Rightarrow E \subset \bigcup_{n \geq 1} E_n$$

and therefore

$$\bigcup_{E \in \operatorname{Lat} T, E \neq H} E = \bigcup_{n \geq 1} E_n, \quad C^T = H \Big\backslash \Big( \bigcup_{n \geq 1} E_n \Big). \tag{11}$$

2. Put $n_1 = 1$ and define inductively $n_2, n_3, \ldots$ such that $n_{k+1}$ is the least integer with $E_{n_{k+1}} \not\subset E_{n_k}$. If this process stops after a finite number of steps, i.e. if for some index $k$ holds $E_n \subset E_{n_k}$, $\forall n$, then we can take for our $x$ the vector $x_{n_k}$. Otherwise set

$$x = \sum_{k \geq 1} c_k x_{n_k}; \quad c_k > 0; \quad \sum_{k > 1} c_k \| x_{n_k} \| < \infty,$$

arranging that for every $k$, $k \geq 2$, holds

$$c_k \| \Delta_k x_{n_k} \| > \sum_{i \geq k+1} c_i \| \Delta_k x_{n_i} \|, \tag{12}$$

where $\Delta_k$ denotes the orthogonal projection onto the subspace $E_{n_k} \ominus E_{n_{k-1}}$. From (12) it follows that $\Delta_k x \neq \mathbb{O}$, $k \geq 2$, and therefore $x \notin \bigcup_{k \geq 1} E_{n_k} = \bigcup_{n \geq 1} E_n$. Using formula (11) applied to the restriction $T_0 \overset{\text{def}}{=} T|(\text{clos} \bigcup_{n\geq 1} E_n)$ we conclude that $x \in C^{T_0}$.

3. If $\lambda$ is a complex number of sufficiently large modulus then $\lambda I - A$ is invertible and $(\lambda I - A)E$ a closed $T$-invariant subspace. If $(\lambda I - A)E \subset E$ then also $AE \subset E$; in the opposite case we must have $(\lambda I - A)E \supset E$ so that $R(\lambda, A)E \subset E$ and by the Lemma on Components (cf. above Sect. 3) again $AE \subset E$. □

Let us now turn to operators of class $C_0$. Since they are decomposable it follows at once that a necessary condition for unicellularity is that the spectrum $\sigma(T)$ reduces to one single point, $\sigma(T) = \{\lambda\}, |\lambda| \leq 1$.

**Theorem on Unicellular $C_0$-Operators.** *Let $T$ be a unicellular operator of class $C_0$, $T \in L(H)$. Then*

1. *The spectrum $\sigma(T)$ consists of just one point $\lambda$, $|\lambda| \leq 1$. If $\dim H < \infty$ then $|\lambda| < 1$ and*

$$m_T = b_\lambda^k, \quad k \geq 1.$$

*If $\dim H = \infty$ then $|\lambda| = 1$ and*

$$m_T = \exp\left(-a_T \frac{\lambda + z}{\lambda - z}\right), \quad a_T > 0.$$

2. Lat $T = \{\text{Ker}\, \varphi(T) : \varphi$ *is an inner divisor of* $m_T\}$.

*Proof.* 1. If $\dim H < \infty$ then $\lambda$ is an eigenvalue of $T$ and we must have $|\lambda| < 1$, since $(s)\text{-lim}\, T^n = \mathbb{O}$.

Conversely, if $|\lambda| < 1$ then $m_T = b_\lambda^k$ for some $k$. This shows that $(T - \lambda I)^k = \mathbb{O}$ so that in particular $T^k$ must be a linear combination of lower powers $T^i$, $0 \leq i < k$. Therefore

$$\bigvee(T^n x : n \geq 0) = \mathscr{L}(T^i x : 0 \leq i < k)$$

for any $x$, $x \in H$ ($\mathscr{L}(\cdot) \overset{\text{def}}{=}$ *linear (algebraic) hull* of $(\cdot)$). Take now for $x$ a cyclic vector (cyclic vectors do exist by the foregoing lemma). It follows that $\dim H \leq k$.

2. The subspaces $\text{Ker}\, \varphi(T)$ ($\varphi$ an inner divisor of $m_T$) are $T$-invariant and by Statement 1 (of the present theorem) and by the Properties (e)–(f) of $C_0$-operators (cf. above Sect. 5) they generate a "maximal chain" of subspaces in the sense that they are linearly ordered and

$$\dim(\bigvee(\text{Ker}\, \varphi'(T) : \varphi \text{ divides } \varphi') \ominus \bigvee(\text{Ker}\, \varphi'(T) : \varphi' \neq \varphi, \varphi' \text{ divides } \varphi)) \leq 1$$

for any inner divisor $\varphi$ (from Statement 1 of the theorem it also follows that this dimension is 1 if $\dim H < \infty$ and 0 if $\dim H = \infty$). $T$ being unicellular we now realize that each $T$-invariant subspace $E$ must coincide with $\text{Ker}\, \varphi(T)$ for a suitable $\varphi$ (it is clear that $\varphi$ can be choosen to be then $m_{T|E}$). □

Which operators in the class $C_0$ are unicellular? It is easy to see that the necessary condition of a degenerating spectrum $\sigma(T) = \{\lambda\}$ cannot be sufficient. This observation can be made already for finite dimensional operators $T$ ($\dim H < \infty$) for which we know from elementary linear algebra a number of equivalent conditions for unicellularity: $\dim \operatorname{Ker}(T - \lambda I) = 1$ ("simple spectrum"); $C^T \neq \varnothing$; the characteristic polynomial of $T$ coincides with the minimal polynomial (cf. e.g. Gantmaher [1], Lyubič-Glazman [1]); the resolvent has maximal possible growth near the spectrum. It turns out that also in the general case the unicellularity of operators of class $C_0$ can be expressed in terms of analogous properties; more exactly we have the following noteworthy result.

**Unicellularity Criterion.** *Let $T$ be an operator of class $C_0$ in the Hilbert space $H$, $\sigma(T) = \{\lambda\}$, $|\lambda| = 1$.*

I. *The following statements are equivalent*:

1. *$T$ is unicellular.*
2. $\operatorname{Lat} T = \{\operatorname{Ker} \varphi(T) : \varphi$ *is an inner divisor of* $m_T\}$.
3. $C^T \neq \varnothing$ (*i.e.* $\mu_T = 1$).

II. *If we in addition assume that* $\dim(I - T^*T)H < \infty$ *and if we identify $T$ with its functional model $P_K S|K$, $K = H^2(E) \ominus \Theta H^2(E)$, then the Statements* 1–3 *are equivalent to each of the following ones*:

4. $m_T = d_T$ *where* $d_T = \det \Theta$.
5. $\underline{\lim}_{r \uparrow 1} (1 - r) \log |\Theta(r\lambda)^{-1}|_E = -\log \det(TT^*)$.
6. $\underline{\lim}_{r \uparrow 1} (1 - r) \log |R(r\lambda, T)|_K = -\log \det(TT^*)$.

Before proving this theorem let us discuss its meaning and further indicate a few complements. Let us at once remark that the Statements 2–6, necessary and sufficient for the unicellularity of the operaor, are expressed in quite different form: Statement 2 in the language of the $H^\infty$-calculus, Statement 3 in terms of the multiplicity of the spectrum, Statements 4–5 make use of a knowledge of the functional model (the characteristic function), finally Statement 6 employs only the traditional apparatus of spectral theory – the resolvent and the determinant. What concerns the existence (and definition) of the determinants entering in Statements 5–6 one can find the relevant comments in connection with the completeness theorem for the model operator (Sects. 5, Completeness Criterion, II). It is convenient to give a different form of Statements 5–6, which does not stipulate explicitly the assumption of a degenerate spectrum $\sigma(T) = \{\lambda\}$ and thereby somewhat strengthen the implications $5 \Rightarrow 1$ and $6 \Rightarrow 1$. This strengthening will follow from the following lemmata of an independent interest.

**Lemma Estimating the Resolvent.** *Let $T = PS|K$ be a model operator, with $P = P_K$, $K = H^2(E) \ominus \Theta H^2(E)$, $\Theta$ an inner $L(E)$-valued functions ($E$ being a Hilbert space). If $\lambda \in \mathbb{D} \backslash \sigma(T)$ then*

$$\tfrac{1}{2}|\Theta(\lambda)^{-1}|_E \leq 1 + (1 - |\lambda|)|R(\lambda, T)|_K \leq 2 + |\Theta(\lambda)^{-1}|_E.$$

*Proof*. By formula (9) in Lecture III (p. 76)

$$R(\lambda, T) = P_K \mathbf{u}|K, \quad u = \frac{\Theta - \Theta(\lambda)}{z - \lambda \mathbb{1}} \Theta(\lambda)^{-1}.$$

On the circle $\mathbb{T}$ the values of the operator function $u$ can be estimated as follows:

$$|u(\zeta)|_E \leq \frac{|\Theta(\lambda)\Theta(\lambda)^{-1}|_E}{1 - |\lambda|} + \frac{1}{1 - |\lambda|} = \frac{1 + |\Theta(\lambda)^{-1}|_E}{1 - |\lambda|}, \quad \zeta \in \mathbb{T},$$

and thus a fortiori the norm of the resolvent $|R(\lambda, T)|_K$ has the same upper bound $(1 - |\lambda|)^{-1} \cdot (1 + |\Theta(\lambda)^{-1}|_E)$.

In order to prove the left inequality we will estimate from below the operator

$$R(\lambda, T)(I - \bar{\lambda}T) = (1 - |\lambda|^2)R(\lambda, T) + \bar{\lambda}I. \tag{13}$$

As a test function we pick

$$x_e \overset{\text{def}}{=} P_K(\mathbb{1} - \bar{\lambda}z)^{-1}e = (\mathbb{1} - \bar{\lambda}z)^{-1}(I - \Theta\Theta(\lambda)^*)e, \quad e \in E,$$

whose norm was computed in Lecture III (formula (5)):

$$\|x_e\|^2 = (1 - \|\Theta(\lambda)^*e\|^2)(1 - |\lambda|^2)^{-1} \le (1 - |\lambda|^2)^{-1}, \tag{14}$$

In the same Lecture (Sect. 7, Theorem on the Spectrum) it was established that $\mathbf{u}\Theta H^2(E) \subset \Theta H^2(E)$ so that

$$R(\lambda, T)(I - \bar{\lambda}T)x_e = -\frac{|\lambda|}{\lambda} P_K \frac{\Theta - \Theta(\lambda)}{b_\lambda} \Theta(\lambda)^{-1} x_e$$

$$= -\frac{|\lambda|}{\lambda} P_K \frac{\Theta - \Theta(\lambda)}{b_\lambda} \Theta(\lambda)^{-1} (\mathbb{1} - \bar{\lambda}z)^{-1} e,$$

where (as always) $b_\lambda = \dfrac{|\lambda|}{\lambda} \dfrac{\lambda\mathbb{1} - z}{\mathbb{1} - \bar{\lambda}z}$. Let us now use the Projection Lemma (Lecture II, Sect. 4), with $P_K = P_\Theta$:

$$\begin{aligned}
\|R(\lambda, T)(I - \bar{\lambda}T)x_e\| &= \left\| P_K \frac{\Theta - \Theta(\lambda)}{z - \lambda\mathbb{1}} \Theta(\lambda)^{-1} e \right\| \\
&= \left\| (I - P_+)\Theta^* \frac{\Theta - \Theta(\lambda)}{z - \lambda\mathbb{1}} \Theta(\lambda)^{-1} e \right\| \\
&= \|(I - P_+)\bar{z}(\mathbb{1} - \lambda\bar{z})^{-1}(I - \Theta^*\Theta(\lambda))\Theta(\lambda)^{-1} e\| \\
&= \|(\mathbb{1} - \lambda\bar{z})^{-1}(I - \Theta^*\Theta(\lambda))\Theta(\lambda)^{-1} e\| \\
&\ge \|\Theta(\lambda)^{-1} e\|_E (1 - |\lambda|^2)^{-1/2} - (1 - |\lambda|^2)^{-1/2} \|e\|_E.
\end{aligned}$$

By the formulae (13) and (14) we get

$$(1 - |\lambda|^2)|R(\lambda, T)|_K \ge |\Theta(\lambda)^{-1}|_E - 1 - |\lambda|,$$

which proves the desired inequality. □

**Lemma on the Exponential Type.** *Let $\Theta$ be an inner $L(E)$-valued function admitting a scalar multiple and let m be the minimal one among the scalar multiples ($\equiv$ the minimal annihilator of the corresponding model operator). If m is singular and $\mu^s$ denotes the representing measure of m then*

1. $\varlimsup_{r \uparrow 1} (1 - r) \log |\Theta(r\lambda)^{-1}|_E = 2\mu^s(\{\lambda\}), \lambda \in \mathbb{T}$.
2. $\varlimsup_{r \uparrow 1} (1 - r) \log \max_{|\zeta| = r} |\Theta(\zeta)^{-1}|_E = 2 \max\{\mu^s(\{\lambda\}) : \lambda \in \mathbb{T}\}$.

*Proof.* Let $\Omega$ be an inner function such that (cf. the Theorem on the Scalar Multiple, Lecture III, Sect. 6)

$$\Theta\Omega = m \cdot I.$$

From this identity we infer that the operators $\Theta(\zeta)$, $\zeta \in \mathbb{D}$, are invertible and further get the estimate

$$2c \overset{\text{def}}{=} \varlimsup_{r\uparrow 1} (1-r)\log|\Theta(r\lambda)^{-1}|_E \leq \varlimsup_{r\uparrow 1} (1-r)\log\frac{1}{|m(r\lambda)|}.$$

Let us first check that

$$\varlimsup_{r\uparrow 1} (1-r)\log\frac{1}{|m(r\lambda)|} = 2\mu^s(\{\lambda\}). \tag{15}$$

For this again it suffices to remark that

$$\begin{aligned}\varlimsup_{r\uparrow 1} (1-r)\log\frac{1}{|m(r\lambda)|} &= \varlimsup_{r\uparrow 1} (1-r)\int_{\mathbb{T}} \operatorname{Re}\frac{\zeta + r\lambda}{\zeta - r\lambda}\, d\mu^s(\zeta) \\ &= \varlimsup_{r\uparrow 1} (1-r)\int_{\mathbb{T}} \frac{1-r^2}{|\zeta - r\lambda|^2}\, d\mu^s(\zeta) \\ &= 2\mu^s(\{\lambda\}) + \varlimsup_{r\uparrow 1} (1-r)\int_{\mathbb{T}\setminus\{\lambda\}} \frac{1-r^2}{|\zeta - r\lambda|^2}\, d\mu^s(\zeta)\end{aligned}$$

and that the last limit is zero. This can very easily be seen by Lebesgue's theorem on dominating convergence applied to the functions $(1+r)(1-r)^2|\zeta - r\lambda|^{-2}$, $r \to 1$.

Thus $2c \leq 2\mu^s(\{\lambda\})$. Assume that $2c < 2\mu^s(\{\lambda\})$ and set $c' = \frac{1}{2}(\mu^s(\{\lambda\}) - c)$. Let us check that the analytic function $\Omega'$,

$$\Omega' \overset{\text{def}}{=} \Omega \exp\left\{c'\frac{\lambda + z}{\lambda - z}\right\}$$

is bounded in the set $\mathbb{D}$. From this will then follow that $m' \overset{\text{def}}{=} m\exp\left\{c'\frac{\lambda + z}{\lambda - z}\right\}$ is a scalar multiple of $\Theta$ and so, by the minimality of $m$, $c' = 0$. Contradiction!

The function $\Omega'$ can be estimated as follows:

$$|\Omega'(\zeta)|_E \leq \exp\left(c'\frac{1-|\zeta|^2}{|\lambda - \zeta|^2}\right) \leq \exp\left(\frac{2c'}{|\lambda - \zeta|}\right), \quad \zeta \in \mathbb{D},$$

$$|\Omega'(\zeta)|_E = 1, \text{ a.e. } \zeta \in \mathbb{T},$$

$$\begin{aligned}|\Omega'(r\lambda)|_E &= |\Omega(r\lambda)|_E \exp\left\{c'\frac{1+r}{1-r}\right\} = |\Theta(r\lambda)^{-1}|_E |m(r\lambda)| \exp\left\{c'\frac{1+r}{1-r}\right\} \\ &\leq \text{const}\cdot\exp\left\{\frac{\mu^s(\{\lambda\}) + c}{1-r} - \mu^s(\{\lambda\})\frac{1+r}{1-r} + \frac{1}{2}(\mu^s(\{\lambda\}) - c)\frac{1+r}{1-r}\right\} \\ &= \text{const}\exp\frac{\mu^s(\{\lambda\}) + c}{2} \overset{\text{def}}{=} C, \; 0 \leq r < 1.\end{aligned}$$

Therefore we can apply to $\Omega'$ the Phragmén-Lindelöf Theorem (Lecture I, Sect. 7) according to which $\Omega'$ is bounded in $\mathbb{D}$, by the constant $C + e^{2c'}$. (In the literal sense of the word, in Lecture I it was question of scalar, not operator valued functions; we have first to apply the theorem quoted to the scalar functions $\zeta \mapsto (\Omega'(\zeta)x, y)_E (x, y \in E)$ and then take the sup with respect to $x$ and $y$ in the estimate obtained:

$$|\Omega'(\zeta)|_E = \sup\{|(\Omega'(\zeta)x, y)| : \|x\|_E \leq 1, \|y\|_E \leq 1\}.)$$

The proof of Statement 1 is complete.

Statement 2 follows readily from Statement 1 once we have verified that

$$\overline{\lim_{r\uparrow 1}}\,(1-r)\log\max_{|\zeta|=r}\frac{1}{|m(\zeta)|} = 2\max\{\mu^s(\{\lambda\}) : \lambda\in\mathbb{T}\}.$$

We know already (formula (15)) that these quantities are connected by the inequality "$\geq$". To prove the converse let us write $\mu^s$ as a sum of a continuous measure $\mu_c$ and a discrete (purely point) measure $\mu_d$ and let $A$ and $\varepsilon$ be given positive numbers. Let $\mu_d = \mu' + \mu''$ where $\mu'$ is a finite sum of atoms (point masses) while as $\mu''\mathbb{T} < \varepsilon$. Let $\zeta \in \mathbb{D}$, $|\zeta| = r$, and denote by $l_{\zeta,A}$ the arc of the circumference $\mathbb{T}$ with center at the point $\zeta|\zeta|^{-1}$ spanning a chord of length $2A(1-r)$ (Fig. 1).

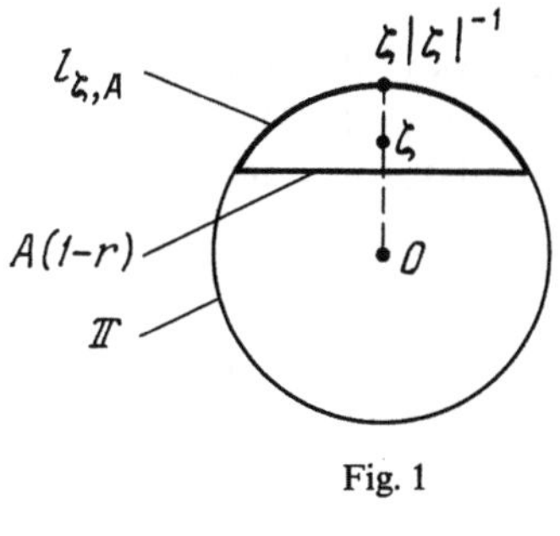


Fig. 1

Then

$$\int_{\mathbb{T}}\frac{1-r^2}{|\zeta-t|^2}d\mu^s(t) \leq \frac{1-r^2}{(1-r)^2}\mu^s l_{\zeta,A} + \int_{\mathbb{T}\setminus l_{\zeta,A}}\frac{1-r^2}{A^2(1-r)^2}d\mu^s(t)$$

$$\leq \frac{1+r}{1-r}(\mu_c l_{\zeta,A} + \mu' l_{\zeta,A} + \varepsilon) + \frac{1+r}{A^2(1-r)}\mu^s\mathbb{T}.$$

For a fixed $A$ and $1-r$ sufficiently small the arc $l_{\zeta,A}$ will contain at most one point carrying an atom of $\mu'$ while $\mu_c l_{\zeta,A}$ can be taken arbitrarily small (uniformly in $\zeta$, $|\zeta| = r$). Consequently

$$\overline{\lim_{r\uparrow 1}}\,(1-r)\max_{|\zeta|=r}\int_{\mathbb{T}}\frac{1-r^2}{|\zeta-t|^2}d\mu^s(t)$$

$$\leq 2(\varepsilon + \max\{\mu^s(\{\lambda\}) : \lambda\in\mathbb{T}\}) + 2\mu^s\mathbb{T}/A^2.$$

Letting $\varepsilon$ tend to 0 and $A$ to infinity we get the desired inequality.

Finally, if $\lambda$ is a point of the circle $\mathbb{T}$ such that

$$\mu^s(\{\lambda\}) = \max\{\mu^s(\{\lambda\}) : \lambda \in \mathbb{T}\},$$

then by what we have said

$$\begin{aligned} 2\mu^s(\{\lambda\}) &= \overline{\lim_{r\uparrow 1}}\,(1-r)\log|\Theta(r\lambda)^{-1}|_E \\ &\le \overline{\lim_{r\uparrow 1}}\,(1-r)\log\max_{|\zeta|=r}|\Theta(\zeta)^{-1}|_E \\ &\le \overline{\lim_{r\uparrow 1}}\,(1-r)\log\max_{|\zeta|=r}\frac{1}{|m(\zeta)|} = 2\mu^s(\{\lambda\}). \ \square \end{aligned}$$

**Corollary.** *Let $T = P_K S|K$ be a model operator, with $K = H^2(E) \ominus \Theta H^2(E)$, $\Theta$ an inner function, assuming that* $\dim E < \infty$. *If $\sigma(T) \subset \mathbb{T}$ then each of the following two conditions entails the unicellularity of $T$ (so that in particular the spectrum of $T$ reduces to one point)*

$$\overline{\lim_{r\uparrow 1}}\,(1-r)\log\max_{|\zeta|=r}|\Theta(\zeta)^{-1}|_E = -\log\det(TT^*),$$

$$\overline{\lim_{r\uparrow 1}}\,(1-r)\log\max_{|\zeta|=r}|R(\zeta, T)|_K = -\log\det(TT^*).$$

In fact, from the lemmata follows $1^0$ that these upper limits coincide and $2^0$ that their common value equals $2\max\{\mu^s(\{\lambda\}) : \lambda \in \mathbb{T}\}$, where $\mu^s$ is the representing measure of the minimal annihilator $m = m_T$. On the other hand (cf. above formula (8))

$$\det TT^* = |\det\Theta(0)|^2,$$

so that we can use the inequality $|\det\Theta| \le m$, resulting from the fact that the inner function $\det\Theta$ is divisible by $m$ (Corollary 2 of the Theorem of the Scalar Multiple, Lecture III, Sect. 6). Thus each of the conditions entails

$$\begin{aligned} \mu^s\mathbb{T} &= \log\frac{1}{|m(0)|} \le \log\frac{1}{|\det\Theta(0)|} \\ &= -\frac{1}{2}\log\det TT^* = \max\{\mu^s(\{\lambda\}) : \lambda \in \mathbb{T}\}, \end{aligned}$$

implying that $\mu^s$ must be a $\delta$-measure at some point of the circle $\mathbb{T}$. The rest follows from the Unicellularity Criterion. $\square$

*Proof of the Unicellularity Criterion.* I. The implications $1 \Rightarrow 2$ and $1 \Rightarrow 3$ have already been proved (the Theorem and the Lemma on Unicellular Operators). The implication $2 \Rightarrow 1$ is immediate because $m_T = \exp\left(-a\dfrac{\lambda + z}{\lambda - z}\right)$ and thus the family of all inner divisors of $m_T$ $\left(\text{that is the family } \exp\left(-a'\dfrac{\lambda + z}{\lambda - z}\right),\ 0 \le a' \le a\right)$ is linearly ordered (with respect to divisibility).

The implication $3 \Rightarrow 2$ is non-trivial and is contained in the Theorem on the Simplicity of the Spectrum (cf. above Sect. 5).

II. By the Theorem on the Simplicity of the Spectrum $4 \Leftrightarrow 3$. The Statements 5 and 6 are equivalent to Statement 4, because by the above lemmata both upper limits equal $2\mu^s(\{\lambda\})$ and we have always

$$2\mu^s(\{\lambda\}) \leq \log|\det \Theta(0)|^{-2} = -\log\det(TT^*)$$

(where we used that the minimal annihilator $m$ divides $\det \Theta$); further the functions $m$ and $\det \Theta$ coincide iff $m(0) = \det \Theta(0)$. □

We conclude this Section by briefly taking up for discussion yet another question related to unicellular operators. As we have already noted the spectrum of an unicellular operator must by necessity be connected, because otherwise the operator could be split up into a sum of non-trivial spectral components with the aid of the Riesz-Dunford calculus and it is clear that the corresponding spectral subspaces are incomparable with respect to set theoretic inclusion. If $T$ admits a richer calculus with the aid of which one can single out spectral subspaces with arbitrary small spectra (see the remarks at the end of Sect. 6) then one can say more, namely, that in such a case the spectrum $\sigma(T)$ is indeed a one point set. For example, this is clearly so for operators in finite dimensional spaces, for operators of class $C_0$, for operators $T$ with a "non-quasianalytic" growth of powers $(\sum_{n\in\mathbb{Z}} |\log|T^n||/(n^2+1) < \infty)$ etc; cf. the survey by Nikol'skiĭ [4]. At one time it was even thought that the degeneracy of the spectrum is a necessary condition for an operator to be unicellular, until Foiaş and Williams constructed the following counter-example.

**Foiaş-Williams Example.** Let $\mathbf{T}_\Theta = P_K S|K$, $K = H^2 \ominus \Theta H^2$, $\Theta = \exp\dfrac{z+\mathbb{1}}{z-\mathbb{1}}$, be a unicellular model operator. Let us pick a function $\varphi$ in $H^\infty$ with the following properties:

(i) the limit $\lim_{r\to 1-0} \varphi(r)$ does not exist;

(ii) there exists a sequence of polynomials $\{p_n\}_{n\geq 1}$ such that $\sup_{n\geq 1} \|p_n \circ \varphi\|_\infty < \infty$; $\lim_n p_n(\varphi(\zeta)) = \zeta$, $\forall \zeta \in \mathbb{D}$.

Once such a function $\varphi$ has been found then it is clear that *the operator* $T \stackrel{\text{def}}{=} \varphi(\mathbf{T}_\Theta)$ *is unicellular, but its spectrum*

$$\sigma(T) = \left\{\zeta : \zeta \in \mathbb{C},\ \inf_{\mathbb{D}}(|\Theta| + |\varphi - \zeta|) = 0\right\}$$

*does not reduce to a point* (it contains all partial limits $\lim_n \varphi(r_n)$, $0 \leq r_n < 1$, $r_n \to 1$). That $T$ is unicellular follows from the fact that each of the operators $T$, $\mathbf{T}_\Theta$ is the weak limit of a sequence of polynomials in the other. In fact,

$$\mathbf{T}_\Theta = \lim_n p_n(T) \left( = \lim_n (p_n \circ \varphi)(\mathbf{T}_\Theta)\right)$$

in the sense of weak convergence, by virtue of the above property (ii) and Statement 1 of the Convergence Lemma, Lecture II. The analogous approximation for $T$ is trivial and in fact contained in the discussion of formula (3), Lecture III.

It remains to construct a function $\varphi$ satisfying (i) and (ii). We are going to use some basic facts on conformal mappings for which we refer the reader to the textbooks by Markuševič [1] and Carathéodory [1]. Let $G$ be a domain obtained from the square $0 \le x, y \le 1$ by deleting either the quadrangles

$$\frac{1}{4n+1} \le x \le \frac{1}{4n}, \quad \frac{1}{n} \le y \le 1 \quad (n = 2, 3, \ldots),$$

$$\frac{1}{4n+3} \le x \le \frac{1}{4n+2}, \quad 0 \le y \le 1 - \frac{1}{n} \quad (n = 2, 3, \ldots),$$

or the analogous triangles (as in Fig. 2, where we discern an unmistakable relative of the hippopotamus of Littlewood [1], p. 46).

Let $G_n$ be a domain in the complement of which there are only left over $n$ teeth (to the right) of the set $\mathbb{C} \setminus G$ and which is slightly expanded in comparison with the set $G$ (the corresponding portions of the set $G$ are drawn with dotted lines) but in such a way that $G_n$ is contained in a $\frac{1}{n}$-neighborhood of $G$. Let $\varphi_0$ and $\varphi_n$ be conformal maps of the disk $\mathbb{D}$ onto the domains $G$ and $G_n$ respectively with the usual normalization at the origin ($\varphi_0(0) = \varphi_n(0)$, $n \ge 1$, $\varphi_0'(0) > 0$, $\varphi_n'(0) > 0$). Then by Carathéodory's convergence theorem (Markuševič [1], vol. 3, p. 33–34, or a simpler variant Walsh [1], p. 32)

$$\lim_n \varphi_n^{-1}(w) = \varphi_0^{-1}(w), \quad w \in G, \tag{16}$$

where $\varphi_n^{-1}$ and $\varphi_0^{-1}$ are the inverse maps in question. If at the point $\zeta$ on the circumference $\mathbb{T}$ the limit $\lim_{r \to 1} \varphi_0(r\zeta) \stackrel{\text{def}}{=} w$ exists then $w$ lies on the "good" part of the boundary $\partial G$ (i.e. $\operatorname{Im} w \ne 0$; here we use the theorem on the correspondence of the boundaries under conformal mapping) but then $\varphi_0$ will also map some neighborhood of $\zeta$ into the "good" part of $\partial G$ (continuity of $\varphi_0$ and $\varphi_0^{-1}$ up to the boundary at good points). If this holds true at each point $\zeta$, $\zeta \in \mathbb{T}$, then the corresponding neighborhoods constitute a covering of $\mathbb{T}$. Taking a finite

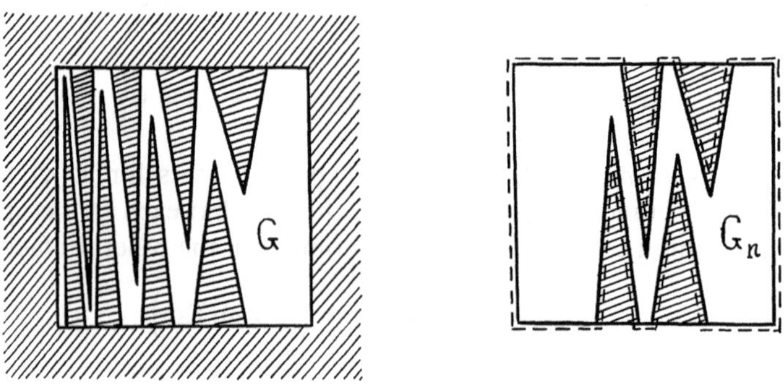


Fig. 2

sub-cover we get a contradiction: part (in fact the "good" part) of the boundary $\partial G$ will not be covered by the image $\varphi_0(\partial\mathbb{D})$. Hence there must be a point $\zeta_0$, $\zeta_0 \in \mathbb{T}$, such that $\lim_{r\to 1} \varphi_0(r\zeta_0)$ does not exist.

Set $\varphi = \varphi_0(\zeta_0 z)$ and let us check that $\varphi$ is the function we are looking for. Property (i) is already settled. Property (ii) is (again by the theorem on the correspondence of boundaries) equivalent to the existence of polynomials $p_n$ approximating $\varphi^{-1}$ pointwise:

$$\sup\{|p_n(w)| : w \in G, n \geq 1\} < \infty,$$

$$\lim_n p_n(w) = \varphi^{-1}(w), \quad w \in G.$$

By virtue of (16) it is sufficient to approximate uniformly and with arbitrary accuracy each function $\varphi_n^{-1}$, $n = 1, 2, \ldots$, with polynomials. As $\varphi_n^{-1}$ is regular in some neighborhood of the set clos $G$ we have

$$\varphi_n^{-1}(w) = \frac{1}{2\pi i} \int_\gamma \frac{\varphi_n^{-1}(t)}{w - t}\, dt, \quad w \in \operatorname{clos} G, \tag{17}$$

where $\gamma$ is a contour encircling clos $G$. It is clear that the Riemann sums for the Cauchy integral (17) approximate $\varphi_n^{-1}$ uniformly in $G$. It remains to approximate each fraction $\frac{1}{t - w}$, $t \in \gamma$, with polynomials, in the $\sup_G$-metric. We have to verify by some method or other that $\frac{1}{t - w} \in \operatorname{clos}_C \mathscr{P}_A$ for $t \in \mathbb{C} \backslash \operatorname{clos} G$; here $C = C(\operatorname{clos} G)$. One proof runs as follows: the set of numbers $t$ such that this approximation is possible is open (the method of "removal of poles" popular in spectral theory too; cf. the Lemma on Components, Sect. 3). Another one is: let $\mu$ be a finite Borel (signed) measure orthogonal to the polynomials ($\int p\, d\mu = 0$, $\forall p \in \mathscr{P}_A$) and concentrated on clos $G$. Then the integral

$$\int \frac{d\mu(w)}{t - w} \tag{18}$$

vanishes for large $t$ (develop into a series in powers of $w/t$). But since (18) is an analytic function of the variable $t$, $t \in \mathbb{C} \backslash \operatorname{clos} G$, this function must vanish throughout $\mathbb{C} \backslash \operatorname{clos} G$. Thus the continuous linear functional $f \mapsto \int f\, d\mu$ (in $C(\operatorname{clos} G)$), vanishing on polynomials, must also vanish on the functions $\frac{1}{t - w}$, $t \in \gamma$. Therefore (Hahn-Banach theorem)

$$\frac{1}{t - w} \in \operatorname{clos}_C \mathscr{P}_A, \quad t \in \gamma. \ \square$$

**8. Concluding Remarks.** The formulae for the invariant subspaces of the model operator derived in Sect. 1 (scalar case) and Sect. 4 (general case) have in essence been known since the time of the proof of the Lax-Halmos Theorem (cf. Lecture I, Sect. 6). From a general point of view it would be natural to connect the invariant subspaces with general factorizations $\Theta = \Theta_1 \Theta_2$ of the function $\Theta$ as a product of contracting factors $\Theta_i$, $i = 1, 2$, especially in the case of the "complete model" of Sz.-Nagy and Foiaş, where the characteristic function is not assumed to be left inner. However, not all contracting factorizations (but only the "regular" ones) correspond to elements of Lat $\mathbf{T}_\Theta$ and the struggle with this difficulty occupies that many pages in the books by Sz.-Nagy and Foiaş [1] and Brodskiĭ [6]. No lesser inconvenience is caused by the necessity to divide for the "nondegeneracy" of the representation of a $\mathbf{T}_\Theta$-invariant subspace $E = \vartheta H^2(D) \ominus \Theta H^2(D)$ (in the sense of Sect. 4), if we insist upon that the function $\vartheta^*\Theta$ should be a "pure" characteristic function for the operator $\mathbf{T}_\Theta | E$ (i.e. that $\vartheta^*\Theta d \not\equiv \text{const}\ \forall d \in D$, $d \neq \mathbb{0}$). It is not hard to see that this property is equivalent to the condition of the operator $\vartheta^*\Theta$ being a strict contraction: $\|(\vartheta^*\Theta)(0)d\| < \|d\|$ for $d \in D$, $d \neq \mathbb{0}$; cf. Sz.-Nagy and Foiaş [1]. Leaving our discussion of factorizations let us further remark that to decompositions of the spectrum (or decompositions of the space $K_\Theta$ into invariant subspaces) there likewise correspond similar decompositions of the function $\Theta$ as a product of (left inner) factors; a few words on such decompositions were said in Sect. 5. We will turn to the study of these decompositions in Lecture VI–VIII (implicitly) and in Lecture IX–X.

Almost all results on spectral synthesis up to this date, as well as a detailed bibliography, can be found in the papers Markus [4] and Nikol'skiĭ [2, 4, 7], already referred to, and in the collection "99 Problems", cf. Nikol'skiĭ-Havin-Hruščev [1]. Besides the general remarks on this topic made in Sect. 1 and 5 we will here add just a little bit more. Let us point out once more that, with the aid of the theory of Sz.-Nagy and Foiaş [3] set forth in the series of Statements (a)–(g) in Sect. 5, for operators of class $C_0$ it was possible to by-pass the explicit formulae for the root spaces and their hulls in Sect. 1, which were used in the proof of the synthesis of the "one dimensional" model operator $\mathbf{T}_\Theta$. On some points our treatment differs from the one of these authors [3, 1]. The explicit formulae (the scalar ones in Sect. 1 and the vector ones in Sect. 4) might be useful beyond the limits of the class $C_0$ or in more subtle questions than the problem of spectral synthesis (cf. Lecture VI–VII).

It should be noted that all classical tests for the completeness of an operator, expressed in "metrical" terms (behavior of the resolvent, the range of the quadratic form etc.; all this is in the spirit of perturbation theory), in fact lead automatically to the inheritance of this completeness property to all restrictions $T|E$ of the operator to arbitrary invariant subspaces $E$, in other words, to the possibility of spectral synthesis. To this category belong the well-known tests by Keldyž, Lidskiĭ, Macaev and others about which one can read in the book Gohberg-Kreĭn [1]. From the point of view of synthesis these criteria are analysed in Markus's paper [1]. The completeness criteria for $C_0$-operators given in the present Lecture can be found in the papers by Sz.-Nagy and Foiaş [3] (Completeness Criterion, I) and Gohberg-Kreĭn [3] (Completeness Criterion, II) but the proofs of these authors are different. It should further be remarked that already much earlier similar completeness criteria for dissipative operators (formulated in terms of traces, and not determinants) were established in the paper by Livšic [2]. The language, in which all these completeness criteria are expressed, has its origin in the case of self-adjoint or unitary operators where the orthogonality of the eigenfunctions allows one to write the completeness condition as an "equation of closedness" (Parseval's identity).

The second proof of the synthesizability of a $C_0$-operator (p. 96) is, as already noted, in some sense preferable to the first one; it was apparently first employed in Vasyunin's paper [3].

The isolation and the study of spectral subspaces of linear operators in terms of the local resolvent (10) has its origin in the work of Dunford [1] and Bishop [1]. The identification of such subspaces with spectral subspaces as well as the theory of decomposable operators is largely due to Foiaş; cf. Dunford-Schwartz [3], Colojoară-Foiaş [1], Nikol'skiĭ [4]. Roughly at the same time when the first results on decomposable operators appeared (cf. Foiaş [3]), Lyubič and Mačaev [1] introduced the notion of "$S$-operator", a notion which is very close to decomposability and perhaps even equivalent to it. There (perhaps at the first time) attention is turned to the rôle of the Levinson-Sjöberg condition

$$\int_0^\delta \log\log\max\{|R(\lambda, T)| : \operatorname{dist}(\lambda, \sigma(T)) \geq t\}\, dt < \infty,$$

guaranteeing the function $|R(\lambda, T)|$ to be an analytical majorant, provided the spectrum $\sigma(T)$ is situated on a smooth curve (i.e. the set of all functions $f$ analytic in a neighborhood of $\sigma(T)$ and satisfying the inequality $|f(\lambda)| \leq |R(\lambda, T)|$, $\forall \lambda \notin \sigma(T)$, is compact).

The Riesz-Dunford calculus and the Lemma on the Spectrum of a Restriction are standard attributes of spectral theory. The spectral subspaces of $C_0$-operators (in the form $\operatorname{Ker} m_\sigma(T)$) as well as the decomposability of such transformations is treated in Foiaş [2] and Ginzburg [1, 2]. It would be interesting to have an explicit formula for the corresponding divisor $\Theta_\sigma$ of the characteristic function $\Theta$ in terms of the parameters of the Potapov-Ginzburg canonical factorization of this function. In Lecture IX we will study the properties of the "subspace valued" function $\sigma \mapsto \operatorname{Ker} m_\sigma(T)$ as a kind of "spectral capacity", reminding of the spectral measure of spectral (or even normal) operators.

The geometric language, which nowadays generally is accepted for the definition of a "Jordan cell" and which is also used in this Lecture, was introduced in Brodskiĭ [1]. The Unicellularity Theorem for the model operator in the space $H^2 \ominus \Theta H^2$ is proved in Sz.-Nagy and Foiaş [3]. The method of proof, relying on an explicit description of all invariant subspaces, for a long time lead only to partial results, as long as the authors (see Sz.-Nagy and Foiaş [7, 8]) were not in possession of the fundamental Theorem on the Simplicity of the Spectrum (cf. Sect. 5). Besides, the corresponding test for unicellularity for operators with a finite rank (or nuclear) defect operator (Part II of our Unicellularity Criterion, p. 104) can be deduced from previous results by Brodskiĭ and Kisilevskiĭ (cf., e.g., the paper [1] by these authors), concerning dissipative operators $\left(\frac{1}{2i}(T - T^*) \geq \mathbb{O}\right)$. The same method is set forth in the book Gohberg-Kreĭn [2] where one can find detailed bibliographical references; bibliographical references are likewise given in the survey by Nikol'skiĭ [4]. The exposition of the present Lecture differs from the papers referred to above in some details. First of all these works are founded on a general unicellularity criterion valid for any completely non-unitary contraction $T$: an equivalent condition for unicellularity is that the set of all "regular" left divisors of the characteristic function $\Theta_T$ of $T$ should be linearly ordered (in the sense of divisibility). Indeed, the results on unicellularity established in Sect. 7 of this Lecture were basically deduced from the Theorem on the Simplicity of the Spectrum by Sz.-Nagy and Foiaş, which depends on subtle features of the calculus admitted by the operator. This theorem will be proved in Appendix 1. We also implicitly study there the possibility of decomposing operators of class $C_0$ as sums of unicellular ones (in a similar way as in Jordan's theorem one splits up matrices or finite dimensional operators).

The derivation from the Unicellularity Theorem (Sect. 3) of the corresponding facts for the operator of (indefinite) integration $J$ is due to Sarason [5] and the corresponding proof of Titchmarsh's Convolution Theorem to Kalisch [1, 2]. The study of the operator $J$ has a long history which, as we have already recorded, has its origin in Gel'fand's problem [1]. The first proof of the unicellularity was in principle obtained already in 1949 when Agmon [1], using a result on the zeros of entire functions, proved the linearity of the lattice $\operatorname{Lat}\{T_\lambda : \lambda > 0\}$ in the spaces $L^p(0,a)$, $1 \le p < \infty$ and $C_{[0,a]}$ for the semigroup of right shifts $T_\lambda : (T_\lambda g)(x) \overset{\text{def}}{=} g(x-\lambda)$ if $x \ge \lambda$, $(T_\lambda g)(x) \overset{\text{def}}{=} 0$ if $x < \lambda$. Indeed this lattice coincides with $\operatorname{Lat} J$: as for each polynomial without constant term $p(J)f = f * q$, where $q = \sum_{n\ge 1} \hat{p}(n) z^{n-1}/(n-1)!$,

$$(f * q)(x) = \int_0^x f(t) q(x-t)\, dt, \quad 0 \le x \le a,$$

upon approximating with polynomials the $\delta$-function $\delta_\lambda$ concentrated at the point $\lambda$, $0 \le \lambda \le a$, we get $\operatorname{Lat} J \subset \operatorname{Lat}\{T_\lambda : \lambda > 0\}$; the opposite inclusion follows if we approximate the Lebesgue measure $dt$ with combinations of $\delta$-measures. Agmon's paper however did not get due attention and the unicellularity of $J$ was rediscovered 8 years later by Brodskiĭ [2], Sahnovič [1] and Donoghue [1], using the theory of characteristic functions (first two papers) or Titchmarsh's Convolution Theorem (third paper). Later "elementary" proofs were published (Brodskiĭ [3], Kalisch [1, 2], Schreiber [1]) and also a considerable number of generalization connected with variations of the space ($W_2^p(0,a)$, $C_0^\infty[0,a]$, $L^p(d\mu)$) as well as with a change of the operator $J$ for allied *Volterra operators* $K$,

$$Kf(x) = \int_0^x k(x,t) f(t)\, dt, \quad 0 \le x \le a.$$

Concerning this (rather numerous) group of works see the survey by Nikol'skiĭ [4] and the book Gohberg-Kreĭn [2]. There one can find information on other types of unicellular operators too.

The Lemma on Unicellular Operators (Sect. 7) comprises results of several authors (Statement 1 is due to Gohberg-Kreĭn [2] and to Rosenthal [1], Statement 2 to Brodskiĭ [6], Statement 3 to Rosenthal [1] and to Nikol'skiĭ [9]) but here everything is expressed somewhat differently.

The Theorem on Unicellular $C_0$-Operators is due to Sz.-Nagy and Foiaş [3].

The Unicellularity Criterion in Sect. 7 has already been discussed at length but it remains to say a few words about the two accompanying lemmata. The Lemma Estimating the Resolvent is proved in Gohberg-Kreĭn [4] in a different way; the Lemma on Exponential Type in the present form perhaps does not appear in the literature. Its Corollary (p. 108) is adopted from Gohberg-Kreĭn [2].

The Foiaş-Williams Example and some other problems lead to the interesting question when the two weakly closed algebras $\mathfrak{R}(T_\Theta)$ and $\mathfrak{R}(\varphi(T_\Theta))$, generated by the operators $I$ and $\mathbf{T}_\Theta$, respectively $I$ and $\varphi(\mathbf{T}_\Theta)$, with $\varphi \in H^\infty$, coincide. Clearly this is so if $\varphi$ is a weak generator of the algebra $H^\infty$ (i.e. $H^\infty = \mathfrak{R}(z) = \mathfrak{R}(\varphi)$) and in fact this argument is used in the example. A complete characterization of weak generators $\varphi$ of the algebra $H^\infty$ was found by Sarason [3] in terms of the (necessarily conformal) image $\varphi(\mathbb{D})$. It is however unknown if it is necessary to pass from $\mathfrak{R}(\varphi(\mathbf{T}_\Theta))$ to $\mathfrak{R}(\varphi)$; cf. Nikol'skiĭ [6].

(Remark (1983): B. Solomyak (unpublished) has resently constructed an example of an operator $T = \varphi(\mathbf{T}_\Theta)$ such that $\mathfrak{R}(\mathbf{T}_\Theta) = \mathfrak{R}(T)$ but $\mathfrak{R}(\varphi') \neq H^\infty$ for any $\varphi'$ with $T = \varphi'(\mathbf{T}_\Theta)$.)

# Lecture V. The Triangular Form of the Truncated Shift

Decompositions of the model operator into spectral subspaces, which in one way or the other are analogous to the Spectral Theorem, will be the object of the next few Lectures. As we will see, the possibility of such a decomposition depends on special (and rather restrictive) requirements imposed on the operator. The classical theory of matrices indicates however that it can be of considerable use already to have a transformation of the operator in question to "triangular form", i.e. representing it as a matrix operator with respect to a basis gotten by orthogonalization from the corresponding family of eigen and root vectors. We give here a continuous analogue of such a "triangularization" for an arbitrary truncated shift using decreasing chains of invariant subspaces.

In this connection it is appropriate to mention that also the elementary theorem of Schur on triangularization of finite dimensional operators (matrices) $T$ depends on the construction of a chain $E_0 \supset E_1 \supset \cdots \supset E_n$ of invariant subspaces, $\dim E_k = n - k$, $0 \le k \le n$. Picking a basis $\{e_k\}_{k=1}^{n}$, $e_k \in E_{k-1} \ominus E_k$, $\|e_k\| = 1$ ($k = 1, \ldots, n$), we transform the matrix of the operator $T$ into lower-diagonal form. $T$ is then decomposed into a sum

$$T = \mathrm{Diag} + N$$

where Diag is the diagonal part of the matrix and $N$ the strictly upper triangular one (nilpotent, $N^n = \mathbb{O}$). A truncated shift with a pure point spectrum (in other words, with a total family of root subspaces) can be triangularized in complete analogy with this finite dimensional procedure. In the case of a continuous spectrum the choice of a chain of invariant subspaces is highly non-unique but we pick the most natural of these chains. The rôle of the diagonal term Diag will be played by a certain (normal) multiplication operator, which can be made subject to further spectral analysis with the aid of the Spectral Theorem, and similarly the rôle of the strictly triangular term $N$ will be played by a suitable operator of Volterra type.

The triangular representation thus obtained will be used in the study of functions of the model operator and further in the study of the operator of indefinite integration $J$ in the space $L^2(0, 1)$.

**1. Pure Point Spectrum.** Here we consider a model operator $T = T_B$, the characteristic function of which is a Blaschke product. Being interested in orthogonalizing the root vectors of $T$ we have to enumerate the point spectrum $\sigma_p(T)$ repeating each eigenvalue as many times as its algebraic multiplicity (i.e. the dimension of the corresponding root subspace) indicates.

**Malmquist-Walsh Lemma.** *Let $B$ be a Blaschke product and let $\Lambda = \{\lambda_n : n \ge 1\}$ be the sequence of its zeros each counted with multiplicity. Put further*

$$\beta_1 = 1, \quad \beta_k = \prod_{n=1}^{k-1} b_{\lambda_n}, \quad k \geq 2, \quad b_\lambda = \frac{\lambda - z}{\mathbb{1} - \bar{\lambda} z} \cdot \frac{|\lambda|}{\lambda},$$

$$e_\lambda = (1 - |\lambda|^2)^{1/2} (\mathbb{1} - \bar{\lambda} z)^{-1}.$$

*Then* $\{\beta_k e_{\lambda_k} : k \geq 1\}$ *is the orthogonalization of the sequence* $\{g_k : k \geq 1\}$ *of root vectors of the operator* $S^*|K_B$, $K_B = H^2 \ominus \mathbf{B}H^2$, *corresponding to* $\Lambda$ *by the following rule: to a group* $\lambda_n = \cdots = \lambda_{n+p} = \lambda$ *of equal members of* $\Lambda$ *there correspond the functions* $g_n = e_\lambda$, $g_{n+1} = e_\lambda^2, \ldots, g_{n+p} = e_\lambda^{p+1}$, *provided* $\lambda \neq 0$, *and* $g_n = \mathbb{1}$, $g_{n+1} = z, \ldots, g_{n+p} = z^p$, *provided* $\lambda = 0$.

*Proof*. Let $m > k$. Then with a suitable $\alpha$, $\alpha = \alpha_{m,k}$,

$$(\beta_m e_{\lambda_m}, \beta_k e_{\lambda_k}) = \left(\frac{\beta_m}{\beta_k} e_{\lambda_m}, e_{\lambda_k}\right) = \alpha \frac{\beta_m}{\beta_k}(\lambda_k) = 0.$$

Similarly if $m = k$ then $(\beta_m e_{\lambda_m}, \beta_k e_{\lambda_k}) = \|e_{\lambda_k}\|^2 = 1$. There remains to check that

$$\bigvee(g_i : 1 \leq i \leq n) = \bigvee(\beta_i e_{\lambda_i} : 1 \leq i \leq n)$$

for all $n$, $n \in \mathbb{N}$. It is clear that $\beta_i e_{\lambda_i} \in \bigvee(g_k : 1 \leq k \leq n)$ for $1 \leq i \leq n$, because this linear span contains all rational functions with the same poles as the $g_k$ $(1 \leq k \leq n)$ provided the multiplicities of these poles do not exceed the maximal of the multiplicities for the $g_k$, $1 \leq k \leq n$. From the orthogonality of the functions $\beta_i e_{\lambda_i}$ we see that they must be linearly independent and therefore

$$\dim \bigvee(\beta_i e_{\lambda_i} : 1 \leq i \leq n) = \dim \bigvee(g_k : 1 \leq k \leq n).$$

This forces the linear spans to coincide. □

**Triangularization Theorem, I.** *Let* $B$ *be a Blaschke product and let* $\Lambda = \{\lambda_n : n \geq 1\}$ *be the sequence of its zeros counted with their multiplicity. Consider the measure* $\mu_B$ *on* $\mathbb{N}$ *given by* $\mu_B(\{n\}) = \frac{1}{2}(1 - |\lambda_n|^2)$, $n \in \mathbb{N}$. *Then the map* $V_B$:

$$V_B c = \frac{1}{\sqrt{2}} \sum_{n \geq 1} c(n)(1 - |\lambda_n|^2)^{1/2} \beta_n e_{\lambda_n}, \quad c = \{c(n)\}_{n \geq 1},$$

*gives a unitary correspondence between the spaces* $L^2(\mu_B)$ *and* $K_B = H^2 \ominus \mathbf{B}H^2$. *The compression of the shift* $T = P_K S|K_B$ *is then mapped onto the operator*

$$V_B^* T V_B = (I - J_B) M_B, \tag{1}$$

*where* $(M_B c)(n) = \lambda_n c(n)$, $n \in \mathbb{N}$, *is a multiplication operator, while* $J_B$,

$$(J_B c)(n) \stackrel{\text{def}}{=} \sum_{k=1}^{n-1} c(k) |\lambda_k|^{-2} \cdot \frac{\beta_n(0)}{\beta_k(0)} (1 - |\lambda_k|^2)^{1/2} \cdot (1 - |\lambda_n|^2)^{1/2}, \quad n \in \mathbb{N},$$

*is a strictly lower-triangular Hilbert-Schmidt operator.*

*Proof*. In order to simplify the notation let us assume that $B(0) \neq 0$. Denote by $\delta_n = \left\{\delta_{nk} \left(\frac{1 - |\lambda_k|^2}{2}\right)^{-1/2}\right\}_{k \geq 1}$ an orthonormal basis vector of $L^2(\mu_B)$. Let us compute the matrix elements for the operator $V_B^* T V_B$:

$$(V_B^* T V_B \delta_n, \delta_k)_{L^2(\mu_B)} = (T\beta_n e_{\lambda_n}, \beta_k e_{\lambda_k})$$

(we use here the definition of $e_\lambda$, along with the inclusion $\beta_k e_{\lambda_k} \in K_B$)

$$= (1 - |\lambda_n|^2)^{1/2}(1 - |\lambda_k|^2)^{1/2}\left(\beta_n \frac{z}{\mathbb{1} - \bar{\lambda}_n z}, \beta_k \frac{\mathbb{1}}{\mathbb{1} - \bar{\lambda}_k z}\right).$$

If $k \leq n$ the last product equals

$$(1 - |\lambda_n|^2)^{1/2}(1 - |\lambda_k|^2)^{1/2}\left(\frac{\beta_n}{\beta_k} \frac{z}{\mathbb{1} - \bar{\lambda}_n z}, \frac{\mathbb{1}}{\mathbb{1} - \bar{\lambda}_k z}\right),$$

i.e. $\lambda_n$ if $k = n$ and 0 if $k < n$. If however $k > n$ the matrix element in question is

$$(1 - |\lambda_n|^2)^{1/2}(1 - |\lambda_k|^2)^{1/2}\left(\frac{1}{\bar{\lambda}_n} \frac{\mathbb{1}}{\mathbb{1} - \bar{\lambda}_n z} - \frac{\mathbb{1}}{\bar{\lambda}_n}, \frac{\beta_k}{\beta_n} \cdot \frac{\mathbb{1}}{\mathbb{1} - \bar{\lambda}_k z}\right)$$

$$= -\frac{(1 - |\lambda_n|^2)^{1/2} \cdot (1 - |\lambda_k|^2)^{1/2}}{\bar{\lambda}_n} \cdot \frac{\beta_k(0)}{\beta_n(0)}.$$

This gives formula (1). Finally we compute the Hilbert-Schmidt norm of $J_B$:

$$\sum_{n \geq 1} \| J_B \delta_n \|^2_{L^2(\mu_B)} = \sum_{n \geq 1} \sum_{k \geq n+1} \left(\frac{\beta_k(0)}{\beta_n(0)}\right)^2 |\lambda_n|^{-4}(1 - |\lambda_k|^2)(1 - |\lambda_n|^2)$$

$$\leq c \sum_{k,n} (1 - |\lambda_k|^2)(1 - |\lambda_n|^2) < \infty. \square$$

**2. Continuous Singular Spectrum.** Let us now turn to the continuous analogues of the theorem just proved. Set $K = K_s \stackrel{\text{def}}{=} H^2 \ominus sH^2$ where $s$ is a singular inner function, denoting its representing measure by $\mu = \mu^s$, i.e.

$$s(\zeta) = \exp\left(-\int_{\mathbb{T}} \frac{t + \zeta}{t - \zeta} d\mu(t)\right), \quad \zeta \in \mathbb{D}.$$

Choose the "natural" increasing chain of $S^*$-invariant subspaces $\{K_\lambda\}$, $K_\lambda = H^2 \ominus s_\lambda H^2$ where

$$s_\lambda(\zeta) = \exp\left(-\int_{\mathbb{T}} \frac{t + \zeta}{t - \zeta} d\mu_\lambda(t)\right), \quad \zeta \in \mathbb{D},$$

and $\mu_\lambda$ is the projection of the measure $\mu$ onto the arc $(1, \lambda] \stackrel{\text{def}}{=} \{\zeta : \zeta \in \mathbb{T}, 0 < \arg \zeta \leq \arg \lambda\}$: $\mu_\lambda e = \mu(e \cap (1, \lambda])$ for any Borel set $e$, $e \subset \mathbb{T}$. The right analogue of the triangularizing operator $V_B$ is the map $V_s$ from the space $L^2(\mu)$ into $K$ given by the identity

$$(V_s c)(\zeta) = \sqrt{2} \int_{\mathbb{T}} c(\lambda) s_\lambda(\zeta) \frac{\lambda \, d\mu(\lambda)}{\lambda - \zeta}, \quad \zeta \in \mathbb{D}. \tag{2}$$

To this formula we are lead from the intuitively plausible fact that the "diagonal" part of the operator $T = P_K S | K$ is $U = \int_{\mathbb{T}} \lambda \, dP_\lambda$ where $P_\lambda = P_{K_\lambda}$ denotes orthogonal projection onto the subspace $K_\lambda$, $\lambda \in \mathbb{T}$. The canonical spectral representation of the unitary operator $U$ should then be defined by the "measure" $d(P_\lambda \mathbb{1})$, since the projection $P_K \mathbb{1}$ is a cyclic vector for $T$ (this follows trivially, for example, from the identities $z^n = z^n P_K \mathbb{1} + z^n(I - P_K)\mathbb{1}$ and $P_K z^n = T^n P_K \mathbb{1}$, $n \geq 0$). If we now recall that $P_\lambda \mathbb{1} = \mathbb{1} - \overline{s_\lambda(0)} s_\lambda$ (by the Projection Lemma, Lecture II) we arrive precisely at an operator of the type (2).

Before studying the operator $V_s$ let us notice one important property of the family $\{s_\lambda\}$: for fixed $\zeta$, $\zeta \in \mathbb{D}$, the measure $d[s_\lambda(\zeta)e^{-\mu(\lambda)}]$ is absolutely continuous with respect to $\mu$ and at each point of continuity of the latter measure there holds the identity

$$d[s_\lambda(\zeta)e^{-\mu(\lambda)}] = s_\lambda(\zeta)e^{-\mu(\lambda)}\left\{-\frac{2\lambda}{\lambda - \zeta}\,d\mu(\lambda)\right\}, \tag{3}$$

with $\mu(\lambda) \stackrel{\text{def}}{=} \mu_\lambda \mathbb{T}$.

In fact, the desired absolute continuity follows from the identity (where it is assumed that $\mu(\lambda) - \mu(\lambda') > 0$)

$$\frac{s_\lambda(\zeta)e^{-\mu(\lambda)} - s_{\lambda'}(\zeta)e^{-\mu(\lambda')}}{\mu(\lambda) - \mu(\lambda')} = \frac{s_\lambda(\zeta)[e^{-\mu(\lambda)} - e^{-\mu(\lambda')}]}{\mu(\lambda) - \mu(\lambda')} + \frac{e^{-\mu(\lambda')}[s_\lambda(\zeta) - s_{\lambda'}(\zeta)]}{\mu(\lambda) - \mu(\lambda')} \tag{4}$$

and the estimates $|s_\lambda(\zeta)| \leq 1$, $|e^{-[\mu(\lambda)-\mu(\lambda')]} - 1| \leq \mu(\lambda) - \mu(\lambda')$,

$$\begin{aligned} |s_\lambda(\zeta) - s_{\lambda'}(\zeta)| &\leq \left|\exp\left(-\int_{(\lambda',\lambda]} \frac{t+\zeta}{t-\zeta}\,d\mu(t)\right) - 1\right| \\ &\leq \exp\left\{\frac{1+|\zeta|}{1-|\zeta|}\mu((\lambda',\lambda])\right\} - 1 \\ &\leq \text{const}(\mu(\lambda) - \mu(\lambda')) \end{aligned}$$

where the const depends only on $\zeta$ and the total mass $\mu\mathbb{T}$. Formula (3) is obtained from (4) by a passage to the limit (for $\lambda' \to \lambda$), taking into account that the functions $e^{-\mu(\lambda)}$ and $s_{(\cdot)}(\zeta)$ are continuous at each point $\lambda$ free from $\mu$-measure (i.e. $\mu(\{\lambda\}) = 0$) and that

$$\begin{aligned} s_\lambda(\zeta) - s_{\lambda'}(\zeta) &= s_\lambda(\zeta)\left[1 - \exp\int_{(\lambda',\lambda]} \frac{t+\zeta}{t-\zeta}\,d\mu(t)\right] \\ &= s_\lambda(\zeta)\left[-\int_{(\lambda',\lambda]} \frac{t+\zeta}{t-\zeta}\,d\mu(t) + o\left(\int_{(\lambda',\lambda]} \frac{t+\zeta}{t-\zeta}\,d\mu(t)\right)\right]. \end{aligned}$$

In fact, the limit of the right hand side of (4) equals

$$-s_\lambda(\zeta)e^{-\mu(\lambda)} - s_\lambda(\zeta)e^{-\mu(\lambda)}\cdot\frac{\lambda+\zeta}{\lambda-\zeta} = -s_\lambda(\zeta)e^{-\mu(\lambda)}\frac{2\lambda}{\lambda-\zeta}. \;\square$$

In an analogous manner one verifies the somewhat more general formula

$$d[s_\lambda(\zeta)\overline{s_\lambda(\eta)}] = -2(1 - \bar{\eta}\zeta)s_\lambda(\zeta)\overline{s_\lambda(\eta)}(1 - \lambda\bar{\eta})^{-1}(1 - \bar{\lambda}\zeta)^{-1}\,d\mu(\lambda). \tag{5}$$

Let us remark that at discontinuity points of the function $\mu(\cdot)$ it is likewise possible to give an explicit expression for the density of the measure $d[s_\lambda(\zeta)e^{-\mu(\lambda)}]$ with respect to $\mu$. However, we will first treat the continuous part of $\mu$. A separate consideration of the continuous and the discrete parts of $\mu$ is necessary, because of the presence of the jump $\mathbf{s}_{\lambda_-}H^2 \ominus \mathbf{s}_{\lambda_+}H^2$ of the chain of subspaces $\{K_\lambda\}$ at each point $\lambda$ which is an atom for $\mu$. Since the truncated shift $T = P_K S|K$ induces a non-scalar operator (i.e. one which is not a multiple of the identity) in this subspace we are forced to "complete" a posteriori our chain $\{K_\lambda\}$ with invariant subspaces, splitting the jumps $s_{\lambda_-}H^2 \ominus \mathbf{s}_{\lambda_+}H^2$.

**Triangularization Theorem, II.** *Let s be a singular inner function with the continuous representing measure $\mu = \mu^s$. Then the mapping (2) gives a unitary correspondence between the spaces $L^2(\mu)$ and $K = K_s \overset{\text{def}}{=} H^2 \ominus \mathbf{s}H^2$. The shift projection $T = P_K S|K$ is then mapped onto the operator*

$$V_s^* T V_s = (I - J_s)M_s,$$

*where $(M_s c)(\lambda) = \lambda c(\lambda)$, $\lambda \in \mathbb{T}$, is a multiplication operator and $J_s$,*

$$(J_s c)(\lambda) \overset{\text{def}}{=} 2\int_{\mathbb{T}} e^{\mu(t)-\mu(\lambda)} c(t)\, d\mu_\lambda(t), \quad \lambda \in \mathbb{T},$$

*a lower-triangular (Volterra) operator of Hilbert-Schmidt class.*

*Proof.* Set $c_{\lambda',\lambda} = \chi_{(\lambda',\lambda]} e^{-\mu(\cdot)}$. Then in view of (3)

$$\begin{aligned}
(Vc_{\lambda',\lambda})(\zeta) &= \sqrt{2}\int_{(\lambda',\lambda]} e^{-\mu(t)} s_t(\zeta) \frac{t\, d\mu(t)}{t-\zeta} \\
&= -\frac{1}{\sqrt{2}}\int_{(\lambda',\lambda]} d[s_t(\zeta)e^{-\mu(t)}] \\
&= \frac{1}{\sqrt{2}}[s_{\lambda'}(\zeta)e^{-\mu(\lambda')} - s_\lambda(\zeta)e^{-\mu(\lambda)}] \\
&= \frac{1}{\sqrt{2}}[s_{\lambda'}(\zeta)s_{\lambda'}(0) - s_\lambda(\zeta)s_\lambda(0)] \\
&= \frac{1}{\sqrt{2}}(P_{K_\lambda \ominus K_{\lambda'}} \mathbb{1})(\zeta).
\end{aligned}$$

Hence

$$\begin{aligned}
\| Vc_{\lambda',\lambda} \|^2 &= \tfrac{1}{2}\| P_{K_\lambda \ominus K_{\lambda'}} \mathbb{1} \|^2 = \tfrac{1}{2}(P_{K_\lambda \ominus K_{\lambda'}} \mathbb{1}, \mathbb{1}) \\
&= \tfrac{1}{2}(s_{\lambda'}(0)^2 - s_\lambda(0)^2) = \tfrac{1}{2}(e^{-2\mu(\lambda')} - e^{-2\mu(\lambda)}) \\
&= \int_{(\lambda',\lambda]} e^{-2\mu(t)}\, d\mu(t) = \| c_{\lambda',\lambda} \|^2.
\end{aligned}$$

Consequently $V = V_s$ is isometric on the linear hull $\mathscr{L}$ of the function $c_{\lambda,\lambda'}$, because to disjoint functions there correspond images lying in orthogonal subspaces. As $\mathscr{L}$ is dense in $L^2(\mu)$ it follows that $V$ maps $L^2(\mu)$ isometrically into $K$.

Let us next check that $VL^2(\mu) = K$. To this end it suffices to show that $VL^2(\mu)$ is dense in $K$ and for this again that the reproducing kernel $k_\Theta$ of $K$ lies in the image $VL^2(\mu)$. This last thing means that for each $\zeta$, $|\zeta| < 1$, there exists a function $c_\zeta$, $c_\zeta \in L^2(\mu)$ such that

$$Vc_\zeta = P_K(\mathbb{1} - \bar{\zeta}z)^{-1} \stackrel{\text{def}}{=} k_\Theta(\cdot, \zeta) \stackrel{\text{def}}{=} k_\zeta.$$

If there exists such a function $c_\zeta$ then $(Vc)(\zeta) = (Vc, (\mathbb{1} - \bar{\zeta}z)^{-1}) = (Vc, k_\zeta) = (Vc, Vc_\zeta) = (c, c_\zeta)_{L^2(\mu)}$ for all $c$, $c \in L^2(\mu)$. On the other hand,

$$(Vc)(\zeta) = \int_{\mathbb{T}} c(\lambda)\{\sqrt{2}\, s_\lambda(\zeta)(\mathbb{1} - \bar{\lambda}\zeta)^{-1}\}\, d\mu(\lambda)$$

so that we must have $c_\zeta(\lambda) = \sqrt{2}\,\overline{s_\lambda(\zeta)}(\mathbb{1} - \lambda\bar{\zeta})^{-1}$ for $\mu$-a.e. $\lambda$, $\lambda \in \mathbb{T}$. It is however clear that this $c_\zeta$ belongs to $L^2(\mu)$ and formula (5) allows us to readily compute its $V$-image at the point $\eta$, $\eta \in \mathbb{D}$:

$$\begin{aligned}(Vc_\zeta)(\eta) &= 2\int_{\mathbb{T}} \overline{s_\lambda(\zeta)} s_\lambda(\eta)(1 - \lambda\bar{\zeta})^{-1}(1 - \bar{\lambda}\eta)^{-1}\, d\mu(\lambda)\\ &= -(1 - \bar{\zeta}\eta)^{-1}\int_{\mathbb{T}} d(s_\lambda(\eta)\overline{s_\lambda(\zeta)})\\ &= (1 - \bar{\zeta}\eta)^{-1}[s_{1+}(\eta)\overline{s_{1+}(\zeta)} - s_{1-}(\eta)\overline{s_{1-}(\zeta)}]\\ &= (1 - \bar{\zeta}\eta)^{-1}[1 - s(\eta)\overline{s(\zeta)}].\end{aligned}$$

The last expression coincides with $k_\zeta(\eta)$, since (Projection Lemma, Lecture II)

$$\begin{aligned}k_\zeta = P_K(\mathbb{1} - \bar{\zeta}z)^{-1} &= (\mathbb{1} - \bar{\zeta}z)^{-1} - s[P_+\bar{s}(\mathbb{1} - \bar{\zeta}z)^{-1}]\\ &= (\mathbb{1} - \bar{\zeta}z)^{-1} - s\cdot\overline{s(\zeta)}\cdot(\mathbb{1} - \bar{\zeta}z)^{-1}.\end{aligned}$$

As the kernel of the integral operator $J$ is square integrable with respect to the product measure $\mu \times \mu$ there remains only to show that the operator $V = V_s$ indeed triangularizes $T$, i.e. that $V(I - J_s)M_s = TV$. Let $\zeta \in \mathbb{D}$ and $c \in L^2(\mu)$. Then, remarking that $d\mu_\lambda = \chi_{(1,\lambda]}\, d\mu$ and $\chi_{[t,1)}(t) = \chi_{[t,1)}(\lambda)$ and performing a change of the order of integration, we get using (3)

$$\begin{aligned}(VJMc)(\zeta) &= 2\sqrt{2}\int_{\mathbb{T}} s_\lambda(\zeta)\frac{\lambda\, d\mu(\lambda)}{\lambda - \zeta}\left(\int_{\mathbb{T}} e^{\mu(t) - \mu(\lambda)} tc(t)\, d\mu_\lambda(t)\right)\\ &= 2\sqrt{2}\int_{\mathbb{T}} tc(t)e^{\mu(t)}\, d\mu(t)\left(\int_{\mathbb{T}} s_\lambda(\zeta)e^{-\mu(\lambda)}\frac{\lambda}{\lambda - \zeta}\chi_{[t,1)}(\lambda)\, d\mu(\lambda)\right)\\ &= \sqrt{2}\int_{\mathbb{T}} tc(t)e^{\mu(t)}[s_t(\zeta)s_t(0) - s(\zeta)s(0)]\, d\mu(t)\\ &= \sqrt{2}\int_{\mathbb{T}} s_t(\zeta)tc(t)\, d\mu(t) + as(\zeta).\end{aligned}$$

Hence

$$(VMc)(\zeta) - (VJMc)(\zeta) = \sqrt{2}\int_{\mathbb{T}} s_t(\zeta)c(t)\left(\frac{t}{1-\bar{t}\zeta} - t\right)d\mu(t) - as(\zeta)$$
$$= \zeta(Vc)(\zeta) - as(\zeta).$$

Finally taking the projection to the subspace $K$ in this last formula we get the desired relationship $VMc - VJMc = TVc$. □

**3. Atomic Singular Spectrum.** Let us turn to the still missing case of a singular function with an atomic (purely discrete) representing measure. The corresponding space $K$ will be built up by the "jumps" of the chain $\{K_\lambda\}$ (see above p. 120), on each of which the diagonal part of $T$ is scalar. In order to describe explicitly the integral representation of the residual Volterra part we use the natural (and unique!) continuous family of invariant subspaces exhausting such a "jump". Let $\Delta$ be a singular inner function with the (purely discrete) representing measure $\mu = \mu^\Delta$. We enumerate the set $\{t : t \in \mathbb{T},\ \mu(\{t\}) > 0\}$ as a sequence $\{t_k\}_{k\in\mathbb{N}}$, for example, so that the corresponding masses $\mu_k \overset{\text{def}}{=} \mu(\{t_k\})$, $k \geq 1$, fall off monotonously. If the set is finite, in order to have a uniform notation, we complete it by an arbitrary countable infinite set of points $t$, equipping the latter with mass $\mu(\{t\}) = 0$. Let further $\mu_\Delta$ be a measure on the halfaxis $\mathbb{R}_+ = [0, \infty)$ such that $d\mu_\Delta(\lambda) = \mu_{[\lambda]+1}\, d\lambda$ and let us define a function $\Delta_\lambda$ on the disk $\mathbb{D}$ by the formula

$$\Delta_\lambda(\zeta) = \exp\left\{-\sum_{k=1}^{[\lambda]} \mu_k \frac{t_k + \zeta}{t_k - \zeta} - (\lambda - [\lambda])\mu_{[\lambda]+1}\frac{t_{[\lambda]+1} + \zeta}{t_{[\lambda]+1} - \zeta}\right\}.$$

Here $[\lambda]$ is the integer part of $\lambda$, $\lambda \in \mathbb{R}_+$, and by definition $\Delta_0 = \mathbb{1}$.

**Triangularization Theorem, III.** *Let $\Delta$ be a singular inner fuction with the atomic representing measure $\mu^\Delta$. Then the mapping $V_\Delta$,*

$$(V_\Delta c)(\zeta) = \sqrt{2}\int_{\mathbb{R}_+} c(\lambda)\Delta_\lambda(\zeta)(1 - \bar{t}_{[\lambda]+1}\zeta)^{-1}\, d\mu_\Delta(\lambda),\ \zeta \in \mathbb{D},$$

*provides a unitary correspondence between the spaces $L^2(\mu_\Delta)$ and $K = K_\Delta = H^2 \ominus \Delta H^2$. The shift compression $T = P_K S|K$ is thereby mapped onto the operator*

$$V_\Delta^* T V_\Delta = (I - J_\Delta)M_\Delta,$$

*where $(M_\Delta c)(\lambda) = t_{[\lambda]+1}c(\lambda)$, $\lambda \in \mathbb{R}_+$, is a multiplication operator and $J_\Delta$,*

$$(J_\Delta c)(\lambda) = \int_0^\lambda c(t)\frac{\Delta_\lambda(0)}{\Delta_t(0)}\, d\mu_\Delta(t),\ \lambda \in \mathbb{R}_+,$$

*lower-triangular and Hilbert-Schmidt.*

*The proof* of this theorem formally repeats the one of Theorem II. We give all calculations in the important special case when $\mu^\Delta$ consists of one single mass;

without loss of generality we may then take $t_1 = 1$, $\mu_1 = 1$. Consequently

$$L^2(\mu_\Delta) = L^2(0,1), \quad M_\Delta = I,$$

$$\Delta_\lambda(\zeta) = \exp\left(-\lambda\frac{1+\zeta}{1-\zeta}\right), \quad 0 \le \lambda \le 1, \quad \zeta \in \mathbb{D};$$

$$d\mu_\Delta(\lambda) = \chi_{(0,1)}(\lambda)\,d\lambda,$$

$$(V_\Delta c)(\zeta) = \sqrt{2}\int_0^1 c(\lambda)\Delta_\lambda(\zeta)(1-\zeta)^{-1}\,d\lambda, \quad \zeta \in \mathbb{D}, \quad c \in L^2(0,1),$$

$$(J_\Delta c)(\lambda) = 2\int_0^\lambda e^{t-\lambda}c(t)\,dt, \quad 0 \le \lambda \le 1, \quad c \in L^2(0,1).$$

Putting as before $c_{\lambda',\lambda}(t) = \chi_{(\lambda',\lambda]}(t)\Delta_t(0)$ we get

$$\begin{aligned}(V_\Delta c_{\lambda',\lambda})(\zeta) &= \sqrt{2}\int_{\lambda'}^{\lambda} \Delta_t(0)\Delta_t(\zeta)(1-\zeta)^{-1}\,dt \\ &= \frac{1}{\sqrt{2}}(\Delta_{\lambda'}(0)\Delta_{\lambda'}(\zeta) - \Delta_\lambda(0)\Delta_\lambda(\zeta)) \\ &= \frac{1}{\sqrt{2}}(P_{K_\lambda \ominus K_{\lambda'}}\mathbb{1})(\zeta),\end{aligned}$$

with $K_\lambda = H^2 \ominus \Delta_\lambda H^2$, $0 \le \lambda \le 1$. This shows that $V_\Delta$ is isometric. The identity $VL^2(0,1) = K$ follows from the easily verifiable representation for the reproducing kernel of the space $K$:

$$Vc_\zeta = \frac{\mathbb{1} - \overline{\Delta(\zeta)}\Delta}{\mathbb{1} - \bar{\zeta}z}; \quad c_\zeta(t) = \sqrt{2}\cdot\overline{\Delta_t(\zeta)}(1-\bar{\zeta})^{-1}, \quad 0 \le t \le 1.$$

Finally as in the preceding theorem (but taking into account that $M_\Delta = I$): $Vc - VJ_\Delta c = zVc - a\Delta$, whence $V_\Delta^* T V_\Delta = I - J_\Delta$. □

**4. The General Case and Applications.** Let us collect the results obtained in the following proposition.

**Triangularization Theorem.** *Let $\Theta$ be an inner function with the canonical factorization $\Theta = B\cdot\Theta^s$, $B$ a Blaschke product and $\Theta^s$ a singular function. Write further $\Theta^s = s\cdot\Delta$ where $s$ and $\Delta$ are singular functions with the representing measures $\mu^s$ and $\mu^\Delta$ respectively, with $\mu^s$ continuous and $\mu^\Delta$ a pure point measure. Then the transformation $V$,*

$$V(c_B, c_s, c_\Delta) = V_B c_B + B(V_s c_s) + Bs(V_\Delta c_\Delta),$$

*defines a unitary correspondence between the spaces $L^2(\mu_B) \times L^2(\mu_s) \times L^2(\mu_\Delta)$ and $K_\Theta = H^2 \ominus \Theta H^2$. Here $V_B, \mu_B, V_s, V_\Delta, \mu_\Delta$ are defined exactly as in Theorems I–III. The shift compression $T = P_\Theta S|K_\Theta$ is thereby mapped onto the operator*

$$V^*TV = M_B \times M_s \times M_\Delta + J, \tag{6}$$

*where* $M_B$, $M_s$, $M_\Delta$ *were defined in the Theorems I–III and* $J = -(J_B M_B \times J_s M_s \times J_\Delta M_\Delta) + A$ *is a Volterra operator of Hilbert-Schmidt class, with* $A^3 = \mathbb{O}$, $\operatorname{rank} A \leq 3$.

*Proof.* Let us first notice the formula

$$K_\Theta = K_B \oplus \mathbf{B}K_s \oplus \mathbf{Bs}K_\Delta, \tag{7}$$

which at once follows from the fact that multiplication by an inner function is an isometric mapping:

$$K_{\Theta_1 \Theta_2} \ominus K_{\Theta_1} = \Theta_1 H^2 \ominus \Theta_1 \Theta_2 H^2 = \Theta_1 K_{\Theta_2}$$

for arbitrary inner functions $\Theta_1$ and $\Theta_2$. This formula shows that $V$ is unitary for the pair of subspaces in question. It is likewise easy to see that

$$V^*f = (V_B^* P_B f, V_s^* P_s(f\bar{B}), V_\Delta^* P_\Delta(f\overline{Bs})), \quad f \in K_\Theta,$$

and that

$$P_\Theta = P_B + \mathbf{B}P_s\bar{\mathbf{B}} + \mathbf{Bs}P_\Delta\overline{\mathbf{Bs}}.$$

The last identity allows us to write down the formula for the matrix of $T$ in the decomposition (7):

$$T = (T_B \oplus \mathbf{B}T_s\bar{\mathbf{B}} + \mathbf{Bs}T_\Delta\overline{\mathbf{Bs}}) + a, \qquad a = \begin{pmatrix} \mathbb{O} & \mathbb{O} & \mathbb{O} \\ a_{12} & \mathbb{O} & \mathbb{O} \\ a_{13} & a_{23} & \mathbb{O} \end{pmatrix},$$

where $a_{ij}$ is the part of the operator $T$ mapping from the term with index $i$ in (7) into the one with index $j$. Each of the operators $a_{ij}$ has rank one. Let us e.g. compute $a_{12}$. Let $f \in K_B$. Then $\bar{B}f \perp H^2$ and, since $Tf = P_\Theta zf = P_B zf + BP_s(\bar{B}zf) + BsP_\Delta(\overline{Bs}zf)$, we get $a_{12}(f) = BP_s(\bar{B}zf) = B(P_s\mathbb{1})\cdot(\bar{B}zf)^\wedge(0)$. In an analogous way we get the other matrix elements too.

Let us check the identity (6). Let $c = (c_B, c_s, c_\Delta) \in L^2(\mu_B) \times L^2(\mu^s) \times L^2(\mu_\Delta)$. Then

$$\begin{aligned} V^*TVc &= V^*T(V_B c_B \oplus BV_s c_s \oplus BsV_\Delta c_\Delta) \\ &= V^*(T_B V_B c_B \oplus BT_s V_s c_s \oplus BsT_\Delta V_\Delta c_\Delta) + V^*aVc \\ &= (V_B^* T_B V_B c_B, V_s^* T_s V_s c_s, V_\Delta^* T_\Delta V_\Delta c_\Delta) + V^*aVc. \end{aligned}$$

It remains to set $A = V^*aV$, to apply the Triangularization Theorems I–III and to notice that $\sigma_p(J) = \varnothing$. The last thing is trivial because $J$ is a lower-triangular operator in the direct product of three $L^2$-spaces and on the diagonal there sit operators with empty point spectra. □

We can now derive some consequences of the results obtained. With the aid of the Triangularization Theorem we will study functions of the truncated shift and the operator of indefinite integration in $L^2(0, 1)$. The first of these applications is patterned on the following well-known fact in finite dimensional linear algebra: if

$A = \{a_{ij}\}_{1 \le i,j \le n}$ is a lower-triangular matrix then the spectrum of $f(A)$ consists of the numbers $f(a_{ii})$, $1 \le i \le n$. In contrast to the finite dimensional case the representation (6) does not yield right away the complete information on the spectrum $\sigma(u(T))$ for all functions $u$ for which $u(T)$ makes sense. Therefore we will here limit ourselves to continuous $u$ only but in return we shall get not only an elementary description of $\sigma(u(T))$ (not using the Corona Theorem) but also a description of a certain part of it, viz. the Fredholm spectrum, of paramount importance in the theory of linear equations.

Let us begin with some definitions. Let us agree to call *Fredholm spectrum* of the operator $T$ the set of all complex numbers $\lambda$ such that there exists a non-compact normalized sequence $\{x_n\}_{n \ge 1}$ with the property that $\lim_n \|(T - \lambda I)x_n\| = 0$. In view of the weak compactness of the unit ball in Hilbert space it is not hard to see that this is equivalent to the existence of a sequence $\{x_n\}_{n \ge 1}$, $\|x_n\| = 1$, weakly converging to $\mathbb{O}$ and such that $\lim_n \|(T - \lambda I)x_n\| = 0$. The Fredholm spectrum of $T$ will be denoted by $\sigma_F(T)$. From the definition it is clear that $\sigma_F(T + C) = \sigma_F(T)$ for any compact operator $C$.

Below we consider continuous functions on $\mathbb{T}$ ($u \in C(\mathbb{T})$) setting $u(T)f \overset{\text{def}}{=} P_K(uf)$, $f \in K$. This definition coincides in the case $u \in C_A$ with the old one (Calculus Lemma, Lecture III); it allows us to reduce the calculation of $\sigma_F(u(T))$ to the determination of the Fredholm spectrum of an appropriate normal operator in $K$. For operators of the latter type the description of the Fredholm spectrum is readily achieved using the Spectral Theorem: $\sigma_F$ is obtained from $\sigma$ if we omit all isolated eigenvalues of finite multiplicity.

**Corollary 1.** *Let $\Theta$ be an inner function setting $K_\Theta = H^2 \ominus \Theta H^2$ and $T = P_K S | K$. Let further $u$ be a continuous function on $\mathbb{T}$. Then*

$$\sigma_F(u(T)) = u(\sigma(\Theta) \cap \mathbb{T}) = u(\sigma_F(T)).$$

For the proof let us first agree that the harmonic continuation to the disk $\mathbb{D}$ of a function $u$, $u \in C(\mathbb{T})$, will be denoted by the same letter $u$, and also remark that if $u$ is a polynomial in $z$ and $\bar{z}$ then in view of the formula (6) holds

$$u(T) = Vu(M)V^* + C \tag{8}$$

where the operators $M = M_B \times M_s \times M_\Delta$ and $V$ have the same meaning as in (6) and $C$ is compact. If now $u \in C(\mathbb{T})$ there exist polynomials $u_n$, $n \ge 1$, such that $\lim_n \|u_n - u\|_\infty = 0$. Hence

$$\lim_n |u(T) - u_n(T)| = 0, \quad \lim_n |u(M) - u_n(M)| = 0$$

so as before (8) holds true with $C$ compact. Using the remarks made after the definition of $\sigma_F$, the description of the operators $M_B$, $M_s$ and $M_\Delta$ in the Triangularization Theorems I–III and the identity

$$\sigma_F(u(M)) = u(\sigma_F(M)) \tag{9}$$

we get $\sigma_F(u(T)) = \sigma_F(u(M)) = u(\sigma(\Theta) \cap \mathbb{T})$. Let us explain the identity (9). If $\lambda = u(\mu)$, $\mu \in \sigma_F(M)$, then for some non-compact sequence $\{x_n\}_{n \ge 1}$ holds $\|x_n\| = 1$

and $\lim_n \|(M - \mu I)x_n\| = 0$. Then for any polynomial in $z$ and $\bar{z}$ holds $\lim_n \|(p(M) - p(\mu)I)x_n\| = 0$. $u$ being continuous on $\sigma(M)$ (and indeed in clos $\mathbb{D}$), approximating it in the uniform metric by polynomials $p$, we get $\mu \in \sigma_F(u(M))$. Thus $u(\sigma_F(M)) \subset \sigma_F(u(M))$. Conversely, if $\lambda \in \sigma_F(u(M))$ then either $\lambda$ is a non-isolated point of $\sigma(u(M))$ and then there exists a sequence of distinct points $\{\mu_n\}_{n\geq 1}$, $\mu_n \in \sigma(M)$, (which we without loss of generality can take to be convergent) such that $\lambda = \lim_n u(\mu_n) = u(\lim_n \mu_n)$, so that $\lambda \in u(\sigma_F(M))$, or else $\lambda$ is an isolated eigenvalue of infinite multiplicity and then the set $u^{-1}(\lambda)$ will be either infinite or contain an eigenvalue of infinite multiplicity for $M$. In any case $\lambda \in u(\sigma_F(M))$. Together with the previous inclusion this gives the "Fredholm spectral mapping theorem" (9). □

In order to deduce from Corollary 1 a spectral mapping theorem for $u(T)$ there remains to identify the eigenvalues of $u(T)$ and $u(T)^*$.

**Corollary 2.** *Let $T$ be as in Corollary* 1 *and $u$ in $C_A$* (recall that $C_A = C(T) \cap H^\infty$). *Then $\sigma(u(T)) = u(\sigma(T))$.*

Indeed, in view of Corollary 1 it suffices to show that

$$u(\sigma(T) \backslash \sigma_F(T)) \subset \sigma(u(T)), \tag{10}$$

$$u(\sigma(T)) \supset \sigma(u(T)) \backslash \sigma_F(u(T)). \tag{11}$$

Let us check that the right hand side of the last inclusion consists of eigenvalues of finite multiplicity for the operator $u(T)$: if the operator $u(T) - \lambda I$ is injective and $\lambda \notin \sigma_F(u(T))$ then on the one hand $\mathrm{GCD}((u - \lambda \mathbb{1})^i, \Theta) = \mathbb{1}$ so that $\mathrm{Ker}(u(T) - \lambda I)^* = \{\mathbb{0}\}$ (Calculus Lemma, Lecture III) and on the other hand $(u(T) - \lambda I)K$ must be a closed set. This means that $(u(T) - \lambda I)K = K$, i.e. $\lambda \notin \sigma(u(T))$. Let thus $\lambda$ be an eigenvalue of finite multiplicity for $u(T)$. Then the $\mathrm{GCD}(u - \lambda \mathbb{1}, \Theta)$ is a finite Blaschke product (Calculus Lemma, Lecture III) and in particular the functions $\Theta$ and $u - \lambda \mathbb{1}$ must have a common zero in $\mathbb{D}$. Consequently $\lambda \in u(\sigma(T))$ and the inclusion (11) is proved.

If $\mu$ is a zero of the function $\Theta$, $|\mu| < 1$ and $\lambda = u(\mu)$, then $\mathrm{Ker}(u(T) - \lambda I) \neq \{\mathbb{0}\}$ (Calculus Lemma, Lecture III) and $\lambda \in \sigma(u(T))$. This proves the inclusion (10). □

We remark once more that the proof of the Spectral Mapping Theorem given here covers only the case of continuous functions of $T$ but in return it is based on only entirely elementary means, whereas the inclusion of $H^\infty$-functions (in Lecture III) required the involvement of the Corona Theorem.

**Corollary 3.** *Let $T$ be the same operator as in Corollary* 1 *and let $u \in C(\mathbb{T})$. Then $u(T)$ is compact iff*

$$u(\sigma(\Theta) \cap \mathbb{T}) = \{0\}.$$

This follows at once from the identity (8) and the definition of the operator $M = M_B \times M_s \times M_\Delta$, cf. the Triangularization Theorems I–III. □

Concluding this Lecture let us show that the integration operator $J$ in the space $L^2(0,1)$ is a "function" of a suitable truncated shift. Let us recall that

$$(Jc)(x) = \int_0^x c\,dt, \quad 0 < x < 1, \quad c \in L^2(0,1).$$

**Corollary 4.** *Set* $\Delta = \exp\frac{z+1}{z-1}$, $K = H^2 \ominus \Delta H^2$, $T = P_K S|K$. *Then the operators* $(I+J)^{-1}$ *and* $\frac{1}{2}(I+T)$ *are unitarily equivalent.*

For the proof we utilize the notation and the formulae of the Triangularization Theorem III. Let us consider the operator $J_\Delta$ from the proof of that theorem. We have

$$e^{-\lambda}\int_0^\lambda c(t)e^t\,dt - \int_0^\lambda e^{-x}d\left(\int_0^x c(t)e^t\,dt\right) = -\int_0^\lambda e^{-x}\left(\int_0^x c(t)e^t\,dt\right)dx,$$

i.e. $J_\Delta - 2J = -JJ_\Delta$. Hence $(I+J)J_\Delta = 2J$, $J_\Delta = 2J(I+J)^{-1} = 2[I-(I+J)^{-1}]$. (Let us recall that the invertibility of $I+J$ was verified in the Concluding Remarks of Lecture IV. Besides, a slight modification of the computations just given leads at once to the identity $J + I = 2(I + V_\Delta^* T V_\Delta)^{-1}$ which in turn implies that $\sigma(J) = \{0\}$). Replacing $J_\Delta$ by the operator $I - V_\Delta^* T V_\Delta$, which indeed equals it (by Theorem III), we get the desired relation $I + V_\Delta^* T V_\Delta = 2(I+J)^{-1}$. □

Some applications of this corollary were given already in Lecture IV.

## Supplements and Bibliographical Notes

**5. Triangular Representations of More General Operators** can be obtained each time the latter have sufficiently many invariant subspaces. For example, if $A$ is a linear operator in a finite dimensional Hilbert space $H$ and $\mathscr{E}_\lambda$ is the spectral projection corresponding to the point $\lambda$, $\lambda \in \sigma(A)$, then the operator

$$A' \stackrel{\text{def}}{=} A - \sum_{\lambda \in \sigma(T)} \lambda \mathscr{E}_\lambda$$

has zero spectrum. Let $\pi = \{P_k : 0 \le k \le \dim H\}$ be a strictly increasing chain of orthogonal projections onto $A$-invariant subspaces $E_k$, $\dim E_k = k$. If $\Delta P = P_k - P_{k-1}$ then apparently $\Delta P A' \Delta P = 0$ for all $P$ so that

$$\begin{aligned} A' &= \sum_{P\in\pi} A'\Delta P = \sum_{P\in\pi} PA'\Delta P = \sum_{P\in\pi} P(A' - A'^*)\Delta P \\ &= 2i \sum_{P\in\pi} PA'_J \Delta P, \end{aligned} \tag{12}$$

where $A'_J = \frac{1}{2i}(A' - A'^*)$ is the imaginary part of $A'$. This simple construction (Brodskiĭ [5]) serves as the point of departure for numerous generalizations based on an elaborate special machinery, viz. the theory of the "integral of the triangular truncation", which in the finite dimensional case at once leads to the transformation $A'_J \mapsto A'$ cutting off the subdiagonal part of the self-adjoint operator $A'_J$. This theory and its application in the triangularization of operators (mainly compact ones) is the subject of the books of Brodskiĭ [6] and Gohberg-Kreĭn [2]. Also all basic results on these themes are due to these authors. The sum (12) has as its analogue the integral

$$2i\int_{\pi} PA_J\,dP, \tag{13}$$

where for the convergence in the operator norm it is necessary that the "blocks" of the operator $A$ at the jumps of the chain $\pi$ equal zero (if

$$P_+ = \inf\{Q : Q\in\pi,\ Q \geq P\},$$
$$P_- = \sup\{Q : Q\in\pi,\ Q < P\},$$

then $(P_+ - P_-)A_J(P_+ - P_-) = \mathbb{O}$). This last condition is fulfilled (Sahnovič [3]) for Volterra operators (compact operators with zero spectrum) and the integral (13) with respect to a "complete" chain of $A$-invariant subspaces allows one to reconstruct this operator from its imaginary part $A_J$ (Brodskiĭ [5]). If on the contrary the self-adjoint operator $A_J = H$ and a chain $\pi$ are given in advance (and the problem is to find an operator $A$ with imaginary part $H$ and such that $\pi \subset \operatorname{Lat} A$) then the question of the convergence of the integral (13) can be rather difficult. It is known that within the class of all continuous chains ($P_+ = P_-$, $\forall P$) the necessary and sufficient condition for convergence is the complete continuity of $H$ along with the following smallness condition on its eigenvalues $\lambda_n$, taken with their multiplicities and ordered so that the moduli $|\lambda_n|$ decrease: $\sum_{n\geq 1}|\lambda_n|/n < \infty$ (Macaev [1]).

Other approaches to triangular models require an explicit construction of a natural "triangularizing basis" in some $L^2$-space; more exactly one considers as triangular model operator any operator $M$ acting in a space $L^2(E,(0,1))$ of vector functions (with values in the Hilbert space $E$) and "triangular" in the sense that

$$ML^2(E,(a,1)) \subset L^2(E,(a,1)), \quad 0 \leq a \leq 1.$$

Such an operator $M$ can be written in the form

$$(Mf)(x) = \frac{d}{dx}\int_0^x M(x,t)f(t)\,dt, \quad x\in(0,1), \tag{14}$$

$f\in L^2(E,(0,1))$ where $M(\cdot,\cdot)$ is a kernel of Carleman type (for details see Sahnovič [3]), and in supplementary hypotheses concerning the closeness of the modelled operator $A$ to the self-adjoint operators (e.g. if $A_J$ is Hilbert-Schmidt; cf. Sahnovič [4]) also a further (constructive) analysis of the model is possible: decomposition into a sum of a self-adjoint constituent and a lower-triangular Volterra operator etc. All details can be found in the papers quoted in this and the preceding paragraphs. Finally, we point out that a general operator $A$ cannot be unitarily equivalent to an operator of the type (14) (because such an operator necessarily admits a continuous chain of invariant subspaces); but one has such an equivalence for a suitable extension $A \oplus N$ with $N$ normal, cf. Sz.-Nagy and Foiaş [5], and if here all blocks of $A$ equal zero on the jumps of some chain in Lat $A$ then one can take $N = \mathbb{O}$, cf. Sahnovič [3].

**6. Concluding Remarks.** The majority of the results of this Lecture is contained in Ahern-Clark [1]. That paper, as well as sharpenings and extensions obtained later (cf. below), arose under the influence of the general theory of triangularization with respect to a chain of invariant subspaces, concerning which some information was given in the foregoing Sect. 5. In the book Gohberg-Kreĭn [2] this theory especially is applied to Volterra operators, while related ideas have been developed for the model operator (in the sense of Sz.-Nagy and Foiaş) with a rank one defect, i.e. the compressions of the shift operator, in papers by a group of American mathematicians (Ahern-Clark [1], Kriete [1], [2], Lubin [1]). The presence of "explicit" formulae for the triangularizing transformation $V$ and its inverse $V^*$ give these authors ground to term the expansion $V^*TV = M + J_T$ obtained by them "concrete triangular models". In Kriete's paper [1] an integral representation, analogous to our Triangularization Theorem, also was obtained for subspaces of $L^2$ of the form $\bar{K}_{\Theta_1} \oplus K_{\Theta_2}$, i.e. for the collection of functions $f$, $f \in L^2$, which are the boundary values of meromorphic functions in $\mathbb{C}\backslash\mathbb{T}$ and such that $f/\Theta_2 \in H^2_-$ and $f/\bar{\Theta}_2 \in H^2$ (here $H^2_- = L^2 \ominus H^2$).

If we base our study not on the shift operator $S$ but on the "equivalent" semigroup of shifts acting in the space $H^2$ in the halfplane (cf. Lecture XI) then in a special case we get as the "triangularizer" $V$ the classical Fourier transform. In this sense every chain of $T$-invariant subspaces gives rise to its own "Fourier transform", the analogue of the Paley-Wiener theorem (the classical Paley-Wiener theorem for the Fourier transform in the space $L^2(-A, A)$ corresponds to the pair of inner functions $\Theta_1 = e^{iAz}$ and $\Theta_2 = e^{iAz}$) etc. It also turns out that the formulations and likewise the proofs are analogous to the ones given in the present Lecture. An important compound of these generalizations constitutes the following criterion (cf. Kriete [2]) for the completeness of an increasing family of inner functions (which entails the unitarity of the corresponding map $V$). Let $\Theta_t$ $(t \geq 0)$ be singular inner functions with the representing measure $\mu_t$, assuming that $\Theta_t$ divides $\Theta_s$ for $t < s$; define a measure $m_0$ on $[0, \infty)$ by the formula $m_0((a, b]) = \mu_b(\mathbb{T}) - \mu_a(\mathbb{T})$; then $\bigvee(\Theta_t : t \geq 0) = H^2$ iff for $m_0$-a.e. value of $t$ the measures

$$\frac{\mu_{t+\varepsilon} - \mu_{t-\varepsilon}}{\mu_{t+\varepsilon}(\mathbb{T}) - \mu_{t-\varepsilon}(\mathbb{T})}$$

converge weakly to the $\delta$-measure $\delta_t$ as $\varepsilon \to 0$. In particular, if this condition is fulfilled then $\bigvee(P_{\Theta_1}\Theta_t : 0 \leq t \leq 1) = H^2 \ominus \Theta_1 H^2$.

Let us further mention that a separate study of the "continuous" and the "discrete" subspaces $K_s$ and $K_\Delta$ can be bypassed if we take the Spectral Theorem not in the form stated in this Lecture (on p. 120) but in the complete form due to von Neumann, providing a decomposition of the space $K$ as a direct integral

$\int \oplus N_\lambda \, d\mu(\lambda)$. The multiplicity $\dim N_\lambda$ then comes from the presence of jumps of the spectral "measure" mentioned on p. 120 (from root subspaces what concerns the point spectrum). Let us however remark that the models discussed there mix the Jordan cells of the operator $T = PS|K$ corresponding to the point spectrum and apparently it is precisely this circumstance which so far prevents us from getting from the integral representations obtained as corollaries any real results. All the generalizations of the Ahern-Clark theorems mentioned in the preceding paragraph can be found in the papers of the American mathematicians listed above. A generalization to the vectorial case of the space $H^2(E)$, $\dim E < \infty$ (operators with finite defects) was accomplished by L. B. Golinskiĭ in diploma work at the Harkov State University (1974, unpublished).

The Malmquist-Walsh Lemma can be found in the book by Walsh [1]; the functions $\beta_k e_{\lambda_k}$ of that lemma are called Malmquist functions. They are often employed in interpolation theory, cf. Walsh [1], Džrbašyan [1].

The proofs given in the text of the Spectral Mapping Theorem (Corollary 1) and the corresponding result for the Fredholm spectrum (Corollary 2) are perhaps not the most economical ones, nor are they the most fruitful ones. They are however instructive as examples of the use of the Triangularization Theorem in a purely spectral theoretic question; it cannot be excluded that similar considerations could be of use in a proof, independent of the Corona Theorem, of the Spectral Mapping Theorem for the $H^\infty$-calculus (see Lecture III, Sect. 3). If we are lucky this would undoubtedly lead to a new interpretation of the Corona Theorem itself; cf. likewise Lecture III and VIII. Further information on Fredholm spectra of general linear operators can be found in the books by Kato [1] and by Douglas [2]. It is easy to see that for completely non-unitary contractions $T$ holds $\sigma_F(T) \supset \mathbb{T} \cap \sigma(T)$ and that for operators with a resolvent which is meromorphic in $\mathbb{D}$ and with finite dimensional rootsubspaces $E_\lambda$ (thus in particular for $C_0$-operators with $\dim E_\lambda < \infty$) this inclusion becomes an equality. Using the description of the spectrum and the point spectrum for $C_0$-operators (Lecture III) along with the Rough Spectral Mapping Theorem (loc. cit.) one can display a direct argument which proves that $\sigma_F(u(T)) = u(\sigma_F(T))$ $(= u(\sigma(m_T) \cap \mathbb{T}))$ for $u \in C_A$ and $T \in C_0$.

Corollary 3 for functions $u$ analytic in $\mathbb{D}$ will be strengthened in Lecture VIII (where the case $u \in H^\infty$ is considered).

The connection between the compression of the shift with the characteristic function $\exp \dfrac{z+\mathbb{1}}{z-\mathbb{1}}$ and the integral operator $J$ (Corollary 4) was established by Sarason [5] in a way differing from the one of this Lecture. (In passing let us point out that the formula transforming $T$ into $J$, apart from a factor $i$, coincides with the Cayley transform.) Using this connection the same author in [6] finds the commutant of $J$ (in $L^2(0,1)$). In contrast to integration $f \mapsto \int_0^x f$ in $L^1(0,1)$ (cf. Dixmier [1]) this commutant contains operators others than just convolutions of finite measures carried by the interval $[0,1)$.

# Lecture VI. Bases and Interpolation (Statement of the Problem)

In this and the following two Lectures we will give a much deeper analysis of the complete truncated shift operators with a simple spectrum. In particular, we will give an exhausting answer to the question raised in the Introductory Lecture: when does a compression of the shift admit a spectral decomposition similar to the Spectral Theorem? The word "similar" can be interpreted literally, i.e. as "linearly similar"; we are thus seeking criteria for the model operator to be equivalent to a normal operator. It is not hard to see (cf. the "Supplements ..." of this Lecture) that among the model operators with an inner characteristic function $\Theta$ and a resolvent meromorphic in the disk $\mathbb{D}$ the hypothesis of completeness and of the simplicity of the spectrum in fact does not diminish the generality – it is indeed a necessary condition for the operator $\mathbf{T}_\Theta$ to be similar to a normal operator. If one thus beforehand imposes this hypothesis (as we will do) the study of the corresponding spectral decomposition is reduced to the description of unconditional bases of eigenfunctions of the operators $\mathbf{T}_\Theta$ and $\mathbf{T}_\Theta^*$. We will precede our study of families of these eigenfunctions with some general results on bases in Hilbert space, the most important of which being the famous theorem of Bari. Then, proceeding to the eigenfunctions of the truncated shift, we will exhibit the connection between the basis problem and some classical interpolation problems in the spaces $H^2$ and $H^\infty$. This will also be the foundation for the main results of Lecture VI–VIII. The latter again will form the pattern upon which we solve the problem of spectrality (in the sense of Dunford) and of generalized spectrality for the model operator in Lecture IX–X.

**1. Riesz Bases**

**Definition.** Let $\mathfrak{N}$ be an at most countably infinite set, $X$ a Banach space and $\mathscr{X} = \{x_n : n \in \mathfrak{N}\}$ a family of vectors in $X$. $\mathscr{X}$ is called *minimal* (or topologically free) if $x_n \notin \bigvee(x_i : i \in \mathfrak{N} \backslash \{n\})$, $n \in \mathfrak{N}$, and *uniformly minimal* if

$$\delta(\mathscr{X}) \stackrel{\text{def}}{=} \inf\{\operatorname{dist}(x_n/\|x_n\|, \bigvee(x_i : i \in \mathfrak{N}\backslash\{n\})) : n \in \mathfrak{N}\} > 0.$$

The family $\mathscr{X}' = \{x_n' : n \in \mathfrak{N}\}$ of continuous linear functionals on $X$ will be called *biorthogonal* to $\mathscr{X}$ if $\langle x_n, x_k' \rangle = \delta_{n,k}$; $n, k \in \mathfrak{N}$.

It is clear that it would have been more appropriate to speak of the biorthogonal pair $(\mathscr{X}, \mathscr{X}')$ reserving for the family $\mathscr{X}'$ some other name ("dual", "coordinate" etc.); however, in the literature one consequently has adhered to the term used in the definition and we do not want to fight against an established use of words.

**Lemma on Minimal Families.** *Let $X$ be a Banach space and $\mathscr{X} = \{x_n : n \in \mathfrak{N}\}$ a family of elements of $X$.*

1. *$\mathscr{X}$ is topologically free iff $\mathscr{X}$ admits a biorthogonal family $\mathscr{X}'$. The latter is uniquely determined by $\mathscr{X}$ iff $\bigvee(\mathscr{X}) = X$.*

2. *Let $\bigvee(\mathscr{X}) = X$. $\mathscr{X}$ is uniformly minimal iff* $\sup \|x_n\| \cdot \|x'_n\| < \infty$ *and this* sup *coincides with $d(\mathscr{X})^{-1}$; if $\mathscr{X}'$ too is total then $d(\mathscr{X}) = d(\mathscr{X}')$.*

*Proof.* All the assertions of the lemma (including the identity $\|x_n\|^{-1} \cdot \|x'_n\|^{-1} = \operatorname{dist}(x_n\|x_n\|^{-1}, \bigvee(x_i : i \in \mathfrak{N}\backslash\{n\}))$ are straightforward consequences of the Hahn-Banach theorem. □

**Definition.** Let $H$ be a Hilbert space and $\mathscr{X} = \{x_n : n \in \mathfrak{N}\}$ a family of elements of $H$. $\mathscr{X}$ is said to be a *basis equivalent to an orthonormal basis* (or a *Riesz basis*) if there exists an isomorphism $U$ mapping $\mathscr{X}$ onto an orthonormal family $\{Ux_n : n \in \mathfrak{N}\}$; the operator $U$ will be called the *orthogonalizer* of $\mathscr{X}$. The expression "a Riesz basis in(for) $H$" means a Riesz basis $\mathscr{X}$ with the completeness property $\bigvee(\mathscr{X}) = H$.

The word "basis" is of course not chosen at random: every vector $x$ in the hull $\bigvee(\mathscr{X})$ can be expanded in an (unconditionally convergent) Fourier series

$$x = U^{-1}Ux = U^{-1}\left(\sum_{n \in \mathfrak{N}} (Ux, Ux_n)Ux_n\right) = \sum_{n \in \mathfrak{N}} (x, U^*Ux_n)x_n.$$

It is clear that a Riesz basis $\mathscr{X}$ is a minimal family, that the corresponding biorthogonal system $\mathscr{X}'$ is gotten by the formula $x'_n = U^*Ux_n$, $n \in \mathfrak{N}$ (because $(Ux_k, Ux_n) = \delta_{kn}$) and that the Fourier coefficients $\{(x, x'_n)\}_{n \in \mathfrak{N}}$ make up the entire space $l^2 = l^2(\mathfrak{N})$ (by the Riesz-Fischer theorem applied to the orthonormal basis $\{Ux_n : n \in \mathfrak{N}\}$). Thus we have the following

**Lemma on Riesz Bases.** *Let $H$ be a Hilbert space and $\mathscr{X} = \{x_n : n \in \mathfrak{N}\}$ a Riesz basis in $H$. Denote the corresponding biorthogonal system by $\mathscr{X}'$. Then 1. $\mathscr{X}$ is uniformly minimal. 2. $\mathscr{X}'$ is a Riesz basis in $H$. 3. If*

$$J_{\mathscr{X}}x \stackrel{\text{def}}{=} \{(x, x'_n) : n \in \mathfrak{N}\}, \quad x \in H,$$

*then $J_{\mathscr{X}}H = l^2$.* □

Next we establish

**Bari's Theorem.** *Let $\mathscr{X} = \{x_n : n \in \mathfrak{N}\}$ be a topologically free family in the Hilbert space $H$ and let $\mathscr{X}'$ be the corresponding biorthogonal system. The following statements are equivalent:*

1. *$\mathscr{X}$ is a Riesz basis in $H$.*
2. *$\mathscr{X}'$ is total in $H$ and $J_{\mathscr{X}}H = l^2$.*
3. *$\mathscr{X}$ is total in $H$ and $J_{\mathscr{X}}H \subset l^2$, $J_{\mathscr{X}'}H \subset l^2$.*
4. *$\mathscr{X}$ is total in $H$ and the Gram matrices $\Gamma_{\mathscr{X}}$ and $\Gamma_{\mathscr{X}'}$ define continuous transformations from $l^2(\mathfrak{N})$ into itself ($\Gamma_{\mathscr{X}} \stackrel{\text{def}}{=} \{(x_n, x_k) : n, k \in \mathfrak{N}\}$ and analogously for $\Gamma_{\mathscr{X}'}$).*
5. *$\mathscr{X}$ is total in $H$ and the matrix $\Gamma_{\mathscr{X}}$ defines a continuous and invertible transformation in $l^2(\mathfrak{N})$.*

*Proof.* The implications $1 \Rightarrow 2$ and $1 \Rightarrow 3$ are contained in the preceding lemma. In the hypothesis of Statement 2, the transformation $J_{\mathscr{X}}$ will be closed and bijective and therefore defines an isomorphism of $H$ onto $l^2(\mathfrak{N})$ which maps the family $\mathscr{X}$ onto the standard basis of the latter space; therefore $2 \Rightarrow 1$. In order to prove the remaining inclusions ($3 \Rightarrow 4 \Rightarrow 5 \Rightarrow 1$) we will employ the following elementary calculation. Let $a$ be an arbitrary finite vector in $l^2$ and set $V_{\mathscr{X}} a \overset{\text{def}}{=} \sum a_n x_n$. Then

$$\| V_{\mathscr{X}} a \|^2 = \left( \sum a_n x_n, \sum a_k x_k \right) = \sum a_n \bar{a}_k (x_n, x_k) = (\Gamma_{\mathscr{X}} a, a), \tag{1}$$

$$(V_{\mathscr{X}} a, b) = \left( \sum a_n x_n, b \right) = \sum a_n \overline{(b, x_n)} = (a, J_{\mathscr{X}'} b), \tag{2}$$

where $b \in H$.

$3 \Rightarrow 4$, because in the hypothesis of Statement 3, the transformation $J_{\mathscr{X}'}$ is continuous (as it is closed) and so are $V_{\mathscr{X}}$ and $\Gamma_{\mathscr{X}}$ (see (2) and then (1)); analogously for $\Gamma_{\mathscr{X}'}$.

$4 \Rightarrow 5$, because $\Gamma_{\mathscr{X}}$ is invertible at the same time as $V_{\mathscr{X}}$ (see (1)); but $J_{\mathscr{X}} V_{\mathscr{X}} = I$, $V_{\mathscr{X}} J_{\mathscr{X}} = I$ (trivial), the operator $J_{\mathscr{X}}$ being continuous, because of the continuity of $V_{\mathscr{X}'}$ (see (2)), and this again in view of the continuity of $\Gamma_{\mathscr{X}'}$ (see (1)).

$5 \Rightarrow 1$, since $V_{\mathscr{X}}$ is an isomorphism (this follows from (1) and the relation $V_{\mathscr{X}} l^2 = \bigvee(\mathscr{X}) = H$.) □

Thus Bari's Theorem reduces the question of Riesz basis to the solution of the abstract moment (or interpolation) problem

$$J_{\mathscr{X}} x = a, \; a \in l^2,$$

in the space $H$, or to two imbedding theorems

$$J_{\mathscr{X}} H \subset l^2, \; J_{\mathscr{X}'} H \subset l^2.$$

In addition, if $U$ is an orthogonalizer of the family $\mathscr{X}$ then the product

$$|U| \cdot |U^{-1}| \overset{\text{def}}{=} r(\mathscr{X})$$

characterizes the deviation of the basis $\mathscr{X}$ from an orthonormal one and $|U|^{-1}$ and $|U^{-1}|$ are the best constants in the inequality

$$c \cdot \left( \sum |a_n|^2 \right)^{1/2} \le \left\| \sum a_n x_n \right\| \le C \cdot \left( \sum |a_n|^2 \right)^{1/2};$$

$r(\mathscr{X})$ will be referred to as the Riesz constant of the family $\mathscr{X}$. It is clear that this is a consistent definition: the norm is independent of which orthogonalizer we choose. The norm $|J_{\mathscr{X}}^{-1}| = |J_{\mathscr{X}'}|$ will be called the *interpolation constant*. As $J_{\mathscr{X}} x_n$ is an element of the standard orthonormal basis in $l^2(\mathfrak{N})$ there holds the inequality $|J_{\mathscr{X}}^{-1}| \ge 1/d(\mathscr{X})$, provided $\| x_n' \| = 1$, $n \in \mathfrak{N}$. It is clear that $r(\mathscr{X}) = |J_{\mathscr{X}}| \cdot |J_{\mathscr{X}'}|$.

**2. Interpolation.** Let us now return to the compression of the shift operator, that is, the operator $T = PS|K$, taking $K = H^2 \ominus \mathbf{B}H^2$, $P = P_K$, where $B$ is a Blaschke product with simple zeros. In this case $B = \prod_{\lambda \in \sigma_p(T)} b_\lambda$ and all eigenvalues of the operators $T^*$ and $T$ are simple, the corresponding eigenfunctions being given by

$$\begin{aligned} x_\lambda &= (1-|\lambda|^2)^{1/2}(\mathbb{1}-\bar{\lambda}z)^{-1}, \\ y_\lambda &= B\cdot(z-\lambda\mathbb{1})^{-1}, \end{aligned} \tag{3}$$

$\lambda\in\sigma_p(T)$. Recall that $\|x_\lambda\| = 1$. Also $\|y_\lambda\| = \|(z-\lambda\mathbb{1})^{-1}\| = (1-|\lambda|^2)^{-1/2}$. We normalize $y_\lambda$ setting $x'_\lambda = \alpha_\lambda y_\lambda$, $\lambda\in\sigma_p(T)$, in such a way that the families $\mathscr{X}$ and $\mathscr{X}'$ become biorthogonal (with respect to each other). The possibility of such a normalization follows at once from the Livšic-Moeller Theorem (Lecture III) and from the definition of the Riesz spectral projection onto an isolated part of the spectrum (in the case at hand thus a subset of $\sigma_p(T)$ consisting of one single element). However, it is also easy to carry out a direct calculation. Setting $B_\lambda = Bb_\lambda^{-1}$ we can determine the factor $\alpha_\lambda$ as follows:

$$\begin{aligned} (x'_\lambda, x_\lambda) &= \alpha_\lambda(1-|\lambda|^2)^{1/2}(B\cdot(z-\lambda\mathbb{1})^{-1}, (\mathbb{1}-\bar{\lambda}z)^{-1}) \\ &= \alpha_\lambda(1-|\lambda|^2)^{1/2}B_\lambda(\lambda)(\bar{\lambda}\lambda-1)^{-1}\frac{|\lambda|}{\lambda}. \end{aligned}$$

We find thus $\alpha_\lambda = -(1-|\lambda|^2)^{1/2}B_\lambda(\lambda)^{-1}\dfrac{\lambda}{|\lambda|}$ so that

$$x'_\lambda = (1-|\lambda|^2)^{1/2}B_\lambda(\lambda)^{-1}\frac{B_\lambda}{\mathbb{1}-\bar{\lambda}z}, \tag{4}$$

$$\|x'_\lambda\| = |B_\lambda(\lambda)|^{-1}. \tag{5}$$

**Lemma on Uniform Minimality.** *The family* $\mathscr{X} = \{x_\lambda : \lambda\in\sigma_p(T)\}$ *is uniformly minimal iff the following* Carleson condition *is fulfilled*:

$$\delta(\mathscr{X}) \overset{\text{def}}{=} \inf\{|B_\lambda(\lambda)| : \lambda\in\sigma_p(T)\} > 0. \tag{C}$$

*Moreover* $\delta(\mathscr{X}) = d(\mathscr{X})$; *this number will be called the* Carleson constant.

*The Proof* is immediate ($\|x_\lambda\| = 1$, $\|x'_\lambda\| = |B_\lambda(\lambda)|^{-1}$). □

**Newman's Lemma.** *The family $\mathscr{X}$ constitutes a Riesz basis for K iff the condition (C) holds and moreover*

$$\sum_{\lambda\in\sigma_p} |f(\lambda)|^2(1-|\lambda|^2) < \infty, \quad \forall f\in H^2. \tag{CN}$$

*Proof.* That the families $\mathscr{X}$ and $\mathscr{X}'$ are complete was established in Lecture IV. From Bari's Theorem (Statement 3) and the formulae (3) and (4) follows that $\mathscr{X}$ is a Riesz basis iff

$$\sum_{\lambda\in\sigma_p} |(f,x_\lambda)|^2 \equiv \sum_{\lambda\in\sigma_p} |f(\lambda)|^2(1-|\lambda|^2) < \infty, \quad \forall f\in K,$$

$$\begin{aligned} \sum_{\lambda\in\sigma_p} |(f,x'_\lambda)|^2 &\equiv \sum_{\lambda\in\sigma_p} |(f, B(\lambda\mathbb{1}-z)^{-1})|^2(1-|\lambda|^2)|B_\lambda(\lambda)|^{-2} \\ &= \sum_{\lambda\in\sigma_p} |(\overline{zf}B)(\lambda)|^2(1-|\lambda|^2)|B_\lambda(\lambda)|^{-2} < \infty, \quad \forall f\in K, \end{aligned}$$

because $\bar{z}B\bar{f} \in H^2$ (see the Lemma on Lat $S^*$, Lecture II). In the first of these formulae we can let $f$ run through all of $H^2$, because $f \in \mathbf{B}H^2 \Rightarrow f(\lambda) = 0$, $\lambda \in \sigma_p$. Thus taking account of Condition ($C$) we see that the two last inequalities are equivalent to the Carleson-Newman Condition ($CN$).

On the other hand if $\mathscr{X}$ is a Riesz basis then by the Lemma on Uniform Minimality condition ($C$) must be fulfilled and by Bari's Theorem ($CN$) too. □

Thus it remains to investigate the subsets $\sigma_p$ of the disk $\mathbb{D}$ such that Carleson's Condition ($C$) and the Newman Condition ($CN$) are fulfilled. The following remarkable result shows that always ($C$) ⇒ ($CN$).

**Theorem on Bases and Interpolation.** *The following Statements are equivalent*:

1. *$\mathscr{X}$ (or $\mathscr{X}'$) is a Riesz basis in $K = H^2 \ominus \mathbf{B}H^2$.*
2. *The interpolation problem $J_{\mathscr{X}'}f = a$,*

$$J_{\mathscr{X}'}f = \{f(\lambda)(1 - |\lambda|^2)^{1/2} : \lambda \in \sigma_p\},$$

*has a solution if $H^2$ iff $a \in l^2$.*

3. *The condition ($C$) is fulfilled.*

This theorem will be proved in the next Lecture by several methods. Presently we observe only that the constant $|J_{\mathscr{X}'}^{-1}|$ is connected with the interpolation of analytic $H^2$-functions:

$$\|J_{\mathscr{X}'}^{-1}a\| = \inf\{\|f\| : f \in H^2; f(\lambda)(1 - |\lambda|^2)^{1/2} = a(\lambda), \forall \lambda \in \sigma_p\},$$

i.e. $J_{\mathscr{X}'}^{-1}a$ is the function with least norm interpolating the family $a$. The comment on Bari's Theorem shows that $|J_{\mathscr{X}'}^{-1}| \geq 1/\delta(\mathscr{X})$, where $\delta(\mathscr{X})$ is the Carleson constant.

Let us turn to the spectral interpretation of the bases theorems.

**3. Spectral Projections and Unconditional Convergence.** Let us consider the "spectral measure" corresponding to the operator $T_B$, i.e. the set function $\sigma \mapsto \mathscr{E}(\sigma)$ where $\mathscr{E}(\sigma)$ is projection onto the subspace $E_\sigma = \bigvee(x'_\lambda : \lambda \in \sigma)$ parallel to $E_{\mathbb{C}\setminus\sigma} = \bigvee(x'_\lambda : \lambda \notin \sigma)$:

$$\mathscr{E}(\sigma)x = \mathscr{E}(\sigma)\left(\sum_{\lambda \in \sigma_p} a_\lambda x'_\lambda\right) = \sum_{\lambda \in \sigma \cap \sigma_p} a_\lambda x'_\lambda$$

for all vectors $x$ in the linear hull of the family $\mathscr{X}'$. It is clear that for these $x$

$$\mathscr{E}(\sigma)x = \sum_{\lambda \in \sigma \cap \sigma_p} (x, x_\lambda)x'_\lambda, \tag{6}$$

that $\mathscr{E}(\varnothing) = \mathbb{0}$, $\mathscr{E}(\mathbb{D}) = I$ and that $\mathscr{E}$ is an additive set function. Let us assume that the projection $\mathscr{E}(\sigma)$ extends to a continuous mapping from the space $K = K_B$ into itself, for any set $\sigma$. From the Banach-Steinhaus theorem it then follows readily that

$$\sup_{\sigma \in \mathscr{X}} |\mathscr{E}(\sigma)| < +\infty \tag{7}$$

and that

$$x = \lim_{\sigma \in \mathscr{K}} \mathscr{E}(\sigma)x \tag{8}$$

where $\mathscr{K}$ is the directed set of all finite subsets of the point spectrum $\sigma_p = \sigma_p(T)$. Let us further remark that $\mathscr{E}(\sigma)$ is the Riesz spectral projection, provided $\sigma$ is a finite set (the same is likewise true for other isolated parts of the spectrum $\sigma(T)$). The relations (6) and (8) can also be associated with an arbitrary biorthogonal pair $\mathscr{X}$, $\mathscr{X}'$, which leads to the following.

**Definition.** The family $\mathscr{X}'$ is called an *unconditional basis* (of its hull $\bigvee(\mathscr{X}')$) if (8) holds true for all $x$, $x \in \bigvee(\mathscr{X}')$.

It is not hard to see that for an arbitrary topologically free family of vectors (as for instance the family (4)) the property of being an unconditional basis is equivalent to the boundedness of the spectral projections (7). If one of these conditions thus holds true for the family (4) then

$$T = \sum_{\lambda \in \sigma_p(T)} \lambda(\cdot, x_\lambda)x'_\lambda = \int_{\mathbb{D}} z\, d\mathscr{E} \tag{9}$$

in the sense of strong unconditional operator convergence. Formula (9) is a perfect analogue of the Spectral Theorem. We show right away that up to a similarity transformation it coincides with that theorem.

**Köthe-Toeplitz Theorem.** *Let $\mathscr{X} = \{x_n : n \in \mathfrak{N}\}$ be a total normalized minimal family in the Hilbert space $H$. The following statements are equivalent:*

1. *$\mathscr{X}$ is an unconditional basis.*
2. *$\mathscr{X}$ is a Riesz basis.*

*Proof.* As any orthonormal basis is unconditional it is clear that $2 \Rightarrow 1$. Conversely, let $\mathscr{X}$ be an unconditional basis. Then by a well-known theorem on numerical series

$$\sum_{n \in \mathfrak{N}} |(x, x'_n)| \cdot |(x_n, x')| < +\infty \tag{10}$$

for all $x$ and $x'$ in $H$. We claim that (10) entails that $\sum_n |(x, x'_n)|^2 < +\infty, x \in H$. To this end set

$$p_\sigma(x') = \sum_{n \in \sigma} |(x, x'_n)| \cdot |(x_n, x')|, \quad x' \in H, \quad \sigma \in \mathscr{K}.$$

Then the $p_\sigma$ are continuous seminorms in $H$ ($p_\sigma(f + g) \le p_\sigma(f) + p_\sigma(g)$, $p_\sigma(\lambda f) = |\lambda| p_\sigma(f)$) having a finite upper bound in the whole of $H$. By a well-known corollary of the Banach-Steinhaus theorem we get from this

$$p_\sigma(x') \le \text{const}\, \|x'\|, \quad x' \in H, \quad \sigma \in \mathscr{K}.$$

If $\{\varepsilon_n : n \in \sigma\}$ is an arbitrary family of unimodular numbers ($|\varepsilon_n| = 1$) then

$$\left| \sum_{n \in \sigma} \varepsilon_n (x, x'_n)(x_n, x') \right| \le p_\sigma(x') \le \text{const}\, \|x'\|, \quad x' \in H.$$

Hence

$$\left\| \sum_{n \in \sigma} \varepsilon_n (x, x'_n) x_n \right\| \leq \text{const}, \quad \sigma \in \mathscr{K},$$

and there remains to choose the numbers $\varepsilon_n$ such that

$$\sum_{n \in \sigma} \|(x, x'_n) x_n\|^2 \leq \left\| \sum_{n \in \sigma} \varepsilon_n (x, x'_n) x_n \right\|^2.$$

The possibility of such a choice is contained in the elementary inequality $\|x\|^2 + \|y\|^2 \leq \|x + \varepsilon y\|^2$ for $\varepsilon = \operatorname{sign}(x, y) \overset{\text{def}}{=} (x, y)/|(x, y)|$ (with $\varepsilon = 1$ if $(x, y) = 0$). By induction (induction over the number of vectors) this inequality yields the following lemma.

**Lemma of Orlicz.** *If $H$ is a Hilbert space and $\{x_n\}$ a finite family of vectors in $H$ then*

$$\sum_n \|x_n\|^2 \leq \max\left\{ \left\| \sum_n \varepsilon_n x_n \right\|^2 : |\varepsilon_n| = 1, \forall n \right\}. \square$$

Thus $\sum_{n \in \sigma} \|(x, x'_n) x_n\|^2 \leq \text{const}$, $\sigma \in \mathscr{K}$ and therefore $\sum_n |(x, x'_n)|^2 < \infty$, $x \in H$. Because of the symmetry in $x$ and $x'$ in inequality (10) we also get $\sum_n |(x', x_n)|^2 < \infty$, $x' \in H$. Bari's Theorem shows that $\mathfrak{X}$ is a Riesz basis. $\square$

**Corollary.** *Condition* (7) *along with the hypothesis of a simple spectrum is a necessary and sufficient condition for a complete truncated shift to be similar to a normal operator.*

In fact, if $T$ is similar to a normal operator then there exists an isomorphism $V$ such that $V\mathscr{E}(\sigma)V^{-1}$ for any $\sigma$ is an orthogonal projection. This implies (7). The converse assertion follows from the definition of a Riesz basis and the Köthe-Toeplitz Theorem combined with the discussion preceding that theorem. $\square$

## Supplements and Bibliographical Notes

**4. Bases of Subspaces.** In Lectures IX–X we need analogues of the bases theorems in a somewhat more general situation than in Sects. 1–3. In the present Section we therefore give the corresponding statements and in the following one we illustrate them in the case of the multidimensional functional model.

Let $X$ be a Banach space and $\mathfrak{X} = \{X_n : n \in \mathfrak{N}\}$ a family of subspaces of $X$ indexed by the countable set $\mathfrak{N}$. Such a family is called *topologically free* (or *minimal*) if for any $n \in \mathfrak{N}$ there exists a continuous linear operator $\mathscr{E}_n$ such that $\mathscr{E}_n | X_k = \mathbb{O}$, $n \neq k$, and $\mathscr{E}_n | X_n = I$. It is clear that here the operators $\mathscr{E}_k$ are pairwise disjoint projections: $\mathscr{E}_n \mathscr{E}_k = \delta_{nk} \mathscr{E}_n$ ($n$, $k \in \mathfrak{N}$). For convenience we will *assume throughout this Section that the family is total in $X$*, i.e. that $X = \bigvee(\mathfrak{X})$. It is then in particular easy to find a family $\mathfrak{X}'$ which is biorthogonal to $\mathfrak{X}$: $X'_n \overset{\text{def}}{=} \mathscr{E}_n^* X^*$, $n \in \mathfrak{N}$, and $\mathfrak{X}' = \{X'_n : n \in \mathfrak{N}\}$. Here $X^*$ is the dual of the space $X$ and $\mathscr{E}_n^*$ the adjoint of the operator $\mathscr{E}_n$. Having future applications in mind we will refer to $\mathscr{E}_n$ or $\mathscr{E}_n^*$ (as well as to sums of these projections, such as $\mathscr{E}_\sigma \overset{\text{def}}{=} \sum_{n\in\sigma} \mathscr{E}_n$, corresponding to subsets $\sigma$ of the set $\mathfrak{N}$) as *spectral projections*. The definition of an unconditional basis in this situation mimics the one in Sect. 3: the family $\mathscr{X}$ is called an *unconditional basis* if

$$x = \lim_{\sigma\in\mathscr{K}} \mathscr{E}_\sigma x, \quad x\in X,$$

where $\mathscr{K}$ is the directed set of all finite subsets of $\mathfrak{N}$.

Let $\mathscr{X} = \{X_n : n\in\mathfrak{N}\}$ be a given family of subspaces of $X$. A (numerical) function $\mu$ defined on $\mathfrak{N}$ will be called a *multiplier* for the family $\mathscr{X}$ if there exists a continuous linear map $M$ from $\bigvee(\mathscr{X})$ into itself such that $M|X_n = \mu_n I$, $n\in\mathfrak{N}$. *The family of all multipliers* will be denoted by $\operatorname{mult}(\mathscr{X})$. It is clear that the requirement that $\mathscr{X}$ should be minimal is equivalent to the inclusion $\{\chi_\sigma : \sigma\in\mathscr{F}\} \subset \operatorname{mult}(\mathscr{X})$ and that the operator $M_{\chi_\sigma}$ corresponding to $\chi_\sigma$ is $\mathscr{E}_\sigma$. In the following theorem when writing $|\mathscr{E}_\sigma| < \infty$ we mean that $\chi_\sigma\in\operatorname{mult}(\mathscr{X})$.

**Theorem on Unconditional Bases.** *Let $X$ be a Banach space and $\mathscr{X} = \{X_n : n\in\mathfrak{N}\}$ a family of subspaces, with $\bigvee(\mathscr{X}) = X$. The following statements are equivalent:*

1. *$\mathscr{X}$ is an unconditional basis.*
2. $\sup\{|\mathscr{E}_\sigma| : \sigma\in\mathscr{K}\} < \infty$.
3. $\operatorname{mult}(\mathscr{X}) = l^\infty(\mathfrak{N})$.
4. $|\mathscr{E}_\sigma| < \infty, \forall\sigma\subset\mathfrak{N}$.

*Proof*. That 1 and 2 are equivalent results from the Banach-Steinhaus theorem.

$2\Rightarrow 3$. If $\sigma\subset\mathfrak{N}$ and $\{\sigma_n\}_{n\geq 1}$ is an increasing sequence of finite subsets with $\sigma = \bigcup_{n\geq 1}\sigma_n$ then $\mathscr{E}_\sigma x = \lim_n \mathscr{E}_{\sigma_n} x$ for every $x$, $x\in X$ (apply once more the Banach-Steinhaus theorem), that is, $\chi_\sigma\in\operatorname{mult}(\mathscr{X})$ and $|\mathscr{E}_\sigma| \leq c$, where $c = \sup\{|\mathscr{E}_\sigma| : \sigma\in\mathscr{K}\}$. Since $\operatorname{mult}(\mathscr{X})$ is a vector space also all step functions (i.e. linear combinations of characteristic functions) belong to $\operatorname{mult}(\mathscr{X})$. Moreover to any such step function $\mu = \sum_{k=1}^n a_k\chi_{\sigma_k}$ there corresponds the operator $M = \sum_{k=1}^n a_k\mathscr{E}_{\sigma_k}$ and the map $\mu\mapsto M$ is linear. The latter is also continuous. For if we assume that the sets $\sigma_k$ are disjoint and the $a_k$ real, taking the latter in increasing order $a_1 < a_2 < \cdots < a_n$, we can write

$$\mu = a_1\chi_{\sigma_1'} + \sum_{k=2}^n (a_k - a_{k-1})\chi_{\sigma_k'},$$

with $\sigma_k' = \bigcup_{i=k}^n \sigma_i$, $1\leq k\leq n$, whence

$$\begin{aligned}|M| &= |a_1\mathscr{E}_{\sigma_1'} + \sum_{k=2}^n (a_k - a_{k-1})\mathscr{E}_{\sigma_k'}| \\ &\leq c(|a_1| + a_n - a_1) \leq 3c\,\|\mu\|_{l^\infty}.\end{aligned}$$

In the general case (complex coefficients) we get $|M| \leq 6c\cdot\|\mu\|_{l^\infty}$. It is now evident that the map $\mu\mapsto M$ extends by continuity to a linear map from $l^\infty(\mathfrak{N})$

into $M$. Also $\mathscr{E}_n Mx = \mu_n \mathscr{E}_n x$, $n \in \mathfrak{N}$, for every $x$ ($x \in X$) and $\mu$ ($\mu \in l^\infty(\mathfrak{N})$). Thus $l^\infty(\mathfrak{N}) \subset \operatorname{mult}(\mathscr{X})$.

The converse assertion is immediate, because $|\mu_n| = \|Mx\| \cdot \|x\|^{-1} \le |M|$ for $\mu \in \operatorname{mult}(\mathscr{X})$, $n \in \mathfrak{N}$, $x \in X_n$.

$3 \Rightarrow 4$. The map $\operatorname{mult}(\mathscr{X}) \to L(X)$ arising from the definition of a multiplier is linear and closed (check it!). If now $\operatorname{mult}(\mathscr{X}) = l^\infty(\mathfrak{N})$ it must also be continuous, whence

$$|\mathscr{E}_\sigma| \le \text{const} \, \|\chi_\sigma\|_{l^\infty} = \text{const}, \quad \sigma \subset \mathfrak{N}.$$

$4 \Rightarrow 2$ by the following Banach-Steinhaus-like arguments. If Statement 2 were not true we could easily verify that for every $\sigma' \in \mathscr{K}$

$$\sup\{|\mathscr{E}_\sigma| : \sigma \in \mathscr{K}, \sigma \cap \sigma' = \varnothing\} = \infty,$$

$$\sup\{|\mathscr{E}_\sigma|(I - \mathscr{E}_{\sigma'})X| : \sigma \in \mathscr{K}\} = \infty.$$

Hence,

$$\sup\{|\mathscr{E}_\sigma|(I - \mathscr{E}_{\sigma'})X| : \sigma \in \mathscr{K}, \sigma \cap \sigma' = \varnothing\} = \infty.$$

Further, the norm of any operator in $X$ is "almost attained" on elements of a subspace $\mathscr{E}_\tau X$ provided $\tau$ ($\tau \in \mathscr{K}$) is sufficiently large. Thus, it is possible to choose (by an induction argument) a sequence of finite sets $\sigma_1, \tau_1, \sigma_2, \tau_2, \dots$ and a sequence of vectors $x_k$ such that:

$$\sigma_k \subset \tau_k; \quad \tau_i \cap \tau_j = \varnothing \quad \text{for } i \ne j;$$

$$x_k \in \mathscr{E}_{\tau_k} X, \quad \|x_k\| = 1, \quad \|\mathscr{E}_{\sigma_k} x_k\| \ge k.$$

Let now $\sigma = \bigcup_{k \ge 1} \sigma_k$. Then

$$\|\mathscr{E}_\sigma x_k\| = \|\mathscr{E}_\sigma \mathscr{E}_{\tau_k} x_k\| = \|\mathscr{E}_{\sigma_k} x_k\| \ge k$$

for every $k$, $k \ge 1$. This contradicts Statement 4. □

*Remark*. If one (and therefore all) of the statements of the theorem is (are) fulfilled for the family $\{X_n : n \in \mathfrak{N}\}$ then the corresponding biorthogonal family $\{X_n' : n \in \mathfrak{N}\}$ constitutes an unconditional basis in the hull $\bigvee(X_n' : n \in \mathfrak{N})$; this hull coincides with $X^*$ if (e.g.) $X$ is reflexive.

A most convenient language for the applications is provided by the operator

$$\mathscr{J}_{\mathscr{X}} x = \{\mathscr{E}_n x : n \in \mathfrak{N}\},$$

and the analogous operator $\mathscr{J}_{\mathscr{X}'}$. If $X$ is a Hilbert space we likewise consider the operator

$$J_{\mathscr{X}} x = \{P_n x : n \in \mathfrak{N}\},$$

where $P_n$ denotes the orthogonal projection onto the subspace $X_n$, $n \in \mathfrak{N}$, as well as the *Gram operator* $\Gamma_{\mathscr{X}}$,

$$\Gamma_{\mathscr{X}} x = \left\{\sum_{k \in \mathfrak{N}} P_n x_k : n \in \mathfrak{N}\right\},$$

defined for finite vectors in the space

$$l^2(\{X_n\}) \stackrel{\text{def}}{=} \left\{x : x = \{x_k\}_{k \in \mathfrak{N}}, \, x_k \in X_k, \, \sum_{k \in \mathfrak{N}} \|x_k\|^2 < \infty \right\};$$

$l^2(\{X_n\})$ will be considered as a subspace of $l^2(X)$. For instance, one sees at once that if $\dim X_n = 1, n \in \mathfrak{N}$ (i.e. the case of a family of vectors, not of subspaces) then the relation $\operatorname{mult}(\mathscr{X}) = l^\infty(\mathfrak{N})$ is equivalent to the requirement of $\mathscr{J}_{\mathscr{X}} X$ to be an ideal space (Banach lattice), i.e.

$$x_n \in X_n, \quad \|x_n\| \le \|\mathscr{E}_n x\|, \quad n \in \mathfrak{N} \Rightarrow \exists y (y \in X) : \mathscr{E}_n y = x_n, \quad n \in \mathfrak{N}.$$

If $X$ is a Hilbert space then in the general case (as well as in the one dimensional case $\dim X_n \equiv 1$, examined in Sects. 1–3) it is useful to have the notion of a *Riesz basis*: this is a total (in $X$) family of subspaces $\{X_n : n \in \mathfrak{N}\}$ which by a suitable invertible operator $V$ (acting from $X$ onto $X$ or from $X$ onto another Hilbert space) is transformed into an orthogonal family: $VX_n \perp VX_k$ for $n \ne k$; $V$ is the *orthogonalizer* of $\{X_n\}$.

**Theorem on Riesz Bases.** *Let $X$ be a Hilbert space and $\mathscr{X} = \{X_n : n \in \mathfrak{N}\}$ a family of subspaces, topologically free and total in $X$, with a total family of spectral projections $\{\mathscr{E}_n : n \in \mathfrak{N}\}$.*

*The following statements are equivalent:*

1. *$\mathscr{X}$ is an unconditional basis.*
2. *$\mathscr{J}_{\mathscr{X}} X \subset l^2(X)$, $\mathscr{J}_{\mathscr{X}'} X \subset l^2(X)$.*
3. *$\mathscr{J}_{\mathscr{X}} X = l^2(\{X_n\})$.*
4. *$\mathscr{X}$ is a Riesz basis.*
5. *$\mathscr{J}_{\mathscr{X}} X \subset l^2(X)$, $J_{\mathscr{X}} X \subset l^2(X)$.*
6. *$\Gamma_{\mathscr{X}}$ extends to a continuous and invertible transformation from the space $l^2(\{X_n\})$ onto itself.*

*The proof* has several general features in common with the arguments of Sects. 1–3. Indeed, the implication $1 \Rightarrow 2$ follows from the Theorem on Unconditional Bases and from the Lemma of Orlicz.

$2 \Rightarrow 3$. By the closed graph theorem the imbeddings $\mathscr{J}_{\mathscr{X}}$ and $\mathscr{J}_{\mathscr{X}'}$ must be continuous. Further for finite families $\mathscr{J}_{\mathscr{X}} x$ (and therefore also for arbitrary ones) holds:

$$\begin{aligned} |(x, x')| &= \Big|\sum_{n \in \mathfrak{N}} (\mathscr{E}_n x, \mathscr{E}_n^* x')\Big| \\ &\le \sum_{n \in \mathfrak{N}} \|\mathscr{E}_n x\| \cdot \|\mathscr{E}_n^* x'\| \\ &\le \|\mathscr{J}_{\mathscr{X}} x\|_{l^2(X)} \|\mathscr{J}_{\mathscr{X}'} x'\|_{l^2(X)} \\ &\le |\mathscr{J}_{\mathscr{X}'}| \cdot \|\mathscr{J}_{\mathscr{X}} x\|_{l^2(X)} \|x'\| \end{aligned}$$

and therefore $\|x\| \le |\mathscr{J}_{\mathscr{X}'}| \cdot \|\mathscr{J}_{\mathscr{X}} x\|_{l^2(X)}$ so that $\mathscr{J}_{\mathscr{X}}$ is an isomorphism onto its image. It is however clear that $\mathscr{J}_{\mathscr{X}} X \subset l^2(\{X_n\})$ and that $\mathscr{J}_{\mathscr{X}} X$ contains the dense set of all finite families in $l^2(\{X_n\})$. Therefore $\mathscr{J}_{\mathscr{X}} X = l^2(\{X_n\})$.

$3 \Rightarrow 4$. In the hypothesis of Statement 3, the operator $\mathcal{J}_{\mathfrak{X}}$ gives an isomorphism between the spaces $X$ and $l^2(\{X_n\})$ (apply the closed graph theorem) mapping the family $\mathfrak{X}$ onto the orthogonal family $\mathcal{J}_{\mathfrak{X}}\mathfrak{X}$ (in $l^2(\{X_n\})$).

$4 \Rightarrow 5$. If $V$ denotes the orthogonalizer of $\mathfrak{X}$ then $\{V\mathscr{E}_n V^{-1} : n \in \mathfrak{N}\}$ is a disjoint family of orthogonal projections and therefore

$$\sum_{n\in\mathfrak{N}} \|\mathscr{E}_n x\|^2 \leq |V^{-1}|^2 \sum_{n\in\mathfrak{N}} \|V\mathscr{E}_n V^{-1} V x\|^2 = |V^{-1}|^2 \cdot \|Vx\|^2$$

so that $\mathcal{J}_{\mathfrak{X}} X \subset l^2(\{X_n\})$. In fact, $\mathcal{J}_{\mathfrak{X}}$ is even an isomorphism, so that the continuity of $J_X$ follows from the following obvious chain of equalities

$$(J_{\mathfrak{X}} x, \mathcal{J}_{\mathfrak{X}} y) = \sum_{n\in\mathfrak{N}} (P_n x, \mathscr{E}_n y) = \sum_{n\in\mathfrak{N}} (x, \mathscr{E}_n y) = (x, y), \tag{11}$$

where we can take the family $\mathcal{J}_{\mathfrak{X}} y$ to be finite.

$5 \Rightarrow 6$. The formulae (11) (on the dense set of vectors $x$ and $y$ with $J_{\mathfrak{X}} x$ and $\mathcal{J}_{\mathfrak{X}} y$ finite) show that the inclusions in Statement 5 (which automatically are continuous by the closed graph theorem) imply

$$\mathcal{J}_{\mathfrak{X}}^* J_{\mathfrak{X}} = I_X, \tag{12}$$

as well as

$$J_{\mathfrak{X}} \mathcal{J}_{\mathfrak{X}}^* = I_{l^2(\{X_n\})} \tag{13}$$

(the operators $\mathcal{J}_{\mathfrak{X}}^*$ and $J_{\mathfrak{X}}^*$ are first defined as the formal adjoints of $\mathcal{J}_{\mathfrak{X}}$ and $J_{\mathfrak{X}}$ respectively on the dense subset of finite families in $l^2(\{X_n\})$ so that $\mathcal{J}_{\mathfrak{X}}^*\{x_n : n \in \mathfrak{N}\} = \sum_{n\in\mathfrak{N}} \mathscr{E}_n^* n_n$, $J_{\mathfrak{X}}^*\{y_n : n \in \mathfrak{N}\} = \sum_{n\in\mathfrak{N}} y_n$ and $(\mathcal{J}_{\mathfrak{X}}^* x, J_{\mathfrak{X}}^* y) = \sum_{n\in\mathfrak{N}} (\mathscr{E}_n^* x_n, y_n) = (x, y)$). Therefore $J_{\mathfrak{X}}$ is an isomorphism; the identity

$$(J_{\mathfrak{X}}^* x, J_{\mathfrak{X}}^* y) = \sum_{n,k\in\mathfrak{N}} (x_n, y_k) = (\Gamma_{\mathfrak{X}} x, y) \tag{14}$$

shows now that the Gram operator $\Gamma_{\mathfrak{X}}$ too is an isomorphism from the space $l^2(\{X_n\})$ onto itself.

$6 \Rightarrow 1$ by the following argument: (14) and (12), (13), applied in turn, provide an isomorphic mapping (with the aid of the transformation $\mathcal{J}_{\mathfrak{X}}$) of the family $\mathfrak{X}$ onto the natural coordinate basis of the space $l^2(\{X_n\})$. Since the latter basis is unconditional the same must be true for the preimage $\mathfrak{X}$. □

*Remark.* From the proof of this theorem it is not hard to derive also other combinations of properties of the operators $\mathcal{J}_{\mathfrak{X}}$ and $\Gamma_{\mathfrak{X}}$ etc. which likewise are equivalent to the requirement of $\mathfrak{X}$ to be an unconditional basis. Here are some of them:

*the Gram operators $\Gamma_{\mathfrak{X}}$ and $\Gamma_{\mathfrak{X}'}$ are continuous on the spaces $l^2(\{X_n\})$ and $l^2(\{X'_n\})$ respectively;*

*the norms $\|\cdot\|_X$ and $\|x\|_1 = (\sum_{n\in\mathfrak{N}} \|\mathscr{E}_n x\|^2)^{1/2}$ are equivalent:*

*the norms $\|\cdot\|_X$ and $\|x\|_2 = (\sum_{n\in\mathfrak{N}} \|P_n x\|^2)^{1/2}$ are equivalent;*

*the operator $U$, $Ux \stackrel{\text{def}}{=} \sum_{n\in\mathfrak{N}} P_n x$, extends from the dense subset of vectors $x$, for which the series in question converges, to an isomorphism of the space $X$ onto itself;*

$$J_{\mathscr{X}}X = l^2(\{X_n\}); \tag{15}$$

*every family of vectors* $\{x_n : n \in \mathfrak{N}\}$, $x_n \in X_n$ $(n \in \mathfrak{N})$ *is an unconditional basis in its hull* $\bigvee(x_n : n \in \mathfrak{N})$;

$\mathscr{J}_{\mathscr{X}}X$ *has the ideal property, i.e. from* $x \in X$ *and* $y_n \in X_n$, $\|y_n\| \le \|\mathscr{E}_n x\|$ $(n \in \mathfrak{N})$ *it follows that* $\{y_n : n \in \mathfrak{N}\} \in \mathscr{J}_{\mathscr{X}}X$.

It turns out that for the applications to the model operator (as in the cases examined in Sects. 1–3) the Statements 2 and 5 of the theorem are especially useful (imbedding of the spaces of "coefficients" into an $l^2$-space) and so is Statement 3 along with formula (15), which all as before have an interpolation character. Also the ideal property of the space $\mathscr{J}_{\mathscr{X}}X$ and the identity $\operatorname{mult}(\mathscr{X}) = l^\infty(\mathfrak{N})$ are connected with interpolation, but a different kind of interpolation, namely operator interpolation. In fact, for unconditional bases in Hilbert space there holds an even stronger result than the previous identity. Namely we have the following.

**Corollary.** *Let* $X$ *be a Hilbert space and* $\mathscr{X} = \{X_n : n \in \mathfrak{N}\}$ *an unconditional basis in* $X$. *Then any bounded collection of operators* $a = \{a_n : n \in \mathfrak{N}\}$ *acting in the subspaces of the family* $\mathscr{X}$ $(a_n \in L(X_n)$, $\sup_{n\in\mathfrak{N}}|a_n| < \infty)$, *has a continuous continuation to the entire space* $X$: *there exists an operator* $A$, $A \in L(X)$ *such that* $AX_n \subset X_n$, $A|X_n = a_n$ $(n \in \mathfrak{N})$. (The set of all such operators will be denoted by $\operatorname{Mult}(\mathscr{X})$.)

For the proof we set

$$Ax = \sum_{n\in\mathfrak{N}} a_n \mathscr{E}_n x, \quad x \in X.$$

By the Theorem on Riesz Bases the convergence of this series is determined only by the condition $\sum_{n\in\mathfrak{N}} \|a_n \mathscr{E}_n x\|^2 < \infty$, which evidently is fulfilled (by the same theorem) for any $x$, $x \in X$. The continuity and other properties needed of the operator $A$ too are evident. □

Lectures VIII–X will be devoted to the study of the operator and interpolation problems connected with this corollary.

**5. Bases of Eigenspaces.** Consider a contraction $T$ in a Hilbert space $H$ and assume that

a) every point of the set $\sigma(T) \cap \mathbb{D}$ is an isolated (in $\sigma(T)$) eigenvalue for $T$;

b) $H = \bigvee(\operatorname{Ker}(T - \lambda I) : \lambda \in \sigma(T) \cap \mathbb{D})$.

In these hypotheses each point $\lambda$ of the set $\sigma(T) \cap \mathbb{D}$ is a *pole of order one for the resolvent* $R(\cdot, T)$. In fact, if $\mathscr{E}_n$ is the Riesz spectral projection corresponding to the singleton $\{\lambda\}$ then

$$H = \mathscr{E}_\lambda H + (I - \mathscr{E}_\lambda)H \tag{16}$$

and it is clear that $E_\lambda \overset{\text{def}}{=} \operatorname{Ker}(T - \mu I) \subset (I - \mathscr{E}_\lambda)H$ for $\mu \ne \lambda$ (cf. Lecture IV), whence $\mathscr{E}_\lambda H = \operatorname{Ker}(T - \lambda I)$ so that $R(\zeta, T) = (\zeta - \lambda)^{-1}\mathscr{E}_\lambda + r$ where $r = \mathbb{O}$ on $\mathscr{E}_\lambda H$ and $r = R(\zeta, T)$ on $(I - \mathscr{E}_\lambda)H$. In particular, it follows that the adjoint operator $T^*$ too satisfies a condition analogous to a) and that $(\lambda I - T)^*H$ is a closed set so that

$$E_\lambda^\perp = (\lambda I - T)^* H = (I - \mathscr{E}_\lambda^*)H.$$

**Corollary 1.** *The operator* $A \overset{\text{def}}{=} P_{E_\lambda^\perp}(\lambda I - T)|E_\lambda^\perp$, *standing on the diagonal of the matrix for the operator* $\lambda I - T$ *with respect to the decomposition* $H = E_\lambda \oplus E_\lambda^\perp$, *that is,*

$$(\lambda I - T) \cong \begin{pmatrix} (\lambda I - T)|E_\lambda & * \\ 0 & A \end{pmatrix},$$

*is invertible.*

In fact, the adjoint operator $A^* = (\lambda I - T)^*|E_\lambda^\perp$ has this property. □

Let us now pass to the functional model unitarily equivalent to the operator $T$. We can of course assume that it coincides with $T$ so that

$$T = PS_E|K, \quad K = H^2(E) \ominus \boldsymbol{\Theta} H^2(E),$$

where $P = P_K$ and $\Theta$ is an *inner* $L(E)$-valued function. The eigenspaces of the operators $T^*$ and $T$ were found in Lecture III, Sect. 7:

$$\begin{aligned} E_\lambda &= (\mathbb{1} - \bar{\lambda} z)^{-1}\, {}_\lambda\Delta E, \\ E'_\lambda &= {}_\lambda\Theta(\mathbb{1} - \bar{\lambda} z)^{-1} \Delta_\lambda E, \end{aligned} \tag{17}$$

where ${}_\lambda\Delta$ is orthogonal projection onto the subspace $\operatorname{Ker} \Theta(\lambda)^*$ and $\Delta_\lambda$ orthogonal projection onto the subspace $\operatorname{Ker} \Theta(\lambda)$, $\lambda \in \sigma_p(T)$. There we also wrote down the factorizations of the function corresponding to these subspaces by the Lax-Halmos Theorem:

$$\Theta = \{{}_\lambda\Delta b_\lambda + (I - {}_\lambda\Delta)\}\Theta_\lambda, \tag{18}$$

$$\Theta = {}_\lambda\Theta\{\Delta_\lambda b_\lambda + (I - \Delta_\lambda)\}, \tag{19}$$

where ${}_\lambda\Theta$ and $\Theta_\lambda$ are inner operator functions.

**Corollary 2.** $\Theta_\lambda(\lambda)$ *and* ${}_\lambda\Theta(\lambda)$ *are invertible operators.*

In fact,

$$K = \{H^2(E) \ominus \boldsymbol{\vartheta}_\lambda H^2(E)\} \oplus \{\boldsymbol{\vartheta}_\lambda H^2(E) \ominus \boldsymbol{\Theta} H^2(E)\}$$

where

$$\vartheta_\lambda = {}_\lambda\Delta b_\lambda + (I - {}_\lambda\Delta), \quad H^2(E) \ominus \boldsymbol{\vartheta}_\lambda H^2(E) = \operatorname{Ker}(\lambda I - T)^* = E_\lambda,$$

and by Corollary 1

$$\bar{\lambda} \notin \sigma(P_{E_\lambda^\perp} T^* | E_\lambda^\perp).$$

It is however clear that the operator

$$P_{E_\lambda^\perp} T | E_\lambda^\perp = P_{E_\lambda^\perp} S_E | E_\lambda^\perp$$

is unitarily equivalent to the operator $T_\lambda \overset{\text{def}}{=} P_\lambda S_E | K_\lambda$ where $K_\lambda = H^2(E) \ominus$

$\Theta_\lambda H^2(E)$ (regarding this see Lecture IV, Sect. 4). As $\lambda \notin \sigma(T_\lambda)$ we see that $\Theta_\lambda(\lambda)$ is an invertible operator in $E$ (Lecture III, Sect. 7). In an analogous way one verifies that ${}_\lambda\Theta(\lambda)$ is invertible. □

Now we make the last preparations for the bases theorem for eigenspaces of the model operator $T$ of the type under consideration, *assuming that the adjoint $T^*$ is complete too* (which is equivalent to the requirement of totalness for the family of projections $\{\mathscr{E}_\lambda : \lambda \in \sigma_p(T)\}$ used in the general theorems of the preceding Section).

**Lemma on Eigenprojections.** *Let $P_\lambda$ be the orthogonal projection and $\mathscr{E}_\lambda$ the spectral projection onto the eigenspace $E_\lambda$ of the operator $T^* = S_E^*|K$. Then*

$$P_\lambda f = \frac{1-|\lambda|^2}{\mathbb{1} - \bar{\lambda} z}\, {}_\lambda\Delta f(\lambda), \quad f \in K, \tag{20}$$

$$\mathscr{E}_\lambda f = \frac{|\lambda|^2 - 1}{|\lambda|(\mathbb{1} - \bar{\lambda} z)} ({}_\lambda\Theta(\lambda)^*)^{-1} \Delta_\lambda \{(\Theta^* f)(1/\bar{\lambda})\}, \quad f \in K, \tag{21}$$

*where $(\Theta^* f)(1/\bar{\lambda})$ denotes the value at the point $1/\bar{\lambda}$ of the natural continuation to the exterior of $\mathbb{D}$ of the function $\Theta^* f$ $(\Theta^* f \in H_-^2(E) \overset{\text{def}}{=} L^2(E) \ominus H^2(E))$.*

*Proof.* It is plain that $P_\lambda f = f$ for functions $f$ in the subspace (17) and that $P_\lambda f \in E_\lambda$, $(f, P_\lambda f) = (1 - |\lambda|^2) \| {}_\lambda\Delta f(\lambda) \|^2 \geq 0$ for any function $f$. Therefore $P_\lambda = P_{E_\lambda}$.

The spectral projection $\mathscr{E}_\lambda^T$ onto the eigenspace $\operatorname{Ker}(T - \lambda I)$ of the operator $T$, $T = P_K S_E | K$, can be found by Riesz's formula (cf. Lecture IV, Sect. 2), integrating the resolvent $R(\zeta, T)$ around a suitable contour $\gamma$ containing exactly one point of the spectrum (namely the point $\lambda$):

$$\mathscr{E}_\lambda^T = \frac{1}{2\pi i} \int_\gamma R(\zeta, T)\, d\zeta.$$

It is however clear that this integral equals the residue $\operatorname{res}_\lambda R(\cdot, T)$ of the resolvent at the point $\lambda$; more precisely, if

$$R(\zeta, T) = \frac{1}{\zeta - \lambda} \operatorname{res}_\lambda R(\cdot, T) + \cdots,$$

the dots indicating a regular function of the spectral parameter $\zeta$, then $\mathscr{E}_\lambda^T = \operatorname{res}_\lambda R(\cdot, T)$. Use now the explicit formula (9) from Lecture III for the resolvent $R(\cdot, T)$; this yields:

$$\mathscr{E}_\lambda^T = \operatorname{res}_\lambda P_K \mathbf{u} \Theta^{-1}, \quad \text{where } u = \frac{\Theta - \Theta(\lambda)}{z - \lambda \mathbb{1}}.$$

Making use of the fact that $u$ is a regular function we get from formula (19) and Corollary 2

$$\mathscr{E}_\lambda^T = P_K \operatorname{res}_\lambda \mathbf{u} \Theta^{-1} = -(1 - |\lambda|^2) P_K \mathbf{u}(\lambda) \Delta_\lambda \cdot \frac{\lambda}{|\lambda|}\, {}_\lambda\Theta(\lambda)^{-1}$$

(since $\Theta(\lambda)\Delta_\lambda = \mathbb{0}$)

$$= -\frac{\lambda}{|\lambda|}(1 - |\lambda|^2)P_K\frac{\Theta}{z - \lambda\cdot\mathbb{1}}\Delta_\lambda\cdot{}_\lambda\Theta(\lambda)^{-1}.$$

It follows that

$$\mathscr{E}_\lambda = (\mathscr{E}_\lambda^T)^* = -\frac{|\lambda|}{\lambda}(1 - |\lambda|^2)P_K{}_\lambda\Theta(\lambda)^{*-1}\Delta_\lambda\frac{\Theta^*}{\bar{z} - \bar{\lambda}\mathbb{1}}. \tag{22}$$

In order to transform this formula into the expression (21) we first notice that from the inclusion $\Theta^* f \in H^2_-(E)(f \in K)$ (Lemma on Lat $S_E^*$, Lecture II) it follows that

$$P_+\frac{\Theta^*}{\bar{z} - \bar{\lambda}\mathbb{1}}f = P_+\frac{z}{\mathbb{1} - \bar{\lambda}z}\Theta^* f = \frac{1}{\bar{\lambda}}(\Theta^* f)\left(\frac{1}{\bar{\lambda}}\right)(\mathbb{1} - \bar{\lambda}z)^{-1}.$$

If we apply to this function the operator ${}_\lambda\Theta(\lambda)^{*-1}.\ \Delta_\lambda$ then we are already in the space $K$, because

$${}_\lambda\Theta(\lambda)^*{}_\lambda\Delta = \Delta_\lambda\Theta_\lambda(\lambda)^* \tag{23}$$

and functions of the form $(\mathbb{1} - \bar{\lambda}z)^{-1}e$, $e \in {}_\lambda\Delta E$, belong to $K$. From this and from (22) we get (21).

In order to prove the identity (23) let us multiply (18) and (19) by ${}_\lambda\Delta$ and $\Delta_\lambda$:

$${}_\lambda\Delta_\lambda\Theta(\zeta)b_\lambda(\zeta)\Delta_\lambda = {}_\lambda\Delta\Theta(\zeta)\Delta_\lambda = {}_\lambda\Delta b_\lambda(\zeta)\Theta_\lambda(\zeta)\Delta_\lambda$$

for all $\zeta, \zeta \in \mathbb{D}$. Dividing this identity by $b_\lambda(\zeta)$ and setting $\zeta = \lambda$ we get ${}_\lambda\Delta_\lambda\Theta(\lambda)\Delta_\lambda = {}_\lambda\Delta\Theta_\lambda(\lambda)\Delta_\lambda$. Moreover $\Theta(\lambda)\Delta_\lambda = \mathbb{0}$, ${}_\lambda\Delta\Theta(\lambda) = \mathbb{0}$ so that

$$(I - {}_\lambda\Delta)\Theta_\lambda(\lambda)\Delta_\lambda = \mathbb{0}, \quad {}_\lambda\Delta_\lambda\Theta(\lambda)(I - \Delta_\lambda) = \mathbb{0}.$$

Therefore finally ${}_\lambda\Delta_\lambda\Theta(\lambda) = \Theta_\lambda(\lambda)\Delta_\lambda$ or, passing to the adjoint operators, ${}_\lambda\Theta(\lambda)^*{}_\lambda\Delta = \Delta_\lambda\Theta_\lambda(\lambda)^*$. □

*Remark.* It is not hard to see that in place of the factor ${}_\lambda\Theta(\lambda)^{*-1}\Delta_\lambda$ in formula (21) one can also use ${}_\lambda\Delta\Theta_\lambda(\lambda)^{*-1}$ or (in symmetric form) ${}_\lambda\Delta_\lambda\Theta(\lambda)^{*-1}\Delta_\lambda$, ${}_\lambda\Delta\Theta_\lambda(\lambda)^{*-1}\Delta_\lambda$.

**Theorem on Riesz Bases, II.** *Let $T = \mathbf{T}_\Theta$ be a model operator in the space $K_\Theta = H^2(E) \ominus \Theta H^2(E)$ which together with $T^*$ satisfies the above conditions* a)–b). *Let $\mathfrak{X} = \{E_\lambda : \lambda \in \sigma_p(T)\}$ be the family of eigenspaces of the operator $T^*$ and let ${}_\lambda\Delta$, $\Delta_\lambda$, ${}_\lambda\Theta$ have the same meaning as in the formulae* (18) *and* (19). *The following statements are equivalent:*

1. *$\mathfrak{X}$ is a Riesz basis for $K_\Theta$.*
2. $$\sum_{\lambda \in \sigma_p(T)} (1 - |\lambda|^2)\|{}_\lambda\Delta f(\lambda)\|_E^2 < \infty, \quad \forall f \in H^2(E),$$
$$\sum_{\lambda \in \sigma_p(T)} (1 - |\lambda|^2)\|\Delta_\lambda f(\bar{\lambda})\|_E^2 < \infty, \quad \forall f \in H^2(E), \tag{24}$$
$$\sup\{|\Delta_\lambda({}_\lambda\Theta(\lambda))^{-1}{}_\lambda\Delta| : \lambda \in \sigma_p(T)\} < \infty.$$

3. $\Delta H^2(E) = l^2(\{_\lambda\Delta E\}_{\lambda\in\sigma_p})$ *where*

$$\Delta f \stackrel{\text{def}}{=} \{(1-|\lambda|^2)^{1/2}{}_\lambda\Delta f(\lambda) : \lambda\in\sigma_p(T)\}.$$

4. *The operator V,*

$$Vf = \sum_{\lambda\in\sigma_p(T)} \frac{1-|\lambda|^2}{\mathbb{1}-\bar{\lambda}z}\,{}_\lambda\Delta f(\lambda),$$

*is continuous and continuously invertible in* $K_\Theta$.

5. *There exist positive constants c and C such that*

$$c\|f\|^2_{H^2(E)} \le \sum_{\lambda\in\sigma_p(T)} (1-|\lambda|^2)\|{}_\lambda\Delta f(\lambda)\|^2_E$$
$$\le C\|f\|^2_{H^2(E)},\ \forall f\in K_\Theta.$$

6. *The Gram matrix*

$$\left\{\frac{(1-|\lambda|^2)^{1/2}(1-|\mu|^2)^{1/2}}{1-\bar{\lambda}\mu}\,{}_\lambda\Delta_\mu\Delta\right\}_{\lambda,\mu\in\sigma_p(T)}$$

*generates a bounded invertible operator in the space* $l^2(\{_\lambda\Delta E\}_{\lambda\in\sigma_p(T)})$.

*Proof.* $1\Leftrightarrow 2$. By the Theorem on Riesz Bases and the Lemma on Eigenprojections it is clear that Statement 1 is equivalent to the two imbeddings:

$$\sum_{\lambda\in\sigma_p(T)} (1-|\lambda|^2)\|{}_\lambda\Delta f(\lambda)\|^2_E < \infty,\quad \forall f\in K_\Theta,$$
$$\sum_{\lambda\in\sigma_p(T)} (1-|\lambda|^2)\left\|({}_\lambda\Theta(\lambda)^*)^{-1}\Delta_\lambda\left\{(\Theta^*f)\left(\frac{1}{\bar{\lambda}}\right)\right\}\right\|^2_E < \infty,\quad \forall f\in K_\Theta. \tag{25}$$

The first of these imbeddings coincides with the first one of Statement 2, because ${}_\lambda\Delta f(\lambda) = \mathbb{0}$ for $f\in\Theta H^2(E)$, and the second one follows from the remaining two conditions of that statement, because $g \stackrel{\text{def}}{=} \Theta^*f\left(\frac{1}{z}\right)\in H^2(E)$ and

$$\|({}_\lambda\Theta(\lambda)^*)^{-1}\Delta_\lambda g(\bar{\lambda})\|_E \le |({}_\lambda\Theta(\lambda)^*)^{-1}\Delta_\lambda|\,\|\Delta_\lambda g(\bar{\lambda})\|_E.$$

Conversely, as $\Delta_\lambda g(1/\bar{\lambda}) = \mathbb{0}$ for every $g$, $g\in\Theta^*H^2_-(E)$, we get from (25)

$$\sum_{\lambda\in\sigma_p(T)} (1-|\lambda|^2)\|({}_\lambda\Theta(\lambda)^*)^{-1}\Delta_\lambda g(1/\bar{\lambda})\|^2_E < \infty$$

for all $g$, $g\in H^2_-(E)$. The left hand side of this inequality can indeed by virtue of the closed graph theorem be estimated by const $\|g\|^2_{H^2_-(E)}$. Passing to the disk $\mathbb{D}$ $\left(g\mapsto g\left(\frac{1}{z}\right)\right)$ and setting $g = (\mathbb{1}-\lambda z)^{-1}e$, $e\in E$, we find $\|({}_\lambda\Theta(\lambda)^*)^{-1}\Delta_\lambda e\|_E \le$ const $\|e\|_E$ from which (24) follows as well as

$$\sum(1-|\lambda|^2)\|\Delta_\lambda g(\bar{\lambda})\|^2_E < +\infty,\quad g\in H^2(E).$$

$1\Leftrightarrow 3$ by virtue of the remark to the Theorem on Riesz Bases (formula (15)) and in view of the formula (20) for the orthogonal projection onto the eigenspace.

$1 \Leftrightarrow 4$, $1 \Leftrightarrow 5$ by virtue of the same remark.

$1 \Leftrightarrow 6$ by the Theorem on Riesz Bases and by formula (20). □

*Remark.* From the proof of the equivalence $1 \Leftrightarrow 2$ we see that

$$|\mathscr{E}_\lambda| = |\Delta_\lambda({}_\lambda\Theta(\lambda))^{-1}| = |\Delta_\lambda({}_\lambda\Theta(\lambda))^{-1}{}_\lambda\Delta|, \quad \lambda \in \sigma_p(T),$$

that is, the condition (24) coincides with the condition of uniform minimality of the family $\mathscr{X}$:

$$\sup\{|\mathscr{E}_\lambda| : \lambda \in \sigma_p(T)\} < +\infty.$$

This remark shows that an operator in the theorem just established, whose spectrum in addition satisfies the Carleson Condition $(C)$, *is linearly similar to a normal operator* (i.e. the family of its eigenspaces constitutes a Riesz basis). In fact, Condition $(C)$ guarantees the uniform minimality of the family (because then by the Theorem on Bases and Interpolation the family $\{(\mathbb{1} - \bar{\lambda}z)^{-1}E : \lambda \in \sigma_p(T)\}$ too is uniformly minimal) as well as both inclusions in Statement 2 of the Theorem on Riesz Bases, II: if $\{e_n\}_{n\geq 1}$ is an orthonormal basis in $E$ then

$$\begin{aligned} f \in H^2(E) \Rightarrow & \sum_{\lambda \in \sigma_p(T)} (1 - |\lambda|^2) \, \|f(\lambda)\|_E^2 \\ & = \sum_{\lambda} (1 - |\lambda|^2) \sum_{n \geq 1} |(f(\lambda), e_n)|^2 \\ & = \sum_{n\geq 1} \sum_{\lambda} (1 - |\lambda|^2) |(f(\lambda), e_n)|^2 \leq \sum_{n \geq 1} \text{const} \, \|(f(\cdot), e_n)\|_{H^2}^2 \\ & = \text{const} \, \|f\|_{H^2(E)}^2. \ \square \end{aligned}$$

In the following Lectures the results of this and the foregoing Section will be subject to a more detailed analysis.

**6. Concluding Remarks.** There is a general principle connecting the solvability of interpolation problems (or – as one says more often – moment problems) with the properties of the biorthogonal expansions

$$x \sim \sum_{n \in \mathfrak{N}} \mathscr{E}_n x \tag{26}$$

associated with a total, topologically free family of subspaces $\mathscr{X}$. The discussion of general such connections will be postponed until Lecture VIII but the theorems proved here already point to one important special case – the unconditional convergence of the expansion (26) being equivalent to the solvability of the problem of "free" interpolation $\mathscr{J}_{\mathscr{X}} x = a$ (or $J_{\mathscr{X}} x = a$), where the datum $a$ runs through a suitable class of functions on the set $\mathfrak{N}$ having the following "ideal" (lattice) property: together with every family $a = \{a_n : n \in \mathfrak{N}\}$ it contains also every family $\{\mu_n a_n : n \in \mathfrak{N}\}$, $\sup_n |\mu_n| < \infty$, and, in the Hilbert space case, also every family of vectors subordinated to the family $a: b_n \in X_n$, $\|b_n\| \leq \|a_n\|$, $n \in \mathfrak{N}$. As a corollary of the last fact we have the isomorphism of all unconditional bases of a Hilbert space (with the same collection of dimensions $\dim X_n$, $n \in \mathfrak{N}$): any two such bases can be mapped onto each other by a suitable isomorphism of the space $X$. In general Banach spaces this property is lost. More exactly, if any two unconditional bases (of one dimensional subspaces) of a space $X$ with an unconditional basis are isomorohic then $X$ itself must be isomorphic to one of the spaces $l^2, l^1, c_0$ (Lindenstrauss-Zippin [1]). The property mentioned remains however in force (Mityagin [1]) for a large class of locally convex spaces – for the so-called "centers of Hilbert scales", among which many useful spaces in mathematical analysis are found (for instance, all the classical spaces of analytic functions obtained by imposing bounds on their growth near the boundary of the domain of definition). In such situations this duality between unconditional summability and free interpolation also can be fruitful for both subjects; see e.g. Dragilev-Zaharyuta-Korobeĭnik [1]. In the Banach space case the ideal property of the "coefficient space" $\mathscr{J}_{\mathfrak{X}}X$ alone is not sufficient for giving a classification up to isomorphism of the corresponding bases, not even in the previous case of the rational fractions $\{(\mathbb{1} - \bar{\lambda}z)^{-1} : \lambda \in \sigma\}$, $\sigma \subset \mathbb{D}$, in the usual spaces of functions analytic in the disk $\mathbb{D}$ considered in Analysis. There is an interesting paper by Vinogradov [1] devoted to the spaces $l^p_A \stackrel{\text{def}}{=} \{f : f = \sum_{n\geq 0} \hat{f}(n)z^n, \sum_{n\geq 0} |\hat{f}(n)|^p < \infty\}$ and their analogues on the halfaxis (and in the halfplane). Bases of subspaces of a Banach space, any two one dimensional sections of which are isomorphic, were considered in Nikol'skiĭ [10].

The Gram operators or matrices, employed in the theorems of this Lecture, have for a long time been in use in geometric problems of Hilbert space as well as in the study of the accuracy of biorthogonal expansions (for instance, with their application in numerical calculations in mind). Referring for these matters to the books Ahieser-Glazman [1] and Mihlin [1] we remark only that a family $\mathfrak{X} = \{x_n : n \in \mathfrak{N}\}$ of vectors in the Hilbert space $H$ is up to a unitary equivalence defined by its Gram matrix $\{(x_n, x_k) : n, k \in \mathfrak{N}\}$ and that in the case of a finite index set $\mathfrak{N}$ the distance $\operatorname{dist}(x_i, \bigvee(x_i : i \in \mathfrak{N}, i \neq n)$ (and thus, for a suitable $n$, also the constant of uniform minimality of $\mathfrak{X}$) can be expressed as a quotient of Gram determinants:

$$\det \Gamma_{\mathfrak{X}} / \det \Gamma_{\mathfrak{X} \setminus \{x_n\}}.$$

We say further a few words on those interpolation problems for analytic functions which appear in the theorems of this Lecture. The simultaneous study of questions of interpolation and of expansion in series is inherent already in the whole classical theory of interpolation by analytic functions. However, the aims of the latter theory, as compared with the theory of "free" interpolation, were initially somewhat different: the classical formulation of the problem consists of reconstructing the function $f$ from its moments $\{(f, x'_n) : n \in \mathfrak{N}\}$ (from the values at the points of a given sets, from the values of its derivatives etc.) whereas the problem of free interpolation is closer to value distribution theory, its object being to find families of linear functionals $\mathfrak{X}' = \{x'_n : n \in \mathfrak{N}\}$ for which the values $J_{\mathfrak{X}} f$ for $f$ in a given class are almost arbitrary (possessing a large degree of freedom, satisfying only a natural requirement of "ideal type"). It is therefore no surprise that the classical theory in general deals with non-minimal families (the problems and series of Newton, Abel-Gončarov etc., cf. for instance Walsh [1], Evgrafov [1]), and if they happen to be minimal then the corresponding biorthogonal families (or even pseudo-biorthogonal, as in the case of the Malmquist-Walsh series (see Walsh [1])) are sometimes prescribed even in advance and are not chosen in the (quotient) space appropriate for the problem. On the contrary, the "free" interpolation problems are suited for a description of minimal families $\mathscr{X}$ to which there correspond Fourier series summable in an in advance given sense (for instance, unconditionally convergent ones). Besides interpolation in the Hardy space $H^2$ there are known many other cases when the problem of free interpolation (i.e. the description of those subsets $\sigma$ of $\mathbb{D}$ for which the restriction of the given space of analytic functions gives rise to an ideal space of functions on $\sigma$) has a complete solution. Some of these will be considered in the "Supplements" of Lecture VIII and other problems of free interpolation (involving values of derivatives) will be solved in Lecture IX–X, but the majority of the results known up to the present moment fall outside the boundaries of this book; regarding this circle of ideas, besides the works already mentioned, we refer to the comprehensive survey by Vinogradov and Havin [1], [2] and the discussion in Lecture VII–X.

Our last general remark concerns the operator theoretic (spectral) meaning of the theorems of this Lecture. It has already been noted that in the hypotheses a)–b) of Sect. 5 they guarantee that the model operator $T$ will be linearly similar to a normal operator. As a matter of fact if we assume that $T \in C_{00}$ (i.e., that $T$ maps onto a functional model with a unitary characteristic function $\Theta = \Theta_T$) and that the resolvent $R(\cdot, T)$ is meromorphic in the disk $\mathbb{D}$ then *these conditions are necessary*. Indeed, if $V$ is an isomorphism of our space mapping $T$ onto a normal operator $VTV^{-1}$ then the latter must also be a contraction (its spectrum being contained in clos $\mathbb{D}$) with the resolvent $VR(\cdot, T)V^{-1}$ meromorphic in $\mathbb{D}$. As moreover

$$(s)\text{-}\lim_n (VTV^{-1})^n = (s)\text{-}\lim_n VT^nV^{-1} = \mathbb{O},$$

the operator $VTV^{-1}$ will be completely non-unitary and so must have a total family of eigenspaces. The same is then true for the operator $T$. □

The Köthe-Toeplitz Theorem and its "subspaces" analogue (the Theorem on Riesz Bases) can also be stated in the following way: If $\mathscr{E}(\cdot)$ is a projection valued measure (that is, $\mathscr{E}(\sigma)$ is a projection in the Hilbert space $H$, $\sigma \subset \mathbb{C}$, $\sigma \in \mathscr{B}$) concentrated on a countable set $\sigma_p$ (that is, $\mathscr{E}(\mathbb{C} \backslash \sigma_p) = \mathbb{O}$) and if

$$\sup\{|\mathscr{E}(\sigma)| : \sigma \in \mathscr{B}\} < \infty$$

then there exists an isomorphism $V$ from $H$ onto itself such that $V\mathscr{E}(\sigma)V^{-1}$ is an orthogonal projection, $\forall \sigma$. If in this formulation we drop the requirement of $\sigma_p$ to be countable we likewise will get a correct result, viz. Wermer's theorem [2]. Let us also point out the geometric meaning of the norm $|\mathscr{E}(\sigma)|$. If $L$ and $L'$ are subspaces of the Hilbert space $H$ we can define the (least) angle $\langle L, L' \rangle$ between them by the formula

$$\cos\langle L, L' \rangle = \sup\left\{\frac{|(l, l')|}{\|l\| \cdot \|l'\|} : l \in L,\, l' \in L'\right\};$$

further if $L \cap L' = \{\mathbb{0}\}$ we define on the algebraic sum $L + L'$ of these subspaces the "spectral projection" $\mathscr{E} = \mathscr{E}(L, L')$ by the formula $\mathscr{E}(l + l') = l$; there holds $|\mathscr{E}(L, L')|^{-1} = \sin\langle L, L'\rangle$. Thus many constructions of this Lecture (e.g. the Theorem on Unconditional Bases) can be reinterpreted geometrically in terms of the angle between the subspaces $\mathscr{E}(\sigma)H$ $(= L)$ and $\mathscr{E}(\mathbb{C}\backslash\sigma)H$ $(= L')$. The analysis and the discussion of the basis problem for the model operators $\mathbf{T}_\Theta$ (and in particular of Carleson's Condition $(C)$ and its analogue (24), especially in the case $\dim E < \infty$) will be the object of the following Lectures.

And at last some bibliographical notes. The main facts concerning minimal families were known already to Banach [1]; in particular, every basis in a Banach space is a minimal family. Bari's Theorem in proved in Bari [1] and its analogue for bases of subspaces in Nikol'skiĭ-Pavlov [1] and Nikol'skiĭ [10] (at least in the form it was used in this Lecture). The Theorem on Unconditional Bases (in the case $\dim X_n = 1$) is due to Lorch [1] and to Grinblyum [1], [2]. Regarding bases of subspaces see also the book by Gohberg and Kreĭn [1], chap. VI, where likewise many results on bases of eigenvectors of linear operators are presented in a systematic fashion. The Lemma of Orlicz is proved in Orlicz [1] and the Köthe-Toeplitz Theorem in Köthe-Toeplitz [1]; the latter was later rediscovered by Lorch [1] and by Gel'fand [2].

The Carleson Condition $(C)$ will play a prominent rôle in all of the following treatment. It was introduced by Newman [1] where there is given a lemma equivalent to our Newman's Lemma. Carleson [1] proved the imbedding theorem $(C) \Rightarrow (CN)$ while the implication $2 \Leftrightarrow 3$ of the Theorem on Bases and Interpolation was established in Shapiro-Shields [1, 2]. The implication $1 \Leftrightarrow 2$ (or $1 \Leftrightarrow 3$) together with an explicit statement of the connection between the problems of bases of rational functions and interpolation appeared in Nikol'skiĭ-Pavlov [1, 2]. In a different context the duality between bases and interpolation was consciously exploited already in work by Carleman [2], Banach [2], Paley-Wiener [1] (theorem 39) and other mathematicians too; for a systematic presentation see Nikol'skiĭ [10] and further the "Supplements" to Lecture VII–X. The results of Sect. 5 are taken from Nikol'skiĭ-Pavlov [1, 2]; cf. also Nikol'skiĭ [10].

# Lecture VII. Bases and Interpolation (Solution)

In order to complete the proof of the Theorem on Bases and Interpolation (Lecture III) it remains to verify that $(C) \Rightarrow (CN)$. This is part of Carleson's imbedding theorem describing all Borel measures $\nu$ on the disk $\mathbb{D}$ such that

$$f \in H^2 \Rightarrow f \in L^2(\nu);$$

such measures will be called *Carleson measures.*

**1. Carleson Measures.** It is somewhat easier to deal with harmonic functions than with analytic ones; more exactly we will consider the harmonic continuation $\mathscr{H}f$,

$$(\mathscr{H}f)(rt) = \sum_{n \in \mathbb{Z}} \hat{f}(n) r^{|n|} t^n, \quad t \in \mathbb{T}, \quad 0 \leq r < 1,$$

of the function $f, f \in L^2(\mathbb{T})$, to the disk $\mathbb{D}$. As

$$\sum_{n \in \mathbb{Z}} r^{|n|} \zeta^n = (1 - r^2)|1 - r\zeta|^{-2} \stackrel{\text{def}}{=} P(r\zeta), \quad \zeta \in \mathbb{T}$$

(Poisson kernel) we have

$$(\mathscr{H}f)(x) = \int_{\mathbb{T}} f(\zeta) P(x\bar{\zeta})\, dm(\zeta), \quad x \in \mathbb{D}.$$

**Imbedding Theorem.** *Let $\nu$ be a Borel measure on the disk $\mathbb{D}$. The following statements are equivalent:*

1. $\mathscr{H} L^2(\mathbb{T}) \subset L^2(\nu)$;
2. $H^2 \subset L^2(\nu)$;
3. $\sup\{(1 - |\zeta|^2)^{1/2} \|(\mathbb{1} - \bar{\zeta} z)^{-1}\|_{L^2(\nu)} : \zeta \in \mathbb{D}\} \stackrel{\text{def}}{=} a < \infty$;
4. $\sup\{(1 - |\zeta|^2)^{1/2} \|(\mathbb{1} - \bar{\zeta} z)^{-1}\|_{L^2(\nu)} : \zeta \in \operatorname{supp} \nu\} \stackrel{\text{def}}{=} C < \infty$.

*Moreover $C \leq a |\mathscr{H}| \leq 4\sqrt{2} C$.*

Before proving this theorem let us remark that the imbedding in question has to be interpreted in the natural sense: every function in $\mathscr{H} L^2(\mathbb{T})$ (in particular in $H^2$) is $\nu$-measurable and the corresponding equivalence class is contained in $L^2(\nu)$. It is clear that $1 \Rightarrow 2$ ($\mathscr{H} | H^2 = I$), $3 \Rightarrow 4$ and $2 \Rightarrow 3$ (we can apply the closed graph theorem, since convergence in $L^2(\mathbb{T})$ implies pointwise convergence of the $\mathscr{H}$ images; therefore $a \leq |\mathscr{H}|_{H^2 \to L^2(\nu)}$). The most important of the implications, namely $4 \Rightarrow 1$, will be derived from the following criterion for the continuity of an integral operator.

**Vinogradov-Seničkin Test.** *Let $Z$ be a measurable space and $k$ a non-negative measurable function on $Z \times Z$. If*

$$\int_Z k(s,t)k(s,x)\,ds \le c(k(t,x)+k(x,t)) \quad \textit{for a.e. } (t,x)\in Z\times Z,$$

*then*

$$Q \overset{\text{def}}{=} \iint_{Z\times Z} k(s,t)g(s)g(t)\,ds\,dt \le 2c$$

*for any function* $g$, $g \ge 0$, $\|g\|_{L^2(Z)} \le 1$, *for which* $Q < \infty$.

*Proof*. $Q^2 = \left(\int_Z g(s)\left(\int_Z k(s,t)g(t)\right)ds\right)^2$

$$\le \|g\|^2_{L^2(Z)}\cdot\left\|\int_Z k(s,t)g(t)\,dt\right\|^2_{L^2(Z)}$$

$$\le \int_Z ds\left(\iint_{Z\times Z} k(s,t)k(s,x)g(t)g(x)\,dt\,dx\right)$$

$$= \iint_{Z\times Z} g(t)g(x)\left(\int_Z k(s,t)k(s,x)\,ds\right)dt\,dx$$

$$\le 2c\iint_{Z\times Z} g(t)g(x)k(t,x)\,dt\,dx = 2cQ. \ \square$$

*Proof of the Imbedding Theorem.* It remains to be shown that $4 \Rightarrow 1$; to this end we invoke the test operator $\mathscr{H}^*$, adjoint to $\mathscr{H}$ (so far only formally). It is clear that

$$(\mathscr{H}^*g)(\zeta) = \int_{\mathbb{D}} g(x)P(x\bar\zeta)\,d\nu(x), \quad \zeta\in\mathbb{T}.$$

Hence

$$\int_{\mathbb{T}} |(\mathscr{H}^*g)(\zeta)|^2\,dm(\zeta) = \iint_{\mathbb{D}\times\mathbb{D}} g(s)\overline{g(t)}\left(\int_{\mathbb{T}} P(s\bar\zeta)P(t\bar\zeta)\,dm(\zeta)\right)d\nu(s)\,d\nu(t)$$

$$= \iint_{\mathbb{D}\times\mathbb{D}} g(s)\overline{g(t)}P(\bar s t)\,d\nu(s)\,d\nu(t)$$

$$\le 4\iint_{\mathbb{D}\times\mathbb{D}} k(s,t)|g(s)|\cdot|g(t)|\,d\nu(s)\,d\nu(t),$$

where $k(s,t) = P(\bar s t)(1+|st|)^{-1}$ for $|s|\ge|t|$, $k(s,t)=0$ for $|s|<|t|$. Let us now remark that

$$|1-x\bar t| = |1-s\bar t+\bar t s - x\bar t|$$
$$\le |1-s\bar t| + |\bar t|\,|s-x|$$
$$\le |1-s\bar t| + |s-x|$$
$$\le |1-s\bar t| + |1-s\bar x|,$$

so that

$$|1-s\bar{t}|^{-1}|1-s\bar{x}|^{-1} \le |1-x\bar{t}|^{-1}\{|1-s\bar{t}|^{-1}+|1-s\bar{x}|^{-1}\}$$

and thus finally

$$|1-s\bar{t}|^{-2}|1-s\bar{x}|^{-2} \le 2|1-x\bar{t}|^{-2}\{|1-s\bar{t}|^{-2}+|1-s\bar{x}|^{-2}\}.$$

Next we estimate the iterated kernel for $|t| \ge |x|$, $t, x \in \operatorname{supp} \nu$:

$$\begin{aligned}\int_{\mathbb{D}} k(s,t)k(s,x)\,d\nu(s) &= \int_{|s|\ge|t|} \frac{1-|st|}{|1-s\bar{t}|^2}\cdot\frac{1-|sx|}{|1-s\bar{x}|^2}\,d\nu(s)\\ &\le \frac{2}{|1-x\bar{t}|^2}\left\{\int_{|s|\ge|t|}\frac{(1-|st|)(1-|sx|)}{|1-s\bar{t}|^2}\,d\nu(s)\right.\\ &\quad\left.+\int_{|s|\ge|t|}\frac{(1-|st|)(1-|sx|)}{|1-s\bar{x}|^2}\,d\nu(s)\right\}\\ &\le \frac{2(1-|xt|)}{|1-x\bar{t}|^2}\left\{\int_{|s|\ge|t|}\frac{1-|t|^2}{|1-s\bar{t}|^2}\,d\nu(s)+\int_{|s|\ge|x|}\frac{1-|x|^2}{|1-s\bar{x}|^2}\,d\nu(s)\right\}\\ &\le 4C^2k(t,x).\end{aligned}$$

Applying the Vinogradov-Seničkin Test we finally get:

$$\|\mathscr{H}^*g\|_2^2 \le 32C^2\|g\|_{L^2(\nu)}^2. \ \square$$

*Remark.* We have only used the finiteness of the upper bound $c$,

$$c^2 \overset{\text{def}}{=} \sup\left\{\int_{|\zeta|\le|t|<1}\frac{1-|\zeta|^2}{|1-\bar{\zeta}t|^2}\,d\nu(t) : \zeta\in\operatorname{supp}\nu\right\}.$$

It is clear that $c \le C$ and that these constants are connected by the further inequalities

$$C \le |\mathscr{H}| \le 4\sqrt{2}c.$$

We have also the following interesting geometric characterization of Carleson measures in terms of their "densities" with respect to Lebesgue measure.

**Lemma on Carleson Measures.** *Let $\nu$ be a Borel measure on the disk $\mathbb{D}$ and set $D(\zeta,l)=\{t\in\mathbb{C}:|t-\zeta|<l\}$ and*

$$A \overset{\text{def}}{=} \sup\{l^{-1}\nu(\mathbb{D}\cap D(\zeta,l)) : l>0, \zeta\in\mathbb{T}\}.$$

*Then*

$$\frac{1}{147}A \le C^2 \le 6A$$

*where $C$ is the upper bound in Statement 4 of the Imbedding Theorem.*

*Proof.* Let $\zeta\in\mathbb{T}$, $0\le r\le 1$ and $\lambda=r\zeta$, $\lambda\in\operatorname{supp}\nu$. Set

$$e_x^\lambda=\left\{t\in\mathbb{D}:\frac{1-|\lambda|^2}{|1-\bar{\lambda}t|^2}>x\right\},\quad x>0.$$

As

$$e_x^\lambda = \left\{t \in \mathbb{D} : |1 - \bar{\lambda} t| < \sqrt{\frac{1 - r^2}{x}}\right\}$$

$$= \left\{t \in \mathbb{D} : |(\zeta - t) + t(1 - r)| < \sqrt{\frac{1 - r^2}{x}}\right\},$$

we have

$$\mathbb{D} \cap D\left(\zeta, \sqrt{\frac{1 - r^2}{x}} - (1 - r)\right) \subset e_x^\lambda \subset \mathbb{D} \cap D\left(\zeta, \sqrt{\frac{1 - r^2}{x}} + (1 - r)\right). \quad (1)$$

The integrals in the definition of the constant $C^2$ (cf. the Imbedding Theorem) can be taken with respect to the distribution $x \mapsto \nu e_x^\lambda$ of the integrand (cf. e.g. Natanson [1]):

$$\int_{\mathbb{D}} \frac{1 - |\lambda|^2}{|1 - \bar{\lambda} t|^2} d\nu(t) = \int_0^\infty (\nu e_x^\lambda)\, dx = \int_0^{(1+r)(1-r)^{-1}} (\nu e_x^\lambda)\, dx,$$

since $e_x^\lambda = \varnothing$ for $x \geq (1 + r)(1 - r)^{-1}$. As the distribution is a monotone function we get using the first inclusion in (1) with $\alpha = 1/7$

$$\begin{aligned} C^2 &\geq \int_0^{(1+r)(1-r)^{-1}} (\nu e_x^\lambda)\, dx \geq \alpha^2 \frac{1 + r}{1 - r} \nu e^\lambda_{\alpha^2(1+r)/(1-r)} \\ &\geq (49(1 - r))^{-1} \nu(\mathbb{D} \cap D(\zeta, 6(1 - r))). \end{aligned} \quad (2)$$

Let $\zeta \in \mathbb{T}$, $0 < l \leq 1$; set $\Delta_{i,n} \stackrel{\text{def}}{=} \{\xi \in \mathbb{D} : 1 - |\xi| \leq 2^{-n} l,\ 5 \cdot 2^{-n-1} l \cdot (i - 1) \leq \arg \bar{\zeta} \xi < 5 \cdot 2^{-n-1} l i\}$ for $n \geq 1$, $i = -2^{n-1} + 1, \ldots, 2^{n-1}$ and $\Delta_0 = \mathbb{D}$. A "square" $\Delta_{i,n}$ (or $\Delta_0$) will be termed black if it contains a point $\lambda$ in $D(\zeta, l) \cap \operatorname{supp} \nu$ with $1 - |\lambda| \geq 2^{-n-1} l$. It is clear that the union of all black "squares" $\Delta_{i,n}$ or $\Delta_0$ contains the set $D(\zeta, l) \cap \operatorname{supp} \nu$ and that this union also is the sum of a family $F$ of pairwise disjoint black "squares". On the other hand $\operatorname{diam} \Delta_{i,n} \leq 3 \cdot 2^{-n} l$, so that if $\lambda_{i,n} \in \Delta_{i,n} \cap \operatorname{supp} \nu$ denotes the point in the definition of a black square then

$$\Delta_{i,n} \subset D(\zeta_{i,n}, 6(1 - |\lambda_{i,n}|)), \quad \zeta_{i,n} \stackrel{\text{def}}{=} \lambda_{i,n} |\lambda_{i,n}|^{-1}.$$

From this and from (2) we infer that

$$\begin{aligned} \nu(\mathbb{D} \cap D(\zeta, l)) &\leq \sum_F \nu D(\zeta_{i,n}, 6(1 - |\lambda_{i,n}|)) \\ &\leq 49 C^2 \sum_F (1 - |\lambda_{i,n}|) \leq 49 C^2 \sum_F 2^{-2} l \leq 49 C^2 l. \end{aligned}$$

If $l \geq 1$ then the intersection $\mathbb{D} \cap D(\zeta, l)$ is covered by three disks of the form $D(\xi, l)$, $\xi \in \mathbb{T}$, so that $\nu(\mathbb{D} \cap D(\zeta, l)) \leq 147 C^2 l$.

The rightmost of the inequalities we are out to prove follows at once from the second inclusion (1):

$$\int_{\mathbb{D}} \frac{1 - |\lambda|^2}{|1 - \bar{\lambda} t|^2} d\nu(t) = \int_0^{(1+r)(1-r)^{-1}} (\nu e_x^\lambda)\, dx$$

$$\le \int_0^{(1+r)(1-r)^{-1}} \nu\left(\mathbb{D} \cap D\left(\zeta, \sqrt{\frac{1-r^2}{x}} + (1-r)\right)\right) dx$$

$$\le A \int_0^{(1+r)(1-r)^{-1}} \left(\sqrt{\frac{1-r^2}{x}} + (1-r)\right) dx$$

$$= A\left(\sqrt{1-r^2}\cdot 2\cdot \sqrt{\frac{1+r}{1-r}} + (1+r)\right) = 3A(1+r) \le 6A. \ \square$$

*Remark.* In order to prove that $\nu$ is a Carleson measure (without an estimate of the constant) it is of course sufficient to check that

$$\sup\{l^{-1}\nu(\mathbb{D} \cap D(\zeta, l)) : 0 < l \le \varepsilon, \zeta \in \mathbb{T}\} < \infty$$

for some $\varepsilon$, $\varepsilon > 0$, or the validity of an analogous estimate for all sufficiently small "rectangles" with a fixed ratio of sides:

$$\sup\{l^{-1}Q(\zeta, l) : \zeta \in \mathbb{T}, 0 < l \le \varepsilon\} < \infty, \quad \varepsilon > 0,$$

where $Q(\zeta, l) = \{t : 1 - l \le |t| < 1; |\arg t - \arg \zeta| \le rl\}$, $r > 0$.

**2. Proof of the Theorem on Bases and Interpolation.** Let $\sigma = \sigma_p(T)$ be a *Carleson set* (i.e. $\sigma$ satisfies Condition (C)) and $\nu = \nu_\sigma$ the measure defined on subsets $e$ of $\mathbb{D}$ by the formula

$$\nu_\sigma e = \sum_{\lambda \in \sigma \cap e} (1 - |\lambda|^2). \tag{3}$$

As $|B_\lambda(\lambda)| \ge \delta$, $\lambda \in \sigma$, where

$$B_\lambda = \prod_{\mu \in \sigma \setminus \{\lambda\}} b_\mu, \quad b_\mu = \frac{|\mu|}{\mu} \frac{\mu\mathbb{1} - z}{\mathbb{1} - \bar{\mu} z},$$

and $\delta$ is the Carleson constant, we get

$$2\log\delta \le \log|B_\lambda(\lambda)|^2 = \sum_{\mu \in \sigma \setminus \{\lambda\}} \log\left|\frac{\mu - \lambda}{1 - \bar{\lambda}\mu}\right|^2$$

$$= \sum_{\mu \in \sigma \setminus \{\lambda\}} \log\left(1 - \frac{(1 - |\lambda|^2)(1 - |\mu|^2)}{|1 - \bar{\lambda}\mu|^2}\right)$$

$$\le - \sum_{\mu \in \sigma \setminus \{\lambda\}} \frac{(1 - |\lambda|^2)(1 - |\mu|^2)}{|1 - \bar{\lambda}\mu|^2}.$$

This gives $C^2 \le 1 + 2\log(1/\delta)$, $C$ being the number in Statement 4 of the Imbedding Theorem. Consequently

$$f \in H^2 \Rightarrow \sum_{\lambda \in \sigma} (1 - |\lambda|^2)|f(\lambda)|^2 \le 32\left(1 + 2\log\frac{1}{\delta}\right) \|f\|_2^2. \ \square$$

Let us point out an estimate for the orthogonalizer $J_{\mathscr{X}'}$ of the family $\mathscr{X}' = \{x'_\lambda : \lambda \in \sigma_p(T)\}$ of the eigenfunctions of the operator $T = T_B$ (cf. (4), Lecture VI), which is contained in the previous computations:

$$|J_{\mathscr{X}'}| \le 4\sqrt{2}\left(1 + 2\log\frac{1}{\delta}\right)^{1/2}.$$

From the proof of Newman's Lemma (Lecture VI, Sect. 2) it follows that the interpolation constant $|J_{\mathscr{X}}^{-1}| = |\mathscr{J}_{\mathscr{X}'}|$ does not exceed $\delta^{-1}|J_{\mathscr{X}'}|$. Let us further recall that $\delta^{-1} = d(\mathscr{X})^{-1} = d(\mathscr{X}') = \sup\{\|x_\lambda\| \cdot \|x'_\lambda\| : \lambda \in \sigma_p\}$ is the constant of uniform minimality for the families $\mathscr{X}$ and $\mathscr{X}'$.

It can be proved that the above mentioned estimates have the "right" growth order for $\delta \to 0$. Namely, for $0 < \delta < 1$ there exists a Carleson set $\sigma$, $\sigma \subset \mathbb{D}$ such that $\delta(\mathscr{X}') = \delta(\mathscr{X}) = \delta$, $\mathscr{X}' = \{x'_\lambda : \lambda \in \sigma\}$ and $|J_{\mathscr{X}'}^{-1}| \ge \alpha\delta^{-1}\log^{\frac{1}{2}}\frac{1}{\delta}$, with an absolute constant $\alpha$, $\alpha > 0$ (see Vinogradov-Havin [1]).

**3. Analysis of Carleson's Condition (C).** The geometric nature of Carleson sets means roughly speaking that such a set is sufficiently thinned out in the disk $\mathbb{D}$: either it will very fast accumulate on the boundary $\mathbb{T}$ of the disk or else, if the speed of this accumulation would slow down, be "well scattered" in circular belts concentric with $\mathbb{T}$. In this Section we give some results on $C$-sets which hopefully will confirm such a conception of the structure of these sets.

Let $\sigma$ be a subset of the disk $\mathbb{D}$, with $\sum_{\lambda\in\sigma}(1 - |\lambda|) < \infty$, and set $B = \prod_{\lambda\in\sigma} b_\lambda$, $B_\lambda = Bb_\lambda^{-1}$, assuming that

$$\inf\{|B_\lambda(\lambda)| : \lambda \in \sigma\} = \delta > 0; \tag{C}$$

in this case (that is, if the Carleson condition is fulfilled) we write $\sigma \in (C)$. As $|b_\lambda| \le 1$, we have also

$$\inf\left\{\left|\frac{\lambda - \mu}{1 - \bar{\lambda}\mu}\right| : \lambda, \mu \in \sigma, \lambda \ne \mu\right\} \ge \delta. \tag{R}$$

The above "sparseness condition" means that each point of the set $\sigma$ is "sufficiently far away" from the remaining parts of the set. This assertion can be made precise in the *hyperbolic metric* $\rho$,

$$\rho(\lambda, \mu) \overset{\text{def}}{=} \log\frac{1 + |b_\lambda(\mu)|}{1 - |b_\lambda(\mu)|}.$$

(In fact, this formula defines a metric. In order to verify the triangle inequality $\rho(\lambda, \mu) \le \rho(\lambda, \nu) + \rho(\nu, \mu)$ we first have to check the conformal invariance of $\rho$ (i.e. that $\zeta \in \mathbb{D}$, $\lambda' = b_\zeta(\lambda)$, $\mu' = b_\zeta(\mu) \Rightarrow |b_{\lambda'}(\mu')| = |b_\lambda(\mu)|$) and afterwards set $\nu = 0$; in that case the inequality in question takes the form $|b_\lambda(\mu)| \le (|\lambda| + |\mu|)(1 + |\lambda\mu|)^{-1}$ or $1 - |b_\lambda(\mu)|^2 = \dfrac{(1 - |\lambda|^2)(1 - |\mu|^2)}{|1 - \bar{\lambda}\mu|^2} \ge \dfrac{1 + |\lambda\mu|^2 - |\lambda|^2 - |\mu|^2}{(1 + |\lambda\mu|)^2}$. The last inequality is trivial to verify. □)

Thus for condition (C) to be fulfilled it is necessary and for condition (R) necessary and sufficient that the hyperbolic disks $\left\{\zeta : \rho(\zeta, \lambda) < \frac{1}{2}\log\frac{1 + \delta}{1 - \delta}\right\}$, $\lambda \in \sigma$, do not intersect. We can also use ordinary Euclidean disks

$$D(\lambda, r) \overset{\text{def}}{=} \{\zeta : |\lambda - \zeta| < r\},$$

if we remark that

$$\begin{aligned}
|\zeta - \lambda| < \frac{\delta}{2}(1 - |\lambda|) \Rightarrow |b_\lambda(\zeta)| &< \frac{\delta}{2} \\
&\Rightarrow \rho(\lambda, \zeta) < \log\frac{1 + \delta/2}{1 - \delta/2} \\
&= \frac{1}{2}\log\left(\frac{1 + \delta/2}{1 - \delta/2}\right)^2 \\
&= \frac{1}{2}\log\left(\frac{1 + \delta + \delta^2/4}{1 - \delta + \delta^2/4}\right) < \frac{1}{2}\log\frac{1 + \delta}{1 - \delta},
\end{aligned}$$

$$\begin{aligned}
\rho(\lambda, \zeta) < \log\frac{1 + \delta}{1 - \delta} &\Rightarrow |b_\lambda(\zeta)| < \delta \\
&\Rightarrow |\lambda - \zeta| < \delta\frac{2}{1 - \delta}(1 - |\lambda|),
\end{aligned}$$

that is

$$(R) \Rightarrow D\left(\lambda, \frac{\delta}{2}(1 - |\lambda|)\right) \cap D\left(\mu, \frac{\delta}{2}(1 - |\mu|)\right) = \varnothing, \quad \lambda, \mu \in \sigma, \lambda \neq \mu; \tag{4}$$

$$D\left(\lambda, \frac{\delta}{1 - \delta}(1 - |\lambda|)\right) \cap D\left(\mu, \frac{\delta}{1 - \delta}(1 - |\mu|)\right) = \varnothing \Rightarrow (R). \tag{5}$$

Let us now write down some further conditions which characterise various degrees of thinness of the set $\sigma$ and which we in what follows will compare with condition $(C)$: we write $\sigma \in (CN)$ if the measure (3) satisfies any (all) of the conditions of the Imbedding Theorem of Carleson (i.e. condition $(CN)$ of Lecture VI is fulfilled).

*Newman's Condition:*

$$\sup\left\{\frac{1 - |\lambda|}{1 - |\mu|} : \lambda \in \sigma \backslash \{\mu\}, |\lambda| \geq |\mu|; \mu \in \sigma\right\} < 1. \tag{N}$$

*Weak Newman Condition:*

$$\sup\left\{\frac{1}{1 - |\mu|}\sum_{\substack{\lambda \in \sigma\backslash\{\mu\} \\ |\lambda| \geq |\mu|}} (1 - |\lambda|) : \mu \in \sigma\right\} < \infty. \tag{wN}$$

*Stolz Condition:* there exists a finite subset $e$ of the circumference $\mathbb{T}$ such that

$$\sup\left\{\frac{\operatorname{dist}(\lambda, e)}{\operatorname{dist}(\lambda, \mathbb{T})} : \lambda \in \sigma\right\} < \infty. \tag{S}$$

*Uniform Frostman Condition:*

$$\sup\left\{\sum_{\lambda\in\sigma}\frac{1-|\lambda|}{|\zeta-\lambda|}:\zeta\in\mathbb{T}\right\}<\infty. \tag{F}$$

It is clear that if two sets $\sigma$ and $\sigma'$ differ only by a finite set then if one of them fulfils some of the conditions $(C)$, $(R)$, $(wN)$, $(S)$, $(F)$ so does the other.

The rest of this Section will be devoted to the proof of the following implications, which together give a good picture of the Carleson sets of the disk $\mathbb{D}$:

$$C\Rightarrow R,$$

$$N\Rightarrow C,$$

$$R\ \&\ CN\Rightarrow C,$$

$$R\ \&\ S\Rightarrow R\ \&\ wN\Rightarrow C,$$

$$F\Rightarrow CN,$$

$$\sum_{\lambda\in\sigma}(1-|\lambda|)<\infty\Rightarrow\forall\lambda\exists\lambda'(|\lambda'|=|\lambda|):\sigma'\stackrel{\text{def}}{=}\{\lambda':\lambda\in\sigma\}\in(C).$$

Let us remark at once that $C\Rightarrow R$ (already proved) and that

$$R\ \&\ CN\Rightarrow C.$$

In fact, from the identity

$$\left|\frac{1-\bar{\lambda}\mu}{\lambda-\mu}\right|^2=1+\frac{(1-|\lambda|^2)(1-|\mu|^2)}{|\lambda-\mu|^2},\quad(\lambda\neq\mu),$$

it follows that

$$\log\prod_{\mu\in\sigma\setminus\{\lambda\}}\left|\frac{1-\lambda\bar{\mu}}{\lambda-\mu}\right|^2=\sum_{\mu\in\sigma\setminus\{\lambda\}}\log\left(1+\frac{(1-|\lambda|^2)(1-|\mu|^2)}{|\lambda-\mu|^2}\right)$$

$$\leq\sum_{\mu\in\sigma\setminus\{\lambda\}}\frac{(1-|\lambda|^2)(1-|\mu|^2)}{|\lambda-\mu|^2}\leq\frac{C^2}{\delta^2},$$

where $\delta$ is the number in condition $(R)$ and $C$ the constant in Carleson's Imbedding Theorem ($\nu=\nu_\sigma$). □

**Corollary.** *If $\sigma\in(CN)$ then $\sigma$ is a finite union of Carleson sets.*

In fact, it suffices just to check the possibility of decomposing the set $\sigma$ into a finite union of sparse sets and this follows readily from the geometric meaning of the Carleson property of the measure $\nu_\sigma$ (cf. Sect. 1): each "square" $Q_{nk}$ (cf. Fig. 3),

$$Q_{nk}:\begin{array}{l}1-2^{-n}\leq|\zeta|\leq1-2^{-n-1}\\ 2\pi k2^{-n}\leq\arg\zeta\leq2\pi(k+1)2^{-n},\end{array}$$

$k=0,\dots,2^n-1$; $n\geq0$, contains not more than $m=[10A]$ points of the set $\sigma$ (where $A$ is the constant in the Lemma on Carleson Measures), and we get the desired decomposition $\sigma = \bigcup_{i=1}^{4m} \sigma_i$ subsequently defining the sets $\sigma_i$, $i = 1, \ldots, m$, by the following rule.

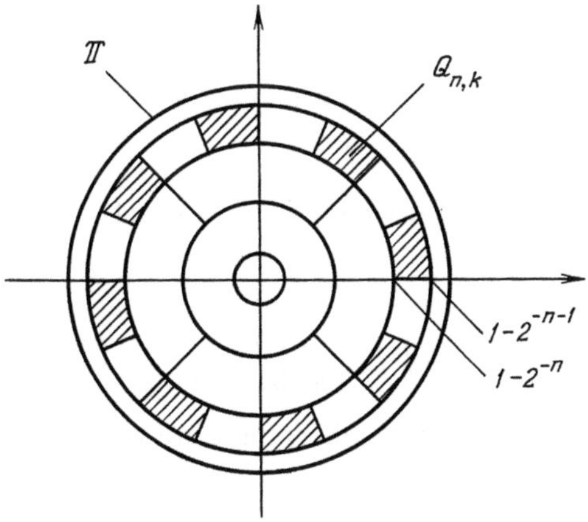


Fig. 3

To $\sigma_1$ we count one point from each non-empty intersection $\sigma \cap Q_{n,k}$ ($n$ and $k$ even) and to $\sigma_2$ one of the remaining points etc.; the sets $\sigma_i$, $i = m+1, \ldots, 2m$, are obtained in an analogous manner with squares $Q_{n,k}$ with $n$ even, $k$ odd; $\sigma_i$, $i = 2m+1, \ldots, 3m$, with $n$ odd and $k$ even; $\sigma_i$, $i = 3m+1, \ldots, 4m$, with $n$ odd and $k$ odd. The preceding comments on the property $(R)$ show that all the sets $\sigma_i$ have that property. □

It is clear the Carleson's condition entails the *Blaschke Condition*:

$$C \Rightarrow \sum_{\lambda \in \sigma} (1 - |\lambda|) < \infty.$$

Therefore the last of the implications under discussion (Naftalevič's Lemma, cf. below) shows that in general the condition $\sigma \in (C)$ does not impose any restrictions on the set of the moduli of the points of $\sigma$ (except the trivial Blaschke condition), but may be reached by suitably distributing the arguments. However, for sets contained in suitable Stolz angles (Condition $(S)$) this is not the case. The remaining ones of the above implications constitute the contents of the following four lemmata.

**Kabaĭla-Newman Lemma.** *Let $\sigma$ be a subset of the disk $\mathbb{D}$ which accumulates only on the circumference $\mathbb{T}$ and put*

$$\gamma \overset{\text{def}}{=} \lim_{\mu \in \sigma, |\mu| \to 1} \sup \left\{ \frac{1 - |\lambda|}{1 - |\mu|} : \lambda \in \sigma \backslash \{\mu\}, |\lambda| \geq |\mu| \right\}.$$

*If $\gamma < 1$ then $\sigma \in (C)$. If $\sigma \subset [0, 1]$ and $\sigma \in (C)$ then $\gamma < 1$.*

*Proof*. We use again the simple but important identity

$$\left|\frac{\lambda-\mu}{1-\bar{\lambda}\mu}\right|^2 = 1 - \frac{(1-|\lambda|^2)(1-|\mu|^2)}{|1-\bar{\lambda}\mu|^2}, \quad \lambda, \mu \in \mathbb{D},$$

from which follows that $|b_\mu(\lambda)| \geq |b_{|\mu|}(|\lambda|)|$. If $\gamma < 1$ we can without loss of generality assume that

$$\lambda, \mu \in \sigma, \lambda \neq \mu \Rightarrow |\lambda| \neq |\mu|$$

and that with a suitable $d$, $0 < d < 1$,

$$\lambda, \mu \in \sigma, |\lambda| < |\mu| \Rightarrow (1-|\mu|) \leq d(1-|\lambda|).$$

Therefore for $\lambda \in \sigma$ holds

$$|B_\lambda(\lambda)| = \prod_{\mu \in \sigma\setminus\{\lambda\}} |b_\mu(\lambda)| \geq \prod_{|\mu|<|\lambda|} \frac{|\lambda|-|\mu|}{1-|\lambda\mu|} \cdot \prod_{|\mu|>|\lambda|} \frac{|\mu|-|\lambda|}{1-|\mu\lambda|}.$$

If $|\mu| > |\lambda|$ and if between these numbers there are yet $m$ ($m \geq 0$) numbers of the set $\{|t| : t \in \sigma\}$ then

$$\begin{aligned} 1-|\mu| &\leq d^{m+1}(1-|\lambda|), \\ |\mu|-|\lambda| &\geq (1-d^{m+1})(1-|\lambda|), \qquad (6) \\ 1-|\lambda\mu| = 1-|\lambda|+|\lambda|(1-|\mu|) &\leq (1+d^{m+1})(1-|\lambda|). \end{aligned}$$

Hence

$$\prod_{\mu\in\sigma, |\mu|>|\lambda|} \frac{|\mu|-|\lambda|}{1-|\lambda\mu|} \geq \prod_{m\geq 0} \frac{1-d^{m+1}}{1+d^{m+1}}.$$

In an analogous manner we find $\prod_{|\mu|<|\lambda|} \geq \prod_{m=0}^{N-1} \frac{1-d^{m+1}}{1+d^{m+1}}$, where $N = \operatorname{card}\{\mu : \mu \in \sigma, |\mu| < |\lambda|\}$, so finally $|B_\lambda(\lambda)| \geq \left(\prod_{m\geq 0} \frac{1-d^{m+1}}{1+d^{m+1}}\right)^2$. The converse statement follows from the inequalities (with $\lambda, \mu \in \sigma$, $\mu > \lambda$)

$$\begin{aligned} 1 - \frac{1-\mu}{1-\lambda} = \frac{\mu-\lambda}{1-\lambda} &\geq \frac{\mu-\lambda}{1-\lambda\mu} \\ &= b_\mu(\lambda) \geq |B_\lambda(\lambda)| \\ &\geq \inf\{|B_\lambda(\lambda)| : \lambda \in \sigma\} > 0. \ \square \end{aligned}$$

**Vinogradov's Lemma.** 1. *$R$ & $S \Rightarrow R$ & $wN$.*
2. *Every* ($wN$)*-set is a finite union of* ($N$)*-sets and vice versa.*
3. *$R$ & $wN \Rightarrow C$.*

*Proof*. Statement 3 follows from Statement 2, as $N \Rightarrow C \Rightarrow CN$, as $(\sigma_i \in (CN) \Rightarrow \bigcup_{i=1}^n \sigma_i \in (CN))$ and as $R$ & $CN \Rightarrow$ C. Statement 2 is in one direction trivial (any finite sum of $N$-sets is a $wN$-set, in view of inequality (6)) and in the other direction it follows by the following reasoning: let $\lambda, \mu \in \sigma$, $\lambda \neq \mu$, $|\mu| \leq |\lambda|$

and $\operatorname{card}\{\zeta : \zeta \in \sigma \backslash \{\mu\}, |\mu| \le |\zeta| \le |\lambda|\} \ge m$ ($m \in \mathbb{N}$); if $c \overset{\text{def}}{=} \sup\{\cdots\}$ is the least upper bound in condition $(wN)$ then

$$c \ge \frac{1}{1-|\mu|} \sum_{\substack{\zeta \in \sigma \backslash \{\mu\} \\ |\lambda| \ge |\zeta| \ge |\mu|}} (1-|\zeta|) \ge \frac{m(1-|\lambda|)}{1-|\mu|}, \quad \frac{1-|\lambda|}{1-|\mu|} \le \frac{c}{m}, \tag{7}$$

so if $c/m < 1$ then $\sigma$ can be divided into a union of not more than $m$ Newman sets.

There remains Statement 1. Let $c \overset{\text{def}}{=} \sup\{\cdots\}$ be the least upper bound in the definition of an (S)-set and let $\delta$ be the number in the definition of a sparse set, writing $D_\lambda = D(\lambda, \delta/2(1-|\lambda|))$, $\lambda \in \sigma$. Further, let $\mu \in \sigma$, $\lambda \in \sigma \backslash \{\mu\}$ with $|\mu| < |\lambda|$ and let $\zeta$ be a point of the set $e$ (cf. the definition of an $S$-set) such that $\operatorname{dist}(\lambda, e) = |\lambda - \zeta|$. Then

$$D_\lambda \subset D(\zeta, 2c(1-|\mu|)),$$

because of the following estimates for a point $t$, $t \in D_\lambda$:

$$|\zeta - t| \le |\zeta - \lambda| + |t - \lambda| \le c(1-|\lambda|) + \frac{\delta}{2}(1-|\lambda|) \le \left(c + \frac{1}{2}\right)(1-|\mu|).$$

As the disks $D_\lambda$, $\lambda \in \sigma \backslash \{\mu\}$, $|\lambda| \ge |\mu|$, are pairwise disjoint (cf. formula (4)) it follows that the sum of their areas cannot exceed the area of the set $\bigcup_{\zeta \in e} D(\zeta, 2c(1-|\mu|))$:

$$\frac{\delta^2}{4} \sum_{\substack{\lambda \in \sigma \backslash \{\mu\} \\ |\lambda| \ge |\mu|}} (1-|\lambda|)^2 \le 4c^2(1-|\mu|)^2 \cdot \operatorname{card} e.$$

Applying the reasoning (7) once more (with the numbers $(1-|\lambda|)^2$ instead of $1-|\lambda|$) we conclude that $\sigma$ is a finite union of $(N)$-sets and therefore we must have $\sigma \in (wN)$. □

**Vinogradov-Goluzina Lemma.**

$$\mathrm{F} \Rightarrow \sup_{\zeta \in \sigma} \sum_{\lambda \in \sigma} \frac{1-|\lambda|}{|1-\bar{\lambda}\zeta|} < \infty \Rightarrow \sigma \in (CN)$$

*Proof*. Let us linearize Frostman's condition, that is, putting

$$f_a = \sum_{\lambda \in \sigma} a_\lambda \frac{1-|\lambda|}{1-\bar{\lambda}z}, \quad a \in l^\infty(\sigma),$$

we rewrite it in the form

$$\alpha \overset{\text{def}}{=} \sup\{\|f_a\|_\infty : \|a\|_{l^\infty(\sigma)} \le 1\} < \infty.$$

It is then clear that $\beta \le \alpha$ where

$$\beta = \sup\left\{\sup_{\zeta \in \sigma} |f_a(\zeta)| : \|a\|_{l^\infty(\sigma)} \le 1\right\} = \sup_{\zeta \in \sigma} \sum_{\lambda \in \sigma} \frac{1-|\lambda|}{|1-\bar{\lambda}\zeta|}.$$

If now the condition $\beta < \infty$ is fulfilled then for any $\zeta$, $\zeta \in \sigma$,

$$\sum_{\lambda \in \sigma} \frac{(1 - |\lambda|^2)(1 - |\zeta|^2)}{|1 - \bar{\lambda}\zeta|^2} \leq 4 \sum_{\lambda \in \sigma} \frac{1 - |\lambda|}{|1 - \bar{\lambda}\zeta|} \cdot \frac{1 - |\zeta|}{1 - |\lambda| \cdot |\zeta|} \leq 4\beta. \square$$

**Naftalevič's Lemma.** *Let $\sigma \subset \mathbb{D}$, $\sum_{\lambda \in \sigma}(1 - |\lambda|) < \infty$. There exists a real valued function $\varphi$ on $\sigma$ such that $\sigma' \in (C)$ if*

$$\sigma' = \{\lambda' : \lambda' = |\lambda| e^{i\varphi(\lambda)}, \lambda \in \sigma\}.$$

*Proof*. Let us enumerate the set $\sigma$ in such a manner that $\sigma = \{\lambda_n : n \geq 1\}$ and $r_n \overset{\text{def}}{=} |\lambda_n| \leq r_{n+1}$, $n \geq 1$. Set

$$\varphi(\lambda_n) = \varphi_n \overset{\text{def}}{=} \sum_{k \geq n} (1 - r_k), \quad n \geq 1.$$

Let us check the sparseness condition $(R)$: if the integer $N$ is sufficiently large and $m > n > N$ then

$$|\lambda_n' - \lambda_m'| \geq \frac{1}{2}|\varphi_n - \varphi_m| = \frac{1}{2} \sum_{k=n}^{m-1} (1 - r_k) \geq \frac{1}{2}(1 - r_n)$$

and therefore (cf. (5)) with a suitable $\delta$, $\delta > 0$, condition $(R)$ is fulfilled for the set $\{\lambda_n' : n > N\}$.

Let us show that $\nu_{\sigma'}$ is a Carleson measure. By the Lemma on Carleson Measures and the remark after that lemma it suffices to estimate the $\nu_{\sigma'}$-mass of the "rectangle" $Q(\zeta, l) = \{t : 1 - l \leq |t| \leq 1, |\arg\zeta - \arg t| \leq l/2\}, 0 < l < 1, \zeta \in \mathbb{T}$. Let $n_1$ and $n_2$ be integers such that

$$\{n : \lambda_n' \in Q(\zeta, l)\} = \{n : n_1 \leq n \leq n_2\}.$$

Then (if $n_2 = \infty$ we put formally $\varphi_{n_2} = 0$, $r_{n_2} = 1$)

$$\begin{aligned} \sum_{\lambda_n' \in Q(\zeta, l)} (1 - r_n^2) &\leq 2 \sum_{\lambda_n' \in Q(\zeta, l)} (1 - r_n) \\ &= 2(\varphi_{n_1} - \varphi_{n_2}) + 2(1 - r_{n_2}) \leq 4l. \square \end{aligned}$$

## Supplements and Bibliographical Notes

**4. Carleson Series.** As we have already noticed condition $(C)$ is for $\dim E > 1$ not any longer necessary for the eigenspaces of a model operator $T$ to be an unconditional basis for the space $H^2(E) \ominus \Theta H^2(E)$. As a (counter-)example we may take the direct (orthogonal) sum $T_1 \oplus T_2$ of two operators both having Carleson spectra but the latter being so close to each other that their union does not any longer fulfil condition $(C)$. This example indicates the possibility to distinguish "series" of eigenvalues satisfying the Carleson condition, provided the eigenspaces form a Riesz basis and the angular parameters ${}_{\lambda}\Delta$ of the series of subspaces $(\mathbb{1} - \bar{\lambda}z)^{-1}{}_{\lambda}\Delta E$ under consideration are "not too much dispersed". The converse of this argument is also plausible: the asymptotic orthogonality of the "angular directors" ${}_{\lambda}\Delta$ of the Carleson series must entail that the whole system is a basis.

In this Section we are going to modify somewhat our notation: if $\sigma$ and $\sigma'$ are countable subsets of the disk $\mathbb{D}$ then $\Gamma(\sigma, \sigma')$ will be the matrix (operator)

$$\{\Gamma_{\lambda,\mu} : \lambda \in \sigma, \mu \in \sigma'\} \stackrel{\text{def}}{=} \left\{ \frac{(1-|\lambda|^2)^{1/2}(1-|\mu|^2)^{1/2}}{1-\bar{\lambda}\mu} : \lambda \in \sigma, \mu \in \sigma' \right\},$$

so that in particular $\Gamma(\sigma) \stackrel{\text{def}}{=} \Gamma(\sigma, \sigma)$ stands for the Gram matrix (operator) of the family of functions $\{(1-|\lambda|^2)^{1/2}(\mathbb{1}-\bar{\lambda}z)^{-1} : \lambda \in \sigma\}$ in the space $H^2$. The requirement of this family to be a basis is equivalent to Carleson's condition; from the estimates of the previous Lecture it follows that if $\sigma$ is a Carleson set then there exist constants $\underline{\gamma}$ and $\bar{\gamma}$ such that $\underline{\gamma} I \le \Gamma(\sigma) \le \bar{\gamma} I$ and vice versa, and similarly the condition $\sigma \in (\overline{CN})$ is equivalent the continuity of the operator $\Gamma(\sigma)$ (i.e. $\Gamma(\sigma) \le \bar{\gamma} I$); also if $\delta = \delta(\sigma)$ is the lower bound in condition $(C)$ then

$$\underline{\gamma}^{-1} \le \frac{32}{\delta^2}\left(1 + 2\log\frac{1}{\delta}\right), \quad \bar{\gamma} \le 32\left(1 + 2\log\frac{1}{\delta}\right).$$

The lower and the upper bounds of the Gram matrix (operator) characterize the deviation of the family under investigation from an orthogonal one. Selecting from the family of eigenspaces $\mathscr{X} = \{E_\lambda : \lambda \in \sigma_p\}$, $E_\lambda = \frac{{}_\lambda\Delta E}{\mathbb{1}-\bar{\lambda}z}$, of the operator $\mathbf{T}_\Theta^*$ a "well-behaved" part $\mathscr{X}(\sigma) \stackrel{\text{def}}{=} \{E_\lambda : \lambda \in \sigma\}$, we can take into account not only the geometry of $\sigma$ but also this deviation. Let $\Delta$ be an orthogonal projection in $E$ and let $0 < \underline{\gamma} < \bar{\gamma} < \infty$. The family $\mathscr{X}(\sigma)$ will be called a *$C(\Delta, \underline{\gamma}, \bar{\gamma})$-Carleson series* if 1) $\lim_{\lambda \in \sigma, |\lambda| \to 1} |\Delta - {}_\lambda\Delta| = 0$; 2) $\underline{\gamma} I \le \Gamma(\sigma) \le \bar{\gamma} I$ (so that in particular $\sigma$ satisfies the Carleson condition). The word "series" will also be used in a wider sense, for distinguishing a certain subset (subsequence) of the point spectrum.

**Series Theorem.** *Let $T = \mathbf{T}_\Theta$ be the compression of the shift onto the subspace $K_\Theta = H^2(E) \ominus \Theta H^2(E)$, satisfying the conditions* a)–b) *of Lecture VI, Sect.* 5 *(i.e. the eigenspaces of $T$ and $T^*$ are total in $K_\Theta$ and the resolvent $R(\cdot, T)$ is meromorphic in the disk $\mathbb{D}$).*

1. *If the family of eigenspaces* $\mathscr{X} = \left\{ \frac{{}_\lambda\Delta E}{\mathbb{1}-\bar{\lambda}z} : \lambda \in \sigma_p(T) \right\}$ *is the union of $N$ $C(\Delta^{(i)}, \underline{\gamma}_i, \bar{\gamma}_i)$-Carleson series, $1 \le i \le N$, and if further*

$$\underline{\gamma}_i > \sum_{j \ne i} \sqrt{\bar{\gamma}_i \bar{\gamma}_j}\,|\Delta^{(i)}\Delta^{(j)}|, \quad i = 1, 2, \dots, N, \tag{8}$$

*then $\mathscr{X}$ is a Riesz basis.*

2. *Let $\delta = \{\delta_\lambda : \lambda \in \sigma\}$ be a family of projection operators in E generating the eigenspaces $E(\delta_\lambda) = (\mathbb{1}-\bar{\lambda}z)^{-1}\delta_\lambda E$; $\mathbb{0} < \delta_\lambda \le {}_\lambda\Delta$. If the family $\delta$ is relatively compact in $L(E)$ then the scalar and the vectorial imbedding theorems for the set $\sigma$ are equivalent:*

$$\sum_{\lambda \in \sigma} (1-|\lambda|^2)\|\delta_\lambda f(\lambda)\|_E^2 < \infty, \quad \forall f \in H^2(E) \Leftrightarrow \sigma \in (CN).$$

3. *If* $\dim E < \infty$ *and $\mathscr{X}$ is a uniformly minimal family then the point spectrum $\sigma_p(T)$ is the union of not more than* $\dim E$ *sparse (satisfying the condition* (R)*) sets $\sigma_i$,*

$i = 1, \ldots, n$; $n \le \dim E$. *In this decomposition we can take account of the points of* $\sigma_p(T)$ *with their multiplicity, that is, we can achieve that*

$$\dim E_\lambda \equiv \sum_{i=1}^{n} \chi_{\sigma_i}(\lambda), \quad \lambda \in \sigma_p(T).$$

Before proving this theorem let us discuss some connections between and some consequences of the various statements. The simplest case when condition (8) is fulfilled is when the series $C(\Delta^{(i)}, \underline{\gamma_i}, \overline{\gamma}_i)$ are *asymptotically orthogonal*, i.e. the condition $\Delta^{(i)}\Delta^{(j)} = \mathbb{O}$, $i \neq j$.

**Corollary 1.** *If* $\mathscr{X}$ *can be written as the finite union of asymptotically orthogonal Carleson series then* $\mathscr{X}$ *is a Riesz basis.* □

The Statements 2 and 3 of the Series Theorem are partial converses of Statement 1, because the imbedding theorem as well as the condition of uniform minimality is a necessary condition for $\mathscr{X}$ to be a basis (cf. the Theorem on Riesz Bases, II; Lecture VI). In the finite dimensional case ($\dim E < \infty$) the premise of Statement 2 is fulfilled in particular if $\delta_\lambda = {}_\lambda\Delta$, $\sigma = \sigma_p(T)$ so that we get.

**Corollary 2.** *If* $\dim E < \infty$ *and* $\mathscr{X}$ *is a Riesz basis then* $\sigma_p(T)$ *is the union of not more than* $\dim E$ *Carleson sets (and in general not a fewer number either).* □

**Corollary 3.** *If* $\Delta = \{{}_\lambda\Delta : \lambda \in \sigma_p(T)\}$ *is a relatively compact family of projection operators and* $\mathscr{X}$ *uniformly minimal then* $\mathscr{X}$ *is a Riesz basis iff* $\sigma_p(T)$ *is a finite union of Carleson sets.* □

Other connections between the geometric properties of the family $\mathscr{X}$ and the location of the spectrum, as well as some quantitative refinements of the results obtained, will be given after the proof of the theorem. Besides, some such refinements are already contained in the following lemmata preparing the proof.

**Hadamard's Lemma.** *Let* $H$ *be a Hilbert space,* $n$ *a natural number and* $A = \{A_{ij}\}_{1 \le i,j \le n}$ *a bounded operator in the direct product* $\mathscr{H} = \sum_{i=1}^{n} \oplus H$,

$$Ax = \left\{\sum_{j=1}^{n} A_{ij}x_j\right\}_{1 \le i \le n}, \quad x = \{x_j\}_{1 \le j \le n} \in \mathscr{H},$$

*where* $A_{ij} \in L(H)$, $1 \le i, j \le n$. *If the operators* $A_{ii}$, $1 \le i \le n$, *are invertible in* $H$ *and*

$$|A_{ii}^{-1}|^{-1} > \sum_{\substack{1 \le j \le n \\ j \neq i}} |A_{ij}|, \quad i = 1, 2, \ldots, n,$$

*then* $A$ *is invertible in* $\mathscr{H}$.

*Proof*. Let us equip $\mathscr{H}$ with another norm $|\cdot|$, equivalent to the Euclidean one, putting

$$|x| = \max_{1 \le j \le n} \|x_j\|_H, \quad x = \{x_j\}_{1 \le j \le n}.$$

Let $\operatorname{Diag} A = \{A_{ij}\delta_{ij}\}_{1 \le i,j \le n}$ be the diagonal part of $A$; it is clear that $\operatorname{Diag} A$ is invertible in $\mathscr{H}$ and that

$$B_{ij} = \begin{cases} A_{ii}^{-1} A_{ij}, & i \neq j \\ \mathbb{O}, & i = j \end{cases}$$

where $B = (\operatorname{Diag} A)^{-1}(A - \operatorname{Diag} A)$. The norm of $B$ in the space $(\mathscr{H}, |\cdot|)$ is estimated in the obvious way:

$$\begin{aligned} |Bx| &= \max_{1 \le i \le n} \left\| \sum_{j=1}^{n} B_{ij} x_j \right\|_H \\ &\le \max_{1 \le i \le n} \left( \max_{1 \le j \le n} \|x_j\|_H \right) \sum_{j=1}^{n} |B_{ij}|_H \\ &= |x| \max_{1 \le i \le n} \sum_{j=1}^{n} |B_{ij}|_H . \end{aligned}$$

By the hypothesis of the lemma $|B|_{(\mathscr{H}, |\cdot|)} < 1$ so that $I + B$ is invertible in $\mathscr{H}$ and therefore also

$$(\operatorname{Diag} A)(I + B) = \operatorname{Diag} A + (A - \operatorname{Diag} A) = A. \ \square$$

**Lemma on Angular Directors.** *Let $\sigma$, $\delta$ and $E(\delta_\lambda)$ have the same meaning as in Statement 2 of the Series Theorem and set $\mathscr{X}(\delta) = \{E(\delta_\lambda) : \lambda \in \sigma\}$.*

1. *If $\alpha \stackrel{\text{def}}{=} \sup |\delta_\lambda - \delta_\mu| < 1$ and the Gram operator $\Gamma_{\mathscr{X}(\delta)}$ is continuous (i.e. the vectorial imbedding theorem holds true for $\mathscr{X}(\delta)$) then $\sigma \in (CN)$.*

2. *If $\underline{\gamma} = \inf \Gamma_{\mathscr{X}(\delta)}$, $\overline{\gamma} = \sup \Gamma_{\mathscr{X}(\delta)}$ and $\alpha^2 \le \frac{7}{23} \underline{\gamma}/\overline{\gamma}$, assuming that $\mathscr{X}(\delta)$ is a Riesz basis, then $\sigma \in (C)$ (that is, the Carleson condition is fulfilled).*

*Proof.* 1. We write the Gram matrix $\Gamma_{\mathscr{X}(\delta)} = \{\Gamma_{\lambda\mu} \delta_\mu \delta_\lambda\}$ as a sum of two positive matrices $\Gamma'$ and $\Gamma''$, the first of which being defined by the identities $\Gamma'_{\lambda,\mu} = \Gamma_{\lambda,\mu} \delta_\mu \delta' \delta_\lambda$ $(\lambda, \mu \in \sigma)$ where $\delta' = (\cdot, e)e$ is orthogonal projection in $E$ in the direction $e$, $\|e\| = 1$, $e \in \delta_{\lambda_0} E$, $\lambda_0 \in \sigma$. Then $\Gamma' \le \overline{\gamma} I$. This makes it possible to estimate the Gram operator $\Gamma(\sigma) = \{\Gamma_{\lambda,\mu}\}$: given a numerical family $a$, $a \in l^2(\sigma)$, and the family of unit vectors $e_\lambda = \delta_\lambda e / \|\delta_\lambda e\|$, $\lambda \in \sigma$, we get

$$\begin{aligned} (\Gamma(\sigma) a, a) &= \sum_{\lambda,\mu} \Gamma_{\lambda,\mu} a_\lambda \bar{a}_\mu = \sum_{\lambda,\mu} \Gamma_{\lambda,\mu} a_\lambda \bar{a}_\mu \frac{(e_\lambda, e)(e, e_\mu)}{(e_\lambda, e)(e, e_\mu)} \\ &= (\Gamma' x, x) \le \overline{\gamma} \|x\|^2 \le \overline{\gamma} (1 - \alpha^2)^{-1} \|a\|^2, \end{aligned}$$

where $x = \{x_\lambda : \lambda \in \sigma\}$, $x_\lambda = a_\lambda e_\lambda / (e_\lambda, e)$, because

$$\begin{aligned} |(e_\lambda, e)|^2 &= \|\delta_\lambda e\|^2 = 1 - \|(I - \delta_\lambda) e\|^2 \\ &= 1 - \|(\delta_{\lambda_0} - \delta_\lambda) e\|^2 \ge 1 - \alpha^2, \end{aligned}$$

so that indeed

$$\begin{aligned} \|x\|^2 &= \sum_\lambda \|x_\lambda\|^2 \le (1 - \alpha^2)^{-1} \sum_\lambda |a_\lambda|^2 \\ &= (1 - \alpha^2)^{-1} \|a\|^2. \end{aligned}$$

Thus $|\Gamma(\sigma)| \le \overline{\gamma} (1 - \alpha^2)^{-1}$ and consequently $\sigma \in (CN)$.

2. In an analogous manner we likewise estimate the difference $\Gamma(\sigma)-\Gamma$ where $\Gamma$ is the restriction of the operator $\Gamma_{\mathscr{X}(\delta)}$ to the subspace $l^2(\delta'_\lambda E:\lambda\in\sigma)$, $\delta'_\lambda \overset{\text{def}}{=} (\cdot, e_\lambda)e_\lambda$ (more exactly, this last space is in a natural way isomorphic to the space $l^2$ where we consider both operators $\Gamma(\sigma)$ and $\Gamma$, also $(\Gamma)_{\lambda,\mu}=\Gamma_{\lambda,\mu}(e_\lambda, e_\mu)$). In fact,

$$(\Gamma(\sigma)-\Gamma)_{\lambda,\mu}=\Gamma_{\lambda,\mu}(1-(e_\lambda,e_\mu))=\Gamma_{\lambda,\mu}\{(e,e-e_\mu)+(e-e_\lambda,e_\mu)\}$$

and

$$(e,e-e_\mu)=1-\|\delta_\mu e\|<1-(1-\alpha^2)^{1/2},$$

$$|(e-e_\lambda,e_\mu)|\le\|e-e_\lambda\|\cdot\|e_\mu-e\|+(e,e-e_\lambda)$$

$$\le 2(1-(1-\alpha^2)^{1/2})+(1-(1-\alpha^2)^{1/2}),$$

so that

$$|\Gamma(\sigma)-\Gamma|\le|\Gamma(\sigma)|\cdot\left\{\sup_\mu(e,e-e_\lambda)+\sup_{\lambda,\mu}|(e-e_\lambda,e_\mu)|\right\}$$

$$\le|\Gamma(\sigma)|\cdot 4(1-(1-\alpha^2)^{1/2}).$$

Consequently $\Gamma(\sigma)=\Gamma-(\Gamma-\Gamma(\sigma))=\Gamma(I-\Gamma^{-1}(\Gamma-\Gamma(\sigma)))$ and, as $\Gamma\ge\underline{\gamma}I$,

$$|\Gamma^{-1}(\Gamma-\Gamma(\sigma))|\le\frac{\overline{\gamma}}{\underline{\gamma}(1-\alpha^2)}\cdot 4(1-(1-\alpha^2)^{1/2})<1.$$

Thus the operator $\Gamma(\sigma)$ is an isomorphism of the space $l^2$, that is, $\sigma\in(C)$. □

In the lemma just proved the decomposition of the spectrum in Carleson series depends on crude metrical conditions concerning the nearness of the "angular directors" ${}_\lambda\Delta$ for different $\lambda$ but does not use linear-algebraic relations in the space $E$. The following lemma is free from this defect.

**Vasyunin's Lemma.** *Let $\{\lambda_n\}_{n\ge1}$ be a sequence of points of the disk $\mathbb{D}$, let $e_n\in E$, $\|e_n\|=1$, $n\ge1$, and assume that the family*

$$\mathscr{X}=\{(1-|\lambda_n|^2)^{1/2}(\mathbb{1}-\bar\lambda_n z)^{-1}e_n : n\ge1\}$$

*is uniformly minimal in the space $H^2(E)$, with the constant of uniform minimality $\varepsilon\overset{\text{def}}{=}d(\mathscr{X})$. Then any disk*

$$D_n(r)\overset{\text{def}}{=}\{\zeta\in\mathbb{C}:|b_{\lambda_n}(\zeta)|\le r\},\quad n\ge1,$$

*does not contain more than* $\dim E$ *points of the sequence $\{\lambda_k\}_{k\ge1}$, provided only $r<\varepsilon^2/(4\dim E)$.*

*Proof*. Let $\mathscr{E}_n$ be "skew" projections corresponding to the family $\mathscr{X}=\{x_n\}_{n\ge1}$ ($\mathscr{E}_n=(\cdot,x'_n)x_n$, $n\ge1$, where $\{x'_n\}_{n\ge1}$ is a family biorthogonal to $\mathscr{X}$); then $\varepsilon^{-1}=\sup\{|\mathscr{E}_n|:n\ge1\}$. Assuming that the lemma is not true we pick an index $i$ such that

$$\operatorname{card}M\ge\dim E+1,\quad M\overset{\text{def}}{=}\{k\in\mathbb{N}:|b_{\lambda_i}(\lambda_k)|\le r\}.$$

Let $n$, $n \in M$, be such that the angular coefficient $e_n$ can be expressed linearly in terms of the remaining $e_k$, $k \in M \backslash \{n\}$. As $D_n(2r) \supset D_i(r)$ (because $\lambda_n \in D_i(r)$) we have $M \subset \{k \in \mathbb{N} : |b_{\lambda_n}(\lambda_k)| \le 2r\}$. Let us now make a unitary transformation

$$f \overset{U}{\mapsto} \frac{(1 - |\lambda_n|^2)^{1/2}}{\mathbb{1} - \bar{\lambda}_n z} f\left(\frac{\lambda_n \mathbb{1} - z}{\mathbb{1} - \bar{\lambda}_n z}\right)$$

of the space $H^2(E)$ onto itself; we get $\tilde{x}_k \overset{\text{def}}{=} Ux_k = \alpha_k(1 - \bar{\mu}_k z)^{-1} e_k$, $k \in M$, with suitable $\alpha_k$ and $\mu_k = (\lambda_k - \lambda_n)(1 - \bar{\lambda}_n \lambda_k)^{-1}$, $k \in M$. It is clear that $\mu_n = 0$, $|\mu_k| \le 2r$ for $k \in M$ and that $\tilde{x}_n \equiv \text{const} \overset{\text{def}}{=} e$. Also $e = \sum_{k \in N} a_k e_k$ with $N \subset M \backslash \{n\}$, $\operatorname{card} N \le \dim E$, $a_k \in \mathbb{C}$. Set

$$x = \sum_{k \in K} a_k \tilde{x}_k$$

and let us show that in these hypotheses the functions $x$ and $\tilde{x}_n$ are closer to each other than permitted by our estimate of the norm $|\mathscr{E}_n|$ ($|\mathscr{E}_n| \le 1/\varepsilon$).

**Sublemma.** *If $e \in \mathscr{L}(e_k, k \in N)$ and $\mu_k \in \mathbb{C}$, $|\mu_k| \le \rho$ for $k \in N$, if $\mathscr{E}$ and $\mathscr{E}^k$ ($k \in N$) are spectral projections corresponding to the linearly independent system of functions $e \cdot \mathbb{1}$ and $\tilde{x}_k = (1 - |\mu_k|^2)^{1/2} \cdot (\mathbb{1} - \bar{\mu}_k z)^{-1} e_k$ ($k \in N$) and if the number $\rho$ ($0 < \rho < 1$) is so small that*

$$\frac{1}{\rho \operatorname{card} N} \ge \max\{|\mathscr{E}^k| : k \in N\} \overset{\text{def}}{=} \frac{1}{d},$$

*then*

$$|\mathscr{E}| \ge \frac{d - \rho \operatorname{card} N}{\rho \operatorname{card} N}.$$

Let us conclude the proof of Vasyunin's Lemma. In that situation $|\mathscr{E}| \le |\mathscr{E}_n|$, $|\mathscr{E}^k| \le |\mathscr{E}_k|$, $k \in N$, so that $d \ge \varepsilon$; if $\rho = 2r$ then $(\rho \operatorname{card} N)^{-1} \ge (2r \dim E)^{-1} > 2/\varepsilon^2 > 1/d$ and by the Sublemma

$$|\mathscr{E}_n| \ge |\mathscr{E}| \ge \frac{d - 2r \dim E}{2r \dim E} > \frac{\varepsilon - \varepsilon^2/2}{\varepsilon^2/2} = \frac{2 - \varepsilon}{\varepsilon} \ge \frac{1}{\varepsilon},$$

which is impossible. □

*Proof of the Sublemma.* We can assume that $\|e\| = \|e_k\| = 1$, writing $e = \sum_{k \in N} (1 - |\mu_k|^2)^{1/2} \cdot a_k e_k$. Set $x = \sum_{k \in N} a_k \tilde{x}_k$, $x_0 = e \cdot \mathbb{1}$. Then

$$\begin{aligned}\|x - x_0\|_{H^2(E)} &= \left\| \sum_{k \in N} a_k(\tilde{x}_k - (1 - |\mu_k|^2)^{1/2} e_k \cdot \mathbb{1}) \right\|_{H^2(E)} \\ &\le \sum_{k \in N} |a_k| \cdot |\mu_k| = \sum_{k \in N} |\mu_k| \cdot \|\mathscr{E}^k x\|_{H^2(E)} \le \frac{\rho \|x\|}{d} \operatorname{card} N.\end{aligned}$$

Moreover

$$\|x\| \le \|x - x_0\| + \|x_0\| \le 1 + \rho \|x\| \cdot d^{-1} \operatorname{card} N$$

whence

$$|\mathscr{E}| \geq \frac{\|\mathscr{E}(x-x_0)\|}{\|x-x_0\|} = \frac{\|x_0\|}{\|x-x_0\|} \geq \frac{d-\rho\,\mathrm{card}\,N}{\rho\,\mathrm{card}\,N}. \ \square$$

*Proof of the Series Theorem.* 1. Let $\mathscr{X}_1, \ldots, \mathscr{X}_N$ be the Carleson series in the theorem and $\sigma_1, \ldots, \sigma_N$ the corresponding parts of the spectrum $\sigma = \sigma_p(T)$ of our operator $T$. The idea is to replace $\Gamma_{\mathscr{X}}$ by $\Gamma(\sigma)$ and to see that this replacement does not too much "spoil" the properties of the Gram operator. What concerns the operator $\Gamma(\sigma)$ it can be given by a block matrix $\Gamma(\sigma) = \{\Gamma(\sigma_i, \sigma_j)\}_{1 \leq i,j \leq N}$ with small off-diagonal elements. In fact, if $a, b \in l^2$ then (let us recall that $\Gamma_{\lambda,\mu}$ are the matrix elements of $\Gamma(\sigma_i, \sigma_j)$ and that $x_\lambda = (1-|\lambda|^2)^{1/2}(\mathbb{1}-\bar{\lambda}z)^{-1}$)

$$\begin{aligned}|(\Gamma(\sigma_i,\sigma_j)a,b)_{l^2}| &= \Big|\sum_{\mu\in\sigma_i,\lambda\in\sigma_j}\Gamma_{\lambda,\mu}a_\mu\bar{b}_\lambda\Big| = \Big|\Big(\sum_{\mu\in\sigma_j}a_\mu x_\mu, \sum_{\lambda\in\sigma_i}b_\lambda x_\lambda\Big)_{H^2}\Big| \\ &\leq \Big\|\sum_{\mu\in\sigma_j}a_\mu x_\mu\Big\|_{H^2}\Big\|\sum_{\lambda\in\sigma_i}b_\lambda x_\lambda\Big\|_{H^2} = (\Gamma(\sigma_j)a,a)^{1/2}(\Gamma(\sigma_i)b,b)^{1/2} \\ &\leq \sqrt{\bar{\gamma}_i\bar{\gamma}_j}\,\|a\|_{l^2}\|b\|_{l^2},\end{aligned}$$

so that $|\Gamma(\sigma_i,\sigma_j)| \leq \sqrt{\bar{\gamma}_i\bar{\gamma}_j}$, $1 \leq i,j \leq N$, whereas $|\Gamma(\sigma_i)^{-1}|^{-1} \geq \underline{\gamma}_i$, $1 \leq i \leq N$. Let us return to the Gram operator $\Gamma_{\mathscr{X}}$ and write it in matrix form

$$\Gamma_{\mathscr{X}} = \{\Gamma(\mathscr{X}_i, \mathscr{X}_j)\}_{1 \leq i,j \leq N},$$

$$\Gamma(\mathscr{X}_i, \mathscr{X}_j) \stackrel{\text{def}}{=} \left\{\frac{(1-|\lambda|^2)^{1/2}(1-|\mu|^2)^{1/2}}{1-\bar{\lambda}\mu}\,{}_\lambda\Delta_\mu\Delta : \lambda\in\sigma_i,\ \mu\in\sigma_j\right\}.$$

The last transformation maps the space $l^2({}_\mu\Delta E : \mu\in\sigma_j) \stackrel{\text{def}}{=} l_j^2$ into $l^2({}_\lambda\Delta E : \lambda\in\sigma_i) \stackrel{\text{def}}{=} l_i^2$. Without loss of generality we may assume that the projections ${}_\lambda\Delta$, $\lambda\in\sigma_i$, are $\varepsilon$-close to the projection operator $\Delta^{(i)}$ (we achieve this by throwing away a suitable finite part of the spectrum); $\varepsilon$ is a sufficiently small positive number. Now it is easy to estimate the operators $\Gamma(\mathscr{X}_i, \mathscr{X}_j)$ in terms of $\Gamma(\sigma_i, \sigma_j)$. If $a \in l_j^2$, $b \in l_i^2$ then

$$|(\Gamma(\mathscr{X}_i,\mathscr{X}_j)a,b)| = \Big|\sum_{\substack{\lambda\in\sigma_i\\ \mu\in\sigma_j}}\Gamma_{\lambda\mu}({}_\lambda\Delta_\mu\Delta a_\mu, b_\lambda)\Big| \leq$$

(let us write ${}_\lambda\Delta$ and ${}_\mu\Delta$ as $\Delta^{(i)} + ({}_\lambda\Delta - \Delta^{(i)})$ and $\Delta^{(j)} + ({}_\mu\Delta - \Delta^{(j)})$, respectively, and use the fact that the norm of the Hilbert "replication" of the operator $\Gamma(\sigma_i,\sigma_j)$ is the same of $\Gamma(\sigma_i,\sigma_j)$ itself:

$$\begin{aligned}\Big|\sum_{\lambda,\mu}\Gamma_{\lambda,\mu}(u_\mu, v_\lambda)\Big| &= \Big|\sum_n\sum_{\lambda,\mu}\Gamma_{\lambda,\mu}\hat{u}_\mu(n)\overline{\hat{v}_\lambda(n)}\Big| \\ &\leq |\Gamma(\sigma_i,\sigma_j)|\sum_n\Big(\sum_\mu|\hat{u}_\mu(n)|^2\Big)^{1/2}\Big(\sum_\lambda|\hat{v}_\lambda(n)|^2\Big)^{1/2} \\ &\leq |\Gamma(\sigma_i,\sigma_j)|\,\|u\|\cdot\|v\|\end{aligned}$$

where $\hat{u}_\mu(n)$ and $\hat{v}_\lambda(n)$ are the Fourier coefficients of $u_\mu$ and $v_\lambda$ in some orthogonal basis in the space $E$)

$$\leq |\Gamma(\sigma_i,\sigma_j)| \left\{ |\Delta^{(i)}\Delta^{(j)}| + \sup_{\lambda\in\sigma_i} |(\Delta^{(i)} - {}_\lambda\Delta)\Delta^{(j)}| \right.$$

$$\left. + \sup_{\mu\in\sigma_j} |\Delta^{(i)}(\Delta^{(j)} - {}_\mu\Delta)| + \sup_{\mu\in\sigma_i} |\Delta^{(i)} - {}_\lambda\Delta| \sup_{\mu\in\sigma_j} |\Delta^{(j)} - {}_\mu\Delta| \right\} \|a\|\,\|b\|$$

$$\leq \sqrt{\bar{\gamma}_i\bar{\gamma}_j}\,\{|\Delta^{(i)}\Delta^{(j)}| + 3\varepsilon\}\,\|a\|\,\|b\|.$$

As the operator $\Gamma(\mathscr{X}_i,\mathscr{X}_j)$ admits the same lower bound as $\Gamma(\sigma_i)$:

$$(\Gamma(\mathscr{X}_i,\mathscr{X}_i)a,a) = \sum_{\lambda,\mu\in\sigma_i} \Gamma_{\lambda,\mu}({}_\mu\Delta a_\mu, {}_\lambda\Delta a_\lambda)_E$$

$$= \sum_n \sum_{\lambda,\mu\in\sigma_i} \Gamma_{\lambda\mu}\hat{a}_\mu(n)\overline{\hat{a}_\lambda(n)} \geq \sum_n \underline{\gamma}_i \sum_{\lambda\in\sigma_i} |\hat{a}_\lambda(n)|^2 = \underline{\gamma}_i\|a\|^2,$$

the matrix $\Gamma(\mathscr{X})$ has a dominant diagonal (cf. condition (8)):

$$|\Gamma(\mathscr{X}_i,\mathscr{X}_i)^{-1}|^{-1} \geq \underline{\gamma}_i > \sum_{j\neq i} |\Gamma(\mathscr{X}_i,\mathscr{X}_j)|; \quad i = 1,\ldots,N,$$

and so by Hadamard's Lemma must give an isomorphism from $l^2({}_\lambda\Delta E : \lambda\in\sigma)$ onto itself.

2. This statement has already been proved: the implication $\Leftarrow$ was established in Lecture VI (in fact at the end of Sect. 5 of that Lecture) and the converse is contained in the Lemma on Angular Directors (in view of the existence of a finite 1-net for the family of projections $\delta$).

3. In each of the subspaces ${}_\lambda\Delta E$, $\lambda\in\sigma_p(T)$, let us pick an orthonormal basis $e_\lambda^1, \ldots, e_\lambda^{k(\lambda)}$ and let us arrange the family of functions $\{(1-|\lambda|^2)^{1/2}(\mathbb{1}-\bar{\lambda}z)^{-1}e_\lambda^i : \lambda\in\sigma_p(T),\ 1\leq i\leq k(\lambda)\}$ in the form of a sequence $x_n = (1-|\lambda_n|^2)^{1/2}(\mathbb{1} - \bar{\lambda}_n z)^{-1}e_n$, $n\in\mathbb{N}$. It is clear that $\mathscr{X}' = \{x_n\}_{n\geq 1}$ is uniformly minimal and that its constant of uniform minimality coincides with the one for the original family $\mathscr{X}$. Apply Vasyunin's Lemma to the sequence $\mathscr{X}'$ taking $r$ to be a number such that

$$\operatorname{card}\{k : \lambda_k\in D_n(r)\} \leq \dim E, \quad n\geq 1.$$

Let $G_1, G_2, \ldots$ be the connected components of the set $\bigcup_{n\geq 1} D_n(r/(4\dim E))$ and set

$$\Lambda_s = \{k : \lambda_k\in G_s\}, \quad s\in\mathbb{N}.$$

Then $\operatorname{card}\Lambda_s \leq \dim E$, because if the opposite were true we could find a connected chain of not more than $\dim E + 1$ disks $D_n(r/(4\dim E))$ containing at least $\dim E + 1$ elements of the sequence $\{\lambda_k\}_{k\geq 1}$. All these disks are contained in a disk $D_n(r)$ whose center coincides with one of the points chosen. This contradicts Vasyunin's Lemma (the choice of $r$). There remains to pick the series with the sparseness condition: $\sigma_i = \{\lambda_k : k\in M_i\}$ where $M_1$ is such that $M_1\cap\Lambda_s$ contains exactly one point, $\forall s$; $M_2$ such that $M_2\cap(\Lambda_s\backslash M_1)$ contains exactly one point, $\forall s$ with $\Lambda_s\backslash M_1 \neq \varnothing$ etc. (the desired sparseness follows from $G_i\cap G_j = \varnothing$, $i\neq j$). □

In conclusion we remark that for the general model operator $\mathbf{T}_\Theta$ (complete and with a resolvent which is meromorphic in $\mathbb{D}$) one has not succeeded in attaining in all questions the same degree of clarity which goes with the model operator with a one dimensional defect ($\dim E = 1$). In particular, it is not clear if for $1 < \dim E < \infty$ the uniform minimality is equivalent to the family of eigenspaces (of vectors) of $\mathbf{T}_\Theta$ being a basis. On the basis of the results obtained one can only establish the following

**Corollary 4.** *Let $T = \mathbf{T}_\Theta$ be the operator in the Series Theorem, assuming $\dim E < \infty$. If the spectrum $\sigma_p(T)$ satisfies the Stolz Condition (S) or the Uniform Frostman Condition (F) then from the uniform minimality of the family $\mathscr{X} = \{E_\lambda : \lambda \in \sigma_p(T)\}$ of the eigenspaces of T it follows that $\mathscr{X}$ must be a Riesz basis.*

In fact, the uniform minimality of $\mathscr{X}$ shows that $\sigma_p(T) = \bigcup_{i=1}^n \sigma_i$ with $\sigma_i \in (R)$, $1 \leq i \leq n$. By Condition (S) or (F) then each $\nu_{\sigma_i}$, $1 \leq i \leq n$, must be a Carleson measure and therefore also the measure $\nu_{\sigma_p(T)}$. By the Theorem on Riesz Bases, II (Lecture VI) $\mathscr{X}$ must be a Riesz basis. □

**5. Remarks on Imbedding Theorems.** The "scalar" part of this Lecture (as the theory of free interpolation in general) has its origin in the celebrated paper by Carleson [1], which contains the implication $(C) \Rightarrow (CN)$. The general imbedding theorem $H^2 \subset L^2(\nu)$ is likewise due to Carleson [2] but the later and more transparent method of proof developed in the text has been communicated to us by Vinogradov. It has also turned out to be advantageous to use the Poisson kernel (and not the Cauchy kernel) and (above all!) the Vinogradov-Seničkin Test, for the proof of the continuity of the corresponding integral operator. This criterion (presented in the symmetrized form due to Havin) takes properly care of the main pecularities of the kernel of the imbedding operator – it decreases when deviating from the diagonal and its convolution with itself helps to manifest this decrease, but also from the quantitative point of view Vinogradov's method gives considerable advantages (it gives the best hitherto known – and possibly even the absolutely best – constant in the Imbedding Theorem). In particular, after a small modification the same method permits us to deduce an estimate of the embedding operator (orthogonalizer) $J_{\mathscr{X}'}$ having the "right order" for the case of "almost orthogonal" bases ($\delta \to 1$) too, namely $|J_{\mathscr{X}'}| \leq 1 + \text{const}(1 - \delta)^{1/2}$, $2^{-1} \leq \delta < 1$ (and $\lesssim \delta^{-1}(1 + \text{const}(1 - \delta)^{1/2})$, $2^{-1} \leq \delta < 1$, for the interpolating operator). A somewhat different treatment of Carleson's Imbedding Theorem is given in the survey article by Vinogradov-Havin [1, 2]. However, in other respects we rely on that paper. The Vinogradov-Seničkin Test appeared first in Nikol'skiĭ [8]. The Lemma on Carleson Measures in its essentials ($A < \infty \Rightarrow H^2 \subset L^2(\nu)$) was known to Carleson [1] but in the quantitative form given in the text (by the way, completely elementary and independent of the Imbedding Theorem) it is due to Vinogradov. The geometric characterization of Carleson measures obtained in this lemma is useful not only in the study of point measures connected with interpolation and bases but also in other situations. For instance, from it one gets at once the following (known) imbeddings:

$$f \in H^2 \Rightarrow \int_\gamma |f|^2\, ds < \infty$$

($\gamma$, any convex curve contained in the disk $\mathbb{D}$),

$$f \in H^2 \Rightarrow \int_0^1 |f|^2\, dx < \infty,$$

the second one clearly being equivalent to the boundedness in $l^2$ of the *Hilbert form*:

$$\int_0^1 |f|^2\, dx = \int_0^1 \sum_{n\geq 0} \hat{f}(n)x^n \left( \sum_{k\geq 0} \overline{\hat{f}(k)} x^k\, dx \right) = \sum_{n,k\geq 0} \frac{\hat{f}(k)\overline{\hat{f}(n)}}{n+k+1}.$$

The Imbedding Theorem of Carleson induces other imbedding theorems, either as immediate consequences (Rosenbaum [1], [2], Šamoyan [1], Vasyunin [1]) or in the form of a pattern to be imitated (Oleĭnik [1]). For instance, one can imagine imbedding theorems with derivatives $f^{(i)}$ in place of the functions $f$ themselves; since the corresponding estimates are as before connected with interpolation (however with multiple interpolation invoking not only the values of the functions at the nodes of interpolation but also of their derivatives; for details cf. Lectures IX–X), let us briefly consider one way of getting them, as outlined in the papers Šamojan [1] and Vasyunin [1]. Let $X$ be a Banach space of functions analytic in the disk $\mathbb{D}$ such that the functional $f \mapsto f(\lambda)$ ($\lambda \in \mathbb{D}$) is continuous in $X$, assuming further that

$$f \in X,\quad f(\lambda) = 0 \quad (\lambda \in \mathbb{D}) \Rightarrow f/b_\lambda \in X.$$

Let $0 \leq r < 1$ and $f \in X$; set

$$\psi_f(r) = \inf\{ \| f^* \|_X^2 : |f(\zeta)| \leq |f^*(\zeta)| \quad \text{for} \quad |\zeta| \leq r;$$
$$f^*(\zeta) \neq 0 \quad \text{for} \quad |\zeta| \leq r\}.$$

It is clear that $\psi_f$ is a increasing positive function.

**Conditional Imbedding Theorem.** *Let $\mu$ be a Borel measure on the disk $\mathbb{D}$, with $X \subset L^2(\mu)$, and let $\alpha$ and $\beta$ be measures on $[0,1]$ such that*

$$\int_0^1 \int_0^1 \frac{\psi_f(r + x(1-r))}{(1-x)^2}\, d\alpha(r)\, d\beta(x) < \infty, \quad f \in X. \tag{9}$$

*If $\beta_n = \int_0^1 x^{2n}\, d\beta(x)$, $n \geq 0$; $\gamma(R) = \int_R^1 d\alpha(r)$, $0 \leq R < 1$, then for any $f$ ($f \in X$) holds*

$$\int_{\mathbb{D}} \gamma(|\zeta|) \sum_{n\geq 0} \beta_n \left| \frac{f^{(n)}(\zeta)}{n!} \right|^2 (1 - |\zeta|)^{2n}\, d\mu(\zeta) \leq \text{const}\, \| f \|_X^2. \tag{10}$$

*Example 1.* If the space $X$ "admits division" by Blaschke products, that is, $\sup_{0\leq r<1} \psi_f(r) < \infty$, $\forall f \in X$, one can take as $\alpha$ the measure assigning mass one to the point 1 and as $\beta$ the measure $d\beta(x) = (1 - x)^{1+\varepsilon}\,dx$, with $\varepsilon > 0$. Then

$$\int_{\mathbb{D}} d\mu(\zeta)\left(\sum_{n\geq 0}\left|\frac{f^{(n)}(\zeta)}{n!}\right|^2 (1 - |\zeta|)^{2n}\frac{1}{n^{2+\varepsilon}}\right) \leq \text{const}\,\|f\|_X^2, \tag{11}$$

still assuming that $X \subset L^2(\mu)$. Division is possible, as we know, in all spaces $X = H^p$, $1 \leq p \leq \infty$, in the spaces

$$l_A^2(w_n) \overset{\text{def}}{=} \left\{f : f = \sum_{n\geq 0} \hat{f}(n)z^n,\ \sum_{n\geq 0} |\hat{f}(n)|^2 w_n^2 < \infty\right\}$$

provided $w_n \leq w_{n+1}$, $n \geq 0$ (cf. Lecture VIII, Sect. 4) and in some other cases.

If $X = H^2$ we can in this theorem as $\mu$ notably take Carleson measures.

However, in a result of type (11) we cannot hope for such a great generality as in the above imbedding theorem, because in general the factor $(1 - |\zeta|)^{-n}$ is significantly less then the ratio $|\phi_{\zeta,n}|_X/|\phi_\zeta|_X$ between the norms of the functionals $\phi_{\zeta,n}(f) = (1/n!)f^{(n)}(\zeta)$ and $\phi_\zeta(f) = f(\zeta)$. This circumstance justifies to some extent the presence of the decreasing weight $\gamma$ in inequality (10). In estimating the function $\psi_f$ one may try to divide by the Blaschke product $B_r = \prod_{|\lambda|\leq r} b_\lambda^{k_f(\lambda)}$ (taken with respect to the zeros of $f$ in the disk $\{\zeta : |\zeta| \leq r\}$ counted with their multiplicity $k_f$), $0 \leq r < 1$, but this leads only to $\psi_f(r) \geq \text{const}(\sum_{|\lambda|\leq r, k_f(\lambda)>0} (1 - |\lambda|))^2$, $0 \leq r < 1$. Some illustrations of this gives

*Example* 2. Let $X = L_A^2(\mathbb{D}) \overset{\text{def}}{=} \{f : f$ analytic in $\mathbb{D}$, $\int_{\mathbb{D}} |f|^2\,dx\,dy \overset{\text{def}}{=} \|f\|^2 < \infty\}$ be the famous Bergman space in the disk $\mathbb{D}$. In this case we know (Oleĭnik [1]) the necessary and sufficient conditions for the imbedding $X \subset L^2(\mu)$; they coincide with the conditions for $X^{(n)} \subset L^2((\mathbb{1} - |z|)^{2n}\,d\mu)$, where $X^{(n)} = \{f^{(n)} : f \in X\}$. From the Conditional Imbedding Theorem we get at once

$$L^2(\mathbb{D})^{(n)} \subset L^2\left((\mathbb{1} - |z|)^{2n+2}\left(\log\frac{1}{\mathbb{1} - |z|}\right)^{-2} d\mu\right),$$

because by a result of Džrbašjan [2] we have $\|f/B_r\|^2 \leq \text{const}\dfrac{\log(1 - r)^{-1}}{(1 - r)^2}\|f\|^2$.

(That this is rather crude compared with Oleĭnik's very exact results (just mentioned) is due to the rather rough estimate from below of the norm $\|B_r g\|$.)

More information on imbedding theorems of the Carleson type (connected with interpolation) can be found in the survey article by Vinogradov and Havin [1], [2].

**6. Concluding Remarks.** The proof of the implications $2 \Leftrightarrow 3$ of the Theorem on Bases and Interpolation was first obtained (Shapiro-Shields [1], Hoffman [1]) using estimates for the Gram matrix obtained with the aid of the well-known *Schur test: if*

$$\int |k(x, y)|p(y)\,dy \leq \alpha p(x), \qquad \int |k(x, y)|p(x)\,dx \leq \beta p(y)$$

*then* $|T_k|_{L^2\to L^2} \leq \sqrt{\alpha\beta}$. However, this condition cannot be applied immediately to the imbedding operator in question (or the Gram matrix $\Gamma_{\mathscr{X}}$, $\mathscr{X} = \{(1-|\lambda|^2)^{1/2}(\mathbb{1}-\bar{\lambda}z)^{-1} : \lambda \in \sigma_p(T)\}$) and in order to get a faster decrease of the off-diagonal terms these authors considered the modified family $\mathscr{X}'$ of the root vectors of the operator $S^*$, $\mathscr{X}' = \{(1-|\lambda|^2)^{3/2}(\mathbb{1}-\bar{\lambda}z)^{-2} : \lambda \in \sigma_p(T)\}$. (This incidentally corresponds to a problem of interpolation of derivatives:

$$(1-|\lambda|^2)^{3/2} f'(\lambda) = c_\lambda, \quad \lambda \in \sigma_p(T),$$

where $c = \{c_\lambda\} \in l^2(\sigma_p)$ and the unknown function $f$ has to be in $H^2$.)

The analysis of the various conditions imposed on the spectrum $\sigma_p(T)$ in the theory of interpolation is here on the whole carried out in the spirit of the paper Vinogradov-Havin [1]. The Corollary (of the Imbedding Theorem) on the decomposition of a set satisfying the Newman-Carleson condition into a finite sum of Carleson sets was without doubt known to the specialists very early; a formal reference can be found in McKenna [1]. The Kabaĭla-Newman Lemma was proved in Kabaĭla [1] and Newman [1]; it guarantees an exponential decrease of the elements $\Gamma_{n,n+m}$ of the matrix $\{\Gamma_{ij}\} = \{(x_{\lambda_i}, x_{\lambda_j})\}$ far off from the diagonal ($m \to \infty$). This case allows an "abstract" formulation and proof (Chalmers [1]) connected with the moment problem

$$(x, x_n) = a_n, \quad n \geq 1,$$

in the Hilbert space $H$. The conditions analogous to the condition ($N$) then take the form

$$\|x_n\| / \|x_{n+1}\| \leq r < 1, \quad |(x_n, x_k)| \leq c \min(\|x_n\|^2, \|x_k\|^2).$$

If $r < (1+2c)^{-1}$ it follows readily from Bari's Theorem that $J_{\mathscr{X}} H = l^2$, where $J_{\mathscr{X}} x = \{(x, x_n)\|x_n\|^{-1} : n \geq 1\}$. In Chalmers [1], as an abstract counterpart of the Stolz Condition ($S$), there are listed the estimates

$$\|x_n\|^2 \leq |(x_n, x_{n+1})| \quad \text{and} \quad |(x_m, x_n)| \leq c \cdot \min(\|x_n\|^2, \|x_m\|^2).$$

Together with the uniform minimality (the "Carleson condition") they imply "Newman's condition" (with the constants $r = (1 - d(\mathscr{X})^2)^{1/2}$ and $c$); for details see the paper Chalmers [1] just quoted where also some applications of these general results to Hilbert spaces of analytic functions (with reproducing kernels) other than the space $H^2$ are outlined.

Vinogradov's Lemma is proved in Vinogradov [3] but is also contained in the papers Nikol'skiĭ-Pavlov [1] and Vinogradov-Havin [1].

The Vinogradov-Goluzina Lemma is proved in Vinogradov-Havin [1]. The Uniform Frostman Condition was introduced earlier by Goluzina in Vinogradov-Goluzina-Havin [1] in connection with the study of multipliers and interpolation in spaces of integrals of the Cauchy-Stieltjes type; Frostman himself used an analogous pointwise condition (without the requirement of uniformity) in the theory of cluster sets of analytic functions (see Collingwood-Lohwater [1]). The proof of the lemma given in the text differs from the proof in Vinogradov-Havin [1]; if we have only the Condition ($F$) in mind then the desired result ($\sigma \in (CN)$) can be obtained in yet another way: as

$$\alpha \overset{\text{def}}{=} \sup\{\|f_a\|_\infty : \|a\|_\infty \le 1\} < \infty,$$

we have

$$|f_a'(\zeta)| \le \alpha \|a\|_\infty (1 - |\zeta|)^{-1},$$

or, otherwise,

$$\sum_{\lambda \in \sigma} |\lambda| \frac{(1 - |\lambda|^2)(1 - |\zeta|^2)}{|1 - \bar{\lambda}\zeta|^2} \le 4\alpha, \quad |\zeta| < 1. \square$$

Naftalevič's Lemma is borrowed from Naftalevič [1], where however the proof is rather complicated (pp. 14–15 and 17–19). The argument in the text is due to Vasyunin. There the radii $r_n$ are taken in increasing order in order to shorten the computations; the same distribution of arguments works also in the general case. Hruščev has found another and also very simple proof of the same lemma based on a decomposition of the disk into dyadic "squares". It is also worth while to mention that the Uniform Frostman Condition $(F)$ – in contrast to the Carleson Condition $(C)$ – puts some auxiliary bounds (besides the obvious Blaschke condition) on the set of moduli $\{|\lambda| : \lambda \in \sigma\}$ of a set $\sigma$ satisfying this condition. For instance, it is easily seen that $\sigma \in (F)$ entails $\sum_{\lambda \in \sigma} (1 - |\lambda|) \log(1 - |\lambda|)^{-1} < \infty$.

The comments on Sect. 4 of this Lecture ("Carleson series") will be preceded by a general remark on the advantages of the functional model compared with the classical approach to the study of spectral decompositions. The very first investigations on spectral decomposition were based on the Riesz formula

$$\mathscr{E}(\sigma) = \frac{1}{2\pi i} \int_\gamma R(\lambda, T)\, d\lambda$$

for the spectral projection onto an isolated portion of the spectrum. The criteria of "spectrality" (for instance, in the sense of Dunford) obtained with the aid of this formula – usually on the basis of an extremely refined analysis – depend on uniform estimates of the resolvent $R(\lambda, T)$ near the spectrum (on special systems of contours cutting the spectrum and contracting to it) and have as a rule the form of characterizations of the "density" of the spectrum of the operator $T$ under consideration. The history and the achievements of this direction is amply covered in the monographs Dunford-Schwartz [2, 3] and Gohberg-Kreĭn [1]. The first essential discovery after the resolvent method (which is still successfully applied in many problems) was the realization of the paramount rôle of calculi more powerful than the calculi resulting from estimates of the resolvent. The main example, which we have already encountered, is still von Neumann's inequality

$$|\phi(T)| \le \|\phi\|_\infty, \quad \phi \in \mathscr{P}_A,$$

valid for contractions $T$ in Hilbert space. If the spectrum of the contraction $T$ is "not too dense" then the weak completion of this calculus is far better capable of decomposing the spectrum (and in particular leads quicker to the "spectrality" of the operator, in one sense or the other) than the previous resolvent method;

regarding this theme consult the "Supplements" of Lecture IV and VIII. It is however clear that also such examples of the "spectrality" of an operator depend on rather crude geometric characteristics (on the spectrum, the root spaces etc.) and can be exact only for comparatively large classes of operators for which all these characteristics coincide. (For example, no natural calculus can decompose Jordan cells corresponding to just one point in the spectrum.)

The second major finding was the possibility of employing functional models in spectral theory. The beginnings of this direction can of course be traced back to Hilbert, v. Neumann and Stone – in the framework of normal operators –, regarding which a few words were said in the Intr. Lecture. For classes of operators not displaying any particular properties with respect to the natural involution $T \mapsto T^*$ one for a long time did not succeed in assigning suitable models. Only more recently one began to this end to use the shift operator $S$ in $H^2$; the initiators of this approach were the mathematicians of the Odessa school of operator theory (Livšic, Potapov, Brodskiĭ) but with particularly strong force the advantages of this new direction really imposed themselves after the introduction of the functional model of Sz.-Nagy and Foiaş. It turned out to be possible not only to employ in the problems of spectral theory the entire "scalar" theory of the Hardy classes but also to manifest with the aid of this model new inner structures describing the properties of the operator under study. In this and the previous Lectures such an auxiliary structure is provided by the "angular directors" ${}_{\lambda}\Delta$ of the eigenspaces of the truncated shift. The second Theorem on Riesz Bases (Lecture VI) and the Series Theorem rest on an analysis of orthogonal projections (and with their aid also spectral projections) onto the eigenspaces of the operator $T$, describing them in terms of the characteristic function $\Theta_T$. This makes it possible not only to get clearly perceivable necessary and sufficient conditions for the spectrality of the operator but also to employ the new "angular" parameters ${}_{\lambda}\Delta E$, taking into account (jointly with the "spectral" part $(\mathbb{1} - \bar{\lambda} z)^{-1}$) the geometry of the family of eigenspaces. Therefore we can express the hope that the effective criteria of bases resulting from these theorems will prove to be more accurate than any previous ones; some support for such a belief can be found, for instance, in the papers Pavlov [2, 3] and Ivanov-Pavlov [1].

As we have already noticed, the vectorial Carleson condition (concerning this condition see the Theorem on Riesz Bases (Lecture VI) as well as the remark to the Lemma on Eigenprojections (same Lecture))

$$\sup_{\lambda \in \sigma_p(T)} |\Delta_\lambda({}_{\lambda}\Theta(\lambda))^{-1}| = \sup_{\lambda \in \sigma_p(T)} |\Theta_\lambda(\lambda))^{-1}{}_{\lambda}\Delta| < \infty \tag{12}$$

is equivalent to the uniform minimality of the family $\mathscr{X}$ ($\sup_\lambda |\mathscr{E}_\lambda| < \infty$), i.e. a uniform estimate from below of the angles between the subspaces

$$\frac{1}{1-\bar{\lambda} z}\,{}_{\lambda}\Delta E \quad \text{and} \quad \bigvee\left(\frac{1}{\mathbb{1}-\bar{\mu} z}\,{}_{\mu}\Delta E : \mu \neq \lambda,\, \mu \in \sigma_p\right), \quad \lambda \in \sigma_p.$$

In this formulation the "angular" nature of the projections ${}_{\lambda}\Delta$ is apparent: worst is the case when ${}_{\lambda}\Delta = \text{const}$, $\lambda \in \sigma_p$, when (12) goes over into the Carleson

Condition $(C)$, best the case of an orthogonal family $\{_\lambda\Delta : \lambda \in \sigma_p\}$. The last of these possibilities is realized for a normal operator $T$ (when there is no doubt that the family of eigenspaces gives a basis) while the first possibility occurs in the case of the "one dimensional" model ($\dim E = 1$). It is pertinent to mention here too that the Carleson Condition $\sigma \in (C)$ is necessary and sufficient for the class of all contraction $T$ with a given (algebraically simple) point spectrum $\sigma_p(T) = \sigma$ to consist of scalar operators in the sense of Dunford (i.e., operators similar to a normal operator). This was first shown by Kacnel'son [1]. On the other hand, there exist operators of class $C_0$ with simple point spectra such that the family of eigenvectors is uniformly minimal but not a basis (Nikol'skiĭ-Pavlov [2]).

The Series Theorem is proved in the papers Nikol'skiĭ-Pavlov [1] (Statements 1 and 2) and Vasyunin [4] (Statement 3), however in a somewhat different form.

Hadamard's Lemma is established in Hadamard [1] for $\dim H = 1$ and is given for $\dim H < \infty$ in Gantmaher's treatise [1]. Not all of the known proofs are suitable for the generalization to the case $\dim H = \infty$; in the text there is used a variant of a reasoning by Parodi [1], p. 15–16.

The Lemma on Angular Directors is contained in Nikol'skiĭ-Pavlov [1] but in a somewhat less exact form. Besides, even its new formulation is not quite perfect – the bounds for the "dispersion" of the angular parameter $_\lambda\Delta$ (the bounds for the number $\alpha$ in the statement of the lemma), securing a Carleson spectrum, clearly are not exact. It would have been more adequate corresponding to the essence of the matter to have a proof directly using the language of the Imbedding Theorem, not the Gram matrices.

Vasyunin's Lemma is taken over from Vasyunin [4].

# Lecture VIII. Operator Interpolation and the Commutant

The construction and the study of a rich functional calculus is one way of getting spectral resolutions or their analogues. In the classical case of a self-adjoint (or normal) operator $A$ the calculus $f \mapsto f(A)$ defined on the algebra of all bounded functions and, in particular, on the characteristic functions $\chi_\sigma$ generates a projection valued (spectral) measure $\sigma \mapsto \mathscr{E}(\sigma) \stackrel{\text{def}}{=} \chi_\sigma(A)$, the study of which leads to the Spectral Theorem. This measure "separates" the spectrum of $A$ sufficiently well so that, for instance, if $A$ is complete we obtain an unconditionally convergent expansion into eigenvectors. For more general transformations having a total and topologically free family $\mathscr{X}$ of eigen (or root) spaces one can always define an analogue of the spectral measure, in a similar manner as this is provided by formula (6), Lecture VI, but in the first instance only on the (dense) linear hull of the family $\mathscr{X}$. If moreover the eigenvalues of the operator are isolated points of the spectrum then this "measure" can be considered as a part of the calculus (the "analytical" Riesz-Dunford calculus). For the compressions of the shift operator $T = PS|K$ we have to use the calculus constructed in Lecture III beyond which (as will be shown) we cannot go. The Sz.-Nagy and Foiaş calculus, it is true, makes it possible to construct the operator $f(T)$ only for analytic functions $f$ ($f \in H^\infty$) but this deficiency is compensated by the fact that in many cases $f(T)$ depends only on $f|\sigma(T)$ or even $f|\sigma_p(T)$. Such functions already can separate the spectrum of $T$ in the desired way.

A principal element of a new approach to the study of spectral resolutions is to obtain an exact estimate for the norm $|f(T)|$ which then in the case at hand allows one to write down the maximal family of functions of the operator $T$, that is, the weakly closed algebra $\mathscr{R}(T)$ generated by $T$ and $I$, and even (formally a more general set) the commutant $\{T\}' \stackrel{\text{def}}{=} \{A : AT = TA\}$ of $T$. In this Lecture (as well as in Lecture IX and X) we will see that our main result, viz. $\{T\}' = \{f(T) : f \in H^\infty\}$ (Sarason's Theorem), has an interpolation character. Here the word "interpolation" can be interpreted in the function theoretic sense as well as in operator sense. In order to clarify this circumstance we first consider a truncated shift $T$ with simple point spectrum. Thus we consider again bases of eigenvectors and it is then natural to apply the following theorem by Lorch and Grinblyum: the family $\mathscr{X} = \{x_\lambda : \lambda \in \sigma\}$ constitutes an unconditional basis for $\bigvee(\mathscr{X})$ iff all multipliers (i.e. bounded transformations of the type $\mu : x_\lambda \mapsto \mu_\lambda x_\lambda$, $\lambda \in \sigma$) come from arbitrary bounded functions on the set $\sigma$. If $Tx_\lambda = \lambda x_\lambda$ we then arrive at a special case of the following problem on operator interpolation: let $T$ be an operator in the space $X$ and $\mathscr{X} = \{X_n : n \in \mathfrak{N}\}$ a family of invariant subspaces; it is required to determine those collections of operators $\{a_n : n \in \mathfrak{N}\}$, $a_n \in \{T|X_n\}'$, $\sup\{|a_n| : n \in \mathfrak{N}\} < \infty$ ("interpolation data") which extend to a global transformation $A$ commuting with $T : A|X_n = a_n$ ($n \in \mathfrak{N}$) and $A \in \{T\}'$. If $\mathscr{X} = \{x_\lambda : \lambda \in \sigma_p(T)\}$ is the family of eigenfunctions of a compression of the shift operator then $A = f(T)$, $f \in H^\infty$ (Sarason's Theorem) so everything is reduced to the interpolation of the multipliers $\mu$ by the values $f(\lambda)$, $\lambda \in \sigma_p(T)$, of bounded analytic functions $f$. Indeed, in exactly the same manner we can study not only unconditionally convergent expansions but more generally the question of the dependence of the geometry of a family of vectors (subspaces) $\mathscr{X}$ on the size of the corresponding space of multipliers.

**1. Interpolation by Bounded Analytic Functions.** Since we have already found a criterion for the family $\mathscr{X} = \{x_\lambda : \lambda \in \sigma_p(T)\}$, $T = \mathbf{T}_B$ to be a basis (Lecture VI–VII, the Theorem on Bases and Interpolation), we will use it to derive the following theorem.

**Carleson's Theorem.** *Let $\sigma \subset \mathbb{D}$. Then*

$$H^\infty | \sigma = l^\infty(\sigma) \Leftrightarrow \sigma \in (C).$$

Before proving this result let us give the exact formulations of the theorems of Lorch-Grinblyum and Sarason alluded to above.

Let $X$ be a Banach space and $\mathscr{X} = \{x_n : n \in \mathfrak{N}\}$ a total (in $X$) family of vectors. Recall (cf. Lecture VI) that a complex valued function $\mu = \{\mu_n : n \in \mathfrak{N}\}$ is called a multiplier for the family $\mathscr{X}$ if the operator $M$ defined on the linear hull of $\mathscr{X}$ by the formulae $Mx_n = \mu_n x_n$, $n \in \mathfrak{N}$, extends to a continuous linear operator on $X$ (for which we retain the same letter $M$ and the appellation multiplier); $\operatorname{mult}(\mathscr{X})$ is the set of all multipliers.

**Lorch-Grinblyum Theorem.** *$\mathscr{X}$ is an unconditional basis in $X \Leftrightarrow \operatorname{mult}(\mathscr{X}) = l^\infty(\mathfrak{N})$.*

This is a special case of the Theorem on Unconditional Bases in Lecture VI, Sect. 4.

**Sarason's Theorem.** *Let $\Theta$ be an inner function and set $K = H^2 \ominus \Theta H^2$, $T = P_K S | K$. If the operator $A$ defined in $K$ commutes with $T$ ($AT = TA$) then there exists a function $a$, $a \in H^\infty$, "interpolating" $A$:*

$$A = a(T) \text{ and } |A| = \|a\|_\infty. \tag{1}$$

Let us use these results to prove Carleson's Theorem.

*Proof of Carleson's Theorem.* It is not hard to see that the Blaschke condition $\sum_{\lambda \in \sigma} (1 - |\lambda|) < \infty$ does not decrease the generality of the situation under study and therefore we can form the Blaschke product $\prod_{\lambda \in \sigma} b_\lambda$. Let $K = H^2 \ominus \mathbf{B} H^2$ and let $\mathscr{X}' = \{x'_\lambda : \lambda \in \sigma\}$ be the family of eigenfunctions of the operator $T = P_K S | K$, $\mathscr{X}$ being the corresponding biorthogonal family. As

$$(f(T)x'_\lambda, x_\mu) = (f x'_\lambda, x_\mu) = f(\lambda)\delta_{\lambda\mu} \quad (\forall \lambda, \mu \in \sigma, f \in H^\infty)$$

we have $f(T)x'_\lambda = f(\lambda)x'_\lambda$ $(\lambda \in \sigma)$ so that $\operatorname{mult}(\mathscr{X}) \supset \{f | \sigma : f \in H^\infty\}$. Now Sarason's Theorem shows that the two sets connected by this inclusion must coincide. The rest follows by combining the Theorem on Bases and Interpolation (Carleson's Condition $(C)$ implies that $\mathscr{X}'$ is an (unconditional) basis, and vice versa) and the Lorch-Grinblyum Theorem (the latter property is equivalent to $l^\infty(\sigma) = \text{mult}(\mathscr{X}')$). $\square$

*Remark.* In fact, the latter proof contains an estimate for the interpolating functions. Indeed, let $\sigma$ be a Carleson set, $\sigma \in \mathbb{D}$, set $\mathscr{X}' = \{x'_\lambda : \lambda \in \sigma\}$ and $\delta = \delta(\mathscr{X}') = \delta(\mathscr{X}) > 0$ (see Lecture VI, Sect. 1.2), and let further $\mu \in l^\infty(\sigma)$. Let us denote by $M$ the corresponding multiplier of the family $\mathscr{X}'$ ($Mx'_\lambda = \mu(\lambda)x'_\lambda$, $\lambda \in \sigma$) and by $M_0$ the diagonal operator on $l^2(\sigma): (M_0\alpha)(\lambda) = \mu(\lambda)\alpha(\lambda)$, $\alpha \in l^2(\sigma)$. Using the bounds for the orthogonalizer $J_{\mathscr{X}'}$ and its inverse $J_{\mathscr{X}'}^{-1}$ (see the remarks at the end of Sect. 2, Lecture VII) we get

$$\begin{aligned} |M| = |J_{\mathscr{X}'}^{-1} M_0 \mathscr{J}_{\mathscr{X}'}| &\le |J_{\mathscr{X}'}^{-1}| \cdot \|\mu\|_{l^\infty(\sigma)} \cdot |\mathscr{J}_{\mathscr{X}'}| \\ &\le 32\delta^{-1}(1 + 2\log\delta^{-1}) \|\mu\|_{l^\infty(\sigma)}. \end{aligned}$$

Now, if $f$ is a function "interpolating $M$" (in the sense of Sarason's Theorem) then

$$f \in H^\infty, f|\sigma = \mu \quad \text{and} \quad \|f\|_\infty = |M| \le 32\delta^{-1}(1 + 2\log\delta^{-1}) \|\mu\|_{l^\infty(\sigma)}. \ \square$$

It is possible to show that this estimate has the "right" growth order for $\delta \to 0$ (see Vinogradov-Havin [1]).

Let us now deduce from Sarason's Theorem two more results (classical) on interpolation by bounded analytic functions.

*The Carathéodory-Fejér Interpolation Problem* consists in finding those collections of complex numbers $(c_0, \dots, c_n)$ for which there exist a function $f$ analytic in $\mathbb{D}$ such that

a) $f \in H^\infty$, $\|f\|_\infty \le 1$;

b) $f = \sum_{k \ge 0} \hat{f}(k) z^k$, $\hat{f}(k) = c_k$ for $0 \le k \le n$ (another variant of property a): $\operatorname{Re} f \ge \mathbb{0}$ in $\mathbb{D}$).

Set $\Theta = z^{n+1}$, $K = H^2 \ominus \Theta H^2$. Then $K$ consists of all polynomials of degree not exceeding $n$ and $Tz^k \overset{\text{def}}{=} P_K S z^k = z^{k+1}$ for $0 \le k < n$, $Tz^n = \mathbb{0}$. The commutant $\{T\}'$ consists of those operators $A$ for which $AT^k = T^k A$, $k \ge 0$, and, in particular, $Az^k = T^k A\mathbb{1}$, $0 \le k \le n$. Thus an operator $A$ commuting with $T$ is uniquely defined by the function $\sum_{k=0}^n c_k z^k \overset{\text{def}}{=} A\mathbb{1}$ and its matrix in the basis $\{z^k\}_{k=0}^n$ takes the form

$$A \sim \begin{pmatrix} c_0 & & & & \\ c_1 & c_0 & & & \mathbb{0} \\ \cdot & \cdot & \cdot & & \\ \cdot & \cdot & \cdot & \cdot & \\ \cdot & \cdot & \cdot & & \cdot \\ c_n & c_{n-1} & \cdot & \dots & c_0 \end{pmatrix}. \tag{2}$$

Moreover

$$\begin{aligned} f(T) = A, \quad f \in H^\infty &\Leftrightarrow f(T)\mathbb{1} = A\mathbb{1} \Leftrightarrow P_K f = A\mathbb{1} \\ &\Leftrightarrow \sum_{k=0}^n \hat{f}(k) z^k = \sum_{k=0}^n c_k z^k. \end{aligned}$$

It follows that the Carathéodory-Fejér problem with given $(c_0, \dots, c_n)$ is solvable iff the matrix (2) defines a contraction operator ($|A| \le 1$). □

From this we get also a criterion for the solvability in the class $\{f : \operatorname{Re} f \ge 0\}$: $\operatorname{Re} A \ge \mathbb{0}$.

*The Nevanlinna-Pick Interpolation Problem* consists of finding those functions $w$ on a finite subset $\sigma$ of the disk $\mathbb{D}$ such that there exist a function $f$ with the properties:

a) $f \in H^\infty$, $\|f\|_\infty \le 1$;

b) $f|\sigma = w$.

Let us set

$$\Theta = B \stackrel{\text{def}}{=} \prod_{\lambda \in \sigma} b_\lambda, \quad K = H^2 \ominus \mathbf{B}H^2, \quad T = P_K S|K.$$

As all the eigenvalues of the (finite dimensional) operator $T$ are simple, every operator $A$ commuting with $T$ too must have the same eigenfunctions $x'_\lambda$, $\lambda \in \sigma : Ax'_\lambda = w(\lambda)x'_\lambda$, $\lambda \in \sigma$. It is now clear that

$$f(T) = A, \quad f \in H^\infty \Leftrightarrow f|\sigma = w.$$

Therefore the solvability of the Nevanlinna-Pick problem is equivalent to the condition $|A| \le 1$. Replacing this by the condition $|A^*| \le 1$ and noting that $A^* x_\lambda = \overline{w(\lambda)} x_\lambda$, $\lambda \in \sigma$, where the $x_\lambda = (\mathbb{1} - \bar{\lambda} z)^{-1}$ are the eigenfunctions of $T^*$, we get a necessary and sufficient condition for the solvability in the form $(x, x) - (A^* x, A^* x) \ge 0$, $x \in K$, or, translating this into the coordinate language for the basis $\{x_\lambda : \lambda \in \sigma\}$, in the form

$$\sum_{\lambda,\mu} c(\lambda)\overline{c(\mu)}(x_\lambda, x_\mu) - \sum_{\lambda,\mu} c(\lambda)\overline{c(\mu)} w(\lambda)\overline{w(\mu)}(x_\lambda, x_\mu)$$
$$= \sum_{\lambda,\mu} c(\lambda)\overline{c(\mu)} \frac{1 - \overline{w(\lambda)} w(\mu)}{1 - \bar{\lambda}\mu} \ge 0,$$

for any function $c$ on the set $\sigma$. Thus the Nevanlinna-Pick problem is solvable iff the matrix

$$\{(1 - \overline{w(\lambda)} w(\mu))(1 - \bar{\lambda}\mu)^{-1}\}_{\lambda,\mu \in \sigma}$$

defines a non-negative operator in $\mathbb{C}^n$ ($n = \operatorname{card} \sigma$). □

From this it is easy to derive a criterion for solvability in the class $\{f : \operatorname{Re} f \ge \mathbb{0}\}$ too.

**2. The Proof of Sarason's Theorem** will be divided into two independent propositions – the description of so-called "Hankel operators" and the expression of the functions of the model operator in terms of the latter. The last result will be useful also in some other questions touched upon in this Lecture. A *Hankel operator* is a continuous linear operator acting from $H^2$ into $H^2_-$ such that

$$P_- z A p = A z p, \quad p \in \mathscr{P}_A. \tag{3}$$

Clearly the matrix of a Hankel operator $A$ (with respect to the standard bases

$\{z^n\}_{n\geq 0}$ and $\{\bar{z}^k\}_{k\geq 1}$) has the following structure: $(Az^n, \bar{z}^k) = (A\mathbb{1}, \bar{z}^{n+k})$, that is, the matrix elements are constant on the diagonals perpendicular to the main diagonal. Conversely, for such matrices the relation (3) always holds true.

**Nehari's Theorem.** *$A$, $A \in L(H^2, H^2_-)$, is a Hankel operator iff there exists a function $f$, $f \in L^\infty(\mathbb{T})$, such that $A = H_f$, where*

$$H_f g \overset{\text{def}}{=} P_- fg, \quad g \in H^2.$$

*Moreocer* $\|A\| = \operatorname{dist}(f, H^\infty)$.

*Proof.* The sufficiency is trivial (i.e. the operator $H_f$ is continuous and (3) holds true). Let us therefore prove the necessity. From (3) we get $P_- z^n A\mathbb{1} = Az^n$, $n \geq 0$, and $P_- pA\mathbb{1} = Ap$, $p \in \mathscr{P}_A$. If $q \in \mathscr{P}_A$, $q(0) = 0$, then $\bar{q} \in H^2_-$ and we have

$$(Ap, \bar{q}) = (P_- pA\mathbb{1}, \bar{q}) = (pA\mathbb{1}, \bar{q}) = \int_{\mathbb{T}} pq \cdot A\mathbb{1}\, dm. \tag{4}$$

Let us now notice that the set $\{\varphi\psi : \|\varphi\|_{H^2} \leq 1, \|\psi\|_{H^2} \leq 1\}$ coincides with the unit ball of the space $H^1$ (the inclusion $\subset$ is trivial whereas $\supset$ follows from the representation $g = g_1 g_2, g_1 = \sqrt{g^e}, g_2 = g^i \sqrt{g^e}; g \in H^1$) and so the products $pq$ run through a set dense in the unit ball of the space $H^1_0 \overset{\text{def}}{=} \{g : g \in H^1, g(0) = 0\}$ when $p$, $q \in \mathscr{P}_A$, $\|p\|_2 \leq 1$, $\|q\|_2 \leq 1$, $q(0) = 0$. From this and formula (4) follows that

$$\|A\| = \sup_{p,q} |(Ap, \bar{q})| = \sup_{p,q} \left| \int_{\mathbb{T}} pqA\mathbb{1}\, dm \right| = \sup_{\substack{\|g\|_1 \leq 1 \\ g(0)=0}} \left| \int_{\mathbb{T}} gA\mathbb{1}\, dm \right|$$

and, in particular, that the functional

$$g \mapsto \int_{\mathbb{T}} gA\mathbb{1}\, dm, \quad g \in \mathscr{P}_A, \quad g(0) = 0, \tag{5}$$

extends to a continuous linear functional on $H^1_0$. Therefore there exists a function $f$, $f \in L^\infty(\mathbb{T})$, such that

$$\int_{\mathbb{T}} A\mathbb{1} \cdot g\, dm = \int_{\mathbb{T}} fg\, dm, \quad g \in \mathscr{P}_A, \quad g(0) = 0, \tag{6}$$

and thus, in view of (4), $(Ap, \bar{q}) = (H_f p, \bar{q})$ for $p$, $q \in \mathscr{P}_A$, $q(0) = 0$. It remains to make the observation that in view of the Hahn-Banach theorem the norm of the functional (5) (i.e. of (6)) on $H^1_0$ equals the distance of the generating function $f$, $f \in L^\infty(\mathbb{T})$, from the annihilator of this subspace (in the standard duality $\langle f, g \rangle = \int_{\mathbb{T}} fg\, dm$ between $L^\infty(\mathbb{T})$ and $L^1(\mathbb{T})$), i.e. from the space $H^\infty$, $H^\infty \subset L^\infty(\mathbb{T})$. □

Let now $\Theta$ be an inner function and set $K_\Theta = H^2 \ominus \Theta H^2$, $T = P_\Theta S | K_\Theta$. Let $\{T\}'$ be the commutant of $T$.

**Lemma.** *Let $A$ be an operator in the space $K_\Theta$ and consider $A_* \overset{\text{def}}{=} \Theta^* A P_\Theta$; this is an operator mapping $H^2$ into $H^2_-$. Then*

$$A \in \{T\}' \Leftrightarrow A_* \text{ is a Hankel operator.}$$

*Proof*. The relation $A \in \{T\}'$ means that

$$P_\Theta \mathbf{z} A P_\Theta = A P_\Theta \mathbf{z},$$

which is the same as $A_*$ being Hankel:

$$\Theta^* P_\Theta \mathbf{z} A P_\Theta \equiv P_- \mathbf{z} A_* = A_* \mathbf{z} \equiv \Theta^* A P_\Theta \mathbf{z}. \ \square$$

It remains to say just a few words more in order to conclude the proof of Sarason's Theorem: if $A \in \{T\}'$ there exists a function $f$, $f \in L^\infty(\mathbb{T})$, such that $\Theta^* A P_\Theta = H_f$. This shows in particular that $P_- fg = \mathbb{O}, \forall g \in \Theta H^2$, i.e. $f\Theta \in H^\infty$ and $f = \bar{\Theta} h$, $h \in H$. Futhermore we get

$$\Theta^* Ag = P_- \bar{\Theta} hg, \quad Ag = \Theta P_- \bar{\Theta} hg = P_\Theta hg = h(T)g \quad (g \in K_\Theta)$$

and it is likewise clear that in this formula one can replace $h$ by any function in the coset $h + \Theta H^\infty$. The existence of an extremal function $a$ such that $a(T) = A$ and $\|a\|_\infty = \inf\{\|h + \Theta g\|_\infty : g \in H^\infty\} = \operatorname{dist}(f, H^\infty) = |H_f| = |A|$ is obtained by minimizing the norm in the coset $h + \Theta H^\infty$: if $|A| = \lim_n \|h + \Theta g_n\|_\infty$ then one may of course assume that the sequence $\{g_n\}_{n \ge 1}$ converges in $\mathbb{D}$ to a function $g$, $g \in H^\infty$, and then $|A| = \|h + \Theta g\|_\infty$. $\square$

**Corollary.** *We have*

$$|P_\Theta \mathbf{h} | K_\Theta| = |P_+ \bar{\mathbf{h}} | K_\Theta| = \|h\|_{H^\infty / \Theta H^\infty} = \|\bar{\Theta} h\|_{L^\infty / H^\infty}$$
$$= \operatorname{dist}(\bar{\Theta} h, H^\infty). \ \square$$

**3. Compact Operators in $\{\mathbf{T}_\Theta\}'$.** Since the commutant $\{\mathbf{T}_\Theta\}'$ of the operator $\mathbf{T}_\Theta = P_K S | K$ consists of precisely the functions $f(\mathbf{T}_\Theta)$, $f \in H^\infty$, the problem of finding all compact operators commuting with $T = \mathbf{T}_\Theta$ is reduced to the study of the set

$$\{f : f \in H^\infty, f(T) \in \mathfrak{S}_\infty\}$$

where $\mathfrak{S}_\infty$ denotes the set of all compact (completely continuous) operators in $K = K_\Theta$.

**Hartman-Sarason Theorem.** *Let $\Theta$ be an inner function and set $K = K_\Theta = H^2 \ominus \Theta H^2$, $T = P_\Theta S | K$. Let further $f \in H^\infty$. The following statements are equivalent:*

1. $f(T) \in \mathfrak{S}_\infty$;
2. $\lim_n |T^n f(T)| = 0$;
3. $f\bar{\Theta} \in \bigvee_{L^\infty}(\bar{z}^n H^\infty : n \ge 0)$;
4. $f\bar{\Theta} \in H^\infty + C(\mathbb{T})$.

The equivalence of the last two statements means essentially that the algebraic sum $H^\infty + C(\mathbb{T})$ is closed in $L^\infty = L^\infty(\mathbb{T})$. At any rate, we reduce the proof of the implication $3 \Leftrightarrow 4$ to this closure property, if we remark that

$$\bigvee_{L^\infty}(\bar{z}^n H^\infty : n \ge 0) = \operatorname{clos}(H^\infty + C(\mathbb{T})).$$

This again is almost trivial:

$$H^\infty + C(\mathbb{T}) = H^\infty + \operatorname{clos}\mathscr{P} \subset \operatorname{clos}(H^\infty + \mathscr{P})$$
$$= \bigvee\nolimits_{L^\infty}(\bar{z}^n H^\infty : n \geq 0),$$

where $\mathscr{P}$ stands for the set of all polynomials in $z$ and $\bar{z}$, and

$$\bigvee\nolimits_{L^\infty}(\bar{z}^n H^\infty : n \geq 0) = \operatorname{clos} \bigcup_{n \geq 0} \bar{z}^n H^\infty \subset \operatorname{clos}(H^\infty + C(\mathbb{T})).$$

Let us further remark that once having the proof that the set $H^\infty + C(\mathbb{T})$ is closed we also get that it is a subalgebra of the algebra $L^\infty$ (since $H^\infty + \mathscr{P}$ is a subalgebra). Let us verify that the sum $H^\infty + C(\mathbb{T})$ indeed is closed.

**Lemma on $H^\infty + C(\mathbb{T})$.** *We have*

$$\operatorname{clos}_{L^\infty}(H^\infty + C(\mathbb{T})) = H^\infty + C(\mathbb{T}).$$

*The proof* makes use of the following elementary remark: if $f \in H^\infty$ and $f_r \stackrel{\text{def}}{=} f(rz)$, $0 \leq r < 1$, then $\|f_r\|_\infty \leq \|f\|_\infty$, and if $f \in C(\mathbb{T})$, $f_r \stackrel{\text{def}}{=} (\mathscr{H}f)(rz)$ (cf. p. 152), then $\lim_{r \to 1} \|f - f_r\|_\infty = 0$. These properties of the Poisson means lead to the identity (where $C_A \stackrel{\text{def}}{=} H^\infty \cap C(\mathbb{T})$)

$$\operatorname{dist}(f, H^\infty) = \operatorname{dist}(f, C_A), \tag{7}$$

for any function $f$ continuous on the circumference $\mathbb{T}$, for the inequality $\operatorname{dist}(f, H^\infty) \leq \operatorname{dist}(f, C_A)$ is trivial and its converse is a consequence of the estimate (for $f \in H^\infty$)

$$\|f - h\|_\infty \geq \lim_{r \to 1} \|(f - h)_r\|_\infty$$
$$\geq \lim_{r \to 1} (\|f - h_r\|_\infty - \|f - f_r\|_\infty)$$
$$= \lim_{r \to 1} \|f - h_r\|_\infty \geq \operatorname{dist}(f, C_A).$$

The inequality (7) thus established implies that the natural imbedding of the quotient space $C(\mathbb{T})/C_A$ into $L^\infty/H^\infty$ is an isomorphism so that the corresponding image ($c$, say) is closed in $L^\infty/H^\infty$. If $\pi$ is the natural projection from $L^\infty$ onto $L^\infty/H^\infty$, it is clear that $\pi^{-1}C = H^\infty + C(\mathbb{T})$ and so $H^\infty + C(\mathbb{T})$ must be a closed subspace of $L^\infty$. □

**Kronecker's Lemma.** *If $F \in L^\infty(\mathbb{T})$ then*

$$\operatorname{rank} H_F < \infty \Leftrightarrow P_- F \text{ is a rational function.}$$

*Proof*. We have

$$\dim P_- \mathbf{F} H^2 = \dim \mathscr{L}(P_- F z^n : n \geq 0) = \dim \mathscr{L}(P_- z^n P_- F : n \geq 0)$$
$$= \dim \mathscr{L}(S^{*n}\overline{P_- F} : n \geq 0)$$

and the finiteness of the last dimension is equivalent to the rationality of $\overline{P_- F}$ (complex conjugate) by the Lemma on Finite Dimensional Subspaces (Lecture II). □

*Proof of the Hartman-Sarason Theorem.* $1 \Rightarrow 2$, since $\lim_n T^n x = \mathbb{O}$ for any $x$, $x \in K$, this convergence being uniform on the (relatively) compact $f(T)$-image of the unit ball $\mathscr{D}$ of the space $K$:

$$0 = \lim_n \sup_{x \in \mathscr{D}} \| T^n f(T) x \| = \lim_n | T^n f(T) |.$$

$2 \Rightarrow 3$, since by the Corollary of Nehari's Theorem

$$| T^n f(T) | = \operatorname{dist}(z^n f \bar{\Theta}, H^\infty) = \operatorname{dist}(f \bar{\Theta}, \bar{z}^n H^\infty).$$

$3 \Rightarrow 1$, since from Statement 3 follows that there exist functions $h_n, h_n \in \bar{z}^n H^\infty$,

$$\lim_n \| f \bar{\Theta} - h_n \|_\infty = 0$$

and from this again

$$\lim_n | H_{f\bar{\Theta}} - H_{h_n} | = 0.$$

The operators $H_{h_n}$, $n \geq 1$, have finite rank by Kronecker's Lemma so that $H_{f\bar{\Theta}} \in \mathfrak{S}_\infty$. Since $f(T) = \Theta P_- \bar{\Theta} \mathbf{f} | K$ we obtain

$$f(T) = \Theta H_{\bar{\Theta} f} \in \mathfrak{S}_\infty.$$

The implication $3 \Leftrightarrow 4$ follows from the Lemma on $H^\infty + C(\mathbb{T})$ and the remarks preceding it. □

Let us point out once more that the Hankel operator $H_{\bar{\Theta} f}$ gives up to the unitary (in $L^2$) factor $\Theta$ the operator $f(T)$:

$$\Theta H_{\bar{\Theta} f} = f(T) \oplus \mathbb{O}.$$

Some auxiliary considerations on the operators $f(T)$, $f \in H^\infty$, will be found in the "Concluding Remarks" of this Lecture as well as in Sect. 6 below.

## Supplements and Bibliographical Notes

**4. The Multiplier Method and the Operator Calculus.** If we wish to study a given family $\mathscr{X} = \{X_n : n \in \mathfrak{N}\}$ of subspaces of a (Banach) space $X$ then a suitable tool for this is often provided by the description of the multiplier algebras $\operatorname{mult} \mathscr{X}$ and $\operatorname{Mult} \mathscr{X}$ (cf. Lecture VI, Sect. 4). Thus, for instance, the Lorch-Grinblyum Theorem says that $\mathscr{X}$ is an unconditional basis iff $\operatorname{mult} \mathscr{X} = l^\infty$ (then we also have $\operatorname{Mult} \mathscr{X} = l^\infty(L(X_n) : n \in \mathfrak{N})$). Another extremal case is the minimality property; it is equivalent to the condition that $\operatorname{mult} \mathscr{X}$ contains all finite numerical families on the set $\mathscr{N}$ (i.e. families equal to zero outside a finite set), and the reasoning required for the proof of this fact is essentially contained in the proof of the Lorch-Grinblyum Theorem. In general, the more multipliers, the more "regularly" the family $\mathscr{X}$ is built (and vice versa). In the following Section we will prove a general result in this direction.

The most important case for us will be when there is singled out in the algebra mult $\mathscr{X}$ (or Mult $\mathscr{X}$) one operator ($T$, say) and notably the case when $\mathscr{X}$ is the family of eigenspaces of $T$: $\mathscr{X} = \{X_\lambda : \lambda \in \sigma_p(T)\}$, $X_\lambda = \operatorname{Ker}(\lambda I - T)$ (or, for instance, the case $\sigma(T|X_n) = \{\lambda_n\}$, $\dim X_n < \infty$, $\forall n$: then the $X_n$ are the root spaces of our operator). One attempts to construct the multipliers of the family $\mathscr{X}$ as "functions" of $T$, taking this word in a as wide sense as possible. In dealing with the eigenspaces of $T$, then the multipliers $\mu$ obtained by this method are exactly those functions which are interpolated (in the sense $f(\lambda) = \mu(\lambda)$, $\lambda \in \sigma_p(T)$) by the elements of the function algebra acting on $T$. Of course, in order not to loose any part of the information one needs a result of the type of Sarason's Theorem, so that every multiplier is generated by some function of $T$. Such results will be discussed at the end of the present Section but first we give yet another general remark on the rôle of the operator calculi.

A rich operator calculus for $T$ is a kind of a spectral theorem. Let $A$ be a (topological) algebra of functions containing the polynomials and $\phi$ a (continuous) homomorphism from $A$ into $L(X)$ mapping the function $z$ onto the operator $T$ and the function $\mathbb{1}$ onto the identity map $I$. Notation: $\varphi(f) = f(A)$, $f \in A$. If $A = C(\sigma)$, $\sigma$ a compact subset of the plane $\mathbb{C}$, then, considering the "scalar spectral measures" $\mu_{x,x'}$ ($x \in X$, $x' \in X^*$) representing the functionals

$$f \mapsto (f(T)x, x'), \quad f \in C(\sigma),$$

and assuming that $X$ is (say) a reflexive Banach space, it is not hard to see that the operator $T$ equipped with the $A$-calculus is *scalar in the sense of Dunford*. By this we mean that there exists a countably additive projection valued Borel measure $\mathscr{E}_T$ commuting with $T$ such that

$$T = \int_\sigma z \, d\mathscr{E}_T;$$

also the integral $x = \int_\sigma d\mathscr{E}_T x$, $x \in X$, gives then an unconditionally convergent resolution of $X$ into "quasi-eigen spectral subspaces". In Hilbert space, as we have already noted, such operators $T$ are similar to normal ones and the measures $\mathscr{E}_T$ similar to orthogonal measures (Wermer [2]). If $A$ is a symmetric function algebra ($f \in A \Rightarrow \bar{f} \in A$) on the set $\sigma$, $\sigma \subset \mathbb{C}$, then the presence of an $A$-calculus for the operator $T$ can still be interpreted in terms of a resolution of the space into "spectral" components, summable by a method connected with the particular algebra $A$. For example, if $A = C^{(n)}(\mathbb{T})$ then from the continuity of the functional $f \mapsto (f(T)x, x')$, $f \in A$ ($x \in X$, $x' \in X^*$), there follows the existence of an operator measure $\mathscr{E}$ and operators $\mathscr{E}_i$, $\mathscr{E}_i \in L(X)$, $0 \le i < n$, such that

$$f(T) = \sum_{i=0}^{n-1} \mathscr{E}_i f^{(i)}(1) + \int_{\mathbb{T}} f^{(n)} \, d\mathscr{E}, \quad f \in C^{(n)}(\mathbb{T}).$$

If this formula is interpreted as the result of an $n$-fold partial integration, we then get the above "weakly summable" resolution of the space $X$. If $T$ admits an even weaker calculus ($A \subset C^\infty(\mathbb{T})$, for instance), then there arises a spectral distribution of higher order (instead of a spectral measure, as in the simple case $A = C(\mathbb{T})$).

The situation further complicates if we pass from symmetric algebras $A$ to algebras of analytic functions; often one can construct only such a calculus for a given operator $T$ (as, for instance, for general contractions; cf. Lecture III). In that case, as we have already noticed, we can only hope that: a) $f(T)$ depends only on "part of the values" of $f$ (more exactly on $f|\sigma$, $f'|\sigma$ etc. where $\sigma$ is a suitable subset of $\mathbb{C}$, as small as possible); b) the algebra $A|\sigma$ is sufficiently rich (symmetric; contains many idempotents etc.). The second of these desiderata constitutes the general interpolation problem connected with the calculus (with the multiplier, if $T$ is complete). Let us illustrate the preceding discussion on some examples.

The example which comes closest is the problem of unconditional bases of eigenspaces of contraction operators in Hilbert space.

*Example* 1. If $T$ is a linear contraction in the Hilbert space $H$ and the portion of the point spectrum $\sigma_p(T)\cap\mathbb{D}$ satisfies the Carleson condition, then the eigenspaces $\{E_\lambda\}$, $E_\lambda=\operatorname{Ker}(T-\lambda I)$, $\lambda\in\sigma_p(T)$, form a Riesz basis in its closed linear hull.

The sufficiency of Carleson's condition for the family of eigenspaces to be a basis however requires only the estimate

$$|p(T)|\le \text{const}\,\|p\|_\infty,\quad p\in\mathscr{P}_A. \tag{8}$$

Indeed, if for the operator $T$ in the Banach space $X$ there holds such an estimate then any family of eigenspaces $\{X_\lambda:\lambda\in\sigma\}$, $X_\lambda=\operatorname{Ker}(T-\lambda I)$, corresponding to a Carleson set ($\sigma\in(C)$), generates an unconditional basis: the calculus (8) for the restriction $T|\bigvee(X_\lambda:\lambda\in\sigma)$ extends to the whole algebra $H^\infty$. An application of Carleson's Theorem and the Theorem on Unconditional Bases concludes the reasoning. □

*Example* 2. Let $X$ be a Banach space of functions analytic in the disk $\mathbb{D}$, assuming that the functionals

$$\varphi_\lambda: f\mapsto f(\lambda),$$

$\lambda\in\mathbb{D}$, are continuous in $X$. If the set of polynomials $\mathscr{P}_A$ is dense in $X$ then every continuous linear functional $\varphi$ in $X$ can be identified with its Taylor series $\sum_{n\ge0}\varphi(z^n)z^n$ and the duality between $X$ and its dual $X^*$ can be written "à la Cauchy": $\langle f,\varphi\rangle=\sum_{n\ge0}\hat f(n)\hat\varphi(n)$ (at least if $f\in\mathscr{P}_A$), where we have set $\hat\varphi(n)=\varphi(z^n)$, $n\ge0$. To the functional $\varphi_\lambda$ there corresponds the series $\sum_{n\ge0}\lambda^n z^n=(\mathbb{1}-\lambda z)^{-1}$. Finally, let $M(X)\overset{\text{def}}{=}\{f: f \text{ analytic in } \mathbb{D}; g\in X\Rightarrow fg\in X\}$ be the space of all multipliers on $X$. *Proposition*: if $M(X)=H^\infty$ and the set $\sigma$, $\sigma\subset\mathbb{D}$, satisfies the Carleson condition then the family $\mathscr{X}\overset{\text{def}}{=}\{(\mathbb{1}-\lambda z)^{-1}:\lambda\in\sigma\}$ forms an unconditional basis for its $X^*$-closed linear hull and (this is the same thing) the space $X|\sigma$ has the ideal property.

Indeed, both of these facts are trivial: $\operatorname{mult}(\mathscr{X})=l^\infty$, because the operator $\mathbf{f}^*$ conjugate to multiplication by $f$, $f\in M(X)$, is a multiplier for $\mathscr{X}$ ($\mathbf{f}^*(\mathbb{1}-\lambda z)^{-1}=f(\lambda)(\mathbb{1}-\lambda z)^{-1}$, $\lambda\in\mathbb{D}$) and moreover $M(X)|\sigma=H^\infty|\sigma=l^\infty$; the ideal property of $X|\sigma$ being self-evident: if the function $\lambda\mapsto g(\lambda)$, $\lambda\in\sigma$, ($g\in X$) belongs to $X|\sigma$ so does the function $\lambda \mapsto f(\lambda)g(\lambda)$, $\lambda \in \sigma$ ($f \in M(X)$), and the factors $f|\sigma$ run through all of $l^\infty$. $\square$

Continuing the discussion of this example, let us notice that the inclusion $M(X) \subset H^\infty$ follows from the other assumptions on the space $X$ ($f \in M(X)$, $\lambda \in \mathbb{D} \Rightarrow |f^n(\lambda)| \leq |\varphi_\lambda| \|f^n|_{M(X)}\| \mathbb{1} \|_X \leq \text{const} \|f\|^n_{M(X)}$, $n \geq 1 \Rightarrow |f(\lambda)| \leq \|f\|_{M(X)}$) and that the ideal property of $X|\sigma$ signifies the solvability of the following interpolation problem:

$$g \in X; \quad |a(\lambda)| \leq |g(\lambda)|,$$
$$\lambda \in \sigma \Rightarrow \exists g_a \in X : g_a(\lambda) = a(\lambda), \quad \lambda \in \sigma.$$

Sometimes one can describe the space of "interpolation data" $X|\sigma$ in a more effective way. For instance, if $X$ is a Hilbert space then (in the hypotheses of Example 2) the normalized family $\{(\mathbb{1} - \bar{\lambda}z)^{-1} \|(\mathbb{1} - \bar{\lambda}z)^{-1}\|_{X^*}^{-1} : \lambda \in \sigma\}$ will be a Riesz basis and the corresponding space of coefficients is therefore $l^2$, that is, we must have

$$X|\sigma = l^2(|\varphi_\lambda|^{-1}) \overset{\text{def}}{=} \left\{a : \sum_{\lambda \in \sigma} |a(\lambda)|^2 |\varphi_\lambda|^{-2} < \infty\right\}.$$

Among the Hilbert spaces $X$ such that $M(X) = H^\infty$ we have all spaces of the form

$$L^2_A(\mu) = \left\{f : f \text{ analytic in } \mathbb{D} \text{ and } \int_{\mathbb{D}} |f|^2 \, d\mu < \infty\right\}$$

(for which the hypotheses of Example 2 are satisfied) or of the form

$$l^2_A(w_n) = \left\{f : f \text{ analytic in } \mathbb{D} \text{ and } \sum_{n \geq 0} |\hat{f}(n)|^2 w_n^2 < \infty\right\},$$

the last ones however under the monotonicity condition $w_n \downarrow$. For $L^2_A$ this is obvious and in the $l^2_A$-case it follows from the fact that the shift $\mathbf{z} : f \mapsto zf$ is a completely non-unitary contraction operator in $l^2_A(w_n)$, thus admitting a $H^\infty$-calculus. Let us notice that for the conjugate space $l^2_A(w_n^{-1})$ (in the Cauchy duality) this fact can be formulated as follows:

$$h \in H^\infty, \quad \varphi \in l^2_A(w_n^{-1}) \Rightarrow P_+ \bar{h}\varphi \in l^2_A(w_n^{-1})$$

where the passage to the circumference $\mathbb{T}$ and the application of the projection $P_+$ is legitimate, since $l^2_A(w_n^{-1}) \subset H^2$ for $w_n \downarrow$. $\square$

The situation is especially simple for *the classical Hardy spaces* $H^p$, $1 \leq p < \infty$. In that case we use the following corollary of the factorization theorem: if $f = f^i \cdot f^e$ with $f^i$ inner and $f^e$ outer then $f \in H^p \Leftrightarrow f^e \in H^p \Leftrightarrow (f^e)^{p/2} \in H^2$ (cf. Appendix 2). From this follows that the sets of moduli $\{|f|\sigma| : f \in H^p\}$ and $\{|g|\sigma|^{2/p} : g \in H^2\}$ majorize each other, in the sense that every function of the first set does not exceed some function of the second set and vice versa. Thus the condition $(C)$ not only allows us to establish the ideal property of $H^p|\sigma$ but also to compute this space (it consists of all functions $a$ such that $\sum_{\lambda\in\sigma}(1-|\lambda|^2)|a(\lambda)|^p < \infty$). It is, incidentally, clear that the condition $\sigma\in(C)$ is necessary for $H^p|\sigma$ to be an ideal, because this property entails in particular the existence of a family of functions $\{f_\lambda:\lambda\in\sigma\}$ with

$$\sup_{\lambda\in\sigma}\|f_\lambda\|_p<\infty,$$

$$f_\lambda(\mu)=\delta_{\lambda\mu}(1-|\lambda|^2)^{-1/p}\text{ for }\lambda,\quad \mu\in\sigma$$

(the uniform minimality of the functionals $\{\varphi_\lambda:\lambda\in\sigma\}$) and then $f_\lambda=B_\lambda g_\lambda$, $\|f_\lambda\|_p=\|g_\lambda\|_p$, $B_\lambda=\prod_{\mu\in\sigma\setminus\{\lambda\}}b_\mu$,

$$\begin{aligned}(1-|\lambda|^2)^{-1/p}&=|f_\lambda(\lambda)|=|B_\lambda(\lambda)|\cdot|g_\lambda(\lambda)|\\&\le|B_\lambda(\lambda)|\cdot\text{const}\,\|g_\lambda\|_p(1-|\lambda|^2)^{-1/p}\end{aligned}$$

so that indeed $\inf_{\lambda\in\sigma}|B_\lambda(\lambda)|>0$. (This reasoning carries over to any space $X$ with the "divisibility" property: $f\in X, f|\sigma\equiv\mathbb{0}\Rightarrow\|f/B_\sigma\|_X\le\text{const}\,\|f\|_X$, where $B_\sigma=\prod_{\lambda\in\sigma}b_\lambda$, with const independent of $f$ and $\sigma$.) Thus we have established the following result essentially due to Shapiro-Shields [1]. (Here we have omitted the argument giving the duality between $H^p$ and $H^{p'}$, $1<p<\infty$, between $H^1$ and $L^\infty/\overline{H_0^\infty}$ and between $L^1/\overline{H_0^1}$ and $H^\infty$ (the bar – stands for complex conjugation; $X_0$ is the set of functions in $X$ vanishing at the origin); cf. Duren [1].)

**Corollary.** *Let* $\sigma\subset\mathbb{D}$, $\sum_{\lambda\in\sigma}(1-|\lambda|)<\infty$, $1\le p\le\infty$ *and* $\mathscr{X}=\{(\mathbb{1}-\bar\lambda z)^{-1}:\lambda\in\sigma\}$. *The following statements are equivalent:*

1. $\mathscr{X}$ *is an unconditional basis in* $\bigvee_{H^{p'}}(\mathscr{X})$ *if* $1<p<\infty$, *in* $\bigvee_{L^\infty/\overline{H_0^\infty}}(\mathscr{X})$ *if* $p=1$, *in* $\bigvee_{L^1/\overline{H_0^1}}(\mathscr{X})$ *if* $p=\infty$.
2. $H^p|\sigma$ *has the ideal property.*
3. $H^p|\sigma=\{a:\{a(\lambda)(1-|\lambda|^2)^{1/p}:\lambda\in\sigma\}\in l^p\}$.
4. $X$ *is uniformly minimal (in the corresponding space:* $H^{p'}$ *if* $1<p<\infty$, $L^\infty/\overline{H_0^\infty}$ *if* $p=1$, $L^1/\overline{H_0^1}$ *if* $p=\infty$).
5. $\sigma\in(C)$. □

*Brief discussion of interpolation formulae.* We have deduced the last Corollary from Carleson's Theorem by very simple and general reasonings. As a matter of fact, the Corollary not only establishes "free interpolation" itself but can also be used to produce *a formula* for an interpolating operator in the case $1<p<\infty$. To see this, let us first remark that the space conjugate to $\bigvee_{H^{p'}}(\mathscr{X})$ can be naturally identified with the quotient space $H^p/\mathbf{B}H^p$ because

$$\bigvee_{H^{p'}}(\mathscr{X})^\perp=\{f\in H^p:\langle f,(\mathbb{1}-\bar\lambda z)^{-1}\rangle=0,\lambda\in\sigma\}=\mathbf{B}H^p.$$

But in the case $1<p<\infty$ the subspace $\mathbf{B}H^p$ is complemented in $H^p$, since the standard projections $\mathbf{B}P_+\bar{\mathbf{B}}$ and $P_B=I-\mathbf{B}P_+\bar{\mathbf{B}}=\mathbf{B}P_-\bar{\mathbf{B}}$ are bounded on $H^p$ (by the well-known theorem of Riesz, see e.g. Duren [1]) and hence

$$H^p=P_BH^p+(I-P_B)H^p=K_B^p+\mathbf{B}H^p,$$

where $K_B^p \overset{\text{def}}{=} P_B H^p = \bigvee(\mathscr{X})$. It follows that the map $x' \to x' + \mathbf{B}H^p$ is an isomorphism of $K_B^p$ onto $H^p/\mathbf{B}H^p$. Now it is easy to check the following.

**Corollary.** *Let $1 < p < \infty$ and $\sigma \in (C)$. Then the operator* In,

$$(\mathrm{In})a \overset{\text{def}}{=} \sum_{\lambda \in \sigma} a(\lambda) \frac{1 - |\lambda|^2}{\mathbb{1} - \bar{\lambda} z} \frac{B_\lambda}{B_\lambda(\lambda)},$$

*where as usual $B_\lambda = \prod_{\mu \in \sigma \setminus \{\lambda\}} b_\mu$, is bounded as a mapping from the space $l^p_{1/p}(\sigma) \overset{\text{def}}{=} \{a : \{a(\lambda)(1 - |\lambda|^2)^{1/p}\}_{\lambda \in \sigma} \in l^p\}$ into $K_B^p$, and*

$$(\mathrm{In})a | \sigma = a, \quad \forall a \in l^p_{1/p}(\sigma).$$

In fact, let $\dot{\mathscr{X}}' = \{\dot{x}'_\lambda : \lambda \in \sigma\}$ be the family biorthogonal to $\mathscr{X}$. Then it follows from the previous Corollary (see the Statements 1 and 3) that for each $a \in l^p_{1/p}(\sigma)$ the series $\sum_{\lambda \in \sigma} a(\lambda)\dot{x}'_\lambda$ is unconditionally convergent (in the dual space $H^p/\mathbf{B}H^p$) and that the operator $(\dot{\mathrm{In}})a \overset{\text{def}}{=} \sum_{\lambda \in \sigma} a(\lambda)\dot{x}'_\lambda$ is bounded from $l^p_{1/p}(\sigma)$ into $H^p/\mathbf{B}H^p$. However, it is easy to see that $\dot{x}'_\lambda = x'_\lambda + \mathbf{B}H^p$, with $x'_\lambda = (1 - |\lambda|^2)(\mathbb{1} - \bar{\lambda} z)^{-1} B_\lambda(\lambda)^{-1} B_\lambda$. By the remarks preceding the formulation of the Corollary we infer the desired properties of In. □

It can be verified (Airapetyan [1]) that the assumption $p \neq 1, \infty$ is essential for the validity of the last Corollary, because for $p = 1$ even $K_B^1 | \sigma = l_1^1(\sigma)$ implies $\operatorname{card} \sigma < \infty$ and for $p = \infty$ $K_B^\infty | \sigma = l^\infty(\sigma)$ implies $\sigma \in (C)$ & $(F)$ (concerning this conditions, see Lecture VII, Sect. 3). Nevertheless for $p = 1, \infty$ there exist linear interpolating operators different from In. The first of them was constructed by Carleson and P. Beurling [1] for the space $H^\infty$. Recently a simple and transparent formula of this kind was found by Peter Jones [3] (for an elegant refinement of Jones's construction see the paper by Vinogradov-Gorin-Hruščev [1]). Unfortunately, the Jones operator has its range "too far" from the space $K_B$ and, as a consequence, the interpolating functions produced by it are transcendental even for finite Blaschke products $B$. Vinogradov [5] contrived the following remarkable modification of Jones's construction free from the last defect.

**Jones-Vinogradov Theorem.** *Let $\sigma \subset \mathbb{D}$, $\sigma \in (C)$ and set $\delta = \inf\{|B_\lambda(\lambda)| : \lambda \in \sigma\}$ (the Carleson constant for $\sigma$). Let further*

$$C^\lambda = \prod_{\substack{\mu \\ \mu \in \sigma \setminus \{\lambda\}, |\mu| \geq |\lambda|}} c_\mu, \quad c_\mu \overset{\text{def}}{=} b_{\mu/\sqrt{|\mu|}}(z\sqrt{|\mu|}) = \frac{\sqrt{|\mu|}}{\mu} \cdot \frac{\mu - |\mu| z}{\mathbb{1} - \bar{\mu} z}.$$

*Then the operator $JV$,*

$$JVa \overset{\text{def}}{=} \sum_{\lambda \in \sigma} a(\lambda) \left( \frac{1 - |\lambda|^2}{\mathbb{1} - \bar{\lambda} z} \right)^2 \frac{B_\lambda}{B_\lambda(\lambda)} \cdot \frac{C^\lambda}{C^\lambda(\lambda)},$$

*is interpolating for $\sigma$ ($JVa | \sigma \equiv a, \forall a$) and for every $p$, $1 \leq p \leq \infty$, $JV$ is a bounded linear mapping from $l^p_{1/p}(\sigma)$ into $K_{B^2}^p$ ($B = \prod_{\lambda \in \sigma} b_\lambda$) with*

$$\| JV \|_{l^p_{1/p}(\sigma) \to H^p} \leq 8\delta^{-2}.$$

*Proof.* It is clear that the terms of the series defining $JVa$ are contained in $K^{\infty}_{B^2}$ and, on the other hand, $JVa|\sigma \equiv a$ if this series converges in $\mathbb{D}$.

Let us estimate the sum of the series. For this purpose we enumerate the set $\sigma$, $\sigma = \{\lambda_k\}_{k\geq 1}$, in such a way that $|\lambda_k| \leq |\lambda_{k+1}|$, $k \geq 1$, and note that for $\zeta \in \mathbb{D}$

$$1 \geq |c_\lambda(\zeta)|^2 = |b_{\lambda/\sqrt{|\lambda|}}(\zeta\sqrt{|\lambda|})|^2 = 1 - \frac{(1-|\lambda|)(1-|\lambda||\zeta|^2)}{|1-\bar{\lambda}\zeta|^2} \geq 0.$$

Hence

$$\begin{aligned}\sum_{\lambda\in\sigma} \frac{(1-|\lambda|^2)^2}{|1-\bar{\lambda}\zeta|^2}|C^\lambda(\zeta)| &\leq 4\sum_{\lambda\in\sigma}\frac{(1-|\lambda|)(1-|\lambda||\zeta|^2)}{|1-\bar{\lambda}\zeta|^2}|C^\lambda(\zeta)| \\ &= 4\sum_{\lambda\in\sigma}(1-|c_\lambda(\zeta)|^2) \prod_{\mu\in\sigma\setminus\{\lambda\},|\mu|\geq|\lambda|} |c_\mu(\zeta)| \\ &\leq 8\sum_{n\geq 1}(1-|c_{\lambda_n}(\zeta)|)\prod_{k\geq n+1}|c_{\lambda_k}(\zeta)| \\ &= 8\sum_{n\geq 1}\left(\prod_{k\geq n+1}|c_{\lambda_k}(\zeta)| - \prod_{k\geq n}|c_{\lambda_k}(\zeta)|\right) \leq 8.\end{aligned}$$

Moreover, for $|\zeta| \leq |\mu| \leq 1$ we have $|\mu - \zeta|\mu||^2 - |\mu||\zeta-\mu|^2 = |\mu|^2 + |\zeta|^2|\mu|^2 - 2\operatorname{Re}\mu\bar{\zeta}|\mu| - |\mu||\zeta|^2 - |\mu|^3 + 2\operatorname{Re}\mu\bar{\zeta}|\mu| = |\mu|(1-|\mu|)(|\mu|-|\zeta|^2) \geq 0$, and in view of the definitions of $c_\mu$ and $b_\mu$

$$|c_\mu(\zeta)| \geq |b_\mu(\zeta)| \quad \text{for} \quad |\zeta| \leq |\mu| \leq 1.$$

Thus $|C^\lambda(\lambda)| \geq \prod_{\mu\in\sigma\setminus\{\lambda\},|\mu|\geq|\lambda|}|b_\mu(\lambda)| \geq |B_\lambda(\lambda)| \geq \delta$. Together with the previous estimate of the sum this gives

$$\begin{aligned}|(JVa)(\zeta)| &\leq \|a\|_{l^\infty(\sigma)}\delta^{-2}\sum_{\lambda\in\sigma}\frac{(1-|\lambda|^2)^2}{|1-\bar{\lambda}\zeta|^2}|C^\lambda(\zeta)| \\ &\leq 8\delta^{-2}\|a\|_{l^\infty(\sigma)}.\end{aligned}$$

An estimate of $JV$ for $p = 1$ is immediate

$$\|JVa\|_{H^1} \leq \delta^{-2}\sum_{\lambda\in\sigma}|a(\lambda)|\left\|\left(\frac{1-|\lambda|^2}{1-\bar{\lambda}z}\right)^2\right\|_{H^1} = \delta^{-2}\|a\|_{l^1_1(\sigma)}.$$

As a result, the operator $JV: l^p_{1/p}(\sigma) \to H^p$ is bounded for $p = 1$ and $p = \infty$. By the famous Riesz-Thorin convexity theorem (see e.g. Reed-Simon [2]) this operator is bounded for each $p$, $1 \leq p \leq \infty$, and $|JV|_{l^p_{1/p}(\sigma)\to H^p} \leq 8\delta^{-2}$. □

It is worth while mentioning that the operator $JV$ has some further interesting properties (see Vinogradov [5]).

**The Vector Valued Case.** Concluding this Section let us remark that the criteria of spectrality (or "basisness") obtained using calculi and interpolation can be exact only within the class of all operators admitting a calculus of prescribed size or else for special extremal operators of the type of the compressions of the scalar shift operator. It is a different thing if we, let us say, instead of an abstract contraction in Hilbert space pass to its functional model, taking the word "calculus" in a wider sense, using additional analytical structures made available by the model. In such a situation the following analogue of Sarason's Theorem in particular is of importance.

**Commutant Theorem.** *Let $E$ and $E'$ be Hilbert spaces and $\Theta$ a left inner function (with values in $H^\infty(L(E', E))$ and set*

$$K = K_\Theta \overset{\text{def}}{=} H^2(E) \ominus \Theta H^2(E'), \quad T = P_K S_E | K.$$

*The commutant of $T$ (i.e. the set $\{A : AT = TA\}$) consists of precisely those operators $A$ which are defined by the formulae*

$$Ax = P_K F x, \quad x \in K,$$

*where $F \in H^\infty(L(E))$ and $F\Theta \in \Theta H^\infty(L(E'))$. Also*

$$|A| = |P_K F| = \inf\{\|F + \Theta h\|_\infty : h \in H^\infty(L(E, E'))\}.$$

*The proof* is similar to the proof of Sarason's Theorem, at least if $\Theta$ is an inner function. Limiting ourselves to this case (the general case is in the book by Sz.-Nagy and Foias [1]) let us indicate the argument required. First of all it is immediate that the lemma used in the proof of Sarason's Theorem carries over (along with its proof). It remains to establish the vector analogue of Nehari's Theorem, which is contained in the following

**Page's Theorem.** *Let $E$ be a Hilbert space and $A$ an operator from $H^2(E)$ into $H^2_-(E)$ such that*

$$P_- z A p = A z p, \quad p \in H^2(E),$$

*Then there exists a function $f$, $f \in L^\infty(L(E))$, such that $A = H_f$,*

$$H_f g \overset{\text{def}}{=} P_-(fg), \quad g \in H^2(E).$$

*Moreover* $|A| = \operatorname{dist}(f, H^\infty(L(E))$.

*Outline of the proof.* The restriction of the operator $A$ to the constants (i.e. the functions $e \cdot \mathbb{1}$, $e \in E$) defines a strong $H^2_-$-operator function $A(\cdot)$ with values in $L(E)$:

$$A(\zeta)e = (Ae)(\zeta), \quad |\zeta| > 1, \quad e \in E.$$

Also $A(z^n e) = P_-(z^n A(\cdot)e) = P_-(A(\cdot)(z^n e))$, $n \geq 0$, $e \in E$, so that

$$Ap = P_- A(\cdot) p$$

for any polynomial $p$, $p \in H^2(E)$. Let now $\mathfrak{S}_2$ be the ideal of all Hilbert-Schmidt operators in $E$ and $\mathfrak{S}_1$ the ideal of nuclear operators (regarding these notions see Appendix 4, 23 and the textbook by Ahiezer-Glazman [1]). If $g_1 \in H^2(\mathfrak{S}_2)$ and $g_2 \in H^2_0(\mathfrak{S}_2)$ then $g_1(\cdot)e \in H^2(E)$ and $g_2(\cdot)^* e \in H^2_-(E)$ for any vector $e$, $e \in E$. Consequently, taking $g_1$ and $g_2$ to be polynomials we get in particular

$$(Ag_1(\cdot)e, g_2(\cdot)^* e) = (A(\cdot) g_1(\cdot) e, g_2(\cdot)^* e),$$

so that (picking an orthonormal basis $\{e_n\}_{n\geq 1}$ in $E$)

$$\int_{\mathbb{T}} \mathrm{Trace}(A(\zeta)g_1(\zeta)g_2(\zeta))\,dm(\zeta) = \int_{\mathbb{T}} \mathrm{Trace}(g_2(\zeta)A(\zeta)g_1(\zeta))\,dm(\zeta)$$

$$= \sum_{n\geq 1} \int_{\mathbb{T}} (A(\zeta)g_1(\zeta)e_n, g_2(\zeta)^* e_n)_E\,dm(\zeta)$$

$$= \sum_{n\geq 1} (Ag_1(\cdot)e_n, g_2(\cdot)^* e_n),$$

$$\left|\int_{\mathbb{T}} \mathrm{Trace}(A(\cdot)g_1 g_2)\,dm\right| \leq \sum_{n\geq 1} |(Ag_1(\cdot)e_n, g_2(\cdot)^* e_n)|$$

$$\leq |A| \left(\sum_{n\geq 1} \|g_1(\cdot)e_n\|^2_{H^2(E)}\right)^{1/2} \left(\sum_{n\geq 1} \|g_2(\cdot)^* e_n\|^2_{H^2_-(E)}\right)^{1/2}$$

$$= |A|\, \|g_1\|_{H^2(\mathfrak{S}_2)} \|g_2\|_{H^2(\mathfrak{S}_2)}.$$

As $g_1$ and $g_2$ run through dense subsets of the unit balls of $H^2(\mathfrak{S}_2)$ and $H^2_0(\mathfrak{S}_2)$ respectively, the products $g_1 g_2$ will generate a dense set in the unit ball of the space $H^1_0(\mathfrak{S}_1)$ (Sarason [6]; this result coincides in essence with the theorem on factorizations of the type $F^*F = G$ mentioned in Lecture I, Sect. 7). The form

$$\langle F, G\rangle = \int_{\mathbb{T}} \mathrm{Trace}(F(\zeta)G(\zeta))\,dm(\zeta)$$

realizes the duality between the spaces $L^\infty(L(E))$ and $L^1(\mathfrak{S}_1)$, for which the annihilator of $H^1_0(\mathfrak{S}_1)$ coincides with $H^\infty(L(E))$. Therefore, there exists a function $f \in L^\infty(L(E))$ such that

$$\int_{\mathbb{T}} \mathrm{Trace}(A(\zeta)g(\zeta))\,dm(\zeta) = \int_{\mathbb{T}} \mathrm{Trace}(f(\zeta)g(\zeta))\,dm(\zeta),\; g \in H^1_0(\mathfrak{S}_1),$$

so in particular $(Ap, q) = (fp, q)$, $p \in H^2(E)$, $q \in H^2_-(E)$, i.e. $A = H_f$. To compute the norm $|A|$ we have to invoke (as in the proof of Nehari's Theorem) the Hahn-Banach theorem. □

Continuing the proof of the Commutant Theorem we apply Page's Theorem to the Hankel operator $A_* = \mathbf{\Theta}^* A P_{\Theta}$. We have $A_* = H_f$ and $\mathbf{f}\mathbf{\Theta} H^2(E) \subset H^2(E)$ since $H_f \mathbf{\Theta} H^2(E) = \{\mathbb{0}\}$. This gives

$$A = \mathbf{\Theta} A_* | K_\Theta = \mathbf{\Theta} P_- \mathbf{\Theta}^* \mathbf{\Theta} \mathbf{f} = P_\Theta \mathbf{\Theta} \mathbf{f}$$

where the function $F = \Theta f$ indeed satisfies the desiderata of the theorem: $\mathbf{F}H^2(E) \subset H^2(E)$ (as $\mathbf{\Theta} P_- \mathbf{f} H^2(E) = (\mathbf{\Theta} H_f) H^2(E) = AP_\Theta H^2(E) \subset H^2(E)$ and $\mathbf{\Theta} P_+ \mathbf{f} H^2(E) \subset H^2(E)$) and $F\mathbf{\Theta} H^2(E) = \mathbf{\Theta}\mathbf{f}\mathbf{\Theta} H^2(E) \subset \mathbf{\Theta} H^2(E)$. □

If we know a criterion for the eigenspaces of $T$ (or $T^*$) (cf. Lecture VI–VII) to be a basis we can also solve the following interpolation problem (regarding the notation ${}_\lambda\Delta$ cf. Lecture VI, Sect. 5 and Lecture VII, Sect. 4)

$$F(\lambda)^* | {}_\lambda\Delta E = a_\lambda, \quad \lambda \in \sigma_p(T), \tag{10}$$

where $a_\lambda$ is a linear operator in ${}_\lambda\Delta E$,

$$\sup_\lambda |a_\lambda| < \infty,$$

the function $F$ being as in the Commutant Theorem ($F \in H^\infty(L(E))$, $F\Theta \in \Theta H^\infty(L(E'))$). Conversely, if the problem (10) has a solution satisfying the last two conditions for any bounded collection of complex numbers $\{a_\lambda\}$ (more exactly, collection of "diagonal" operators $\{a_\lambda I\}$, $\sup_\lambda |a_\lambda| < \infty$), then the family of eigenspaces for $T$ must give an unconditional basis.

In some special cases it is possible to return anew to "scalar functions" of $T$ but these will not any longer belong to $H^\infty$, cf. Sz.-Nagy and Foiaş [1], [2], [10].

**5. Summation Bases.** The property of the family $\mathscr{X}$ to be a $V$-basis with respect to some method of summation $V$ can likewise be expressed in the language of the algebra mult $\mathscr{X}$. For example, one sees at once that the existence of a finite-rowed summation method is equivalent to the presence of a directed family of finite rank operators in mult $\mathscr{X}$ converging to $I$ (in the strong or in the weak operator topology). Let us state a result which is sufficient in many concrete cases. Let $V = \{v_n : n \geq 1\}$ be a family of functions defined, piecewise continuous and bounded on the segment $[0, 1)$, with $\lim_{t\to 1-0} v_n(t) = 1$, $\forall n \in \mathfrak{N}$. We say that *the series* $\sum_{n\geq 1} x_n$ *is summable by the method* $V$ if the series $\sum_{n\geq 1} v_n(t)x_n$ converges for all $t$, $0 \leq t < 1$, and if, in addition, $\lim_{t\to 0} \sum_{n\geq 1} v_n(t)x_n$ exists. The family $\mathscr{X} = \{x_n : n \in \mathfrak{N}\}$ of vectors in the Banach space $X$ will be called a *V-basis* (in the closure of its linear hull) if for any $x$, $x \in \bigvee(\mathscr{X})$, there exists a unique collection of scalars $\{a_n : n \in \mathfrak{N}\}$ such that $x = (V)\text{-}\sum_{n\geq 1} a_n x_n$.

**Theorem on Summation Bases.** *Let $X$ be a Banach space, let $\mathscr{X} = \{x_n : n \geq 1\}$ be a total family in $X$ with the total biorthogonal system $\mathscr{X}'$ and let $V$ be a summation method with the property that the series $\sum_{n\geq 1} v_n(t)(x, x'_n)x_n$, for any $x$, $x \in X$, converges uniformly on compact subsets of the interval* $[0, 1)$. *The following Statements are equivalent*:

1. *$\mathscr{X}$ is a $V$-basis.*
2. *$\hat{V} \subset \operatorname{mult} X$ where $\hat{V} = \{\mu : \mu = \{\mu_n\}_{n\geq 1}, \ \mu_n = \int_0^1 v_n \, d\mathscr{M}, \ n \geq 1; \ \mathscr{M}$ any finite measure on* $[0, 1)\}$.

*Proof.* Let us first show that $1 \Rightarrow 2$. To this end we remark that for each $t$ the operator

$$V(t) : x \mapsto \sum_{n\geq 1} v_n(t)(x, x'_n)x_n$$

is linear and continuous and moreover holds $\sup_{0\leq t<1} |V(t)| < \infty$. In fact, in view of Statement 1 it is sufficient to establish that $\sup_{0\leq t\leq 1-\delta} |V(t)| < \infty$ for any $\delta$, $\delta > 0$. But this follows from the Banach-Steinhaus theorem and the incompatability of the following two propositions a)–b): a) $\sup_n |(V(t_n)x, x')| = \infty$; $0 \leq t_n \leq 1 - \delta$, $x \in X$, $x' \in X^*$, b) $|\sum_{1\leq k\leq N} v_k(t)(x, x'_k)(x_k, x') - (V(t)x, x')| < \varepsilon$ for all $t$, $0 \leq t \leq 1 - \delta$, and $N$ sufficiently large. Since the function $t \mapsto V(t)$, $0 \leq t < 1$, is strongly measurable and $\sup_{0\leq t<1} |V(t)| < \infty$ the operator $\mu$, $\mu =$

$\int_0^1 V(t)\,d\mathcal{M}(t)$, must be continuous in $X$, for any (finite) measure $\mathcal{M}$. It is clear that

$$\mu x_n = \mu_n x_n; \quad \mu_n \stackrel{\text{def}}{=} \int_0^1 v_n(t)\,d\mathcal{M}(t), \quad n \geq 1. \tag{14}$$

In order to get the implication $2 \Rightarrow 1$ it is again sufficient to establish that $\sup_{0 \leq t < 1} |V(t)| < \infty$. From Statement 2 and the closed graph theorem it follows that the transformation $\mathcal{M} \mapsto \mu$, assigning to the measure $\mathcal{M}$ the multiplier $\mu$ (a bounded operator in $X$) by formula (14), must be continuous. Consequently $|\mu| \leq \text{const Var}\, M$. But $V(t) = \mu_{\delta(t)}$, where $\delta(t)$ denotes mass one at the point $t$, $0 \leq t < 1$. Hence $\sup_{0 \leq t < 1} |V(t)| < \infty$. □

**Corollary 1.** 1. *$\mathfrak{X}$ is a basis in $\bigvee(\mathfrak{X})$ (i.e. with respect to ordinary convergence of series) iff* bv $\subset$ mult $\mathfrak{X}$ *where* bv *is the set of all sequences $\{\mu_n\}_{n \geq 1}$ of bounded variation ($\sum_{n \geq 1} |\mu_{n+1} - \mu_n| < \infty$).*

2. *$\mathfrak{X}$ is a Cesaro basis in $\bigvee(\mathfrak{X})$ (i.e. with respect to the method of arithmetic means) iff every bounded positive convex sequence is in* mult $\mathfrak{X}$.

Let us verify, for instance, Statement 1. To this end we pick a sequence of numbers $\{t_n\}_{n \geq 1}$ such that $0 \leq t_n < 1$, $t_n < t_{n+1}$, $\lim_n t_n = 1$ and set $v_n(t) = \begin{cases} 0, & 0 \leq t \leq t_n \\ 1, & t_n < t < 1 \end{cases}$. Since the propositions: i) $\mu \in \text{bv}$, ii) there exists a number $c$ and a finite measure $\mathcal{M}$ on $[0, 1)$ with $\mu_n = c + \mathcal{M}([t_n, 1))$, $n \geq 1$, apparently are equivalent, we get the desired inclusion (using of course the Theorem on Summation Bases). □

Some general results on $V$-basis were essentially mentioned already in Sect. 4, because the $A$-calculus for the operator $T$ leads to a special summation of the "spectral resolutions". Here we list a few more examples of the use of calculi as well as examples of interpolation problems arising from the connection with $V$-bases.

*Example* 1. Let $T$ be a complete, polynomially bounded operator in the Banach space $X$ (i.e. $|p(T)| \leq \text{const} \|p\|_\infty$, $\forall p \in \mathcal{P}_A$) having root spaces $X_\lambda$ of height $k(\lambda)$, $\lambda \in \sigma_p(T)$, $\sigma_p(T) \subset \mathbb{D}$. If, in addition, $\sum_{\lambda \in \sigma_p(T)} k(\lambda)(1 - |\lambda|) < \infty$ then there exists a finite-rowed method $V$ summing all (formal) Fourier series $\sum_{\lambda \in \sigma_p(T)} \mathcal{E}_\lambda x$ to the vectors $x$ generating them.

In view of the completeness of the family $\mathfrak{X} = \{X_\lambda : \lambda \in \sigma_p\}$ and the inclusion $\sigma_p \subset \mathbb{D}$ the map $p \mapsto p(T)$, $p \in \mathcal{P}_A$, extends to the algebra $H^\infty$ with preservation of the estimate of the norm. The desired method is supplied by the operators $B_n(T)$, $B_n \stackrel{\text{def}}{=} \prod_{\lambda \in \sigma_p, |\lambda| \geq 1 - 1/n} b_\lambda^{k(\lambda)}$, $n \geq 1$. In fact, as $B_n(T)\mathcal{E}_\lambda x = \mathcal{E}_\lambda B_n(T)x$ we have $B_n(T)x \sim \sum_{\lambda \in \sigma_p} B_n(T)\mathcal{E}_\lambda x$. This series contains only a finite number of nonzero terms and its sum equals $B_n(T)x$ (notice that $B_n(T)x \in \bigvee(X_\lambda : \lambda \in \sigma_p, |\lambda| < 1 - 1/n)$ because of $B_1(T) = \mathbb{O}$). Finally $|B_n(T)| \leq \text{const}$, $n \geq 1$, the convergence $B_n(T)x \to x$ being obvious on the total subset $\bigcup_{\lambda \in \sigma_p} X_\lambda$. □

A special case of this proposition was used already in Lecture IV, Sect. 5.

*Example* 2. Let $T$ be an operator in the Banach space $X$ with $\sup_{n\geq 0}|T^n| < \infty$ and assume also that the sequence of its eigenvalues $\sigma = \{\lambda_n\}_{n\geq 1}$, $|\lambda_n| \leq |\lambda_{n+1}|$, $\sigma \subset \mathbb{D}$, lies in a Stolz angle and satisfies the sparseness condition $(R)$ (for these notions consult Lecture VII). Then the sequence $\{X_{\lambda_n}\}_{n\geq 1}$, $X_{\lambda_n} = \operatorname{Ker}(T - \lambda_n I)$, $n \geq 1$, is a basis in the closure of its linear hull.

In fact, Vinogradov [1] has shown that in the preceding hypotheses one has $l_A^1|\sigma = \mathrm{bv}$ so that, in view of $f(T)|X_{\lambda_n} = f(\lambda_n)I$, $n \geq 1$, everything follows from the Theorem on Summation Bases. □

Besides ordinary bases ($V$-convergence = convergence) one quite frequently uses Abel bases ($v_n(t) = t^n$ in the definition of the $V$-method). They can be described in the spirit of Corollary 1 (the algebra mult $\mathscr{X}$ has to contain all "positive definite" sequences). In this direction we give one result concretizing what has been said.

**Corollary 2.** *Let $T$ be a complete contraction of class $C_0$ with the minimal annihilator $B = \prod_{n\geq 1} b_{\lambda_n}$, $\sigma_p(T) = \{\lambda_n : n \geq 1\}$; $\lambda_n \neq \lambda_k$ for $n \neq k$. If*

$$\sup_{n\geq 1} t^n |B_{\lambda_n}(\lambda_n)|^{-1} < \infty, \quad t \in (0,1), \tag{15}$$

*and if there exist functions $f_t$, $f_t \in H^\infty$, $0 < t < 1$, with*

$$f_t(\lambda_n) = t^n, \quad n \geq 1, \quad \overline{\lim_{t\to 1}}\, \|f_t\|_\infty < \infty, \tag{16}$$

*then the spectral expansion of $T$ is summable by Abel's method, i.e. the series $\sum_{n\geq 1} t^n \mathscr{E}_{\lambda_n}$ converge for all $t$, $0 \leq t < 1$ (uniformly on compact subsets of $[0,1)$) and*

$$\lim_{t\to 1} \sum_{n\geq 1} t^n \mathscr{E}_{\lambda_n} x = x, \ \forall x.$$

*If $T$ is a model operator, $T = P_K S|K$, $K = H^2 \ominus \mathbf{B}H^2$, then the conditions* (15) *and* (16) *are even necessary.*

In fact, condition (16) means that

$$f_t(T)\mathscr{E}_{\lambda_n} = t^n \mathscr{E}_{\lambda_n}, \quad n \geq 1, \tag{17}$$

and (15) that $\sup_{n\geq 0} t^n |\mathscr{E}_{\lambda_n}| < \infty$, $t \in (0,1)$ and – together with (17) – that $f_t(T) = \sum_{n\geq 0} t^n \mathscr{E}_{\lambda_n}$. As the convergence $x = \lim_{t\to 1} f_t(T)x$ is trivial on the total set of eigenvectors and as $\overline{\lim}_{t\to 1} |f_t(T)| \leq \overline{\lim}_{t\to 1} \|f_t\|_\infty < \infty$ we see that $\{\mathscr{E}_{\lambda_n}\}_{n\geq 1}$ indeed is an Abel basis.

Conversely, if the series $\sum_{n\geq 1} t^n \mathscr{E}_{\lambda_n} x$ converges for all $x$ then $\sup_{n\geq 1} t^n |\mathscr{E}_{\lambda_n}| < \infty$ and so the necessity of the conditions (15) and (16) (in the "model" case) will follow from Sarason's Theorem along with the formula $|\mathscr{E}_{\lambda_n}| = |B_{\lambda_n}(\lambda_n)|^{-1}$ (Lecture VI). □

*Example* 3. Let us consider the model operator $T = P_K S|K$, $K = H^2 \ominus \mathbf{B}H^2$ where $B = \prod_{n\geq 1} b_{\lambda_n}$ and the eigenvalues make up one of the following sequences:

$$\lambda_n = 1 - 2e^{-n^\alpha}, \quad 0 < \alpha < 1;$$
$$\lambda_n = 1 - 2e^{-(\log n)^\alpha}, \quad \alpha > 1;$$
$$\lambda_n = 1 - n^{-\alpha}, \quad \alpha > 1.$$

One can show (Vasyunin [3]) that in the first two cases the series of eigenvectors are summable by Abel's method (one takes $f_t = t^{\left(\log \frac{2}{1-z}\right)^{1/\alpha}}$ and $f_t = t^{\exp\left(\log \frac{2}{1-z}\right)^{1/\alpha}}$ respectively), while this is not the case in the third one (condition (15) is not fulfilled).

Other examples show that the conditions (15) and (16) are independent of each other (i.e. (15) $\not\Rightarrow$ (16), (16) $\not\Rightarrow$ (15)).

All this indicates that the summability of series of rational fractions by one method or other is not in the first place connected with ideal properties but depends on a more refined description of the space of restrictions $X|\sigma$ of the given class of functions $X$ to a given set $\sigma$. In this context various sufficient conditions for interpolation of the type $l^q(w_n) \subset H^p|\{\lambda_n\}$ etc. can be useful; cf. Snyder [1], Taylor-Williams [1], Švedenko [1], Garnett [1].

**6. Hankel Operators and Angles Between Subspaces.** As was noted in Sect. 3, studying functions of a model operator we in fact study certain Hankel operators. Now we will establish yet some other important properties of Hankel operators and use them to describe bases of reproducing kernels in the model spaces $K_\Theta$.

The following classical problem on nonharmonic Fourier series constitutes the principal special case of the problem solved in this Section: given a collection of points $\{\lambda_n\}_{n\geq 1}$ of the complex plane and a positive number $a$, to decide whether the family of exponentials $\{e^{i\lambda_n x}\}_{n\geq 1}$ is an unconditional basis in the space $L^2(0, a)$ or not. Until present times many deep facts have been accumulated connected with the geometry of families of exponentials on a finite interval (one has found necessary and/or sufficient conditions for having a basis, for minimality, for completeness etc. in $L^2(0, a)$). However, the clearest and most systematic study of this circle of ideas, nonharmonic Fourier analysis, is still perhaps the book Paley-Wiener [1]. The ideas contained there are even now far from exhausted although the majority of the results proved by these authors were outshined years ago. Also one of the basic underlying principles of that book (cf. chap. VI and VII, in particular theorem 37) clearly lies behind the main result of this Section: for the family $\{e^{i\lambda_n x}\}_{n\geq 1}$ to be a basis it is sufficient that it is close to a basis. At the time of Paley-Wiener the only known class of bases was in essence the class of orthogonal bases and, estimating in one way or the other the deviation of the family $\{e^{i\lambda_n x}\}_{n\geq 1}$ from an orthogonal one, they obtained their fundamental results on bases of exponentials. We are going to measure the deviation of the family $\{e^{i\lambda_n x}|[0, a)\}_{n\geq 1}$ from the family $\{e^{i\lambda_n x}|[0, \infty)\}_{n\geq 1}$. Here the latter family differs only by a conformal map followed by a Fourier transformation from the well-known family of rational fractions $\{(\mathbb{1} - \mu_n z)^{-1} : n \geq 1\}$ in the Hardy space $H^2$. In Lecture XI we will in detail consider all these transformations. Right now we remark just that the exponentials *on the segment* are gotten from the exponentials *on the halfaxis* by orthogonal projection (multiplication by the characteristic function $\chi_{[0,a)}$) and that after a passage from the space $L^2(0,\infty)$ to functions analytic in $\mathbb{D}$ (the space $H^2$) the subspace $\chi_{(0,a)}L^2(0,\infty)$ is replaced by the model space $H^2 \ominus \exp\left\{a\frac{z+1}{z-1}\right\}H^2$ and the projection $f \mapsto \chi_{(0,a)}f$ by the projection $f \to P_\Theta f$, $\Theta = \exp\left\{a\frac{z+1}{z-1}\right\}$. Carrying out the corresponding calculation we get criteria for the family $\{e^{i\lambda_n x}|[0,a)\}_{n\geq 1}$ to be an unconditional basis (under the auxiliary hypothesis that $\inf_{n\geq 1}\operatorname{Im}\lambda_n > -\infty$) in terms of the closeness of the inner function $q = e^{iaz}$ to the Blaschke product (for the halfspace) constructed with the zeros $\{\lambda_n\}_{n\geq 1}$. The adequate translation of these conditions into the language of the frequencies $\{\lambda_n\}_{n\geq 1}$ requires independent work; this will be done in Lecture XI only in one but a most significant case (Kadec's $\frac{1}{4}$-Theorem). But now, as already told, we turn to the study of the geometry of the reproducing kernel $\{P_\Theta(\mathbb{1} - \bar{\lambda}z)^{-1} : \lambda \in \sigma\}$ for an arbitrary inner function $\Theta$ and an arbitrary set $\sigma$, $\sigma \subset \mathbb{D}$. However, not all details of the following investigation are necessary for the analysis of the problem for the exponentials.

In this Section (as well as in Lecture IX–X) it will be necessary to use certain geometric notions: the angle between two subspaces and the projection onto one subspace parallel to another; a particular case was encountered already in connection with our study of "spectral projections". Here are the general definitions. Let $L$ and $M$ be subspaces of a Hilbert space, with $L \cap M = \{\mathbb{0}\}$. Then we define on the sum $L + M$ the operator

$$\mathscr{P}_{L\|M} : l + m \mapsto l, \quad l \in L, \quad m \in M;$$

this is the *projection* ($\mathscr{P}^2_{L\|M} = \mathscr{P}_{L\|M}$) *onto L parallel to M* ($\mathscr{P}|L = I, \mathscr{P}|M = \mathbb{0}$). By the closed graph theorem this operator is continuous iff $L + M = \operatorname{clos}(L + M)$. *The angle* $\langle L, M\rangle$ *between L and M* is defined by the conditions

$$\langle L, M\rangle \in \left[0, \frac{\pi}{2}\right], \quad \cos\langle L, M\rangle = \sup_{x\in L, y\in M} \frac{|(x,y)|}{\|x\|\cdot\|y\|}.$$

It follows readily from the definition that

$$\cos\langle L, M\rangle = \sup_{x\in L} \frac{\|P_M x\|}{\|x\|} = |P_M P_L|.$$

$$\sin\langle L, M\rangle = \inf_{x\in L} \frac{\|(I - P_M)x\|}{\|x\|} = |\mathscr{P}_{L\|M}|^{-1},$$

where $P_M$ and $P_L$ are the corresponding orthogonal projections.

The rôle of Hankel operators in the evaluation of the angle between subspaces depends on the following celebrated

**Helson-Szegö Theorem.** *Let $\mu$ be a finite Borel measure on the circumference $\mathbb{T}$ and let $\mathscr{P}$ be the set of all trigonometrical polynomials, setting $\mathscr{P}_A = \mathscr{P} \cap H^2$, $\mathscr{P}_- = \mathscr{P} \cap H^2_-$. The following statements are equivalent.*

1. *The operator $P_+|\mathscr{P}$ can be continued to a continuous transformation from $L^2(\mu)$ into $L^2(\mu)$.*

2. *The angle between the subspaces $\operatorname{clos}_{L^2(\mu)}\mathscr{P}_A$ and $\operatorname{clos}_{L^2(\mu)}\mathscr{P}_-$ is positive.*

3. *The family $\{z^n : n\in\mathbb{Z}\}$ constitutes a basis in $L^2(\mu)$.*

4. *$d\mu = |h|^2\,dm$ where $h$ is an outer function in $H^2$ and $|H_{\bar{h}/h}| = \operatorname{dist}(\bar{h}/h, H^\infty) < 1$.*

5. *$\mu$ is absolutely continuous with respect to Lebesgue measure $m$ and there exist bounded realvalued functions $u$ and $v$ such that*

$$d\mu/dm = e^{u+\tilde{v}}, \quad \|v\|_\infty < \pi/2, \tag{HS}$$

*where $\tilde{v}$ is the function conjugate to $v$ (i.e. $\tilde{v} = \operatorname{Im} f$, $v = \operatorname{Re} f$ for some $f$ in $H^2$ and $\tilde{v}(0) = 0$).*

Prior to the proof let us remark that the family $\{z^n : n\in\mathbb{Z}\}$ forms a basis in $L^2(\mu)$ iff (by definition) each function $f\in L^2(\mu)$ has a unique representation $f = \sum_{k\in\mathbb{Z}} a_k z^k \stackrel{\text{def}}{=} \lim_n P_{n,n}f$, $P_{n,n}f \stackrel{\text{def}}{=} \sum_{k=-n}^{n} a_k z^k$ and that the coordinate functionals $a_k = a_k(f)$ then are continuous. Obviously, $a_k(f) = \hat{f}(k)$ for all trigonometrical polynomials $f$ and for all $k\in\mathbb{Z}$ and it is further easy to check that $z^{m-n}P_{n,n}\cdot\bar{z}^{m-n}P_{m,m}f = P_{n,m}f \stackrel{\text{def}}{=} \sum_{k=-n}^{m} a_k z^k$ for $m\ge n$, $f\in L^2(\mu)$. Therefore we have $f = \lim_{m,n} P_{m,n}f$, $f\in L^2(\mu)$ (in the basis case).

*Proof of the theorem.* As $P_+|\mathscr{P}$ also is the projection on the vector space $\mathscr{P}_A$ parallel to $\mathscr{P}_-$ we have $1\Leftrightarrow 2$ (in view of the remarks preceding the proof). What concerns Statement 3, it is clear that it is equivalent to the previous ones, because of the following facts: the operation $\mathbf{z}$ of multiplication by the independent variable is an isometry in $L^2(\mu)$, the formulae

$$P_{n,m}f = z^{m+1}(I-P_+)\bar{z}^{(n+m+1)}P_+z^n f, \quad f\in\mathscr{P}, \quad P_+f = \lim_n z^n P_{n,n}\bar{z}^n f,$$

for the projection $P_+$ and the partial sum $P_{n,m}f = \sum_{k=-n}^{m}\hat{f}(k)z^k$ of the Fourier series of the polynomial $f$, $f\in\mathscr{P}$, and, lastly, the Banach-Steinhaus theorem, according to which $\{z^n : n\in\mathbb{Z}\}$ is a basis in $L^2(\mu)$ iff $\sup_{n,m\ge 0}|P_{n,m}| < \infty$.

1, 2, 3 $\Rightarrow$ 4. From Statement 3 it follows that $\bigcap_{n\ge 0}\mathbf{z}^n \operatorname{clos}_{L^2(\mu)}\mathscr{P}_A = \{\mathbb{0}\}$, and thus that $\mu$ is absolutely continuous with respect to the Lebesgue measure $m$ (Corollary of the Szegö-Kolmogorov Theorem, Lecture I, Sect. 7). Let $\mu = w\,dm$. Further $\mathbb{1}\notin \operatorname{clos}_{L^2(\mu)}\mathbf{z}\mathscr{P}_A$ because of the minimality of the family $\{z^n : n\ge 0\}$. By the Szegö-Kolmogorov Theorem this entails $\log w\in L^1(\mathbb{T})$. (By the way, it is easy to see that in terms of the density $w$ the necessary and sufficient condition for minimality is $1/w\in L^1(\mathbb{T})$ (Kolmogorov)). Let $h$ be an outer function in $H^2$ with modulus $w^{1/2}$. Let further $f\in\mathscr{P}_A$ and $g\in\mathscr{P}_-$. Then

$$(f,g)_{L^2(\mu)} = \int_{\mathbb{T}} f\bar{g}\,d\mu = \int_{\mathbb{T}} fh\bar{g}\bar{h}\frac{\bar{h}}{h}\,dm = (H_{\bar{h}/h}(fh), g\bar{h})_{L^2(\mathbb{T})}. \tag{18}$$

Since $h$ is outer we have $\operatorname{clos}_{L^2(\mathbb{T})} h\mathscr{P}_A = H^2$, $\operatorname{clos}_{L^2(\mathbb{T})}\bar{h}\mathscr{P}_- = H^2_-$ and moreover holds $\|fh\|_{H^2} = \|f\|_{L^2(\mu)}$, $f\in\mathscr{P}_A$; $\|g\bar{h}\|_{H^2_-} = \|g\|_{L^2(\mu)}$, $g\in\mathscr{P}_-$. From this and from (18) we infer that

$$\cos_{L^2(\mu)}\langle \mathscr{P}_A, \mathscr{P}_- \rangle = |H_{\bar{h}/h}|,$$

and, the angle between the subspaces in question being positive, Statement 4 therefore follows.

It is clear that we thereby also have established $4 \Rightarrow 2$.

There remains the equivalence of 4 and 5. If $g \in H^\infty$ and $\left\| \frac{\bar{h}}{h} - g \right\|_\infty < 1$ then $\|1 - (hg)/\bar{h}\|_\infty < 1$ so that $\sup|\arg(hg)/\bar{h}| < \pi/2$. It is likewise clear that $\arg(hg)/\bar{h} = \arg h^2 g$ and that $h^2 g \in H^1$. Hence the function $h^2 g$, being the Poisson integral of its boundary values, maps the disk $\mathbb{D}$ into the same sector as $h^2 g|\mathbb{T}$. Moreover, we have for the standard branch of the logarithm $\sup_{\mathbb{D}}|\operatorname{Im}\log(h^2 g)| = \sup_{\mathbb{D}}|\arg h^2 g| < \pi/2$. Setting $v = -\operatorname{Im}\log(h^2 g) = -\arg h^2 g$ we get $h^2 g = e^{\tilde{v}+c-iv}$, $|h|^2 = e^{u+\tilde{v}}$, with $u = c - \log|g|$, $c = \text{const}$. It remains to make the observation that $\log|g|$ is a bounded function, because $\inf_{\mathbb{T}}|g| \geq 1 - \|\bar{h}/h - g\|_\infty > 0$. Thus $4 \Rightarrow 5$.

Conversely, if $h$ is an outer function and $|h|^2 = e^{u+\tilde{v}}$, $\|v\|_\infty < \pi/2$ then $h^2 = \exp\{u + \tilde{v} + i(\tilde{u} - v) + ic\}$. Set $g = e^{-(u+i\tilde{u})-ic}$. Then $hg/\bar{h} = e^{-u-iv}$ so that for small positive $\varepsilon$ the function $\varepsilon hg/\bar{h}$ will take its values in the set shaded in Fig. 4 and $\|\mathbb{1} - \varepsilon hg/\bar{h}\|_\infty = \|\bar{h}/h - \varepsilon g\|_\infty < 1$. □

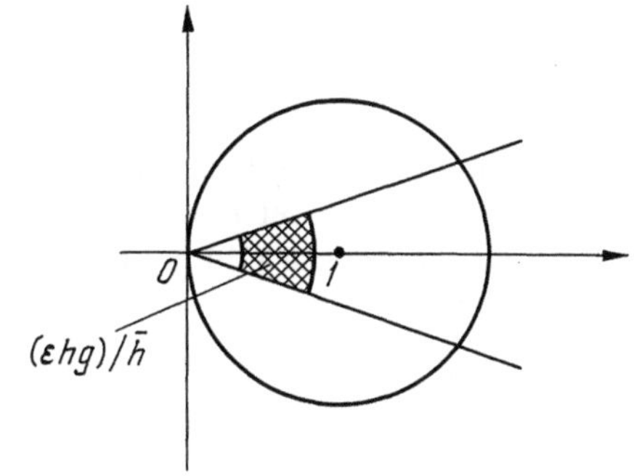


Fig. 4

**Corollary.** *If $h$ is an outer function in $H^2$ then*

$$\operatorname{dist}(\bar{h}/h, H^\infty) < 1 \Leftrightarrow \operatorname{dist}(h/\bar{h}, H^\infty) < 1, \quad h^{-1} \in H^2. \ \square$$

We now use these considerations for the study of the geometry of the reproducing kernel of the model subspace $K_\Theta$, $K = H^2 \ominus \Theta H^2$, $\Theta$ an inner function. As was noted in Lecture V, this kernel is given by

$$k_\Theta(\zeta, \lambda) = (P_\Theta(\mathbb{1} - \bar{\lambda}z)^{-1})(\zeta) = \frac{1 - \Theta(\zeta)\overline{\Theta(\lambda)}}{1 - \bar{\lambda}\zeta},$$

$\lambda, \zeta \in \mathbb{D}$. We are interested in the question of unconditional bases of the form

$$\{k_\Theta(\cdot, \lambda) : \lambda \in \sigma\} \tag{19}$$

where $\sigma$ is a subset of the disk $\mathbb{D}$. A special case of this question ($\Theta = \prod_{\lambda\in\sigma} b_\lambda$) was considered already in Lecture VI–VII. It is not hard to see that for the minimality of the family (19) the Blaschke condition

$$\sum_{\lambda\in\sigma}(1-|\lambda|)<\infty \tag{20}$$

must be necessary (if $T$ is a linear operator and $T\mathscr{X}$ is a minimal family then $\mathscr{X}$ too must be minimal; here $\mathscr{X}=\{(\mathbb{1}-\bar{\lambda}z)^{-1}:\lambda\in\sigma\}$, $T=P_\Theta$). *In what follows we assume this condition to be fulfilled.*

A somewhat more special[1] fact is *the necessity of the Carleson condition* ($C$) for the family (19) to be an unconditional basis for its hull. To see this let us remark that $\|k_\Theta(\cdot,\lambda)\|^2=\dfrac{1-|\Theta(\lambda)|^2}{1-|\lambda|^2}$ and that the Gram matrix for the (normalized) family (19) takes the form

$$\Gamma_{\lambda,\mu}\overset{\text{def}}{=}\frac{(1-|\lambda|^2)^{1/2}(1-|\mu|^2)^{1/2}}{1-\bar{\lambda}\mu}\cdot\frac{1-\Theta(\mu)\overline{\Theta(\lambda)}}{(1-|\Theta(\lambda)|^2)^{1/2}(1-|\Theta(\mu)|^2)^{1/2}},\quad \lambda,\mu\in\sigma.$$

From this we see that

$$\frac{(1-|\lambda|^2)^{1/2}(1-|\mu|^2)^{1/2}}{|1-\bar{\lambda}\mu|}\le|\Gamma_{\lambda,\mu}|,\quad \lambda,\mu\in\sigma,$$

so that the "sparseness" of the family (19) (i.e. the fact that $\sup_{\lambda\neq\mu}|\Gamma_{\lambda,\mu}|<1$) entails the sparseness condition ($R$) for the set $\sigma$ (see Lecture VII). In an analogous manner we get the imbedding theorem: if (19) is an unconditional basis then $\sup_{\lambda\in\sigma}\sum_{\mu\in\sigma}|\Gamma_{\lambda,\mu}|^2<\infty$ so that

$$\sup_{\lambda\in\sigma}\sum_{\mu\in\sigma}\frac{(1-|\lambda|^2)(1-|\mu|^2)}{|1-\bar{\lambda}\mu|^2}<\infty,$$

and $\nu_\sigma=\mu^B$, $B=\prod_{\lambda\in\sigma}b_\lambda$, is a Carleson measure ($\sigma\in(CN)$; cf. Lecture VII). Thus $\sigma\in(C)$. □

All this allows us to consider the family (19) as a "small" perturbation of the family of rational functions $\{(\mathbb{1}-\bar{\lambda}z)^{-1}:\lambda\in\sigma\}$. In a special case when these families are uniformly close to each other, i.e. if

$$\sup_{\lambda\in\sigma}|\Theta(\lambda)|<1 \tag{21}$$

(or, equivalently, if $\|k_\Theta(\cdot,\lambda)\|_2\asymp\|(\mathbb{1}-\bar{\lambda}z)^{-1}\|_2$, $\lambda\in\sigma$) the preceding conclusion can be simplified in an essential way: already the uniform minimality of the family (19) entails the uniform minimality of the family of "pure" fractions $\{(\mathbb{1}-\bar{\lambda}z)^{-1}:\lambda\in\sigma\}$ and thus the Carleson condition for the set $\sigma$.

[1] "Special" in the sense that in contrast to minimality the property of uniform minimality is not necessarily possessed by the family $\mathscr{X}$ if it is possessed by a linear deformation $T\mathscr{X}$.

In the hypothesis of (20) and (21) we will now seek the necessary and sufficient conditions for the family (19) to be: a) an unconditional basis in $K_\Theta$, b) an unconditional basis for its hull. Let us remark that the last property is equivalent to the interpolation identity

$$\left\{\left\{\frac{(1-|\lambda|^2)^{1/2}}{(1-|\Theta(\lambda)|^2)^{1/2}} f(\lambda) : \lambda \in \sigma\right\} : f \in K_\Theta\right\} = l^2$$

and that an elementary sufficient condition for this is condition $(C)$ ($\delta \overset{\text{def}}{=} \inf_{\lambda\in\sigma} |B_\lambda(\lambda)| > 0$, $B_\lambda = \prod_{\mu\in\sigma\setminus\{\lambda\}} b_\mu$) along with the assumption of a small "uniform perturbation": $\sup_{\lambda\in\sigma} |\Theta(\lambda)| < \varepsilon$ (with a corresponding $\varepsilon = \varepsilon(\delta)$). First however we have to do some more preparatory work; the additional information gained in this way has also an independent interest.

**Lemma on Close Subspaces.** *Let $H$ be a Hilbert space and let $L$ and $M$ be (closed) subspaces.*

*The following statements are equivalent*: 1. $\operatorname{Ker}(P_L|M) = \{\mathbb{0}\}$; 2. $M \cap L^\perp = \{\mathbb{0}\}$; 3. $\operatorname{clos}(L + M^\perp) = H$; 4. $\operatorname{clos} P_M L = M$.

*The following statements are equivalent*: 1. *$P_L|M$ is an isomorphism onto its image*; 2. $P_M L = M$; 3. $H = M^\perp + L$; 4. $|P_{L\perp} P_M| < 1$.

*The following statements are equivalent*: 1. *$P_L|M$ is an isomorphism from $M$ onto $L$*; 2. $P_M L = M$, $L \cap M^\perp = \{\mathbb{0}\}$; 3. $H = M^\perp + L$, $\operatorname{clos}(M + L^\perp) = H$; 4. $|P_{L\perp} P_M| < 1$, $|P_{M\perp} P_L| < 1$.

*Proof.* The four first statements coincide in a trivial way once we remark that $(P_L : M \to L)^* = (P_M : L \to M)$ (since $x \in L$, $y \in M \Rightarrow (P_L y, x) = (y, x) = (y, P_M x)$).

The next series is based on Banach's isomorphism theorem (the operator $A$ is an isomorphic imbedding of one Banach space into another iff the adjoint operator is onto), which implies $1 \Leftrightarrow 2$. It is also clear that $3 \Rightarrow 2$; conversely if $P_M L = M$ then $H = M^\perp + L$ since $H = M^\perp + M$ and always

$$\begin{aligned} M^\perp + P_M L &= \bigcup_{l\in L} (M^\perp + l - P_{M\perp} l) \\ &= \bigcup_{l\in L} (M^\perp + l) = M^\perp + L. \end{aligned} \tag{22}$$

The operator $P_L|M$ is an isomorphism onto its image iff $\operatorname{Ker}(P_L|M) = \{\mathbb{0}\}$ and $\operatorname{clos} P_L M = P_L M$. The first relation is equivalent to the linear independence of the subspaces $M$ and $L^\perp$ ($M \cap L^\perp = \{\mathbb{0}\}$) and the second one to the sum $L^\perp + P_L M = L^\perp + M$ being closed (this last identity was established above). Finally, in view of the closed graph theorem the two last conditions together are equivalent to the continuity of the projection $\mathscr{P}_{L\perp\|M}$, i.e. the inequality

$$\cos\langle L^\perp, M\rangle = |P_{L\perp} P_M| < 1.$$

The last series of statements is contained in the union of the two preceding ones. □

**Theorem on Close Subspaces.** *Let $\Theta$ and $\Theta'$ be inner functions and set $K_\Theta = H^2 \ominus \Theta H^2$, $P_\Theta$ being the orthogonal projection onto $K_\Theta$. The following state-*

*ments are equivalent*:

1. $P_\Theta|K_{\Theta'}$ *is an isomorphism from* $K_{\Theta'}$ *onto* $K_\Theta$.
2. $\operatorname{dist}(\Theta\bar{\Theta}', H^\infty) < 1, \operatorname{dist}(\Theta'\bar{\Theta}, H^\infty) < 1$.
3. *There exists an outer function* $\psi$, $\psi \in H^\infty$, *with*

$$\|\Theta\bar{\Theta}' - \psi\|_\infty < 1.$$

4. $\inf_{c \in \mathbb{R}, h=\bar{h} \in L^\infty} \|\arg\Theta\bar{\Theta}' - \tilde{h} - c\|_\infty < \pi/2$ *where* $\tilde{h}$ *is the function conjugate to* $h$.

**Corollary.** *The Statements* 1–4 *are true if* $\|\Theta - \Theta'\|_\infty < 1$. □

Let us remark right away that for the applications to the problem of bases of exponentials $\{e^{i\lambda_n x}\}_{n\geq 1}$ in $L^2(0, a)$ as discussed at the beginning of this Section expecially the equivalence of Statements 1 and 4 will be of great use; cf. also Corollary 1 and 2 on p. 205–206 and Sect. 2 of Lecture XI.

*Proof of the Theorem.* The implication $1 \Leftrightarrow 2$ follows from the lemma with the same name if we set $L = K_\Theta$, $M = K_{\Theta'}$, $\mathbf{T}_{\Theta'} = P_{\Theta'}S|K_{\Theta'}$ noting that.

$$|P_\Theta^\perp|K_{\Theta'}| = |\Theta P_+\bar{\Theta}|K_{\Theta'}| = |P_+\bar{\Theta}|K_{\Theta'}| = |\Theta(\mathbf{T}_{\Theta'})|$$

and that $|\Theta(\mathbf{T}_{\Theta'})| = \operatorname{dist}(\Theta\bar{\Theta}', H^\infty)$ (the Corollary of Nehari's Theorem).

$2 \Rightarrow 3$. Consider $\varphi$ and $\psi$ in $H^\infty$ with $\|\Theta\bar{\Theta}' - \psi\|_\infty < 1$ and $\|\Theta'\bar{\Theta} - \varphi\|_\infty < 1$. Then $\|\mathbb{1} - \bar{\Theta}\Theta'\psi\|_\infty < 1$ and $\|\mathbb{1} - \bar{\Theta}'\Theta\varphi\|_\infty < 1$ so the values of $\bar{\Theta}\Theta'\psi$ and $\bar{\Theta}'\Theta\varphi$ lie in a suitable disk $\{\zeta : |\zeta - 1| < r\}, 0 < r < 1$. Therefore the values, taken on the circle $\mathbb{T}$ by the product $\varphi\psi$ of the functions, must lie in the region $\Omega \stackrel{\text{def}}{=} \{\zeta : (1-r)^2 \leq |\zeta| \leq (1+r)^2, |\arg\zeta| \leq \pi - \varepsilon\}$, provided the number $\varepsilon$ is sufficiently small. This gives $(\varphi\psi)(\mathbb{D}) \subset \Omega$, because for each $\mu$, $\mu \notin \Omega$, there exists a polynomial $p$ with $p(\mu) = 1$, $\max_\Omega |p| < 1$ so that the assumption of the existence of a point $\zeta$, $\zeta \in \mathbb{D}$, with $(\varphi\psi)(\zeta) = \mu$ would contradict the maximum principle: $1 = |p(\varphi\psi)(\zeta)| \leq \sup_{\mathbb{T}} |p(\varphi\psi)| \leq \sup_\Omega |p| < 1$. Thus $\inf_{\mathbb{D}} |\varphi\psi| > 0$ and $\varphi\psi$ has to be outer. But then $\psi$ is outer too.

$3 \Rightarrow 4$. If $\psi$ is the function in Statement 3 we have $\|\mathbb{1} - \bar{\Theta}\Theta'\psi\|_\infty < 1$ and $\|\arg\Theta\bar{\Theta}' - \arg\psi\|_\infty < \pi/2$ and in addition holds $\log|\psi| \in L^\infty(\mathbb{T})$, in view of the inequalities $|\psi(\zeta)| \geq |\Theta\bar{\Theta}'(\zeta)| - \|\Theta\bar{\Theta}' - \psi\|_\infty = 1 - \|\Theta\bar{\Theta}' - \psi\|_\infty > 0, \zeta \in \mathbb{T}$.

$4 \Rightarrow 2$. Let $c \in \mathbb{R}$ and let $h$ be a real function, $h \in L^\infty(\mathbb{T})$ and $\|\arg\Theta\bar{\Theta}' - \tilde{h} - c\|_\infty < \pi/2$. Then the values of $\bar{\Theta}\Theta'\varphi$ ($\varphi \stackrel{\text{def}}{=} e^{h+i\tilde{h}+ic}$) lie in a set of the type indicated in Fig. 4 (p. 199) so for small $\varepsilon$, $\varepsilon > 0$, there holds $\|\mathbb{1} - \varepsilon\bar{\Theta}\Theta' \cdot \varphi\|_\infty = \|\Theta\bar{\Theta}' - \varepsilon\varphi\|_\infty < 1$. Because of the symmetry of Statement 4 with respect to $\Theta$ and $\Theta'$ this gives also the second one of the inequalities desired. □

From the Lemma on Close Subspaces one can also derive other conditions for the closeness of the model spaces $K_\Theta$, which employ the language of Hankel and Toeplitz operators. To this end we need the following results for such operators, of an independent interest.

a) *If* $f \in L^\infty$ *and the operator* $H_f$ *attains its norm* (*i.e. there exists a function* $g$, $g \in H^2$; $g \neq \mathbb{0}$, *with* $\|H_f g\| = |H_f| \cdot \|g\|$) *then the set* $f + H^\infty$ *contains a unique element of least norm and the latter is of the form*

$$\lambda \frac{\bar{h}}{hv},$$

*where* $\lambda =$ const, $h$ *is an outer function in* $H^2$ *and* $v$ *an inner one with* $v(0) = 0$.

In fact, without loss of generality we can assume that $|H_f| = 1$ and $\|g\| = 1$. Let thus $F \in f + H^\infty$, $\|F\|_\infty = 1$. Then $H_F = H_f$ and

$$1 = \|H_f g\| = \|P_- Fg\| \leq \|Fg\| = \|g\| = 1,$$

so that $|F| = 1$ a.e. on $\mathbb{T}$ and $Fg \in H^2_-$, i.e. $H_f g = Fg$ or $F = H_f g/g$. Since further $|H_f g| = |g|$ a.e. we get the requested representation for $F$ (with $h = g^e$ (the outer part of $g$) and $v = g^i \cdot \overline{(H_f g)^i}$). □

*Remark.* From this proof we see that if $f = \varphi\bar{\Theta}$, $\varphi \in H^\infty$, $\Theta$ an inner function, then the extremal vector $g$ belongs to the space $K_\Theta = H^2 \ominus \Theta H^2$ (since the operator $H_f$ acts as zero on $\Theta H^2$) and further if $F = \varphi\bar{\Theta} + k$, $k \in H^\infty$, then $(\varphi + \Theta k)g \in \Theta H^2_-$. Thus the function $(\varphi + \Theta k)|\varphi(\mathbf{T}_\Theta)|^{-1}$ is inner, $(\varphi + \Theta k)(\mathbf{T}_\Theta) = \varphi(\mathbf{T}_\Theta)$ and moreover it is the ratio of two functions in $K_\Theta$.

b) *If* $f \in L^\infty$ *and* $\operatorname{dist}(f, H^\infty) < 1$, $\operatorname{dist}(\bar{z}f, H^\infty) = 1$ *then* $H_f$ *attains its norm.*

In fact, by the assumptions imposed $|H_{\bar{z}f}| = 1$, whereas $\operatorname{dist}(\bar{z}f, \bar{z}H^\infty) = \operatorname{dist}(\bar{z}f + c\bar{z}, H^\infty) < 1$ for a suitable $c$, i.e. $|H_{\bar{z}f + c\bar{z}}| < 1$. This gives what is required: if the operator $A$ does not attain its norm and $C$ is completely continuous (in our case: $A = H_{\bar{z}f}$, $C = H_{c\bar{z}}$) then $|A + C| \geq |A|$ (in fact, then $A^*A$ does not attain its norm either and by the Spectral Theorem there exist vectors $x_n$ with pairwise disjoint spectra such that $\|x_n\| = 1$, $\lim_n \|A^*Ax_n\| = |A^*A|$; since $x_n \xrightarrow{\text{weakly}} \mathbb{0}$ we have $|A^*| \cdot |A| = |A^*A| = \lim_n \|A^*Ax_n + fA^*Cx_n\|$, which gives the desired conclusion). □

c) *Let* $f \in L^\infty$; *consider the Toeplitz operator* $T_f$ *setting*

$$T_f g = P_+ fg, \; g \in H^2.$$

*If* $f$ *is unimodular* ($|f| = 1$ *a.e. on* $\mathbb{T}$) *then* $T_f$ *is left invertible (is an isomorphism onto its image) iff*

$$\operatorname{dist}(f, H^\infty) < 1$$

*In the same hypothesis,* $T_f$ *is an isomorphism of* $H^2$ *onto itself iff*

$$\operatorname{dist}(f, H^\infty) < 1, \quad \operatorname{dist}(\bar{f}, H^\infty) < 1.$$

Both assertions follow from the relations $T_f^* = T_{\bar{f}}$ ($\forall f \in L^\infty$) and $\|x\|^2 = \|T_f x\|^2 + \|H_f x\|^2$, $\forall x \in H^2$ (in the hypothesis $|f| = 1$ a.e.). □

d) *Let* $f \in L^\infty$ *and assume that* $T_f$ *is left invertible. Then the following statements are equivalent*: 1. $T_f$ *is invertible.* 2. $\operatorname{Ker} T_{\bar{z}f} \neq \{\mathbb{0}\}$. 3. $T_{\bar{z}f}$ *is not left invertible.*

In fact, $T_{\bar{z}f} = S^*T_f$ which gives $1 \Rightarrow 2$. It remains to verify that $3 \Rightarrow 1$: if $\inf_{\|x\|=1} \|T_{\bar{z}f}x\| = 0$ then $\inf_{\|x\|=1} \|T_f x - (T_f x)^\wedge(0)\mathbb{1}\| = 0$ and therefore, since $\inf_{\|x\|=1} \|T_f x\| > 0$, we conclude that $\mathbb{1} \in \operatorname{clos} T_f H^2 = T_f H^2$. But then $T_f H^2 = H^2$, in view of $\mathscr{P}_A \subset T_f H^2$ (the last inclusion follows from the identities $\mathbb{1} = T_f \bar{z}^n z^n T_f^{-1}\mathbb{1} = S^{*n} T_f z^n T_f^{-1}\mathbb{1}$, $n \geq 0$). □

e) *If* $f \in L^\infty$ *and* $\|\mathbb{1} - f\|_\infty < 1$ *then* $T_f$ *is invertible.* □

f) *If* $f \in L^\infty$ *and* $\psi \in H^\infty$ *then* $T_{\bar{\psi} f} = T_{\bar{\psi}} T_f = T_\psi^* T_f$.

Immediate verification: $T_{\bar{\psi} f} x = P_+ \bar{\psi} f x = P_+ \bar{\psi} P_+ f x = T_{\bar{\psi}} T_f x$, $x \in H^2$. □

Before using these results for the study of the reciprocal situation of subspaces of type $K_\Theta$ let us point out that the Statements a)–b) contain essentially the following facts.

**Adamyan-Arov-Kreĭn Theorem.** *If* $f \in L^\infty$ *and* $\operatorname{dist}(f, H^\infty) < 1$ *then the set* $f + H^\infty$ *contains an unimodular function.*

*Proof.* Set $H_0^\infty = \{f : f \in H^\infty, \ f(0) = 0\}$. The continuous function $c \mapsto \operatorname{dist}(f + c\mathbb{1}, H_0^\infty)$, $c \in \mathbb{C}$, takes on both values which are larger than 1 (obvious) and smaller than 1 (by the hypothesis). Therefore for some $c$ we must have $\operatorname{dist}(f + c\mathbb{1}, H_0^\infty) = 1$. Applying Statement b) to the function $f + c\mathbb{1}$ and afterwards Statement a) to the function $\bar{z}(f + c\mathbb{1})$ we conclude that the set $\bar{z}(f + c\mathbb{1}) + H^\infty$ contains a unimodular element. This gives the result. □

**Corollary 1.** *Let* $\Theta$ *be an inner function and let* $\varphi \in H^\infty$, *assuming that* $|\varphi(\mathbf{T}_\Theta)| < 1$. *Then there exists an inner function* $\psi$ *such that* $\psi(\mathbf{T}_\Theta) = \varphi(\mathbf{T}_\Theta)$ *(i.e.* $\varphi - \psi \in \Theta H^\infty$*) and this function is the ratio of two functions in* $K_{z\Theta}$.

Indeed, $|\varphi(\mathbf{T}_\Theta)| = \operatorname{dist}(\varphi\bar{\Theta}, H^\infty) < 1$ and the existence of $\psi$ follows from the theorem. Use now the remark to Statement a), applying it to the function $\bar{z}(\varphi\bar{\Theta} + c\mathbb{1}) = (\varphi + c\Theta)\bar{z}\bar{\Theta}$ ($c$ being the constant in the proof of the theorem). We get then an inner function $\psi = \varphi + c\Theta + k\Theta z$ satisfying all the required conditions. □

**Corollary 2 (Nevanlinna's Theorem).** *If* $B = \prod_{\lambda \in \sigma} b_\lambda$ *is the Blaschke product with the set of zeros* $\sigma$, $\sigma \subset \mathbb{D}$, *and if* $\varphi \in H^\infty$ *with* $\operatorname{dist}(\varphi\bar{B}, H^\infty) < 1$, *then there exists an inner function* $\psi$ *such that* $\psi|\sigma = \varphi|\sigma$. *In particular this is so (with any* $B$*) if* $\|\varphi\|_\infty < 1$. □

**Corollary 3 (Schur's Theorem).** *If* $\varphi \in H^\infty$, $\|\varphi\|_\infty \le 1$, *then there exist finite Blaschke products* $\psi_n$, $n \ge 1$, *such that* $\varphi(\zeta) = \lim_n \psi_n(\zeta)$, $\forall \zeta \in \mathbb{D}$.

In proving this corollary we may assume that $\|\varphi\|_\infty < 1$ (as $\lim_{r \to 1-0} \varphi(r\zeta) = \varphi(\zeta)$ and $\|\varphi(rz)\|_\infty < 1$, $r < 1$, if $\varphi \not\equiv \text{const}$). By Corollary 1 we can find inner functions $\psi_n$, $n \ge 1$, with $\varphi - \psi_n \in \mathbf{z}^n H^\infty$. It is then clear that $\psi_n \xrightarrow{\text{weak } H^2} \varphi$ for $n \to \infty$ and therefore also $\lim_n \psi_n(\zeta) = \varphi(\zeta)$, $\forall \zeta \in \mathbb{D}$. The functions $\psi_n$ are quotients of elements of the subspaces $H^2 \ominus \mathbf{z}^{n+1} H^2$, i.e. ratios of polynomials. Therefore they must be finite Blaschke products, $n \ge 1$. □

Let us now turn to estimating the closeness of subspaces of the form $K_\Theta$.

**Complement 1** to the Theorem on Close Subspaces. Statements 1–4 can also be expressed in the following equivalent forms:

5. $|\Theta(\mathbf{T}_{\Theta'})| < 1$, $|\Theta'(\mathbf{T}_\Theta)| < 1$.
6. *The Toeplitz operator* $T_{\Theta\bar{\Theta}'} \stackrel{\text{def}}{=} P_+ \Theta\bar{\Theta}' | H^2$ *is invertible.*
7. $K_\Theta + \Theta' H^2 = H^2$, $K_{\Theta'} + \Theta H^2 = H^2$.

8. $\operatorname{dist}(\Theta\bar{\Theta}', H^\infty) < 1$, $\Theta' H^2 \cap \Theta H^2_- = \{\mathbb{0}\}$.
9. $\operatorname{dist}(\Theta\bar{\Theta}', H^\infty) < 1$, $\operatorname{dist}(\bar{z}\Theta\bar{\Theta}', H^\infty) = 1$.
10. *There exists an outer function h such that*

$$h \in H^2, \quad |h|^2 \in (HS), \quad \Theta\bar{\Theta}' = \lambda \frac{\bar{h}}{h} \quad (\lambda = \text{const}).$$

In fact, Statement 5 coincides with Statement 2 (the Corollary of Sarason's Theorem) and Statement 6 is equivalent to Statement 5, because of Statements c) and d).

It is likewise clear that $1 \Leftrightarrow 7$ and that the implication $6 \Leftrightarrow 9$ is contained in the Statements c) and d) preceding the theorem.

In Statement 8 we are told that the operator $P_\Theta | K_{\Theta'}$ is an isomorphism onto its image and that the adjoint operator $P_{\Theta'} | K_\Theta$ is injective (i.e. the same as in Statement 1).

$10 \Rightarrow 2$ by the Helson-Szegö Theorem.

$9 \Rightarrow 10$. Using Statements a) and b) we conclude that $\bar{z}\Theta\bar{\Theta}' = \lambda \dfrac{\bar{h}}{hv}$ where $\lambda = \text{const}$, $h$ outer ($h \in H^2$) and $v$ inner, with $v(0) = 0$. In view of the symmetry of Statement 9 with respect to $\Theta$ and $\Theta'$ (this is seen, for example, from $9 \Leftrightarrow 6$) we further have $\bar{z}\Theta'\bar{\Theta} = \lambda_1 \dfrac{\bar{h}_1}{h_1 v_1}$ with analogous $h_1$ and $v_1$. This gives $\bar{z}^2 h h_1 v v_1 = \lambda\lambda_1 \overline{h h_1}$ so that both members must be constant. Therefore $\bar{z}v = \text{const}$ and $\bar{z}v_1 = \text{const}$. The rest (i.e. the inclusion $|h|^2 \in (HS)$) follows from the Helson-Szegö Theorem. □

**Complement 2.** The same arguments give also the equivalence of the analogues of Statements 2–4, 6, 9, 10 gotten by replacing $\Theta\bar{\Theta}'$ by an arbitrary function $f$, $|f| = 1$ a.e. It is also easy to indicate the changes that are needed in the Statements 2, 5–9 so that they become equivalent to the left invertability of $P_\Theta | K_{\Theta'}$ (for example, instead of Statement 5 one only has to impose $|\Theta(\mathbf{T}_{\Theta'})| < 1$ etc.).

Returning to the bases of values of the reproducing kernel of the space $K_\Theta$ we get the following result.

**Corollary 1.** *Let $\Theta$ be an inner function and $B = \prod_{\lambda \in \sigma} b_\lambda$ a Blaschke product with simple zeros ($\sigma \subset \mathbb{D}$). Assume that*

$$\sup_{\lambda \in \sigma} |\Theta(\lambda)| < 1.$$

*The following statements are equivalent*:

1. *The family* $\left\{ \dfrac{\mathbb{1} - \Theta\overline{\Theta(\lambda)}}{\mathbb{1} - \bar{\lambda}z} : \lambda \in \sigma \right\}$ *is an unconditional basis in* $K_\Theta$.
2. $\sigma \in (C)$ *and the operator* $P_\Theta | K_B$ *defines an isomorphism from* $K_B$ *onto* $K_\Theta$.
3. $\sigma \in (C)$ *and* $|\Theta(\mathbf{T}_B)| < 1$, $|B(\mathbf{T}_\Theta)| < 1$.
4. $\sigma \in (C)$ *and* $\operatorname{dist}(\Theta\bar{B}, H^\infty) < 1$, $\operatorname{dist}(B\bar{\Theta}, H^\infty) < 1$.

5. $\sigma \in (C)$ *and there exists an outer function* $h$ *such that*

$$h \in H^2, \quad |h|^2 \in (HS), \quad \Theta\bar{B} = \lambda\frac{\bar{h}}{h} \quad (\lambda = \text{const}).$$

6. $\sigma \in (C)$ *and there exists an outer function* $\psi$ *with* $\|\Theta\bar{B} - \psi\|_\infty < 1$.
7. $\sigma \in (C)$, $\inf_{c \in \mathbb{R}, h=\tilde{h} \in L^\infty} \|\arg \Theta\bar{B} - \tilde{h} - c\|_\infty < \pi/2$.

In fact, $1 \Rightarrow 2$, since the implication $1 \Rightarrow (\sigma \in (C))$ has been established in the comments preceding the Lemma on Close Subspaces and further it is clear that the operator $P_\Theta|K_B$, mapping the Riesz basis $\{(1 - |\lambda|^2)^{1/2}(\mathbb{1} - \bar{\lambda}z)^{-1} : \lambda \in \sigma\}$ onto the Riesz basis $\{(1 - |\lambda|^2)^{1/2} k_\Theta(\cdot, \lambda) : \lambda \in \sigma\}$, must be an isomorphism. Conversely, $2 \Rightarrow 1$ in a trivial way. The rest follows from the Theorem on Close Subspaces and Complement 1. □

As illustration we will only consider one comparatively simple case, viz. the case of the singular function $\Theta = \exp\left\{a\frac{z + \mathbb{1}}{z - \mathbb{1}}\right\}$, $a > 0$; then (as will be shown) the condition $\sup_{\lambda \in \sigma} |\Theta(\lambda)| < 1$ is not essential. After conformal transplantation to the halfplane this special case also provides the solution of the problem of bases of exponentials in the space $L^2(0, a)$ and therefore it will be deferred to Lecture XI. In a more general situation the study of the geometry of the reproducing kernel $k_\Theta$ (in particular the question of bases $\{k_\Theta(\cdot, \lambda) : \lambda \in \sigma\}$) is obstructed by serious analytical difficulties, which will be discussed in the following Section and in corresponding places in Lecture XI.

Applying Complement 2 of the theorem we get the analogue of Corollary 1 for bases of hulls.

**Corollary 2.** *Let the hypotheses of Corollary 1 be fulfilled. Then the following statements are equivalent*:

1. *The family* $\left\{\frac{\mathbb{1} - \Theta\overline{\Theta(\lambda)}}{\mathbb{1} - \bar{\lambda}z} : \lambda \in \sigma\right\}$ *is an unconditional basis in the closure of its linear hull.*
2. $\sigma \in (C)$ *and* $P_\Theta|K_B$ *maps* $K_B$ *isomorphically into* $K_\Theta$.
3. $\sigma \in (C)$, $|\Theta(\mathbf{T}_B)| < 1$.
4. $\sigma \in (C)$, $\operatorname{dist}(\Theta\bar{B}, H^\infty) < 1$. □

**7. Concluding Remarks.** Functional calculi as a method of spectral analysis is the object of an extensive literature of which, as a complement to what has already been said, we mention only the monographs Dunford-Schwartz [3], Colojoara-Foiaş [1] and Maslov [1]. A not less important and popular object of study is provided by the multipliers of families of vectors (subspaces). They are used not only in the study of the geometry of these families but also in purely analytic questions (e.g. multipliers of Fourier series or integrals, in the solution of the corresponding convolution equations etc.). The possible structure of the multiplier algebras themselves is investigated in the papers of Yamazaki [1]. What concerns the analysis of spectral resolutions with the aid of the multiplier method then in the general situation (that is, for operators far from shifts, differentiation etc., for which one uses multipliers of Fourier series (transformations) a hindering factor was for a long time the absence of non-trivial calculi. The results presented in this Lecture owe their existence to v. Neumann's inequality ($|p(T)| \le \max_{|\zeta| \le |T|} |p(\zeta)|$, $p \in \mathscr{P}_A$) and the subsequent $H^\infty$-calculus and to Carleson's Theorem on free interpolation in $H^\infty$. This combination of ideas, developed afterwards by many authors, can first be found in Kacnel'son [1]; in that paper there is in particular shown that a contraction operator with a complete family of eigenspaces and a Carleson point spectrum must be scalar in Dunford's sense (Example 1 of Sect. 4). More general results will be obtained in Lecture IX.

As just noted, Carleson's Theorem along with its generalizations (Lecture IX) lies at the heart of our study of the spectral properties of contraction operators. It was first proved by Carleson [1]; numerous comments and problems connected with it can be found in Duren [1] and in Vinogradov-Havin [1], [2]. The scheme of reasoning of this Lecture (the Lorch-Grinblyum Theorem + Sarason's Theorem + the Theorem on Bases and Interpolation ⇒ Carleson's Theorem) is proposed in Nikol'skiĭ [8]. The applicability of the same scheme in other situations has been discussed in detail in the foregoing Sections of this Lecture.

The original references in connection with the Lorch-Grinblyum Theorem are Lorch [1] and Grinblyum [1], [2].

Sarason's Theorem was proved by that author in [6], which paper also contains the applications given in Sect. 1 to the classical interpolation theorems. The connection between the commutant $\{\mathbf{T}_\Theta\}'$ and Hankel operators too can be traced to Sarason [6] but there this connection does not play the same rôle as in Sect. 2 of this Lecture, where we essentially show that Sarason's Theorem is an immediate consequence of Nehari's Theorem. This fact is of particular importance for the vector variant of Sarason's Theorem (viz. the Commutant Theorem), the original proof of which is noted for being very complicated (Sz.-Nagy and Foiaş [1], [6]). Let us further remark that Page [1] derived his theorem (Sect. 4) from the Commutant Theorem but the opposite route used in Sect. 4 appears much simpler and more natural. Yet another proof of Page's Theorem is contained in Adamyan-Arov-Kreĭn [1].

General information on Hankel operators (matrices) can be found in the book by Grenander-Szegö [1] and in Appendix 4. Nehari's Theorem was proved by that author in [1].

The Hartman-Sarason Theorem is contained in the papers Hartman [1] and Sarason [6], in the first of these in fact at once for general Hankel operators $H_g$, $g \in L^\infty$, and not only for the operators

$$f(\mathbf{T}_\Theta)P_\Theta = \Theta H_{\Theta^* f}. \tag{23}$$

Later on one has also considered a general notion of "Hankel operator with respect to an isometry $U$" ($U^*A = AU$) (Rosenblum [1]) and the analogue of Hartman's Theorem for the latter was obtained in Page's paper [1] and then – in a much simpler manner – in Bonsall-Power [1]. The study of this question in Sect.

3 is close to the last paper; it is also applicable in the case of general Hankel operators. Let us in passing notice that the set of products $\Theta\bar{f}$, where $\Theta$ is an inner function and $f \in H^\infty$, is dense in $L^\infty(\mathbb{T})$. This is a special case of a theorem by Douglas and Rudin (see Appendix 2, Sect. 30 and the survey by Havin [1]). It is likewise appropriate to mention the rather recent theorem of Chang and Marshall describing all subalgebras of $L^\infty(\mathbb{T})$ containing $H^\infty$ (each of these is generated by the inner functions contained in the algebra in question, along with their complex conjugates); we refer to Sarason [9], to Appendix 4 and to the references mentioned there. The Lemma on (the closedness of) $H^\infty + C(\mathbb{T})$ is due to Sarason [6]; other proofs can be found in Douglas [2], Rudin [3] and Sarason [6]. The proof in this Lecture is due to Zal'cman but can in essence be found in Adamyan-Arov-Kreĭn [2].

The identity (23) shows that the operators $f(\mathbf{T}_\Theta)$ and $H_{\Theta^* f}$ must have the same "smoothness", i.e. they are at the same time continuous, nuclear, Hilbert-Schmidt ($\mathfrak{S}_2$ operators) etc. The last condition is easily expressed in terms of the symbol:

$$H_f \in \mathfrak{S}_2 \Leftrightarrow \sum_{n\geq 1} n|\hat{f}(-n)|^2 < \infty$$

$$\left(\sum_{n\geq 0} \|H_f z^n\|_2^2 = \sum_{n\geq 0} \sum_{k\leq -n-1} |\hat{f}(k)|^2 = \sum_{k\geq 1} k|\hat{f}(-k)|^2\right).$$

Concerning membership in other symmetrical normed ideals see Kronecker [1] (where Kronecker's Lemma is first established), Clark [1], Koosis [2], Dym-Shapiro [1]. Recently Peller [2]–[5] found a complete description of Hankel operators in the Schatten-von Neumann class $\mathfrak{S}_p$, $0 < p < \infty$; see Appendix 4, Sect. 23, and for the proof and other details Appendix 5. Regarding vectorical generatizations (i.e. theorems on operators in $\{\mathbf{T}_\Theta\}'$ and on Hankel operators in the space $H^2(E)$) see also (besides the papers already mentioned) Muhly [1] where as a corollary there is obtained the following criterion for the compactness of a contraction $T$: its characteristic function $\Theta_T$ (after a suitable "normalization", since it is defined by $T$ only up to a constant unitary factor) must admit analytic continuation into a neighborhood of clos $\mathbb{D}$, and all the operators $\Theta_T(\zeta) - \zeta I$, $\zeta \in \mathbb{D}$, have to be compact.

There are several extremal problems connected with essentially the same type of questions (i.e. the study of spectral properties of Hankel operators, their "smoothness" etc.). One of these, which has been touched upon in Sect. 6, is the search for a function $a$, $a \in f + H^\infty$, such that $\|a\|_\infty = \inf_{h\in H^\infty} \|f + h\|_\infty$. The main results in this direction are due to Adamyan, Arov and Kreĭn; see their papers [1]–[5] and also the paper Carleson-Jacobs [1]. A short survey of results obtained here can likewise be found in Sarason's lectures [9] and in Appendix 4.

Sect. 4 was written in the spirit of Nikol'skiĭ [10] where one can find other details too.

Sect. 5 follows in its essentials Nikol'skiĭ [2]. Corollary 2 and Examples 1 and 3 are due to Vasyunin [3].

The treatment of Sect. 6 was inspired by the papers by Pavlov [1] and Hruščev [1] from where we borrow notably the idea of studying isomorphisms generated by projections and where (apart from a conformal transplantation from the halfplane to the disk; cf. for more details Lecture IX) Riesz bases consisting of fractions $k_\Theta(z,\lambda) = (\mathbb{1} - \Theta\overline{\Theta(\lambda)})(\mathbb{1} - \bar{\lambda}z)^{-1}$, $\lambda \in \sigma$; $\Theta = \exp\left(a\frac{z+1}{z-1}\right)$, $a > 0$, are considered. The comments and supplements to this pivotal case will be deferred to Lecture XI. Here we give some general remarks on the contents of Sect. 6.

Concerning angles between subspaces and the geometry connected with this see Gohberg-Kreĭn [1].

The Helson-Szegö Theorem was proved by these authors in [1]; our presentation is close to Sarason [9]. There is yet another condition equivalent to the Statement 1–5 of that theorem: the measure $\mu$ is absolutely continuous with respect to Lebesgue measure $m$ and its density $w = d\mu/dm$ satisfies

$$\sup\left(\frac{1}{mI}\int_I w\,dm\right)\left(\frac{1}{mI}\int_I \frac{1}{w}\,dm\right) < \infty,$$

where the sup extends over all arcs (intervals) $I$, $I \subset \mathbb{T}$ (Hunt-Muckenhoupt-Wheeden [1]). The analogue of this condition in the $L^p$-norm leads to a theorem for $L^p(\mu)$ which is similar to the Helson-Szegö Theorem (cf. loc. cit.). See also Koosis [4], Sawyer [1], Carleson-Jones [1], Treĭl' [1] for some new interesting results about this matter, as well as a series of the papers by Arosena-Cotlar-Sadosky (we mention just one of them [1]).

Our presentation of the problem of bases of reproducing kernels follows the paper Nikol'skiĭ [11].

The properties (a)–(e) of Hankel operators set forth in Sect. 6 and the equivalence of the Statements 2, 3, 5–10 of the Theorem on Close Subspaces resulting from them are part of a theory developed by Hartman, Wintner, Widom, Devinatz, Rosenblum, Coburn, Douglas, Adamyan, Arov, Kreĭn and others. The reader will find much additional material as well as references in Appendix 4, in Douglas's book [2] and in Sarason's lectures [9]. The latters brilliant exposition is to a large extent used in the derivation of a)–e) and the Theorem on Close Subspaces. What concerns the individual statements let us remark that the properties a) and b) of Hankel operators (in a more general situation) were established by Adamyan, Arov and Kreĭn, property c) by Widom and Devinatz (also the implication $6 \Leftrightarrow 10$ of Complement 1 to the Theorem on Close Subspaces is due to them), finally the implication $3 \Leftrightarrow 6$ by Douglas; for details we refer to the three sources just mentioned.

It should be noted that the proof of a part of the Theorem on Close Subspaces, namely implication $2 \Rightarrow 3$, can (following Peller-Hruščev [1]) be simplified with the aid of above mentioned property c) as follows. Let $f \in L^\infty$, $|f| = 1$ a.e., $\operatorname{dist}(f, H^\infty) < 1$, $\operatorname{dist}(\bar{f}, H^\infty) < 1$ and let $h \in H^\infty$, $\|\bar{f} - h\|_\infty < 1$. Then the operator $T_{fh} = T_f T_h$ is invertible in view of $\|1 - fh\|_\infty = \|\bar{f} - h\|_\infty < 1$ and $T_f$ is also invertible by the property c). It follows that $T_h$ is invertible, that is $h^{-1} \in H^\infty$, and hence the function $h$ is outer. □

The Adamyan-Arov-Kreĭn Theorem (in a considerably more general form) is contained in the paper [2] by these authors and Corollary 2 and 3 – in a somewhat different version compared with our's – in [4]. Nevanlinna's Theorem [2] (cf. also Walsh [1]) as a matter of fact says more than proved in Corollary 2: if $\varphi \in H^\infty$ and if there exist two *different* functions $\psi_1$ and $\psi_2$ such that $\|\psi_i\|_\infty \leq 1$ and $\varphi|\sigma = \psi_i|\sigma\,(i = 1, 2)$ then there exists also an inner function $\psi$ coinciding with $\varphi$ on $\sigma$. This result likewise follows from the general results of Adamyan-Arov-Kreĭn [2], [3], [4] (cf. also Kreĭn-Nudel'man [1]), which in fact provide a complete parametrization of all inner functions $\psi$ such that $\varphi - \psi \in \Theta H^\infty$ ($\Theta$, a given inner function). For details see Sect. 165. 18–19 of Appendix 4. For a discussion of Schur's Theorem see Lecture II, Sect. 5. Some strengthenings of Nevanlinna's Theorem were found by Jerison [1].

Turning to the main discussion at issue let us mention some important questions where the answer is not known; first of all: does there exist an unconditional basis of the type $\{k_\Theta(\cdot, \lambda) : \lambda \in \sigma\}$ in every subspace $K_\Theta$? which unconditional bases $\mathscr{X}_\sigma \overset{\text{def}}{=} \{k_\Theta(\cdot, \lambda) : \lambda \in \sigma\}$ (in its hull) can be extended to a basis of the same kind $\mathscr{X}_{\sigma'}$, $\sigma' \supset \sigma$, in the whole space $K_\Theta$? for which inner functions $\Theta$ does there exist a Blaschke product $B$ connected with $\Theta$ as in Corollary 1 (p. 205–206)? In particular, is it true that for every inner function $\Theta$ there exist Blaschke products whose zeros form a Carleson interpolation set, and such that $\|\Theta - B\|_\infty < 1$? It is easy to see that there exist functions $\Theta$ for which none of the Blaschke products of the form $B_\tau \overset{\text{def}}{=} \dfrac{\Theta - \tau \mathbb{1}}{\mathbb{1} - \bar{\tau}\Theta}$, $|\tau| < 1$, (compare Frostman's Theorem, Lecture II) has a Carleson zero set (remark by Jones; see his paper mentioned in remark 6), below). These questions touch upon a large circle of ideas but we shall not enter into their discussion, limiting ourselves to enumerating the following facts, which can be found (together with more additional information) in Hruščev-Nikol'skiĭ-Pavlov [1]:

1) if the operator $P_\Theta|K_B$ is an isomorphism onto its image then we must have $\sup_{\lambda \in \sigma} |\Theta(\lambda)| < 1$ ($\sigma = B^{-1}(0)$);

2) the condition $\lim_n \operatorname{dist}(\lambda_n, \sigma(\Theta)) = 0$ is necessary for the minimality of the family $\{k_\Theta(\cdot, \lambda_n) : n \geq 1\}$;

3) there exists an unconditional basis $\{k_\Theta(\cdot, \lambda_n) : n \geq 1\}$ in $K_\Theta$, $\Theta = \exp\left(a\dfrac{z+1}{z-1}\right)$, $a > 0$, with $\lim_n |\Theta(\lambda_n)| = 1$ (cf. Lecture XI);

4) there exists a family of the same type which is an unconditional basis for its hull but with $\lim_n |\Theta(\lambda_n)| = 0$.

5) if $\{k_\Theta(\cdot, \lambda) : \lambda \in \sigma\}$ is an unconditional basis in $K_\Theta$ then it is necessary that $\sum_{\lambda \in \sigma}(1 - |\lambda|^2)(1 - |\Theta(\lambda)|^2)^{-1} < \infty$ and that the operator $f \mapsto P_B G f$, $f \in K_\Theta$, gives an isomorphism from $K_\Theta$ onto $K_B$ (here $G \in H^2$, $G(\lambda) = (1 - |\Theta(\lambda)|^2)^{-1/2}$, $\lambda \in \sigma$); these conditions (together with Carleson's Condition $(C)$) are indeed also sufficient.

6) Jones [1] showed that every inner function $\Theta$ can be approximated (in the norm $\|\cdot\|_\infty$) arbitrarily well by ratios of Blaschke products the zeros of which form an interpolation set; in particular, we have $\|\Theta - B_1/B_2\|_\infty < 1$ for suitable such

$B_1$ and $B_2$, which shows that in the space $K_{\Theta B_2} = K_\Theta \vee K_{B_2}$ there by necessity must exist an unconditional basis of values of the reproducing kernel (and one has such a basis in $K_{B_2}$ too, $K_{\Theta B_2}$ being a "quasi-direct sum" of $K_\Theta$ and $K_{B_2}$; $K_\Theta \cap K_{B_2} = \{\mathbb{O}\}$ and $\operatorname{clos}(K_\Theta + K_{B_2}) = K_{\Theta B_2}$).

7) If $\bigvee(k_\Theta(\cdot,\lambda): \lambda \in \sigma) \neq K_\Theta$ then $k_\Theta(\cdot,\mu) \notin \bigvee(k_\Theta(\cdot,\lambda): \lambda \in \sigma)$ for every $\mu$, $\mu \in \mathbb{D}\backslash\sigma$. (In fact, there exists $f \in K_\Theta$, $f \not\equiv 0$ such that $f \perp k_\Theta(\cdot,\lambda)$, $\lambda \in \sigma$ or equivalently $f(\lambda) = 0$, $\lambda \in \sigma$; but $g \stackrel{\text{def}}{=} f/b_\mu^k = P_+ \bar{b}_\mu^k f \in K_\Theta$ where $k = k_g(\mu)$ is the multiplicity of the zero at $\mu$, and hence $g \not\equiv 0$, $g(\lambda) = 0$, $\lambda \in \sigma$ and $g(\mu) \neq 0$. $\square$).

Some more comments and examples can be found in Lecture XI.

# Lecture IX. Generalized Spectrality and Interpolation of Germs of Analytic Functions

In this Lecture we describe those cases when a "spectral decomposition" of the model operator $\mathbf{T}_\Theta$ is possible, taking these words in a broader sense than in Lectures VI–VIII. In place of eigenspaces we will consider arbitrary spectral (and even more general) subspaces and the corresponding parts of the spectrum $\sigma(\mathbf{T}_\Theta)$ will be drawn from a specially chosen $\sigma$-algebra of Borel sets. We must of course maintain sufficiently tight connections with the classical expansion as provided by the Spectral Theorem for normal operators. The necessary language will be borrowed from the theory of spectral operators in the sense of Dunford, concerning which some notions will be developed in the "Concluding Notes" of this Lecture.

**1. Generalized Spectrality.** This introductory Section of Lecture IX sets out by explaining part of the motivation leading to the problem of generalized spectrality. We define scalar operators (in the sense of Dunford) as those operators which display the closest analogy with the normal ones: for them holds (by definition) the unconditionally convergent spectral expansion

$$A = \int_{\sigma(A)} \lambda \, d\mathscr{E}_\lambda$$

generated by the spectral measure $\mathscr{E}$ of $A$. In Hilbert space the scalar operators are just the operators which are linearly similar to some normal operator (Wermer); among the complete operators these again are precisely the operators whose eigenspaces define an unconditional (Riesz) basis in the space in question and for which thus holds the expansion

$$A = \sum_{\lambda \in \sigma_p(A)} \lambda \mathscr{E}(\{\lambda\}).$$

For the model operator $\mathbf{T}_\Theta$ a criterion for scalarity was obtained in Lectures VI–VII. It consists of the requirement that $\Theta$ is a Blaschke product ($\Theta = B$), that $k(\lambda) \leq 1$ (all eigenvalues are simple!) and finally that the Carleson condition

$$\inf_{\lambda \in \sigma_p(T)} |B_\lambda(\lambda)| > 0$$

should hold true; here $B_\lambda \overset{\text{def}}{=} Bb_\lambda^{-k(\lambda)}$.

The study of spectral operators in the sense of Dunford gives in the case of the functional model rise to a considerably more general problem. If the operator is complete, then the question is to find conditions on the spectrum $\sigma_p(T)$ which are equivalent to the decomposition of the space $K_B$ into an unconditionally convergent series of rootspaces. Without the completeness condition it becomes necessary to invoke generalized spectral subspaces, to a study of which we now turn.

By the Riesz-Nevanlinna theorem the inner function $\Theta$ factorizes as a Blaschke product $B$ times a singular function $S: \Theta = BS$, $B$ being defined by the divisor of the zeros $k$ (an integer valued function counting the multiplicity), subject to the condition $\sum_{\lambda \in \mathbb{D}}(1 - |\lambda|)k(\lambda) < \infty$. $S$ is defined by a measure $\mu = \mu^s$, distributed on $\mathbb{T}$, singular with respect to Lebesgue measure $m$ and representing $S$ in the sense that

$$S = \exp\left\{-\int_{\mathbb{T}} \frac{\zeta + z}{\zeta - z} d\mu^s(\zeta)\right\}.$$

If it is necessary to emphasize the dependence of $B$ and $S$ on the parameters defining these functions we write $B(k, \cdot)$ and $S(\mu, \cdot)$.

Let us set $\mu^\Theta = \mu^s + \mu^B$ where $\mu^B = \frac{1}{2}\sum_{\lambda \in \mathbb{D}}(1 - |\lambda|^2)k(\lambda)\delta_\lambda$, $\delta_\lambda$ being the Dirac measure at the point $\lambda$. If $e$ is a Borel set on clos $\mathbb{D}$ and $\mu$ a Borel measure we set $(\chi_e\mu)(e') \stackrel{\text{def}}{=} \mu(e \cap e') \forall e' \in \mathscr{B}$. (The algebra of all Borel sets will be denoted by $\mathscr{B}$; $\chi_e$ is the characteristic function of the set $e$.) Let us also write $ce = \operatorname{clos} \mathbb{D} \backslash e$, $e^* = \{\bar{\zeta} : \zeta \in e\}$. It is clear that the inner function $\Theta_e \stackrel{\text{def}}{=} \Theta(\chi_e\mu^\Theta, \cdot)$ is a divisor of $\Theta$. The corresponding subspace $K_e = K_{\Theta_e}$ is "pseudo-spectral" for the operator $T^* = \mathbf{T}_\Theta^*$ in the sense that

$$T^*K_e \subset K_e, \quad \sigma(T^*|K_e) \subset \operatorname{clos} e^*,$$

$$T^*K_{ce} \subset K_{ce}, \quad \sigma(T^*|K_{ce}) \subset \operatorname{clos}(ce^*), \quad K_e \vee K_{ce} = K;$$

the subspace $\tilde{K}_e = K \ominus K_e$ is "pseudo-spectral" for $T$:

$$T\tilde{K}_e \subset \tilde{K}_e, \quad \sigma(T|\tilde{K}_e) \subset \operatorname{clos} e,$$

$$T\tilde{K}_{ce} \subset \tilde{K}_{ce}, \quad \sigma(T|\tilde{K}_{ce}) \subset \operatorname{clos} ce, \quad \tilde{K}_e \vee \tilde{K}_{ce} = K.$$

Let us define spectral projections: Let $L$ and $M$ subspaces of $K$ such that: $L \cap M = \{\mathbb{0}\}$, $L \vee M = K$; then on the dense subset $L + M$ of $K$ there is defined the projection onto $L$ parallel to $M$, which we denote by $\mathscr{P}_{L\|M}$. The spectral projections of $T$ are the projections $\mathscr{P}_e \stackrel{\text{def}}{=} \mathscr{P}_{K_e\|K_{ce}}$, $e \in \mathscr{B}$. The requirement of spectrality of $T$ imposes that the operators $\mathscr{P}_e$ should be continuous with uniformly bounded norms: $\sup_{e \in B} |\mathscr{P}_e| < \infty$.

If $\mu^\Theta$ is a pure point measure concentrated on the set $\sigma$ then the spectrality of $T$ is still equivalent to the possibility of decomposing the space $K$ into an unconditionally convergent sum of "quasi-rootspaces", that is, spectral subspaces with one point spectra:

$$K = \sum_{\lambda \in \sigma} K_\lambda.$$

The question to which this Lecture is directly devoted is the search of conditions to be imposed on the characteristic function $\Theta$ guaranteeing the "generalized spectrality" of the operator $\mathbf{T}_\Theta$. The generalization in question is that we require the existence and the boundedness of the spectral measure $e \mapsto \mathscr{P}_e$ not necessarily on the whole algebra but fix a certain subalgebra $\mathscr{B}_0$ with which we connect the notion of "$\mathscr{B}_0$-spectrality": $\sup_{e \in \mathscr{B}_0} |\mathscr{P}_e| < \infty$. The algebra $\mathscr{B}_0$

considered below will be generated by a certain (countably infinite) decomposition of the spectrum $\sigma(T)$ so that we are always dealing with bases of spectral subspaces. In this connection we need again the following geometric terminology.

If there is given a family of subspaces $\{K_n\}_{n\in\mathfrak{N}}$ then set $K_n' = \bigvee(K_s : s \neq n)$ and let $\mathscr{P}_n$ be the projection onto $K_n$ parallel to $K_n'$, which operator is defined on $K_n + K_n'$. The family of orthogonal complements $\{K_n'^{\perp}\}_{n\in\mathfrak{N}}$ is then biorthogonal to $\{K_n\}_{n\in\mathfrak{N}}$. The angle between two subspaces is defined by the conditions:

$$\langle L, M\rangle \in [0, \pi/2], \quad \cos\langle L, M\rangle = \sup_{x\in L, y\in M} |(x, y)|/\|x\|\cdot\|y\|.$$

It follows immediately from the definition that

$$\cos\langle L, M\rangle = \sup_{x\in L} \|P_M x\|/\|x\| = |P_M P_L|;$$

$$\sin\langle L, M\rangle = \inf_{x\in L} \|(I - P_M)x\|/\|x\| = |\mathscr{P}_{L\|M}|^{-1}.$$

The family $\{K_n\}_{n\in\mathfrak{N}}$ will be called *minimal* if $|\mathscr{P}_n| < \infty$, $\forall n$, $n\in\mathfrak{N}$; *uniformly* minimal if $\sup_{n\in\mathfrak{N}} |\mathscr{P}_n| < \infty$. By the above expression for the sine between two subspaces the condition of the minimality of the family $\{K_n\}_{n\in\mathfrak{N}}$ can be put in the form $\sin\langle K_n, K_n'\rangle > 0$, $n\in\mathfrak{N}$, and that of the uniform minimality becomes $\inf_{n\in\mathfrak{N}} \sin\langle K_n, K_n'\rangle > 0$. A total family $\{K_n\}_{n\in\mathfrak{N}}$ constitutes an unconditional basis iff the angles between the appropriate subspaces are uniformly separated from zero. Therefore a substantial part of Lectures IX and X will be devoted to estimates of the angle between subspaces of the form $K_\Theta$.

Below we show that a necessary and sufficient condition for $\mathscr{B}_0$-spectrality is the generalized Carleson condition $(CV)$. In particular $\mathbf{T}_\Theta$ is spectral in Dunford's sense iff the measure $\mu_\Theta$ has no continuous component and in addition condition $(CV)$ is fulfilled, where each factor $\vartheta_n$ (cf. p. 217) has a one point spectrum. This condition is to a considerable extent independent from and does not reduce to combinations of conditions known in the theory of interpolation (cf. Lecture VII and Lecture X).

**2. Non-Classical Interpolation in $H^\infty$ and Bases.** The general scheme for using "operator interpolation" in the problem of bases leads in the case of the model operator to the class $H^\infty$. This scheme (cf. Lectures VI and VIII) is based on the Theorems of Sarason (Lecture VIII) and Lorch-Grinblyum (Lecture VI) but in contrast to Lecture VIII it will now be necessary to invoke also the complete (operator) form of the last theorem (the Corollary of the Theorem on Riesz Bases, Sect. 4, Lecture VI). A sufficient supply of multipliers of the family $\{K_n\}_{n\in\mathfrak{N}}$ of $\mathbf{T}_\Theta$-invariant subspaces of the Hilbert space $K = K_\Theta$ will as before be looked for among the functions $f(\mathbf{T}_\Theta)$ of the operator $\mathbf{T}_\Theta$. It turns out that the following simple corollaries of the Calculus Lemma (Lecture III) will be essential; here $T = \mathbf{T}_\Theta$, $f\in H^\infty$, the $\vartheta_n$ are divisors of $\Theta$, $K_n = K_{\vartheta_n}$ and the $\mu_n$ complex numbers:

$$f(T)^*\mathbf{K}_\vartheta = \{\mathbb{O}\} \Leftrightarrow f\in\vartheta H^\infty; \tag{1}$$

$$(f(T)^*|K_n = \bar{\mu}_n I,\ n\geq 1) \Leftrightarrow ((f - \mu_n\mathbb{1})\in\vartheta_n H^\infty,\ n\geq 1); \tag{2}$$

$$f(T)^*|K_1 \vee K_2 = \mathscr{P}_{K_1\|K_2} \Leftrightarrow f \in \vartheta_2 H^\infty, \quad (f - \mathbb{1}) \in \vartheta_1 H^\infty. \tag{3}$$

These statements render the following definitions most natural.

**Definition.** We say that the function $f_\mu$, $f_\mu \in H^\infty$, *interpolates the family* $\mu = \{\mu_n\}_{n\in\mathfrak{N}}$, $\mu \in l^\infty$, *with respect to the family of inner functions* $\{\vartheta_n\}_{n\in\mathfrak{N}}$, if

$$(f_\mu - \mu_n \mathbb{1}) \in \vartheta_n H^\infty, \quad n \in \mathfrak{N}. \tag{4}$$

Let us mention that in the case $\vartheta_\lambda = b_\lambda$ or $\vartheta_\lambda = b_\lambda^{k(\lambda)}$, $\lambda \in \sigma_p$, interpolation with respect to the family $\{\vartheta_\lambda\}$ reduces to the usual notion ($f(\lambda) = \mu_\lambda$, $\lambda \in \sigma_p$) or to multiple interpolation ($f(\lambda) = \mu_\lambda$, $f^{(k)}(\lambda) = 0$, $1 \le k < k(\lambda)$, $\lambda \in \sigma_p$).

We have essentially already shown (the Theorem on Unconditional Bases + formula (2)) that the solvability of the interpolation problem (4) for all data in $l^\infty$ entails that $\{K_{\vartheta_n}\}$ is an unconditional basis of subspaces. The proof of the converse statement uses Sarason's Theorem on interpolation, and connecting this theorem with the previous results we get the following.

**Theorem on Unconditional Bases, II.** *Let* $\Theta$ *and* $\vartheta_n$, $n \in \mathfrak{N}$, *be inner functions* $\Theta = \prod_{n\in\mathfrak{N}} \vartheta_n$, *and set* $K = H^2 \ominus \Theta H^2$, $K_n = H^2 \ominus \vartheta_n H^2$. *The following statements are equivalent.*

1. *The family* $\{K_n\}_{n\in\mathfrak{N}}$ *is an unconditional basis (Riesz basis) in* $K$.
2. *Every family* $\mu$ *in* $l^\infty$ *can be interpolated with respect to the family* $\{\vartheta_n\}_{n\in\mathfrak{N}}$ *by a suitable function* $f = f_\mu$ *in* $H^\infty$.
3. *Every idempotent family (a characteristic function of a subset of* $\mathfrak{N}$*) can be interpolated with respect to* $\{\vartheta_n\}_{n\in\mathfrak{N}}$ *by a function in* $H^\infty$.

*Proof*. The relation (2) shows that from the solvability of the interpolation problem (4) it follows that $\bar{\mu} \in \operatorname{mult}(\{K_n\}_{n\in\mathfrak{N}})$. Conversely, if $\bar{\mu}$ is a multiplier for $\{K_n\}_{n\in\mathfrak{N}}$ then the corresponding operator $\overline{M}$ commutes with $T^*$ and so by Sarason's Theorem it must have the form $f(T)^*$, $f \in H^\infty$, that is, (once more by (2)) the problem (4) must be solvable. Thus Statements 2 and 3 mean respectively that $\operatorname{mult}(\{K_n\}_{n\in\mathfrak{N}}) = l^\infty$ and that all characteristic functions are multipliers. Moreover each of the Statements 2 and 3 entails the completeness of our family in $K_\Theta : \bigvee(K_{\vartheta_n} : n \in \mathfrak{N}) = K_{\mathrm{LCM}\{\vartheta_n\}}$ (Corollary of Theorem *BH*, Lecture I). Also the (weaker) Statement 3 entails that the functions $\vartheta_n$ are relatively prime (for any $n$ and $m$ there exists a function $f$ such that $f = \vartheta_n g$ and $\mathbb{1} - f = \vartheta_m h$, $g, h \in H^\infty$, that is, $\mathbb{1} = \vartheta_n g + \vartheta_m h$. Thus $\mathrm{LCM}\{\vartheta_n\} = \Theta$.

It remains to apply the Theorem on Unconditional Bases (Lecture VI) to the family $\{K_n\}_{n\in\mathfrak{N}}$. □

*Remark.* Any of the statements of the theorem entails

$$\sup \inf \| f_\mu \|_\infty < \infty,$$

where the sup is taken over the unit ball of $l^\infty(\mathfrak{N})$ ($\|\mu\|_{l^\infty} \le 1$) and the inf over all functions $f_\mu$ interpolating the element $\mu$ with respect to the family $\{\vartheta_n\}_{n\in\mathfrak{N}}$. To see this it suffices to remark that $\inf \| f_\mu \| = |\overline{M}|$ where $\overline{M}$ is the multiplier corresponding to the family $\{\bar{\mu}_n\}$. But this is a consequence of Sarason's Theorem, because $f_\mu$ runs through precisely all functions interpolating the operator $\overline{M}$: $\overline{M} = f_\mu(T)^*$. □

**3. The Rôle of the Uniform Minimality** in the problem of generalized spectrality is as before of central importance (exactly as in Lectures VI–VIII in the study of a simple point spectrum). As we have already noticed, we consider the $\sigma$-algebra $\mathscr{B}_0$ generated by a suitable countably infinite decomposition of the disk $\mathbb{D}$, $\mathbb{D} = \bigcup_{n\in\mathfrak{N}}\sigma_n$. The operator $\mathbf{T}_\Theta^*$ (or $\mathbf{T}_\Theta$) is then $\mathscr{B}_0$-spectral, provided the family $\{K_n\}_{n\in\mathfrak{N}}$, $K_n \overset{\text{def}}{=} K_{\vartheta_n}$, $\vartheta_n = \Theta_{\sigma_n}$, corresponding to the "atomic" spectral subspaces is an unconditional basis, i.e. when this family is total in $K_\Theta$ and $\sup\{|\mathscr{E}(\sigma)| : \sigma\in\mathscr{B}_0\} < \infty$, here $\mathscr{E}(\sigma) = \mathscr{P}_\sigma$ is the spectral projection on the part $\sigma\cap\sigma(\mathbf{T}_\Theta)$ of the spectrum $\sigma(\mathbf{T}_\Theta)$. We will soon see that the last condition is equivalent to the uniform minimality of $\{K_{\vartheta_n}\}_{n\in\mathfrak{N}}$. We shall use a geometric language, because, as is clear from the definitions, the uniform minimality is reduced to an estimate from below of angles:

$$\inf_n \sin\langle K_n, K_n'\rangle > 0, \tag{5}$$

where $K_n' = \bigvee(K_n : i \neq n)$. Let us first estimate the angle between the subspaces $K_{\Theta_i}$, $i = 1, 2$, by the constant in the Carleson Corona Theorem.

**Theorem on Angles.** *If* $K_i = K_{\Theta_i}$, $\Theta_i$ *being inner functions with* $\inf_{\mathbb{D}}\{|\Theta_1| + |\Theta_2|\} = \delta$, *then*

$$\delta^3/63 \le \sin\langle K_1, K_2\rangle \le 2\delta.$$

*Proof*. Assume that (throughout the disk $\mathbb{D}$) $|\Theta_1| + |\Theta_2| \ge \delta$, $\delta < 1/2$. By the Corona Theorem of Carleson (Lecture III) we may find $f$ in $H^\infty$ such that $f\in\Theta_1 H^\infty$, $(\mathbb{1} - f)\in\Theta_2 H^\infty$ and $\|f\|_\infty \le 63\delta^{-3}$ (taking account of that $|\Theta_1|^2 + |\Theta_2|^2 \le 2$). From formula (3) it follows that $\mathscr{P}_{K_2\|K_1} = f(T)^*$ so that

$$\sin\langle K_1, K_2\rangle = |\mathscr{P}_{K_2\|K_1}|^{-1} \ge \|f\|_\infty^{-1} \ge \delta^3/63.$$

Conversely, let the angle between $K_1$ and $K_2$ be different from zero. Then $\mathscr{P}_{K_1\|K_2}$ exists and is continuous. As already has been noted, this projection commutes with $T^*$, i.e. $\mathscr{P}^*_{K_2\|K_1}$ commutes with $T$ and by Sarason's Theorem there exists a function $f$ in $H^\infty$ such that $f(T) = \mathscr{P}^*_{K_2\|K_1}$ and $\|f\|_\infty = |\mathscr{P}^*_{K_2\|K_1}| = (\sin\langle K_1, K_2\rangle)^{-1}$. According to formula (3) then $f = \Theta_1 h$, $h\in H^\infty$, $\mathbb{1} - f = \Theta_2 g$, $g\in H^\infty$, $\|h\|_\infty = \|f\|_\infty$, $\|g\|_\infty \le 1 + \|f\|_\infty$. Therefore

$$\begin{aligned}\mathbb{1} = \Theta_1 h + \Theta_2 g &\le |\Theta_1||h| + |\Theta_2||g| \le |\Theta_1|\,\|h\|_\infty + |\Theta_2|\,\|g\|_\infty \\ &\le (|\Theta_1| + |\Theta_2|)(1 + \|f\|_\infty) \le 2\|f\|_\infty(|\Theta_1| + |\Theta_2|),\end{aligned}$$

and consequently

$$2(|\Theta_1| + |\Theta_2|) \ge \|f\|_\infty^{-1} = \sin\langle K_1, K_2\rangle. \ \square$$

**Corollary.** *Let* $\Theta = \prod_{n\in\mathfrak{N}}\vartheta_n$, $K = H^2 \ominus \Theta H^2$, $K_n = K_{\vartheta_n}$ *and, if* $\sigma$ *is an arbitrary subset of* $\mathfrak{N}$, *set* $\Theta_\sigma = \prod_{n\in\sigma}\vartheta_n$, $\Theta_\sigma' = \prod_{n\notin\sigma}\vartheta_n$. *Then the family of subspaces*

$\{K_n\}_{n\in\mathfrak{N}}$ *is a Riesz basis for K iff*

$$\inf_{\sigma}\inf_{\mathbb{D}}\{|\Theta_\sigma| + |\Theta'_\sigma|\} > 0. \tag{6}$$

In fact, by the Theorem on Angles condition (6) is equivalent to the condition $\sup_\sigma|\mathscr{P}_\sigma| < \infty$, because $|\mathscr{P}_\sigma|^{-1} = \sin\langle K_\sigma, K'_\sigma\rangle$. Here $\mathscr{P}_\sigma$ is the projection onto $K_\sigma$ parallel to $K'_\sigma$, $\mathscr{P}_\sigma = \sum_{n\in\sigma}\mathscr{P}_n$. As has already been noted, for a total family $\{K_n\}$ in order to be an unconditional basis (Riesz basis) it is necessary and sufficient that $\sup_\sigma|\mathscr{P}_\sigma| < \infty$. Therefore in order to complete the proof it remains to show that (6) implies the completeness of $\{K_n\}$.

But it is clear that $\bigvee(K_n : n\in\mathfrak{N}) = K_{\mathrm{LCM}(\vartheta_n)}$ and that $\mathrm{LCM}\{\vartheta_n\} = \Theta$, provided the functions $\vartheta_n$ are pairwise relatively prime. Let us assume that this is not so, that is, there exist two functions $\vartheta_i$ and $\vartheta_j$ with a common divisor: $\vartheta_i = \vartheta\tilde{\vartheta}_i$, $\vartheta_j = \vartheta\tilde{\vartheta}_j$, $\vartheta \neq \mathbb{1}$. Then

$$\inf_{\sigma}\inf_{\mathbb{D}}\{|\Theta_\sigma| + |\Theta'_\sigma|\} \le \inf_{\mathbb{D}}\{|\vartheta_i| + |\vartheta_j|\} \le 2\inf_{\mathbb{D}}|\vartheta| = 0,$$

which contradicts condition (5). □

The uniform minimality is an immediate property of any Riesz basis. In the general situation this property is of course much weaker than the requirement of being a basis. It is therefore remarkable that in the case of spectral subspaces of the model operator these properties turn out to be equivalent (cf. below). In what follows we will exploit this circumstance in a manifold of ways but regretfully the geometric mechanism of its origin cannot be said to be ultimately understood.

**Vasyunin's Theorem.** *Let* $\Theta = \prod_{n\in\mathfrak{N}}\vartheta_n$, $K_\Theta = H^2 \ominus \Theta H^2$, $K_n = K_{\vartheta_n}$, $n\in\mathfrak{N}$. *The following statements are equivalent*:

1. *The family* $\{K_n\}_{n\in\mathfrak{N}}$ *is a Riesz basis of* $K_\Theta$.
2. *The family* $\{K_n\}_{n\in\mathfrak{N}}$ *is uniformly minimal.*
3. *There exists a number* $\delta$, $\delta > 0$, *such that*

$$|\Theta(\zeta)| \ge \delta \inf_{n\in\mathfrak{N}}|\vartheta_n(\zeta)|, \quad \zeta\in\mathbb{D}. \tag{CV}$$

The theorem shows in particular that there does not exist conditionally convergent expansions generated by a family of $\mathbf{T}^*_\Theta$-invariant (or $\mathbf{T}_\Theta$-invariant) subspaces: if such a family determines a basis with respect to some enumeration (some ordering) then it is by necessity uniformly minimal (Banach [1]) and thus already an unconditional basis.

In the following Lecture we shall study in detail the Carleson-Vasyunin Condition $(CV)$. Here we remark only than in the special case $\Theta = B = \prod_{\lambda\in\sigma_p} b_\lambda$ (a Blaschke product with simple zeros) and $\vartheta_\lambda = b_\lambda$, $\lambda\in\sigma_p$, condition $(CV)$ is equivalent to Carleson's Condition $(C)$. This follows from a confrontation of Vasyunin's Theorem with the Theorem on Bases and Interpolation (Lecture VI) but can also be seen directly. One of the implications $((CV)\Rightarrow(C))$ is trivial because for $\lambda\in\sigma_p$ and $\zeta$ near $\lambda$ Condition $(CV)$ implies

$$|B(\zeta)/b_\lambda(\zeta)| \geq \delta;$$

putting here $\zeta = \lambda$ we get $|B_\lambda(\lambda)| \geq \delta$, $\lambda \in \sigma_p$, i.e. Carleson's Condition $(C)$.

It is worth while to give the opposite reasoning too, although the latter also can be found implicitly in the computations leading to the theorems in Lectures IX–X. If $|B_\lambda(\lambda)| \geq \delta$, $\lambda \in \sigma_p$, and $D_\lambda \overset{\text{def}}{=} \{\mu : |b_\lambda(\lambda)| \leq \delta/3\}$, $\lambda \in \sigma_p$, then for $\mu \in D_\lambda$,

$$|B_\lambda(\mu)| \geq \delta/2.$$

In fact, from the condition $|b_\lambda(\mu)| \leq \delta/3$ follows that

$$|\lambda - \mu| \leq \delta/3 \cdot \frac{1}{1 - \delta/3} \min\{(1 - |\lambda|^2), (1 - |\mu|^2)\}$$
$$< \delta/2 \min\{(1 - |\lambda|^2), (1 - |\mu|^2)\},$$

and hence (by Lagrange's inequality)

$$|B_\lambda(\mu)| \geq |B_\lambda(\lambda)| - |B'_\lambda(\xi)|\,|\lambda - \mu| \geq \delta - 1/2\delta = \delta/2.$$

Here we used that $1 - |\xi|^2 \geq \min\{(1 - |\lambda|^2), (1 - |\mu|^2)\}$ and that

$$|B'_\lambda(\xi)| \leq (1 - |\xi|^2)^{-1}.$$

(Two methods of proof: by Cauchy's formula

$$|f'(\xi)| = \left| \int_{\mathbb{T}} \frac{\bar{t} f(t)}{(1 - \bar{t}\xi)^2} dm(t) \right| \leq \|f\|_\infty (1 - |\xi|^2)^{-1}; \square$$

with the aid of conformal mapping: $\tau = (\xi\mathbb{1} - z)(\mathbb{1} - \bar{\xi}z)^{-1}$, $f \overset{\text{def}}{=} g \circ \tau$ and $|f'(\xi)| = |g'(0)|\,|\tau'(\xi)| \leq \|g\|_\infty |\tau'(\xi)| = \|f\|_\infty (1 - |\xi|^2)^{-1}$. $\square$)

Thus

$$\mu \in \bigcup_{\lambda \in \sigma_p} D_\lambda \Rightarrow |B(\mu)| \geq \delta/2 \inf_{\lambda \in \sigma_p} |b_\lambda(\mu)|.$$

Outside the union of the disks $\mathbb{D}_\lambda$ we use the maximum principle

$$\inf_{\mathbb{D} \setminus \bigcup_\lambda D_\lambda} |B| = \inf_{\lambda \in \sigma_p} \min_{\partial D_\lambda} |B| = \inf_{\lambda \in \sigma_p} \min_{\partial D_\lambda} |b_\lambda|\,|B_\lambda|$$
$$= \inf_{\lambda \in \sigma_p} \min_{\partial D_\lambda} \delta/3 |B_\lambda| \geq \delta/3 \cdot \delta/2 = \delta^2/6.$$

In order to make this argument more convincing, we could in this last computation first have assumed that $B$ is a finite Blaschke product (so that there is also only finitely many disks $D_\lambda$), afterwards passing to the limit: from the partial products to the complete product $B$. $\square$

Yet another source for extracting consequences from Condition $(CV)$ is based on the conformal invariance of that condition. Let us perform a conformal transformation mapping the point $\zeta$ onto the origin 0.

1) If $\vartheta_\lambda(\zeta) = \left( \frac{|\lambda|}{\lambda} \frac{\lambda - \zeta}{1 - \bar{\lambda}\zeta} \right)^{k(\lambda)}$, $\lambda \in \sigma_p$, we can write the condition for having a

basis in the following form:

$$\prod_{\lambda \in \sigma_p} |\lambda(\zeta)|^{k(\lambda)} \geq \delta \inf_{\lambda \in \sigma_p} |\lambda(\zeta)|^{k(\lambda)}, \quad \forall \zeta \in \mathbb{D},$$

where $\lambda(\zeta)$ are the points of the unit disk which are the images of the zeros of the functions $\vartheta_\lambda$ under the given conformal mapping.

2) If $\vartheta_\lambda(\zeta) = \exp\left\{-a_\lambda \frac{\lambda+\zeta}{\lambda-\zeta}\right\}$, $\lambda \in \sigma$, $\sigma \subset \mathbb{T}$, then the condition for being a basis can be rewritten as

$$\sum_{\lambda \in \sigma} a_\lambda(\zeta) \leq A + \sup_{\lambda \in \sigma} a_\lambda(\zeta),$$

where $a_\lambda(\zeta) = \dfrac{a_\lambda(1-|\zeta|^2)}{|\lambda-\zeta|^2}$ is the mass of the singular measure of the inner function gotten from $\vartheta_\lambda$ by the above conformal transformation.

Finally, let us remark that Condition ($CV$) (and thus the condition of $\mathscr{B}_0$-spectrality of the operator $\mathbf{T}_\Theta^* = S^*|K_\Theta$) can be cast in the following form, directly connected with the operator $\mathbf{T}_\Theta^*$.

**Corollary.** *The Statements* 1–3 *of Vasyunin's Theorem are equivalent to the existence of a number* $C$, $C > 0$, *such that*

$$\|R(\lambda, \mathbf{T}_\Theta^*)\| \leq C \sup_{n \in \mathfrak{N}} \|R(\lambda, \mathbf{T}_\Theta^*)|K_n\|, \; \lambda \in \mathbb{D} \backslash \sigma(\mathbf{T}_\Theta^*). \tag{7}$$

In fact, $R(\lambda, \mathbf{T}_\Theta^*)|K_n = R(\lambda, \mathbf{T}_{\vartheta_n}^*)$ for the values of $\lambda$ in question so that by the Lemma Estimating the Resolvent (Lecture IV)

$$\|R(\lambda, \mathbf{T}_\Theta^*)|K_n\| \leq \text{const}(1-|\lambda|)^{-1}|\vartheta_n(\lambda)|^{-1},$$

$$|\Theta(\lambda)|^{-1}(1-|\lambda|)^{-1} \leq \text{const}((1-|\lambda|)^{-1} + \|R(\lambda, \mathbf{T}_\Theta^*)\|).$$

Therefore condition ($CV$) follows from (7).

The converse implication is trivial (and in fact contained in the Lorch-Grinblyum Theorem): if $\{K_n\}_{n \in \mathfrak{N}}$ is a Riesz basis and $A$ an operator leaving invariant the subspaces $K_n$, $n \in \mathfrak{N}$ (in particular thus if $A = R(\lambda, \mathbf{T}_\Theta^*)$, $\lambda \in \mathbb{D} \backslash \sigma(\mathbf{T}_\Theta^*)$) then

$$\|A\| \leq \text{const} \sup_n \|A|K_n\|, \tag{8}$$

where the constant const depends only on the characteristics of the basis $\{K_n\}_{n \in \mathfrak{N}}$ (the norm of the orthogonalizer and its inverse) and not on the operator $A$. □

The reasoning on the previous paragraph once more underlines the inique rôle of the criterion (7) for the family of spectral subspaces $\{K_{\vartheta_n}\}_{n \in \mathfrak{N}}$ to be a basis. The point is that if we in condition (7) replace the set of the simplest rational functions $(\lambda \mathbb{1} - z)^{-1}$, $\lambda \in \mathbb{D} \backslash \sigma(\mathbf{T}_\Theta^*)$ by the set $R(\sigma(\mathbf{T}_\Theta^*))$ of all rational functions (applicable to the operator $\mathbf{T}_\Theta^*$) then the equivalence of the estimate arising in this way, viz.

$$\|\phi(\mathbf{T}_\Theta^*)\| \leq C \sup_{n \in \mathfrak{N}} \|\phi(\mathbf{T}_\Theta^*)|K_n\|, \quad \phi \in R(\sigma(\mathbf{T}_\Theta^*)), \tag{9}$$

and the condition for the family of subspaces $\{K_n\}_{n\in\mathfrak{N}}$ to be a basis becomes a much more elementary fact, and, before turning to the proof of Vasyunin's Theorem, we will explicate this in one special but important case.

Namely, let $\Theta$ be a Blaschke product and let $\vartheta_n$ be pairwise relatively prime partial products of $\Theta$, $n\in\mathfrak{N}$. In the following reasoning we shall only use the fact that $\{K_n\}$ is a total in $K$ family of spectral subspaces of the (some) operator $V$ and that the spectra $\sigma_n \overset{\text{def}}{=} \sigma(V|K_n)$, $n\in\mathfrak{N}$, are isolated (open-closed) parts of the spectrum $\sigma(V)$ ($\overset{\text{def}}{=}$ *property* $(*)$).

If these assumptions are fulfilled then for any finite set $N$, $N\subset\mathfrak{N}$, there exist non-intersecting closed neighborhoods $F_1$ and $F_2$ of the sets $\sigma=\bigcup_{n\in N}\sigma_n$ and $\sigma'=\sigma(V)\backslash\sigma$ respectively. Pick a rational function $\varphi$ with poles in $\hat{\mathbb{C}}\backslash(F_1\cup F_s)$ which with an in advance given accuracy approximates the function $\mathbb{1}$ on the set $F_1$ and the function $\mathbb{0}$ on the set $F_2$ (to this end we can, for instance, take the Riemann sums of Cauchy's integral for the function to be approximated, analytic in a neighborhood of $F_1\cup F_2$). Then (we assume that $F_i$, $i=1,2$, are compact sets with smooth oriented boundaries. By $\mathscr{E}(\sigma)$ we denote the Riesz spectral projection corresponding to the portion $\sigma$ of the spectrum $\sigma(V)$.)

$$\varphi(V)\mathscr{E}(\sigma)=\frac{1}{2\pi i}\int_{\partial F_1}R(\lambda,V)\varphi(\lambda)\,d\lambda.$$

$$|(\varphi(V)-I)\mathscr{E}(\sigma)|\le\frac{1}{2\pi}\max_{\lambda\in\partial F_1}|R(\lambda,V)|\,|\varphi(\lambda)-1|.$$

This estimate and the analogous one for $|\varphi(V)\mathscr{E}(\sigma')|$ show that the operator $\varphi(V)$ approximates arbitrary well the spectral projection $\mathscr{E}(\sigma)$. The analogue of condition (9) for the operator $V$ is now

$$|\varphi(V)|\le C\cdot\sup_n|\varphi(V)|K_n|. \tag{10}$$

But we have already estimated the right hand side of this inequality: for suitable $\varphi$ we thus have $|\varphi(V)|K_n|<1+\varepsilon$ if $n\in N$ and $|\varphi(V)|K_n|<\varepsilon$ if $n\in\mathfrak{N}\backslash N$. Hence $|\mathscr{E}(\sigma)|\le C$ and consequently

$$\sup_{\sigma\subset\mathfrak{N}}|\mathscr{E}(\sigma)|<\infty.$$

Thus condition (9) for the operator $\mathbf{T}_\Theta^*$ (or condition (10) for a general operator $V$ possessing property $(*)$) entails the unconditional convergence of the spectral expansion

$$I=\sum_{n\in\mathfrak{N}}\mathscr{E}(\sigma_n).$$

The converse assertion is trivial and was already mentioned in the deduction of inequality (7). □

Concluding our preliminary discussion of Vasyunin's Theorem let us once more underline that in the general case (i.e. not for the model operator $\mathbf{T}_\Theta$) the gap between the conditions (10) and

$$|R(\lambda, V)| \leq C \sup_{n\in\mathfrak{N}} |R(\lambda, V)|K_n|, \quad \lambda \notin \sigma(V),$$

is very big.

Let us pass to the proof of Vasyunin's Theorem. The Corollary of the Theorem on Angles indicates that we have to start by estimating inner functions near their set of zeros.

**Lemma 1.** *Let* $\{\vartheta_n\}_{n\in\mathfrak{N}}$ *be an arbitrary family of inner functions,* $\Theta = \prod_{n\in\mathfrak{N}} \vartheta_n$, $\Theta_n = \Theta/\vartheta_n$,

$$\inf_{n\in\mathfrak{N}} \inf_{\mathbb{D}} \{|\vartheta_n| + |\Theta_n|\} = \delta > 0,$$

$$E_n = \{\zeta : \zeta \in \mathbb{D}, |\vartheta_n(\zeta)| < \varepsilon\}, \quad E = \bigcup_{n\in\mathfrak{N}} E_n.$$

*Then* $\inf_{\mathbb{D}\setminus E} |\Theta| \geq \varepsilon(\delta - \varepsilon)$.

*Proof*. First let us assume that $\Theta$ is a finite Blaschke product. Then $\Theta^{-1}$ is analytic in a neighborhood of the set clos $\mathbb{D}\setminus E$ and therefore by the maximum principle (taking into account that $|\Theta| = 1$ on $\mathbb{T}$) we get

$$\inf_{\mathbb{D}\setminus E} |\Theta| = \inf_{\partial E} |\Theta| = \inf_{n\in\mathfrak{N}} \inf_{\partial E_n} |\vartheta_n||\Theta| \geq \inf_{n\in\mathfrak{N}} \inf_{\partial E_n} |\vartheta_n|(\delta - |\vartheta_n|) = \varepsilon(\delta - \varepsilon).$$

Next, let $\Theta$ be an arbitrary Blaschke product. Let us fix $\zeta$, $\zeta \in \mathbb{D}\setminus E$, and, given a positive number $\eta$, consider a finite partial product $B$ such that $|\Theta(\zeta)| \geq (1 - \eta)|B(\zeta)|$. The products $\vartheta_n$ are then replaced by corresponding partial products $b_n$ so that $|b_n(\zeta)| \geq |\vartheta_n(\zeta)| \geq \varepsilon$, $|b_n(\zeta)| + |B_n(\zeta)| \geq |\vartheta_n(\zeta)| + |\Theta_n(\zeta)| \geq \delta$ and, by what we just proved, $|B(\zeta)| \geq \varepsilon(\delta - \varepsilon)$. Hence $|\Theta(\zeta)| \geq (1 - \eta)\varepsilon(\delta - \varepsilon)$, which gives $|\Theta(\zeta)| \geq \varepsilon(\delta - \varepsilon)$, $\forall \zeta \in \mathbb{D}\setminus E$.

Finally, let $\Theta$ be an arbitrary inner function. We have now to invoke Frostman's Theorem (on uniform approximation of inner functions by Blaschke products) which was stated in Lecture II, Sect. 5. According to this theorem given positive number $\eta_n$, $n \in \mathfrak{N}$, there exist Blaschke products $b_n$, $n \in \mathfrak{N}$, such that $\|\vartheta_n - b_n\|_\infty \leq \eta_n$, $n \in \mathfrak{N}$. In order to complete the proof it is convenient to renumerate $\mathfrak{N}$ taking $\mathfrak{N} = \mathbb{N}$. Let $\eta = \sum_{n\geq 1} \eta_n < \infty$ and set $h_n = \prod_{1\leq s<n} b_s \cdot \prod_{s\geq n} \vartheta_s$, $h_1 = \Theta$. As $\|h_{n+1} - h_n\|_\infty = \|b_n - \vartheta_n\|_\infty \leq \eta_n$ we get $\|h_{n+m} - h_n\|_\infty \leq \sum_{s=n}^{n+m-1} \eta_s$, that is, the sequence $\{h_n\}_{n\geq 1}$ converges in $H^\infty$. Put $B = \lim_n h_n$. As $\lim_n \prod_{s\geq n} \vartheta_s(\zeta) = 1$, $\forall \zeta \in \mathbb{D}$, clearly $B$ is a Blaschke product and $B = \prod_{n\geq 1} b_n$.

The Blaschke product $B$ thus obtained is close to $\Theta$ : $\|\Theta - B\|_\infty \leq \eta$, as is seen by passage to the limit in the inequalities

$$\|h_{n+1} - h_1\|_\infty \leq \sum_{s=1}^{n} \|h_{s+1} - h_s\|_\infty \leq \sum_{s=1}^{n} \eta_s < \eta.$$

It is likewise clear that $\|\Theta_n - B_n\|_\infty \leq \eta$. Therefore for points $\zeta$ in $\mathbb{D}\setminus E$

$$|b_n(\zeta)| > |\vartheta_n(\zeta)| - \eta \geq \varepsilon - \eta,$$

$$|b_n(\zeta)| + |B_n(\zeta)| \geq |\vartheta_n(\zeta)| - \eta + |\Theta_n(\zeta)| - \eta \geq \delta - 2\eta.$$

Since the lemma is already proved for arbitrary Blaschke products we get

$$|B(\zeta)| \geq (\varepsilon - \eta)[(\delta - 2\eta) - (\varepsilon - \eta)] = (\varepsilon - \eta)(\delta - \varepsilon - \eta),$$

$$|\Theta(\zeta)| \geq |B(\zeta)| - |\Theta(\zeta) - B(\zeta)| \geq (\varepsilon - \eta)(\delta - \varepsilon - \eta) - \eta.$$

Therefore, $\eta$ being arbitrary, we find $|\Theta(\zeta)| \geq \varepsilon(\delta - \varepsilon)$, $\forall \zeta \in \mathbb{D} \backslash E$. □

**Lemma 2.** *Let $\{\vartheta_n\}_{n \in \mathfrak{N}}$ be an arbitrary family of inner functions, $\Theta = \prod_{n \in \mathfrak{N}} \vartheta_n$, $\sigma$ an arbitrary subset of $\mathfrak{N}$, $\Theta_\sigma = \prod_{n \in \sigma} \vartheta_n$, $\Theta'_\sigma = \prod_{n \notin \sigma} \vartheta_n$. Then*

$$\inf_{\sigma} \inf_{\mathbb{D}} \{|\Theta_\sigma| + |\Theta'_\sigma|\} \geq \frac{4}{5} \inf_{n \in \mathfrak{N}} \inf_{\mathbb{D}} \{|\vartheta_n| + |\Theta_n|\}.$$

*Proof*. Set $\delta \stackrel{\text{def}}{=} \inf_{n \in \mathfrak{N}} \inf_{\mathbb{D}} \{|\Theta_n| + |\vartheta_n|\}$.

$$E_n = \{\zeta : \zeta \in \mathbb{D}, |\vartheta_n(\zeta)| < \delta/5\}, \quad E = \bigcup_{n \in \mathfrak{N}} E_n.$$

If $\zeta \in E_n$, $n \in \sigma$, then

$$|\Theta_\sigma(\zeta)| + |\Theta'_\sigma(\zeta)| \geq |\Theta'_\sigma(\zeta)| \geq \delta - |\vartheta_n(\zeta)| \geq \frac{4}{5}\delta;$$

if $n \notin \sigma$, then

$$|\Theta_\sigma(\zeta)| + |\Theta'_\sigma(\zeta)| \geq |\Theta_\sigma(\zeta)| \geq |\Theta_n(\zeta)| \geq \frac{4}{5}\delta.$$

Finally, let $\zeta \notin E$. Applying Lemma 1 (with $\varepsilon = \delta/5$) we get

$$|\Theta(\zeta)| \geq \frac{1}{5}\delta\left(\delta - \frac{1}{5}\delta\right) = \frac{4}{25}\delta^2,$$

$$|\Theta_\sigma(\zeta)| + |\Theta'_\sigma(\zeta)| \geq 2\sqrt{|\Theta_\sigma(\zeta)|\,|\Theta'_\sigma(\zeta)|} = 2\sqrt{|\Theta(\zeta)|} \geq \frac{4}{5}\delta. \ \square$$

**Lemma 3.** *Let $\vartheta_n$ be inner functions, $n \in \mathfrak{N}$, $\Theta = \prod_{n \in \mathfrak{N}} \vartheta_n$, $\Theta_n = \Theta \vartheta_n^{-1}$, $n \in \mathfrak{N}$. Set*

$$\delta = \inf_{n \in \mathfrak{N}} \inf_{\zeta \in \mathbb{D}} \{|\vartheta_n(\zeta)| + |\Theta_n(\zeta)|\}, \quad d = \inf_{\zeta \in \mathbb{D}} \sup_{n \in \mathfrak{N}} |\Theta_n(\zeta)|.$$

*Then $\frac{1}{4}\delta^2 \leq d \leq \delta$.*

*Proof*. Let us set $d_\Theta(\zeta) = \inf_{n \in \mathfrak{N}} |\vartheta_n(\zeta)|$, $\zeta \in \mathbb{D}$. Let $E_n = \{\zeta : \zeta \in \mathbb{D}, |\vartheta_n(\zeta)| < \delta/2\}$, $E = \bigcup_{n \in \mathfrak{N}} E_n$. If $\zeta \in E_n$ then $|\vartheta_n(\zeta)| < \delta/2$ and $|\vartheta_s(\zeta)| \geq |\Theta_n(\zeta)| \geq \delta - |\vartheta_n(\zeta)| \geq \delta/2$, $\forall s$, $s \neq n$, so that $d_\Theta(\zeta) = |\vartheta_n(\zeta)|$ and $|\Theta(\zeta)| = |\Theta_n(\zeta)| \cdot |\vartheta_n(\zeta)| \geq \delta/2 \cdot d_\Theta(\zeta)$, $\forall \zeta \in E$.

If $\zeta \notin E$ then by Lemma 1 (with $\varepsilon = \delta/2$)

$$|\Theta(\zeta)| \geq \delta/2(\delta - \delta/2) = \delta^2/4 \geq \delta^2/4 \cdot d_\Theta(\zeta).$$

Therefore $|\Theta(\zeta)| \geq \frac{1}{4}\delta^2 d_\Theta(\zeta)$, $\forall \zeta$, $\zeta \in \mathbb{D}$, which means that $d \geq \frac{1}{4}\delta^2$.

For the proof of the second inequality we remark that for all $m, n \in \mathfrak{N}$ and all $\zeta$, $\zeta \in \mathbb{D}$, holds $|\vartheta_n(\zeta)| + |\Theta_n(\zeta)| \geq |\Theta_m(\zeta)|$ (if $n \neq m$ this follows from the fact that $\vartheta_n$

divides $\Theta_m$), whence

$$\delta = \inf_{\zeta \in \mathbb{D}} \inf_{n \in \mathfrak{N}} \{|\vartheta_n(\zeta)| + |\Theta_n(\zeta)|\} \geq \inf_{\mathbb{D}} \sup_{m \in \mathfrak{N}} |\Theta_m(\zeta)| = d. \ \square$$

*Remark.* It is clear that the number $d$ defined in the lemma is the best constant in the inequality

$$|\Theta(\zeta)| \geq d \inf_{n \in \mathfrak{N}} |\vartheta_n(\zeta)|, \quad \zeta \in \mathbb{D}.$$

*Proof of Vasyunin's Theorem.* By the Theorem on Angles (cf. also formula (5)) the uniform minimality of the family $\{K_n\}_{n \in \mathfrak{N}}$ is equivalent to the condition $\inf_{n \in \mathfrak{N}} \inf_{\mathbb{D}} \{|\vartheta_n| + |\Theta_n|\} > 0$, while the condition for having a basis takes the form $\inf_{\sigma \subset \mathfrak{N}} \inf_{\mathbb{D}} \{|\Theta_\sigma| + |\Theta'_\sigma|\} > 0$ (by the Corollary of that theorem). But these conditions coincide (Lemma 2) so that $1 \Leftrightarrow 2$.

The equivalence of Conditions 2 and 3 follows from Lemma 3. $\square$

**4. Interpolation of Germs of Analytic Functions.** The results obtained in Sects. 2 and 3 allow us to solve the problem of "free" interpolation by analytic functions and their successive derivatives. The general problem of this type consists of the following. Given a space $X$ of functions regular in the disk $\mathbb{D}$ and given a divisor $k$, that is, a non-negative integer valued function on $\mathbb{D}$ such that the set $\sigma = \{\lambda : k(\lambda) > 0\}$ has no points of accumulation in $\mathbb{D}$. If $\lambda \in \mathbb{D}$ and $f$ is a function regular in a neighborhood of the point $\lambda$ then we denote by $(f)_{k(\lambda)}$ the germ of height $k(\lambda)$ at $\lambda$, i.e. the Taylor polynomial of the corresponding degree

$$(f)_{k(\lambda)} \overset{\text{def}}{=} \sum_{k=0}^{k(\lambda)-1} \frac{f^{(k)}(\lambda)}{k!}(z-\lambda)^k, \quad \text{if} \quad f = \sum_{k \geq 0} \frac{f^{(k)}(\lambda)}{k!} \cdot (z-\lambda)^k.$$

Our problem is now to describe the family of germs $\{f_\lambda : \lambda \in \sigma\}$ consisting precisely of the germs of some function $f$ in $X$, i.e. germs such that there exists $f$, $f \in X$, with

$$(f)_{k(\lambda)} = f_\lambda, \quad \lambda \in \sigma. \tag{11}$$

The "freedom" of such an interpolation (in the spirit of what was said regarding the meaning of this word in Lectures VI–VIII) has to consist of the freedom of the interpolation data $\{f_\lambda : \lambda \in \sigma\}$ from any "global" condition, reminiscent of their origin (11). In particular, this is so if the space of all such data is an ideal: if $\|g_\lambda\| \leq \|f_\lambda\|$, $\lambda \in \sigma$, then the family $\{g_\lambda\}$ has to be among the interpolation data, provided this is the case for $\{f_\lambda\}$. It often happens however that such a counting of the magnitude of the germs $\{f_\lambda\}$ in fact turns out to be "unfree", the reason being that the norm in the space $X$ is weakly coupled to the local properties of the function $f$ belonging to that space (with its values, the values of its derivatives etc.). A more flexible method of describing the interpolation data is then with the aid of the norms $\|a_\lambda f_\lambda\|$ where the $a_\lambda$ are appropriate operators in a natural way associated with the space $X$ and the divisor $k$. The following theorems serves to confirm what we have just said.

**Interpolation Theorem.** *Let $k$ be a divisor satisfying the Blaschke condition $\sum_{\lambda\in\sigma} k(\lambda)(1-|\lambda|)<\infty$ and let $B=\prod_{\lambda\in\sigma} b_\lambda^{k(\lambda)}$ be the corresponding product $(\sigma=\{\lambda : k(\lambda)>0\})$. The following statements are equivalent.*

1. *The Carleson-Vasyunin Condition (CV) is fulfilled*:

$$|B(\zeta)| \geq \operatorname{const} \inf_{\lambda\in\sigma} |b_\lambda(\zeta)|^{k(\lambda)}, \quad \zeta\in\mathbb{D}.$$

2. *If $f^\lambda\in H^\infty$ and $\sup_{\lambda\in\sigma}\|f^\lambda\|_\infty<\infty$ then there exists a function $f, f\in H^\infty$, such that $f-f^\lambda\in b_\lambda^{k(\lambda)}H^\infty$, i.e.*

$$(f)_{k(\lambda)} = (f^\lambda)_{k(\lambda)}, \quad \lambda\in\sigma. \tag{12}$$

3. *Let $P_\lambda$ denote the orthogonal projection onto the subspace $K_\lambda = H^2\ominus \mathbf{b}_\lambda^{k(\lambda)}H^2$, $\lambda\in\sigma$, and let $\{f^\lambda : \lambda\in\sigma\}$ be a family of functions in $H^2$. There exists a function $f, f\in H^2$, solving the interpolation problem (12), iff*

$$\sum_{\lambda\in\sigma} \|P_\lambda f^\lambda\|_{H^2}^2 < \infty, \tag{13}$$

4. *If $f^\lambda\in H^2$ $(\lambda\in\sigma)$ then the problem (12) has a solution $f$ in the class $H^2$ iff*

$$\sum_{\lambda\in\sigma}\left\|\left(\frac{f^\lambda(z-\lambda\mathbb{1})^{k(\lambda)}}{B}\right)_{k(\lambda)}\frac{B}{(z-\lambda\mathbb{1})^{k(\lambda)}}\right\|_{H^2}^2<\infty.$$

*Proof*. The rootspaces $K_\lambda = H^2\ominus \mathbf{b}_\lambda^{k(\lambda)}H^2$, $\lambda\in\sigma$, of the operator $\mathbf{T}_B^* = S^*|K_B$, $K_B = H^2\ominus \mathbf{B}H^2$, form a total minimal family. The corresponding biorthogonal family is the family $\{K'_\lambda\}_{\lambda\in\sigma}$, $K'_\lambda = \mathbf{B}_\lambda H^2\ominus \mathbf{B}H^2$, of rootspaces of the operator $\mathbf{T}_B$ (here $B_\lambda = B\bar{b}^{k(\lambda)}$, $\lambda\in\sigma$), likewise total in $K_B$ (by the Completeness Theorem for the model operator, Lecture IV). Therefore (by the Theorem on Riesz Bases, Lecture VI) $\{K_\lambda\}_{\lambda\in\sigma}$ is a Riesz basis iff any of the maps $f\mapsto\{P_\lambda f : \lambda\in\sigma\}$ and $f\mapsto\{\mathscr{E}_\lambda f : \lambda\in\sigma\}$ sends the space $K_B$ onto $l^2(\{K_\lambda\}_{\lambda\in\sigma})$; here $\mathscr{E}_\lambda$ is the spectral projection onto the subspace $K'_\lambda$, $\lambda\in\sigma$. By Vasyunin's Theorem Statement 1 is equivalent to the same basis condition. In order to prove the implications

$$1\Leftrightarrow 3\Leftrightarrow 4 \tag{14}$$

we have only to remark that $(12)\Rightarrow f-f^\lambda\in\mathbf{b}_\lambda^{k(\lambda)}H^2$ (trivial) and to compute the projection $\mathscr{E}_\lambda P_K$, $K=K_B$. Let us verify that

$$\mathscr{E}_\lambda P_K f = \left(\frac{f(z-\lambda\mathbb{1})^{k(\lambda)}}{B}\right)_{k(\lambda)}\frac{B}{(z-\lambda\mathbb{1})^{k(\lambda)}}, \quad f\in H^2. \tag{15}$$

In fact, if $f\in\mathbf{B}H^2$ then both the left and the right hand side equal zero. Let us now remark that by Corollary 3 of the Theorem on Lat $\mathbf{T}_\Theta$ (Lecture IV)

$$\begin{aligned} K'_\lambda &= \operatorname{Ker}(T_B-\lambda I)^{k(\lambda)} \\ &= \{B(z-\lambda\mathbb{1})^{-k(\lambda)}p : p \text{ a polynomial of degree not exceeding } k(\lambda)-1\}. \end{aligned}$$

Let $f = B(z-\mu\mathbb{1})^{-k(\mu)}p$, $p$ of degree not exceeding $k(\lambda)-1$. Then $f(z-\lambda\mathbb{1})^{k(\lambda)}B^{-1}$ for $\lambda\neq\mu$ is regular at the point $\lambda$ and has there a zero of order $k(\lambda)$, that is,

$$\left(\frac{f(z-\lambda\mathbb{1})^{k(\lambda)}}{B}\right)_{k(\lambda)} = \mathbb{0}.$$

If however $\lambda = \mu$ then the corresponding germ equals $p$ and

$$\left(\frac{f(z-\lambda\mathbb{1})^{k(\lambda)}}{B}\right)_{k(\lambda)}\frac{B}{(z-\lambda\mathbb{1})^{k(\lambda)}} = f.$$

Thus the identity (15) is established (formula (15) can also be proved by "a direct computation" using the resolvent and the Riesz integral representing the projection $\mathscr{E}_\lambda$). And thereby also the relationship (14).

$2 \Rightarrow 1$ by the Theorem on Unconditional Bases, II (taking the $f^\lambda$ to be constant functions, $\lambda \in \sigma$).

$1 \Rightarrow 2$. In this case the families $\{K_k\}_{\lambda\in\sigma}$ and $\{K'_\lambda\}_{\lambda\in\sigma}$ generate unconditional bases and we have to invoke the corollary on operator interpolation (Lecture VI, p. 142). Let $a_\lambda$ be the operator in $K_\lambda$ defined by the formula

$$a_\lambda = P_+\overline{\mathbf{f}^\lambda}\,|K_\lambda. \tag{16}$$

Then $\sup_{\lambda\in\sigma}|a_\lambda| \le \sup_{\lambda\in\sigma}\|f^\lambda\|_\infty < \infty$ and by the corollary mentioned there exists an operator $A$ in $K_B$ interpolating the family $\{a_\lambda\}_{\lambda\in\sigma}$:

$$A|K_\lambda = a_\lambda, \lambda\in\sigma. \tag{17}$$

Since $a_\lambda \mathbf{T}_B^*|K_\lambda = T_B^* a_\lambda$, $\lambda\in\sigma$ also $A\mathbf{T}_B^* = \mathbf{T}_B^* A$; so by Sarason's Theorem (Lecture VIII) there exists a function $f, f\in H^\infty$, such that $A = P_+\overline{\mathbf{f}}|K_B$. From (16) and (17) it follows that $P_+(\overline{\mathbf{f}} - \overline{\mathbf{f}^\lambda})|K_\lambda = \mathbb{0}$, $\lambda\in\sigma$, and by the Calculus Lemma (Lecture III; cf. also relation (1) of the present Lecture reproducing the fact needed now)

$$f - f^\lambda \in b_\lambda^{k(\lambda)} H^\infty, \quad \lambda\in\sigma. \ \square$$

It is pertinent to make a few remarks. Clearly one may in Statement 3 as well as in Statement 4 consider the functions $f^\lambda$ as germs of height $k(\lambda)$ (at the point $\lambda$) and the formula (12) simplifies somewhat: $(f)_{k(\lambda)} = f^\lambda$, $\lambda\in\sigma$. In the case of Statement 2 one cannot perform such a change (there is no projection onto $K_\lambda$) but if we can in some other way estimate the magnitude of the germs of $f^\lambda$, then we get again a correct formulation: for any family of germs $\{f^\lambda\}_{\lambda\in\sigma}$ ($f^\lambda$, a polynomial of degree not exceeding $k(\lambda)-1$) such that

$$\sup_{\lambda\in\sigma}\|f^\lambda\|_{H^\infty/b_\lambda^{k(\lambda)}H^\infty} < \infty,$$

one can find $f$ in $H^\infty$ such that

$$(f)_{k(\lambda)} = f^\lambda, \quad \lambda\in\sigma.$$

This quotient norm is precisely the norm of the operator $a_\lambda$ (formula (16)); for this see Sarason's Theorem (Lecture VIII).

The projection $P_\lambda$ can be expressed in terms of the Malmquist functions (Lecture V): since these functions $l_{\lambda,k}$,

$$l_{\lambda,k} = \frac{(1-|\lambda|^2)^{1/2}}{\mathbb{1}-\bar{\lambda}z} b_\lambda^{k-1}, \quad 1\le k\le k(\lambda), \tag{18}$$

generate an orthonormal basis in the space $K_\lambda$ (the Malmquist-Walsh Lemma, Lecture V) we have

$$P_\lambda = \sum_{k=1}^{k(\lambda)} (\cdot, l_{\lambda,k}) l_{\lambda,k},$$
$$\|P_\lambda f\|^2 = \sum_{k=1}^{k(\lambda)} |(f, l_{\lambda,k})|^2, \quad f \in H^2. \tag{19}$$

The last formula shows that the norm $\|P_\lambda f\|$ is in a rather complicated way expressed in terms of the natural quantities $f(\lambda), f'(\lambda), \ldots$ describing the local (near the point $\lambda$) behavior of $f$. If however the multiplicities $k(\lambda)$, $\lambda \in \sigma$, are uniformly bounded, then this obstacle can be overcome. Below (Sect. 5) we will study this question in greater detail. Here we limit ourselves to the following simple consequence of the Interpolation Theorem.

**Theorem on Multiple Interpolation.** *Let $k$ be a divisor in $\mathbb{D}$ satisfying the Blaschke condition, set $\sigma = \{\lambda \in \mathbb{D} : k(\lambda) > 0\}$ and assume that $n = \sup_{\lambda \in \mathbb{D}} k(\lambda) < \infty$. Put further*

$$If \stackrel{\text{def}}{=} \{(1 - |\lambda|)^{s+1/2} f^{(s)}(\lambda) : 0 \le s < k(\lambda), \lambda \in \sigma\}, \quad f \in H^2.$$

*Then $IH^2 = l^2$ iff $\sigma$ is a Carleson set (i.e. $\sigma \in (C)$).*

*Proof*. Set $B^1 = \prod_{\lambda \in \sigma} b_\lambda$, $B = \prod_{\lambda \in \sigma} b_\lambda^{k(\lambda)}$. We have already shown that (Sect. 3, above)

$$\sigma \in (C) \Leftrightarrow \inf_{\zeta \in \mathbb{D}} \sup_{\lambda \in \sigma} |B_\lambda^1(\zeta)| > 0.$$

Now since

$$|B_\lambda^1(\zeta)|^n \le |B_\lambda(\zeta)| \le |B_\lambda^1(\zeta)|, \quad \zeta \in \mathbb{D},$$

the condition $\sigma \in (C)$ is equivalent to condition $(CV)$ for the divisor $k$. By the Interpolation Theorem it suffices to show that the norms (13) and $\|If\|_{l^2}^2$ are equivalent in $H^2$. Let us compute $P_\lambda f$, using the formulae (19) and (18):

$$\begin{aligned}(f, l_{\lambda,k}) &= (1 - |\lambda|^2)^{1/2} (P_+ \bar{b}_\lambda^{k-1} f)(\lambda) \\ &= (1 - |\lambda|^2)^{1/2} \sum_{j=0}^{k-1} \frac{f^{(j)}(\lambda)}{j!} (P_+ (z - \lambda \mathbb{1})^j \bar{b}_\lambda^{k-1})(\lambda).\end{aligned}$$

The general term of the sum has here as factor in front of $\dfrac{f^{(j)}(\lambda)}{j!}$ the number

$$\begin{aligned}(P_+(z - \lambda \mathbb{1})^j \bar{b}_\lambda^{k-1})(\lambda) &= \left(-\frac{\lambda}{|\lambda|}\right)^{k-1} \frac{1}{2\pi i} \int_{\mathbb{T}} \frac{(1 - \bar{\lambda}\zeta)^{k-1}}{(\zeta - \lambda)^{k-j}} d\zeta \\ &= \left(-\frac{\lambda}{|\lambda|}\right)^{k-1} \frac{1}{(k-j-1)!} [(\mathbb{1} - \bar{\lambda} z)^{k-1}]^{(k-j-1)}(\lambda) \\ &= \left(-\frac{\lambda}{|\lambda|}\right)^{k-1} (-\bar{\lambda})^{k-j-1} \frac{(k-1)!}{j!(k-j-1)!} (1 - |\lambda|^2)^j.\end{aligned}$$

This shows that there exist constants $c_1$ and $c_2$ such that

$$c_1 \sum_{k=1}^{k(\lambda)} (1-|\lambda|^2)^{2k-1} |f^{(k-1)}(\lambda)|^2 \le \|P_\lambda f\|^2$$
$$\le c_2 \sum_{k=1}^{k(\lambda)} (1-|\lambda|^2)^{2k-1} |f^{(k-1)}(\lambda)|^2$$

for all $f$, $f \in H^2$, and all $\lambda$, $\lambda \in \sigma$. □

*Remark.* An intermediate step in the proof was the equivalence

$$\sigma \in (C) \Leftrightarrow k \in (CV)$$

(in the hypothesis $\sup_{\lambda \in \mathbb{D}} k(\lambda) < \infty$; $\sigma = \{\lambda : k(\lambda) > 0\}$).

**5. Splitting and Blocking of Rootspaces.** If the family of subspaces $\{K_\lambda\}_{\lambda \in \sigma}$ forms a Riesz basis in $K_B$ then selecting in every $K_\lambda$ an orthonormal basis $\{x_{\lambda,n}\}_n$ (or a basis "close" to an orthonormal one) we get a family of vectors $\{x_{\lambda,n}\}$ constituting a Riesz basis for $K_B$. In order to have an "operator interpretation" of this family we can, for instance, require that

$$\bigvee(x_{\lambda,n} : 1 \le n \le k) = \operatorname{Ker}(T^* - \bar{\lambda} I)^k, \quad 1 \le k \le k(\lambda) \tag{20}$$

Then $\{x_{\lambda,n}\}$ is the Gram-Schmidt orthogonalization of the basis $\{z^{k-1}(\mathbb{1} - \bar{\lambda} z)^{-k}\}_{k \le k(\lambda)}$, i.e. the Malmquist functions (18). However among the bases (not necessarily orthogonal) of the space $K_B$ having the property (20) there is one corresponding to the classical problem of "multiple interpolation". This is the basis

$$\mathscr{X} = \{x_{\lambda,k} : 0 \le k < k(\lambda), \lambda \in \sigma\},$$
$$x_{\lambda,k} \stackrel{\text{def}}{=} z^k(\mathbb{1} - \bar{\lambda} z)^{-k-1}, \quad 0 \le k < k(\lambda).$$

Then the "interpolation operator" takes the form

$$If = \{(f, x_{\lambda,k}) : 0 \le k < k(\lambda), \lambda \in \sigma\}$$
$$= \left\{\frac{1}{k!} f^{(k)}(\lambda) : 0 \le k < k(\lambda)\right\}, \quad f \in H^2,$$

and our problem consists of clarifying in what circumstances

$$IH^2 = l^2(\|x_{\lambda,k}\|^{-1});$$

here $l^2(w_n) \stackrel{\text{def}}{=} \{\{a_n\} : \sum |a_n|^2 w_n^2 < \infty\}$.

Since we have already found (cf. Sect 3 above) a criterion for the family of rootspaces $\{K_\lambda : \lambda \in \sigma\}$ to be a basis, to solve this problem it remains only to explain a few things (how the angles between the vectors $x_{\lambda,k}\|x_{\lambda,k}\|^{-1}$ and $x_{\lambda,i}\|x_{\lambda,i}\|^{-1}$ behave when $\lambda$, $k$, $i$ vary). In particular, we shall in a quite different way prove the foregoing theorem.

**Theorem on Multiple Interpolation, II.** *Let $k$ be a divisor on the disk $\mathbb{D}$, the values of which are non-negative integers or the symbol $\infty$. Set $\sigma = \{\lambda \in \mathbb{D} : k(\lambda) > 0\}$. The following statements are equivalent:*

1. *The family* $\mathscr{X} = \{x_{\lambda,k} : 0 \le k < k(\lambda),\ \lambda \in \sigma\}$ *is an unconditional basis in the closure of its linear hull.*
2. $IH^2$ *is an ideal space.*
3. $IH^2 = l^2(\|x_{\lambda,k}\|^{-1})$.
4. $\sigma \in (C)$ *and either* $\sigma = \{0\}$ *or* $\sup_{\lambda \in \sigma} k(\lambda) < \infty$.

*Proof*. The equivalence of Statements 1, 2 and 3 is a consequence of the general theorems on bases (Lecture VI).

$1 \Rightarrow 4$. As $\mathscr{X}$ is a basis, clearly $\mathscr{X}_0 = \{(\mathbb{1} - \bar{\lambda}z)^{-1} : \lambda \in \sigma\}$ too is a basis so that $\sigma$ must be a Carleson set and in particular it thus cannot have points of accumulation in $\mathbb{D}$. Let $\mathscr{E}_{\lambda,n}$ be projection in the subspace $K_{\lambda,n+1} = \bigvee(x_{\lambda,k} : 0 \le k < n+1\}$ onto the direction $x_{\lambda,n}$ parallel to $K_{\lambda,n}$. Then $\mathscr{E}_{\lambda,n} = (\cdot, f_{\lambda,n})x_{\lambda,n}$ and $|\mathscr{E}_{\lambda,n}| = \|f_{\lambda,n}\| \cdot \|x_{\lambda,n}\|$ if $f_{\lambda,n} \in K_{\lambda,n+1}$. These conditions define the function $f_{\lambda,n}$ uniquely and it is clear that apart from a constant factor this is the Malmquist function $f_{\lambda,n} = cb_\lambda^n(\mathbb{1} - \bar{\lambda}z)^{-1}$. The formula $\mathscr{E}_{\lambda,n}x_{\lambda,n} = x_{\lambda,n}$ shows that $(x_{\lambda,n}, f_{\lambda,n}) = 1$ and $1 = |c|(1 - |\lambda|^2)^{-n-1}$. Thus $|\mathscr{E}_{\lambda,n}| = \|x_{\lambda,n}\| \cdot (1 - |\lambda|^2)^{n+1/2}$.

In computing the norm $\|x_{\lambda,n}\|^2$, in order to get the answer in the form of a polynomial in $|\lambda|^2$, we perform a change of variable of integration $w = \dfrac{\lambda - \zeta}{1 - \bar{\lambda}\zeta}$:

$$\|x_{\lambda,n}\|^2 = \int_{\mathbb{T}} \frac{dm}{|\mathbb{1} - \bar{\lambda}z|^{2n+2}} = \frac{1}{(1 - |\lambda|^2)^{2n+1}} \cdot \int_{\mathbb{T}} |\mathbb{1} - \bar{\lambda}z|^{2n}\, dm.$$

It is clear that if we expand $|\mathbb{1} - \bar{\lambda}z|^{2n}$ into powers of $z$, then the integral equals the constant term, while the remaining ones are zero. Since $(\mathbb{1} - \bar{\lambda}z)^n = \sum_{s=0}^n \binom{n}{s} \times (-\bar{\lambda}z)^s$ the coefficient in question will be $\sum_{s=0}^n \binom{n}{s}^2 |\lambda|^{2s}$, that is,

$$\|x_{\lambda,n}\|^2 = \frac{1}{(1 - |\lambda|^2)^{2n+1}} \sum_{s=0}^{n} \binom{n}{s}^2 |\lambda|^{2s}.$$

Thus $|\mathscr{E}_{\lambda,n}| = (\sum_{s=0}^n \binom{n}{s}^2 |\lambda|^{2s})^{1/2}$ and, as $\sup\{|\mathscr{E}_{\lambda,n}| : 0 \le n < k(\lambda), \lambda \in \sigma\} < \infty$ ($\mathscr{X}$ is an unconditional basis) we see that

$$\sup_{\lambda \in \sigma} k(\lambda)|\lambda| < \infty. \tag{21}$$

If $k(0) = \infty$ then $k(\lambda) = 0$ for all other $\lambda$, $\lambda \in \mathbb{D}$, because already $\{z^n\}_{n \ge 0}$ is a basis. Assume thus that $k(0) < \infty$. If $\sum_\lambda k(\lambda)(1 - |\lambda|) = \infty$, then the family $\mathscr{X}$ cannot be a basis since it is not even minimal. Therefore $\sum_\lambda k(\lambda)(1 - |\lambda|) < \infty$ and, in particular, $\sigma$ cannot have 0 as a point of accumulation. Now (21) gives $\sup_{\lambda \in \sigma} k(\lambda) < \infty$.

$4 \Rightarrow 1$. By Vasyunin's Theorem the family $\{K_\lambda : \lambda \in \sigma\}$ must be a basis for $\bigvee(\mathscr{X})$. Therefore it suffices to show that there exists a uniform family of isomorphisms $\{\mathscr{O}_\lambda : \lambda \in \sigma\}$ (i.e. we have $\sup_\lambda |\mathscr{O}_\lambda||\mathscr{O}_\lambda^{-1}| < \infty$) mapping $\mathscr{X} \cap K_\lambda$ onto an orthogonal basis in $K_\lambda$. For this orthogonal basis we take the Malmquist functions $l_{\lambda,i+1}$ (the orthogonalization of the set $x_{\lambda,0}, \ldots, x_{\lambda,k(\lambda)-1}$) and set $\mathscr{O}_\lambda l_{\lambda,i+1} \overset{\text{def}}{=} x_{\lambda,i}/\|x_{\lambda,i}\|$, $0 \le i < k(\lambda)$. Then

$$|\mathscr{O}_\lambda|^2 \le |\mathscr{O}_\lambda|_2^2 \overset{\text{def}}{=} \sum_{i=0}^{k(\lambda)-1} \|x_{\lambda,i}\|^2/\|x_{\lambda,i}\|^2 = k(\lambda).$$

It remains to estimate $|\det \mathcal{O}_\lambda|$. The matrix of the operator $\mathcal{O}_\lambda$ is triangular with diagonal elements

$$|(\mathcal{O}_\lambda l_{\lambda,i+1}, l_{\lambda,i+1})| = 1/i! \cdot 1/\|x_{\lambda,i}\| \cdot |l^{(i)}_{\lambda,i+1}|$$
$$= 1/\|x_{\lambda,i}\| \cdot (1-|\lambda|^2)^{-i-1/2}$$

such that

$$1 \le |(\mathcal{O}_\lambda l_{\lambda,i+1}, l_{\lambda,i+1})^{-1}| \le \left(\sum_{s=0}^{i} \binom{i}{s}^2\right)^{1/2},$$

$i = 0, \ldots, k(\lambda) - 1$. □

This theorem shows that if $\sup_\lambda k(\lambda) = \infty$ no free interpolation problem of the form $If = a$ (i.e. a problem with an ideal space of data) is solvable, a single exception being the case

$$If = \{f^{(n)}(0) : n \ge 0\}.$$

If the family of rootspaces $\{K_\lambda\}_{\lambda \in \sigma}$ is not a basis then the simplest way to correct this inconvenience is to try to lump together "sufficiently close" rootspaces. After such a unification into blocks the new family will have more chances to be a basis. The two theorems to be established below confirm this point of view. Let us point out beforehand that these theorems are close in spirit to the theorems on Carleson series in Lecture VII. In the proof of the first of these theorems one can use a sufficient condition for uniform minimality of the family of subspaces $K_{\vartheta_n} = H^2 \ominus \vartheta_n H^2$, which follows from the estimates in Lecture X: set $\Theta = \prod_{n \in \mathfrak{N}} \vartheta_n$, $\vartheta_n = \prod_\lambda b_\lambda^{k_n(\lambda)}$, and $B_n = \Theta \vartheta_n^{-1}$; then the condition says that $\inf_n \prod_\lambda |B_n(\lambda)|^{k_n(\lambda)} > 0$. However, it is simpler to use an elementary criterion for uniform minimality, which follows immediately from the Theorem on Angles: $\inf_n \inf_{\mathbb{D}} \{|\vartheta_n| + |B_n|\} > 0$.

**Blocking Theorem.** *Let $\{\Lambda^j\}$ be a collection of Carleson subsets of the unit disk, denoting by $k^j$ the characteristic function of $\Lambda^j$, $1 \le j \le N$, and writing $k = \sum_j k^j$, $B = \prod_\lambda b_\lambda^{k(\lambda)}$. Then there exists a family of spectral subspaces $\{K_n\}$, $\dim K_n \le N$, which forms a basis for $K_B$.*

*Proof.* Each of the sets $\Lambda^j$ is sparse (cf. Lecture VII) so that the pairwise distances between the points of each set are bounded away from zero. Here we have the hyperbolic distance in mind, defined by the formula $\operatorname{th} \dfrac{\rho(\lambda,\mu)}{2} = \left|\dfrac{\lambda-\mu}{1-\bar\lambda\mu}\right|$. Let us set $\rho_0 = \min_j \inf_{\lambda,\mu \in \Lambda^j, \lambda \ne \mu} \rho(\lambda,\mu)$ (if it happens that $\operatorname{th} \rho_0/2 > 1/2$ we replace $\rho_0$ by $2 \operatorname{arcth} \frac{1}{2}$) and $\rho = \rho_0/32N$. Each point $\lambda$, $\lambda \in \Lambda^j$, $j \le N$, will be surrounded by a disk with radius $\rho$, $D_\lambda = \{\zeta : \rho(\lambda,\zeta) \le \rho\}$. Let $\Lambda$ be the set of pairs $\{\{\lambda, j\} : \lambda \in \Lambda^j\}$; we divide it into subsets $\Lambda_n$ corresponding to the decomposition of the set $\bigcup D_\lambda$ into connected components $G_n$. Let us show that the number of elements in $\Lambda_n$ does not exceed $N$.

Assume the contrary. Then one can find a subset $\Lambda'_n$, $\Lambda'_n \subset \Lambda_n$, consisting of $N+1$ elements such that the union of the corresponding disks is a connected set. Therefore the distance between the points $\lambda$, $\{\lambda, j\} \in \Lambda'_n$, cannot exceed $2N\rho$. As

$2N\rho = \frac{1}{16}\rho_0 < \rho_0$, for any two pairs $\{\lambda, i\} \in \Lambda'_n$ and $\{\mu, j\} \in \Lambda'_n$ we must have $i \neq j$ and this contradicts the fact that $\Lambda'_n$ consists of $N+1$ elements. Consequently there are not more than $N$ elements in any of the subsets $\Lambda_n$.

Set $\vartheta_n = \prod_{\lambda \in G_n} b_\lambda^{k(\lambda)}$, $B_n = B \cdot \vartheta_n^{-1}$, $K_n = H^2 \ominus \vartheta_n H^2$, $\sigma = \bigcup_j \Lambda^j$, $G = \bigcup_n G_n$. Let us show that the family $\{K_n\}$ is a basis in $K_B$. To this end it suffices to verify that $\{K_n\}$ is indeed uniformly minimal, i.e. that

$$\inf_n \inf_{\mathbb{D}} \{|\vartheta_n| + |B_n|\} > 0.$$

If $\zeta \in \mathbb{D} \backslash G$ then $\rho(\zeta, \sigma) \geq \rho$ and

$$|\vartheta_n(\zeta)| = \prod_{\lambda \in G_n} \left| \operatorname{th} \frac{\rho(\zeta, \lambda)}{2} \right|^{k(\lambda)} \geq \left( \operatorname{th} \frac{\rho}{2} \right)^N.$$

If $\zeta \in G$ then there exists a $\lambda$, $\lambda \in \sigma$, such that $\rho(\zeta, \lambda) < \rho$. Assume that $\lambda \in G_n$. For any $m$, $m \neq n : \rho(\zeta, G_m \cap \sigma) \geq \rho$; therefore $|\vartheta_m(\zeta)| \geq (\operatorname{th} \frac{\rho}{2})^N$. In order to complete the proof it remains to estimate $|B_n(\zeta)|$.

Set $k_n^j = k^j \chi_{\mathbb{D} \backslash G_n}$; then $B_n = \prod_{j=1}^N B(k_n^j, \cdot)$ (the symbol $B(k, \cdot)$ was defined on p. 213). Consider separately the two cases 1) $G_n \cap \Lambda^j \neq \varnothing$ and 2) $G_n \cap \Lambda^j = \varnothing$.

1) Let $G \cap \Lambda^j \neq \varnothing$ and assume that the $\Lambda^j$ satisfy a $\delta$-Carleson condition (meaning that the infimum in condition (C) equals $\delta$). Then they are "$\delta$-sparse", i.e. $\operatorname{th} \rho_0/2 < \delta$. Now if $\{\mu, i\} \in \Lambda_n$, as we have already remarked, $\rho(\lambda, \mu) \leq 2N\rho < \rho_0/16$ and therefore $\operatorname{th} \frac{1}{2}\rho(\lambda, \zeta) \leq \operatorname{th} \rho_0/32 < \frac{1}{4}\delta$ (since $\rho_0$ was chosen such that $\operatorname{th} \rho_0/2 \leq \frac{1}{2}$). This allows us to conclude (cf. p. 218) that from the given condition $|B(k_n^j, \lambda)| \geq \delta$ follows $|B(k_n^j, \zeta)| \geq \delta/2$.

2) Let $G_n \cap \Lambda^j = \varnothing$ and assume again that the sets $\Lambda^j$ satisfy a $\delta$-Carleson condition. Then if $\lambda \in \Lambda^j$ and $|b_\lambda(\zeta)| \leq \delta/4$ we must have $|B_\lambda(k^j, \zeta)| \geq \delta/2$, i.e.

$$\inf_{\lambda \in \Lambda^j} \inf_{\zeta \in \mathbb{D}} \{|b_\lambda(\zeta)| + |B_\lambda(k^j, \zeta)|\} \geq \delta/4.$$

As $\rho(\zeta, \Lambda^j) > \rho$ we have $\sup_{\lambda \in \Lambda^j} |b_\lambda(\zeta)| \geq \operatorname{th} \rho/2 \stackrel{\text{def}}{=} \varepsilon$. By Lemma 1 to Vasyunin's Theorem then holds the estimate $|B(k^j, \zeta)| \geq \varepsilon(\frac{\delta}{4} - \varepsilon)$. Finally, since $\varepsilon = \operatorname{th} \rho/2 = \operatorname{th} \rho_0/64N < \frac{1}{4}\delta$, clearly $\varepsilon(\frac{\delta}{4} - \varepsilon) > 0$.

Thus in both cases we can estimate $|B(k_n^j, \zeta)|$ by a constant independent of $n, j$ and $\zeta$. □

**Splitting Theorem.** *Assume that the family of subspaces $\{K_n\}$, $K_n = H^2 \ominus \vartheta_n H^2$, is a basis for $K_\Theta$, $\Theta = \prod \vartheta_n$, and that $\dim K_n \leq N$. Then $\Theta$ is a Blaschke product which can be split into a product of $N$ factors with simple zeros, each satisfying a Carleson condition.*

*Proof*. As $\dim K_n \leq N$ then $\vartheta_n$ is a finite Blaschke product. Pick in each of the spaces $K_n$ an eigenvector of $T^*$, say, $x_n = \dfrac{1}{\mathbb{1} - \bar{\lambda}_n z}$. By hypothesis these eigenvectors form a basis for the closure of their linear hull so that the family $\{\lambda_n\}$ must satisfy the Carleson condition. Let us now set $\vartheta'_n = \vartheta_n b_{\lambda_n}^{-1}$; $K'_n = H^2 \ominus \vartheta'_n H^2$. Then $\dim K'_n \leq N - 1$. The proof is completed by induction. □

**6. Spectrality and $\mathscr{B}_0$-Spectrality.** Let us first show that a model operator with a characteristic function "in general position" as a matter of fact cannot be spectral in the classical sense.

**Foiaş's Theorem.** *If the model operator $\mathbf{T}_\Theta$ with the characteristic function $\Theta$ is spectral in Dunford's sense then its representing measure $\mu^\Theta$ must be a pure point measure (no continuous component).*

*The proof* of this theorem essentially blows down to the following statement: if $\mu$ is a continuous singular measure on the circle $\mathbb{T}$ then we can find relatively singular measure $\mu'$ and $\mu''$ such that $\mu' + \mu'' = \mu$ and $\sin\langle K', K''\rangle = 0$ where $K' = H^2 \ominus S(\mu', \cdot)H^2$ and $K'' = H^2 \ominus S(\mu'', \cdot)H^2$.

For the proof we divide $\mathbb{T}$ into intervals $\tilde{\Delta}_n$ such that $\mu(\tilde{\Delta}_n) > 0$, $n = 1, 2, \ldots$. Since $\mu$ is singular we can find intervals $\Delta_n$, $\Delta_n \subset \tilde{\Delta}_n$, such that $\mu(\Delta_n) \geq 2\pi n m(\Delta_n)$. As $\mu$ is continuous, we can divide each $\Delta_n$ into two pieces $\Delta_n'$ and $\Delta_n''$ such that $\mu(\Delta_n') = \mu(\Delta_n'') = \frac{1}{2}\mu(\Delta_n)$. Set $\mu_n' = \chi_{\Delta_n'}\mu$, $\mu_n'' = \chi_{\Delta_n''}\mu$; the corresponding inner functions will be denoted by $s_n'$ and $s_n''$ and the subspaces by $K_n'$ and $K_n''$. Put $\mu' = \sum_n \mu_n'$, $\mu'' = \mu - \mu'$; it is clear that $\mu'' \geq \sum_n \mu_n''$. To complete the proof it suffices to show that

$$\inf_n \inf_{\zeta \in \mathbb{D}} \{|S(\mu_n', \zeta)| + |S(\mu_n'', \zeta)|\} = 0.$$

Let $\zeta_n$ be the middle point of the arc $\Delta_n$. Then

$$|S(\mu_n', r\zeta_n)| = \exp\left\{-\int_{\Delta_n'} \frac{1 - r^2}{|\xi - r\zeta_n|^2} d\mu(\xi)\right\},$$

$$|\xi - r\zeta_n| \leq |\xi - \zeta_n| + |\zeta_n - r\zeta_n| \leq \pi m(\Delta_n) + (1 - r).$$

For $r = 1 - \pi m(\Delta_n)$ we get the estimate $|\xi - r\zeta_n| \leq 2\pi m(\Delta_n)$ and consequently

$$\int_{\Delta_n'} \frac{1 - r^2}{|\xi - r\zeta_n|^2} d\mu(\xi) \geq \frac{1}{4\pi^2} \frac{1 - r}{m(\Delta_n)^2} \mu(\Delta_n') = \frac{1}{4\pi m(\Delta_n)} \frac{\mu(\Delta_n)}{2} \geq \frac{n}{4},$$

i.e. $|S(\mu_n', r\zeta_n)| \leq \exp(-\frac{n}{4})$. In exactly the same way $|S(\mu_n'', r\zeta_n| \leq \exp(-\frac{n}{4})$ and therefore

$$\inf_n \inf_{\zeta \in \mathbb{D}} \{|S(\mu_n', \zeta)| + |S(\mu_n'', \zeta)|\} \leq 2 \inf_n \exp\left(-\frac{n}{4}\right) = 0. \square$$

In the remainder of this Section we consider a model operator $T$ whose characteristic function is an arbitrary singular inner function $S$, i.e. its generating measure $\mu = \mu^s$ may contain a continuous as well as a discrete component.

Let $\mathbb{T}_\delta$ denote the set of all points of the circle $\mathbb{T}$ such that $\inf_{0<r<1} |S(r\zeta)| \geq \delta$ and let $\mathscr{B}_\delta$ be the $\sigma$-algebra generated by those open subsets of $\mathbb{T}$ for which the endpoints of every connected component belong to $\mathbb{T}_\delta$.

**Theorem on $\mathscr{B}_0$-Spectrality.** 1. *The operator $\mathbf{T}_S$ is $\mathscr{B}_0$ spectral.*

2. *For every $\gamma$, $\gamma < 1$, one can find a number $\delta$, $\delta > 0$, such $m\mathbb{T}_\delta \geq \gamma$.*

*Proof*. 1. Let $\Delta \in \mathscr{B}_\delta$ and put $\Delta' = \mathbb{T}\backslash\Delta$, $S_\Delta = S(\chi_\Delta\mu, \cdot)$, $S'_\Delta = S(\chi_{\Delta'}\mu, \cdot)$. By the Theorem on Angles $\mathbf{T}_S$ will be $\mathscr{B}_\delta$ spectral iff

$$\inf_{\Delta \in \mathscr{B}_\delta} \inf_{\mathbb{D}} \{|S_\Delta| + |S'_\Delta|\} > 0.$$

If $\zeta \in \mathbb{T}_\delta$ then

$$\inf_{0<r<1} \{|S_\Delta(r\zeta)| + |S'_\Delta(r\zeta)|\} \geq 2 \inf_{0<r<1} |S(r\zeta)| \geq 2\delta.$$

Let $\zeta \in \mathbb{T}\backslash\mathbb{T}_\delta$ and let $l$ be the contiguity interval of $\mathbb{T}_\delta$ containing $\zeta$ ($\mathbb{T}_\delta$ is a closed set). Two cases are possible: either $l \subset \Delta$ or else $l \subset \Delta'$. Indeed, this property is by definition shared by any open subset in $\mathscr{B}_\delta$ and therefore is inherited by the whole $\sigma$-algebra.

If $l \subset \Delta$ then $|S'_\Delta(\xi)| \geq |S'_l(\xi)|$, $\xi \in \mathbb{D}$; if $l \subset \Delta'$ then $|S_\Delta(\xi)| \geq |S'_l(\xi)|$, $\xi \in \mathbb{D}$; that is, in any case $|S_\Delta(\xi)| + |S'_\Delta(\xi)| \geq |S'_l(\xi)|$, $\xi \in \mathbb{D}$. Thus, for the proof of the theorem it remains to check that $\inf_{\zeta \in \mathbb{T}\backslash\mathbb{T}_\delta} \inf_{0<r<1} |S'_l(r\zeta)| > 0$. Let us fix $\zeta$, $\zeta \in l$. Let $\xi_0$ be the middle of the interval $l'$ and let $\xi_1$ and $\xi_2$ denote its endpoints. If the point $\xi$, $\xi \in l'$, lies between the points $\xi_0$ and $\xi_i$ then $|\xi - r\zeta| \geq |\xi - r\xi_i|$, $i = 1, 2$; therefore

$$|S'_l(r\zeta)| = \exp\left\{-\int_{l'} \frac{1-r^2}{|\xi - r\zeta|^2}\, d\mu(\xi)\right\} \geq \min_{i=1,2} |S'_l(r\xi_i)| \geq \delta,$$

because $\xi_i \in \mathbb{T}_\delta$.

2. Let

$$\mathbb{T}_0 = \bigcup_{\delta>0} \mathbb{T}_\delta = \left\{\zeta : \inf_{0<r<1} |S(r\zeta)| > 0\right\}.$$

Assume that at the point $\zeta$, $\zeta \in \mathbb{T}$, the limit $\lim_{r\to 1} |S(r\zeta)| = 1$ exists. Then $\zeta \in \mathbb{T}_0$. But the set of these points has full measure, $m(\mathbb{T}_0) = 1$, i.e. $\lim_{\delta\to 0} m(\mathbb{T}_\delta) = 1$. □

Thus there exist arbitrary fine decompositions of $\mathbb{T}$ generating a bounded family of spectral projections.

**7. Concluding Remarks.** The main result of Dunford's theory consists of the following. Every spectral operator $T$ has the form $T = \int z\, d\mathscr{E} + N$ where $\mathscr{E}$ is a suitable projection valued Borel measure, which commutes with $T$, further $NT = TN$ and $\sigma(N) = \{0\}$. All applications of this theorem are connected with spectrality criteria based on special estimates of the resolvent, somehow reminiscent of the estimate (7). Operators $T$ scalar in the sense of Dunford (i.e. those for which $N = \mathbb{O}$) are characterized by the possibility of applying to them any function continuous on the spectrum $\sigma(T)$, and some traces of this theorem can be found in Lecture VIII, Sect. 4. A practically inexhaustible source of information on spectral operators is the book Dunford-Schwartz [3].

Almost all of the material of this Lecture is taken over from Vasyunin's paper [2]; the general Interpolation Theorem (Sect. 4) is established in Nikol'skiĭ [10] and the Theorems on Multiple Interpolation (I and II) in Vasyunin [4]. Let us also remark that these theorems admit an extension similar to the results on interpolation in Lecture VIII: condition $(CV)$ is (as in the theorems of Sect. 4) sufficient for the interpolation of germs in any space $X$ with $M(X) = H^\infty$. If $X = H^p$ the exact formulation can be found in Nikol'skiĭ [10]. After the publication of the Russian edition of this book the papers Vinogradov-Rukšin [1] and Rukšin [1] appeared which contain proofs of Interpolation Theorem in the spirit of the "pure" function theory. In the first of these papers it is also proved that a) condition $(CV)$ is not only sufficient but also necessary for the possibility of free interpolation of germs in the space $H^p$, $1 \leq p \leq \infty$, that b) for $1 < p < \infty$ there always exists a linear interpolating operator and that c) for $p = 1, \infty$ such an operator exists iff $\sup_{\lambda \in \sigma} k(\lambda) < \infty$.

As already has been noted, the functional model allows us to distinguish not only spectral but also "pseudo-spectral" subspaces $K_e$, corresponding to arbitrary Borel subsets $e$ of the closed disk clos $\mathbb{D}$. The correspondence $e \mapsto K_e$ is a "capacity in the sense of Apostol" (cf. Colojoara-Foiaş [1]), i.e. it is a geometric variant of the spectral measure (on the level of subspaces, not projections). It would be of importance to know how to integrate with respect to such a "measure", not only with respect to the $\sigma$-algebras $\mathscr{B}_0$, generated by a countable decomposition of the spectrum (as was done in the present Lecture); this would allow us to get integral representations for the operator $\mathbf{T}_\Theta$ analogous to Dunford's representation.

Vector and operator analogues of some of the results of this Lecture can be found in Nikol'skiĭ [10].

# Lecture X. Analysis of the Carleson-Vasyunin Condition

The Carleson condition $(C)$ imposed on the Blaschke product $B = \prod_{\lambda \in \sigma} b_\lambda$ ($\sigma \subset \mathbb{D}$) with simple zeros

$$\inf_{\lambda \in \sigma} |B_\lambda(\lambda)| > 0, \tag{C}$$

was analyzed in great detail in Lecture VII. The power of this analysis, which helps us to visualize rather completely the geometric nature of Carleson sets, depends to a large extent on the local character of condition $(C)$ – it is expressed exclusively in terms of the local parameters of $B$, more exactly, the values of its derivative $|B'(\lambda)|$ $(= (1 - |\lambda|^2)^{-1} |B_\lambda(\lambda)|, \lambda \in \sigma)$ on the spectrum $\sigma(B) = \sigma$. Not less important is the circumstance that condition $(C)$ by definition is expressed in terms of the representing measure $\mu^B$ of $B$ and a fixed kernel on the product $\sigma \times \sigma$:

$$\inf_{\lambda \in \sigma} 2 \int_{\mathbb{D} \setminus \{\lambda\}} \frac{\log \left| \dfrac{\zeta - \lambda}{1 - \bar{\zeta}\lambda} \right|}{1 - |\zeta|^2} d\mu^B(\zeta) > -\infty.$$

This Lecture will be devoted to a similar study of the Carleson-Vasyunin condition $(CV)$, as we have seen, corresponding to the generalized $\mathscr{B}_0$-spectrality of the model operator $\mathbf{T}_\Theta$. We will examine several important special cases ($\Theta = B$, $\Theta$ a singular function with a purely discrete representing measure) and get visible necessary and/or sufficient conditions for $\Theta \in (CV)$ but we cannot hope to achieve the same clarity as in the case of the Carleson condition $(C)$. The principal obstruction turns out to be the non-local character of condition $(CV)$. This is clearly out of question for functions $\Theta$ possessing a singular factor but already in the case of a Blaschke product $\Theta = B = \prod_{\lambda \in \sigma} b_\lambda^{k(\lambda)}$ with unbounded multiplicities condition $(CV)$ cannot be reduced to combinations of conditions imposed on the germs of the function at the points of the spectrum $\sigma$. At any rate in all natural conjectures of this kind we produce corresponding counter-examples (Sect. 5 below), the acquaintance with which ought to bring an essential contribution towards the understanding of the mechanism governing the condition $(CV)$.

**1. An Estimate for the Angle in Terms of Representing Measures.** We are concerned with the angle between subspaces $K_{\Theta_i} = H^2 \ominus \Theta_i H^2, i = 1, 2$, in terms of which (as has already repeatedly been noticed) the norms of the corresponding projections $\mathscr{P}_{K_{\Theta_1} \| K_{\Theta_2}}$ can be expressed. In order to shorten the notation we will leave out part of the multiple level indices writing, for instance, $K_{\Theta_i} = K_i$, $\mu^{\Theta}i = \mu_i$, etc., each time when this will not lead to any confusion. The main result of this Section is the following inequality.

**Theorem on Angles, II.** *Let* $\Theta_i$. $i = 1, 2$, *be inner functions with the representing* *measures* $\mu_i$, $i = 1, 2$. *Then*

$$\sin\langle K_1, K_2\rangle \geq \exp\left\{4\int_{\mathbb{D}}\int_{\mathbb{D}}\frac{\log\left|\dfrac{\zeta-\xi}{1-\bar{\zeta}\xi}\right|}{(1-|\zeta|^2)(1-|\xi|^2)}\,d\mu_1(\zeta)\,d\mu_2(\xi)\right\}. \tag{1}$$

Let us discuss this inequality. Let us denote by $F(\Theta_1, \Theta_2)$ the expression in the right hand side and consider it as a function of the arguments $\Theta_1$ and $\Theta_2$. From the definition we see immediately that $F$ is symmetric ($F(\Theta_1, \Theta_2) = F(\Theta_2, \Theta_1)$) and multiplicative ($F(\Theta_1, \Theta_2\Theta_3) = F(\Theta_1, \Theta_2)F(\Theta_1, \Theta_3)$) The last assertion follows from an evident property of the representing measures: $\mu^{\Theta_2\Theta_3} = \mu^{\Theta_2} + \mu^{\Theta_3}$.

In special cases $F$ takes on a substantially simpler form. If the $B_i$ are Blaschke products defined by the divisors $k_i$, $i = 1, 2$, then

$$\begin{aligned} F(B_1, B_2) &= \exp\left\{\sum_{\zeta\in\mathbb{D},\,\xi\in\mathbb{D}} k_1(\zeta)k_2(\xi)\log\left|\frac{\zeta-\xi}{1-\bar{\zeta}\xi}\right|\right\} \\ &= \prod_{\zeta\in\mathbb{D},\,\xi\in\mathbb{D}}\left|\frac{\zeta-\xi}{1-\bar{\zeta}\xi}\right|^{k_1(\zeta)k_2(\xi)} \\ &= \prod_{\xi\in\mathbb{D}}|B_1(\xi)|^{k_2(\xi)} \\ &= \prod_{\zeta\in\mathbb{D}}|B_2(\zeta)|^{k_1(\zeta)}. \end{aligned}$$

This formula shows that the estimate (1) is not too sharp at least for functions with a more or less arbitrary distribution of the spectra. For example, assume that $k_i \leq 1$, $i = 1, 2$, that the product $B = B_1 B_2$ satisfies the Carleson condition $(C)$ and that the zeros of the first factor are dispersed so well that they give to the product $B_\lambda(\lambda)$, $\lambda\in\sigma_1 = \sigma(B_1)$, a contribution sufficiently close to one and are situated so close to the corresponding points of $\sigma_2 = \sigma(B_2)$ that $|B_2(\lambda)| \sim \delta \stackrel{\text{def}}{=} \inf_{\lambda\in\sigma_1\cup\sigma_2}|B_\lambda(\lambda)|$ for $\lambda\in\sigma_1$. Then inequality (1) estimates the angle between $K_{B_1}$ and $K_{B_2}$ by the number $F(B_1, B_2) \sim \delta^N$, $N = \operatorname{card}\sigma_1$, while from the estimate of the orthogonalizer of the corresponding family of eigenfunctions (Lecture VII, p. 156) we know that the angle in question is not less than const $\delta\,(\log 1/\delta)^{-1/2}$. The function $F(\Theta_1, \Theta_2)$ is however expressed in a very simple manner in terms of the representing measures $\mu_i$, $i = 1, 2$, and gives therefore in many cases a decent estimate of the angle.

Let now $\Theta_1 = S_1$ be a singular inner function. The function under the integral sign has on the circumference $|\zeta| = 1$ the form $-\frac{1}{2}|\zeta-\xi|^{-2}$. Indeed, if $|\zeta|\to 1$, then

$$\begin{aligned} \frac{\log\left|\dfrac{\zeta-\xi}{1-\zeta\xi}\right|}{(1-|\zeta|^2)(1-|\xi|^2)} &= \frac{\dfrac{1}{2}\log\left[1-\dfrac{(1-|\zeta|^2)(1-|\xi|^2)}{|1-\bar{\zeta}\xi|^2}\right]}{(1-|\zeta|^2)(1-|\xi|^2)} \sim -\frac{1}{2}\frac{1}{|1-\bar{\zeta}\xi|^2} \\ &= -\frac{1}{2}\frac{1}{|\zeta-\xi|^2}. \end{aligned}$$

Thus

$$F(S_1,\Theta_2)=\exp\left\{-2\int_{\mathbb{T}}\int_{\overline{\mathbb{D}}}\frac{d\mu_1(\zeta)\,d\mu_2(\xi)}{|\zeta-\xi|^2}\right\}.$$

In the special case $\Theta_2=B_2$ we get

$$\begin{aligned}F(S_1,B_2)&=\exp\left\{-\sum_{\xi\in\mathbb{D}}k_2(\xi)(1-|\xi|^2)\int_{\mathbb{T}}\frac{d\mu_1(\zeta)}{|\zeta-\xi|^2}\right\}\\&=\prod_{\xi\in\mathbb{D}}|S_1(\xi)|^{k_2(\xi)},\end{aligned}$$

since

$$\begin{aligned}|S_1(\xi)|&=\exp\left\{-\operatorname{Re}\int_{\mathbb{T}}\frac{\zeta+\xi}{\zeta-\xi}\,d\mu_1(\zeta)\right\}\\&=\exp\left\{-(1-|\xi|^2)\int_{\mathbb{T}}\frac{d\mu_1(\zeta)}{|\zeta-\xi|^2}\right\}.\end{aligned}$$

Consequently, for an arbitrary $\Theta_1$,

$$\begin{aligned}F(\Theta_1,B_2)&=F(B_1,B_2)F(S_1,B_2)\\&=\prod_{\xi\in\mathbb{D}}|B_1(\xi)|^{k_2(\xi)}\prod_{\xi\in\mathbb{D}}|S_1(\xi)|^{k_2(\xi)}\\&=\prod_{\xi\in\mathbb{D}}|\Theta_1(\xi)|^{k_2(\xi)}.\end{aligned}$$

The proof of the Theorem on Angles will be divided into several steps. We prove first inequality (1) for Blaschke products and then for singular functions with discrete representing measures and finally with the aid of these two cases for arbitrary inner functions. To accomplish this plan we need some properties of angles between subspaces and a special type of continuity of the function $F$, which constitute the contents of the following four lemmata.

**Lemma on the Semi-Continuity of the Angle.** *Let $X$ be any Banach space and $X_n$ a sequence of subspaces of $X$ ($n=0,1,\ldots$). Then*

$$|\mathscr{P}_{(\underline{\lim}_n X_n)\|X_0}|\le\underline{\lim}_n|\mathscr{P}_{X_n\|X_0}|,$$

$$|\mathscr{P}_{X_0\|(\underline{\lim}_n X_n)}|\le\underline{\lim}_n|\mathscr{P}_{X_0\|X_n}|.$$

*Proof.* If $x\in\underline{\lim}_n X_n$ then there exist elements $x_n$, $x_n\in X_n$, approximating $x:\lim_n\|x-x_n\|=0$. Let $x'\in X_0$. Then

$$\begin{aligned}\|x\|&=\lim_n\|x_n\|=\lim_n\|\mathscr{P}_{X_n\|X_0}(x_n+x')\|\\&\le\underline{\lim}_n|\mathscr{P}_{X_n\|X_0}|\cdot\|x_n+x'\|\\&=\|x+x'\|\cdot\underline{\lim}_n|\mathscr{P}_{X_n\|X_0}|.\ \square\end{aligned}$$

**Lemma on the Continuity of the Angle.** *Let $K$ be an arbitrary subspace of $H^2$ and set $K_i = H^2 \ominus \Theta_i H^2$, the $\Theta_i$ being inner functions, $i = 1, 2$. Then*

$$|\sin\langle K_1, K\rangle - \sin\langle K_2, K\rangle| \leq \|\Theta_1 - \Theta_2\|_\infty.$$

*Proof*. Set $P_i = P_{K_i}$. Then

$$\begin{aligned}
\sin\langle K_2, K\rangle &= \inf_{x\in K} \frac{\|(I-P_2)x\|_2}{\|x\|_2} = \inf_{x\in K}\frac{\|P_+\bar{\Theta}_2 x\|_2}{\|x\|_2} \\
&\leq \inf_{x\in K}\frac{\|P_+(\bar{\Theta}_1 - \bar{\Theta}_2)x\|_2 + \|P_+\bar{\Theta}_1 x\|_2}{\|x\|_2} \\
&\leq \inf_{x\in K}\left\{\|\Theta_1 - \Theta_2\|_\infty + \frac{\|P_+\bar{\Theta}_1 x\|_2}{\|x\|_2}\right\} \\
&= \|\Theta_1 - \Theta_2\|_\infty + \inf_{x\in K}\frac{\|(I-P_1)x\|_2}{\|x\|_2} \\
&= \|\Theta_1 - \Theta_2\|_\infty + \sin\langle K_1, K\rangle. \; \square
\end{aligned}$$

**Lemma on the Submultiplicativity of the Angle.** *Let $\Theta$, $\Theta_1$ and $\Theta_2$ be inner functions and write $K = H^2 \ominus \Theta H^2$, $K_i = H^2 \ominus \Theta_i H^2$, $i = 1, 2$, $K_{12} = H^2 \ominus \Theta_1\Theta_2 H^2$. Then*

$$\sin\langle K_{12}, K\rangle \geq \sin\langle K_1, K\rangle \sin\langle K_2, K\rangle.$$

*Proof*. Without loss of generality we can take $\sin\langle K_i, K\rangle > 0$, for otherwise everything is trivial. Let us first also assume that $\Theta_1$ and $\Theta_2$ are relatively prime so that $K_{12} = K_1 \vee K_2$.

Let $T_i$ be the compression of the shift onto the subspace $K_i \vee K$ ($T_i x = P_{K_i\vee K} Sx$) and $T$ the compression onto $K_{12} \vee K$ ($Tx = P_{K_{12}\vee K}Sx$). Since $\sin\langle K_i, K\rangle > 0$ we have continuous projections $\mathscr{P}_{K\|K_i}$. By Sarason's Theorem we can find functions $f_i$, $f_i \in H^\infty$, such that $\mathscr{P}_{K\|K_i} = f_i(T_i)^*$ and $\|f_i\|_\infty = |\mathscr{P}_{K\|K_i}| = \sin^{-1}\langle K_i, K\rangle$. By formula (3) in Lecture IX $f_i \in \Theta_i H^\infty$, $(\mathbb{1} - f_i) \in \Theta H^\infty$. Set $f = f_1 f_2$. Then $f \in \Theta_1\Theta_2 H^\infty$ and $(\mathbb{1} - f) = [(\mathbb{1} - f_1) + f_1(\mathbb{1} - f_2)] \in \Theta H^\infty$, that is, again by the same formula (3), Lecture IX, $f(T)^* = \mathscr{P}_{K\|K_{12}}$ and consequently

$$\begin{aligned}
\sin\langle K_{12}, K\rangle &= |\mathscr{P}_{K\|K_{12}}|^{-1} \geq \|f\|_\infty^{-1} \geq \|f_1\|_\infty^{-1}\|f_2\|_\infty^{-1} \\
&= \sin\langle K_1, K\rangle \sin\langle K_2, K\rangle.
\end{aligned}$$

Next let us consider the general case. Using Frostman's Theorem (cf. Lecture II, p. 59) pick a sequence of Blaschke products $\{B_n\}_{n\geq 1}$ uniformly converging to $\Theta_1$. Moving slightly the zeros we can arrange that the $B_n$ ($n = 1, 2, \ldots$) are relatively prime to $\Theta_2$. Then by the previous proof

$$\sin\langle K_{B_n\Theta_2}, K\rangle \geq \sin\langle K_{B_n}, K\rangle \cdot \sin\langle K_2, K\rangle.$$

It remains just to apply the Lemmata on the Semi-Continuity and the Continuity of the Angle:

$$\sin\langle K_{12}, K\rangle \geq \overline{\lim_n} \sin\langle K_{B_n\Theta_2}, K\rangle \geq \overline{\lim_n} \sin\langle K_{B_n}, K\rangle \cdot \sin\langle K_2, K\rangle$$

$$= \sin\langle K_1, K\rangle \cdot \sin\langle K_2, K\rangle. \square$$

**Lemma on the Continuity of $F$.** 1. *If $\Theta$ and $\Theta_n$ are inner functions ($n \geq 1$) and if $\Theta_n$ divides $\Theta_{n+1}$, $n \geq 1$, then $\lim_n F(\Theta, \Theta_n) = F(\Theta, \Theta')$ where $\Theta' = \lim_n \Theta_n$.*

2. *Set*

$$s_\zeta = \exp\left\{-a\frac{\zeta + z}{\zeta - z}\right\}, \quad s_\xi = \exp\left\{-b\frac{\xi + z}{\xi - z}\right\},$$

$$B_\tau = (s_\zeta - e^{-\tau}\mathbb{1})(\mathbb{1} - e^{-\tau}s_\zeta)^{-1}, \quad \tau > 0.$$

*Then $\lim_{\tau\to\infty} F(B_\tau, s_\xi) = F(s_\zeta, s_\xi)$.*

*Proof*. 1. The representing measures $\mu_n = \mu^{\Theta_n}$ form in increasing sequence: $\mu_{n+1} \geq \mu_n$, $n \geq 1$. Therefore one can pass to the limit under the sign of integration in the integral representing $-\log F(\Theta, \Theta_n)$. As $\mu^{\Theta'}e = \lim_n \mu_n e$ for every Borel set $e$, $e \subset \operatorname{clos}\mathbb{D}$, the desired result follows.

2. It is not hard to see that $B_\tau$ is a Blaschke product with simple zeros at the points $\lambda_n = \dfrac{\tau - a + 2\pi i n}{\tau + a + 2\pi i n}$ so that

$$1 - |\lambda_n|^2 = \frac{4a\tau}{(\tau + a)^2 + 4\pi^2 n^2},$$

$$|\xi - \lambda_n|^2 = 4\frac{\tau^2 \sin^2\beta + (a\cos\beta - 2\pi n \sin\beta)^2}{(\tau + a)^2 + 4\pi^2 n^2},$$

with $2\beta = \arg\zeta - \arg\xi$.

We use the identity

$$\sum_{n=-\infty}^{\infty} \frac{1}{(2\pi n + x)^2 + y^2} = \frac{\operatorname{sh} y}{2y(\operatorname{ch} y - \cos x)}$$

for the computation of the quantity $F(B_\tau, s_\xi)$:

$$F(B_\tau, s_\xi) = \prod_{n=-\infty}^{\infty} |s_\xi(\lambda_n)| = \exp\left\{-b\sum_{n=-\infty}^{\infty} \frac{1 - |\lambda_n|^2}{|\xi - \lambda_n|^2}\right\}$$

$$= \exp\left\{-ab\tau \sum_{n=-\infty}^{\infty} \frac{1}{(2\pi n \sin\beta - s\cos\beta)^2 + \tau^2\sin^2\beta}\right\}$$

$$= \exp\left\{-\frac{ab\tau}{\sin^2\beta}\frac{\operatorname{sh}\tau}{2\tau[\operatorname{ch}\tau - \cos(a\cot\beta)]}\right\}.$$

From this relation follows that

$$\lim_{\tau\to\infty} F(B_\tau, s_\xi) = \exp\left\{-\frac{ab}{2\sin^2\beta}\right\} = \exp\left\{-\frac{2ab}{|\zeta - \xi|^2}\right\} = F(s_\zeta, s_\xi). \square$$

*Proof of the Theorem on Angles.*

*Step 1. If $\Theta$ is an inner function and $B$ is a Blaschke product then* $\sin\langle K_\Theta, K_B\rangle \geq F(\Theta, B)$.

In fact, for an individual Blaschke factor $b_\lambda$ the angle $\langle K_\Theta, K_\lambda\rangle$, $K_\lambda = K_{b_\lambda}$, is easily found:

$$\sin\langle K_\Theta, K_\lambda\rangle = \inf_{x\in K_\lambda} \frac{\|(I - P_\Theta)x\|_2}{\|x\|_2} = \frac{\|P_+\bar{\Theta}(\mathbb{1} - \bar{\lambda}z)^{-1}\|_2}{\|(\mathbb{1} - \bar{\lambda}z)^{-1}\|_2}$$

$$= \frac{\|\Theta(\lambda)(\mathbb{1} - \bar{\lambda}z)^{-1}\|_2}{\|(\mathbb{1} - \bar{\lambda}z)^{-1}\|_2} = |\Theta(\lambda)|.$$

Put $B_r = \prod_{|\lambda|<r} \left(\frac{|\lambda|}{\lambda}\frac{\lambda\mathbb{1} - z}{\mathbb{1} - \bar{\lambda}z}\right)^{k(\lambda)}$, $K_r = H^2 \ominus \mathbf{B}_r H^2$. $B_r$ is a finite Blaschke product. Therefore by the Lemma on the Submultiplicativity of the Angle

$$\sin\langle K_\Theta, K_r\rangle \geq \prod_{|\lambda|<r} (\sin\langle K_\Theta, K_r\rangle)^{k(\lambda)}$$

$$= \prod_{|\lambda|<r} |\Theta(\lambda)|^{k(\lambda)} = F(\Theta, B_r).$$

We can now use the monotonicity of $F$ $(F(\Theta, B_r) \geq F(\Theta, B))$ along with the Lemma on the Semi-Continuity of the Angle (it is clear that $K_B = \underline{\lim}_{r\to 1} K_r$, cf. Lecture II):

$$\sin\langle K_\Theta, K_B\rangle \geq \overline{\lim_{r\to 1}} \sin\langle K_\Theta, K_r\rangle \geq F(\Theta, B). \ \square$$

Here it is worthwhile to point out that the indirect path for the proof of the estimate (1) chosen is connected with the impossibility of a "weak" passage to the limit in that inequality: the function $\sin\langle\cdot,\cdot\rangle$ is upper semi-continuous (by the corresponding lemma) and the function $F(\cdot,\cdot)$ too has this property $(F(\Theta', \Theta) \geq \overline{\lim}_n F(\Theta_n, \Theta)$ if the inner functions $\Theta_n$ converge weakly to the inner function $\Theta')$, but for the passage to the limit in view one needs semi-continuity of the opposite sign (which, as is easily seen, cannot be maintained). Therefore using the explicit calculations of the second statement of the Lemma on the Continuity of $F$ we reduce the remaining part of the proof to the case of singular functions with purely discrete representing measures.

Let $\Theta_i = S_i B_i$ where $S_i$ is a singular function and $B_i$ a Blaschke product, $i = 1$, 2. We have just proved that

$$\sin\langle K_{B_1}, K_{\Theta_2}\rangle \geq F(B_1, \Theta_2), \quad \sin\langle K_{S_1}, K_{B_2}\rangle \geq F(S_1, B_1).$$

Therefore in order to complete the proof of the theorem it suffices, using the Lemma on the Submultiplicativity of the Angle and the Lemma on the Multiplicativity of $F$, to show that for any pair of singular inner functions holds the estimate

$$\sin\langle K_{S_1}, K_{S_2}\rangle \geq F(S_1, S_2). \tag{2}$$

*Step 2. $\mu^{S_1}$ is a purely discrete measure.*

If $S_1 = s_\zeta$, $S_2 = s_\xi$ then the desired estimate holds true, because by the Lemma on the Continuity of $F: \lim_{\tau\to\infty} F(B_\tau, S_2) = F(S_1, S_2)$, and by the Lemma on the Continuity of the Angle: $\lim_{\tau\to\infty} \sin\langle K_{B_\tau}, K_{S_2}\rangle = \sin\langle K_{S_1}, K_{S_2}\rangle$, since $B_\tau$ converges uniformly to $S_1$.

Upon applying the Lemma on the Submultiplicativity of the Angle we get the estimate (2) with $S_1 = s_\zeta$ and $S_2$ a finite product of factors of the type $s_\xi$.

Let now $S_1 = s_\zeta$, $S_2 = S(\mu, \cdot)$, $\mu$ being an arbitrary singular measure on the circumference $\mathbb{T}$. If $F(S_1, S_2) = 0$ then everything is trivial. If $F(S_1, S_2) > 0$ then the integral $\int_{\mathbb{T}} \frac{d\mu(\xi)}{|\zeta - \xi|^2}$ must converge and therefore for any $n$ we can find a number $\delta_n$, $\delta_n > 0$, such that $\int_{|\zeta-\xi|<\delta_n} \frac{d\mu(\xi)}{|\zeta - \xi|^2} \le \frac{1}{n}$. We can likewise find a discrete measure $\mu_n$ with finite support approximating $\mu$ such that $S(\mu_n, \cdot)$ converges uniformly to $S(\mu, \cdot)$ on any compact subset of $\mathbb{D}$ and such that there holds the inequality

$$\left| \int_{|\zeta-\xi|\ge\delta_n} \frac{d\mu(\xi)}{|\zeta - \xi|^2} - \int_{\mathbb{T}} \frac{d\mu_n(\xi)}{|\zeta - \xi|^2} \right| \le \frac{1}{n}.$$

Then on the one hand $\lim\limits_{n\to\infty} \int_{\mathbb{T}} \frac{d\mu_n(\xi)}{|\zeta - \xi|^2} = \int_{\mathbb{T}} \frac{d\mu(\xi)}{|\zeta - \xi|^2}$, i.e. $\lim_{n\to\infty} F(S_1, S(\mu_n, \cdot)) = F(S_1, S_2)$, and on the other hand $S(\mu_n, \cdot) \xrightarrow{\text{weak}} S(\mu, \cdot)$. Therefore by the Lemma on the Semi-Continuity of the Angle inequality (2) holds also for $S_1 = s_\zeta$ and $S_2$ an arbitrary singular inner function. From the submultiplicativity of the angle follows now that it remains in force if $S_1$ is generated by a discrete measure with a finite set of masses.

Finally, let $S_1 = S(\nu, \cdot)$, $S_2 = S(\mu, \cdot)$ with $\mu$ arbitrary singular and $\nu$ purely discrete. We approximate $\nu$ with a sequence $\nu_n$, gotten from $\nu$ upon omitting all masses less than $\frac{1}{n}$. As $F(S(\nu_n, \cdot), S_2) \ge F(S_1, S_2)$ and $S(\nu_n, \cdot) \xrightarrow{\text{weak}} S(\nu, \cdot)$, by the Lemma on the Semi-Continuity of the Angle, (2) must be true also in this case. □

*Step 3. The measures $\mu^{S_i}$ $(i = 1, 2)$ are continuous.*

Let $S_1 = S(\nu, \cdot)$, $S_2 = S(\mu, \cdot)$, $\mu$ and $\nu$ being continuous singular measures on the unit circle.

Put $g(\zeta) = \int_{\mathbb{T}} \frac{d\mu(\xi)}{|\zeta - \xi|^2}$, $\zeta \in \mathbb{T}$. If $g \notin L^1(\nu)$ then $F(S_1, S_2) = 0$ and there is nothing to prove. If $g \in L^1(\nu)$ then it is not hard to find a sequence $\{\nu_n\}_{n\ge1}$ of point measures with finite supports such that $\lim_n \nu_n = \nu$ (in the sense of weak convergence of measures) and $\lim_n \int_{\mathbb{T}} g\, d\nu_n = \int_{\mathbb{T}} g\, d\nu$. Then $F(S(\nu_n, \cdot), S_2) \to F(S_1, S_2)$. As in all the preceding cases, the functions $S(\nu_n, \cdot)$ will converge uniformly to $S_1$ on compact subsets of $\mathbb{D}$ so that $\underline{\lim}_n K_{S(\nu_n, \cdot)} = K_{S_1}$ (Lecture II) and by the Lemma on the Semi-Continuity of the Angle, inequality (2) holds for the inner functions $S_1$ and $S_2$, i.e. for arbitrary singular inner functions with continuous representing measures. □

Using once more the multiplicativity of $F$ and the Lemma on the Submultiplicativity of the Angle we conclude that (2) hold for an arbitrary pair of singular inner functions. □

Using the Theorems on the Angle I and II we now get bilateral estimates in two important special cases.

**Corollary 1.** *Let* $\Theta_i = b_i^{k_i}$, $b_i = \dfrac{\lambda_i \mathbb{1} - z}{\mathbb{1} - \bar{\lambda} z}$, $i = 1, 2$, *and set* $\varepsilon = \dfrac{1}{4} \cdot \sin\langle K_1, K_2 \rangle$.

*Then*

$$\exp\left(-\frac{\log^2 \varepsilon}{2\varepsilon}\right) \leq \left|\frac{\lambda_1 - \lambda_2}{1 - \bar{\lambda}_1 \lambda_2}\right|^{k_1 k_2} \leq 4\varepsilon.$$

Indeed, the rightmost inequality is just the estimate of the Theorem on Angles, II. For the proof of the left inequality let us connect the points $\lambda_1$ and $\lambda_2$ by a (geodesic) line whose parametric representation has the form

$$\lambda(t) = \frac{\lambda_1(1 - \bar{\lambda}_1 \lambda_2) + t(\lambda_2 - \lambda_1)}{1 - \bar{\lambda}_1 \lambda_2 + t\bar{\lambda}_1(\lambda_2 - \lambda_1)}, \quad 0 \leq t \leq 1,$$

$$\lambda(0) = \lambda_1, \quad \lambda(1) = \lambda_2,$$

$$b_1(\lambda(t)) = \frac{\lambda_1 - \lambda_2}{1 - \bar{\lambda}_1 \lambda_2} t, \quad b_2(\lambda(t)) = \frac{(\lambda_2 - \lambda_1)(1 - \bar{\lambda}_1 \lambda_2)(1 - t)}{|1 - \bar{\lambda}_1 \lambda_2|^2 - t|\lambda_1 - \lambda_2|^2}.$$

When the parameter $t$ runs from 0 to 1, the quantity $|b_1(\lambda(t))|$ increases continuously from 0 to $|b_1(\lambda_2)|$ while $|b_2(\lambda(t))|$ decreases from $|b_1(\lambda_2)|$ to 0. Thus we can find a value of the parameter, $\tau$, say, with $0 < \tau < 1$, such that $|b_1(\lambda(\tau))|^{k_1} = |b_2(\lambda(\tau))|^{k_2} \stackrel{\text{def}}{=} \delta$. Then

$$\delta = \frac{1}{2}\{|b_1(\lambda(\tau))|^{k_1} + |b_2(\lambda(\tau))|^{k_2}\}$$

$$\geq \frac{1}{2} \inf_{\mathbb{D}} \{|b_1|^{k_1} + |b_2|^{k_2}\} \geq \frac{1}{4} \sin\langle K_1, K_2 \rangle = \varepsilon,$$

that is, using the Theorem on Angles (Lecture IX) we get the inequality $\delta \geq \varepsilon$.

As the point $\lambda(t)$ for each $t$ lies on the geodesic connecting $\lambda_1$ and $\lambda_2$, the triangle inequality reduces to the identity $|b_1(\lambda_2)| = \dfrac{|b_1(\lambda(t))| + |b_2(\lambda(t))|}{1 + |b_1(\lambda(t))|\,|b_2(\lambda(t))|}$ (this can also easily be found by a direct computation) and, in particular, for $t = \tau$ we have $|b_1(\lambda_2)| = \dfrac{\delta^{1/k_1} + \delta^{1/k_2}}{1 + \delta^{1/k_1}\delta^{1/k_2}}$. Therefore

$$\log \frac{1}{|b_1(\lambda_2)|^{k_1 k_2}} = k_1 k_2 \log \frac{1 + \delta^{1/k_1 + 1/k_2}}{\delta^{1/k_1} + \delta^{1/k_2}}$$

$$= k_1 k_2 \log\left[1 + \frac{(1 - \delta^{1/k_1})(1 - \delta^{1/k_2})}{\delta^{1/k_1} + \delta^{1/k_2}}\right]$$

$$\leq k_1 k_2 \frac{(1 - \delta^{1/k_1})(1 - \delta^{1/k_2})}{\delta^{1/k_1} + \delta^{1/k_2}} \leq \frac{\log^2 \delta}{2\delta},$$

because $\delta^{1/k} \geq \delta$ and $k(1 - \delta^{1/k}) \leq \log 1/\delta$.

As the function $\delta \to \frac{1}{\delta}\log^2\delta$ is steadily decreasing on the segment $(0, 1)$, then $\log\frac{1}{|b_1(\lambda_2)|^{k_1k_2}} \le \log^2\varepsilon/2\varepsilon$, i.e.

$$|b_1(\lambda_2)|^{k_1k_2} = \left|\frac{\lambda_1-\lambda_2}{1-\bar{\lambda}_1\lambda_2}\right|^{k_1k_2} \ge \exp\left(-\frac{\log^2\varepsilon}{2\varepsilon}\right). \square$$

Let us give the analogous corollary for singular inner functions.

**Corollary 2.** *Let* $\Theta_i = s_{\zeta_i}$, $s_{\zeta_i} = \exp\left\{-a_i\frac{\zeta_i+z}{\zeta_i-z}\right\}$, $i = 1, 2$. *Then*

$$\exp\left\{-\frac{2a_1a_2}{|\zeta_1-\zeta_2|^2}\right\} \le \sin\langle K_1, K_2\rangle \le 4\exp\left\{-\frac{2\sqrt{a_1a_2}}{|\zeta_1-\zeta_2|}\right\}.$$

In fact, the left inequality is the estimate from the Theorem on Angles, II. For the proof of the right inequality we locate, as in the previous corollary, a point $\zeta_0$ on the geodesic connecting $\zeta_1$ and $\zeta_2$ such that $|\Theta_1(\zeta_0)| = |\Theta_2(\zeta_0)|$. In this case it is not necessary to enter into the cumbersome calculations of the previous proof, because we can write down explicitely the desired point:

$$\zeta_0 = \frac{a_1\zeta_2 + a_2\zeta_1}{a_1 + a_2 + \sqrt{a_1a_2}|\zeta_1-\zeta_2|},$$

$$|s_{\zeta_1}(\zeta_0)| = |s_{\zeta_2}(\zeta_0)| = \exp\left\{-\frac{2\sqrt{a_1a_2}}{|\zeta_1-\zeta_2|}\right\}.$$

Therefore by the Theorem on Angles (Lecture IX)

$$\begin{aligned}\sin\langle K_1, K_2\rangle &\le 2\inf_{\mathbb{D}}\{|\Theta_1| + |\Theta_2|\}\\ &\le 2\{|\Theta_1(\zeta_0)| + |\Theta_2(\zeta_0)|\}\\ &= 4\exp\left\{-\frac{2\sqrt{a_1a_2}}{|\zeta_1-\zeta_2|}\right\}. \square\end{aligned}$$

**2. Bases of Rootspaces.** Now we will study in greater detail the conditions for the family of rootspaces of the operator $T_B$, whose characteristic function is a Blaschke product $B$, to be a basis. During the computations we shall find necessary and/or sufficient conditions for $B \in (CV)$ which are suitable for the applications. The multiplicity of the zeros of $B$ is then only restricted by the (necessary) Blaschke condition.

Vasyunin's Theorem (Lecture IX) gives the following criterion for the family of subspaces $K_\lambda$, $K_\lambda = H^2 \ominus \mathbf{b}_\lambda^{k(\lambda)}H^2$, to be a basis: the exists a number $\delta$, $\delta > 0$, such that

$$|B(k,\zeta)| \ge \delta\inf_\lambda|b_\lambda(\lambda)|^{k(\lambda)}, \quad \zeta \in \mathbb{D}.$$

In this Section we display a more traditional local condition (of the type of

Carleson's condition), which uses only the values of the partial products at the zeros of the characteristic function.

**Corollary 3.** *For the family of subspaces* $\{K_\lambda\}$ *to be a basis in the space* $K = H^2 \ominus \mathbf{B}H^2$, $B = \prod_\lambda b^{k(\lambda)}$, *it is*

1. *necessary that* $\inf_\lambda |B_\lambda(\lambda)| > 0$ *and* $\inf_{\lambda,\mu;\lambda\neq\mu} |b_\lambda(\mu)|^{k(\lambda)k(\mu)} > 0$.
2. *sufficient that* $\inf_\lambda |B_\lambda(\lambda)|^{k(\lambda)} > 0$.

For the proof we use the notation $K'_\lambda = \bigvee(K_\mu : \mu \neq \lambda)$; we then have

$$\sin\langle K_\lambda, K'_\lambda\rangle = \inf_{x\in K_\lambda} \frac{\|P_+ \bar{B}_\lambda x\|_2}{\|x\|_2} \leq \frac{\|P_+ \bar{B}_\lambda(\mathbb{1} - \bar{\lambda}z)^{-1}\|_2}{\|(\mathbb{1} - \bar{\lambda}z)^{-1}\|_2} = |B_\lambda(\lambda)|.$$

Therefore, if the family $\{K_\lambda\}$ is a basis, then $0 < \inf_\lambda \sin\langle K_\lambda, K'_\lambda\rangle \leq \inf_\lambda |B_\lambda(\lambda)|$ and a fortiori $\inf_{\lambda,\mu,\lambda\neq\mu} \sin\langle K_\lambda, K_\mu\rangle > 0$, that is, in view of Corollary 1 $\inf_{\lambda,\mu,\lambda\neq\mu} |b_\lambda(\mu)|^{k(\lambda)k(\mu)} > 0$. This establishes the necessity in Condition 1.

Next let us assume that Condition 2 is fulfilled. From the Theorem on Angles, II we get the estimate $\inf_\lambda \sin\langle K_\lambda, K'_\lambda\rangle > 0$, because $K'_\lambda = H^2 \ominus \mathbf{B}_\lambda H^2$. Thus $\{K_\lambda\}$ is uniformly minimal and thus by Vasyunin's Theorem (Lecture IX) a basis. □

*Remark.* Now we obtain once more that if $\sup_\lambda k(\lambda) < \infty$ then $\{K_\lambda\}$ is a Riesz basis iff the Carleson condition is fulfilled: $\inf_\lambda |B_\lambda(\lambda)| > 0$.

Let us remark that, generally speaking, this necessary (sufficient) condition is not sufficient (necessary) (cf. Sect. 5 infra) and in the general case there is no hope of finding basis criteria in terms of only the spectrum $\sigma(T)$ and the multiplicity function $k$. It is however a mollifying circumstance that the "margin" between the Conditions 1–2 of Corollary 3 is rather unessential, as will be seen on the counter-examples in Sect. 5. Moreover, in the interesting special case when the zeros of $B$ satisfy the Stolz requirement (Lecture VII), these conditions simply coincide.

**3. Stolzian Spectrum.** Let us start with a radial distribution of the zeros. A separate study of this case is justified by that it is so much more transparent compared with the general Stolzian spectrum.

**Corollary 4.** *Let* $B$ *be a Blaschke product with* $B^{-1}(0) = \{\lambda_n\}$, $\lambda_{n+1} > \lambda_n > 0$, $k_n = k(\lambda_n)$, $n \geq 1$. *The following statements are equivalent*:

1. $\{K_n\}_{n\geq 1}$ *is a Riesz basis in* $K_B$; $K_n \stackrel{\text{def}}{=} H^2 \ominus \mathbf{b}_{\lambda_n}^{k_n} H^2$;
2. $\inf_n |B_n(\lambda_n)|^{k_n} > 0$;
3. $\inf_n |b_n(\lambda_{n+1})|^{k_n k_{n+1}} > 0$;
4. *there exist constants* $c_1$ *and* $c_2$, $0 < c_1 < 1$, $c_2 > 0$, *such that*

$$\frac{1 - \lambda_{n+1}}{1 - \lambda_n} \leq \min\left\{c_1, \frac{c_2}{k_n k_{n+1}}\right\}.$$

Beginning the proof let us remark that the implications $2 \Rightarrow 1 \Rightarrow 3$ follow from Corollary 3. Let us therefore prove the implications $3 \Rightarrow 4 \Rightarrow 2$.

$3 \Rightarrow 4$. Set $\inf_n |b_n(\lambda_{n+1})|^{k_n k_{n+1}} = \delta$, $c_1 = 1 - \delta$, $c_2 = \log 1/\delta$, $\beta_n = \dfrac{1 - \lambda_{n+1}}{1 - \lambda_n}$.
Then $1 - \beta_n = 1 - \dfrac{1 - \lambda_{n+1}}{1 - \lambda_n} = \dfrac{\lambda_{n+1} - \lambda_n}{1 - \lambda_n} \geq \dfrac{\lambda_{n+1} - \lambda_n}{1 - \lambda_{n+1}\lambda_n} \geq \delta^{1/k_n k_{n+1}} \geq \delta$, i.e. on the one hand $\beta_n \leq 1 - \delta = c_1$ and on the other hand

$$\beta_n \leq 1 - \delta^{1/k_n k_{n+1}} \leq \frac{1}{k_n k_{n+1}} \log \frac{1}{\delta} = \frac{c_2}{k_n k_{n+1}}.$$

$4 \Rightarrow 2$. Set $\alpha_{ns} = \prod_{i=0}^{s-1} \beta_{n+i}$. Then $1 - \lambda_{n+s} = \alpha_{ns}(1 - \lambda_n)$. Let us break up the product $|B_n(\lambda_n)|^{k_n}$ into two factors:

$$B' = \prod_{i=1}^{n-1} |b_i(\lambda_n)|^{k_i k_n}, \quad B'' = \prod_{i=n+1}^{\infty} |b_i(\lambda_n)|^{k_i k_n}.$$

As

$$\begin{aligned} \frac{1 - \lambda_{n+s}\lambda_n}{\lambda_{n+s} - \lambda_n} &= \frac{(1 - \lambda_n) + \lambda_n(1 - \lambda_{n+s})}{(1 - \lambda_n) - (1 - \lambda_{n+s})} \\ &= \frac{1 + \lambda_n \alpha_{ns}}{1 - \alpha_{ns}} \leq \frac{1 + \alpha_{ns}}{1 - \alpha_{ns}} \\ &= 1 + \frac{2\alpha_{ns}}{1 - \alpha_{ns}} \leq 1 + \frac{2\alpha_{ns}}{1 - c_1}, \end{aligned}$$

we get for $B''$ the estimate

$$\begin{aligned} \log 1/B'' &= \sum_{s=1}^{\infty} \log \left( \frac{1 - \lambda_{n+s}\lambda_n}{\lambda_{n+s} - \lambda_n} \right)^{k_{n+s} k_s} \\ &\leq \sum_{s=1}^{\infty} k_n k_{n+s} \log \left( 1 + \frac{2\alpha_{ns}}{1 - c_1} \right) \\ &\leq \frac{2}{1 - c_1} \sum_{s=1}^{\infty} k_n k_{n+s} \alpha_{ns}. \end{aligned}$$

For $s = 1$ holds $\alpha_{n1} = \beta_n \leq c_2/k_n k_{n+1}$, i.e. $k_n k_{n+1} \alpha_{n1} \leq c_2$. For $s \geq 2$

$$\alpha_{ns} = \beta_n \beta_{n+s-1} \prod_{i=n+1}^{n+s-2} \beta_i \leq \frac{c_2}{k_n k_{n+1}} \frac{c_2}{k_{n+s-1} k_{n+s}} c_1^{s-2} \leq \frac{c_2^2 c_1^{s-2}}{k_n k_{n+s}},$$

i.e. $k_n k_{n+s} \alpha_{ns} \leq c_2^2 c_1^{s-2}$. Thus

$$\sum_{s=1}^{\infty} k_n k_{n+s} \alpha_{ns} \leq c_2 + c_2^2 \sum_{s=2}^{\infty} c_1^{s-2} = c_2 + \frac{c_2^2}{1 - c_1} \equiv c_3.$$

and $\log 1/B'' \leq \dfrac{2c_3}{1 - c_1} \equiv c_4$. In exactly the same way we get the estimate $\log 1/B' \leq c_4$. Therefore $|B_n(\lambda_n)|^{k_n} = B'B'' \geq \exp(-2c_4)$. □

There is a simple geometrical principle why results similar to the ones obtained for points on a ray also hold for points in an angle. Namely that the angle in a neighborhood of its vertex constitutes a strip of bounded width in the hyperbolic metric. The "sparseness" condition $\inf_{\lambda,\mu,\lambda\neq\mu}\left|\frac{\lambda-\mu}{1-\bar{\lambda}\mu}\right| > 0$ means that in this metric the points are uniformly separated so that such a set, thus contained in a strip, can contain only a finite number of radial sequences. We can therefore hope for a not more than multiple "deterioration" of the lower estimate of the complete Blaschke product, compared with what was done for points on a ray.

**Corollary 5.** *Let $B$ be a Blaschke product with $B^{-1}(0) = \{\lambda_n\}$, $|\lambda_{n+1}| \geq |\lambda_n|$, $k_n = k(\lambda_n)$, $n \geq 1$, and assume that the Stolz condition (S) is fulfilled: there exist a finite subset $e$ of $\mathbb{T}$ such that*

$$\sup_n (1-|\lambda_n|)^{-1}\operatorname{dist}(\lambda_n, e) < \infty.$$

*The following statements are equivalent:*

1. *The family $\{K_n\}_{n\geq 1}$ is a Riesz basis in $K_B$; $K_n = H^2 \ominus \mathbf{b}_{\lambda_n}^{k_n} H^2$;*
2. $\inf_n |B_n(\lambda_n)|^{k_n} > 0$;
3. $\inf_{i,j,i\neq j} |b_i(\lambda_j)|^{k_i k_j} > 0$;
4. *If $\alpha_{ns} \overset{\text{def}}{=} \dfrac{1-|\lambda_{n+s}|}{1-|\lambda_n|}$ then $\sup_{n,s}(k_n k_{n+s}\alpha_{ns}) < \infty$, and there exists an $N$ such that* $\sup_n \alpha_{nN} < 1$, $\inf\{|b_i(\lambda_j)|^{k_i k_j} : 0 < |i-j| < N\} = \delta > 0$.

In proving this corollary we can use the same scheme as in the proof of Corollary 4, so we shall dwell only on those changes that are necessary for the proof in question. Without loss of generality we can assume that the zeros of $B$ accumulate at the point $\zeta = 1$, because the general case differs from this one only by the presence of a larger number of points of accumulation for the spectrum $\{\lambda_n\}$. If we divide the spectrum into a finite union of sets, $\{\lambda_n\} = \bigcup_{\zeta\in e}\sigma_\zeta$, each of which having a unique point of accumulation ($\mathbb{T}\cap\operatorname{clos}\sigma_\zeta \subset \{\zeta\}$, $\zeta\in e$), and set $B^\zeta = \prod_{\lambda_n\in\sigma_\zeta} b_{\lambda_n}^{k_n}$, $\zeta\in e$, then $B = \prod_{\zeta\in e} B^\zeta$ and the spaces $K_{B^\zeta}$ and $\bigvee(K_{B^s} : s\in e\backslash\{\zeta\})$ meet at a non-zero angle ($\forall\zeta\in e$). Therefore the finite family of subspaces $\{K_{B^\zeta} : \zeta\in e\}$ forms a basis for $K_B$, so henceforth one can consider only the case of a single point of accumulation. Set $\zeta = 1$ and

$$c \overset{\text{def}}{=} \sup_n \frac{|1-\lambda_n|}{1-|\lambda_n|} < \infty. \tag{3}$$

$3 \Rightarrow 4$. We know that for sparse sequences (that is, sequences such that $\inf_{i\neq j}|b_i(\lambda_j)| > 0$) satisfying a Stolz condition the second of the conditions 4 holds true (Lecture VII).

For the proof of the first condition we estimate the expression $1 - |b_i(\lambda_j)|^2$, $i = j + s$. On the one hand

$$\begin{aligned} 1-|b_i(\lambda_j)|^2 &= \frac{(1-|\lambda_i|^2)(1-|\lambda_j|^2)}{|1-\bar{\lambda}_i\lambda_j|^2} \\ &\geq \frac{(1-|\lambda_i|)(1-|\lambda_j|)}{[|1-\lambda_i|+|1-\lambda_j|]^2} \end{aligned}$$

$$\geq \frac{1}{c^2} \frac{(1-|\lambda_i|)(1-|\lambda_j|)}{[(1-|\lambda_i|)+(1-|\lambda_j|)]^2}$$

$$= \frac{\alpha_{js}}{c^2(\alpha_{js}+1)^2} \geq \frac{\alpha_{js}}{4c^2}.$$

Here we used the Stolz condition (3) and an enumeration such that $|\lambda_{n+1}| \geq |\lambda_n|$ and $\alpha_{ns} \leq 1$.

On the other hand if $\inf |b_i(\lambda_j)|^{k_i k_j} \stackrel{\text{def}}{=} \delta$ then

$$1 - |b_i(\lambda_j)|^2 \leq 1 - \delta^{2/k_i k_j} \leq \frac{2}{k_i k_j} \log \frac{1}{\delta}.$$

Consequently $k_j k_{j+s} \alpha_{js} \leq 8c^2 \log 1/\delta$.

$4 \Rightarrow 2$. Let $N$ be a natural number such that $c_1 \stackrel{\text{def}}{=} \sup_n \alpha_{nN} < 1$ and take $s \geq N$. Set $c_2 \stackrel{\text{def}}{=} \sup_{i,j} k_i k_{i+j} \alpha_{ij}$. Then

$$\left| \frac{1-\lambda_{n+s}\lambda_n}{\lambda_{n+s}-\lambda_n} \right|^2 - 1 = \frac{(1-|\lambda_n|^2)(1-|\lambda_{n+s}|^2)}{|\lambda_{n+s}-\lambda_n|^2}$$

$$\leq \frac{4(1-|\lambda_n|)(1-|\lambda_{n+s}|)}{[(1-|\lambda_n|)-(1-|\lambda_{n+s}|)]^2}$$

$$= \frac{4\alpha_{ns}}{(1-\alpha_{ns})^2},$$

i.e.

$$\left| \frac{1-\lambda_{n+s}\lambda_n}{\lambda_{n+s}-\lambda_n} \right| \leq \left[ 1 + \frac{4\alpha_{ns}}{(1-\alpha_{ns})^2} \right]^{1/2} = \frac{1+\alpha_{ns}}{1-\alpha_{ns}} \leq 1 + \frac{2\alpha_{ns}}{1-c_1},$$

since $\alpha_{ns} \leq \alpha_{nN} \leq c_1$.

We split up the product $|B_n(\lambda_n)|^{k_n}$ into three factors

$$B' = \prod_{i=1}^{n-N} |b_i(\lambda_n)|^{k_i k_n}, \quad B'' = \prod_{i=n+N}^{\infty} |b_i(\lambda_n)|^{k_i k_n},$$

$$B''' = |B_n(\lambda_n)|^{k_n} (B' B'')^{-1};$$

then $B'''$ will contain $2(N-1)$ elementary factors so that $B''' \geq \delta^{2(N-1)}$. For $B''$ we get, as in the proof of Corollary 4, by the inequality above the estimate

$$\log \frac{1}{B''} \leq \frac{2}{1-c_1} \sum_{s=N}^{\infty} k_n k_{n+s} \alpha_{ns}.$$

We set $\beta_n = \dfrac{1-|\lambda_{n+1}|}{1-|\lambda_n|}$, $\beta_n = \alpha_{n1} \leq \dfrac{c_2}{k_n k_{n+1}}$. Following the proof of Corollary 4 we estimate $\alpha_{ns}$ as a product of factors $\beta_n$. However, now instead of the inequality $\beta_n \leq c_1$ we have at our disposal only the estimate $\alpha_{nN} \leq c_1$. Therefore we collect the $\beta_n$ in groups of $N$ factors.

If $s = jN + i + 2, j \geq 0, 0 \leq i \leq N-1$ then

$$\alpha_{n-1,s} = \beta_{n-1}\alpha_{nN}\alpha_{n+N,N}\alpha_{n+2N,N}\cdots\alpha_{n+(j-1)N,N}\beta_{n+jN}\beta_{n+jN+1}\cdots\beta_{n+jN+i-1}\beta_{n+jN+i}$$

$$\leq \frac{c_2}{k_{n-1}} c_1^j c_2^j \frac{c_2}{k_{n+jN+i+1}} \leq \frac{c_1^j c_2^{N+1}}{k_{n-1}k_{n-1+s}}.$$

Thus $k_n k_{n+s}\alpha_{ns} \leq c_1^j c_2^{N+1}$ whence

$$\sum_{s=N}^{\infty} k_n k_{n+s}\alpha_{ns} = k_n k_{n+N}\alpha_{nN} + k_n k_{n+N+1}\alpha_{n,N+1} + \sum_{j=1}^{\infty}\sum_{i=0}^{N-1} k_n k_{n+s}\alpha_{ns}$$

$$\leq 2c_2 + \sum_{j=1}^{\infty} N c_1^j c_2^{N+1}$$

$$= 2c_2 + \frac{N c_1 c_2^{N+1}}{1-c_2} \equiv c_3$$

and $\log 1/B'' \leq \dfrac{2c_3}{1-c_1} \equiv c_4$. Analogously $\log 1/B' \leq c_4$ and consequently $|B_n(\lambda_n)|^{k_n} \geq \delta^{2(N-1)}\exp(-2c_4)$. □

**4. Singular Discrete Spectrum.** Let us consider the question when the "quasi-root" subspaces $\{K_\zeta\}$, $K_\zeta = H^2 \ominus \mathbf{s}_\zeta H^2$, $s_\zeta = \exp\left\{-\mu(\zeta)\dfrac{\zeta+z}{\zeta-z}\right\}$, $\sum_\zeta \mu(\zeta) < \infty$, $S = \prod_\zeta s_\zeta$, form a basis in the corresponding closed linear hull, i.e., in the space $K_S$, $K_S = H^2 \ominus \mathbf{S}H^2$. We have the following analogue of Corollary 3.

**Corollary 6.** *For the family* $\{K_\zeta\}$ *to be a Riesz basis in* $K_S$ *it is*

1. *necessary that* $\sup\limits_{\zeta,\xi;\zeta\neq\xi} \dfrac{\mu(\zeta)\mu(\xi)}{|\zeta-\xi|^2} < \infty$,
2. *sufficient that* $\sup\limits_{\zeta} \sum\limits_{\xi,\xi\neq\zeta} \dfrac{\mu(\zeta)\mu(\xi)}{|\zeta-\xi|^2} < \infty$.

In fact, if $\{K_\zeta\}$ is a basis then $\inf_{\zeta,\xi;\zeta\neq\xi}\sin\langle K_\zeta, K_\xi\rangle > 0$ and by Corollary 2 $\sup\limits_{\zeta,\xi;\zeta\neq\xi} \dfrac{\mu(\zeta)\mu(\xi)}{|\zeta-\xi|^2} < \infty$.

By the Theorem on Angles, II

$$\sin\langle K_\zeta, K_\zeta'\rangle \geq \exp\left\{-2\mu(\zeta)\sum_{\xi,\xi\neq\zeta}\frac{\mu(\xi)}{|\zeta-\xi|^2}\right\},$$

and therefore $2 \Rightarrow \inf_\zeta\langle K_\zeta, K_\zeta'\rangle > 0$, i.e. $K_\zeta$ is uniformly minimal. By Vasyunin's Theorem (Lecture IX) it forms a Riesz basis.□

*Example.* The family $\{K_\zeta\}$ can be a Riesz basis also in cases when the spectrum $\sigma(S)$ coincides with the entire unit circle.

Let us call a point $\zeta$ on the circle $\mathbb{T}$ rational if $\frac{1}{2\pi}\arg\zeta = p(\zeta)/q(\zeta)$ where $p(\zeta)$ and $q(\zeta)$ are natural numbers. Let there correspond to each rational point $\zeta$ a number $\alpha(\zeta)$, $\alpha(\zeta) > 0$, in such a way that $\sum\alpha(\zeta) < \infty$. Set $\mu(\zeta) = \alpha(\zeta)q(\zeta)^{-2}$.

Then $\{K_\zeta\}$ forms a Riesz basis. In fact, as $|\zeta-\xi| \geq \frac{2}{\pi}|\arg\zeta - \arg\xi| \pmod{2\pi}$, we have

$$\sup_{\zeta}\sum_{\xi,\xi\neq\zeta}\frac{\mu(\zeta)\mu(\xi)}{|\zeta-\xi|^2}\leq\frac{1}{16}\sup_{\zeta}\sum_{\xi,\xi\neq\zeta}\frac{\alpha(\zeta)}{q(\zeta)^2}\frac{\alpha(\xi)}{q(\xi)^2}\left|\frac{p(\zeta)}{q(\zeta)}-\frac{p(\xi)}{q(\zeta)}\right|^{-2}$$

$$=\frac{1}{16}\sup_{\zeta}\sum_{\xi,\xi\neq\zeta}\frac{\alpha(\zeta)\alpha(\xi)}{|p(\zeta)q(\xi)-p(\xi)q(\zeta)|^2}$$

$$\leq\frac{1}{16}\sup_{\zeta}\sum_{\xi}\alpha(\zeta)\alpha(\xi)\leq\left[\frac{1}{4}\sum_{\zeta}\alpha(\zeta)\right]^2<\infty.$$

It remains only to make an appeal to Corollary 6. □

**5. Counterexamples.** In this Section we construct counter-examples showing that the necessary conditions in Corollaries 3 and 6 for being a basis are not sufficient and that the sufficient ones are not necessary. All proofs and other details can be found in Vasyunin's paper [2].

**Example 1.** A family of subspaces $\{K_\zeta\}$, $K_\zeta = H^2 \ominus s_\zeta H^2$, $s_\zeta = \exp\left\{-\mu(\zeta)\times\frac{\zeta+z}{\zeta-z}\right\}$, $|\zeta|=1$, which is not a basis for the subspace $K_\Theta$, $\Theta=\prod_{\mu(\zeta)>0}s_\zeta$, but for which holds

$$\sup_{\zeta\neq\xi}\mu(\zeta)\mu(\xi)|\zeta-\xi|^{-2}<\infty. \tag{4}$$

*Plan of the construction*: For given positive numbers $\varepsilon_n$, $\varepsilon_n\downarrow 0$ $(n\geq 0)$, one constructs functions $\mu_n$ on the circumference $\mathbb{T}$ with finite supports such that

$$\sup_{\zeta\neq\xi}\mu_n(\zeta)\mu_n(\xi)|\zeta-\xi|^{-2}\leq\varepsilon_0,$$

$$\inf_{\mu'+\mu''=\mu_n}\inf_{\mathbb{D}}\{|\Theta(\mu',\zeta)|+|\Theta(\mu'',\zeta)|\}\leq\varepsilon_n,\quad n\geq 1$$

(here and in the sequel the outer inf is taken with respect to all decompositions of the measure $\mu_n\overset{\text{def}}{=}\sum_{\zeta\in\mathbb{T}}\mu_n(\zeta)\delta_\zeta$ into two positive constituents which are singular with respect to each other); thereafter one picks by induction numbers $w_n$, $|w_n|<1$, such that the function $\mu$,

$$\mu(\zeta)=\sum_{n\geq 1}\frac{1-|w_n|^2}{|\tau_{w_n}\zeta-w_n|^2}\mu_n(\tau_{w_n}\zeta),\quad \zeta\in\mathbb{T}\left(\tau_w\zeta\overset{\text{def}}{=}\frac{\zeta-w}{1-w\zeta}\right)$$

fulfills condition (4), although on the other hand

$$\inf_{\mu'+\mu''=\mu}\inf_{\zeta\in\mathbb{D}}\{|\Theta(\mu',\zeta)|+|\Theta(\mu'',\zeta)|\}\leq\inf_{n\geq 1}\inf_{\mu'+\mu''=\mu_n}\inf_{\zeta\in\mathbb{D}}\{|\Theta(\mu',\zeta)|+|\Theta(\mu'',\zeta)|\}$$

$$\leq\inf_{n\geq 1}\varepsilon_n=0.$$

By the Corollary of the Theorem on Angles, I (Lecture IX) the family $\{K_\zeta\}_{\mu(\zeta)>0}$ then will not be a basis for its closed linear hull. □

**Example 2.** A family $\{K_\lambda\}$, $K_\lambda = H^2 \ominus \mathbf{b}_\lambda^{k(\lambda)}\mathbf{H}^2$, $\lambda \in \mathbb{D}$, which is not a basis in the space $K_B$, $B = \prod_\lambda b_\lambda^{k(\lambda)}$, but for which holds

$$\inf_\lambda |B_\lambda(\lambda)| > 0, \quad \inf_{\lambda \neq \mu} |b_\mu(\lambda)|^{k(\lambda)k(\mu)} > 0. \tag{5}$$

*Plan of the construction*: For given positive numbers $\delta$, $\delta < 1$, and $\varepsilon_n$, $\varepsilon_n \downarrow 0$ $(n \geq 1)$, one constructs finite Blaschke products $B(k_n, \cdot)$ such that

$$\inf_\lambda |B_\lambda(k_n, \lambda)| \geq \delta, \quad \inf_{\lambda \neq \mu} |b_\mu(\lambda)|^{k_n(\lambda)k_n(\mu)} \geq \delta,$$

$$\inf_{k'+k''=k_n} \inf_{\zeta \in \mathbb{D}} \{|B(k', \zeta)| + |B(k'', \zeta)|\} \leq \varepsilon_n;$$

thereafter one picks by induction numbers $w_n$, $|w_n| < 1$, such that the function $k(\lambda) \overset{\text{def}}{=} \sum_{n \geq 1} k_n(\tau_{w_n}\lambda)$, $\lambda \in \mathbb{D}$, has the property (5), although on the other hand

$$\inf_{k'+k''=k} \inf_{\zeta \in \mathbb{D}} \{|B(k', \zeta)| + |B(k'', \zeta)|\} \leq \inf_{n \geq 1} \inf_{k'+k''=k_n} \inf_{\zeta \in \mathbb{D}} \{|B(k', \zeta)| + |B(k'', \zeta)|\}$$

$$\leq \inf_{n \geq 1} \varepsilon_n = 0.$$

By the very same Corollary of the Theorem on Angles, I (Lecture IX) the family $\{K_\lambda\}$, corresponding to the function $k$, satisfies the desired requirements. □

**Example 3.** A family $\{K_\lambda\}$, $K_\lambda = H^2 \ominus \mathbf{b}_\lambda^{k(\lambda)}H^2$, which is a Riesz basis in $K_B$, $B = \prod_\lambda b_\lambda^{k(\lambda)}$, but for which $\inf_\lambda |B_\lambda(k, \lambda)|^{k(\lambda)} = 0$.

*Plan of the construction*: There exist a number $\delta$, $\delta > 0$, such that for arbitrary $\varepsilon_n$, $\varepsilon_n \downarrow 0$, one can find finite Blaschke products $B(k_n, \cdot)$ with the properties $\inf_\lambda |B_\lambda(k_n, \lambda)|^{k_n(\lambda)} \leq \varepsilon_n$, $|B(k_n, \zeta)| \geq \delta \inf_\lambda |b_\lambda^{k_n(\lambda)}(\zeta)|$, $\zeta \in \mathbb{D}$; thereafter by induction one constructs numbers $w_n$, $|w_n| < 1$, such that the function $k(\lambda) \overset{\text{def}}{=} \sum_{n \geq 1} k_n(\tau_{w_n}, \lambda)$, $\lambda \in \mathbb{D}$, has the property $|B(k, \zeta)| \geq \frac{1}{2}\delta \inf_\lambda |b_\lambda^{k(\lambda)}(\zeta)|$, $\zeta \in \mathbb{D}$, although on the other hand $\inf_\lambda |B_\lambda(k, \lambda)|^{k(\lambda)} \leq \inf_{n \geq 1} \inf_\lambda |B_\lambda(k_n, \lambda)|^{k_n(\lambda)} \leq \inf_{n \geq 1} \varepsilon_n = 0$. By Vasyunin's Theorem (Lecture IX) the family $\{K_\lambda\}$, corresponding to the function $k$, has the desired properties. □

**Example 4.** A family $\{K_\zeta\}$, $K_\zeta = H^2 \ominus \mathbf{s}_\zeta H^2$, $s_\zeta = \exp\left\{-\mu(\zeta)\dfrac{\zeta + z}{\zeta - z}\right\}$, which is a Riesz basis in the space $K_\Theta$, $\Theta = \prod_{\mu(\zeta)>0} s_\zeta$, but such that $\sup_{\zeta \in \mathbb{T}} \sum_{\xi, \xi \neq \zeta} \dfrac{\mu(\zeta)\mu(\xi)}{|\zeta - \xi|^2} = \infty$.

*Plan of the construction*: The desired function $\mu$ is obtained from the identities $\mu(1) = 1$, $\mu(\zeta_n) = (1 + 4\pi n^2)^{-1}$ for $\zeta_n = (2\pi ni - 1)(2\pi ni + 1)^{-1}$ and $\mu(\zeta) = 0$ for all other $\zeta$, $\zeta \in \mathbb{T}$. That $\{K_\zeta\}$ forms a Riesz basis follows again from Vasyunin's Theorem ($|\Theta(\zeta)| \geq \delta \inf_n |s_{\zeta_n}(\zeta)|$, $\zeta \in \mathbb{D}$, with a suitable $\delta$, $\delta > 0$). □

**6. Concluding Remarks.** The entire material of this Lecture (except for some inessential complements) is taken over from Vasyunin's paper [2]. Analogous results can of course be obtained also for the halfplane where they are connected with spectral (or generalized spectral) expansions for dissipative operators; for this see "Supplements and Bibliographical Notes" of Lectures VIII and IX. Let us likewise notice that an estimate from below of the angle between the rootspaces $\mathrm{Ker}(S^* - \bar{\lambda}I)^{k(\lambda)}$, $\lambda = \lambda_1, \lambda_2$, analogous to half of Corollary 1 of the Theorem on Angles, II, can be found in the early work of Markus [2] (the dissipative case) and Kacnel'son [1], and in the case of eigenspaces ($k(\lambda_1) = k(\lambda_2) = 1$) already in Mukminov [1] and Glazman [1]. In these papers it is question of general contraction (or dissipative) operators but the estimates obtained are in one direction only. These estimates follow of course from the estimates of the present Lecture (for $k(\lambda) \equiv 1$ from the results of Lecture VIII), as the projection

$$\mathscr{P} = \mathscr{P}_{K_{\lambda_1} \| K_{\lambda_2}}, K_{\lambda_i} = \mathrm{Ker}(T - \lambda_i I)^{k(\lambda_i)}, \quad i = 1, 2,$$

onto a rootspace of a contraction operator $T$ is as before given by the formula

$$\mathscr{P} = u(T)$$

where $u \in H^\infty$, $u \in b_{\lambda_2}^{k(\lambda_2)} H^\infty$, $(\mathbb{1} - u) \in b_{\lambda_1}^{k(\lambda_1)} H^\infty$, with

$$|\mathscr{P}| \leq \|u\|_\infty .$$

The papers quoted further contain conditions (derived from estimates of the angles) for the family of (root) eigenspaces to be a basis; cf. also Gohberg-Kreĭn [1]. These conditions however concede to those which can be obtained by a straight-forward application of the Sz.-Nagy and Foiaş calculus in the case of an arbitrary contraction operator.

Thus as far as our estimates from above for the spectral projections (i.e. from below for the angles between the subspaces) are concerned they can be immediately transfered with the help of the functional calculus of Sz.-Nagy and Foiaş to arbitrary contractive (dissipative) operators. In particular, if we in the propositions of this Lecture everywhere only keep this half of the estimates (conclusions) then we get theorems valid for all contractions. As in the problem of bases of eigenspaces (cf. "Supplements and Bibliographical Notes" to Lecture VII) a further improvement (i.e. suitable estimates from below for the spectral projections) is possible only for the functional model – using those auxiliary "angular parameters" it displays. There are so far (1982) no results of this type in the literature.

# Lecture XI. On the Line and in the Halfplane

All the previous Lectures were devoted to the study of the semigroup $\{\mathcal{S}^n\}_{n\geq 0}$ generated by the shift operator $\mathcal{S}$ in the space $L^2(\mathbb{T})$. Also the continuous analogue of this theory, connected with the space $L^2(\mathbb{R})$, is useful, in particular in scattering theory and in the study of stationary stochastic processes. The semigroup of shifts in $L^2(\mathbb{R})$ is interesting also from the point of view of function theory and we shall reduce the study of it to results already established for the operator $\mathcal{S}$, limiting ourselves however only to the proof of Lax's fundamental theorem on the description of invariant subspaces and to a transplantation of a small number of the results of Lecture I–VIII into the language of the $\mathbb{R}$-theory. Concerning the details of the theory of semigroups of contractions obtained by projecting the shifts onto coinvariant subspaces, the reader is referred to the books and papers listed in the "Concluding Remarks".

**1. The Invariant Subspaces** of the semigroup alluded to above will be found on the basis of the Beurling-Helson Theorem (Lecture I); the connection between the "spectral" and the "translation" representation will be implemented by the Fourier transform. In order to clarify the situation we have to invoke the important theorem of Paley-Wiener, which will be proved in this Section.

In this Lecture the letter $w$ will stand for the identity map $w(\eta) = \eta$ of the upper halfplane $\Pi$ onto itself,

$$\Pi \stackrel{\text{def}}{=} \{\eta : \eta \in \mathbb{C}, \operatorname{Im} \eta > 0\},$$

and also for the identity map $w(x) = x$ from the real line $\mathbb{R}$ onto itself. It should be clear from the context which one we have in mind. $\mathbb{R}_+$ denotes the right halfaxis $[0, \infty)$.

If $t \in \mathbb{R}$ then $S_t$ is the operator shifting the function $f$, defined on $\mathbb{R}$, $t$ units to the right: $(S_t f)(x) = f(x - t)$, $x \in \mathbb{R}$. We get information on the operators $e^{itw}$ and $S_t$ studying one special semigroup in $L^2(\mathbb{T})$.

**Sz.-Nagy and Foiaş Lemma.** *Set* $u_t = \exp\left\{t\left(\frac{z+\mathbb{1}}{z-\mathbb{1}}\right)\right\}$, $t \geq 0$ *and let* $E$ *be a subspace of* $L^2(\mathbb{T})$. *Then*

$$u_t(\mathcal{S})E \subset E,\ \forall t \geq 0 \Leftrightarrow \mathcal{S}E \subset E.$$

*Proof.* We begin the proof with an elementary remark on the operator $u_t(\mathcal{S})$: it is well-defined on $L^2(\mathbb{T})$, because the function $u_t$ is bounded, measurable and defined a.e. with respect to the spectral measure of $\mathcal{S}$. By definition we can write $u_t(\mathcal{S}) = \mathbf{u}_t$, that is, $u_t(\mathcal{S})$ is multiplication by the function $u_t$. The implication $\Leftarrow$ follows from the fact that this function is the limit of its Abel means in the sense of bounded convergence a.e.: $u_t = \lim_{r\to 1-0} u_t^r$ a.e. on $\mathbb{T}$ and $\|u_t^r\|_\infty \le \|u_t\|_\infty$, where $u_t^r(\zeta) \stackrel{\text{def}}{=} u_t(r\zeta)$, $\zeta \in \mathbb{T}$, $0 \le r < 1$. Consequently $(s)$-$\lim_{r\to 1-0} u_t^r(\mathcal{S}) = u_t(\mathcal{S})$, so that it suffices to check that $E$ is invariant for the operators $u_t^r(\mathcal{S})$, $0 \le r < 1$. But this is obvious, because $u_t^r(\mathcal{S})$ is the sum of an absolutely convergent series in powers of $\mathcal{S}$.

By an analogous reasoning, to establish the converse implication $\Rightarrow$ it suffices to remark that the "independent variable" $z$ is the bounded pointwise limit of the functions $w_t \stackrel{\text{def}}{=} [u_t - \mathbb{1}(1-t)]\cdot[u_t - \mathbb{1}(1+t)]^{-1}$ as $t \to +0$. The desired property of the family $\{w_t\}_{t\ge 0}$ follows from an elementary computation:

$$w_t(\zeta) = \frac{1 + t\alpha + o(t\alpha) - 1 + t}{1 + t\alpha + o(t\alpha) - 1 - t} = \frac{1 + \alpha + \alpha\cdot o(1)}{\alpha - 1 + \alpha\cdot o(1)} = \zeta + o(1), \text{ for } t \to +0,$$

where $|\zeta| = 1$, $\zeta \ne 1$, $\alpha = \dfrac{\zeta+1}{\zeta-1}$. $\square$

*Remark.* It is clear that if we in this lemma everywhere replace $\mathcal{S}$ by $\mathcal{S}^* = \mathcal{S}^{-1}$ and $u_t(\mathcal{S})$ by $u_t(\mathcal{S}^*) = u_{-t}(\mathcal{S})$ then the resulting statement is still true.

The formula defining the operator $u_t(\mathcal{S})$ points to the necessity of making a change of variables $z \stackrel{\omega}{\to} i\dfrac{\mathbb{1}+z}{\mathbb{1}-z}$, sending the punctured circumference $\mathbb{T}\backslash\{1\}$ onto the line $\mathbb{R}$. Under this map the Lebesgue measure $m$ goes over into the measure $\dfrac{dx}{\pi(1+x^2)}$ on $\mathbb{R}$; the inverse map is given by the formula $\omega^{-1}(x) = \dfrac{x-i}{x+i}$, $x \in \mathbb{R}$. Thus the operator $U$,

$$(Uf)(x) = \frac{1}{\sqrt{\pi}}\frac{1}{x+i} f\left(\frac{x-i}{x+i}\right), \; x \in \mathbb{R},$$

defines an isometric transformation from $L^2(\mathbb{T})$ onto $L^2(\mathbb{R})$. It is clear that the operator $u_t(\mathcal{S})$ then goes over into the multiplication operator $e^{itw}$.

Thus we end up with the semigroup $\{e^{itw}\}_{t\ge 0}$. Let us now compute the image $UH^2$. To this end let us first notice that the function $\omega$ not only carries $\mathbb{T}\backslash\{1\}$ onto $\mathbb{R}$ but also induces a conformal map of the disk $\mathbb{D}$ onto the halfplane

$$\Pi = \{\eta : \eta \in \mathbb{C}, \operatorname{Im}\eta > 0\}.$$

The conformity is preserved up to the boundary $\mathbb{T}$ so that any Stolz angle in $\mathbb{D}$ is mapped onto a subset of $\Pi$ locally represented as an angle – the intersection with a small neighborhood of the corresponding point $x$ of the real axis contains the intersection of this neighborhood with an angle $\left|\arg\frac{1}{i}(\eta - x)\right| < \alpha$ $(\alpha < \pi/2)$ and is contained in another set of this type. Therefore, if $f$ is analytic in $\mathbb{D}$, admitting angular boundary values on some subset of $\mathbb{T}$ (cf. Appendix 2), then $f \circ \omega^{-1}$ too admits (corresponding) boundary values on the $\omega$-image of the set in question. In particular, when speaking on functions of the form $Uf$, $f \in H^2$, we can consider them as functions in the halfplane $\Pi$ or identify them with their boundary values $(Uf)_{\mathbb{R}}$; clearly $Uf_{\mathbb{T}} = (Uf)_{\mathbb{R}}$ (in the obvious notation).

**Lemma on the Image of $H^2$.** *The following sets coincide.*

1. $UH^2$.
2. $\bigvee_{L^2(\mathbb{R})}\left(\frac{1}{w-\mu\mathbb{1}} : \operatorname{Im}\mu < 0\right)$.
3. *the set $H^2_\Pi$ of all functions $g$ analytic in $\Pi$ such that*

$$\sup_{y>0} \int_{\mathbb{R}} |g(x+iy)|^2\, dx < \infty. \tag{1}$$

*Proof.* The sets 1 and 2 coincide, because $U$ is isometric and

$$U(\mathbb{1} - \bar{\lambda}z)^{-1} = \frac{1}{\sqrt{\pi}} \frac{1}{1-\bar{\lambda}} \frac{1}{w - \overline{\omega(\lambda)}}, \quad \lambda \in \mathbb{D}, \tag{2}$$

and

$$\bigvee_{L^2(\mathbb{T})}((\mathbb{1} - \bar{\lambda}z)^{-1} : \lambda \in \mathbb{D}) = H^2.$$

Let $g$ be a function analytic in $\Pi$ and $f$ one in $\mathbb{D}$ such that $Uf = g$. The pre-image of the line $\operatorname{Im}\eta = y$ under the map $\omega$ is a circle $C_y$ tangent to $\mathbb{T}$ at the point 1. Computing the Jacobian of the map $\omega^{-1}$ on this line $(2i(w+i)^{-2})$ we see that

$$\int_{\mathbb{R}} |g(x+iy)|^2\, dx = \frac{1}{2\pi} \int_{C_y} |f(\xi)|^2 |d\xi|$$

(integration with respect to arc length on $C_y$). There remains to be shown that the condition

$$\sup_{y>0} \int_{C_y} |f(\xi)|^2 |d\xi| < \infty \tag{3}$$

singles out among the functions regular in $\mathbb{D}$ precisely the Hardy class $H^2$.

If $f \in H^2$ then (3) holds true, because arc length on $C_y$ defines a Carleson measure $\nu_y$ on the disk $\mathbb{D}$ and the constants $A = A(\nu_y)$ in the Lemma on Carleson Measures are uniformly bounded (trivial; cf. Lecture VII). (Of course, a reference to this lemma in such a simple situation is not very natural: but we have in mind a very particular case of the lemma, which was noted in Lecture VII, Sect. 5 (p. 171) and a direct proof of which (independent of the Carleson Imbedding Theorem) can be found in the textbooks in Function Theory.)

The converse inclusion is somewhat more delicate. For its proof we have to show that

$$g \in H^2_\Pi \Rightarrow \sup\{|g(\eta)| : \operatorname{Im}\eta \geq \varepsilon\} < \infty, \forall \varepsilon > 0. \tag{4}$$

Once this is done we can conclude that the restriction of $f$ to the interior int $C_y$ of the circle $C_y$ belongs to the class[1] $H^2$, because then $f = 2i\sqrt{\pi}(\mathbb{1} - z)^{-1}(g \circ \omega)$

[1] Since it is question of the disks int $C_y$, $y > 0$, the modification of all notions connected with the class $H^2$ (including the basis definition) does not cause any difficulties and can be obtained by a linear change of variable.

and we can apply Corollary 3 of the Beurling-Helson Theorem (Lecture I): the function $g \circ \omega$ is bounded in $\operatorname{int} C_y$ and the function $\mathbb{1} - z$ is outer there while $f \in L^2(C_y)$. Now we can apply once more the Lemma on Carleson Measures (Lecture VII), this time with the disk $\operatorname{int} C_y$ and the measure $v^r$ equal to arc length on the circle $\mathbb{T}_r \overset{\text{def}}{=} \{\zeta : |\zeta| = r\}, r < 1$. Since (again) the constants $A = A(v^r)$ in that lemma are uniformly bounded (they are to be taken with respect to any circle $C_y$ containing $\mathbb{T}_r$; cf. Fig. 5), condition (3) shows that $\sup_{0 \le r < 1} \int_{\mathbb{T}_r} |f(\zeta)|^2 |d\zeta| < \infty$, that is, $f \in H^2$.

There remains the proof of the implication (4). If $\operatorname{Im} \eta > 0$ then by the mean value theorem

$$\begin{aligned}
|g(\eta)| &= \frac{1}{\pi(\operatorname{Im}\eta)^2} \left| \iint_{|\eta - (x+iy)| < \operatorname{Im}\eta} g(x+iy)\,dx\,dy \right| \\
&\le \frac{1}{\pi(\operatorname{Im}\eta)^2} (\sqrt{\pi}\operatorname{Im}\eta) \left( \iint_{|\eta - (x+iy)| < \operatorname{Im}\eta} |g(x+iy)|^2\,dx\,dy \right)^{1/2} \\
&\le \frac{1}{\sqrt{\pi}\operatorname{Im}\eta} \left( \int_0^{2\operatorname{Im}\eta} dy \int_{\mathbb{R}} |g(x+iy)|^2\,dx \right)^{1/2} \\
&\le \left( \frac{2}{\pi} \frac{\text{const}}{\operatorname{Im}\eta} \right)^{1/2},
\end{aligned}$$

where the const stands for the left hand side of inequality (1). □

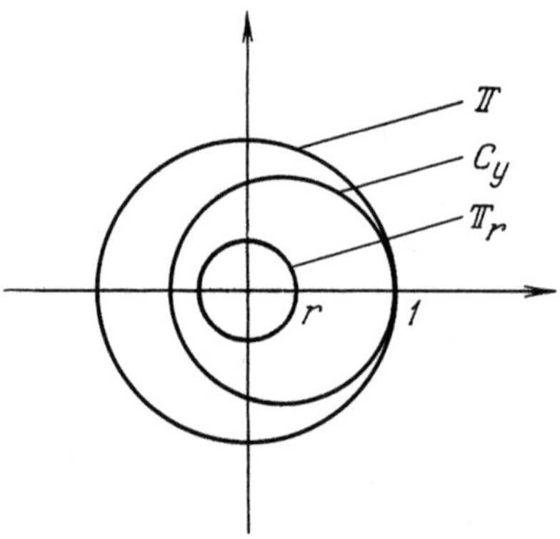


Fig. 5

**Corollary (Paley-Wiener Theorem).** *$H^2_\Pi = \mathscr{F}^{-1} L^2(\mathscr{R}_+)$ where $\mathscr{F}^{-1}$ is the inverse Fourier transform $((\mathscr{F}^{-1}f)(x) = 1/\sqrt{2\pi} \int e^{itx} f(t)\,dt$ for functions $f$ in $L^1(\mathbb{R}) \cap L^2(\mathbb{R})$) and $L^2(\mathbb{R}_+) = \chi_{\mathbb{R}_+} L^2(\mathbb{R})$.*

In fact, since $\mathscr{F}^{-1}$ is unitary, it is sufficient to show that a suitable fundamental subset of $L^2(\mathbb{R}_+)$ is mapped onto a fundamental subset of $H^2_\Pi$. As such a subset we take $A = \{e^{i\lambda w}\chi_{\mathbb{R}_+} : \operatorname{Im}\lambda > 0\}$; the linear hull of this set is dense in $L^2(\mathbb{R}_+)$ because the classical Fourier transform is injective: from $f \in L^2(\mathbb{R}_+)$, $f \perp e^{i\lambda w}\chi_{\mathbb{R}_+}$ for $\lambda =$

$i + t$, $t \in \mathbb{R}$ it follows that $\mathscr{F}(fe^{-w}\chi_{\mathbb{R}_+}) \equiv 0$ and hence $f \equiv 0$. As

$$\mathscr{F}^{-1}e^{i\lambda w}\chi_{\mathbb{R}_+} = \frac{i}{\sqrt{2\pi}}(w + \lambda\mathbb{1})^{-1} \tag{5}$$

for $\operatorname{Im}\lambda > 0$, we see by the Lemma on the Image of $H^2$ that $\bigvee_{L^2(\mathbb{R})}(\mathscr{F}^{-1}A) = H^2_\Pi$. $\square$

Now we can get a preliminary result from the work done and state the "spectral form" of Lax's Theorem.

**Lax's Theorem.** *Let $E$ be a subspace of $L^2(\mathbb{R})$.*

1. *If $e^{itw}E \subset E$, $\forall t \in \mathbb{R}$, then $E = \chi_e L^2(\mathbb{R})$ where $e$ is a suitable measurable subset of $\mathbb{R}$.*
2. *If $e^{itw}E \subset E$, $t \geq 0$, but $e^{itw}E \not\subset E$ for some (and therefore all) $t$, $t < 0$, then there exists a measurable function $q$ on $\mathbb{R}$, $|q| = 1$ a.e., such that $E = qH^2_\Pi$.*

*The proof* follows at once from the Beurling-Helson Theorem (applied to the subspace $U^{-1}E$), the Sz.-Nagy and Foiaş Lemma and its remark, along with the Lemma on the Image of $H^2$. $\square$

Let us now perform a Fourier transformation, so that we can deal with the right shifts $S_t$, $t \geq 0$. If $\mathscr{F}$ is the Fourier transform in $L^2(\mathbb{R})$, it is easy to see that $\mathscr{F}e^{itw} = S_t\mathscr{F}$. The map $\mathscr{F}U$ maps as before $L^2(\mathbb{T})$ isometrically onto $L^2(\mathbb{R})$, the image of $H^2$ being $\chi_{\mathbb{R}_+}L^2(\mathbb{R}) = L^2(\mathbb{R}_+)$ (the Paley-Wiener Theorem and the Lemma on the Image of $H^2$). This gives the "translation" form of Lax's Theorem.

**Lax's Theorem.** *Let $E$ be a subspace of $L^2(\mathbb{R})$.*

1. *If $S_tE \subset E$, $\forall t \in \mathbb{R}$, then $E = \mathscr{F}\chi_e L^2(\mathbb{R})$ where $e$ is a measurable subset of $\mathbb{R}$.*
2. *If $S_tE \subset E$, $t \geq 0$, but $S_tE \not\subset E$ for some (and thereby all) $t$, $t < 0$, then there exists a measurable function $q$ on $\mathbb{R}$, $|q| = 1$ a.e., such that $E = \mathscr{F}qH^2_\Pi$.* $\square$

Lax's Theorem allows us to get the analogues of all the results of Lecture I for the semigroups $\{e^{itw}\}_{t\geq 0}$ and $\{S_t\}_{t\geq 0}$. The proofs can be patterned on the $\mathscr{S}$-theory of Lecture I or alternatively obtained by a formal reduction to the operator $\mathscr{S}$ with the aid of the transformations $U$ and $\mathscr{F}U$. Inner functions in the halfplane $\Pi$ are now defined as preimages of inner functions in the disk (with respect to the transformation $f \mapsto f \circ \omega$), i.e. as bounded analytic functions in the half-plane (the collection of all these functions $\stackrel{\text{def}}{=} H^\infty_\Pi$) whose boundary values are unimodular. Outer functions are defined analogously and the property for $f$, $f \in H^2_\Pi$, to be outer is equivalent to $\bigvee(e^{itw}f : t \geq 0) = H^2_\Pi$.

**Corollary.** 1. *If $f \in H^2_\Pi$ and $f = 0$ a.e. on some subset of positive Lebesgue measure then $f = \mathbb{0}$.*

2. *If $f \in H^2_\Pi$ then $\bigvee(e^{itw}f : t \geq 0) = f^iH^2_\Pi$ where $f^i$ is the inner part of $f$.*

3. *If $f \in L^2(\mathbb{R})$ then $\bigvee(e^{itw}f : t \geq 0) = L^2(\mathbb{R}) \Leftrightarrow \left(f \neq 0 \text{ a.e. on } \mathbb{R} \text{ and } \int_{\mathbb{R}} \frac{\log|f|}{1 + x^2}\,dx = -\infty\right)$.*

4. *If* $f \in L^2(\mathbb{R})$ *then* $\bigvee(S_t f : t \in \mathbb{R}) = L^2(\mathbb{R}) \Leftrightarrow \mathscr{F}f \neq 0$ *a.e. on* $\mathbb{R}$.

5. *If* $f \in L^2(\mathbb{R})$ *then* $\bigvee(S_t f : t \geq 0) = L^2(\mathbb{R}) \Leftrightarrow \left(\mathscr{F}f \neq 0\right.$ *a.e. on* $\mathbb{R}$ *and* $\left.\int_{\mathbb{R}} \frac{\log|\mathscr{F}f|}{1+x^2}dx = -\infty\right)$. □

Let us finally point out a trivial fact, in a way converse to Lax's Theorem: the subspaces encountered in that theorem are invariant for the corresponding groups (semigroups) of operators for an arbitrary choice of the "parameters" $e$ and $q$. These parameters are determined uniquely (in the obvious sense) by the subspace.

**2. Bases of Exponentials.** The known results on the geometry of families of rational fractions $\{(\mathbb{1} - \bar{\lambda}z)^{-1} : \lambda \in \sigma\}$ in $H^2$ can easily be converted into analogous information on other families of functions – the fractions $(w - \bar{\mu}\mathbb{1})^{-1}$, $\operatorname{Im}\mu > 0$, in $H^2_\Pi$; the exponentials $e^{-i\bar{\lambda}x}$, $\operatorname{Im}\lambda > 0$, in $L^2(0,\infty)$; the powers $x^\nu$, $\operatorname{Re}\nu > -1/2$ in $L^2(0,1)$, etc. The passage from one family to another is implemented by unitary transformations:

$$(1 - |\lambda|^2)^{1/2}(\mathbb{1} - \bar{\lambda}z)^{-1} \mapsto c(\mu)(\pi^{-1}\operatorname{Im}\mu)^{1/2}(w - \bar{\mu}\mathbb{1})^{-1},$$

$\mu = \omega(\lambda)$, $\lambda \in \mathbb{D}$, $c(\mu) = |\mu + i|(\mu + i)^{-1}$ (the transformation $U$),

$$(1 - |\lambda|^2)^{1/2}(\mathbb{1} - \bar{\lambda}z)^{-1} \mapsto -ic(\mu)(2\operatorname{Im}\mu)^{1/2}e^{-i\bar{\mu}w}\chi_{\mathbb{R}_+}, \tag{6}$$

$\mu = \omega(\lambda)$, $\lambda \in \mathbb{D}$ (the transformation $\mathscr{F}U$),

$$(2\operatorname{Im}\mu)^{1/2}e^{-i\bar{\mu}w}\chi_{\mathbb{R}_+} \mapsto (2\operatorname{Re}\nu + 1)^{1/2}x^\nu, \tag{7}$$

$\nu = i\bar{\mu} - 1/2$, $\operatorname{Im}\mu > 0$ (the transformation $f \mapsto f(\log 1/x)x^{-1/2}$ from the space $L^2(\mathbb{R}_+)$ into the space $L^2(0,1)$).

Therefore one can in the study of these new families limit oneself just to translating theorems on functions regular in $\mathbb{D}$ into the language of the new parameters ($\lambda \mapsto \mu$, $\mu \mapsto \nu$ etc.) Let us carry out this task for the map $\lambda \mapsto \mu = \omega(\lambda)$.

The *Blaschke condition* in the halfplane (necessary and sufficient for the family $\{(w - \bar{\mu}\mathbb{1})^{-1} : \mu \in \tau\}$, $\tau \subset \Pi$, not to be total in $H^2_\Pi$ or the same thing for the family $\{e^{-i\bar{\mu}w}\chi_{\mathbb{R}_+} : \mu \in \tau\}$ in $L^2(\mathbb{R}_+)$) reads:

$$\sum_{\mu \in \tau} \frac{\operatorname{Im}\mu}{1 + |\mu|^2} < \infty.$$

In fact

$$1 - |\omega^{-1}(\mu)|^2 = \frac{|\mu + i|^2 - |\mu - i|^2}{|\mu + i|^2} = \frac{4\operatorname{Im}\mu}{|\mu + i|^2} \tag{8}$$

and $|\mu|^2 + 1 \leq |\mu + i| \leq 2(|\mu|^2 + 1)$ for $\operatorname{Im}\mu > 0$. The corresponding condition for the divisor $k$:

$$\sum_{\mu \in \Pi} k(\mu)\operatorname{Im}\mu(1 + |\mu|^2)^{-1} < \infty.$$

The *Blaschke product* in the halfplane is the function $B = B(k, \cdot)$ corresponding to the divisor $k$ (or a set $\tau$, $\tau \subset \Pi$, then $k = \chi_\tau$), satisfying the Blaschke condition, given by the formula

$$B = \prod_{\operatorname{Im}\mu > 0} \left( \frac{w - \mu}{w - \bar{\mu}} \cdot \frac{|1 + \mu^2|}{1 + \mu^2} \right)^{k(\mu)}.$$

The *Carleson Condition* (*C*) for a Blaschke product with simple zeros ($\tau$ being the set of these zeros) takes in the halfplane the form

$$\inf_{\mu' \in \tau} \prod_{\mu \in \tau \setminus \{\mu'\}} \left| \frac{\mu' - \mu}{\mu' - \bar{\mu}} \right| > 0. \tag{C}$$

Likewise there will arise no difficulties in translating to the case of the halfplane those geometric properties of a set which imply that it is a Carleson set: the sparseness condition (*R*) and the property of the corresponding measure to be a Carleson measure (cf. Lecture VII). Let us give the necessary computations:

The *sparseness condition* (*R*) amounts formally speaking to

$$\sup_{\mu \neq \mu';\, \mu, \mu' \in \tau} (4 \operatorname{Im}\mu \cdot \operatorname{Im}\mu') |(e^{-i\bar{\mu}w}, e^{-i\bar{\mu}'w})_{L^2(\mathbb{R}_+)}|^2 < 1,$$

i.e. to

$$\sup_{\mu \neq \mu'} \frac{4 \operatorname{Im}\mu \cdot \operatorname{Im}\mu'}{|\bar{\mu} - \mu'|^2} < 1,$$

or equivalently

$$\inf_{\mu \neq \mu'} \left| \frac{\mu' - \mu}{\mu' - \bar{\mu}} \right| > 0. \tag{R}$$

As in the case of the disk $\mathbb{D}$ this condition is equivalent to the requirement of the disks

$$D_\mu(\varepsilon) \overset{\text{def}}{=} \left\{ \eta : \left| \frac{\eta - \mu}{\eta - \bar{\mu}} \right| < \varepsilon \right\}, \mu \in \tau,$$

to be disjoint for sufficiently small $\varepsilon$, $\varepsilon > 0$ (one can use the conformal invariance of this condition so there is no need of reproducing the argument of Lecture VII). The following "Euclidean" version is much more transparent: condition (*R*) for the subset $\tau$ of the halfplane $\Pi$ amounts to requiring that the disks

$$D(\mu, \delta \operatorname{Im}\mu) = \{\eta : |\eta - \mu| < \delta \operatorname{Im}\mu\}, \mu \in \tau,$$

are pairwise nonintersecting for small $\delta$, $\delta > 0$,

$$D(\mu, \delta \operatorname{Im}\mu) \cap D(\mu', \delta \operatorname{Im}\mu') = \varnothing, \mu \neq \mu';\ \mu, \mu' \in \tau. \tag{9}$$

To see this it suffices to notice the inclusions

$$D(\mu, \delta \operatorname{Im}\mu) \subset D_\mu(\delta)$$

(trivial in view of $|\eta - \bar{\mu}| > \operatorname{Im}\mu$) and

$$D_\mu(\delta) \subset D\left(\mu, \frac{2\delta}{1-\delta}\operatorname{Im}\mu\right)$$

$$\left(\text{in view of } \left|\frac{\eta-\mu}{\eta-\bar{\mu}}\right| < \delta \Rightarrow \left|1 - \frac{\mu-\bar{\mu}}{\eta-\bar{\mu}}\right| < \delta \Rightarrow \frac{2\operatorname{Im}\mu}{|\eta-\bar{\mu}|} > 1-\delta\right). \square$$

The *condition* (*CN*) *for the measure*

$$\nu_\sigma = \frac{1}{2}\sum_{\lambda\in\sigma}(1-|\lambda|^2)\delta_\lambda$$

*to be a Carleson measure* with respect to the space $H^2$ (here $\delta_\lambda$ is the delta-measure concentrated at the point $\lambda$) is likewise readily transplanted to the halfplane: from the Lemma on the Image of $H^2$ and inequality (8) we see that the inclusion needed for the interpolation takes the form

$$g\in H^2_\Pi \Rightarrow \sum_{\mu\in\tau}\operatorname{Im}\mu|g(\mu)|^2 < \infty. \qquad (CN)$$

(This is self-evident, because the invariant form of this imbedding – part of Bari's Theorem (Lecture VI) – says that $\sum_{\mu\in\tau}|(g,e_\mu)|^2 < \infty$, $\forall g\in H^2_\Pi$, where $e_\mu = (\pi^{-1}\operatorname{Im}\mu)^{1/2}(w-\bar{\mu}\mathbb{1})^{-1}$.) A necessary and sufficient condition for the validity of this imbedding theorem is gotten most simply by returning to the disk $\mathbb{D}$ and applying the Lemma on Carleson Measures (Lecture VII). Since then the test disks $D(\zeta,l)$ in this lemma go over into similar disks (or the exteriors of such disks if $1\in D(\zeta,l)$) with centers on $\mathbb{R}$ and the arc length of $\mathbb{T}\cap D(\zeta,l)$ equals

$$\frac{1}{\pi}\int_{\omega(\mathbb{T}\cap D(\zeta,l))}\frac{dx}{1+x^2},$$

we see that the condition

$$\sup_{\zeta\in\mathbb{T},\, l>0} l^{-1}\nu_\sigma(D(\zeta,l)) < \infty$$

is transformed into the condition

$$\sum_{\mu\in\tau\cap Q_{\eta,r}}\frac{\operatorname{Im}\mu}{1+|\mu|^2} \le \text{const}\int_{\mathbb{R}\cap Q_{\eta,r}}\frac{dx}{1+x^2};\ \eta\in\mathbb{R},\ r>0, \qquad (10)$$

where $Q_{\eta,r}$ runs through the set of all disks $D(\eta,r)$ or their complements $\mathbb{C}\backslash D(\eta,r)$; $\eta\in\mathbb{R}$, $r>0$. This criterion for $\nu_\tau = \sum_{\mu\in\tau}\delta_\mu\operatorname{Im}\mu$ to be a Carleson measure can be simplified if we replace the invariant measure of the set

$$\mathbb{R}\cap D(\eta,r)$$

by the Euclidean measure. Namely *condition* (10) *is equivalent to the condition*

$$\sum_{\mu\in\tau\cap D(\eta,r)}\operatorname{Im}\mu \le \text{const}\cdot r;\ \eta\in\mathbb{R},\ r>0. \qquad (11)$$

Moreover, from the Imbedding Theorem of Carleson for the space $H^2$ (Lecture VII, p. 151) it follows at once that if $\nu$ is a Borel measure in $\Pi$ then

$$H^2_\Pi \subset L^2(\nu) \Leftrightarrow \sup_{\zeta\in\mathbb{D}} \left\| U\frac{(1-|\zeta|^2)^{1/2}}{1-\bar{\zeta}z} \right\|_{L^2(\nu)} < \infty.$$

In other words, that $\nu$ is a Carleson measure in $\Pi$ is equivalent to the condition

$$\sup_{\operatorname{Im}\mu>0} \int_\Pi \frac{\operatorname{Im}\mu}{|w-\bar{\mu}|^2}\,d\nu < \infty. \tag{12}$$

It is now immediate to verify that this last property of $\nu$ is equivalent to the estimate

$$\nu(D(\eta,r)) \le Ar;\ \eta\in\mathbb{R},\ r>0.$$

*Let us prove this proposition.* Let $a$ be the left hand side of (12) and write $\mu = \eta + ir$. Then $|w-\bar{\mu}|^2 < 4r^2$ for $|w-\eta|<r$, $w\in\Pi$, whence

$$\frac{1}{4r^2}\nu(D(\eta,r)) \le \int_\Pi \frac{d\nu}{|w-\bar{\mu}|^2},$$

i.e. $\nu(D(\eta,r)) \le 4ar$.

Conversely, let

$$\nu(D(\eta,r)) \le Ar$$

for all $\eta\in\mathbb{R}$, $r>0$. Computing the integral in terms of the distribution function we get (in the previous notation)

$$\begin{aligned}
\int_\Pi \frac{\operatorname{Im}\mu}{|w-\mu|^2}\,d\nu &= r\int_0^\infty \nu\{w: |w-\bar{\mu}|^{-2} > t\}\,dt \\
&= r\int_0^{1/r^2} \nu\{w: |w-\bar{\mu}|^2 < t^{-1}\}\,dt \\
&\le r\int_0^{1/r^2} \nu\{w: |w-\eta|^2 < (t^{-1}-r^2)\}\,dt \\
&\le r\int_0^{1/r^2} A(t^{-1}-r^2)^{1/2}\,dt \\
&\le r\int_0^{1/r^2} At^{-1/2}\,dt = 2A. \ \Box
\end{aligned}$$

In two special cases we have $(R)\Rightarrow(CN)$ – *when the set $\tau$ lies in a strip*

$$\left\{\eta: 0 < \inf_{\mu\in\tau}\operatorname{Im}\mu \le \operatorname{Im}\eta \le \sup_{\mu\in\tau}\operatorname{Im}\mu < \infty\right\}, \tag{13}$$

or when it lies in *an arbitrary Stolz angle*:

$$\sup_{\mu\in\tau}\frac{|\mu-x|}{\operatorname{Im}\mu} < \infty$$

for some real $x$. In fact, in the case of the strip (13) condition $(R)$ means the ordinary separation of points in the plane:

$$\inf_{\mu \neq \mu';\, \mu, \mu' \in \tau} |\mu - \mu'| > 0,$$

or equivalently $D(\mu, \delta) \cap D(\mu', \delta) = \varnothing (\mu \neq \mu';\ \mu,\ \mu' \in \tau)$ for small $\delta$, $\delta > 0$. This implies condition (11): the number of point $\mu$, $\mu \in \tau$, falling into the disk $D(\eta, r)$ (and indeed in the strip $\{\eta - r < \operatorname{Re} \mu < \eta + r\}$) does not exceed const $r$ (compare the area of the rectangle which cuts that strip with the strip (13) with the area of the sum $\bigcup_{\mu \in \tau \cap D(\eta, r)} D(\mu, r)$).

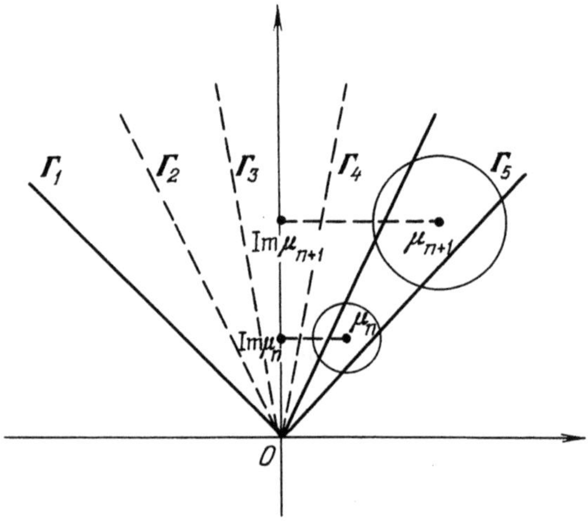


Fig. 6

If $\tau$ lies in a Stolz angle (say, with the vertex at the point $x = 0$) and satisfies condition (9) then this angle can be divided into a sum of angles $\Gamma_1, \ldots, \Gamma_N$ all of equal and so small magnitude that each disk $D(\mu, \delta \operatorname{Im} \mu)$, $\mu \in \Gamma_i$, intersects *both* sides of $\Gamma_i$, $i = 1, \ldots, N$, cf. Fig. 6. Let us enumerate the points $\mu$ of the set $\tau \cap \Gamma_i$ with $\operatorname{Im} \mu$ increasing. By geometric considerations it is clear (cf. Fig. 6) that there exists an $\varepsilon$, $\varepsilon > 0$, such that

$$\operatorname{Im} \mu_{n+1} - \operatorname{Im} \mu_n \geq \varepsilon |\mu_{n+1} - \mu_n| \geq \varepsilon\delta(\operatorname{Im} \mu_{n+1} + \operatorname{Im} \mu_n) \geq \varepsilon\delta \operatorname{Im} \mu_n.$$

Consequently $\operatorname{Im} \mu_{n+1}/\operatorname{Im} \mu_n \geq 1 + \varepsilon\delta$ for all $n$. For such sequences both conditions (12) are trivially fulfilled. They are then also fulfilled for the union $\bigcup_{i=1}^{N} (\Gamma_i \cap \tau) = \tau$. □

*Remark.* Of course, instead of the last proofs we could have referred to the corresponding results in Lecture VII but the geometric reasoning set forth in the text seems to be preferable.

Let us now combine the results on the "transplantation" into the halfplane of the basic characteristics of a set (responsible for the interpolation) with the Theorem on Bases and Interpolation and the Carleson Imbedding Theorem. In the following theorem we deal with the family of exponentials $\{e^{i\lambda w}\chi_{\mathbb{R}_+} : \lambda \in \sigma\}$, $\sigma \subset \Pi$, in the space $L^2(\mathbb{R}_+)$ but the Fourier transform or the logarithmic change of variable (7), mentionned at the beginning of this Section, allow us to describe in exactly the same way bases and interpolation for the rational fractions $(w - \mu\mathbb{1})^{-1}$, $\operatorname{Im}\mu < 0$ (in the space $H^2_\Pi$) or the powers $x^\nu$, $2\operatorname{Re}\nu > -1$ (in the space $L^2(0,1)$).

**Theorem on Exponentials.** *Let $\tau$ be a subset of the upper halfplane $\Pi$. The following statements are equivalent.*

1. *The family $\{(2\operatorname{Im}\mu)^{1/2}e^{i\mu w}\chi_{\mathbb{R}_+} : \mu \in \tau\}$ forms a Riesz basis for its closed linear hull (in the space $L^2(\mathbb{R}_+)$)*
2. *$JH^2 = l^2$ where $Jf = \{(\operatorname{Im}\mu)^{1/2}f(\mu) : \mu \in \tau\}$.*
3. *$H^\infty_\Pi|\tau = l^\infty$.*
4. *$\tau \in (C)$.*
5. *$\tau \in (R)$, $\tau \in (CN)$.* □

Passing to the study of families of exponentials on a *finite interval*, in the space $L^2(0,a)$, $a > 0$, we will adhere to the point of view expressed in Lecture VIII, Sect. 6; namely, we will consider such families as "perturbations" (projections) of the family $e^{i\lambda w}\chi_{\mathbb{R}_+}$ in the space $L^2(\mathbb{R}_+)$. This approach excludes from the discussion those families $e^{i\mu w}\chi_{(0,a)}$ which do not possess a "pre-image" in $L^2(\mathbb{R}_+)$, i.e., families with $\operatorname{Im}\mu \le 0$. Some general remarks on the problems arising here will be found in the following Sect. 3. In the present Section we study only the property for a family of exponentials to be a basis (in the whole space $L^2(0,a)$ or in its closed linear hull).

First of all let us nail down that the space $L^2(0,a)$ will be considered as imbedded in $L^2(\mathbb{R})$ (the natural continuation by zero on $\mathbb{R}\backslash(0,a)$ identifies $L^2(0,a)$ and $\chi_{(0,a)}L^2(\mathbb{R})$) so that $L^2(0,a) = L^2(\mathbb{R}_+) \ominus S_aL^2(\mathbb{R}_+) = \mathscr{F}(H^2_\Pi \ominus e^{iaw}H^2_\Pi)$. The pre-image of the subspace $e^{iaw}H^2_\Pi$ under the map $U$ (cf. Sect. 1, above) is the space $\Theta H^2$, $\Theta = \exp\left(a\dfrac{z+\mathbb{1}}{z-\mathbb{1}}\right)$, so that $L^2(0,a) = \mathscr{F}UK_\Theta$, $K_\Theta = H^2 \ominus \Theta H^2$. The orthogonal projection $P_\Theta$ is then mapped onto the projection $\chi_{(0,a)}$ onto the subspace $L^2(0,a)$ and we can use all our information on bases of values of the reproducing kernel $k_\Theta$, as obtained in Lecture VIII, Sect. 6. To the functions $k_\Theta(\cdot,\lambda) = (\mathbb{1} - \Theta\overline{\Theta(\lambda)})(\mathbb{1} - \bar{\lambda}z)^{-1}$ there correspond now the functions $\chi_{(0,a)}e^{-i\bar{\mu}w}$, $\mu = \omega(\lambda)$ (up to a numerical factor which can be easily found with the aid of (6)).

In the space $L^2(0,a)$ there exist isomorphisms sending exponentials into exponentials. Among these isomorphisms we have the following ones: $f \mapsto fe^{itw}$ ($t \in \mathbb{C}$; this transformation maps the family $\{e^{i\mu w} : \mu \in \tau\}$ onto the family $\{e^{i\mu w} : \mu \in \tau + t\}$); $f(x) \mapsto f(a-x)$ (this transformation maps, apart from a constant factor, the family $\{e^{i\mu w} : \mu \in \tau\}$ onto the family $\{e^{i\mu w} : \mu \in (-\tau)\}$); $f \mapsto \bar{f}$ (complex conjugation; $\{e^{i\mu w} : \mu \in \tau\}$ gets replaced by $\{e^{i\mu w} : \mu \in (-\bar{\tau})\}$). All these transformations preserve the geometric properties of interest for us: the first two since they are linear isomorphisms of $L^2(0,a)$ onto itself, and the last one since it maps any fundamental set in $L^2(0,a)$ onto a fundamental set and the Gram matrix $\{(f_\alpha, f_\beta)\}$ of any family $\{f_\alpha\}$ onto its transpose (for the rôle of Gram matrices, see Lecture VI). Therefore, without destroying the property of the family $\{e^{i\mu w}\chi_{(0,a)} : \mu \in \tau\}$ to be an unconditional basis in $L^2(0,a)$ (or a basis for its own closed linear hull) one can replace $\tau$ by any translate or by a reflexion in the coordinate axes.

Let us further recall that (among other things) we established in Lecture VIII (in terms of the kernel $k_\Theta$) the necessity of the Carleson condition $(C)$ for the family of exponentials $\{e^{i\mu w} : \mu \in \tau\}$, $\tau \subset \Pi$, to be a basis for the closure of its linear hull; in particular, the Blaschke condition is necessary.

**Theorem on Exponentials in $L^2(0,a)$.** *Let $\tau \subset \mathbb{C}$ and assume that*

$$\alpha_\tau \stackrel{\text{def}}{=} \inf \operatorname{Im} \mu > -\infty.$$

*The following statements are equivalent:*

1. $\{e^{i\mu w}|[0,a) : \mu \in \tau\}$ *is an unconditional basis for* $L^2(0,a)$.

2. *If* $\mathscr{I}_\alpha f \stackrel{\text{def}}{=} \{(\operatorname{Im}\mu)^{1/2}(\mathscr{F}f)(\mu) : \mu \in \tau + i\alpha\}$ *then* $\mathscr{I}_\alpha L^2(0,a) = l^2$ *and* $\|f\|_{L^2(0,a)} \asymp \|\mathscr{I}_\alpha f\|_{l^2}$, $f \in L^2(0,a)$, *for any (some)* $\alpha$, $\alpha > -\alpha_\tau$.

3. $\tau + i\alpha \in (C)$, $\inf\{\|\arg(e^{-ia w}B^\alpha) - \tilde{h} - c\|_\infty : h = \bar{h} \in L^\infty(\mathbb{R}), c \in \mathbb{R}\} < \pi/2$ *for any (some)* $\alpha$, $\alpha > -\alpha_\tau$; *here* $B^\alpha$ *is the Blaschke product formed with the zeros* $\tau + i\alpha$.

4. $\tau + i\alpha \in (C)$, $\operatorname{dist}(e^{iaw}, B^\alpha H^\infty_\Pi) < 1$, $\operatorname{dist}(B^\alpha, e^{iaw}H^\infty_\Pi) < 1$ *for any (some)* $\alpha$, $\alpha > -\alpha_\tau$.

*Proof.* Taking account of the discussion preceding the theorem and Corollary 1 of the Theorem on Close Subspaces and its Complements (p. 205–206), it remains only to remark that $\inf\{\operatorname{Im}\mu : \mu \in \tau + i\alpha\} > 0$ for $\alpha > -\alpha_\tau$ so that

$$\|e^{i\mu w}|[0,a)\|^2_{L^2(0,a)} = \frac{1}{2\operatorname{Im}\mu}(1 - e^{-2a\operatorname{Im}\mu}) \asymp \frac{1}{\operatorname{Im}\mu}$$

for $\mu \in \tau + i\alpha$. The equivalence of the Statements 1 and 2 now follows from Bari's Theorem (Lecture VI), since the natural orthogonalizer

$$f \mapsto \{\|e^{i\mu w}|(0,a)\|^{-1}_{L^2(0,a)}(f, e^{i\mu w})_{L^2(0,a)} : \mu \in \tau + i\alpha\}$$

is (in the sense of a twosided estimate) equivalent to the operator $\mathscr{I}_\alpha$. □

Of course, also other statements (than the ones employed here) of the Theorem on Close Subspace, Complement 1 of that theorem and its Corollary 1 can be useful. Below we shall outline one interesting application of the Theorem on Exponentials which is based on the implications 1 ⇔ 3. Let us further point out that if we wish to get a criterion for an unconditional basis of exponentials *in their closed linear hull* one has to replace Statements 2 and 4 by the following ones:

2′. $\mathscr{I}_\alpha L^2(0,a) = l^2$.

4′. $\tau + i\alpha \in (C)$, $\operatorname{dist}(e^{iaw}, B^\alpha H^\infty_\Pi) < 1$ *for any (some)* $\alpha$, $\alpha > -\alpha_\tau$.

A crude sufficient condtion for Statement 4 to be true is

$$\tau + i\alpha \in (C), \ \|e^{iaw} - B^\alpha\|_\infty < 1, \tag{14}$$

and for Statement 4′ the same Carleson condition $(C)$ together with the assumption of a *sufficiently small* upper bound

$$\sup\{|e^{ia\mu}| : \mu \in \tau + i\alpha\} = \exp\left\{-a\cdot \inf_{\mu\in\tau+i\alpha} \operatorname{Im}\mu\right\}$$

(this "smallness" depends only on the Carleson constant of the set $\tau + i\alpha$; this constant is lost if $\alpha$ grows). See Lecture VIII, Sect. 6.

**Corollary 1.** *Let $\tau \subset \Pi$ and assume that for any $\alpha$, $\alpha > 0$, $\{\mu : \mu \in \tau, \operatorname{Im}\mu < \alpha\}$ is a finite set. Then the family of exponentials $\{e^{i\mu w}\chi_{(0,a)} : \mu \in \tau\}$ is an unconditional basis in its closed linear hull iff $\tau \in (C)$. If $\tau \in (C)$ then this hull does not coincide with $L^2(0, a)$.*

Let us give only an outline of the proof. As it was mentioned before this Corollary there exists a positive function $c = c(\delta)$ such that every family $\{e^{i\mu w}|[0, a] : \mu \in \tau\}$ with the Carleson constant $\delta$ is an unconditional basis in its closed linear hull, provided $\inf\{\operatorname{Im}\mu : \mu \in \tau\} \geq c$.

Let us suppose now that we have a family $\tau \in (C)$ (with the Carleson constant $\delta_0$, say) such that the subsets $\{\mu : \mu \in \tau, \operatorname{Im}\mu < \alpha\}$ are finite for every $\alpha > 0$. Then $\tau$ is a uniformly separated (sparse) set (i.e. (9) holds for some $\delta$) and the corresponding measure is a Carleson measure (i.e. (11) holds for some constant $\beta_0$, say). Fix a number $\beta_1 > \beta_0$. Then there exists a constant $\delta_1$, $0 < \delta_1 < \delta$, such that every set $\sigma$ satisfying the $\delta$-condition (9) and the $\beta_1$-condition (11) is a Carleson set with the constant $\delta_1$ (see Lecture VII, Sect. 3). Put now $\sigma' = \{\mu : \mu \in \tau, \operatorname{Im}\mu \geq c(\delta_1)\}$ and choose on the line $\operatorname{Im}\mu = c(\delta_1)$ an infinite set $\sigma''$ so sparse that, for $\sigma = \sigma' \cup \sigma''$, the $\delta$-condition (9) and the $\beta_1$-condition (11) hold. Therefore the family $\{e^{i\mu w}\chi_{(0,a)} : \mu \in \sigma\}$ forms an unconditional basis in its closed linear hull and hence $\{e^{i\mu w}\chi_{(0,a)} : \mu \in \tau\}$ too; moreover $\operatorname{codim} \bigvee \{e^{i\mu w}\chi_{(0,a)} : \mu \in \tau\} = \infty$ (see remark 7) at the end of Sect. 7, Lecture VIII). □

Another extremal case is provided by a set $\tau$ lying in a strip. The classical case, viz. $a = 2\pi$, $\tau = \mathbb{Z}$ = the set of integers, leads in a trivial way (sic!) to a basis: the Blaschke product of $\mathbb{Z} + i\alpha$ is

$$B^\alpha = \frac{\sin \pi(w - i\alpha)}{\sin \pi(w + i\alpha)}$$

and

$$\| B^\alpha + e^{2\pi i w}\|_{H^\infty_\Pi} = e^{-\pi\alpha} \left\| \frac{\sin 2\pi w}{\sin \pi(w + i\alpha)} \right\|_{L^\infty(\mathbb{R})} = o(e^{-\pi\alpha}),$$

whence

$$\arg B^\alpha(x) = 2\pi x + \pi + c_\alpha(x), \tag{15}$$

$|c_\alpha(x)| \leq \text{const}\, e^{-\pi\alpha}$. Therefore Statement 4 is fulfilled. In a similar way we can also consider "small perturbations" of $\mathbb{Z}$. Let us set

$$\tau = \{n + \delta_n : n \in \mathbb{Z}\},\ \delta_n \in \mathbb{R}.$$

**Corollary 2 (Kadec's 1/4-Theorem).** *If*

$$\sup_{n\in\mathbb{Z}} |\delta_n| \overset{\text{def}}{=} \delta < 1/4,$$

*then* $\{e^{i\mu w}\chi_{(0,2\pi)} : \mu\in\tau\}$ *is an unconditional basis in* $L^2(0,2\pi)$.

In fact, for any $x$, $x\in\mathbb{R}$, the function

$$s\mapsto \arg\frac{x-(s+i\alpha)}{x-(s-i\alpha)} = 2\arg(x-(s+i\alpha))$$

decreases as $s$ grows, $s\in\mathbb{R}$ ($\alpha > 0$). Therefore the largest value of the modulus of the difference

$$A = \left|\sum_{n\in\mathbb{Z}}\left\{\arg\frac{x-(n+i\alpha)}{x-(n-i\alpha)} - \arg\frac{x-(n+\delta_n+i\alpha)}{x-(n+\delta_n-i\alpha)}\right\}\right|$$

is assumed when all terms have the same sign and take their extremal value (under the side condition $|\delta_n| \le \delta$, $n\in\mathbb{Z}$). Consequently

$$A \le \left|\sum_{n\in\mathbb{Z}}\left\{\arg\frac{x-(n+i\alpha)}{x-(n-i\alpha)} - \arg\frac{x-(n+\delta+i\alpha)}{x-(n+\delta-i\alpha)}\right\}\right|.$$

Since the products consisting of the normalizing factors $|\mu_n^2+1|(\mu_n^2+1)^{-1}$ for the sets $\tau + i\alpha$ and $\mathbb{Z} + i\alpha$ ($\alpha > 0$) converge it follows that

$$\begin{aligned}|\arg B^\alpha_{\mathbb{Z}}(x)\overline{B^\alpha_\tau(x)} - \text{const}| &= A \le |\arg B^\alpha_{\mathbb{Z}}(x)\overline{B^\alpha_{\mathbb{Z}}(x-\delta)}|\\ &= |2\pi\delta + c_\alpha(x) - c_\alpha(x-\delta)| < \pi/2,\end{aligned}$$

this in view of (15), with $\alpha$ sufficiently large. □

**3. Concluding Remarks.** There are also several other ways (besides the one taken here) to connect the spaces $H^2$ and $H^2_\Pi$; cf. Hoffman [1], Sarason [2], Duren [1]. Against the background of the developments on the space $H^2$ in Lecture I–X it was natural to choose the shortest (even if not the most elementary) variant reducing the theory of $H^2$-functions in the half plane to the analogous theory in the disk $\mathbb{D}$. Let us emphasize once more that the use in this connection of the estimate of the integral along a curve by the integral along the circumference (known as Gabriel's theorem; Gabriel [1]) is considerably simpler than the Imbedding Theorem by Carleson.

The Sz.-Nagy and Foiaş Lemma can be found e.g. in the book [1] by these authors. One can also study the invariant subspaces of the semigroups $\{e^{itw}\}_{t\ge 0}$ and $\{S_t\}_{t\ge 0}$ in $L^2(\mathbb{R})$ without first transforming to the case of the shift operators $\mathcal{S}$ and $S$. Such an investigation is undertaken e.g. in the books Duren [1], Lax-Phillips [1], Helson [1–3], Dym-MacKean [1].

The original proof of Lax's Theorem is contained in the paper Lax [1] and is based on purely function theoretic ideas. As in the case of the Beurling-Helson theorem there exist several independent proofs. Halmos [1] replaced several of Lax's "analytical" arguments by more elementary Hilbert space methods and generalized somewhat the results in Lax's paper [1] considering also the space of vector functions $H^2(E)$, $\dim E \leq \infty$ (in [1] the presentation is adapted to the case $\dim E < \infty$). An essentially different approach was given by Helson and Lowdenslager [1]. These authors start with the analysis of the group $\lambda \overset{V_\lambda}{\mapsto} \int_{\mathbb{R}} e^{it\lambda}\, dP_t$ where $P_t$ is orthogonal projection onto $e^{itw}E$, $t \in \mathbb{R}$ ($E \subset L^2(\mathbb{R})$, $e^{itw}E \subset E$, $t \geq 0$) which satisfies the so-called Weyl commutation relations: $e^{itw}V_\lambda = e^{it\lambda}V_\lambda e^{itw}$. Then one uses the uniqueness theorem of Mackey (in a rather special form) from which the simultaneous unitary equivalence of the pairs $\{e^{itw}, V_\lambda\}$ and $\{e^{itw}, S_\lambda\}$ follows, at any rate, if the $e^{itw}$-invariant subspace $E$ does not reduce the first of these groups. Next one shows that such subspaces $E$ are in one-to-one correspondence with "cocycles", that is, with families $\{A_t\}$ of measurable unimodular functions on $\mathbb{R}$ satisfying the equation $A_{t+u} - A_t S_t A_u = \mathbb{0}$, $\forall t, u \in \mathbb{R}$. To get Lax's Theorem from here it remains to establish that each such cocycle is a "coboundary" in the sense that $A_t = qS_tq^{-1}$, $t \in \mathbb{R}$, where $|q| = 1$ a.e. on $\mathbb{R}$.

The proof of Lax's Theorem outlined here is not simple but compared with other ones it has the advantage that it gives considerable information on translation invariant subspaces on locally compact groups other than $\mathbb{R}$, reducing the problem to the study of cocycles. The details can be found in the paper Helson-Lowdenslager [1] and in the books Helson [1–3], Gamelin [1].

In Lax's Theorem there are absorbed several important theorems historically preceding it: its first statement is due to Wiener and this is also the case for Statement 4 of the Corollary; Statements 3 and 5 of the same Corollary are special cases of a theorem by Szegö-Kolmogorov-Kreĭn, often employed in the theory of stationary stochastic processes; cf., for example, Wiener [1], Rozanov [2], Dym-McKean [1].

The Paley-Wiener Theorem was obtained by these authors in [1] and is set forth in many textbooks on function theory; cf., for example, Levin [1], Hoffman [1], Duren [1].

Rational fractions in the halfplane, exponentials on the halfaxis $\mathbb{R}_+$ and other families equivalent to them were studied rather early. The modern epoch of this study, as already observed, started with the appearance of the book by Paley and Wiener [1]. The fact that the Blaschke condition in the half-plane is a criterion for non-completeness for the family $\{e^{i\mu w} : \mu \in \tau\}$ in $L^2(\mathbb{R}_+)$ was established by Szàsz [1]; cf. likewise Paley-Wiener [1], Ahieser [1]. The Theorem on Exponentials in the form presented in Sect. 2 is from the paper Nikol'skiĭ-Pavlov [1]. However, that paper does not contain the deciphering of the Carleson Condition ($C$) for the halfplane in terms of the Sparseness Condition ($R$) and the imbedding theorem ($CN$). These conditions look in the halfplane $\Pi$ much more "geometrical" than in the disk $\mathbb{D}$ (cf. the formulae (9) and (11)). It is likewise clear that exactly as in the case of the Theorem on Bases and Interpolation and the Carleson Imbedding Theorem one can "transplant" from the disk into the halfplane also all results on multiple and generalized interpolation in Lecture IX and X. There is a rather extensive literature devoted to direct proofs of theorems of this type, cf., for example, Džrbašyan's paper [3], which contains additional bibliographical references. The study of bases and interpolation in non-Hilbert metrics (for example, $H^p$, $\mathscr{F}L^p(\mathbb{R}_+)$, etc.) is harder to reduce to the analogous problems for spaces of functions analytic in the disk, so in this case for the most part "direct" methods, specific for the half-plane, are used; cf. Gurariĭ-Macaev [1] (bases of powers $x^\mu$ in the spaces $L^p(0,1)$ and $C_{[0,1]}$), Vinogradov [1], Džrbašyan [3].

Exponentials on a segment have been studied intensively since time immemorial and all that was known in 1934 was covered in Paley and Wiener's book [1] already quoted. They proved in particular that if $\delta \stackrel{\text{def}}{=} \sup_{n\in\mathbb{Z}}|n-\mu_n| < 1/\pi^2$ then the family $\{e^{i\mu_n w}\chi_{(0,2\pi)} : n\in\mathbb{Z}\}$ is a Riesz basis in $L^2(0,2\pi)$. Subsequently there appeared many papers (cf. e.g. the bibliography in Kacnel'son [2], likewise devoted to that question) where the constant $1/\pi^2$ is replaced by larger values. Kadec [1] showed that one can take the value 1/4 for this constant (this is Corollary 2 of the Theorem on Exponentials in $L^2(0,a)$). Before that it was long known (Ingham's example, cf. Levinson [1], p. 48) that for $\delta \geq 1/4$ the family of exponentials need not be a basis (in fact, if $\mu_n = n + \delta\operatorname{sign} n$, $n\in\mathbb{Z}$, and $\delta > 1/4$, this family is not total in $L^2(0,2\pi)$ and for $\delta \leq -1/4$ not minimal). In the paper by Levin [2] and the subsequent papers by Kacnel'son [2] and Avdonin [1] (and some other ones; cf. the bibliography of these papers) there are given sufficient conditions for a basis in the same terms as Paley and Wiener used to establish totalness or minimality in $L^2(0,a)$ of this family of exponentials, i.e. in terms of the entire function $F$ whose zero set is $\{\mu_n : n\in\mathbb{Z}\}$. The restrictions imposed on $F$ require the function $F/(w+i)$ to be square summable on $\mathbb{R}$ and $|F|$ to have slow and regular growth (decrease) on suitable lines parallel to $\mathbb{R}$. The papers Pavlov [1] and Hruščev [1] contain a complete solution of the *problem of Riesz basis* of exponentials $\{e^{i\mu w}\chi_{(0,a)} : \mu\in\tau\}$ in the space $L^2(0,a)$. The difference, as compared with the theorem established in Sect. 2, is that there it is assumed in advance that the set $\tau$ lies in a strip, as this entails the condition $0 < \inf_{\mu\in\tau}\|e^{i\mu w}\|_{L^2(0,a)} \leq \sup_{\mu\in\tau}\|e^{i\mu w}\|_{L^2(0,a)} < \infty$. Their solution is likewise stated in the language of the function $F$ and amounts to the following. First one remarks that if the family in question is total and minimal in $L^2(0,a)$ then $\tau$ is the set of (all) zeros of an entire function $G$ such that: a) $G|\Pi\in H^2_\Pi$; b) $|G(\eta)| \leq \text{const}\, e^{a|\eta|}$; c) $\overline{\lim}_{y\to+\infty} y^{-1}\log|G(iy)| = 0$; $\overline{\lim}_{y\to+\infty} y^{-1}\log|G(-iy)| = a$. It follows from a) that $(1+|w|^2)^{-1}\log|G|\in L^1(\mathbb{R})$. In particular $G$ belongs to the socalled "Cartwright class" (cf. Levinson [1], Levin [1]); therefore its zero set must be "symmetrically distributed" in the sense that $\lim_{r\to\infty} r^{-1}n_\pm(r) = a/2\pi$, $\sup_{r>0}|\sum_{\mu\in\tau,|\mu|\leq r} 1/\mu| < \infty$ (here $n_\pm(r) = \operatorname{card}\{\mu : \mu\in\tau,\ \operatorname{Re}\mu \gtrless 0,\ |\mu|\leq r\}$). To these two requirements one adds the conditions $0<\inf_{\mu\in\tau}\operatorname{Im}\mu\leq\sup_{\mu\in\tau}\operatorname{Im}\mu<\infty$ and $\inf_{\mu\neq\mu';\mu,\mu'\in\tau}|\mu-\mu'| > 0$ (that is, condition $(R)$, in the strip equivalent to Carleson's Condition $(C)$). A set $\tau$ having all these properties is called *standard* in Hruščev [1]. The basic idea in Pavlov [1] is that the operator $\mathbf{F}P_-\mathbf{F}^{-1}$ provides the projection of the subspace $\mathbf{\Theta}H^2_-$ parallel to the subspace $\mathbf{B}H^2_\Pi$ (here $\Theta = e^{iaw}$, $H^2_- = L^2(\mathbb{R})\ominus H^2_\Pi$, $P_-$ being orthogonal projection in $L^2(\mathbb{R})$ onto the subspace $H^2_-$), i.e. the operator $P_- + (P_B|K_\Theta)^{-1}P_B$ in our notation. First one must check however that (cf. Hruščev [1]) the set $\{u : u\in H^2_\Pi;\ uF, uF^{-1}\in L^2(\mathbb{R})\}$ is dense in $H^2_\Pi$. The condition for the continuity of the operator $\mathbf{F}P_-\mathbf{F}^{-1}$ is likewise precisely the Helson-Szegö Condition (HS) for the function $|F|^2$. Using the theory of the Cartwright class referred to above Pavlov's theorem can be stated as follows (cf. Hruščev [1]): a necessary and sufficient condition for the family $\{e^{i\mu w}\chi_{(0,a)}:\mu\in\tau\}$ to be a Riesz basis in $L^2(0,a)$ is that the shift $\tau+i\alpha$ is a standard set for large $\alpha$, $\alpha>0$, and that the square of the modulus of the corresponding function $F$,

$$F \stackrel{\text{def}}{=} e^{i\frac{a}{2}w} \lim_{r\to\infty} \prod_{|\mu|\le r,\,\mu\in\tau} \left(1-\frac{w}{\mu}\right),$$

satisfies Condition $(HS)$ (more exactly, its analogue on the line $\mathbb{R}$).

The exposition of the problem of bases of exponentials presented in this Lecture is based on Nikol'skiĭ [11] and Hruščev-Nikol'skiĭ-Pavlov [1].

The derivation of Kadec's $\frac{1}{4}$-Theorem from the Theorem on Exponentials in $L^2(0,a)$ is contained in Hruščev [1] but there this is done somewhat differently. In that paper also some other corollaries of the latter theorem are given; for example, if $\tau$ lies in a strip then whether the corresponding family of exponentials is an unconditional basis for $L^2(0,a)$ or not depends only on the set $\operatorname{Re}\tau$. This can also be perceived within the framework of the method by which we in Sect. 2 proved Corollary 2 of the Theorem on Exponentials in $L^2(0,a)$. In particular there exist bases with the property $\lim_n \operatorname{Im}\mu_n = 0$ ($\{\mu_n : n\in\mathbb{Z}\}=\tau$), as mentioned in Lecture VIII, Sect. 7.

All results on bases of exponentials mentioned in this Section can be found in Hruščev-Nikol'skiĭ-Pavlov [1]; the last paper contains also many other results. In particular, it contains proof of the following theorem of Vinogradov's (compare with one of the questions of Sect. 7, Lecture VIII): if $\operatorname{Im}\mu_n>0$, $\lim_n \operatorname{Im}\mu_n = +\infty$ and $\{\mu_n\}_{n\ge 0}$ satisfies the Carleson condition, then for every $a>0$ there exists a subset $\tau$ of the halfplane $\Pi$ such that $\{\mu_n\}_{n\ge 1}\subset\tau$ and such that the family $\{e^{i\mu x}\chi_{(0,a)}:\mu\in\tau\}$ forms an unconditional basis of the space $L^2(0,a)$. Regarding other questions see the Concluding Notes of Lecture VIII.

When speaking of unconditional bases of exponentials in their closed linear hulls (or of the interpolation identity 2′ on p. 262 (after the Theorem on Exponentials in $L^2(0,a)$)) one can give yet another form of the main criterion ("$P_\Theta|K_B$ is an isomorphism onto its image"; cf. Lecture VIII): the norms $\|f\chi_{(0,a)}\|_{L^2(\mathbb{R}_+)}$ and $\|f\|_{L^2(\mathbb{R}_+)}$ are equivalent in $K_B$. In this formulation the problem is very close to the one studied by Schwartz [1]. In this direction an interesting paper by Vol'berg [2] should be mentioned containing an effective necessary and sufficient condition for the norms $\|\cdot\|_{L^2(dm)}$ and $\|\cdot\|_{L^2(wdm)}$ to be equivalent on the space $K_\Theta$. Koosis [2] finds necessary and sufficient conditions on the set of zeros $\tau$ which insure the operator $f\mapsto\chi_{(a,\infty)}f$, $f\in K_B$ ($B=\prod_{\omega(\mu)\in\tau}(b_\mu\circ\omega^{-1})$), to be completely continuous for any $a$, $a>0$. In this connection we refer once more to the papers Dym-Shapiro [1], Hruščev-Vol'berg [1].

Somewhat apart from the problems studied here we have the question of *totalness* in $L^2(0,a)$ of the family of exponentials $\{e^{i\mu w}\chi_{(0,a)}:\mu\in\tau\}$. It is connected with rather hard problems of the theory of entire functions and the best result in this direction is the theorem by Beurling and Malliavin [1] where $\sup\{a:\bigvee(e^{i\mu w}\chi_{(0,a)}:\mu\in\tau)=L^2(0,a)\}$ is evaluated (in terms of the distribution of the set $\tau$).

In conclusion we wish to point out that the connections between function theory in the disk and in the halfplane extend much further than we have been able to trace them in the material of this Lecture. Here we should also have in mind the theory of Lax-Phillips [1] and many concrete problems, which arise through its intermediary in mathematical physics, but which in reality are concerned with the compressions $P_K S_E | K$ of the (multiple) shift operator, and likewise the preservation of the spectral properties of the operator (or semi-group of operators) under the Cayley transform and much more the like. For some of these subjects see Lax-Phillips [1], Sz.-Nagy and Foiaş [1], Brodskiĭ [6], Gohberg-Kreĭn [2].

# Appendix 1. The Theory of Spectral Multiplicity for Operators of Class $C_0$

**1. Contents:** We describe the invariant subspaces of $C_0$-operators with a simple spectrum (i.e. those operators $T$ of class $C_0$ for which $\mu_T = 1$, $\mu_T \stackrel{\text{def}}{=} \inf\{\text{card}\, A : \bigvee(T^n A : n \geq 0) = H\}$; we show further that every operator $T$, $T \in C_0$, is "quasi-similar" (definition below) to a unique operator of the form $\mathbf{T}_{m_1} \oplus \mathbf{T}_{m_2} \oplus \cdots \oplus \mathbf{T}_{m_s}$ where $m_1 = m_T$, $m_i/m_{i+1} \in H^\infty$ $(1 \leq i < s \leq \infty)$ and $\mathbf{T}_m = PS|K_m$, $K_m = H^2 \ominus \mathbf{m}H^2$, is the (one dimensional) model operator. Some of the intermediate results also have an independent interest.

**2. Diagonal Characteristic Function.** Let $E$ be a Hilbert space, $\Theta$ an inner $L(E)$-valued function and set $T = PS|K_\Theta$ where $K_\Theta = H^2(E) \ominus \Theta H^2(E)$. If $\dim E \leq N < \infty$ then we say that $T$ (or any unitarily equivalent operator) belongs to the class $C_0(N)$. For operators of class $C_0(N)$ with a diagonal characteristic function $\Theta = \text{diag}(\vartheta_1, \ldots, \vartheta_N)$ it is easy to estimate from below the spectral multiplicity $\mu_T$:

$$\mu_T \geq \max \text{card} \left\{ \sigma : \sigma \subset \{1, \ldots, N\},\ \bigwedge_{i \in \sigma} \vartheta_i \neq \mathbb{1} \right\}. \tag{1}$$

In fact, if $F$ is a linear submanifold of $K_\Theta$ and if there exists a set $\sigma$, $\sigma \subset \{1, \ldots, N\}$, $\bigwedge_{i \in \sigma} \vartheta_i \stackrel{\text{def}}{=} \vartheta \neq \mathbb{1}$, containing more than $\dim F$ elements then $F$ cannot be $T$-cyclic. In proving this last assertions we may assume that $\sigma = \{1, \ldots, N\}$. If then $K_\Theta = \bigvee(z^n F : n \geq 0)$, i.e. $H^2(E) = \bigvee(z^n F + \Theta H^2(E) : n \geq 0)$, there will exist $F$-valued polynomials $p_{n,j}$ and functions $h_{n,j}$ in $H^2(E)$ such that $\lim_n \| F p_{nj} + \Theta h_{nj} - e_j \mathbb{1} \|_2 = 0$, $1 \leq j \leq N$ (here, as in the Theorem on Outer Functions (Lecture I, p. 21), with which the present reasoning has some features in common, we consider, in order to shorten the notation, $F$ as a strong $H^2$-function, $F(\zeta)f = f(\zeta)$, $f \in F$). Setting $\Delta_n = \det\{F p_{nj} + \Theta h_{nj}\}_{1 \leq j \leq N}$ and taking into account that $\dim F < N$ and that all the functions $\vartheta_j$ are divisible by $\vartheta$ we get $\Delta_n = \det\{F p_{nj}\} + \vartheta \delta_n = \vartheta \delta_n \in \boldsymbol{\vartheta} H^{2/N}$, although (cf. Appendix 2, Sect. 29) $\lim_n \| \Delta_n - \mathbb{1} \|_{H^{2/N}} = 0$. This contradicts Appendix 2, Sect. 27. $\square$

In fact, in (1) we can write "$=$" instead of "$\geq$" (cf. Sec. 16 below).

**3. Quasi-Similarity.** Not only a linear similarity conserves the spectral multiplicity but this holds also true for considerably weaker notions of equivalence of operators. Let $A_i$ be an operator in a Hilbert space $H_i$, $i = 1, 2$; let us say that $A_1$ is a *deformation* of $A_2$ and write $A_1 \prec A_2$ if there exists an operator $X$ from $H_1$ into $H_2$ such that $\text{Ker}\, X = \{\mathbb{0}\}$, $\text{clos}\, X H_1 = H_2$ and $X A_1 = A_2 X$. If $A_1 \prec A_2$ and $A_2 \prec A_1$ we say that the operators $A_1$ and $A_2$ are *quasi-similar*. (An operator $X$ with these properties will be called a *deformation* from $H_1$ into $H_2$.) It is clear that

$$A_1 \prec A_2, A_2 \prec A_3 \Rightarrow A_1 \prec A_3,$$

$$A_1 \prec A_2 \Rightarrow A_2^* \prec A_1^*,$$

$$A_1 \prec A_2 \Rightarrow (\text{the } A_i \text{ are simultaneously in class } C_0 \text{ or not and } m_{A_1} = m_{A_2}); \text{ here } A_i \text{ are contractions,}$$
$$A_1 \prec A_2 \Rightarrow \mu_{A_1} \geq \mu_{A_2},$$

so that quasi-similar operators have the same spectral multiplicity. The main result of this Appendix says that any operator of class $C_0$ is quasi-similar to an operator with a diagonal (possibly infinite dimensional) characteristic function of a special type. Besides other information one can from this formula also deduce a formula for the spectral multiplicity.

**Theorem on the Jordan Model.** *Let $T \in C_0$. Among the operators quasi-similar to $T$ there exists a unique operator of the form $J_{\{m_i\}} \overset{\text{def}}{=} \sum_{i \geq 1} \oplus \mathbf{T}_{m_i}$, where the number of terms equals the spectral multiplicity $\mu_T$, with $m_1 = m_T$; $m_i/m_{i+1} \in H^\infty$, $m_i \neq \text{const}$, $i \geq 1$*

Following Sz.-Nagy and Foiaş we call operators $J_{\{m_i\}}$ of this type *Jordan operators*. The proof of this theorem will be broken up into a series of lemmata; after the proof we consider some consequences of the theorem.

**4. Quasi-Equivalence of Matrices.** For operators of class $C_0(N)$, $N < \infty$, the quasi-similarity to a Jordan model can be displayed explicitly with the aid of suitable matrix transformations. Let $\dim E < \infty$, $\Theta_k \in H^\infty(L(E))$, $k = 1, 2$. The functions $\Theta_1$ and $\Theta_2$ are called *quasi-equivalent* if there exist functions $X$ and $Y$ in $H^\infty(L(E))$ such that $X\Theta_1 = \Theta_2 Y$ and such that the inner parts $(\det X)^i$ and $(\det Y)^i$ of the corresponding determinants are relatively prime to $(\det \Theta_k)^i$, $k = 1, 2$.

**Nordgren-Moore Theorem.** *Let* $\dim E < \infty$. *Then*: 1. *If the inner functions $\Theta_1$ and $\Theta_2$ are quasi-equivalent then the operators $\mathbf{T}_{\Theta_1}$ and $\mathbf{T}_{\Theta_2}$ are quasi-similar.*

2. *Every inner function $\Theta$, $\Theta \in H^\infty(L(E))$, is quasi-equivalent to a unique function* $\operatorname{diag}(m_1, m_2, \ldots, m_N)$, $N = \dim E$, $m_i/m_{i+1} \in H^\infty$, $1 \leq i \leq N-1$, *and* $m_k = \delta_{k-1}/\delta_k$, $1 \leq k \leq N$, *where $\delta_k$ is the* GCD *of the set of all the inner parts of the minors of order $N-k$ of the matrix $\Theta$.*

*The proof* of this theorem will be given only to the extent we use it in the following considerations, namely we prove Statement 1 and a special case of Statement 2 ($N = 2$).

1. Set $K_i = H^2(E) \ominus \mathbf{\Theta}_i H^2(E)$, $i = 1, 2$, and let $X$ and $Y$ be the functions in the definition of the quasi-equivalence, writing $A = P_{\Theta_2}\mathbf{X}|K_1$. As $\mathbf{X}\mathbf{\Theta}_1 H^2(E) = \mathbf{\Theta}_2 \mathbf{Y} H^2(E) \subset \mathbf{\Theta}_2 H^2(E)$, for $f \in K_1$ we get

$$A\mathbf{T}_{\Theta_1} f = P_{\Theta_2} X P_{\Theta_1} zf = P_{\Theta_2} X zf = P_{\Theta_2} zXf = P_{\Theta_2} z P_{\Theta_2} Xf = \mathbf{T}_{\Theta_2} Af.$$

Let us check that the operator $A$ is a deformation: $AK_1 = P_{\Theta_2}\mathbf{X}K_1 = P_{\Theta_2}\mathbf{X}H^2(E) \supset P_{\Theta_2}(\det X)H^2(E)$, where the last subspace is dense in $K_2$, since $(\det X)^i$ and $(\det \Theta_2)^i$ were relatively prime (cf. property (a) of $C_0$-operators in Lecture IV, Sect. 5). That $A$ is one-to-one is proved analogously: if $f \in K_1$ and $Xf \in \mathbf{\Theta}_2 H^2(E)$ then $\Theta_2^* Xf = Y\Theta_1^* f \in H^2(E)$ and thus a fortiori $(\det Y)\Theta_1^* f \in H^2(E)$, i.e. $(\det Y) f \in$

$\Theta_1 H^2(E)$. By the same property (a) and the fact that $(\det Y)^i$ and $(\det \Theta_1)^i$ were relatively prime we get $f \in \Theta_1 H^2(E)$, i.e. $f = \mathbb{0}$.

Thus we have shown that $\mathbf{T}_{\Theta_1} \prec \mathbf{T}_{\Theta_2}$. But quasi-equivalence is an equivalence relation ($X\Theta_1 = \Theta_2 Y \Rightarrow \Theta_1(Y^{co} \det X) = (X^{co} \det Y)\Theta_2$, where $(\dots)^{co}$ stands for the co-matrix). Hence likewise $\mathbf{T}_{\Theta_2} \prec \mathbf{T}_{\Theta_1}$.

2. If $N = 2$ we are dealing with $2 \times 2$ matrices and it is easy to write down *all* pairs of matrices $X$, $Y$ such that $X\Theta = DY$ where $D = \operatorname{diag}(m_1, m_2)$, $m_1 = \det \Theta / \bigwedge_{k,j} \vartheta_{kj}^i$; $m_2 = \bigwedge_{k,j} \vartheta_{kj}^i$, $\Theta = \{\vartheta_{kj}\}$. It is clear that it suffices to consider the case $m_2 = \mathbb{1}$. Then among these pairs $X$, $Y$ there is the family of pairs $X_\alpha$, $Y_\alpha$ $(\alpha \in \mathbb{C})$ given by the identities

$$X_\alpha = \begin{pmatrix} \alpha\vartheta_{22} - \vartheta_{21} & \vartheta_{11} - \alpha\vartheta_{12} \\ \mathbb{1} & \alpha^2 \mathbb{1} \end{pmatrix}, \quad Y_\alpha = \begin{pmatrix} \alpha \mathbb{1} & \mathbb{1} \\ \vartheta_{11} + \alpha^2 \vartheta_{21} & \vartheta_{12} + \alpha\vartheta_{22} \end{pmatrix},$$

with $d_\alpha \overset{\text{def}}{=} \det X_\alpha = \det Y_\alpha = -\vartheta_{11} + \alpha\vartheta_{12} - \alpha^2\vartheta_{21} + \alpha^3\vartheta_{22}$. There remains to determine $\alpha$, $\alpha \in \mathbb{C}$, such that $d_\alpha^i \wedge \det\Theta = \mathbb{1}$. To this end let us set $\varphi_\alpha = d_\alpha^i \wedge (\det \Theta)$ and remark that $\varphi_{\alpha_1} \wedge \varphi_{\alpha_2} \wedge \varphi_{\alpha_3} \wedge \varphi_{\alpha_4} = \mathbb{1}$ for any four pairwise different points $\alpha_k$, $1 \le k \le 4$ (this follows from $\bigwedge_{k,j} \vartheta_{kj} = \mathbb{1}$ and the non-degeneracy of the van der Monde determinant). The argument is concluded by the following elementary.

**Lemma on Relatively Prime Functions.** *Let $\mathscr{D}$ be any family of scalar inner functions and $d_0$ an inner function, setting $\Phi \overset{\text{def}}{=} \{d_0 \wedge d : d \in \mathscr{D}\}$. Assume that there exists a positive integer $n$ such that $\bigwedge_{\varphi \in M} \varphi = \mathbb{1}$ for every subset $M$, $M \subset \Phi$, with* card $M \ge n$. *Then all functions in $\Phi$, except for an at most countable subset, equal $\mathbb{1}$.*

*Proof of the Lemma.* If $n = 1$ the assertion is trivial. Assume therefore that it is fulfilled for all numbers less than $n$. Let $A$ be the set of all subsets of $\Phi$ each consisting of $n-1$ elements and set $\Psi \overset{\text{def}}{=} \{\psi_a : \psi_a = \bigwedge_{\varphi \in a} \varphi,\ a \in A\}$. All the functions in $\Psi$ are relatively prime (cf. the hypothesis of the lemma) and all divide the inner function $d_0$; therefore (cf., for example, Appendix 2, Sect. 22, 23) $\Psi$ contains at most a countable subset $\Psi_0$ of functions different from $\mathbb{1}$. The set $\Phi_0 \overset{\text{def}}{=} \bigcup_{\psi_a \in \Psi_0} a$ is likewise at most countable. If $a \subset \Phi \backslash \Phi_0$, card $a = n-1$, then $\bigwedge_{\varphi \in a} \varphi = \mathbb{1}$; by the induction hypothesis therefore all functions in $\Phi \backslash \Phi_0$, except for a countable subset, equal $\mathbb{1}$. □□

**Corollary.** *Let $T \in C_0(2)$. The following statements are equivalent*: 1. $\mu_T = 1$; 2. $\mu_{T^*} = 1$; 3. $\det \Theta_T = m_T$.

In fact, quasi-similarity preserves the multiplicities $\mu_T$ and $\mu_{T^*}$ and the minimal annihilator and if it is achieved (as in the theorem) at the expense of quasi-equivalence of the characteristic functions then the determinant $\det \Theta_T$ also is preserved. For the canonical (Jordan) matrix $\operatorname{diag}(m_1, m_2)$ the desired equivalence is trivial. □

**5. Lemma.** *If $T \in C_0$ then there exists an operator $T_1$, $T_1 \in C_0(\mu_T)$, such that $T_1 \prec T$.*

*Proof.* Let $U$ be a unitary dilation of the operator $T$ acting in the space $H$ and $L$ a subspace of $H$ with $\dim L = \mu_T$, $\bigvee(T^n L : n \geq 0) = H$. Let us set $N = \bigvee(U^n L : n \geq 0)$, $N_0 = \operatorname{Ker} P_H | N$, $H_1 = N \ominus N_0$, $X = P_H | H_1$, $T_1 = P_{H_1} U | H_1$. Then $TX = P_H U P_H | H_1 = P_H U | H_1 = P_H(P_{H_1} + P_{N_0}) U | H_1 = P_H P_{H_1} U | H_1 = XT_1$.

Let us verify that $X$ has a trivial kernel and dense range: $\operatorname{Ker} X = (\operatorname{Ker} P_H) \cap H_1 = N_0 \cap H_1 = \{\mathbb{0}\}$, $\operatorname{clos} XH_1 = \operatorname{clos} P_H H_1 = \operatorname{clos} P_H N = \operatorname{clos} P_H(\bigvee(U^n L : n \geq 0)) = \bigvee(T^n L : n \geq 0) = H$.

Next let us show that $T_1 \in C_0(\mu_T)$. As $TX = XT_1$ we have $m_T(T_1) = \mathbb{0}$, i.e. $T_1 \in C_0$. It remains to remark that $T_1$ has as unitary dilation the shift of multiplicity not exceeding $\mu_T$: $\dim(N \ominus UN) = \dim(\bigvee(U^n L : n \geq 0) \ominus \bigvee(U^n L : n \geq 1)) \leq \dim L = \mu_T$. □

The rest follows by combining the Dilation Theorem (Intr. Lecture), Sarason's Projection Lemma (Lecture III, Sect. 5) and the Theorem on Invariant Subspaces (Lecture I, Sect. 5). □

**6. Lemma.** *Let $T$ be an operator of class $C_0$ in the Hilbert space $H$ and set $m_x \overset{\text{def}}{=} m_{T|E_x}$, $E_x \overset{\text{def}}{=} \bigvee(T^n x : n \geq 0)$ for $x \in H$. Then the set $A = \{x \in H : m_T \neq m_x\}$ is an $F_\sigma$ set and of the first category in $H$.*

*Proof.* We can assume that $m_T(0) > 0$. Set

$$A_n = \{x \in H : m_x(0) \geq m_T(0) + 1/n\}, \; n \geq 1.$$

Then $A = \bigcup_{n \geq 1} A_n$ and moreover all the $A_n$ are closed. In fact, if $x_k \in A_n$, $\lim_k x_k = x$ then we can assume that $\lim_k m_{x_k}(\zeta) = \varphi(\zeta)$, $|\zeta| < 1$, where $\varphi \in H^\infty$. Then $\varphi(0) \geq m_T(0) + 1/n$, $|\varphi| \leq 1$ in $\mathbb{D}$ and in view of the strong convergence (s)-$\lim_k m_{x_k}(T) = \varphi(T)$ (Lecture III, Sect. 5) holds $\varphi(T)x = \mathbb{0}$. It is clear that then also $\varphi^i(T)x = \mathbb{0}$ (Lecture IV, Sect. 5, Statement (b)), $\varphi^i(0) \geq m_T(0) + 1/n$ whence $m_x(0) \geq m_T(0) + 1/n$.

The sets $A_n$ do not contain inner points. If the sequence $\{x_k\}$ is dense in some ball in the space $H$ then $m_T(0) = \lim_n(m_{x_1} \vee m_{x_2} \vee \dots \vee m_{x_n})(0)$ so that it suffices to prove (for any $n$) the existence of an element $x$ in this ball such that $m_x = m_{x_1} \vee \cdots \vee m_{x_n}$. To this end it is again sufficient to find a convex combination $x_\alpha = \alpha x + (1 - \alpha) y$, $0 \leq \alpha \leq 1$, of two elements $x$ and $y$ with the property that $m_{x_\alpha} = m_x \vee m_y$ and then proceed by induction. It is clear that the functions $m_\alpha \overset{\text{def}}{=} m_{x_\alpha}$ divide the function $m_x \vee m_y$ and that for $\alpha \neq \alpha'$ we have $m_\alpha \vee m_{\alpha'} = m_x \vee m_y$. Finally, apply the Lemma on Relatively Prime Functions to the ratios $m^\alpha = (m_x \vee m_y)/m_\alpha$. □

**7. Lemma.** $J_{\{m_i\}} \prec J_{\{m_i'\}} \Rightarrow J_{\{m_i\}} = J_{\{m_i'\}}$.

*Proof.* It is clear that $m_1 = m_1'$. Assume that $m_i = m_i'$ for $i < j$ and let us show that $m_j = m_j'$ (which will complete the proof). As $J^*_{\{m_i'\}} \prec J^*_{\{m_i\}}$ and for one dimensional model operators the adjoint $\mathbf{T}^*_m$ is unitarily equivalent to $\mathbf{T}_{m^t}$, it is sufficient to establish that $m_j$ divides $m_j'$. If now $m_j$ does not divide $m_j'$ there exists an inner function $\vartheta$, $\vartheta \not\equiv \text{const}$, dividing $m_j$, with $\vartheta \wedge m_j' = \mathbb{1}$. Let us set $K = \sum_{i=1}^{j} \oplus(\mathbf{m}_i/\vartheta H^2 \ominus \mathbf{m}_i H^2)$, $K' = \sum_{i=1}^{j-1} \oplus(\mathbf{m}_i/\vartheta H^2 \ominus \mathbf{m}_i H^2)$ and consider a deformation $X$ mapping $J_{\{m_i'\}}$ into $J_{\{m_i'\}}: J_{\{m_i'\}}X = XJ_{\{m_i\}}$. Set $T = J_{\{m_i\}}|K$, $T' = J_{\{m_i'\}}|K'$. As $K \subset \operatorname{Ker} \vartheta(J_{\{m_i\}})$ then $XK \subset \operatorname{Ker} \vartheta(J_{\{m_i'\}}) \subset K'$. Let $Y = X|K$. Then $T'Y = YT$ and $T^*Y^* = Y^*(T')^*$, with clos $Y^*K' = K$ in view of Ker $Y = \{\mathbb{0}\}$. As $\mu_{(T')^*} = j - 1$ (cf. Sect. 2 above; here we further use the fact that the restriction $\mathbf{T}_{m_i}|(\mathbf{m}_i/\boldsymbol{\vartheta}H^2 \ominus \mathbf{m}_i H^2)$ is unitarily equivalent to $\mathbf{T}_\vartheta$) therefore $\mu_{T^*} \leq j - 1$, contradicting the identity $\mu_{T^*} = j$, which again follows from Sect. 2 and the definition of the operator $T$. □

**8. Lemma.** *If $T \in C_0$, $\mu_T \leq 2$, $\mu_{T^*} = 1$ then $\mu_T = 1$ and $T$ is quasi-similar to the rank one model operator $\mathbf{T}_{m_T}$.*

*Proof.* By Lemma 5 there exist operators $T_i (i = 1, 2)$ such that $T_1 \in C_0(1)$ and $T_2 \in C_0(2)$, with $T_2 \prec T$, $T_1 \prec T^*$. But the class $C_0(1)$ consists only of the model operators $\mathbf{T}_m$ and $T_1^* \in C_0(1)$; hence $T_1^* = \mathbf{T}_m$ for some inner function $m$ and so from $T \prec T_1^*$ we get $m = m_T$. By the Nordgren-Moore Theorem (the case $N = 2$) $T_2$ must be quasi-similar to the model $J_{\{m_1, m_2\}}$, whence $J_{\{m_1, m_2\}} \prec T \prec \mathbf{T}_m$. Finally, by Lemma 7 $J_{\{m_1, m_2\}} = \mathbf{T}_m$. □

**9. Lemma.** *Let $T$ be an operator of class $C_0$ in the Hilbert space $H$, with $\mu_T \leq 2$, $\mu_{T^*} = 1$. If $E \in \operatorname{Lat} T$ and $\mu_{T|E} = 1$, $m_{T|E} = m_T$ then $E = H$.*

*Proof.* By Lemma 8 there exist deformations $X$ and $Y$ such that $\mathbf{T}_m X = XT$, $Y\mathbf{T}_m = TY$ where $m = m_T$ and $\mathbf{T}_m = PS|K_m$, $K_m = H^2 \ominus \mathbf{m}H^2$. As $1 \leq \mu_{(T|E)^*} \leq \mu_{T^*} = 1$ (cf. Lecture II, Sect. 4) we can apply Lemma 8 to $T|E$ too and we get a deformation $Y_E$ such that $Y_E \mathbf{T}_m = (T|E) Y_E$. As $E \subset H$, we can assume that $Y_E$ acts from $K_m$ into $H$ and clos $Y_E K_m = E$.

Let us set $Z = Y_E X$; then clos $ZH$ = clos $Y_E XH$ = clos $Y_E K_m = E$. From the fact that the operators $X$, $Y$ and $Y_E$ intertwine with the operators $T$, $\mathbf{T}_m$ and $T|E$ (in appropriate combinations), we immediately see that $(XY)\mathbf{T}_m = \mathbf{T}_m(XY)$, $(XZY)\mathbf{T}_m = \mathbf{T}_m(XZY)$. By Sarason's Theorem (Lecture VIII) there exist functions $u, v \in H^\infty$ such that $XZY = u(\mathbf{T}_m)$, $XY = v(\mathbf{T}_m)$. Hence $YXZY = Yu(\mathbf{T}_m) = u(T)Y$, i.e. $YXZ = u(T)$, and $YXY = Yv(\mathbf{T}_m) = v(T)Y$, i.e. $YX = v(T)$. The operators $X$, $Y$, $Z$ being injective we have then Ker $u(T) = \{\mathbb{0}\}$ and $u^i \wedge m = \mathbb{1}$ (cf. Lecture IV, Sect. 5, Statement (a)). Consequently clos $u(\mathbf{T}_m)K_m = K_m$ and $H$ = clos $YK_m$ = clos $Yu(\mathbf{T}_m)K_m$ = clos $YXZYK_m$ = clos $v(T)ZH$ = clos $v(T)E \subset E$, i.e. $H = E$. □

**10. Lemma.** *Let $T \in C_0$. Then $\mu_T = 1 \Leftrightarrow \mu_{T^*} = 1$.*

*Proof.* It is clear that it suffices to prove only one of the implications. Assume that $\mu_{T^*} = 1$. By Lemma 6 there exists a vector $x$ such that $m_x = m_T$. Let us verify that $E_x \overset{\text{def}}{=} \bigvee (T^n x : n \geq 0) = H$ (the space where $T$ acts). Let $y \in H$ and set $H_y = E_x \vee E_y$, $T_y = T|H_y$. Then $\mu_{T_y} \leq 2$, $m_{T_y} = m_T$, that is, Lemma 9 is applicable to the subspace $E = E_x$. By that lemma $H_y = E_x$, i.e. $y \in E_x$. □

**11. Lemma.** *Let $T$ be an operator of class $C_0$ in the space $H$. For any $x$, $x \in H$, and any $\varepsilon$, $\varepsilon > 0$, there exist an element $y$, $y \in H$, a subspace $E$, $E \in \operatorname{Lat} T$, and a deformation $X$, $X: K_m \to E_y$, with $m = m_T$, such that:* a) $m_y = m$; b) $\|x - y\| < \varepsilon$; c) $X\mathbf{T}_m = TX$; d) $E \vee E_y = H$; e) $XK_m \cap E = \{\mathbb{0}\}$.

*Proof*. By Lemma 6 there exists a vector $u$ such that $\|u - x\| < \varepsilon/2$, $m_{T^*|E_u^*} = m_{T^*}$ $(\equiv m_T^t)$; here $E_u^* = \bigvee(T^{*n}u : n \geq 0)$. Set $T_u = (T^*|E_u^*)^*$; then $T_u \in C_0$; $\mu_{T_u^*} = 1$, whence $\mu_{T_u} = 1$. Consequently, there exists a cyclic vector $y$ for $T_u$ with $\|y - u\| < \varepsilon/2$. (That the cyclic vectors of $T$ ($T \in C_0, \mu_T = 1$) are dense in $H$ follows by juxtaposition of the Lemmata 6, 9 and 10. Direct proof: if $x \in C^T$ and $\varphi^i \wedge m_T = \mathbb{1}$ then $\operatorname{Ker} \varphi(T)^* = \{\mathbb{0}\}$, i.e. $\varphi(T)x \in C^T$; it is however clear that there exist polynomials $\varphi$ arbitrarily close to a given polynomial $p$ which are also relatively prime to $m_T$. □) As $P_{E_u^*}T^n = T_u^n P_{E_u^*}$ then $E_u^* = \operatorname{clos} P_{E_u^*}E_y$. Let us show that $m_{T_u} = m_{T|E_y} = m_T : m_{T_u} = m_{T_u^*}^t = (m_T^t)^t = m_T$ and, since $y \in C^{T_u}$ and $m_{T|E_y}(T_u)y = P_{E_u^*}m_{T|E_y}(T)y = P_{E_u^*}m_{T|E_y}(T|E_y)y = \mathbb{0}$, we have $m_{T|E_y}(T_u) = \mathbb{0}$, i.e. $m_T$ divides $m_{T|E_y}$, i.e. $m_{T|E_y} = m_T$.

By Lemma 5 we can find an operator $T_1$, $T_1 \in C_0(1)$, such that $T_1 \prec T|E_y$. It is clear that $T_1 = \mathbf{T}_m$ for some inner function $m$ and that there exists a deformation $X$, $X : K_m \to E_y$, with $(T|E_y)X = X\mathbf{T}_m$, i.e. $TX = X\mathbf{T}_m$. Hence $m = m_T$.

Let us set $E = (E_u^*)^\perp$; $E \vee E_y = (E_u^*)^\perp \vee P_{E_u^*}E_y = (E_u^*)^\perp \vee E_u^* = H$. There remains the verification of Statement e). Applying Lemma 5 to the operator $T_u^*$ we get $T_u^* \prec \mathbf{T}_m t$, i.e. $T_u \prec \mathbf{T}_m^* t = \mathbf{T}_m$. Thus there exists a deformation $Y$, $Y : E_u^* \to K_m$, intertwining $\mathbf{T}_m$ and $T_u$: $\mathbf{T}_m Y = YT_u$. Set $W = YP_{E_u^*}X$. Then $\mathbf{T}_m W = \mathbf{T}_m YP_{E_u^*}X = YT_u P_{E_u^*}X = YP_{E_u^*}TX = YP_{E_u^*}X\mathbf{T}_m = W\mathbf{T}_m$. By Sarason's Theorem (Lecture VIII) we can find a function $w$, $w \in H^\infty$ such that $W = w(\mathbf{T}_m)$. As $\operatorname{clos} WK_m = \operatorname{clos} YP_{E_u^*}XK_m = \operatorname{clos} YP_{E_u^*}E_y = \operatorname{clos} YE_u^* = K_m$ we have $w \wedge m_T = \mathbb{1}$. Finally let $h \in E \cap XK_m$, i.e. $h = Xg$, $g \in K_m$. Then $Wg = YP_{E_u^*}h = \mathbb{0}$ so that $g = \mathbb{0}$ and $h = \mathbb{0}$. □

**12. Lemma.** *If* $T \in C_0$ *there exists a Jordan operator* $J_{\{m_i\}}$ *which is a deformation of* $T$.

*Proof*. Let $D$ be a countable dense subset of $H$ (the space where $T$ acts) and let $\{x_n\}_{n \geq 1}$ be a sequence assuming every value in $D$ infinitely many times. By Lemma 11 we find an element $y_1$ with $m_{y_1} = m_T \overset{\text{def}}{=} m_1$,

$$\|x_1 - y_1\| < 1/2, \tag{2}$$

a subspace $E_1 \in \operatorname{Lat} T$ and a deformation $X_1 : K_{m_1} \to E_{y_1}$, such that: 1) $X_1\mathbf{T}_{m_1} = TX_1$; 2) $E_1 \vee E_{y_1} = H$; 3) $X_1K_{m_1} \cap E_1 = \{\mathbb{0}\}$; 4) $|X_1| < 1/2$.

It follows from (2) that $\|(I - P_{E_{y_1}})x_1\| < 1/2$. By property 2) there exist $u_1 \in E_1$ and $v_1 \in E_{y_1}$ with

$$\|x_2 - (u_1 + v_1)\| < 1/8, \tag{3}$$

in $E_1$ a vector $y_2$ with $m_{y_2} = m_{T|E_1} \overset{\text{def}}{=} m_2$,

$$\|y_2 - u_1\| < 1/8, \tag{4}$$

a subspace $E_2 \in \operatorname{Lat}(T|E_1)$ and, finally, a deformation $X_2$, $X_2 : K_{m_2} \to E_{y_2}$, such that $X_2\mathbf{T}_{m_2} = TX_2$, $E_2 \vee E_{y_2} = E_1$, $X_2K_{m_2} \cap E_2 = \{\mathbb{0}\}$, $|X_2| < 1/4$. From (3) and (4) it follows that $\|x_2 - (v_1 + y_2)\| < 1/4$, i.e. $\|(I - P_{E_{y_1} \vee E_{y_2}})x_2\| < 1/4$.

By induction we construct subspaces $E_n \in \operatorname{Lat} T$, $E_{n+1} \subset E_n$, vectors $y_n$, $y_n \in E_{n-1}$, $m_{y_n} = m_{T|E_{n-1}} \overset{\text{def}}{=} m_n$, and deformations $X_n$, $X_n : K_{m_n} \to E_{y_n}$ such that $X_n\mathbf{T}_{m_n} =$

$TX_n$, $E_n \vee E_{y_n} = E_{n-1}$, $X_n K_n \cap E_n = \{\mathbb{0}\}$, $|X_n| < 2^{-n}$, $\|(I - P_{E_{y_1} \vee \cdots \vee E_{y_n}})x_n\| < 2^{-n}$. Set now $K = \sum_{n \geq 1} \oplus K_{m_n}$, $X\{h_n\}_{n \geq 1} = \sum_{n \geq 1} X_n h_n$ for $\{h_n\}_{n \geq 1} \in K$. It is clear that $X$ is a continuous operator and that $XJ_{\{m_n\}} = TX$ where $J_{\{m_n\}} = \sum_{n \geq 1} \oplus \mathbf{T}_{m_n}$. It remains to be shown that $X$ is a deformation. If $X\{h_n\}_{n \geq 1} = \mathbb{0}$ then (setting $h_0 \stackrel{\text{def}}{=} \mathbb{0}$) from $h_n = \mathbb{0}$, $n < j$, it follows that $h_j = \mathbb{0} : X_j h_j = -\sum_{n > j} X_n h_n \in \bigvee(E_{y_n} : n > j) \subset E_j$, so that $h_j = \mathbb{0}$ as $X_j K_{m_n} \cap E_j = \{\mathbb{0}\}$. Hence $\operatorname{Ker} X = \{\mathbb{0}\}$.

$$\operatorname{clos} XK = \bigvee(X_n K_{m_n} : n \geq 1) = \bigvee(E_{y_n} : n \geq 1) \stackrel{\text{def}}{=} H',$$

so that $D \subset H'$ and $H' = H$, since $\|(I - P_{H'})x_n\| < 2^{-n}$. □

**13. Proof of the Theorem on the Jordan Model.** By Lemma 12 there exists a Jordan operator $J = J_{\{m_i\}}$, $J \prec T$; analogously $J' \prec T^*$, i.e. $J \prec T \prec J'^*$. By Lemma 7 $J'^* = J$, i.e. $T$ is quasi-similar to $J$. □

**14. Corollary.** *Let $T \in C_0$. The following statements are equivalent.*

1. $\mu_T = 1$.
2. $\mu_{T^*} = 1$.
3. *$T$ is quasi-similar to $\mathbf{T}_m$, $m = m_T$.*
4. *If $E \in \operatorname{Lat} T$ then there exists a divisor $m_E$ of $m$ such that $E = \operatorname{Ker} m_E(T)$.*
5. *If $E \in \operatorname{Lat} T$, $m_{T|E} = m$ then $E = H$ ($H$, the space where $T$ acts).*

In fact, the implications $1 \Leftrightarrow 2 \Leftrightarrow 3$ have already been established. $1 \mathbin{\&} 2 \Rightarrow 4$: set

$$m_E = m_{T|E} \text{ and } E' = \operatorname{Ker} m_E(T),\ T' = T|E';$$

then by Lemma 9 (applied to the operator $T'$ and the subspace $E$, $E \subset E'$) $E = E'$.

$4 \Rightarrow 5$ in the trivial manner.

$5 \Rightarrow 1$: by Lemma 6 there exists an $x$, $x \in H$, such that $m_{T|E_x} = m_T$; by the hypothesis then $E_x = H$ and consequently $\mu_T = 1$. □

*Remark.* The Theorem on the Simplicity of the Spectrum, formulated in Lecture IV (Sect. 5), is contained in the union of the preceding corollary and the Nordgren-Moore Theorem.

**15. Corollary.** *Let $T \in C_0$, $\mu_T = 1$. The vector $x$ is cyclic for $T$ iff*

$$\varphi(T)x = \mathbb{0},\ \varphi \in H^\infty \Rightarrow \varphi/m_T \in H^\infty. \ \square$$

**16. Corollary.** *In formula* (1) *we can take "$=$" in place of "$\geq$".*

By the Nordgren-Moore Theorem,

$$\mu_T = \min\left\{N - s : \bigwedge_{\tau \subset \{1,\ldots,N\},\, \operatorname{card} \tau = s} \left(\prod_{i \in \tau} \vartheta_i\right) = \mathbb{1}\right\}.$$

It is however not hard to see that this min always coincides with the number max in (1). □

**17. Concluding Remarks.** The simplicity of the spectrum of a $C_0$-operator can be expressed yet in a number of other ways; for example, it is equivalent to the identity $\{T\}' = \{\{T\}'\}'$ ($\{\dots\}'$ is the commutant of the family $\{\dots\}$) or the identity

$$\{T\}' = \{A : \exists u, v \in H^\infty : v^i \wedge m_T = \mathbb{1}, A = u(T)v(T)^{-1}\}.$$

These and almost all other theorems on spectral multiplicity were obtained by Sz.-Nagy and Foiaş, cf. [7, 8]; instead of the word "deformation" there is used in these papers the term "quasi-affine transformation". The notion of quasi-similarity has turned out to be useful in other questions too (for example, it preserves the non-triviality of the lattice Lat$\{T\}'$ and, for $C_0(N)$ operators, this lattice itself; cf. Sz.-Nagy and Foiaş [1]). Regarding the Nordgren-Moore Theorem, see Nordgren [1], Moore-Nordgren [1]. The notion of quasi-similarity generalizes the notion of "quasi-direct sum of operators" in the sense of Brodskiĭ and Kisilevskiĭ, which was used earlier for analogous purposes; cf. Brodskiĭ [6]. For a generalization of the Nordgren-Moore Theorem to general $C_0$-operators see Berkovici-Voiculescu [1], Müller [1].

In connection with the topics of Appendix 1 see further Berkovici, Foiaş and Sz.-Nagy [1], Szücz [1], Sherman [1], Sz.-Nagy [3].

# Appendix 2. Summary of $H^p$ Spaces

**1. Contents:** Definition, Poisson kernel, imbedding of the space $H^p$ into $L^p(\mathbb{T})$, Fatou's theorem, Jensen's inequality, uniqueness theorems for the classes $H^p$, Blaschke products, Herglotz's theorem and singular functions, the multiplicative structure of $H^p$-functions, characterization of outer functions, the classes $H^p$ for $0 < p < 1$, the Douglas-Rudin theorem, the Nevanlinna class.

**2. Definition.** Let $f$ be a function analytic in the disk $\mathbb{D} = \{\zeta \in \mathbb{C} : |\zeta| < 1\}$ and let $0 < p < \infty$. We say that $f$ belongs to the *Hardy class* $H^p$ if

$$\|f\|_p \overset{\text{def}}{=} \sup_{0 \le r < 1} \left( \int_{\mathbb{T}} |f(r\zeta)|^p \, dm(\zeta) \right)^{1/p} < \infty.$$

Obvious modification for $p = \infty$: $\|f\|_\infty \overset{\text{def}}{=} \sup\{|f(\zeta)| : \zeta \in \mathbb{D}\}$.

It is clear that $H^p$ is a vector space, on which for $p \ge 1$ the functional $\|\cdot\|_p$ is a norm, whereas for $0 < p < 1$ the formula $\varrho(f, g) = \|f - g\|_p^p$ gives an invariant metric.

For $p = 2$ we get by Parseval's formula

$$\|f\|_2 = \sup_{0 \le r < 1} \left( \sum_{n \ge 0} |\hat{f}(n)|^2 r^{2n} \right)^{1/2} = \left( \sum_{n \ge 0} |\hat{f}(n)|^2 \right)^{1/2},$$

where $\hat{f}(n)$ are the Taylor coefficients of $f$ ($f = \sum_{n \ge 0} \hat{f}(n) z^n$).

**3. The Poisson Kernel.** This is the function $\xi \to P(\xi) \overset{\text{def}}{=} (1 - |\xi|^2)|1 - \xi|^{-2}$, $\xi \in \mathbb{D}$, or the family of functions $\{P_\zeta\}_{\zeta \in \mathbb{D}}$, $P_\zeta(\xi) \overset{\text{def}}{=} P(\xi\zeta)$, $|\xi| \le 1$. With every analytic function in $\mathbb{D}$ we may associate the family $\{f_\zeta\}_{\zeta \in \mathbb{D}}$, $f_\zeta(\xi) \overset{\text{def}}{=} f(\zeta\xi)$, $|\xi| \le 1$. It is easily seen that $f_{\zeta\xi} = f_\zeta * P_\xi$ for all $\zeta, \xi \in \mathbb{D}$, where convolution $*$ is defined by the formula $(f * g)(\lambda) = \int_{\mathbb{T}} f(t) g(\lambda \bar{t}) \, dm(t)$, $\lambda \in \mathbb{D}$. As

$$\|f * g\|_{L^p(\mathbb{T})} \le \|f\|_{L^p(\mathbb{T})} \|g\|_{L^1(\mathbb{T})}$$

for $p \ge 1$ (well-known and easy to prove), it follows from this and the identity

$$\|P_\zeta\|_{L^1(\mathbb{T})} = \int_{\mathbb{T}} P(\zeta t) \, dm(t) = P(0) = 1$$

that

$$\|f_{r'}\|_{L^p(\mathbb{T})} \le \|f_r\|_{L^p(\mathbb{T})} \text{ for } 0 \le r' \le r < 1, p \ge 1. \tag{1}$$

It is clear that this inequality is true also for functions $f$ holomorphic just in the disk $\{\zeta : |\zeta| \le r\}$. The general case $p > 0$ requires a minor additional argument.

**4. Lemma.** $\|f_{r'}\|_{L^p(\mathbb{T})} \le \|f_r\|_{L^p(\mathbb{T})}$ *for any* $0 \le r' \le r < 1$, $p > 0$, *and any function* $f$ *analytic in the disk* $\mathbb{D}$.

*Proof.* Let (as everywhere in this book) $b_\lambda = |\lambda|\lambda^{-1}\cdot(\lambda - z)(\mathbb{1} - \bar{\lambda}z)^{-1}$, $\lambda \in \mathbb{D}$, and let

$$B = \prod_{\lambda, |\lambda| < r} b_{\lambda/r}\left(\frac{z}{r}\right)^{k_f(\lambda)}, \tag{2}$$

be the finite product extended only over those $\lambda$ for which $|\lambda| < r$, $k_f(\lambda) > 0$ ($k_f(\lambda)$ is the multiplicitiy of the zero of $f$ at the point $\lambda$). As the integral $\int_{\mathbb{T}} |f_r|^p\, dm$ depends continuously on $r$, it suffices to consider the case $f(\zeta) \neq \mathbb{0}$, $|\zeta| = r$. Then $f/B$ does not vanish in the disk $\{\zeta : |\zeta| \leq r\}$ and $|B| \leq 1$ in this disk and $= 1$ on the boundary. Therefore

$$\begin{aligned}\int_{\mathbb{T}} |f_{r'}|^p\, dm &\leq \int_{\mathbb{T}} |(f/B)_{r'}|^p\, dm \\ &= \int_{\mathbb{T}} |((f/B)^p)_{r'}|\, dm \\ &\leq \text{(in view of (1))} \int_{\mathbb{T}} |((f/B)^p)_r|\, dm \\ &= \int_{\mathbb{T}} |f_r|^p\, dm. \ \square\end{aligned}$$

**5. Corollary.** *If* $f \in H^p$, $p > 0$, *then* $\|f\|_p = \lim_{r\to 1-0} \|f_r\|_{L^p(\mathbb{T})}$. $\square$

**6. Theorem.** *Let* $f \in H^p$, $p \geq 1$. *Then*:

a. *There exists a unique "boundary function" $bf$ in $L^p(\mathbb{T})$ such that* $f_r|\mathbb{T} = bf * P_r$, $0 \leq r < 1$.

b. $\|bf\|_{L^p(\mathbb{T})} = \|f\|_p$ *and* $\lim_{r\to 1-0} \|bf - f_r\|_{L^p(\mathbb{T})} = 0$ *for* $p < \infty$.

c. $bH^p = \{F \in L^p(\mathbb{T}) : \hat{F}(n) = 0, n < 0\}$.

*Proof.* Assume first that $p > 1$. In view of the weak compactness of the unit ball of $L^p(\mathbb{T})$ every family $\{f_{r_n}|\mathbb{T}\}$, $r_n \uparrow 1$, has a (weak) limit point. Moreover, all these limit points coincide (because the Fourier series of the function $f_r|\mathbb{T}$ equals $\sum_{k\geq 0} \hat{f}(k) r^k z^k$ and $\lim_n r_n^k = 1$, $\forall k$). If we denote this unique function by $bf$ and compare the Fourier coefficients of $f_r|\mathbb{T}$ and $bf * P_r$, we obtain Statement a.

For $p = 1$ we use in a similar way the weak compactness of the ball in the space of complex Borel measures (i.e., in the dual of the space $C(\mathbb{T})$). We get a measure $bf$ such that $(bf)^\wedge(n) = 0$, $n < 0$, $(bf)^\wedge(n) = \hat{f}(n)$, $n \geq 0$ (in the right hand side of this identity we have Taylor coefficients, in the left hand side Fourier coefficients). By the F. and M. Riesz Theorem (Lecture I, Sect. 5) the measure $bf$ must be absolutely continuous with respect to Lebesgue measure $m$, and Statement a follows anew.

As the family $\{P_r\}_{0\leq r<1}$ is an approximate identity (i.e. $\sup_r \|P_r\|_{L^1(\mathbb{T})} < \infty$, $\lim_{r\to 1-0} \hat{P}_r(n) = 1$, $\forall n \in \mathbb{Z}$), we can for $p < \infty$ use the trivial part of the Banach-Steinhaus theorem (*hint*: the trigonometric polynomials are dense in $L^p(\mathbb{T})$) and we get

$$\lim_{r\to 1-0} \|bf - f_r\|_{L^p(\mathbb{T})} = \lim_{r\to 1-0} \|bf - bf * P_r\|_{L^p(\mathbb{T})} = 0.$$

It follows from this and Corollary 5 that $\|bf\|_{L^p(\mathbb{T})} = \|f\|_p$, $p < \infty$. If $p = \infty$ taking here the exact upper bound with respect to $p$ we get $\|bf\|_{L^\infty(\mathbb{T})} = \|f\|_\infty$. Statement $b$ is thus proved.

From the above it is clear that $bH^p \subset \{F \in L^p(\mathbb{T}) : \hat{F}(n) = 0, n < 0\}$. Conversely, if $F \in \{\dots\}$ let us set $f = \sum_{n \geq 0} \hat{F}(n) z^n$. Then $f_r | \mathbb{T} = F * P_r$, $0 \leq r < 1$. Hence $f \in H^p$. Upon comparing the Fourier coefficients of $F$ and $bf$ we see that $bf = F$. □

**7. Convention.** We agree to identify the functions $bf$ and $f$ ($f \in H^p$) everywhere where this does not lead to any ambiguity. Thus as a complement to Definition 2 we may assert that

$$H^p = \{f \in L^p(\mathbb{T}) : \hat{f}(n) = 0, n < 0\}, \tag{3}$$

and that $f_r = f * P_r$, $0 \leq r < 1$, for $f \in H^p$. Here $p \geq 1$.

**8. Fatou's Theorem.** *Let $\mu$ be a complex Borel measure on the circumference $\mathbb{T}$ and set $f = P * \mu$, that is,*

$$f(\zeta) = \int_{\mathbb{T}} P(\zeta \bar{t}) \, d\mu(t), \; \zeta \in \mathbb{D}.$$

*Let $t$ be a point of $\mathbb{T}$ where $\mu$ is differentiable with respect to the Lebesgue measure $m$. Then $f$ has an angular boundary value at $t$, which equals $d\mu(t)/dm(t)$, in the sense that,*

$$\lim_{\zeta \in St, \zeta \to t} f(\zeta) = d\mu(t)/dm,$$

*where $St$ is an arbitrary Stolz domain with vertex at $t$, i.e. an angle of the form $St = \{\zeta \in \mathbb{D} : |\arg(t - \zeta)| < \pi/2 - \alpha\}$, $\alpha > 0$.*

*Outline of proof.* Without loss of generality we may take $t = 1$ and $\mu(\mathbb{T}) = 0$. The measure $\mu$ admits a primitive $F$, that is, a function of bounded variation on the interval $[-\pi, \pi]$ such that $F(-\pi) = F(\pi)$ and

$$f(\zeta) = \int_{-\pi}^{\pi} P(\zeta e^{-i\tau}) \, dF(\tau), \; \zeta \in \mathbb{D}.$$

Integrating by parts we get for $\zeta \in \mathbb{D}$

$$f(\zeta) = \int_{-\pi}^{\pi} (\partial P_\zeta / \partial \tau)(e^{-i\tau}) F(\tau) \, d\tau,$$

where $\partial P_\zeta / \partial \tau$ is the derivative of the function $\tau \to P_\zeta(e^{i\tau})$. The Poisson kernel being an even function we get

$$\begin{aligned} f(\zeta) &= \frac{1}{2} \int_{-\pi}^{\pi} (\partial P_\zeta / \partial \tau)(e^{-i\tau}) [F(\tau) - F(-\tau)] \, d\tau \\ &= 2\pi \int_{-\pi}^{\pi} [\tau (\partial P_\zeta / \partial \tau)(e^{-i\tau})] \frac{F(\tau) - F(-\tau)}{2\tau} \, d\tau / 2\pi. \end{aligned}$$

One can now verify that for the family of functions $\{k_\zeta\}_{\zeta\in St}$, $k_\zeta(\tau)\overset{\text{def}}{=}\tau(\partial P_\zeta/\partial\tau)(e^{-i\tau})$, and an arbitrary Stolz domain $St$ the following properties hold true: (a) $\sup\{\|k_\zeta\|_{L^1(\mathbb{T})}:\zeta\in St\}<\infty$; (b) $\lim_{\zeta\in St,\zeta\to 1}\int_{\mathbb{T}}k_\zeta\,dm=1$; (c) $\lim_{\zeta\in St,\zeta\to 1}\sup\{|k_\zeta(\tau)|:\alpha\le|\tau|\le\pi\}=0$, $\alpha>0$. From this and from the existence of the limit $\lim_{\tau\to 0}(2\tau)^{-1}[F(\tau)-F(-\tau)]=F'(0)=(2\pi)^{-1}(d\mu/dm)(1)$ it is easy to see that

$$\lim_{\zeta\in St,\zeta\to 1} f(\zeta)=(d\mu/dm)(1).\ \square$$

**9. Corollary.** *If $f\in H^p$, $p\ge 1$ then $f(t)=\lim_{\zeta\in St,\zeta\to t}f(\zeta)$ for a.e. $t$, $t\in\mathbb{T}$, and for every Stoltz domain $St$.*

This is Theorem 6 and Theorem 8 for the measure $d\mu=f\,dm$. □

**10. Jensen's Formula.** *Let $f$ be a function analytic in the disk $\mathbb{D}$, $f(0)\ne 0$, and $r$, $0<r<1$, a number such that $f(\lambda)\ne 0$ for $|\lambda|=r$ (i.e. $k_f(\lambda)\equiv 0$ for $|\lambda|=r$). Then*

$$\log|f(0)|+\sum_{\lambda,|\lambda|<r}k_f(\lambda)\log(r/|\lambda|)=\int_{\mathbb{T}}\log|f_r|\,dm. \tag{4}$$

*Proof.* If $B$ is the product (2) then $\log|f/B|$ is harmonic in an open disk containing $\{\zeta:|\zeta|\le r\}$ and $|B(\zeta)|=1$ for $|\zeta|=r$. Therefore

$$\begin{aligned}\int_{\mathbb{T}}\log|f_r|\,dm&=\int_{\mathbb{T}}\log|(f/B)_r|\,dm\\&=\log|(f/B)(0)|\\&=\log|f(0)|+\sum_{\lambda,|\lambda|<r}k_f(\lambda)\log(r/|\lambda|).\ \square\end{aligned}$$

**11. Corollary** (the Blaschke condition). *If $f$ is analytic in $\mathbb{D}$, $f\not\equiv 0$, and $\sup_{0\le r<1}\int_{\mathbb{T}}\log|f_r|\,dm<\infty$ then*

$$\sum_{\lambda\in\mathbb{D}}k_f(\lambda)(1-|\lambda|)<\infty. \tag{B}$$

In fact, if necessary passing to the function $f\cdot z^{-k_f(0)}$, we may assume that $f(0)\ne 0$. By Jensen's formula $\sum_{\lambda,|\lambda|<r}k_f(\lambda)\log(r/|\lambda|)\le\text{const}$, $0<r<1$. Thus for arbitary $R$, $r$, $0\le R<r<1$.

$$\sum_{\lambda,|\lambda|<R}k_f(\lambda)\log(r/|\lambda|)\le\text{const}.$$

Making now $r\to 1$ and afterwards $R\to 1$ we obtain $\sum_{\lambda\in\mathbb{D}}k_f(\lambda)\cdot\log(1/|\lambda|)<\infty$. □

**12. Corollary.** *If $f\in H^p$, $p>0$, and $f\not\equiv 0$ then condition (B) is fulfilled.*

In addition to Corollary 11 one just has to invoke the inequality between the arithmetical and the geometrical means: $p\int_{\mathbb{T}}\log|f_r|\,dm\le\log\int_{\mathbb{T}}|f_r|^p\,dm$, $0\le r<1$. □

**13. Lemma on Blaschke Products.** *Let $k$ be an integer-valued function in the disk $\mathbb{D}$, $k\ge 0$ and $\sum_{\lambda\in\mathbb{D}}k(\lambda)(1-|\lambda|)<\infty$ (i.e., $k$ is a divisor satisfying the*

*Blaschke condition). Then the product*

$$B(k,\cdot)=\prod_{\lambda\in\mathbb{D}} b_\lambda^{k(\lambda)} \stackrel{\text{def}}{=} \lim_{r\to 1}\prod_{|\lambda|<r} b_\lambda^{k(\lambda)}$$

*converges uniformly on compact subsets of the disk* $\mathbb{D}$, $|B|\le 1$ *on* $\mathbb{D}$, $|B|=1$ *a.e. on* $\mathbb{T}$ *and* $k_B\equiv k$.

*Proof.* Set $B^r=\prod_{|\lambda|<r} b_\lambda^{k(\lambda)}$, $0\le r<1$ (a finite Blaschke product for each $r$). Then for $0\le r'<r<1$

$$\begin{aligned}\|B^r-B^{r'}\|_2^2 &= 2-2\operatorname{Re}(B^r,B^{r'}) \text{ (the scalar product in } H^2\text{, i.e. in } L^2(\mathbb{T}))\\ &= 2-2\operatorname{Re}\int_{\mathbb{T}} B^r\overline{B^{r'}}\,dm\\ &= 2-2\operatorname{Re}\int_{\mathbb{T}} B^r/B^{r'}\,dm\\ &= 2-2\prod_{r'\le|\lambda|<r}|\lambda|^{k(\lambda)}.\end{aligned} \tag{5}$$

Hence $\lim_{r,r'\to 1}\|B^r-B^{r'}\|_2=0$ and therefore there exists a function $B$ of class $H^2$ such that $\lim_{r\to 1}\|B-B^r\|_2=0$. It is clear that $|B|=1$ a.e. on $\mathbb{T}$ and that $B=\lim_{r\to 1}B^r$ uniformly on compact subsets of $\mathbb{D}$. As $|B^r(\zeta)|\le 1$ for $|\zeta|<1$ we must have $|B(\zeta)|\le 1$ for $|\zeta|<1$. The same formula (5) also shows that $\lim_{r\to 1}\|B/B^r-1\|_2=0$ and therefore $\lim_{r\to 1}(B/B^r)(\zeta)=1$, $\zeta\in\mathbb{D}$. This yields $k_B=k$. □

**14. Lemma.** *Let* $f\in H^p$, $p>0$, $f\not\equiv 0$. *Then* $f=B\cdot F$ *where* $B=B(k_f,\cdot)$, $F\in H^p$ *and* $F(\zeta)\ne 0$, $\zeta\in\mathbb{D}$.

*Proof.* Let $0\le r$, $r'<1$ and let $B^r=\prod_{|\lambda|<r} b_\lambda^{k_f(\lambda)}$ be a "partial" Blaschke product of $f$. From Corollary 5 we obtain

$$\|(f/B^r)_{r'}\|_{L^p(\mathbb{T})}\le\|f/B^r\|_p=\|f\|_p,$$

and subsequently for $r\to 1$ (cf. likewise 12 and 13) $\|(f/B)_{r'}\|_{L^p(\mathbb{T})}\le\|f\|_p$. This means that $F\in H^p$ (and that $\|F\|_p\le\|f\|_p$ and, the converse inequality being trivial, $\|F\|_p=\|f\|_p$). □

**15. Corollary.** *If* $f\in H^p$, $p>0$, *then there exist functions* $f_1,f_2$ *of class* $H^p$ *such that* $f=f_1+f_2$ *and* $f_k(\zeta)\ne 0$, $\zeta\in\mathbb{D}$; $k=1,2$.

Indeed, take $f_1=1/2\cdot(B+\mathbb{1})F$, $f_2=1/2\cdot(B-\mathbb{1})F$. □

**16. Corollary.** *Each function* $f$ *of class* $H^p$, $p>0$, *takes on a.e. point of the circumference* $\mathbb{T}$ *angular boundary values (denoted by* $bf$*);* $bf\in L^p(\mathbb{T})$, $\|f\|_p=\|bf\|_{L^p(\mathbb{T})}$ *and* $\lim_{r\to 1}\|bf-f_r\|_{L^p(\mathbb{T})}=0$ *for* $p<\infty$.

In fact, in view of 14 we may assume that $f(\zeta)\ne 0$, $\zeta\in\mathbb{D}$. Then $f^p\in H^1$; the existence of the angular boundary values $bf$, the containment $bf\in L^p(\mathbb{T})$ and the relationship $\|f\|_p=\|bf\|_{L^p(\mathbb{T})}$ now follow from 6 and 9. As in Sect. 7 we make the convention not to distinguish between $bf$ and $f$. For the proof of

$\lim_{r\to 1}\|f-f_r\|_{L^p(\mathbb{T})}=0$ we pick an integer $n$ such that $np>1$. Then

$$\int_{\mathbb{T}}|f-f_r|^p\,dm=\int_{\mathbb{T}}|(f^{1/n}-f_r^{1/n})\sum_{k=0}^{n-1}f^{k/n}\cdot f_r^{(n-1-k)/n}|^p$$
$$\leq\left(\int_{\mathbb{T}}|f^{1/n}-f_r^{1/n}|^{pn}\right)^{1/n}$$
$$\cdot\left(\int_{\mathbb{T}}\left|\sum_{k=0}^{n-1}f^{k/n}f_r^{(n-1-k)/n}\right|^{pn/(n-1)}\right)^{(n-1)/n}$$

As $f^{1/n}\in H^{pn}$ then in view of 6 it suffices to verify that the second of the last two integrals is bounded for $r\to 1$. To this end let us remark that $(\sum_{k=0}^{n-1}a_k)^\alpha\leq c_{n,\alpha}\sum_{k=0}^{n-1}a_k^\alpha$ $(a_k\geq 0,\alpha>0)$ so it suffices to estimate any of the terms using Hölder's inequality:

$$\int_{\mathbb{T}}|f|^{pk/(n-1)}|f_r|^{(n-1-k)/(n-1)}\,dm$$
$$\leq\left(\int_{\mathbb{T}}|f|^p\right)^{k/(n-1)}\left(\int_{\mathbb{T}}|f_r|^p\right)^{(n-1-k)/(n-1)}$$
$$\leq\int_{\mathbb{T}}|f|^p.\ \square$$

**17. Corollary (Jensen's Inequality).** *If $f\in H^p$, $p>0$, then*

$$\log|f(0)|\leq\int_{\mathbb{T}}|f|\,dm.$$

*In particular, if $f\not\equiv 0$ then $\log|f|\in L^1(\mathbb{T})$.*

In fact, it follows from formula (4) that

$$\log|f(0)|\leq\int_{\mathbb{T}}\log|f_r|\,dm\leq\int_{\mathbb{T}}\log(|f_r|+\varepsilon)\,dm$$

for arbitrary $r$, $0\leq r<1$, and $\varepsilon$, $\varepsilon>0$. Again it follows from 16 that

$$\lim_{r\to 1}\|\log(|f|+\varepsilon)-\log(|f_r|+\varepsilon)\|_{L^1(\mathbb{T})}=0.$$

But then $\log|f(0)|\leq\int_{\mathbb{T}}\log(|f|+\varepsilon)\,dm$. Finally make $\varepsilon\to 0$. $\square$

**18. Remark.** Exactly in the same way we get from the formula $\int_{\mathbb{T}}P(\zeta\bar{t})\times\log|f_r(t)|\,dm(t)=\log|(f/B)(\zeta)|$, $\zeta\in\mathbb{D}$ (cf. the proof of Jensen's formula, Sect. 10) and the inequality $|B(\zeta)|\leq 1$ the more general Jensen's inequality:

$$\log|f(\zeta)|\leq\int_{\mathbb{T}}P(\zeta\bar{t})\log|f(t)|\,dm(t),\ \zeta\in\mathbb{D},\ f\in H^p,\ p>0.$$

**19. Lemma.** *Let $f\in L^p(\mathbb{T})$, $\log|f|\in L^1(\mathbb{T})$. Then the function $f_e$*

$$f_e\overset{\text{def}}{=}\exp\left(\int_{\mathbb{T}}\frac{\zeta+z}{\zeta-z}\log|f(\zeta)|\,dm(\zeta)\right),$$

*belongs to the class $H^p$ and $|f_e| = |f|$ a.e. on $\mathbb{T}$. If $g \in H^q$, $q > 0$, and $|g| \le |f|$ on $\mathbb{T}$ then $|g| \le |f_e|$ on $\mathbb{D}$.*

*Proof.* The inequality between the means (the geometric and the arithmetic ones) gives

$$\begin{aligned}\int_{\mathbb{T}} |(f_e)_r|^p\,dm &= \int_{\mathbb{T}} \exp\left(\int_{\mathbb{T}} P(r\zeta\bar{t})\log|f(t)|^p\,dm(t)\right)dm(\zeta)\\ &\le \int_{\mathbb{T}}\left(\int_{\mathbb{T}} P(r\zeta\bar{t})|f(t)|^p\,dm(t)\right)dm(\zeta)\\ &= \int_{\mathbb{T}} |f(t)|^p\,dm(t)\end{aligned}$$

for all $r$, $0 \le r < 1$, so that $f_e \in H^p$. The case $p = \infty$ requires an obvious modification. The equality $|f_e| = |f|$ a.e. on $\mathbb{T}$ follows from Theorem 8 applied to the measure $d\mu = \log|f|\,dm$.

If $g \in H^q$, $q > 0$, and $\zeta \in \mathbb{D}$ then by Jensen's inequality 18 we obtain

$$\begin{aligned}\log|g(\zeta)| &\le \int_{\mathbb{T}} P(\zeta\bar{t})\log|g(t)|\,dm(t)\\ &\le \int_{\mathbb{T}} P(\zeta\bar{t})\log|f(t)|\,dm(t)\\ &= \log|f_e(\zeta)|. \square\end{aligned}$$

**20. Corollary.** *If $f \in H^p$, $p > 0$, and $f|\mathbb{T} \in L^q(\mathbb{T})$, $p < q \le \infty$, then $f \in H^q$.* □

**21. Herglotz's Theorem.** *Let $u$ be a function harmonic in the disk $\mathbb{D}$, with $u \ge 0$. Then there exists a unique Borel measure $\mu$ on the circle $\mathbb{T}$ such that $u = P * \mu$. Also $\mu \ge 0$.*

*Proof.* By hypothesis

$$u(0) = \int_{\mathbb{T}} u_r\,dm = \int_{\mathbb{T}} |u_r|\,dm,\ 0 \le r < 1,$$

so that the family of measures $\{u_{r_n}\,dm : r_n \uparrow 1\}$ is weakly relatively compact in the space of all measures. Let $\mu$ be a limit point of these measures. It is clear that $\mu \ge 0$ and that it follows from the formula

$$u(r\zeta) = \int_{\mathbb{T}} P(\zeta\bar{t})u_r(t)\,dm(t),\ \zeta \in \mathbb{D},$$

that

$$u(\zeta) = \int_{\mathbb{T}} P(\zeta\bar{t})\,d\mu(t),\ \zeta \in \mathbb{D}.$$

Uniqueness of the representation: if $u = P * \mu'$ then $\mathbb{0} = P * \nu$, $\nu = \mu - \mu'$, whence

$(P * \nu)(r\zeta) = \sum_{n \in \mathbb{Z}} r^{|n|}\zeta^n \hat{\nu}(n) \equiv 0, \zeta \in \mathbb{T}, 0 < r < 1$ ($\hat{\nu}(n)$ are the Fourier coefficients of $\nu$). It follows now from the uniqueness theorem for Fourier series that $\hat{\nu}(n) = 0$, $n \in \mathbb{Z}$, and therefore (the theorem of Weierstrass) that $\nu = \mathbb{0}$. □

**22. Corollary (Structure of Singular Inner Functions).** *Let $v \in H^p, p > 0, |v| = 1$ a.e. on $\mathbb{T}$, $v(\zeta) \neq 0$ for all $\zeta, \zeta \in \mathbb{D}$. Then there exists a unique measure $\mu$ on the circle $\mathbb{T}$ such that*

$$v = \lambda \exp\left(-\int_{\mathbb{T}} \frac{\zeta + z}{\zeta - z} d\mu(\zeta)\right), \lambda = v(0)/|v(0)|.$$

*This measure is non-negative and singular with respect to the Lebesgue measure m.*

In fact, let us apply Theorem 21 to the function $u = -\log|v|$ ($u \geq 0$ in $\mathbb{D}$, in view of 20 and 6). This gives the desired measure $\mu$, $u = P * \mu$. Decomposing the measure $\mu$ into its absolutely continuous and singular parts, $d\mu = (d\mu/dm) \times dm + d\mu_s$. and applying Theorem 8, we get $d\mu/dm = \mathbb{0}$, i.e. $\mu = \mu_s$. The functions $\int (\zeta + z)/(\zeta - z)\, d\mu(\zeta)$ and $-\log v$ have the same real parts, so that they differ by a purely imaginary constant only. □

**23. Riesz-Nevanlinna Theorem.** *If $f \in H^p$, $p > 0$, then $f = \lambda B v f_e$, where $|\lambda| = 1$, $B = B(k_f, \cdot)$ and $v$ is the function in Sect. 22 (a singular inner function).*

*The proof* is contained in the union of 12, 13, 17, 19, 22. □

**24. Lemma (Structure of Outer Functions).** *Let $p, q, r > 0, f \in H^p$. The following statements are equivalent.*

1. $f = \lambda f_e, |\lambda| = 1$.
2. $\log|f(\zeta)| = \int_{\mathbb{T}} P(\zeta\bar{t}) \log|f(t)|\, dm(t)$ *for some (all) $\zeta, \zeta \in \mathbb{D}$.*
3. *If $g \in H^q$, $g/f \in L^r(\mathbb{T})$ then $g/f \in H^r$.*

*If $p = q = r = 2$ then Statements 1–3 are also equivalent to the following one.*

4. *$f$ is an outer function (in the sense of Lecture I, Sect. 2).*

*Proof.* 1 ⇒ 2. Trivial.

2 ⇒ 1. Let us apply Theorem 23: if $|f(\zeta)| = |f_e(\zeta)|$ for some $\zeta$, $\zeta \in \mathbb{D}$, then $|B(\zeta)| \cdot |v(\zeta)| = 1$ and (in view of $|B| \leq 1, |v| \leq 1$) $|B(\zeta)| = |v(\zeta)| = 1$. Recalling the definition of the functions $B$ and $v$ (cf. 13, 22), we conclude that $B = \mathbb{1}$ and $\mu = \mathbb{0}$, i.e. $v = \mathbb{1}$.

1 ⇒ 3. Set $g = \lambda_1 g_e v_1 B_1$ (in the sense of Theorem 23). Then $g/f = (\lambda_1 \lambda^{-1})(g_e/f_e) v_1 B_1 = \lambda_2 (g/f)_e v_1 B_1 \in H^r$ (cf. 19).

3 ⇒ 1. Let us use Theorem 23: $f = \lambda f_e v B$. Put $g = (\min(|f|, 1))_e$. Then $g \in H^\infty$ (cf. 19) and $g/f = (g/f)_e \lambda^{-1} B^{-1} v^{-1}$ so that $g/f \in L^\infty(\mathbb{T})$ and therefore also $g/f \in H^r$. By Theorem 23 $g/f = \lambda_1 (g/f)_e B_1 v_1$, whence $B B_1 v v_1 = \text{const}$. As we have already noted this implies $B \equiv \text{const}$, $v \equiv \text{const}$.

Thus we have shown that the Statements 3 for various collections $p, q, r$ all are equivalent. The equivalence of Statement 4 and Statement 3 for $p = q = r = 2$ was proved in Lecture I, Sect. 2. □

So, if $f \in H^2$ we have for the inner and the outer factors of $f$ the formulae

$f^i = \lambda Bv$ and $f^e = \mu f_e$ ($|\lambda| = |\mu| = 1$) respectively, where we used the notation of Theorem 23.

It is not hard to see also that, given a function $f \in H^p$ ($p > 0$), each of the following requirements implies the equality $f = \lambda f_e$: a) $f^{-1} \in H^\infty$; b) $\operatorname{Re} f(\zeta) > 0$, $\zeta \in \mathbb{D}$; c) $f(\zeta) \neq 0$, $\zeta \in \mathbb{D}$ and $\sup\{|\arg f(\zeta)| : \zeta \in \mathbb{D}\} < \infty$, where $\arg f$ is a continuous branch of the argument of $f$.

**25. Some Auxiliary Results** for the classes $H^p$, $p > 0$ (mainly as metric spaces) are listed in Sect. 26–29.

**26. Theorem.** *The space $H^p$, $p > 0$, is complete.*

*The proof* of this fact is very standard for spaces with sup-metrics and for $p \geq 1$ it follows in general immediately from formula (3). Let us begin by establishing the continuity of the imbedding $H^p \subset A(\mathbb{D})$ into the space of all analytic functions in $\mathbb{D}$ with the topology of uniform convergence on compact subsets. This follows at once from the following estimate:

$$|f(\zeta)| \leq \|f\|_p (1 - |\zeta|^2)^{-1/p}; \; \zeta \in \mathbb{D}, \; f \in H^p.$$

Proof of the estimate: it follows from 13 and 14 that

$$\begin{aligned} |f(\zeta)| &= |B(\zeta)F(\zeta)| \leq |F(\zeta)| \\ &= |F(\zeta)^{p/2}|^{2/p} \leq (\|F^{p/2}\|_2 (1 - |\zeta|^2)^{-1/2})^{2/p} \\ &= \|F\|_p (1 - |\zeta|^2)^{-1/p} \\ &= \|f\|_p (1 - |\zeta|^2)^{-1/p}. \end{aligned}$$

Let now $\{f_n\}_{n \geq 1}$ be a Cauchy sequence in the space $H^p$, $p > 0$. By the preceding discussion $\lim_n f_n = f$ then exists uniformly on compact subsets of $\mathbb{D}$. Let $\varepsilon > 0$ and pick a number $N$ such that $n, k > N \Rightarrow \|f_n - f_k\|_p < \varepsilon$. Then $\|(f_n)_r - (f_k)_r\|_{L^p(\mathbb{T})} < \varepsilon$, $0 < r < 1$. Let $k \to \infty$ for $r$ fixed. Then $\|(f_n)_r - f_r\|_{L^p(\mathbb{T})} \leq \varepsilon$ for $n > N$, that is, $\|f_n - f\|_p \leq \varepsilon$ for $n > N$. It follows that $f \in H^p$ and $\lim_n \|f_n - f\|_p = 0$. □

**27. Corollary.** *Let $\Theta$ be an inner function (i.e. $\Theta \in H^\infty$, $|\Theta| = 1$ a.e. on $\mathbb{T}$), $\Theta \not\equiv$ const, and let $p > 0$. Then* $\operatorname{clos} \Theta H^p = \Theta H^p \neq H^p$.

In fact, if $f_n = \Theta g_n$, $g_n \in H^p$ and $\lim_n \|f_n - f\|_p = 0$ then the sequence $\{g_n\}_{n \geq 1}$ is Cauchy ($\|g_n - g_k\|_p = \|g_n - g_k\|_{L^p(\mathbb{T})} = \|f_n - f_k\|_p$) and therefore there exists a function $g$, $g \in H^p$, such that $\lim_n \|g_n - g\|_p = 0$. It is clear that $f = \Theta g$ and consequently $f \in \Theta H^p$.

Let us check that $\mathbb{1} \notin \Theta H^p$: if $\mathbb{1} = \Theta \cdot g$, $g \in H^p$, then $|g| = 1$ a.e. on $\mathbb{T}$ and in view of 20 $g \in H^\infty$, i.e. $g$ is an inner function. Hence (cf. 24 or Lecture I) $\Theta \equiv$ const, $g \equiv$ const. Contradiction. □

**28. Functions of Bounded Type** (or of bounded characteristic) are by definition meromorphic functions of the form $f/g$, with $f, g \in H^\infty$. The set of all such functions is called the *Nevanlinna class* and noted Nev. *Statement*: if $f, g \in H^p$,

$p > 0$, then $f/g \in \mathrm{Nev}$. In fact, define outer functions $f_1$, $g_1$ by the relations $f_1 = (\min(|f|, 1))^e$, $g_1 = \min(|g|, 1))^e$ and let $f = f^e \cdot f^i$, $g = g^e \cdot g^i$ be the outer-inner factorizations of the functions $f$ and $g$ (cf. Lecture I or Sect. 23). Then $f_1$, $g_1 \in H^\infty$, $f_1/f^e$, $g_1/g^e \in H^\infty$ and

$$f/g = \frac{f_1 \cdot f/f_1}{g_1 \cdot g/g_1} = \frac{f_1 f^i \cdot g_1/g^e}{g_1 g^i \cdot f_1/f^e}. \square$$

**29. The Multilinear Map** $(f_1, \ldots, f_n) \to \prod_{k=1}^n f_k$ from $H^{p_1} \times \ldots \times H^{p_n}$ into $H^p$ is continuous provided $p^{-1} = \sum_{k=1}^n p_k^{-1}$.

*Proof*: repeated application of Hölder's inequality. $\square$

**30. Douglas-Rudin Theorem.** *Let* $\Sigma$ *be the set of all inner functions. Then*

$$L^\infty(\mathbb{T}) = \mathrm{clos}_{L^\infty}\{\bar{\Theta} H^\infty : \Theta \in \Sigma\},$$

$$L^\infty(\mathbb{T}) = \bigvee\nolimits_{L^\infty}\{\Theta_1 \bar{\Theta}_2 : \Theta_1, \Theta_2 \in \Sigma\}.$$

*Proof.* Clearly the first identity is a consequence of the second one. If we agree to denote by $\mathrm{alg}\, M$ the uniformly closed algebra generated by the set $M$, $M \subset L^\infty(\mathbb{T})$, then the second identity affirms that $L^\infty(\mathbb{T}) = \mathrm{alg}\{\Sigma, \bar{\Sigma}\}$. In order to prove this last identity it is sufficient to check that $\chi_\sigma \in \mathrm{alg}\{\Sigma, \bar{\Sigma}\}$ for every Borel set $\sigma$, $\sigma \subset \mathbb{T}$ ($\chi_\sigma$ is the characteristic function). Let $f_n$ be an outer function, with $|f_n| = n\chi_\sigma + 1/n \cdot \chi_{\mathbb{T}\setminus\sigma}$ on the circumference $\mathbb{T}$, $n \geq 1$. Moreover, set $A = \{\zeta \in \mathbb{C} : n^{-1} < |\zeta| < n\}$. It is then clear that $f_n \in H^\infty$, $f_n(\mathbb{D}) \subset A$, $f_n(\mathbb{T}) \subset \partial A$. Set $\varphi_1(w) = w + w^{-1}$, $w \in \mathbb{C}\setminus\{0\}$. Then the image $\varphi_1(A)$ is an ellipse and the conformal map $\omega : \varphi_1(A) \to \mathbb{D}$ is continuous up to the boundary. Therefore the function $\Theta_1 \stackrel{\mathrm{def}}{=} \omega \circ \varphi_1 \circ f_n$ must be inner. The function $\omega^{-1}$ being continuous on $\mathrm{clos}\, \mathbb{D}$, it can be uniformly approximated by polynomials (for instance, by its Fejér polynomials). Consequently $\varphi_1 \circ f_n = f_n + f_n^{-1} \in \mathrm{alg}\{\Theta_1\}$. Doing the same to the function $\varphi_2(w) \equiv w - w^{-1}$ we produce an inner function $\Theta_2$ such that $f_n - f_n^{-1} \in \mathrm{alg}\{\Theta_2\}$. Hence $f_n \in \mathrm{alg}\{\Theta_1, \Theta_2\}$, $|f_n|^2 \in \mathrm{alg}\{\Theta_1, \Theta_2, \bar{\Theta}_1, \bar{\Theta}_2\}$ and

$$\chi_\sigma + n^{-4}\chi_{\mathbb{T}\setminus\sigma} \in \mathrm{alg}\{\Theta_1, \Theta_2, \bar{\Theta}_1, \bar{\Theta}_2\}.$$

Letting $n \to \infty$ we conclude that $\chi_\sigma \in \mathrm{alg}\{\Sigma, \bar{\Sigma}\}$. $\square$

**31. Remark.** In view of Frostman's Theorem (Lecture II, Sect. 5) one can everywhere in Theorem 30 replace the set $\Sigma$ by the set of all Blaschke products.

**32. Lemma on the Algebra** $L^\infty \cap \mathrm{Nev}$. *Let* $f_i \in L^\infty(\mathbb{T}) \cap \mathrm{Nev}$, $1 \leq i \leq k$. *The following statements are equivalent*:

1. $\mathrm{ess\,inf}_{\mathbb{T}} \sum_{i=1}^k |f_i|^2 > 0$.
2. $\exists g_i : g_i \in L^\infty \cap \mathrm{Nev} (1 \leq i \leq k)$, $\sum_{i=1}^k g_i f_i = 1$.
3. $\exists g_i : g_i \in H^\infty$ $(1 \leq i \leq k)$, $\sum_{i=1}^k f_i g_i$ *is an inner function.*
4. $\exists g_i : g_i \in H^\infty$ $(1 \leq i \leq k)$, $\mathrm{ess\,inf}_{\mathbb{T}} |\sum_{i=1}^k g_i f_i| > 0$.

*Proof.* $1 \Rightarrow 4$. By the Douglas-Rudin Theorem (see 30) there exist inner functions $\Theta_i$ and $H^\infty$-functions $h_i$ $(1 \leq i \leq k)$ such that $|\bar{f}_i - \bar{\Theta}_i h_i| < 2^{-1}\delta C^{-1} k^{-1/2}$ a.e. on $\mathbb{T}$ $(1 \leq i \leq k)$, where $\delta = \mathrm{ess\,inf}_{\mathbb{T}} \sum_{i=1}^k |f_i|^2$, $C^2 =$

ess $\sup_{\mathbb{T}} \sum_{i=1}^k |f_i|^2$. Let now $\Theta = \prod_{i=1}^k \Theta_i$, $g_i = \Theta_i \bar{\Theta}_i h_i$, $1 \le i \le k$. Then $g_i \in H^\infty$ and

$$\left|\sum_{i=1}^k g_i f_i\right| = \left|\sum_{i=1}^k \bar{\Theta}_i h_i f_i\right| = \left|\sum_{i=1}^k |f_i|^2 + \sum_{i=1}^k (\bar{\Theta}_i h_i - \bar{f}_i) f_i\right|$$

$$\ge \delta - \sum_{i=1}^k |\bar{\Theta}_i h_i - \bar{f}_i||f_i| \ge \delta - 2^{-1}\delta C^{-1} \cdot C = 2^{-1}\delta.$$

$4 \Rightarrow 3$. It is sufficient to divide the function $\sum_{i=1}^k g_i f_i$ in Statement 4 by the outer function $h$ with $|h| = |\sum_{i=1}^k g_i f_i|$ on $\mathbb{T}$.

$3 \Rightarrow 2$. Easy.

$2 \Rightarrow 1$. Easy.

**33. Concluding Remarks.** The material covered here is classical. All the results and many more particulars connected with the classes $H^p$ can be found in the books Privalov [1], Hoffman [1], Duren [1], Koosis [4], Garnett [2]. There detailed bibliographical references likewise are located.

# Appendix 3. The Corona Problem, the Problem of Sz.-Nagy, Ideals in the Algebra $H^\infty$

**1. Contents:** Proof of the Corona Theorem (formulation in Lecture III, p. 66); estimates from above and from below of the moduli of the solving functions; operator variant of the Corona Theorem; complementing of matrix valued functions to isomorphisms (the problem of Sz.-Nagy); finitely generated ideals of the algebra $H^\infty$; the corona theorem in other Banach algebras.

**2. Carleson's Corona Theorem.** This theorem, formulated in Lecture III and amounting to the solvability of the equation $\sum_{i=1}^n f_i g_i = \mathbb{1}$, $g_i \in H^\infty$, in the sole hypothesis $\inf_{\zeta\in\mathbb{D}} \sum_{i=1}^n |f_i(\zeta)|^2 > 0$, will now be proved and indeed at once for an infinite family of functions $\{f_i : i \in \mathbb{N}\}$. Let us further remark that (cf. also the "Concluding Remarks", Sect. 14 of this Appendix) the estimates obtained below for the solutions $g_i$ are also of special interest.

We consider a vector valued function $\zeta \mapsto f(\zeta) = \{f_i(\zeta)\}_{i\geq 1}$ with values in the space $l^2$ of square summable sequences; the standard norm in $l^2$ will be denoted by $|f(\zeta)| \stackrel{\text{def}}{=} (\sum_{i\geq 1} |f_i(\zeta)|^2)^{1/2}$.

Let $H^\infty(l^2)$ be the space of all $l^2$-valued bounded analytic functions $f$ in the disk $\mathbb{D}$, $\|f\|_{H^\infty(l^2)} = \sup_{\zeta\in\mathbb{D}} |f(\zeta)|$.

**3. Corona Theorem.** *Let* $0 < \delta \leq 1$ *and* $f \in H^\infty(l^2)$. *If*

$$\delta \leq |f(\zeta)| \leq 1, \ \zeta \in \mathbb{D},$$

*there exists a function* $g$, $g \in H^\infty(l^2)$ *such that*

$$\sum_{i\geq 1} f_i g_i = \mathbb{1}, \ \|g\|_{H^\infty(l^2)} \leq 1/\delta + (1/\delta^2).\,[7\sqrt{\log 1/\delta} + 20 \log 1/\delta].$$

**4. Preliminary Remarks Concerning Green's Formula.** A) $\Delta \stackrel{\text{def}}{=} \partial^2/\partial x^2 + \partial^2/\partial y^2 = 4\partial\bar\partial$ where $\partial \stackrel{\text{def}}{=} \frac{1}{2}(\partial/\partial x - i\partial/\partial y)$, $\bar\partial \stackrel{\text{def}}{=} \frac{1}{2}(\partial/\partial x + i\partial/\partial y)$.

B) If $u$ is a smooth function (for instance, of class $C^2$) with compact support and

$$v_0(\zeta) = \frac{1}{\pi} \iint_{\mathbb{C}} \frac{u}{\zeta - z} \tag{1}$$

(all integrals $\iint$ are taken with respect to the planar Lebesgue measure $dx\,dy$), then $v_0 \in C^2$ and $\bar\partial v_0 = u$. (*Hint*: apply Green's (Stokes's) formula to the integral $\bar\partial v_0(\zeta) = \dfrac{1}{2\pi} \displaystyle\iint_{\mathbb{C}} \frac{\partial u/\partial x + i\partial u/\partial y}{\zeta - z}$ and the domain $\{\xi \in \mathbb{C} : |\xi| < R, |\zeta - \xi| > \varepsilon\}$ with large $R > 0$ and small $\varepsilon > 0$, let $\varepsilon \to 0$. □)

C) If $w$ is a smooth ($C^2$) function in the disk clos $\mathbb{D}$, then

$$2\pi w(0) = 2\pi \int_{\mathbb{T}} w\, dm - \iint_{\mathbb{D}} \log(1/|z|)\Delta w \tag{2}$$

(*Hint*: apply Green's formula $\iint_G (u\Delta u_1 - u_1 \Delta u) = \int_{\partial G} (u \cdot \partial u_1/\partial n - u_1 \cdot \partial u/\partial n)\, ds$ to the functions $u = \log(1/|z|)$, $u_1 = w$ and the domain $G = \{\zeta : \varepsilon < |\zeta| < 1\}$ and let $\varepsilon \to 0$; here $\partial/\partial n$ stands for differentiation in the direction of the outer normal; $ds$ is arc length. □)

**5. Proof of the Corona Theorem.** We can assume that $f$ is regular in the closed disk clos $\mathbb{D}$, because once we have got a uniform estimate of the corresponding functions $g_i$, it is easy to pass to the limit using the functions $f(rz)$, $0 < r < 1$.

Set $h_i = \bar{f_i} |f|^{-2}$, $i \geq 1$. It is clear that the functions $h_i$ solve the desired equation but they are not analytic in $\mathbb{D}$. Let us notice that $\bar\partial h_i = (\bar{f_i'} |f|^2 - \bar{f_i}(f, f'))|f|^{-4}$, where $(\cdot, \cdot)$ is the scalar product in $l^2$, and likewise set

$$u_{ik} \stackrel{\text{def}}{=} h_i \bar\partial h_k - h_k \bar\partial h_i = (\bar{f_i} \bar{f_k'} - \bar{f_k} \bar{f_i'})|f|^{-4}.$$

If we can find functions $v_{ik}$ such that $\bar\partial v_{ik} = u_{ik}$, $v_{ik} = -v_{ki}$ (so that for the corresponding matrix function $v$ holds $\bar\partial v = u$ and $v = -v^T$, where $T$ stands for the operation of taking the transpose) then the functions

$$g_i \stackrel{\text{def}}{=} h_i + \sum_{k \geq 1} v_{ik} f_k, \; i \geq 1,$$

will be analytic in $\mathbb{D}$ (as $\bar\partial g_i = \bar\partial h_i + \sum_{k\geq 1} (h_i \bar\partial h_k f_k - \bar\partial h_i h_k f_k) = \bar\partial h_i - \bar\partial h_i \equiv \mathbb{0}$, in view of $\sum_{k \geq 1} f_k \bar\partial h_k = \mathbb{0}$ and $\sum_{k\geq 1} h_k f_k = \mathbb{1}$), with

$$\sum_{i=1}^{n} g_i f_i = \sum_{i=1}^{n} h_i f_i = \mathbb{1}$$

(because of $v = -v^T$) and $|g(\zeta)| \leq |h(\zeta)| + |v(\zeta)||f(\zeta)|$, i.e.

$$\|g\|_{H^\infty(l^2)} \leq \delta^{-1} + \|v\|_{L^\infty(L(l^2))}. \tag{3}$$

Let us estimate the right hand side. The general solution of the equation $\bar\partial v = u$ has the form $v = v_1 + v_0$ where $v_1$ is analytic in $\mathbb{D}$ and $v_0$ is defined by the formula (1). (In this formula $u$ stands for any continuation of the function $u|\text{clos}\,\mathbb{D}$ to a smooth function with compact support in the whole plane.) By Montel's theorem it is clear that the greatest lower bound $\inf_{v_1} \|v_1 + v_0\|_{L^\infty}$ is attained, in fact by an anti-symmetric matrix $v = v_1 + v_0$ (because $\|v_0 + (v_1 - v_1^T)/2\| = \frac{1}{2}\|v_0 + v_1 - (v_1^T + v_0^T)\| \leq \|v_0 + v_1\|$). Let us estimate the norm $\|v\|_{L^\infty(\mathfrak{S}_2)} = \text{ess sup}_{\zeta \in \mathbb{T}} |v(\zeta)|_{\mathfrak{S}_2}$ where $|d|_{\mathfrak{S}_2} \stackrel{\text{def}}{=} (\sum_{i,k\geq 1} |d_{ik}|^2)^{1/2}$ is *the Hilbert-Schmidt norm* of the matrix $d$. Let $\langle \alpha, \beta \rangle = \sum_{i,k \geq 1} \alpha_{ik} \beta_{ik}$ be the *bilinear* form establishing the duality of the space $\mathfrak{S}_2$ (with itself). By the Hahn-Banach theorem

$$\|v\|_{L^\infty(\mathfrak{S}_2)} = \sup\left\{\left|\int_{\mathbb{T}} \langle F, v_0 \rangle \, dm\right| : F \in H^1(\mathfrak{S}_2),\ \|F\|_{H^1(\mathfrak{S}_2)} \leq 1,\ F(0) = \mathbb{0}\right\}.$$

Let us apply (2) to the function $w = \langle F, v_0 \rangle$ (it is clear that a simple passage to the limit extends (2) to such products of smooth functions $v_0$ and functions $F$ in $H^1$). As $\Delta w = 4\partial\bar\partial\langle F, v_0\rangle = 4\partial\langle F, \bar\partial v_0\rangle = 4\partial\langle F, u\rangle = 4\langle F', u\rangle + 4\langle F, \partial u\rangle$ we have

$$\|v\|_{L^\infty(\mathfrak{S}_2)} \le \frac{2}{\pi}\sup\left\{\iint_{\mathbb{D}} (|\langle F', u\rangle| + |\langle F, \partial u\rangle|)\log(1/|z|)\right. \tag{4}$$
$$\left. : \|F\|_{H^1(\gamma_2)} \le 1,\ F(0) = \mathbb{O}\right\}.$$

Let us estimate the integral $I_1 \stackrel{\text{def}}{=} \iint_{\mathbb{D}} |\langle F', u\rangle| \log(1/|z|)$. To this end define the outer function $\varphi$ by the formula $|\varphi(\zeta)|^2 = |F(\zeta)|_{\mathfrak{S}_2}$, $\zeta \in \mathbb{T}$. Then

$$I_1 \le \left(\iint_{\mathbb{D}} \frac{|F'(z)|^2_{\mathfrak{S}_2}}{|\varphi(z)|^2}\log(1/|z|)\right)^{1/2}\left(\iint_{\mathbb{D}} |\varphi(z)|^2 |u(z)|^2_{\mathfrak{S}_2}\cdot\log(1/|z|)\right)^{1/2} \stackrel{\text{def}}{=} I_1'\cdot I_1''.$$

Set $r = F\varphi^{-1}$. Then $|r(\zeta)|_{\mathfrak{S}_2} = |\varphi(\zeta)|$ a.e. on $\mathbb{T}$ and $|r(\zeta)|_{\mathfrak{S}_2} \le |\varphi(\zeta)|$, $\zeta \in \mathbb{D}$. Further

$$|F'(\zeta)|^2_{\mathfrak{S}_2} = |r'(\zeta)\varphi(\zeta) + r(\zeta)\varphi'(\zeta)|^2_{\mathfrak{S}_2}$$
$$\le 2(|r'(\zeta)|^2_{\mathfrak{S}_2}|\varphi(\zeta)|^2 + |r(\zeta)|^2_{\mathfrak{S}_2}|\varphi'(\zeta)|^2)$$

so that

$$(I_1')^2 \le 2\iint_{\mathbb{D}} (|r'(z)|^2_{\mathfrak{S}_2} + |\varphi'(z)|^2)\log(1/|z|) \le 2\pi,$$

because by formula (2) (applied to the function $w = |\varphi|^2$)

$$4\iint_{\mathbb{D}} |\varphi'|^2 \log(1/|z|) = 2\pi\int_{\mathbb{T}} |\varphi|^2\, dm - 2\pi|\varphi(0)|^2 \le 2\pi, \tag{5}$$

and similarly for the second integral.

In order to estimate the second factor (i.e. the integral $(I_1'')^2$) we remark that

$$|u(\zeta)|^2_{\mathfrak{S}_2} = \sum_{i,k\ge 1} |u_{ik}(\zeta)|^2$$
$$= 2|f(\zeta)|^{-8}(|f(\zeta)|^2|f'(\zeta)|^2 - |(f(\zeta), f'(\zeta))|^2)$$

and, on the other hand, $\Delta(\log|f|^2) = 4\bar\partial(f, f')/|f|^2 = 4|f|^{-4}\cdot(|f'|^2|f|^2 - |(f, f')|^2)$. Hence

$$(I_1'')^2 \le \frac{1}{2}\iint_{\mathbb{D}} |\varphi|^2 \frac{\Delta(\log|f|^2)}{|f|^4}\log(1/|z|)$$
$$\le \frac{1}{2\delta^4}\iint_{\mathbb{D}} |\varphi|^2\cdot\Delta(\log|f|^2)\log(1/|z|)$$
$$\le 2\pi e\delta^{-4}\log\delta^{-1}.$$

The last inequality was obtained by application of the following lemma, in which we have to take $a = \log(|f|^2\delta^{-2})$.

**6. Lemma.** *Let $a$ be smooth in* clos $\mathbb{D}$ *(e.g., of class $C^2$), positive and subharmonic (i.e. $\Delta a \ge 0$); set $\|a\|_\infty \stackrel{\text{def}}{=} \max_{\operatorname{clos}\mathbb{D}} |a|$. Then*

$$(\Delta a)\log(1/|z|)\, dx\, dy$$

*is a Carleson measure and the norm of the corresponding imbedding operator* (*cf.* *Lecture VII, Sect.* 1) *is at most*

$$(2\pi e\|a\|_\infty)^{1/2}.$$

Continuing the proof of the theorem let us join the estimates for the integrals $I_1'$ and $I_1''$:

$$I_1 \le 2\pi\delta^{-2}\sqrt{e\log\delta^{-1}}.$$

Next let us estimate the integral $I_2 \stackrel{\text{def}}{=} \iint_{\mathbb{D}} |\langle F, \partial u\rangle| \log(1/|z|)$. To this end let us remark that $\partial u = -2u|f|^{-2}\partial|f|^2 = -2u|f|^{-2}(f', f)$, and further that

$$\begin{aligned}|\partial u|_{\mathfrak{S}_2} &= 2|u|_{\mathfrak{S}_2}|f|^{-2}|(f',f)|\\ &= 2\sqrt{2}|f|^{-6}|(f',f)|(|f|^2|f'|^2 - |(f',f)|^2)^{1/2},\end{aligned}$$

because $|u|^2_{\mathfrak{S}_2} = 2|f|^{-8}(|f|^2|f'|^2 - |(f,f')|^2)$.

Setting $a = \log(2|f|^2 + \delta^2)/3\delta^2$, $\alpha = \delta^2/2|f|^2$ and $\beta = |(f',f)|^2\cdot|f|^{-2}|f'|^{-2}$ we find

$$\Delta a = 4|f'|^2|f|^{-2}(1+\alpha-\beta)(1+\alpha)^{-2}.$$

Hence

$$|\partial u|_{\mathfrak{S}_2} = 2^{-1/2}|f|^{-2}(1+\alpha-\beta)^{-1}(1+\alpha)^2\sqrt{\beta(1-\beta)}\cdot\Delta a = \sqrt{2}\omega(\alpha,\beta)\Delta a,$$

where

$$\omega(\alpha,\beta) = \alpha(1+\alpha)^2(1+\alpha-\beta)^{-1}\sqrt{\beta(1-\beta)}.$$

Since $\sup\{\omega(\alpha,\beta): 0\le\alpha\le 1/2, 0\le\beta\le 1\} = \omega(1/2, 3/4) = 3\sqrt{3}/8$, we have

$$|\partial u|_{\mathfrak{S}_2} \le (3\sqrt{6}/8)\delta^{-2}\Delta a.$$

Therefore

$$\begin{aligned}I_2 &\le \iint_{\mathbb{D}} |F(z)|_{\mathfrak{S}_2}|\partial u(z)|_{\mathfrak{S}_2}\log 1/|z| \le (3\sqrt{6}/8)\delta^{-2}\iint_{\mathbb{D}}|\varphi|^2\Delta a\cdot\log 1/|z|\\ &\le \text{(by Lemma 6) } (3\sqrt{6}/8)\delta^{-2}\cdot 2\pi e\|a\|_\infty \le (3\pi e\sqrt{6}/2)\delta^{-2}\log\delta^{-1}.\end{aligned}$$

Returning to inequalities (3) and (4) we see that

$$\begin{aligned}\|g\|_{H^\infty(l^2)} &\le \delta^{-1} + \delta^{-2}(4\sqrt{e\log\delta^{-1}} + 3e\sqrt{6}\log\delta^{-1})\\ &\le \delta^{-1} + \delta^{-2}(7\sqrt{\log\delta^{-1}} + 20\log\delta^{-1}). \ \square\end{aligned}$$

**7. Proof of Lemma 6.** If $\varphi$ is a polynomial ($\varphi\in\mathscr{P}_A$) we get by differentiation

$$\begin{aligned}\Delta(|\varphi|^2e^{ta}) &= 4|\varphi'|^2e^{ta} + |\varphi|^2\Delta e^{ta} + 8\,\mathrm{Re}\,\bar\partial|\varphi|^2\partial e^{ta}\\ &= 4|\varphi'|^2e^{ta} + |\varphi|^2(4t^2|\partial a|^2 + t\Delta a)e^{ta} + 8te^{ta}\,\mathrm{Re}\,\varphi\,\bar\varphi'\partial a \ge t|\varphi|^2e^{ta}\Delta a\\ &\ge t|\varphi|^2\Delta a.\end{aligned}$$

Therefore

$$\iint_{\mathbb{D}} |\varphi|^2 \Delta a \log 1/|z| \le \inf_{t>0} t^{-1} \iint_{\mathbb{D}} \Delta(|\varphi|^2 e^{ta}) \log 1/|z|$$
$$\le \inf_{t>0} t^{-1} 2\pi \| e^{ta} \|_\infty \| \varphi \|_2^2 = 2\pi e \| a \|_\infty \| \varphi \|_2^2. \square$$

**8. Remark on the Best Constant.** Let us set

$$c(\delta) = \sup\left\{\min\left[\| g \|_{H^\infty(l^2)} : \sum_{i\ge 1} g_i f_i = \mathbb{1}\right] : f \in H^\infty(l^2),\ \delta \le |f(\zeta)| \le 1,\ \forall \zeta \in \mathbb{D}\right\}.$$

Theorem 3 shows that $c(\delta) \le 1/\delta + (1/\delta^2)\,[7\sqrt{\log 1/\delta} + 20 \log 1/\delta]$. In Sect. 12 we shall give a lower bound for $c(\delta)$; for other estimates see also Sect. 13 and 14.

**9. The Operator Valued Corona Problem. Complementing to an Isomorphism.** This problem was briefly touched upon in Lecture III. It concerns a family of operator functions $f_i$, $f_i \in H^\infty(L(E))$, $1 \le i \le n$, and consists in deciding whether the condition

$$\sum_{i=1}^{n} \| f_i(\zeta) e \|^2 \ge \delta^2 \| e \|,\ \forall e \in E,\ \forall \zeta \in \mathbb{D},$$

is sufficient for the existence of functions $g_i$, $g_i \in H^\infty(L(E))$, $1 \le i \le n$, such that

$$\sum_{i=1}^{n} g_i(\zeta) f_i(\zeta) = I_E,\ \zeta \in \mathbb{D},$$

In fact, it is more natural to consider operators acting from $E'$ into $E$ where $E'$ and $E$ are different Hilbert spaces (always separable). Then the corona problem admits the following more economical and more general formulation (Sz.-Nagy): given a function $f$, $f \in H^\infty(L(E', E)) \stackrel{\text{def}}{=} H^\infty(E', E)$, possessing the property

$$\| f(\zeta) e \| \ge \delta \| e \|,\ \forall e \in E'; \tag{6}$$

we ask if there exists a function $g$, $g \in H^\infty(E, E')$ for which

$$g(\zeta) f(\zeta) \equiv I_{E'},\ \zeta \in \mathbb{D}. \tag{7}$$

The original statement of the question becomes a special case of the problem of Sz.-Nagy if we set $E = E' \oplus \cdots \oplus E'$ ($n$ times) and $f = (f_1, \dots, f_n)$ (coordinate notation).

The problem (6)–(7) is sometimes equivalent to another interesting problem (the complementing of an imbedding to an isomorphism). Before stating the latter, we remark that condition (6) itself can hold true only if $\dim E' \le \dim E$. If this is the case, we can without loss of generality assume that $E' \subset E$. The problem of complementing to an isomorphism consists now of describing those functions $f$, $f \in H^\infty(E', E)$, such that there exists an invertible function $F$ with

$$F,\ F^{-1} \in H^\infty(E, E);\ F | E' = f. \tag{8}$$

It is clear that condition (6) (uniform left invertibility) is necessary for such a complementing; below we show that this condition is for $\dim E' < \infty$ indeed sufficient and that it is then also sufficient for the solvability of equation (7). In the general case ($\dim E' = \infty$) there is also another necessary condition for (8):

$$\dim(E \ominus E') = \dim(E \ominus f(\zeta)E'), \, |\zeta| < 1. \tag{9}$$

Example when this last condition fails but (6) nevertheless holds true: $E' \neq E$; $f(\zeta) \equiv V, |\zeta| < 1$, where $V$ is any isomorphism of $E'$ *onto* $E$ ($VE' = E$); by the way, in this example problem (7) has a trivial solution. It is not known whether the conditions (6) and (9) are sufficient for (8).

Let us state two results concerning the problems (7) and (8) which have close analogues (for the algebra $L^\infty \cap \mathrm{Nev}$ instead of $H^\infty$) in Lecture II (see Complementing Lemma).

**10. Tolokonnikov's Lemma.** *Let* $E' \subset E$, $\dim E' < \infty$, $f \in H^\infty(E', E)$. *The following statements are equivalent.*

a. *There exists* $g$, $g \in H^\infty(E, E')$ *such that* $gf = I_{E'}$.

b. *There exists* $F$, $F \in H^\infty(E, E)$ *such that* $F|E' = f$, $F^{-1} \in H^\infty(E, E)$.

This statement is contained among the results of Lecture II, Sect. 4; cf. there the Complementing Lemma and *its sublemma*, where one has to put $m = \mathbb{1}$. In particular, from the proof of this sublemma it follows that $F = fP_{E'} + \Theta P_{E \ominus E'}$, where $\Theta$ is a left inner function, and that $F^{-1} = g + \Theta^{-1}(I - fg)$; thus $\|F\|_\infty \leq \|f\|_\infty + 1$, $\|F^{-1}\|_\infty \leq \|g\|_\infty(1 + \|f\|_\infty) + 1$.

**11. Fuhrmann-Vasyunin Theorem.** *Let* $E$ *and* $E'$ *be Hilbert spaces,* $n \stackrel{\text{def}}{=} \dim E' < \infty$, *and* $f \in H^\infty(E', E)$, $\|f\|_\infty \leq 1$. *If condition* (6) *is fulfilled then there exists a solution* $g$ *of equation* (7) *such that* $\|g\|_\infty \leq \sqrt{n}c(\delta^n)$.

*The proof* in the principal special case $\dim E < \infty$, with which we begin, is based on the Binet-Cauchy formula, already employed for similar purposes in the proof of the Complementing Lemma (Lecture II).

*The case* $\dim E < \infty$. As in the proof of the lemma just mentioned

$$\delta^{2n} \leq \det f(\zeta)^* f(\zeta) = \sum_{\kappa, \operatorname{card} \kappa = n} |\det f_\kappa|^2 \leq 1$$

where $f_\kappa = \pi_\kappa f$, $\pi_\kappa$ the orthogonal projection in $E$ onto the subspace $\bigvee(e_j : j \in \kappa)$, $\kappa \subset \{1, \ldots, \dim E\}$, $\{e_j\}$ an orthonormal basis in $E$. By Theorem 3 there exist functions $g_\kappa$ in $H^\infty$ such that

$$\sum_{\kappa, \operatorname{card} \kappa = n} g_\kappa \det f_\kappa = \mathbb{1}, \quad \left( \sum_{\kappa, \operatorname{card} = n} |g_\kappa|^2 \right)^{1/2} \leq c(\delta^n).$$

The operator function $g$ will be obtained fixing a suitable orthonormal basis $\{e'_i\}$ in $E$ and defining the matrix elements $g_{ij}$ by the formula

$$g_{ij} = \sum_{\kappa; j \in \kappa} g_\kappa f_\kappa^{ij}$$

where $f_\kappa^{ij}$ is the algebraic complement of the element $f_{ij}$ in the matrix for the operator function $f_\kappa$. From elementary properties of determinants (cf., for instance, Gantmaher [1]) it follows that for all $i, l$ $(1 \leq i, l \leq n)$ holds

$$\sum_{1 \leq j \leq \dim E} g_{ij} f_{jl} = \sum_{\kappa, \operatorname{card} \kappa = n} g_\kappa \sum_{j, j \in \kappa} f_\kappa^{ij} f_{jl} = \sum_{\kappa, \operatorname{card} \kappa = n} g_\kappa \cdot \delta_{il} \det f_\kappa = \delta_{il} \mathbb{1},$$

i.e. $gf \equiv I_{E'}$.

In order to estimate $|g(\zeta)|$, $\zeta \in \mathbb{D}$, we consider any vector $a = \sum_j a_j e_j \in E$, $\sum_j |a_j|^2 \leq 1$, and apply anew the Binet-Cauchy formula

$$\left|\sum_j g_{ij} a_j\right| = \left|\sum_\kappa g_\kappa \sum_{j, j \in \kappa} f_\kappa^{ij} a_j\right| \leq \left(\sum_\kappa |g_\kappa|^2\right)^{1/2} \cdot \left(\sum_\kappa \left|\sum_{j, j \in \kappa} f_\kappa^{ij} a_j\right|^2\right)^{1/2}$$
$$\leq c(\delta^n)(\det f_{a,i}^* f_{a,i})^{1/2},$$

where the map $f_{a,i}(\zeta)$ is defined by the identity $f_{a,i}(\zeta) e_k' = f(\zeta) e_k'$ for $1 \leq k \leq n$, $k \neq i$, $f_{a,i}(\zeta) e_i' = a$. By Hadamard's famous inequality (cf., for example, Gantmaher [1], p. 230)

$$\det f_{a,i}^* f_{a,i}(\zeta) \leq \left(\sum_{\substack{k \neq i \\ 1 \leq k \leq n}} \|f(\zeta) e_k'\|^2\right) \|a\|^2 \leq 1,$$

whence

$$\sum_{i=1}^n \left|\sum_j g_{ij} a_j\right|^2 \leq n \cdot c(\delta^n)^2, \quad |g(\zeta)| \leq \sqrt{n} c(\delta^n).$$

*The case of a function $f$ continuous in* clos $\mathbb{D}$, *with*

$$\delta \|e\| < \|f(\zeta) e\| \leq \|e\| \quad (\zeta \in \operatorname{clos} \mathbb{D}, e \in E' \backslash \{\mathbb{0}\}).$$

From the continuity of $f$ and the compactness of the unit ball in $E'$ we get that $\|f(\zeta) e\| \geq (\delta + \varepsilon) \|e\|$, $\forall e \in E'$, for some $\varepsilon$, $\varepsilon > 0$. By the same token, there exist in $E$ orthogonal projections $P_k$, rank $P_k < \infty$, $P_k \uparrow I$, such that $\|P_k f(\zeta) e\| \geq \delta \|e\|$, $\forall e \in E'$, $\forall \zeta \in \mathbb{D}$. We can now utilize the part of the theorem which we have already established and produce functions $g_k$, $g_k \in H^\infty(P_k E, E')$ such that $g_k P_k f \equiv I_{E'}$, $\|g_k\|_\infty \leq \sqrt{n} c(\delta^n)$. Since $E'$ is finite dimensional it follows from Montel's theorem that a suitable subsequence of $\{g_k P_k\}_{k \geq 1}$ converges uniformly on compact subsets of the disk $\mathbb{D}$ to a function $g$ with $gf \equiv I_{E'}$, $g \in H^\infty(E, E')$, $\|g\|_\infty \leq \sqrt{n} c(\delta^n)$.

*The general case* will be derived from the special case just settled. To this end we will show that $f = f^1 \oplus f^2$ where $f^1 \equiv$ const and $f^2$ is a pure contraction, i.e. $\|f^1(\zeta) e\| < \|e\|$, $\forall \zeta \in \mathbb{D}$, $\forall e \in E_2' \backslash \{\mathbb{0}\}$ (we have in mind that $E = E_1 \oplus E_2$, $E' = E_1' \oplus E_2'$ with $f^1(\zeta): E_1' \to E_1$, $f^2(\zeta): E_2' \to E_2$). Once this has been done we can consider the functions $f_r^2 \stackrel{\text{def}}{=} f^2(rz)$, $0 < r < 1$, which obviously are continuous in clos $\mathbb{D}$ and in addition satisfy

$$\delta \|e\| \leq \|f_r^2(\zeta) e\| \leq (1 - \varepsilon) \|e\|, \; e \in E_2'$$

(use again the condition $\dim E' < \infty$), with $\varepsilon = \varepsilon(r) = o(1)$, $r \to 1$. Utilizing the part of the theorem already proved (with the function $(1 - \varepsilon)^{-1} f_r^2$) we find functions $g_r$, $g_r \in H^\infty(E_2, E_2')$, such that $\|g_r\|_\infty \leq \sqrt{n} c(\delta^n)$, $g_r f_r^2 \equiv (1 - \varepsilon) I_{E_2'}$. As before we can pass to the limit using a suitable sequence $\{r_k\}$, $\lim_k r_k = 1$, and we get the desired function $g$, $g = (f^1)^{-1} \oplus \lim_k g_{r_k}$, $gf \equiv I_{E'}$, $\|g\|_\infty \leq \sqrt{n} c(\delta^n)$.

Thus it remains to see that a decomposition with the above properties is possible. This is a very well-known argument (and we have already mentioned it in passing in Lecture IV, p. 112, l. 11–13) but for completeness sake we give the details. If $\zeta_0 \in \mathbb{D}, e \in E'$ and $\|f(\zeta_0)e\| = \|e\| = 1$ then by the maximum principle it follows that the function $(f(\cdot)e, f(\zeta_0)e)$ is identically $\mathbb{1}$ and by the strong Schwarz inequality that $f(\zeta)e = f(\zeta_0)e, \zeta \in \mathbb{D}$. Moreover, if $e'$ is any other vector in $E'$ and $(e, e') = 0$ then for all $\lambda$, $\lambda \in \mathbb{C}$,

$$|\lambda|^2 + \|e'\|^2 \geq \|f(\zeta)(\lambda e + e')\|^2$$
$$= |\lambda|^2 + \|f(\zeta)e'\|^2 + 2 \operatorname{Re} \lambda (f(\zeta)e, f(\zeta)e'),$$

which shows that $(f(\zeta)e, f(\zeta)e') = 0, \forall \zeta \in \mathbb{D}$.

It is now clear that there exists a largest subspace $E_1'$, $E_1' \subset E'$, on which $f(\cdot)$ is constant and isometric. On the complement $E_2' \stackrel{\text{def}}{=} E' \ominus E_1'$ then $f$ is a pure contraction and in addition holds $f(\zeta)E_1' \perp f(\zeta)E_2', \forall \zeta \in \mathbb{D}$. The desired decomposition is thus: $f = f^1 + f^2, f^1 \stackrel{\text{def}}{=} f|E_1', f^2 \stackrel{\text{def}}{=} f|E_2'$. □

**12. A Lower Estimate in the Corona Theorem.** Let us first make a remark concerning the number $\inf \|g\|_\infty$, the least of the norms of the functions $g = \{g_i\}_{i \geq 1}$ solving the equation $\sum_{i \geq 1} f_i g_i = \mathbb{1}$; here $f$, $f \in H^\infty(l^2)$, is fixed. Namely, if we for fixed $\delta$, $q$ $(0 < \delta < q)$ set

$$c(\delta, q) = \sup\{\inf \|g\|_\infty : f \in H^\infty(l^2), \delta \leq |f| \leq q\},$$

then it is easy to read off that $c(\delta, q) = 1/q \cdot c(\delta/q)$.

Next let us show that the estimate of $c(\delta)$ (cf. Sect. 8) contained in Theorem 3 is close to the best one. More exactly, *we prove the inequality*

$$c(\delta) \geq \max(1/\delta, \sqrt{1 - \delta^2}/\delta^2) \tag{10}$$

To this end take $f_1 = (z^n - \delta\mathbb{1})(\mathbb{1} - \delta z^n)^{-1}$, $f_2 = \delta z$, where $n \in \mathbb{N}$, and let us estimate from below the $H^\infty$-norm of the functions $g_1$ and $g_2$ solving the equation $f_1 g_1 + f_2 g_2 = \mathbb{1}$. The function $f_1$ is a Blaschke product (more correctly, up to a unimodular factor) with the zeros $\lambda_i$, $1 \leq i \leq n$ (the roots of the equation $z^n = \delta$). In particular $g_2(\lambda_i) = \delta^{-1}\lambda_i^{-1}$, $1 \leq i \leq n$, and we can estimate $\|g_2\|_\infty$ by the minimum of the norms of the functions interpolating these values at the points $\lambda_i$:

$$\|g_2\|_\infty \geq \inf_{g \in H^\infty} \|g_2 + g_1 g\|_\infty = \operatorname{dist}(g_2 \bar{f}_1, H^\infty)$$
$$= \text{(Hahn-Banach theorem)}$$
$$\sup\left\{\left|\int_{\mathbb{T}} g_2 \bar{f}_1 h \, dm\right| : h \in H^1, \|h\|_1 \leq 1, h(0) = 0\right\}$$
$$= \sup\left\{\left|\frac{1}{2\pi i} \int_{\mathbb{T}} \frac{g_2}{f_1} h \, dz\right| : h \in H^1, \|h\|_1 \leq 1\right\}$$
$$= \text{(residue theorem)}$$
$$\sup\left\{\left|\sum_{i=1}^n \frac{g_2(\lambda_i)h(\lambda_i)}{f_1'(\lambda_i)}\right| : h \in H^1, \|h\|_1 \leq 1\right\}.$$

As the test function $h$ we now choose $h = (1-\delta^2)(\mathbb{1} - \delta z^n)^{-2}$. Since $f_1'(\lambda_i) = n\lambda_i^{n-1}(1-\delta^2)^{-1}$ and $h(\lambda_i) = (1-\delta^2)^{-1}$, $1 \le i \le n$, we get $\|g_2\|_\infty \ge \delta^{-2}$.

On the other hand, for $|\zeta| \le (\delta/n)^{1/n}$ we have $|f_1(\zeta)| \ge (\delta - \delta/n)(1+\delta^2/n)^{-1}$ and for $|\zeta| \ge (\delta/n)^{1/n}$ similarly $|f_2(\zeta)| \ge \delta \cdot (\delta/n)^{1/n}$ so that throughout the disk $\mathbb{D}$ holds

$$1 + \delta^2 \ge |f_1|^2 + |f_2|^2 \ge \delta^2 \min\left\{\left(\frac{n-1}{n+\delta^2}\right)^2, \left(\frac{\delta}{n}\right)^{2/n}\right\} \overset{\text{def}}{=} \delta^2\alpha_n^2$$

Thus $c(\delta\alpha_n, \sqrt{1+\delta^2}) \ge \delta^{-2}$, $n \ge 1$. As $\lim_n \alpha_n = 1$ and further $c(\cdot, q)$ is a monotone decreasing function $(\forall q)$ we find $c(\delta\alpha, \sqrt{1+\delta^2} \ge \delta^{-2}$ for all $\alpha$, $\delta$, $0 < \alpha < 1, 0 < \delta < 1$. Hence

$$\frac{1}{\sqrt{1+\delta^2}} \cdot c\left(\frac{\delta\alpha}{\sqrt{1+\delta^2}}\right) \ge \delta^{-2}$$

or, otherwise, $c(\varepsilon) \ge \alpha/\varepsilon\delta$, where $\varepsilon$ and $\delta$ are connected by the equation $\varepsilon^2 = \delta^2\alpha^2(1+\delta^2)^{-1}$. Choosing $\alpha$ arbitrarily close to 1 we get the inequality

$$c(\varepsilon) \ge \sqrt{1-\varepsilon^2}/\varepsilon^2, 0 < \varepsilon < 1/\sqrt{2}$$

and thus the estimate (10) for all $\delta$, $0 < \delta < 1$. □

**13. Remarks on Other Estimates.** From Theorem 3 it follows that $c(\delta) - 1 = o((1-\delta)^{1/2})$ for $\delta \to 1 - 0$. In fact, one can show that $c(\delta) - 1 \asymp 1 - \delta$ in a neighborhood of $\delta = 1$; cf. Tolokonnikov [2]. In the operator version of the Corona Problem (the problem of Sz.-Nagy, Sect. 9) one can likewise define a best constant in exactly the same way as in Sect. 8 for the scalar case; let us denote this constant by $c_n(\delta)$, $n = \dim E'$, $0 < \delta < 1$. Then $c_n(\delta) \ge \delta^{1-n} \cdot 2/\sqrt{n}(2n)^{3(n-1)/2}$ and in particular $\sup_{n\ge1} c_n(\delta) \ge A\exp(B\delta^{2/3})$ for small $\delta$ ($A$, $B$ being absolute constants); for this cf. Tolokonnikov [2].

**14. Concluding Remarks.** As already mentioned in Lecture III, the original proof of the Corona Theorem, due to Carleson [2], concerned the case $f = (f_1, \ldots, f_n)$ with an estimate $\|g\|_\infty \le \delta^{-A}$ for the solution which depended on $n$ ($A = A(n)$). In 1979 Thomas Wolff gave a new remarkable proof of Carleson's theorem which (with some simplifications due to Varopoulos and Garnett) is set forth in Koosis's book [4], p. 359–376; the estimate for the solution has the form $\|g\|_\infty \le 2\delta^{-1} + 2(n-1)M$ where $M$ depends on $\delta$ only. Based on this proof (improved by Gamelin [2]), the first estimate of the solution not depending on $n$ and providing the possibility to solve the corona problem for $l^2$-valed functions was independently and simultaneously obtained by Tolokonnikov [1] ($\|g\|_\infty \le 25\delta^{-4}$) and Rosenblum [2] ($\|g\|_\infty \le 65\delta^{-4}$). About the same time Tolokonnikov (see his papers [1], [2]) also got an estimate close to Theorem 3: $\|g\|_\infty \le \delta^{-1} + 42\delta^{-2}(\log\delta^{-1})^{1/2}(1/4 + \log\delta^{-1})$. An estimate for the solution having order $\delta^{-2}\log\delta^{-1}$ for $\delta \to 0$ has been obtained by Uchiyama [1].

The proof of Theorem 3, as given in the text, follows Tolokonnikov's paper [2] and differs just in one detail from that paper [2], namely in the estimate of the integral $I_2$ (p. 291, l. $9_{-}$–$8_{-}$) due to Uchiyama [1]. There (Tolokonnikov [2]) Lemma 10 and Theorem 11 also appear; chronologically Tolokonnikov's Lemma is preceded by the Complementing Lemma (Lecture II) from which it was derived in our exposition. The proof of Theorem 11 (modulo some details) is due to Vasyunin, who employed part of a reasoning of Fuhrmann's, who in turn solved the operator corona problem in the special case $E = E' \oplus \cdots \oplus E'$ ($n$ times), $\dim E' < \infty$ (cf. Fuhrmann [1]). The problem of Sz.-Nagy is formulated in the paper [1] by that author; there one can find interesting comments on other aspects of this problem too. The estimate in Sect. 12 is due to Tolokonnikov [1, 2].

Let us mention some other results connected with the Corona Theorem. First of all if we want to estimate the joint size of the infinite family $\{f_i\}_{i\geq 1} \overset{\text{def}}{=} f$, then it is not at all necessary to use the Euclidean norm of the space $l^2$ (as we did in Theorem 3). Putting $|f(\zeta)|_p = (\sum_{i\geq 1} |f_i(\zeta)|^p)^{1/p}$, $\zeta \in \mathbb{D}$, one can ask if from the conditions $f \in H^\infty(l^p)$, $\delta \leq |f(\zeta)|_p \leq 1$ $(\zeta \in \mathbb{D})$ one can deduce the existence of a function $g = \{g_i\}_{i\geq 1}$, $g \in H^\infty(l^{p'})$ $(1/p + 1/p' = 1)$ such that $\sum_{i\geq 1} f_i g_i = \mathbb{1}$. If $p = 2k$, $k \in \mathbb{N}$, *the answer is positive* (remark by Tolokonnikov) and one has indeed an estimate $\|g\|_{H^\infty(l^{p'})} \leq c(\delta^k)$, where $c(\cdot)$ is the constant of Sect. 8. The case $p = \infty$ is analyzed in Uchiyama's paper [1] (one has a positive answer with the estimate $\|g\|_{H^\infty(l^1)} \leq \min(c\delta^{-c}, 1 + c/\log(1-\delta)^{-1})$, $c$ a suitable constant).

Returning to the case $f \in H^\infty(l^2)$ let us mention a result by Arveson [1], Sz.-Nagy and Foiaş [10] and Schubert [1] to the effect that

$$\inf\left\{\|g\|_{H^\infty(l^2)} : \sum_{i\geq 1} f_i g_i = \mathbb{1}\right\} = \inf\{1/\varepsilon : \|T_{\bar f} h\|_{H^2(l^2)} \geq \varepsilon \|h\|_{H^2}, \forall h \in H^2\};$$

here $T_{\bar f} h \overset{\text{def}}{=} \{P_+ \bar f_i h\}_{i\geq 1}$ (a vector valued Toeplitz operator); for the proof see Appendix 4, Sect. 165. 169. In fact, it was precisely an estimate for such Toeplitz operators (and corresponding Hankel operators $H_{\bar f}$; regarding these estimates see Lecture VIII and IX and Appendix 4) that led Tolokonnikov to the first of the statements mentioned in Sect. 13.

The question of the existence of a corona is interesting and important also in other Banach algebras of analytic functions, not only in $H^\infty$. Some interesting results in this direction were recently obtained by Tolokonnikov [3]. Let us indicate some of these. In a first group of results it is question of algebras $X$ of analytic functions in the disk $\mathbb{D}$, distinguished by the fact that the derivatives (some derivatives) $f^{(i)}$ belong to a suitable ideal space of functions $E_i$. Then it is natural to define the class $X(l^2)$ of $l^2$-valued analytic functions such that $|f^{(i)}| \in E_i$ and to solve in this class the problem of the equation $\sum_{i\geq 1} f_i g_i = \mathbb{1}$ and of the estimate of the solution $g$ under the conditions $\|f\|_{X(l^2)} \leq 1$, $\delta \leq |f(\zeta)|$ $(\zeta \in \mathbb{D})$.

*Examples.* a) $X = \{f : f^{(k)} \in H^1, \|f\|_X \overset{\text{def}}{=} \|f^{(k)}\|_E + \|f\|_\infty < \infty\}$, $k \geq 1$, where $E$ is a symmetric (rearrangment invariant) Banach space of functions on the circumference $\mathbb{T}$, either separable or conjugate with respect to the natural duality to an analogous symmetric space; estimate of the solution:

$$\|g\|_{X(l^2)} \leq \text{const} \cdot \delta^{-k-3};$$

b) particular case: $X = H_k^p \stackrel{\text{def}}{=} \{f : f^{(k)} \in H^p\}$, $1 \le p < \infty$, $k \ge 1$; another special case: $X = H^1 \cap \operatorname{Lip}\alpha$, $0 < \alpha < 1$ (in the class $H^1 \cap \operatorname{Lip}\alpha$ Šamoyan has given an explicit formula for the solutions $g_i$ (based on integral representations due to Džrbašžyan) and some estimates for these functions as well);

c) $X = B^s_{pq}$ (a Besov space of analytic functions in $\mathbb{D}$; cf. definition in Appendix 5) for $s > 1/p$ or $s = 1/p$, $q = 1$ (i.e. when $X$ is an algebra with respect to multiplication); estimate of the solution: $\| g \|_{X(l^2)} \le c_{s,p,q} \delta^{-[s]-3}$;

d) $X = H^\infty \cap B^{1/2}_{2,2}$ (the algebra of bounded functions with a finite Dirichlet integral $\iint_{\mathbb{D}} |f'|^2 < \infty$); estimate for the solution: $\| g \|_{X(l^2)} \le \text{const} \cdot \delta^{-3}$ (this answers a question posed in Chow-Protas [1]).

Another group of results of the same paper concern the solvability of the equation $\sum_{i=1}^n f_i g_i = \mathbb{1}$ in the *algebras of multipliers* $M(X)$ of suitable spaces $X$ of analytic functions, $M(X) \stackrel{\text{def}}{=} \{g \in H^\infty : gX \subset X\}$. The estimates of the solutions for these algebras obtained in the paper in question depend as a rule on $n$.

*Examples* a) $M(X)$ for $X = B^s_{pq}$ ($s > 0$, $1 \le p$, $q \le \infty$); estimate for the solution: $\| g \|_{l^2_n(M(X))} \le c_{p,q} n^2 \delta^{-3}$ for $s < 1$ and $\le c_{p,q} n^2 \delta^{-4}$ for $s = 1$;

b) $M(X)$ for $X = \{f : \iint_{\mathbb{D}} (|f'|\psi)^p < \infty\}$, where $\psi = \psi(r) \downarrow 0$ (the function $\psi$ has to decrease as a power for $1 - r \to 0$, for details see Tolokonnikov [3]); estimate for the solution: $\| g \|_{l^2_n(M(X))} \le c_\psi n^2 \delta^{-3}$. The example embraces for $p = 2$ the algebras $M(X)$ for spaces of the type $X = \{f : f = \sum_{k \ge 0} \hat{f}(k) z^k, \sum_{k \ge 0} |\hat{f}(k)|^2 \beta_k^2 < \infty\}$ considered in Shields's paper [1].

A question close to the Corona Theorem is the question of ideals in $H^\infty$ generated by a collection of functions $f_i$, $f_i \in H^\infty$, $i \ge 1$. Let, for example, $f = \{f_i\}_{i \ge 1} \in H^\infty(l^2)$,

$$I(f) \stackrel{\text{def}}{=} \left\{ \sum_{i \ge 1} f_i g_i : g \in H^\infty(l^2) \right\}.$$

If $x \mapsto \alpha(x)$ is a positive function on $\mathbb{R}_+$ then by definition

$$\mathscr{I}^{(\alpha)}(f) = \{h \in H^\infty : |h(\zeta)| \le \text{const}\, \alpha(|f(\zeta)|), \zeta \in \mathbb{D}\}.$$

It is clear that $I(f) \subset \mathscr{I}^{(x)}(f)$ (we have the function $x \mapsto x$ in mind). There are many papers devoted to the characterization of those functions $\alpha$ for which one has the inverse inclusion $\mathscr{I}^{(\alpha)}(f) \subset I(f)$. The record (1981) apparently is hold by Tolokonnikov [2] who proved that a) $\mathscr{I}^{(x^{2+\varepsilon})}(f) \subset I(f)$, $\forall \varepsilon > 0$, $\forall f \in H^\infty(l^2)$; b) there exist functions $f_i \in H^\infty$, $i = 1, 2$, such that the condition $x^2 = o(\alpha(x))$ for $x \to 0$ entails $\mathscr{I}^{(\alpha)}(f) \not\subset I(f)$, where $f = (f_1, f_2)$. Finally, let us mention that the proof of Statement a) only in minor details differs from the above proof of Theorem 3 and that a comprehensive bibliography on this problem of ideals in $H^\infty$ can be located in the same paper.

# Appendix 4. Essays on the Spectral Theory of Hankel and Toeplitz Operators

**1. Contents:** This Appendix gives a short survey of some directions of the spectral theory of Toeplitz and Hankel operators which naturally enter into the general framework of the main body of the book. In the first place it will be question of results close to the spectral theory of functions and the theory of the functional model – on the expression of $s$-numbers of Hankel operators in terms of best approximation, on the connections of this problem with interpolation problems and on a wide circle of problems arising from attempts to describe the spectra of Toeplitz (Wiener-Hopf) operators. The criteria for the Toeplitz operator $T_\varphi$ to be invertible or Fredholm will be stated in terms of the distance $\mathrm{dist}_{\mathrm{BMO}}(\arg\varphi, \widetilde{L}^\infty + \mathbb{R})$ in the metric of the famous space BMO but for each concrete class of symbols the translation of such criteria into recipes suitable for the applications constitutes an interesting independent problem. The solution of several such problems will likewise be given in this Appendix.

By its character this Appendix (together with the corresponding portions of Lectures VIII and XI) gives a *first introduction* to the spectral theory of Hankel and Toeplitz operators. In all cases preference was given to "direct" proofs as compared to "high-brow" ones, even if the latter were connected with very interesting theories. In particular (and this is apparently the most serious gap in this Appendix) we were forced to omit the part of the theory which is based on an analysis of the maximal ideal space of the algebra $H^\infty$ (or its super algebras); the same was the fate of the "generalized Fredholm" property in von Neumann algebras and part of the operator $K$-theory connected with the computations of the spectrum of Toeplitz operators. The inclusion of this material would have lead to a considerable expansion of the size of this Appendix. These and other aspects of the theory not treated here will be discussed briefly in the "Concluding Remarks".

The plan of the Appendix follows the logic of that protion of the subject which we have succeeded in developing on these pages. After a brief description of the problems connected with the *metrical properties of Hankel operators* we pass to the study of *the spectra of Toeplitz operators*. The presentation relies on the following "three pillars"[1] of the theory of Toeplitz operators: these are the *direct methods* developing the factorization technique of Riemann, Hilbert, Wiener-Hopf; further the *localization technique* arising from general considerations on the dependence of singular integrals on the local behavior of the functions as well as various variants of the local theory of Banach algebras (partition of the unity, maximal sets of antisymmetry etc.); finally the *algebraic approach* which is aimed at a study of a Toeplitz operator within the collection of

[1] Some other "pillars" will be mentioned in the "Concluding Remarks".

operators "close to it" (which is generated by a certain *function algebra*) and uses the information that is left over after ignoring the non-commutativity of these operators (this approach is connected with the symbolic calculi and contains as one of its main ingredients various *spectral inclusion theorems*).

Here is a list of the main themata of the Appendix (with the numbers of the Sections where their treatment starts as well as the corresponding page references):

All bibliographical references are collected in the corresponding divisions of the "Concluding Remarks".

**2. This Short Preface** to the Appendix is called for by the need to relate the theory to be set forth below with the remainder of the book. Let us first recall that a Hankel operator is for us a (continuous) operator acting from the space $H^2$ into the space $H^2_- \overset{\text{def}}{=} L^2(\mathbb{T}) \ominus H^2$ according to the rule

$$H_\varphi x = P_- \varphi x, \quad x \in H^2,$$

where $\varphi \in L^\infty(\mathbb{T})$. Nehari's Theorem (Lecture VIII) guarantees that every (continuous) operator $A$, $A: H^2 \to H^2_-$, having a matrix characteristic for Hankel operators ($(Az^n, \bar{z}^m) = (Az^k, \bar{z}^l)$ for $n + m = k + l$ ($n, k \geq 0$, $m, l \geq 1$) or equivalently $P_- zA = Az$) in fact has the form $A = H_\varphi$, $\varphi \in L^\infty$, with $|H_\varphi| = \operatorname{dist}_{L^\infty}(\varphi, H^\infty)$.

Lecture VIII, XI and XII have given us ample occasions to use Hankel operators as a tool in our study – in connection with functions $\varphi(\mathbf{T}_\Theta)$ of the model operator $\mathbf{T}_\Theta$, in connection with the computation of the angle between the "past" $\bigvee(z^k : k < 0)$ and the "future" $\bigvee(z^n : n \geq 0)$ in a weighted metric (the Helson-Szegö Theorem) and in the study of bases of exponentials and of reproducing kernels. There could have been many more occasions, because Hankel operators (and Toeplitz operators too, of which we will say more below) are encountered in a great variety of different mathematical problems – in the theory of integral transformations, in the algebraic problem concerning the separation of the roots of an equation, in the spectral theory of stationary random processes, in problems of statistical estimation, in many questions of function theory (interpolation, best approximation, boundary value problems, ...), in control theory etc., etc. Some of these applications of Hankel and Toeplitz operators will be found in Appendix 5, the remainder in the literature, to be discussed at the end of the present Appendix (cf. the "Concluding Remarks"). Here we mention just one class of Hankel operators, arising in the moment problem for powers:

$$\int_{-1}^{+1} t^n \, d\nu(t) = a_n, \quad n \geq 0,$$

where $\nu$ is a finite (positive) measure on the segment $(-1, 1)$. Already at the very beginning of the study of this problem it was established that for given numbers $a_n$, $a_n \in \mathbb{R}$, the existence of such a measure $\nu$ is equivalent to the positive definiteness of the Hankel matrix $H \stackrel{\text{def}}{=} \{a_{n+k}\}_{n,k \geq 0}$, along with the condition $\lim_n a_n = 0$. The problem of the continuity of the operator $H[\nu]$ (likewise termed a Hankel operator) generated by the matrix $H$ in the space $l^2$ is solved by the following theorem.

**3. Widom's Theorem.** *Let $\nu$ be a finite (positive) measure on the segment $(-1, 1)$. The following statements are equivalent.*

a. *The operator $H[\nu]$ is continuous.*

b. *$\nu$ is a Carleson measure.*

c. $\nu((\alpha, 1)) + \nu((-1, -\alpha)) = O((1-\alpha))$ *for* $\alpha \to 1-0$.

*The proof* amounts essentially to a computation of the matrix of the operator $\mathscr{H}^*\mathscr{H}$ where $\mathscr{H}: H^2 \to L^2(\nu)$ is the imbedding operator from Lecture VIII, Sect. 1:

$$(\mathscr{H}^*\mathscr{H} z^n, z^m)_{H^2} = (\mathscr{H} z^n, \mathscr{H} z^m)_{L^2(\nu)} = \int_{-1}^{1} x^n x^m \, d\nu = a_{n+m}.$$

It is now clear that the continuity of the operator $\mathscr{H}$ and the form $(H[\nu]a, b) = (\mathscr{H}^*\mathscr{H} \sum_{n \geq 0} a_n z^n, \sum_{m \geq 0} b_m z^m)$ (initially defined on the polynomials only) are equivalent. It remains to make an appeal to Carleson's Imbedding Theorem (Lecture VIII). □

**4. The Connection between Hankel, Toeplitz and Model Operators.** If $\Theta$ is an inner function, $K_\Theta = H^2 \ominus \Theta H^2$, $\mathbf{T}_\Theta = P_\Theta S | K$, and if $\varphi \in H^\infty$ then

$$\varphi(\mathbf{T}_\Theta) P_\Theta = \Theta H_{\bar{\Theta}\varphi}. \tag{1}$$

Thanks to this simple formula (and the Lemma in Lecture VIII, Sect. 2) we were able to use Hankel operators in the study of the commutant of the model operator $\mathbf{T}_\Theta$. In Lecture VIII we obtained compactness criteria for the operator $\varphi(\mathbf{T}_\Theta)$ and thus for the operator $H_{\bar{\Theta}\varphi}$. There it was likewise remarked (with the proof however being deferred to Appendix 2) that the set of functions of the form $\bar{\Theta}\varphi$ ($\Theta$ an inner function, $\varphi \in H^\infty$) is dense in the space $L^\infty(\mathbb{T})$ so that the criterion in question comprises indeed "almost all" Hankel operators $H_f$, $f \in L^\infty$. In the present Appendix we will in fact once more get Hartman's Theorem ($H_f \in \mathfrak{S}_\infty \Leftrightarrow f \in H^\infty + C(\mathbb{T})$) but on an entirely different basis (as a consequence of the Adamyan-Arov-Kreĭn Theorem). Let us here further remark that the Hankel operators encountered in (1) are completely characterized by having a non-trivial kernel.

**5. Lemma on** $\operatorname{Ker} H_f$. *Let $f \in L^\infty(\mathbb{T})$. 1. The kernel $\operatorname{Ker} H_f$ is an $S$-invariant subspace. 2. $\operatorname{Ker} H_f \neq \{\mathbb{0}\} \Leftrightarrow$ the function $f$ has the form $f = \bar{\Theta}\varphi$ where $\Theta$ is an inner function and $\varphi \in H^\infty$.*

*The proof* of the first statement is obvious: $H_f x = P_- fx = \mathbb{0} \Rightarrow fx \in H^2 \Rightarrow zfx \in H^2 \Rightarrow H_f zx = \mathbb{0}$.

Half of the second assertion ($\Leftarrow$) too is trivial (and has already been recorded above). The converse implication ($\Rightarrow$) follows from what already has been proved and the Beurling-Helson Theorem (Lecture I): $\operatorname{Ker} H_f = \Theta H^2$, where $\Theta$ is an inner function, i.e. $\mathbf{f}\Theta H^2 \subset H^2$, i.e. $f\Theta \in H^2 \cap L^\infty = H^\infty$. □

**6. The Toeplitz operator** $T_\varphi$ with symbol $\varphi$, $\varphi \in L^\infty$, is defined by the identity

$$T_\varphi x = P_+ \varphi x, \quad x \in H^2.$$

Such operators have also arisen more sporadically in this book (Lectures II, VIII); their study constitutes a large division of spectral theory, in view of their rôle in mathematical physics (where they are encountered in connection with "Wiener-Hopf equations", i.e. the equations $T_\varphi x = y$), in the theory of stochastic processes etc. Within the framework of this book the most important thing is however the impact of the theory of Toeplitz operators on the spectral theory of functions – by the intermediary of Hankel operators (with which they are intimately connected), through the analysis of interesting intermediate Banach algebras between $H^\infty$ and $L^\infty(\mathbb{T})$ ("Douglas algebras"), in connection with the Corona Theorem. The influence of Toeplitz operators reveals itself in yet some other ways too about which we say more in what follows.

The connection with Hankel operators just referred to is very simple and is based on the identity

$$H_\varphi x + T_\varphi x = \varphi x, \quad x \in H^2.$$

Besides this identity let us mention the functions (and likewise "functions" in a wider sense) of the model operator $\mathbf{T}$, $\mathbf{T} = P_K S|K$, $K = H^2 \ominus \Theta H^2$: if $\varphi \in L^\infty$ then the operator $P_K \boldsymbol{\varphi}|K$ can on the one hand be considered as a "section" (compression) of the Toeplitz operator onto the subspace $K_\Theta$ (if $\Theta = z^n$ one gets the standard finite section of the Toeplitz matrix $\{(P_+ \varphi z^k, z^l)\} = \{\hat{\varphi}(l-k)\}_{l,k \geq 0}$ so popular in the "projection methods" for the solution of Wiener-Hopf equations) and on the other hand this operator $P_K \boldsymbol{\varphi} = P_+ \Theta P_- \bar{\Theta} \boldsymbol{\varphi}$ also coincides with the product $H^*_{\bar{\Theta}} H_{\bar{\Theta}\varphi}$. We get an analogous product if we consider the "semi-commutator" of two Toeplitz operators ($f, g \in L^\infty$):

$$T_{fg} - T_f T_g = P_+ fg - P_+ f P_+ g = P_+ f P_- g = H^*_{\bar{f}} \cdot H_g. \tag{2}$$

Also, a Hankel matrix $\{a_{i+j}\}_{1 \geq i,j \geq n}$ goes over into a Toeplitz matrix $\{b_{i+n-j}\}_{1 \leq i,j \leq n}$ if we multiply it by a simple isometrical factor, the matrix $\{\delta_{i,n-j}\}_{1 \leq i,j \leq n}$ (corresponding to a change (permutation) of basis).

The following treatment of Hankel operators will center around the fundamental theorem of Adamyan-Arov-Kreĭn connecting the $s$-numbers of the operator $H_\varphi$ with the approximation of its symbol $\varphi$ by rational functions.

**7. The Singular Numbers** (or $s$-numbers) of a linear operator $A$ acting from one Hilbert space into another are defined by the identity

$$s_n(A) = \inf\{|A - K| : K \in \mathscr{K}_n\}, \quad n \geq 0,$$

where $\mathscr{K}_n$ is the set of all operators of rank not exceeding $n$. It is clear that $s_n(A) = |A|$, that $s_n \downarrow$ and that $\lim_n s_n(A) = 0$ iff $A \in \mathfrak{S}_\infty$ (where $\mathfrak{S}_\infty$ is the class of all compact operators). The well-known theorem of Schmidt (cf., e.g., Reed-Simon [1], theorem VI.17 or Gohberg-Kreĭn [1], chap. 2, theorem 2.1) says that assuming compactness ($A \in \mathfrak{S}_\infty$) the numbers $s_n^2$ are precisely the eigenvalues of $A^*A$ taken in decreasing order (and counted with their multiplicities) and the corresponding eigenvectors of $A^*A$ ($e_n$, say) and $AA^*$ ($e'_n$, say) generate the expansion of $A$ into a Schmidt series

$$Ax = \sum_{n \geq 0} s_n(x, e_n)e'_n.$$

If $A$ is not compact then the positive operator $A^*A$ cannot be compact either. The spectrum of such an operator $A^*A$ consists of the union of the isolate eigenvalues of finite multiplicity and the Fredholm spectrum $\sigma_F(A^*A)$ (likewise termed the essential or continuous spectrum, definition in Lecture V, Sect. 4), cf. the text books referred to above. Setting

$$s_\infty^2 = \max \sigma_F(A^*A)$$

we see that all the points of the set $\sigma(A^*A) \cap (s_\infty^2, \infty)$ are as before (that is, as for compact operators $A$, in which case $s_\infty = 0$) squares of $s$-numbers of the operator $A$. If this set is finite and the sum of the dimensions of the corresponding eigenspaces $\operatorname{Ker}(A^*A - s^2 I)$, $s > s_\infty$, equals $n$, then $s_k(A) = s_\infty$ for $k \geq n$ and the $s$-numbers $s_0(A), \ldots, s_{n-1}(A)$ are determined by the eigenvalues of the operator $A^*A$ situated to the right of the number $s_\infty^2$ (if there are any, that is, if $n > 0$). We have always $s_\infty(A) = \lim_n s_n(A)$.

An arbitrary eigenvector $f$ of the operator $A^*A$ will be termed a *Schmidt vector* (an $s$-Schmidt vector or simply $s$-vector if $A^*Af = s^2 f$, $s > 0$).

In preparation for the Adamjan-Arov-Kreĭn Theorem let us mention two simple properties of $s$-numbers and $s$-vectors.

**8. Lemma.** $s_n(BAC) \leq |B| s_n(A) |C|$, $n \geq 0$, *if the product* $BAC$ *makes sense.*

*The proof* follows at once from the definition of $s$-numbers. □

**9. Lemma.** *If* $A^*Af = s^2 f$, $s > 0$, *then the operator* $\frac{1}{s}A$ *is unitary on the element* $f$, *i.e.*

$$(Af, Ag) = s^2(f, g) \forall g.$$

*The proof* is immediate. □

**10. Adamyan-Arov-Kreĭn Theorem.** *Let* $A = H_\varphi$ *be a Hankel operator* ($\varphi \in L^\infty$) *and denote for* $n \geq 0$ *by* $\mathscr{R}_n$ *the set of all rational functions with poles in the disk* $\mathbb{D}$ *and with the sum of the multiplicities of these poles (for each such function) not exceeding* $n$. *Then*

$$s_n(A) = \inf\{|A - H_\psi| : \operatorname{rank} H_\psi \leq n\} = \operatorname{dist}(\varphi, \mathscr{R}_n + H^\infty).$$

**11. The Comments** on this theorem will start with the first identity. It shows that in order to find the $s$-numbers of a Hankel operator $A$ one can stay within the set Hank of all Hankel operators (in fact, Hank is a closed subspace of the space $L(H^2, H_-^2)$ of all linear operators): the exact lower bound for the norm $|A - K|$ for *all* operators $K$ of rank not exceeding a given number is the same as for the *subset* of such operators which are Hankel ($\text{Hank} \cap \mathscr{K}_n$). It is as if there were in $L(H^2, H_-^2)$ a projection $\pi$ onto the subset Hank, $|\pi| = 1$, which does not increase the rank of an operator. However such a projection does not exist; there exists no bounded projection onto the subspace Hank.

It is further clear that in the very definition of $s$-number as well as in the Adamyan-Arov-Kreĭn Theorem one can *replace* inf by min (because the unit ball of $L(H^2, H_-^2)$ is weakly compact and the subspace Hank is weakly closed).

In trying to understand this identity it is likewise convenient to recall Kronecker's Lemma (Lecture VIII, Sect. 3) describing the Hankel operators of finite rank: rank $H_\varphi < \infty \Leftrightarrow P_-\varphi$ is a rational function. In fact, from the proof of Kronecker's Lemma it follows that $\dim H_\varphi(H^2) = \deg P_-\varphi$ where deg denotes the order (or degree) of the rational function (i.e. the sum of the multiplicities of all of its poles in $\hat{\mathbb{C}}$). In the case rank $H_\varphi < \infty$ it is easy to find an explicit formula for $H_\varphi f$: if $|\lambda| < 1$ then

$$P_-\left(\frac{1}{z-\lambda} f\right) = P_-\left(\frac{f - f(\lambda)}{z-\lambda} + \frac{f(\lambda)}{z-\lambda}\right) = \frac{f(\lambda)}{z-\lambda}, \quad f \in H^2,$$

and thus, in the case of a rational function $P_-\varphi$ with simple poles,

$$P_-\varphi \stackrel{\text{def}}{=} \sum_{k=1}^{n} \frac{c_k}{z-\lambda_k}, \quad |\lambda_k| < 1,$$

we get

$$H_\varphi f = \sum_{k=1}^{n} \frac{c_k f(\lambda_k)}{z-\lambda_k}, \quad f \in H^2.$$

The general formula will contain terms of the form

$$P_-\frac{1}{(z-\lambda)^{k+1}} f = \frac{1}{k!}\frac{d^k}{d\lambda^k}\left[\frac{f(\lambda)}{z-\lambda}\right].$$

The second identity in Theorem 10 follows at once from the preceding discussion in conjunction with Nehari's Theorem (Lecture VIII). To this we may add that (as in Nehari's Theorem as well as in Sarason's Theorem on the Commutant, Lecture VIII) the distance $\operatorname{dist}(\varphi, \mathscr{R}_n + H^\infty) = \inf\{\|\varphi - \psi\|_\infty : \psi \in \mathscr{R}_n + H^\infty\}$ is attained ($= \|\varphi - \psi\|_\infty$ for some function $\psi$ in $\mathscr{R}_n + H^\infty$), by virtue of the weak compactness of the operator ball and the ball in the space $H^\infty$.

Let us pass to the proof of Theorem 10.

**12. Lemma.** *Let $A$ be a Hankel operator and $f$ a $s$-Schmidt function for $A$ ($s > 0$). Then*

$$f\bar{g} - \frac{1}{s^2} Af \cdot \overline{Ag} \in H_-^1, \quad \forall g \in H^2.$$

*The proof* follows from Lemma 9 applied to the function $z^n g$ ($n \geq 0$):

$$(f, z^n g) = \frac{1}{s^2}(Af, Az^n g) = \frac{1}{s^2}(Af, P_- z^n Ag) = \frac{1}{s^2}(Af, z^n Ag). \square$$

**13. Corollary.** *If $A$ is a Hankel operator and $f$ and $g$ $s$-functions ($\neq \mathbb{0}$) for $A$, $s > 0$, then*

$$\frac{Af}{sf} = \frac{Ag}{sg} = \frac{\bar{F}}{z\vartheta F}$$

*where $\vartheta$ is an inner function and $F$ an outer one belonging to $H^2$.*

In fact, using Lemma 12 and interchanging there the rôles of $f$ and $g$ we get $Af/sf = \overline{sg}/\overline{Ag}$, from which it follows that the left hand side is independent of the choice of $f$ among all possible $s$-functions. Also (taking $f = g$) we see that the ratio $Af/sf$ is unimodular. It remains to remark that the $H^2$-functions $sf$ and $\overline{Af}$ have the same modulus (and therefore the same outer factor in the canonical factorization). $\square$

**14. Remark.** If the norm of the operator $A = H_\varphi$ is attained, that is, if $\|Ag\| = |A| \, \|g\|$, then $g$ is an $|A|$-Schmidt function: $((A^*A - |A|^2)g, g) = 0$, whence $A^*Ag = |A|^2 g$. The converse is likewise true. $\square$

**15. Proof of Theorem 10. The Case** $s_n(A) > s_\infty(A)$. It suffices to put $\psi = \varphi - Af/f$ and to check that

$$\operatorname{rank} H_\psi \leq n,$$

where $f$ is an $s$-Schmidt function and $n$ the least number such that $s = s_n(A)$; in this remark we have taken into account that $|A - H_\psi| = |H_{Af/f}| = \operatorname{dist}\left(s\frac{Af}{fs}, H^\infty\right) \leq s$.

Let us first remark that $f \in \operatorname{Ker} H_\psi : H_\psi f = P_-(\varphi f - Af) = Af - Af = \mathbb{0}$. As the subspace $\operatorname{Ker} H_\psi$ is $S$-invariant (Lemma 5) it contains together with $f$ the subspace

$$\bigvee(S^k f : k \geq 0, A^*Af = s^2 f).$$

By the Beurling-Helson Theorem (Lecture I) this subspace must have the form $\Theta H^2$ where $\Theta$ is an inner function (= the GCD of the inner parts $f^i$ of the functions $f, f \in \operatorname{Ker}(A^*A - s^2 I)$). Thus $\Theta H^2 \subset \operatorname{Ker} H_\psi$ and therefore

$$\operatorname{rank} H_\psi \leq \dim(\Theta H^2)^\perp = \deg \Theta.$$

(Here $\deg \Theta = \infty$ if $\Theta$ is not a finite Blaschke product.)

It is now sufficient to show that $\deg \Theta \leq n$. Set $k = \dim \operatorname{Ker}(A^*A - s^2 I)$, $m = \deg \Theta$ and consider the operator $A\Theta = H_{\varphi\Theta}$. If we can show that

$$\dim \operatorname{Ker}(H_{\varphi\Theta}^* H_{\varphi\Theta} - s^2 I) \geq k + m, \tag{3}$$

the proof will be finished, because from this will follow that $s_n(A) = s \leq s_{m+k}(A\Theta)$, $s_{m+k}(A\Theta) \leq |\Theta| s_{m+k}(A) = s_{m+k}(A)$, and, as the multiplicity of the $s$-number $s$ of the operator $A$ by definition equals $k$, we get $m + k \le n + k$, that is, $\deg \Theta = m \le n$.

There remains the proof of (3). To this end we consider a chain of inner divisiors $\Theta_1, \dots, \Theta_m$ of $\Theta$ such that divides $\Theta_{j+1}, \Theta_j \ne \Theta_{j+1}, \Theta_m = \Theta$, and remark that

$$\dim \bigvee(\bar{\Theta}_j \operatorname{Ker}(A^*A - s^2 I) : 1 \le j \le m) \ge k + m.$$

Let us check that the functions belonging to this linear hull are $s$-functions for the operator $A\Theta = H_{\varphi\Theta}$ (and thereby complete the proof). Let us consider in $L^\infty(\mathbb{T})$ the conjugate-linear operator $\mathcal{J}$,

$$\mathcal{J}f \overset{\text{def}}{=} \overline{zf}, \quad f \in L^2(\mathbb{T}).$$

It is not hard to see that

$$\mathcal{J}^2 = I, \quad \mathcal{J}P_+ = P_-\mathcal{J}, \quad \mathcal{J}A = A^*\mathcal{J}, \quad A^*A = (\mathcal{J}A)^2.$$

Hence $(\mathcal{J}A)f \in \operatorname{Ker}(A^*A - s^2 I) \subset \Theta H^2$ if $f \in \operatorname{Ker}(A^*A - s^2 I)$ and thus $Af \in \mathcal{J}\Theta H^2 = \bar{\Theta} H^2_-$. The function $\Theta_j' \overset{\text{def}}{=} \bar{\Theta}_j \Theta$ being a divisor of $\Theta$ we have $\Theta_j' Af \in H^2_-$ and therefore

$$H_{\varphi\Theta}(\bar{\Theta}_j f) = P_-\varphi\Theta\bar{\Theta}_j f = P_-\varphi\Theta_j' f = P_-\Theta_j' P_-\varphi f = \Theta_j' Af,$$

$$H^*_{\varphi\Theta} H_{\varphi\Theta}(\bar{\Theta}_j f) = P_+\overline{\varphi\Theta}(\Theta_j' Af) = P_+\bar{\Theta}_j P_+\bar{\varphi} Af = P_+\bar{\Theta}_j A^*Af = s^2(\bar{\Theta}_j f). \ \square$$

**16. Completion of the Proof: the Case $s_n(A) = s_\infty(A)$.** It is sufficient to consider just an $s$-number $s_n(A)$ such that $s_\infty(A) = s_n(A) < s_{n-1}(A)$ (if already $s_0(A) = s_\infty(A)$ then the assertion of the theorem is trivial). Making thus this assumption let us show that for numbers $\rho$ arbitrarily close to $s_n(A)$, with $s_n(A) < \rho < s_{n-1}(A)$, there exist functions $\psi$, $\psi \in L^\infty$, such that $|H_\psi| \le \rho$, $\operatorname{rank}(A - H_\psi) \le n$.

To this end we want to construct a Hankel operator $B$ with the property

$$B^*B = A^*A + (\cdot, h)h \tag{4}$$

for a suitable $h$ in $H^2$, $h \ne \mathbb{0}$. Such an operator will be found in the form

$$B = \bar{\mathbf{z}}A + (\cdot, h)\bar{z}. \tag{5}$$

The identity (4) is then immediate whereas the requirement $B \in \text{Hank}$ (equivalent to the identity $P_- z^n B\mathbb{1} = Bz^n$, $n \ge 0$) means that

$$h = \alpha + zP_+\overline{z\varphi} = \alpha + zA^*\bar{z}, \quad \alpha \in \mathbb{C}. \tag{6}$$

Our remaining concern is now to find a number $\alpha$ such that

$$\operatorname{Ker}(B^*B - \rho^2 I) \ne \{\mathbb{0}\}.$$

Once this is done formula (5) and the definition of $s$-number immediately lead to the inequalities

$$s_{n+1}(B) \le s_n(A) \le s_n(B) \le s_{n-1}(A) \le s_{n-1}(B).$$

(As a matter of fact $s_{n+1}(B) = s_\infty(B) = s_\infty(A) = s_n(BA)$.) As $s_n(A) < \rho < s_{n-1}(A)$, $\rho$ being an $s$-number of $B$, we get $s_{n+1}(B) < s_n(B) = \rho < s_{n-1}(B)$. By the previous Sect. 15 we see that there exists a function $\psi_0$ such that rank $H_{\psi_0} \le n$, $|B - H_{\psi_0}| = \rho$ (we can take $\psi_0 = f/Bf$ where $f$ is a $\rho$-function for $B$). It remains to make the observation that $P_-\mathbf{z}B = A$ and to set $\psi = z\psi_0 : |A - H_\psi| = |P_-\mathbf{z}B - P_-\mathbf{z}H_{\psi_0}| \le \rho$, rank $H_\psi =$ rank $P_-\mathbf{z}H_{\psi_0} \le n$. This is precisely what we were set out to prove.

Thus we have to find $\alpha$ such that the equation

$$\rho^2 f - B^*Bf = \mathbb{0}$$

has a non-zero solution. We rewrite it as

$$(\rho^2 I - A^*A)f = (f,h)h. \tag{7}$$

Let us set $R = (\rho^2 I - A^*A)^{-1}, r = r(\rho) = (R\mathbb{1}, \mathbb{1})$ and show that in the hypothesis $r \neq 0$ there exists an $\alpha$, $\alpha \in \mathbb{C}$, such that

$$(Rh, h) = 1.$$

Then we can solve the equation (7) setting $f = Rh$ and the proof of the theorem will be over. Let us first notice that the function $\rho \to r(\rho)$ cannot have more than one zero in the segment $(s_n(A), s_{n-1}(A))$, because it is analytic and $dr/d\rho = -2\rho(R^2\mathbb{1}, \mathbb{1}) < 0$.

Let us pass to the search of a number $\alpha$ with $(Rh, h) = 1$. Notation: $a = zA^*\bar{z}$, $h = \alpha\mathbb{1} + a$. We get then the following equation for $\alpha$:

$$|\alpha|^2 r + 2\,\mathrm{Re}\,\alpha(R\mathbb{1}, a) + (Ra, a) = 1,$$

or

$$|\beta r + (a, R\mathbb{1})|^2 = r + |(R\mathbb{1}, a)|^2 - r(Ra, a).$$

Let us check that

$$r + |(R\mathbb{1}, a)|^2 - r(Ra, a) = 1/\rho^2; \tag{8}$$

then we have *all* solutions $\alpha : \alpha = \varepsilon/\rho r - 1/r \cdot (a, R\mathbb{1})$. Attention: the remainder of the proof has the character of a calculation exercise.

Identity (8) can easily be gotten from the formula

$$A'R\mathbb{1} = \rho^2 rRa - \rho^2(Ra, \mathbb{1})R\mathbb{1}, \tag{9}$$

where $A' = S\mathscr{J}A$. In fact, taking here the scalar product with the function $a = A'\mathbb{1}$ we get

$$\begin{aligned} \rho^2 r(Ra, a) - \rho^2|(Ra, \mathbb{1})|^2 &= (A'R\mathbb{1}, A'\mathbb{1}) = (S\mathscr{J}AR\mathbb{1}, S\mathscr{J}A\mathbb{1}) \\ &= (\mathscr{J}AR\mathbb{1}, \mathscr{J}A\mathbb{1}) = (A\mathbb{1}, AR\mathbb{1}) \\ &= (\mathbb{1}, A^*AR\mathbb{1}) = (\mathbb{1}, \rho^2 R\mathbb{1} - \mathbb{1}) \\ &= \rho^2 r - 1, \end{aligned}$$

that is, the desired formula (8).

In order to verify the identity (9) let us apply the operator $\rho^2 - A^*A = R^{-1}$ to both of its members, yielding

$$R^{-1}A'R\mathbb{1} = \rho^2 ra - \rho^2(Ra, \mathbb{1})\mathbb{1},$$

and let us thereupon show that

$$(I - P_0)R^{-1}A'R\mathbb{1} = \rho^2 ra, \tag{10}$$

$$(R^{-1}A'R\mathbb{1}, \mathbb{1}) = -\rho^2(Ra, \mathbb{1}), \tag{11}$$

where $P_0 \stackrel{\text{def}}{=} (\cdot, \mathbb{1})\mathbb{1}$. In order to check (10) recall that $\mathscr{J}A = A^*\mathscr{J}$ and $\mathscr{J}AS = S^*A^*\mathscr{J}$ ($\equiv Az = P_- zA$, which is the Hankel condition) so that

$$A'^2 = S\mathscr{J}AS\mathscr{J}A = SS^*A^*A = (I - P_0)A^*A,$$

or

$$(I - P_0)R^{-1} = (I - P_0)(\rho^2 - A'^2).$$

This gives

$$\begin{aligned}(I - P_0)R^{-1}A'R\mathbb{1} &= (I - P_0)(\rho^2 - A'^2)A'R\mathbb{1} = A'(\rho^2 - A'^2)R\mathbb{1} \\ &= A'(I - P_0)R^{-1}R\mathbb{1} + A'P_0(\rho^2 - A'^2)R\mathbb{1} \\ &= \rho^2 A'P_0R\mathbb{1} = \rho^2 rA'\mathbb{1} = \rho^2 ra.\end{aligned}$$

For the proof of (11) notice that

$$AA' = \mathscr{J}(\mathscr{J}AS)\mathscr{J}A = \mathscr{J}(S^*A^*\mathscr{J})\mathscr{J}A = \mathscr{J}S^*(\rho^2 - R^{-1}),$$

which gives

$$\begin{aligned}(R^{-1}A'R\mathbb{1}, \mathbb{1}) &= -(A^*AA'R\mathbb{1}, \mathbb{1}) = -(\mathscr{J}S^*(\rho^2 - R^{-1})R\mathbb{1}, A\mathbb{1}) \\ &= -\rho^2(\mathscr{J}S^*R\mathbb{1}, A\mathbb{1}) = -\rho^2(\mathscr{J}A\mathbb{1}, S^*R\mathbb{1}) = -\rho^2(a, R\mathbb{1}). \ \square\end{aligned}$$

**17. Corollary.** $s_n(H_\varphi) = \inf\{|H_{B\varphi}| : B$ *a Blaschke product,* $\deg B \leq n\}$.

In fact, $|H_{B\varphi}| = \operatorname{dist}(B\varphi, H^\infty) = \operatorname{dist}(\varphi, \bar{B}H^\infty)$ and the set of functions of the form $h/B$ makes up the set $\mathscr{R}_n + H^\infty$ when $h$ runs through $H^\infty$ and $B$ the set of all Blaschke products with $\deg B \leq n$. $\square$

**18. Consequences for Interpolation.** In a similar way as Nehari's Theorem (through the intermediary of Sarason's Theorem) solved the interpolation problems of Carathéodory and Nevanlinna-Pick (Lecture VIII, Sect. 1), the Adamyan-Arov-Kreĭn Theorem can be used to answer the following more general question.

*Given pairwise disjoint points* $\zeta_1, \ldots, \zeta_m$ *in the disk* $\mathbb{D}$, *germs (polynomials)* $p_1, \ldots, p_m$ *of height* $k_1, \ldots, k_m$ *at these points and an integer* $n$, $n \geq 0$; *to find* $\inf \|f\|_\infty$ *extended over all functions* $f$ *in* $\mathscr{R}_n + H^\infty$ *regular at the points* $\zeta_i$ *and having there a Taylor expansion*

$$f(\zeta) = p_i(\zeta) + a_i(\zeta - \zeta_i)^{k_i + i} + \ldots \tag{12}$$

If $m = 1$, $\zeta_1 = 0$ this is known as the Schur-Takagi problem and for general $m$ and $k_i = 1$ $(1 \leq i \leq m)$ as the Pick-Ahieser problem. In the latter case one gives thus the values $w_i, w_i \in \mathbb{C}$, of the function $f$ at the points $\zeta_i : f(\zeta_i) = w_i, 1 \le i \le m$. If $n = 0$ these problems have been considered in Lecture VIII, Sect. 1.

*The solution* of this problem can be connected with the $s$-numbers of a certain Hankel operator. In fact, let $f \in \mathscr{R}_n + H^\infty$ and assume that we have the expansion (12), $1 \le i \le m$; let in addition $g$ be any function in $H^\infty$ with given germs $p_i$, $1 \le i \le m$, at the points $\zeta_i$ (by the way, this function can be taken in the space $K_B$, $B \stackrel{\text{def}}{=} \prod_{i=1}^m b_{\zeta_i}^{k_i}$ and then it is uniquely defined). Now

$$\frac{f}{B} = \frac{g}{B} + \frac{f-g}{B}$$

and it is clear that $\dfrac{f-g}{B} \in \mathscr{R}_n + H^\infty$ (we can think of the functions in $\mathscr{R}_n + H^\infty$ as meromorphic functions in the disk $\mathbb{D}$ with no more than $n$ poles and in addition bounded near the circumference). From this and from the converse assertion (if $f/B = g/B + h, h \in \mathscr{R}_n + H^\infty$, then $f = g + Bh$ and by a small displacement of the poles of the function $h$ one can attain the identity (12)) combined with Theorem 10 we get *a formula for the lower bound*:

$$\inf \| f \|_\infty = s_n(H_{g\bar{B}}). \ \square$$

**19. The Uniqueness of the Solution** of the preceding problem, that is, the uniqueness of the function $F$, $F \in \mathscr{R}_n + H^\infty$, for which $\|F\|_\infty = \inf \| f \|_\infty$ and condition (12) is fulfilled (the inf is taken over all functions in $\mathscr{R}_n + H^\infty$ satisfying condition (12)), holds true iff there exists a unique solution $x$ realizing $\operatorname{dist}(g\bar{B}, \mathscr{R}_n + H^\infty)$ (in the sense that $\| g\bar{B} - x \|_\infty = \inf\{ \| g\bar{B} - y \|_\infty : y \in \mathscr{R}_n + H^\infty\}$, $x \in \mathscr{R}_n + H^\infty$). For rational functions $g\bar{B}$ (as in Sect. 18) this function $x$ is unique. Moreover, if $\varphi \in L^\infty(\mathbb{T})$ and $s_n(H_\varphi)^2$ is an eigenvalue of the operator $H_\varphi^* H_\varphi$ then there exists a unique $x$ in $\mathscr{R}_n + H^\infty$ with

$$\| \varphi - x \|_\infty = \operatorname{dist}(\varphi, \mathscr{R}_n + H^\infty);$$

in fact $x = \varphi - Af/f$ where $f$ is a corresponding Schmidt function for $H_\varphi$. If $n = 0$ this was proved in Lecture VIII (let us recall that the number $s_0(H_\varphi)^2 = |H_\varphi|^2$ is an eigenvalue for $H_\varphi^* H_\varphi$ iff the norm of $H_\varphi$ is attained; cf. remark 14). Cf. Adamyan-Arov-Kreĭn [5]. There one can also find a general necessary and sufficient condition for uniqueness $\lim_{\rho \to s_n + 0} ((\rho^2 I - H_\varphi^* H_\varphi)^{-1} \mathbb{1}, \mathbb{1}) = \infty$. In contradistinction to the *uniqueness*, the *existence* of a function $x$ does not guarantee the existence of a function $F$ realizing the inf in Sect. 8 (the poles of this unique function $x$ may hit the points $\zeta_i$).

**20. Corollary.** *If* $\operatorname{dist}(\varphi, H^\infty + C(\mathbb{T})) < \operatorname{dist}(\varphi, H^\infty)$ *then there exists precisely one function $x$ in $H^\infty$ such that* $\| \varphi - x \|_\infty = \operatorname{dist}(\varphi, H^\infty)$. *In particular this is so if* $\varphi \in H^\infty + C(\mathbb{T})$ *or if* $\operatorname{dist}(B\varphi, H^\infty) < \operatorname{dist}(\varphi, H^\infty)$ *for some finite Blaschke product.*

This statement too is essentially contained in Lecture VIII: formally speaking, we considered there just the case $B = z$ but the proof extends to any finite Blaschke product $B$. As $\operatorname{dist}(\varphi, H^\infty + C(\mathbb{T})) = \lim_n \operatorname{dist}(\varphi, \bar{z}^n H^\infty)$, the first statement of the corollary likewise is proved. $\square$

**21. The Essential Norm** of an operator $A$ acting from one Hilbert space into another is defined by the formula

$$|A|_{\text{ess}} \stackrel{\text{def}}{=} \operatorname{dist}(A, \mathfrak{S}_\infty) = \inf\{|A+K| : K \in \mathfrak{S}_\infty\}.$$

From Theorem 10 it follows that *for a continuous Hankel operator* $A = H_\varphi$, $\varphi \in L^\infty$, *there holds the formula*

$$|H_\varphi|_{\text{ess}} = \operatorname{dist}(\varphi, H^\infty + C(\mathbb{T})).$$

In fact, for the proof of this formula it suffices to make a general remark on the distance in an arbitrary metric space $X$:

$$\operatorname{dist}\left(x, \operatorname{clos} \bigcup_{n\geq 1} X_n\right) = \operatorname{dist}\left(x, \bigcup_{n\geq 1} X_n\right) = \lim_n \operatorname{dist}(x, X_n),$$

if $x \in X$, $X_n \subset X_{n+1} \subset X$ $(n \geq 1)$. Let us apply this with $X_n = \mathscr{K}_n$ (in the space of operators):

$$|A|_{\text{ess}} = \lim_n s_n(A).$$

and with $X_n = \mathscr{R}_n + H^\infty$ (in $L^\infty(\mathbb{T})$):

$$\operatorname{dist}(\varphi, H^\infty + C(\mathbb{T})) = \lim_n \operatorname{dist}(\varphi, H^\infty + \mathscr{R}_n),$$

in conjunction with Theorem 10. □

**22. Hartman's Theorem.** *Let* $\varphi \in L^\infty$. *Then* $H_\varphi \in \mathfrak{S}_\infty \Leftrightarrow \varphi \in H^\infty + C(\mathbb{T})$.

*Proof*. $H_\varphi \in \mathfrak{S}_\infty \Leftrightarrow |H_\varphi|_{\text{ess}} = 0 \Leftrightarrow \operatorname{dist}(\varphi, H^\infty + C(\mathbb{T})) = 0 \Leftrightarrow \varphi \in H^\infty + C(\mathbb{T})$ (the last implication uses the closedness of the sum $H^\infty + C(\mathbb{T})$, established in Lecture VIII, Sect. 3). □

Another (more elementary) proof, not based on Theorem 10, can be found in Lecture VIII.

**23. The Schatten-von Neumann Classes** $\mathfrak{S}_p$, $1 \leq p < \infty$, already mentioned in Lecture VIII in connection with Hankel operators, are defined by the requirement that $\sum_{n\geq 0} s_n(A)^p < \infty$. A description of the symbols $\varphi$, $\varphi \in L^\infty$, corresponding to Hankel operators in $\mathfrak{S}_p$, is provided by the following theorem.

**Peller's Theorem.** *Let* $\varphi \in L^\infty(\mathbb{T})$, $1 \leq p < \infty$. *The following statements are equivalent.*

1. $H_\varphi \in \mathfrak{S}_p$.
2. $\sum_{n\geq 0} \operatorname{dist}(\varphi, \mathscr{R}_n + H^\infty)^p < \infty$.
3. $\iint_{\mathbb{D}} |(P_-\varphi)''(1/z)|^p (1-|z|)^{2p-2}\, dx\, dy < \infty$.
4. $P_-\varphi \in B^{1/p}_{p,p}$ *(Besov class).*
5. $\varphi \in H^\infty + B^{1/p}_{p,p}$.
6. $\dfrac{(P_-\varphi)(\lambda^{-1}) - (P_-\varphi)(\mu^{-1})}{\lambda - \mu} = \sum_{n\geq 1} u_n(\lambda) v_n(\mu)$ $(\lambda, \mu \in \mathbb{D})$, *where* $u_n$, $v_n \in H^2$, *the systems* $\{u_n\}_{n\geq 1}$, $\{v_n\}_{n\geq 1}$ *being orthogonal and* $\sum_{n\geq 1} (\|u_n\|_{H^2} \|v_n\|_{H^2})^p < \infty$.

All details connected with this theorem (including the definition of the Besov classes $B_{p,p}^{1/p}$) can be found in Appendix 5. Here we just remark that a) the implication $1 \Leftrightarrow 2$ *has already been proved; b) Statement 3 for* $p > 1$ can alternatively be written as

$$\iint_{\mathbb{D}} |(P_- \varphi)'(1/z)|^p (1 - |z|)^{p-2}\, dx\, dy < \infty;$$

c) the class Hank $\cap \mathfrak{S}^2$ admits a very simple description in terms of the Fourier coefficients of the symbol $\varphi$ (as was noticed in Lecture VIII): $\sum_{n \geq 0} \| H_\varphi z^n \|_{H^2}^2 = \sum_{k \geq 1} k |\hat{f}(-k)|^2 < \infty$; d) in Statement 6 for $p = 1$ it is not necessary to require the orthogonality of the system $\{u_n\}_{n \geq 1}$, $\{v_n\}_{n \geq 1}$.

**24. Toeplitz Operators: Matrices, Commutator Identity, Symbols.** A *Toeplitz matrix* $\{a_{ij}\}_{i,j \geq 0}$ (finite or infinite) is a matrix with the diagonals parallel to the main diagonal constant: $a_{i+k,j+k} = a_{ij}$, $\forall k \geq 0$. It is clear that the Toeplitz operator $T_\varphi$ has precisely such a matrix in the standard basis $\{z^n\}_{n \geq 0}$ of the space $H^2: (T_\varphi z^k, z^n) = (\varphi z^k, z^n) = \hat{\varphi}(n-k)$ (the Fourier coefficients of the symbol $\varphi$). The converse is likewise true: an operator with a Toeplitz matrix is Toeplitz. In order to see this it is convenient to mention a commutator relation which is an equivalent condition for the matrix of a given operator to be Toeplitz.

**25. Lemma.** *Let $T$ be a linear operator from the space of polynomials $\mathscr{P}_A$ into $H^2$. The following statement are equivalent.*

1. *The matrix $\{(Tz^k, z^k)\}_{k,n \geq 0}$ of $T$ is Toeplitz.*
2. $S^*TS = T$.

*The proof* of the lemma is almost self-evident, because in order to verify the identity 2 it is sufficient to identity the bilinear forms $(S^*TSf, g)$ and $(Tf, g)$ and indeed just on the basis elements $f = z^k$, $g = z^n$ $(k, n \geq 0)$ but then Statement 2 is reduced to Statement 1: the identities

$$(Tz^k, z^n) = (\bar{z}Tz \cdot z^k, z^n) \quad (k, n \geq 0).$$

$$(Tz^k, z^n) = (Tz^{k+1}, z^{n+1}) \quad (k, n \geq 0)$$

are equivalent. □

**26. Theorem.** *Let $T$ be a continuous linear operator in the space $H^2$ having a Toeplitz matrix in the standard basis. Then there exists a unique function $\varphi$, $\varphi \in L^\infty(\mathbb{T})$, with $T = T_\varphi$; in addition $|T| = \|\varphi\|_\infty$ and*

$$\varphi x = \lim_n \bar{z}^n T P_+ z^n x, \quad x \in L^2(\mathbb{T}). \tag{13}$$

*Proof*. Once the existence of a function $\varphi$ with the property (13) is established, everything is proved:

$$\|\varphi x\| = \lim_n \|\bar{z}^n T P_+ z^n x\| \leq |T|\, \|x\| \Rightarrow \|\varphi\|_\infty \leq |T|,$$

$$P_+ \varphi x = \lim_n P_+ \bar{z}^n T z^n = (\text{Lemma 25}) \lim_n Tx = Tx,$$

so that $T = T_\varphi$; the converse inequality $|T_\varphi| \leq \|\varphi\|_\infty$ is trivial.

To prove (13) we begin with the weak convergence in $L^2(\mathbb{T})$ of the sequence $\{\bar{z}^n T z^n\}_{n\geq 0}$: it is bounded and the sequence of Fourier coefficients $\{(\bar{z}^n T z^n)^\wedge(k)\}_{n\geq 1}$ stabilizes ($k \in \mathbb{Z}$):

$$(\bar{z}^n T z^n, z^k) = (T\mathbb{1}, z^k) \quad \text{for} \quad k \geq 0,$$

$$(\bar{z}^n T z^n, z^k) = (\bar{z}^{n+k} T z^{n+k} \cdot \bar{z}^k, \mathbb{1}) = (T\bar{z}^k, \mathbb{1}) \quad \text{for} \quad n \geq |k|, \quad k < 0.$$

It follows that there exists a function $\varphi$, $\varphi \in L^2(\mathbb{T})$, with $(\bar{z}^n T z^n, z^k) = (\varphi, z^k)$, $k \geq -n$, and thus $\lim_n \|\varphi - \bar{z}^n T z^n\| = 0$. Hence also

$$\lim_n \|\varphi z^k - \bar{z}^n T z^n z^k\| = \lim_n \|\varphi - \bar{z}^{n+k} T z^{n+k}\| = 0, \quad k \in \mathbb{Z}.$$

As $|\bar{z}^n T P_+ z^n| \leq |T|$ and polynomials in $z$ and $\bar{z}$ are dense in $L^2(\mathbb{T})$ everything is proved. □

**27. Corollary.** *Let $\varphi_{ij} \in L^\infty$ (a finite family), $K \in \mathfrak{S}_\infty$. Then*

$$\left\| \sum_i \prod_j \varphi_{ij} \right\|_\infty \leq \left| K + \sum_i \prod_j T_{\varphi_{ij}} \right|.$$

*In particular, every compact Toeplitz operator must be zero.*

In fact, by the weak convergence to $\mathbb{0}$ of the sequence $\{z^n\}_{n\geq 1}$ it follows that $\lim_n \|Kz^n\| = 0$ while formula (13) states that for any function $\varphi$, $\varphi \in L^\infty$, we have the identity $\boldsymbol{\varphi} = \lim_n \bar{\mathbf{z}}^n T_\varphi \cdot P_+ \mathbf{z}^n$ (strong operator convergence in $L^2(\mathbb{T})$). Hence $\prod_j \boldsymbol{\varphi}_{ij} = \prod_j \lim_n \bar{\mathbf{z}}^n T_{\varphi_{ij}} P_+ \mathbf{z}^n = \lim_n \prod_j (\bar{\mathbf{z}}^n T_{\varphi_{ij}} P_+ \mathbf{z}^n) =$ (in view of $\bar{z}^n z^n = 1$) $\lim_n \bar{\mathbf{z}}^n (\prod_j T_{\varphi_{ij}}) P_+ \mathbf{z}^n$ and similarly for the sum $\sum_i \prod_j \boldsymbol{\varphi}_{ij}$. We thus get

$$\left\| \sum_i \prod_j \varphi_{ij} \right\|_\infty = \left| \sum_i \prod_j \boldsymbol{\varphi}_{ij} \right| = \left| \lim_n \bar{\mathbf{z}}^n \left( \sum_i \prod_j T_{\varphi_{ij}} + K \right) P_+ \mathbf{z}^n \right|$$

$$\leq \left| \sum_i \prod_j T_{\varphi_{ij}} + K \right|. \; \square$$

**28. Corollary.** *If $\varphi \in L^\infty(\mathbb{T})$ then $|T_\varphi^n| = \|\varphi^n\|_\infty$, $r(T_\varphi) = \|\varphi\|_\infty$, $r(T)$ denoting the spectral radius of the operator $T$ ($\stackrel{\text{def}}{=} \max\{|\lambda| : \lambda \in \sigma(T)\} = \lim_n \|T^n\|^{1/n}$).*

In fact, the estimate $|T_\varphi^n| \leq \|\varphi\|_\infty^n = \|\varphi^n\|_\infty$ is trivial while the converse inequality is a special case of Corollary 27: $|T_\varphi^n| \geq \|\varphi^n\|_\infty$. □

**29. The Essential Spectrum** $\sigma_{\text{ess}}(T)$ of the operator $T$, acting in the space $H$, is defined as the spectrum of its image in the *Calkin algebra* $L(H)/\mathfrak{S}_\infty$. As is readily seen, the invertibility of the operator $T$, $T \in L(H)$, in the Calkin algebra is equivalent to the existence of operators $U$ and $V$ ($U, V \in L(H)$) such that $TU - I \in \mathfrak{S}_\infty$, $VT - I \in \mathfrak{S}_\infty$. (They are called the right and (respectively) the left regularizers of $A$.) Here is another description, which follows from the Fredholm-Riesz-Schauder theory: $0 \notin \sigma_{\text{ess}}(T) \Leftrightarrow \dim \operatorname{Ker} T < \infty$, $\dim \operatorname{Ker} T^* < \infty$, $TH = \operatorname{clos} TH$. Such operators (thus invertible in $L(H)/\mathfrak{S}_\infty$) are called *Fredholm operators*. Thus

$$\sigma_{\text{ess}}(T) = \{\lambda :^* T - \lambda I \text{ is not Fredholm}\}.$$

It is clear that $\sigma_{\text{ess}}$ is a closed set, $\sigma_{\text{ess}}(T) \subset \sigma(T)$, and it is not hard to see that in fact $\sigma_{\text{ess}}(T) = \sigma_F(T) \cup \sigma_F(T^*)$, where $\sigma_F$ is the Fredholm spectrum (the continuous spectrum according to a different terminology) as defined in Lecture V, Sect. 4.

Now we proceed to the so-called "spectral inclusion theorem" but first we make the convention to denote by Range($\varphi$) the *essential range* of $\varphi$, i.e. the set $\{\lambda \in \mathbb{C}: \operatorname{ess\,inf} |\varphi - \lambda| = 0\}$ and by conv($\cdot$) the convex hull of the set ($\cdot$).

**30. Theorem.** *Let* $\varphi \in L^\infty(\mathbb{T})$. *Then*

1. Range($\varphi$) $\subset \sigma_F(T_\varphi)$.
2. $\sigma(T_\varphi) \subset \operatorname{conv} \operatorname{Range}(\varphi)$.

*The proof* will as before be based on formula (13).

1. If $\operatorname{ess\,inf} |\varphi - \lambda| = 0$ we can find functions $x_k$, $x_k \in L^2$, such that $\|x_k\|_2 = 1$, $\lim_k \|(\varphi - \lambda)x_k\|_2 = 0$. As

$$\lim_n \|(T_\varphi - \lambda I)P_+ z^n x_k\|_2 = \|(\varphi - \lambda)x_k\|_2,$$

there exist integers $n_k$ such that the sequence $\{y_k\}_{k \geq 1}$, $y_k = P_+ z^{n_k} x_k / \|P_+ z^{n_k} x_k\|_2$, is not compact but $\lim_k \|(T_\varphi - \lambda I)y_k\|_2 = 0$. Thus $\lambda \in \sigma_F(T_\varphi)$.

2. Assume that $\lambda \notin \operatorname{conv} \operatorname{Range}(\varphi)$. Since Range($\varphi$) is compact there exists a disk $K$ (with center $a$) containing this set but not the point $\lambda$. As

$$|\lambda - a| > \sup |\operatorname{Range}(\varphi) - a| = \operatorname{ess\,sup} |\varphi - a|$$

it follows from Corollary 28 that the operator $T_{\varphi - a} - (\lambda - a)I = T_\varphi - \lambda I$ is invertible. Hence $\lambda \notin \sigma(T_\varphi)$. □

**31. Corollary.** *If* Range($\varphi$) *is contained in an angle* (*with vertex at the origin* 0) *less than* $\pi$, (*i.e.* $\operatorname{ess\,sup}\{|\arg \varphi(\zeta) - \arg \varphi(\zeta')| : \zeta, \zeta' \in \mathbb{T}\} < \pi$) *and if* $\operatorname{ess\,inf}_{\mathbb{T}} |\varphi| > 0$ *then* $0 \notin \sigma(T_\varphi)$. □

A function $\varphi$ with $\operatorname{ess\,sup}\{|\arg \varphi(\zeta) - \arg \varphi(\zeta')| : \zeta, \zeta' \in \mathbb{T}\} < \pi$ will often be termed *sectorial*.

**32. Corollary.** *If* $\sigma_F(T_\varphi) \in \mathbb{R}$ *then* $\varphi$ *must be real* (*so that* $T_\varphi$ *is a self-adjoint operator*). □

**33. Corollary (Hartman-Wintner Theorem).** *If* $\varphi$ *is a real function in* $L^\infty(\mathbb{T})$ *then*

$$\sigma(T_\varphi) = \operatorname{conv} \operatorname{Range}(\varphi) = [\operatorname{ess\,inf} \varphi, \operatorname{ess\,sup} \varphi].$$

In fact, it follows from Theorem 30 that it is sufficient to prove the implication: if $\lambda \in \mathbb{R}$ and the operator $T_\varphi - \lambda I$ is invertible then $\operatorname{sign}(\varphi - \lambda) \equiv$ const. But the fact that the function $\psi \overset{\text{def}}{=} \varphi - \lambda$ then has constant sign follows from the solvability of the equation $P_+ \psi f = \mathbb{1}$: if $f$ is the solution ($f \in H^2$) then $(\psi f, f z^n) = (\mathbb{1}, f z^n) = 0$, $n \geq 1$, i.e. $\int_{\mathbb{T}} \psi |f|^2 \bar{z}^n \, dm = 0$ for $n \geq 1$ and, $\psi$ being realvalued, for $n \in \mathbb{Z} \backslash \{0\}$ too, i.e. $\psi |f|^2 \equiv$ const; hence $\operatorname{sign} \psi \equiv$ const. □

In the same way we prove the following statement.

**34. Corollary.** *The boundary of the set* Range($\varphi$) *does not contain eigenvalues of* *the operator* $T_\varphi$ *if* $\varphi \not\equiv$ const. *In particular* $\sigma_p(T_\varphi) = \varnothing$ *if* $T_\varphi = T_\varphi^*$ (*i.e.* $\varphi$ *is a real function*).

In fact, as in Theorem 30, by a linear transformation this statement is reduced to the following one: $\varphi \neq \mathbb{O}$, $\operatorname{Im} \varphi \geq 0 \Rightarrow \operatorname{Ker} T_\varphi = \{\mathbb{O}\}$. Let us check the last statement: if $P_+ \varphi f = \mathbb{O}$, $f \in H^2$ ($f \neq \mathbb{O}$), then $0 = (\varphi f, f z^n) = \int_{\mathbb{T}} \varphi |f|^2 \bar{z}^n \, dm$ for $n \geq 0$ and, in particular, $\int_{\mathbb{T}} \operatorname{Im} \varphi |f|^2 \, dm = 0$, i.e. $\operatorname{Im} \varphi \equiv \mathbb{O}$; since $\varphi$ is a real function we now get $\int_{\mathbb{T}} \varphi |f|^2 z^n \, dm = 0$, $n \in \mathbb{Z}$, i.e. $\varphi = \mathbb{O}$. Contradiction. □

**35. A Fundamental Criterion for the Invertibility** of a Toeplitz operator $T_\varphi$, essentially due to Widom, was actually established already in Lecture VIII, where we studied operators $T_\varphi$ with unimodular symbols $\varphi$ ($|\varphi| = 1$ a.e.). Formally speaking, in Lecture VIII these symbols had the form $\varphi = \Theta \bar{\vartheta}$, where $\Theta$, $\vartheta$ are inner functions, but we noticed there that most of the formulations and the proofs formally extended to the general case $|\varphi| = 1$ a.e. To this we have to add just a little (cf. Lemma 37) to get the Widom-Devinatz criterion but first let us mention some necessary conditions for the invertibility of Toeplitz operators with unimodular symbols, to be used in the sequel.

**36. Theorem.** *Let* $\varphi \in L^\infty(\mathbb{T})$, $|\varphi| = 1$ *a.e. on* $\mathbb{T}$. *The following statements are equivalent.*

1. *The operator* $T_\varphi$ *is invertible.*
2. $\operatorname{dist}(\varphi, H^\infty) < 1$, $\operatorname{dist}(\bar{\varphi}, H^\infty) < 1$.
3. *There exists an outer function* $h$ *such that* $\|\varphi - h\|_\infty < 1$.
4. $\varphi = e^{i(u + \tilde{v} + c)}$ *where* $u$ *and* $v$ *are bounded real functions*, $c \in \mathbb{R}$ *and* $\|u\|_\infty < \pi/2$.
5. *There exists an outer function* $h$, $h \in H^2$, *such that* $\varphi = \lambda \bar{h}/h$, $|h|^2 \in (HS)$ (*i.e.* $|h|^2 = e^{u + \tilde{v}}$; $u, v \in L^\infty$, $\|v\|_\infty < \pi/2$), $\lambda =$ const.

*The proof* of this theorem, as already noted, can be found in Lecture VIII: the implications $1 \Leftrightarrow 2$ are contained in Statement c) (Lecture VIII, Sect. 6); the implications $2 \Leftrightarrow 3 \Leftrightarrow 4$ coincide with the same implications of the Theorem on Close Subspaces; the equivalence of Statement 5 with the remaining ones is established in Complement 1 to this theorem (p. 204–205). □

Let us likewise recall (cf. Statement f), Lecture VIII, Sect. 6) that if $\varphi \in L^\infty$, $\psi \in H^\infty$ then $T_\varphi T_\psi = T_{\varphi\psi}$, $T_{\bar{\psi}} T_\varphi = T_{\bar{\psi}\varphi}$.

**37. Lemma** *Let* $\varphi \in L^\infty$. *Then*

$$0 \notin \sigma(T_\varphi) \Leftrightarrow \operatorname*{ess\,inf}_{\mathbb{T}} |\varphi| > 0,\ 0 \notin \sigma(T_{\varphi/|\varphi|})$$

$$\Leftrightarrow \operatorname*{ess\,inf}_{\mathbb{T}} |\varphi| > 0, \quad 0 \notin \sigma(T_{\varphi/\varphi_e}).$$

*Here* $\varphi_e$ *is the* outer function *with* $|\varphi_e| = |\varphi|$ *a.e.*

*Proof*. If $0 \notin \sigma(T_\varphi)$ then by Theorem 30 $\operatorname{ess\,inf}_{\mathbb{T}} |\varphi| > 0$. Let $h$ be an outer function with $|h| = |\varphi|^{-1/2}$. Then

$$T_{\varphi/|\varphi|} = T_{\bar{h}\varphi h} = T_h^* T_\varphi T_h. \tag{14}$$

But the multiplication operator $T_h$ is invertible ($h$ is outer, $\operatorname{ess\,inf}_{\mathbb{T}}|h| > 0$, and therefore $h^{-1} \in H^\infty$, $T_{h^{-1}} = (T_h)^{-1}$). Consequently $0 \notin \sigma(T_{\varphi/|\varphi|})$.

The converse implication likewise follows from formula (14).

The second way of reducing the invertibility to the case of a unimodular symbol, namely now $\varphi/\varphi_e$, is settled in an analogous manner, because

$$T_\varphi = T_{\varphi/\varphi_e} T_{\varphi_e}. \square \tag{15}$$

**38. Corollary (Widom-Devinatz Theorem).** *Let $\varphi \in L^\infty$. The following statements are equivalent.*

1. $0 \notin \sigma(T_\varphi)$.

2. $\operatorname{ess\,inf}_{\mathbb{T}}|\varphi| > 0$; $\varphi = |\varphi| e^{i(u+\tilde{v}+c)}$ *where $u, v$ are bounded real functions, $c \in \mathbb{R}$, and $\|u\|_\infty < \pi/2$.*

3. $\varphi = g e^{v+iu}$ *where $g$, $1/g \in H^\infty$ and $u$, $v$ are bounded real functions with $\|u\|_\infty < \pi/2$.*

4. $\operatorname{ess\,inf}_{\mathbb{T}}|\varphi| > 0$ *and there exists a function $h$ ($h, 1/h \in H^\infty$) such that the product $h\varphi$ is sectorial.*

In fact, the implications $1 \Leftrightarrow 2$ follow at once from 36 and 37.

$1 \Rightarrow 3$. From the invertibility of $T_\varphi$ and from 36, 37 it follows that $\varphi = \varphi_e \exp\{iu + i\tilde{v} + ic\} = g\exp\{-v + iu\}$, with $\|u\|_\infty < \pi/2$ and $g = \varphi_e \exp\{v + i\tilde{v} + ic\}$, $\varphi_e$ being the outer function with $|\varphi_e| = |\varphi|$. It is clear that $g$, $1/g \in H^\infty$.

$3 \Rightarrow 4$. Take $h = 1/g$.

$4 \Rightarrow 1$. $T_\varphi = T_{h\varphi} T_{h^{-1}} = T_{h\varphi}(T_h)^{-1}$ and it suffices to apply Corollary 31.

**39. Comments.** Thus the assumption that $\varphi$ is sectorial is in an essential way connected with the computation of the spectrum of the Toeplitz operator $T_\varphi$: together with the obvious condition $\operatorname{ess\,inf}_{\mathbb{T}}|\varphi| > 0$ it is not only sufficient for the invertibility of $T_\varphi$ (Corollary 31) but up to distinguishing an analytical invertible factor of $T_\varphi = T_{\varphi h}(T_h)^{-1}$ also necessary. Later on in Section 80 we will prove that likewise the requirement for $\varphi$ to be *locally* sectorial is a sufficient condition for invertibility. Let us further point out that from 38 we get once more the inclusion $\sigma(T_\varphi) \subset \operatorname{conv}\operatorname{Range}(\varphi)$: if $\lambda \notin \operatorname{conv}\operatorname{Range}(\varphi)$ then for a suitable $\alpha$, $\alpha \in \mathbb{R}$, holds $\|\arg e^{i\alpha}(\varphi - \lambda)\|_\infty < \pi/2$, so that $\lambda \notin \sigma(T_\varphi)$ by Statement 2 of Corollary 38.

**40. The Left and Right Spectra** of an operator $A$, $A \in L(H)$, as well as the onesided essential spectra, arise in a natural way from problems connected with the solvability of the equations $Ax - \lambda x = y$ and (on the other hand) give rise to a certain classification of the spectrum $\sigma(A)$. Definitions:

$$\sigma^l(A) = \{\lambda \in \mathbb{C} : B(A - \lambda I) \neq I, \forall B \in L(H)\},$$

$$\sigma^r(A) = \{\lambda \in \mathbb{C} : (A - \lambda I)B \neq I, \forall B \in L(H)\},$$

$$\sigma^l_{\mathrm{ess}}(A) = \{\lambda \in \mathbb{C} : B(A - \lambda I) - I \notin \mathfrak{S}_\infty, \forall B \in L(H)\},$$

$$\sigma^r_{\mathrm{ess}}(A) = \{\lambda \in \mathbb{C} : (A - \lambda I)B - I \notin \mathfrak{S}_\infty, \forall B \in L(H)\}.$$

From standard results on compact operators it follows that (say) the essential left invertibility of $A$ ($0 \notin \sigma^l_{\mathrm{ess}}(A)$) is equivalent to the existence of a subspace $E$,

$E \subset H$, and a number $\varepsilon$, $\varepsilon > 0$, such that

$$\operatorname{codim} E < \infty; \quad \|Ax\| \geq \varepsilon \|x\|, \quad \forall x \in E. \tag{16}$$

It is further easy to see that the left and right spectra are interchanged if we pass to the adjoint operator and that always $\sigma(A) = \sigma^l(A) \cup \sigma^r(A)$, $\sigma_{\text{ess}}(A) = \sigma^l_{\text{ess}}(A) \cup \sigma^r_{\text{ess}}(A)$, $\sigma_F(A) \subset \sigma^l_{\text{ess}}(A)$. Therefore it follows from the essential invertibility (from the left or from the right) of a Toeplitz operator $T_\varphi$ that $\operatorname{ess\,inf}_{\mathbb{T}} |\varphi| > 0$ (Theorem 30). In the following proposition only the formulae for the *left* spectra of $T_\varphi$ are given, the corresponding *right* ones are gotten upon replacing the symbol $\varphi$ by its complex conjugate $\bar{\varphi}$. By the reason just mentioned the invertibility condition $1/\varphi \in L^\infty$ (clearly necessary at any rate!) is included in the premise of the theorem.

**41. Theorem.** *Let* $\varphi \in L^\infty(\mathbb{T})$, $\operatorname{ess\,inf}_{\mathbb{T}} |\varphi| > 0$.

1. $0 \notin \sigma^l(T_\varphi) \Leftrightarrow \operatorname{dist}(\varphi/|\varphi|, H^\infty) < 1 \Leftrightarrow \operatorname{dist}(\varphi/\varphi_e, H^\infty) < 1$.
2. $0 \notin \sigma^l_{\text{ess}}(T_\varphi) \Leftrightarrow \operatorname{dist}(\varphi/|\varphi|, H^\infty + C) < 1 \Leftrightarrow \operatorname{dist}(\varphi/\varphi_e, H^\infty + C) < 1$.

*Here* $\varphi_e$ *is the outer function with* $|\varphi_e| = |\varphi|$ *a.e.*

*Proof*. Statement 1 follows from Statement c), Lecture VIII, Sect. 6, and the present formulae (14) and (15).

Statement 2 too depends on these formulae: they reduce the issue to the essential left invertibility of the operator $T_u$ with $u = \varphi/|\varphi|$ or $u = \varphi/\varphi_e$. From (2) it follows that

$$|I - T_{\bar{u}} T_u|_{\text{ess}} = |H_u^* H_u|_{\text{ess}}.$$

If $\operatorname{dist}(u, H^\infty + C) < 1$ then $|H_u|_{\text{ess}} < 1$ (cf. (21)) and $|H_u^* H_u|_{\text{ess}} < 1$. Therefore there exists a compact operator $K$ such that $|I - T_{\bar{u}} T_u - K| < 1$. Hence $(T_{\bar{u}} T_u + K)^{-1} T_{\bar{u}} T_u = \mathrm{I} - (T_{\bar{u}} T_u + K)^{-1} K$, i.e. $0 \notin \sigma^l_{\text{ess}}(T_u)$.

Conversely, if $0 \notin \sigma^l_{\text{ess}}(T_u)$ then (16) holds true with $A = T_u$. As now $\|H_u x\|_2^2 + \|T_u x\|_2^2 = \|ux\|_2^2 = \|x\|_2^2$, $x \in H^2$, we find $\|H_u x\|_2^2 \leq (1 - \varepsilon^2)\|x\|_2^2$, $x \in E$. Hence $\operatorname{dist}(u, H^\infty + C) = |H_u|_{\text{ess}} \leq |H_u P_E| < 1$. □

**42. The Rôle of the Essential Spectrum and of the Index.** Theorem 38 (or any other combination of the Statements 36, 37) is not always easy to apply in practise, because there exists no simple algorithm for representing the function $\arg \varphi/|\varphi|$ in the form $u + \tilde{v} + c$, $\|u\|_\infty < \pi/2$, even if such a representation is known to exist. Therefore for various concrete classes of functions $\varphi$ it is natural to try to describe the spectrum $\sigma(T_\varphi)$ in a different language (preferably in geometric terms, employing the image of the circumference Range($\varphi$)). Below such a description of $\sigma(T_\varphi)$ will be obtained (on the basis of the Widom-Devinatz Theorem) in the case $\varphi \in H^\infty + C(\mathbb{T})$, $\varphi \in PC$ (the closure of all piecewise continuous functions on $\mathbb{T}$) and in some other cases too.

The geometric description of the spectrum of $T_\varphi$ is as a rule connected with finding criteria for the operator to be Fredholm. And the point is that not only is the Fredholm property a necessary condition for invertibility, the study of the first of these properties usually being simpler than the second one, but that, as

Lemma 43 below indicates, the invertibility of a Fredholm Toeplitz operator $T_\varphi$ depends *only on its index* ind $T_\varphi$,

$$\operatorname{ind} T_\varphi \stackrel{\text{def}}{=} \dim \operatorname{Ker} T_\varphi - \dim \operatorname{Ker} T_\varphi^*.$$

The index of the operator $T_\varphi$ for continuous symbols $\varphi$ is given by the "winding number" of the curve $\varphi(\mathbb{T})$, and for some other classes of symbols by the winding number of appropriate continuous "approximations" to $\varphi(\mathbb{T})$.

Other more profound reasons for the influence of the Fredholm phenomenon on the spectral theory of Toeplitz operators will be seen more clearly later on – the property of $T_\varphi$ being Fredholm (in contradistinction to its invertibility) depends only on local properties of the symbol $\varphi$, indeed it (and not the invertibility of the operator) often reduces to the invertibility of the symbol, and in the algebraic approach to the theory one can even frequently completely disregard the ideal $\mathfrak{S}_\infty$.

**43. Coburn's Lemma.** *If* $\varphi \in L^\infty$, $\varphi \not\equiv \mathbb{0}$, *then either* $\operatorname{Ker} T_\varphi = \{\mathbb{0}\}$ *or* $\operatorname{Ker} T_\varphi^* = \{\mathbb{0}\}$.

*Proof*. If $f$, $g \in H^2$ and $P_+ \varphi f = \mathbb{0}$, $P_+ \bar{\varphi} g = \mathbb{0}$, then $\varphi f = \bar{a}$, $\bar{\varphi} g = \bar{b}$ with $a$, $b \in H_0^2$. Hence $\overline{ga} = fb$ and, as $fb$, $ga \in H^2 \cdot H_0^2 \subset H_0^1$, it follows that $fb = \mathbb{0}$, $ga = \mathbb{0}$. If $f \neq \mathbb{0}$ then $f \neq 0$ a.e. and $b = \mathbb{0}$, i.e. $\bar{\varphi} g = \mathbb{0}$ and $g = 0$ on a set of positive measure. Consequently $g = \mathbb{0}$. □

**44. Corollary.** $\sigma(T_\varphi) = \sigma_{\text{ess}}(T_\varphi) \cup \{\lambda \in \mathbb{C} \setminus \sigma_{\text{ess}}(T_\varphi) : \operatorname{ind}(T_\varphi - \lambda I) \neq 0\}$. □

**45. The General Properties of the Index** $\operatorname{ind} A \stackrel{\text{def}}{=} \dim \operatorname{Ker} A - \dim \operatorname{Ker} A^*$ **of a Fredholm Operator** $A$, which we will need, are the following: $\operatorname{ind}(A + K) = \operatorname{ind} A$ if $K \in \mathfrak{S}_\infty$; $\operatorname{ind} AB = \operatorname{ind} A + \operatorname{ind} B$ provided $B$ is Fredholm too; there exists a number $\varepsilon$, $\varepsilon > 0$, such that if $|B| < \varepsilon$ then $A + B$ is Fredholm, with $\operatorname{ind}(A + B) = \operatorname{ind} A$. For all this we refer to, for instance, Kato [1].

**46. The Class** $PC$ is defined as the closure in $L^\infty(\mathbb{T})$ of the set of all piecewise continuous functions. Thus $\varphi \in PC$ iff the function has right and left limits at *each point* of the circumference (such a function is continuous everywhere except on an at most countably infinite set of points at which it has jumps; the set of points where the modulus of the jump exceeds $\varepsilon$ is finite, for any $\varepsilon > 0$). Let us agree to normalize the functions $\varphi$ of class $PC$ by the condition of *continuity to the right* (this corresponds to the choice of a definite representative among the equivalent functions defining $\varphi$ as an element of the space $L^\infty(\mathbb{T})$). It is then clear that

$$\varphi \in PC \Rightarrow \operatorname{Range}(\varphi) = \operatorname{clos} \varphi(\mathbb{T}).$$

If $\varphi \in PC$ then we define a *connected closed curve* $\varphi_*(\mathbb{T})$ in the plane $\mathbb{C}$ by connecting the points $\varphi(e^{it-0}) \stackrel{\text{def}}{=} \lim_{\tau \to t-0} \varphi(e^{i\tau})$ and $\varphi(e^{it})$, $0 < t \leq 2\pi$, by the straight line segment $[\varphi(e^{it-0}), \varphi(e^{it})]$. The natural parametrization of the curve $\varphi_*(\mathbb{T})$ (the path with support $\varphi_*(\mathbb{T})$) will be denoted by $\varphi_*$. If $\varphi \in C(\mathbb{T})$ then $\varphi_* = \varphi$.

The *winding number* (or the index with respect to the origin) of a closed continuous path $\alpha$ ($\alpha: \mathbb{T} \to \mathbb{C}\backslash\{0\}$) is as usual defined as the integer

$$\operatorname{wind}\alpha \overset{\text{def}}{=} \frac{1}{2\pi i}\int_{\alpha(\mathbb{T})} \frac{dz}{z}.$$

In picturesque language $2\pi\cdot\operatorname{wind}\alpha$ is the change of the argument of the point $\alpha(e^{it})$ when $t$ runs from 0 to $2\pi$, i.e. $\operatorname{wind}\alpha$ is the number of turns of this point around the origin. It is not hard to see that $\operatorname{wind} z^n = n$ ($n \in \mathbb{Z}$) and that $\operatorname{wind}\alpha\beta = \operatorname{wind}\alpha + \operatorname{wind}\beta$. It is clear that $\operatorname{wind}\alpha = 0$ characterizes those continuous functions which possess on $\mathbb{T}$ a continuous branch of the argument (in fact this statement reduces to the definition of $\operatorname{wind}\alpha$): $\operatorname{wind}\alpha = 0 \Leftrightarrow$ there exists a function $\beta$, $\beta \in C(\mathbb{T})$ such that $\alpha/|\alpha| = e^{i\beta}$).

**47. Widom-Devinatz Theorem.** *Let* $\varphi \in PC$. *Then*

$$\sigma_{\text{ess}}(T_\varphi) = \varphi_*(T_\varphi),$$
$$\sigma(T_\varphi) = \varphi_*(\mathbb{T}) \cup \{\lambda \in \mathbb{C}\backslash\varphi_*(\mathbb{T}) : \operatorname{wind}(\varphi_* - \lambda) \neq 0\},$$
$$\operatorname{ind}(T_\varphi - \lambda I) = -\operatorname{wind}(\varphi_* - \lambda), \quad \lambda \in \mathbb{C}\backslash\varphi_*(\mathbb{T}).$$

*Proof*. Assume first that $0 \notin \varphi_*(\mathbb{T})$, i.e. $0 \notin \varphi(\mathbb{T})$ and none of the segments connecting the jumps of the function $\varphi$ passes through the point 0. It is then clear that $\varphi/|\varphi| \in PC$ and that one can pick a branch $\psi$ of the argument of $\varphi$ such that $\psi \in PC$ and the absolute values of all jumps of the (real) function $\psi$ are less than $\pi$ (the corresponding segments do not pass through the point 0), except for at most one jump arising from the winding number $\operatorname{wind}\varphi_*$. Fixing such a function $\psi$ we see that $\varphi = |\varphi|e^{i\psi}$, $\operatorname{wind}\varphi_* = \dfrac{1}{2\pi}(\psi(1) - \psi(e^{i(2\pi-0)}))$.

If $\operatorname{wind}\varphi_* = 0$ then all jumps of $\psi$ are in modulus less than $\pi$ and therefore also less than $\pi - \delta$ for a suitable $\delta > 0$. It is clear that for a given $\varepsilon$, $\varepsilon > 0$, we can find open arcs $I_k$ ($k = 1, \ldots, n$) on the circle $\mathbb{T}$ such that $\mathbb{T}\backslash\bigcup_{k=1}^n I_k$ is a finite set and $\sup\{|\psi(\zeta) - \psi(\zeta')| : \zeta, \zeta' \in I_k\} < \varepsilon$ ($k = 1, \ldots, n$). From this and from the preceding assertion it follows that there exists a continuous function $\omega$ with $\|\psi - \omega\|_\infty < \dfrac{\pi - \delta}{2}$. We now approximate the function $\omega$ by a real trigonometric polynomial $p$ ($\|\omega - p\|_\infty < \delta/3$) and set $v = \tilde{p}$. Then $\psi = (\psi - p) + \tilde{v} + \hat{p}(0)$ and $\|\psi - p\|_\infty < \pi/2$. Therefore by Corollary 38 $0 \notin \sigma(T_\varphi)$.

Thus we have established the inclusion $\sigma(T_\varphi) \subset \varphi_*(\mathbb{T}) \cup \{\lambda : \operatorname{wind}(\varphi_* - \lambda) \neq 0\}$.

Assume next that $n \overset{\text{def}}{=} \operatorname{wind}\varphi_* \neq 0$ (and as before $0 \notin \varphi_*(\mathbb{T})$). Then the argument variation of the function $z^n\varphi = |\varphi|z^n e^{i\psi}$ equals zero and so, by what we just have proved, $T_{\varphi z^n}$ is an invertible operator. Using formula (2) we get

$$\begin{aligned} T_{\varphi z^n} - T_\varphi T_{z^n} &= H^*_{\bar{\varphi}} H_{z^n}, \\ T_{\varphi z^n} - T_{z^n} T_\varphi &= H^*_{\bar{z}^n} H_\varphi. \end{aligned} \tag{17}$$

As $\operatorname{rank} H_{z^n} < \infty$, $\operatorname{rank} H^*_{\bar{z}^n} < \infty$ the operators $T_{z^n}(T_{\varphi z^n})^{-1}$ and $(T_{\varphi z^n})^{-1}T_{z^n}$ are right and (respectively) left regularizers of $T_\varphi$. Hence $0 \notin \sigma_{\text{ess}}(T_\varphi)$; this proves the inclusion $\sigma_{\text{ess}}(T_\varphi) \subset \varphi_*(\mathbb{T})$.

Let us compute the index. From formula (17) and the properties of the index 45 it follows that

$$0 = \operatorname{ind} T_{\varphi z^n} = \operatorname{ind} T_\varphi T_{z^n} = \operatorname{ind} T_\varphi + \operatorname{ind} T_{z^n} = \operatorname{ind} T_\varphi - n,$$

i.e. $\operatorname{ind} T_\varphi = -\operatorname{wind} \varphi_*$, whence also $\{\lambda \in \mathbb{C} \backslash \varphi_*(\mathbb{T}) : \operatorname{wind}(\varphi_* - \lambda) \neq 0\} \subset \sigma(T_\varphi)$.

It remains to see that $\varphi_*(\mathbb{T}) \subset \sigma_{\text{ess}}(T_\varphi)$. The inclusion $\operatorname{clos} \varphi(\mathbb{T}) \subset \sigma_{\text{ess}}(T_\varphi)$ is contained in Theorem 30. Therefore from the assumption $0 \in \varphi_*(\mathbb{T}) \backslash \sigma_{\text{ess}} \bigvee T_\varphi)$ it follows that $\operatorname{dist}(0, \varphi(\mathbb{T})) > 0$, i.e. the point 0 must belong to a (finite) set of open segments connecting the jumps of $\varphi$. Make the point $\lambda$ run through a circle $\{\lambda : |\lambda| = \varepsilon\}$ of sufficiently small radius. When this point intersects any of the segments in question the index of the operator $T - \lambda I$ changes (cf. the preceding paragraph), which contradicts the stability of the index for small perturbations (cf. 45). Thus $\varphi_*(\mathbb{T}) \backslash \sigma_{\text{ess}}(T_\varphi) = \varnothing$. □

**48. The Class** $H^\infty + C(\mathbb{T})$ plays an essential rôle in the spectral theory of Toeplitz and Hankel operators not merely because of Hartman's Theorem (cf. 22). It is also connected in an essential way with the description of those Toeplitz operators which are unitary mod $\mathfrak{S}_\infty$ (cf. 54) and with the compactness of the semicommutators $T_\varphi T_\psi - T_{\varphi\psi}$ (cf. 130, 133, 144) and with other problems. Of equally great importance is the fact that $H^\infty + C$ is the least closed subalgebra of $L^\infty$ containing $H^\infty$ and also (certain) non-analytic functions (cf. 60).

The spectra of operators $T_\varphi$ with symbols in $H^\infty + C(\mathbb{T})$ will be described in a similar way as is done for symbols belonging to the subalgebras $C(\mathbb{T})$ and $H^\infty$ generating $H^\infty + C$. The result for $C(\mathbb{T})$ is contained in 47 (if $\varphi \in C(\mathbb{T})$ then $0 \in \sigma(T_\varphi) \Leftrightarrow 1/\varphi \in C(\mathbb{T})$ and $\operatorname{wind} \varphi = 0$) and for $H^\infty$ it constitutes the contents of the following proposition. Let us recall that if $\varphi \in H^\infty$ then by $\varphi^i$ and $\varphi^e$ we denote respectively the inner and the outer part of $\varphi$, $\varphi = \varphi^i \varphi^e$.

**49. Lemma.** *Let* $\varphi \in H^\infty$. *Then*

1. $\sigma(T_\varphi) = \operatorname{clos} \varphi(\mathbb{D}) = \{\lambda \in \mathbb{C} : (\varphi - \lambda)^{-1} \notin H^\infty\}$.
2. $\sigma_{\text{ess}}(T_\varphi) = \operatorname{Range}(\varphi) \cup \{\lambda \in \mathbb{C} : \deg(\varphi - \lambda)^i = \infty\}$.
3. *If* $\lambda \notin \sigma_{\text{ess}}(T_\varphi)$ *then* $\operatorname{ind}(T_\varphi - \lambda) = -\deg(\varphi - \lambda)^i = -\lim_{r \to 1-0} \operatorname{wind}(\varphi(r\mathbb{T}) - \lambda)$.

*The proof* follows from Theorem 30: if $\varphi = \varphi^i \varphi^e$ is the inner-outer factorization then from the condition (necessary for the operator to be Freholm) $0 \notin \operatorname{Range}(\varphi)$ (i.e. $\operatorname{ess\,inf}_{\mathbb{T}} |\varphi| > 0$) it follows that $1/\varphi^e \in H^\infty$ and then $T_\varphi H^2 = \varphi^i H^2$ and (by the definition of the degree deg of an inner function) $\operatorname{codim} T_\varphi H^2 = \deg \varphi^i$. From this Statements 1 and 2 follow at once. Statement 3 is trivial. □

**50. Douglas's Theorem.** *Let* $\varphi \in H^\infty + C(\mathbb{T})$. *The following statements are equivalent.*

1. $T_\varphi$ *is Fredholm.*
2. $\varphi^{-1} \in H^\infty + C$.

3. *There exists a number* $\delta$, $\delta > 0$, *such that* $|\mathscr{H}\varphi(\zeta)| \geq \delta$ *for* $1 - \delta < |\zeta| < 1$; *here* $\mathscr{H}\varphi$ *is the harmonic continuation of* $\varphi$ *into the disk* $\mathbb{D}$.

*Proof*. $2 \Rightarrow 1$. From (2) and Hartman's Theorem we get $I - T_\varphi T_{\varphi^{-1}} = H^*_{\bar\varphi} H_{\varphi^{-1}} \in \mathfrak{S}_\infty$, $I - T_{\varphi^{-1}} T_\varphi = H^*_{\bar\varphi^{-1}} H_\varphi \in \mathfrak{S}_\infty$, i.e. $T_\varphi$ is Fredholm.

$1 \Rightarrow 2$. By Theorem 30 $\operatorname{ess\,inf}_{\mathbb{T}} |\varphi| > 0$, so that $\varphi_e^{-1} \in H^\infty$, where $\varphi_e$ is an outer function, $|\varphi_e| \equiv |\varphi|$. Hence $u \overset{\text{def}}{=} \varphi/\varphi_e \in H^\infty + C$ and therefore (Hartman's Theorem) $H_u \in \mathfrak{S}_\infty$. Since $|u| = 1$ a.e. we have

$$H^*_{\bar u} H_{\bar u} T_u = (I - T_u T_{\bar u}) T_u = T_u (I - T_{\bar u} T_u) = T_u H^*_u H_u. \tag{18}$$

Thus the operator $H^*_{\bar u} H_{\bar u} T_u$ is compact while $T_u \overset{\text{def}}{=} T_{\varphi\varphi_e^{-1}} = T_\varphi \cdot (T_{\varphi_e})^{-1}$ is Fredholm along with $T_\varphi$. It is therefore clear that $H_{\bar u} \in \mathfrak{S}_\infty$, i.e. $\bar u = 1/u \in H^\infty + C$. Together with the invertibility in $H^\infty$ of $\varphi_e$ this gives the desired result: $\varphi^{-1} = u^{-1}\varphi_e^{-1} \in H^\infty + C$.

The proof of the implications $2 \Leftrightarrow 3$ is based on the asymptotic multiplicativity of the operator of harmonic continuation $\mathscr{H}$ in the class $H^\infty + C$. If $f \in L^\infty(\mathbb{T})$ we set $f_r(\zeta) = (\mathscr{H}f)(r\zeta)$, $\zeta \in \mathbb{T}$. What we need is now that

$$\lim_{r \to 1} \| f_r g_r - (fg)_r \|_\infty = 0$$

for any pair of functions $f, g$ in $H^\infty + C$. We use this lemma to complete the proof of the theorem, relegating its proper verification to Sect. 52.

$2 \Rightarrow 3$. If $\varphi$, $\varphi^{-1} \in H^\infty + C$ then $|\varphi_r| \leq \|\varphi\|_\infty$, $|(\varphi^{-1})_r| \leq \|\varphi^{-1}\|_\infty$ while by Lemma 52 $\lim_{r\to 1} \|\varphi_r(\varphi^{-1})_r - \mathbb{1}\|_\infty = 0$. Hence Statement 3 follows.

$3 \Rightarrow 2$. Let $0 < \varepsilon < \delta/2$, $n \geq 1$, and pick a function $\psi$ such that $\psi \in H^\infty$, $\|\varphi - z^{-n}\psi\|_\infty < \varepsilon$. By Lemma 52 $\lim_{r \to 1} \|(z^n\varphi)_r - r^n z^n \varphi_r\|_\infty = 0$. From this and Statement 3 it follows that $|\psi(z)| \geq \delta - \varepsilon$ for $1 - \delta_1 < |\zeta| < 1$, with $\delta_1 > 0$. Therefore the inner factor $\psi^i$ of $\psi$ must be a finite Blaschke product (by the Theorem on the Spectrum, Lecture III) and $1/\psi^e \in H^\infty$. Together these inclusions give $1/\psi \in H^\infty + C$. Moreover

$$\|z^n\varphi\psi^{-1} - \mathbb{1}\|_\infty \leq \|\psi^{-1}\|_\infty \cdot \varepsilon \leq \varepsilon(\delta - \varepsilon)^{-1} < 1.$$

The invertibility of $z^n\varphi\psi^{-1}$ now entails the invertibility of $\varphi$. □

**51. Corollary.** *Let* $\varphi \in H^\infty + C$. *Then*

1. $\sigma_{\text{ess}}(T_\varphi) = \{\lambda \in \mathbb{C} : (\varphi - \lambda)^{-1} \notin H^\infty + C\} = \bigcap_{0<\delta<1} \operatorname{clos}\{\mathscr{H}\varphi(\zeta) : \delta < |\zeta| < 1\}$.
2. $\sigma_{\text{ess}}(T_\varphi)$ *is connected.*
3. $\sigma(T_\varphi)$ *is connected.*
4. *If* $\varphi^{-1} \in H^\infty + C$ *then* $\operatorname{ind} T_\varphi = -\lim_{r\to 1} \operatorname{wind} \varphi_r(\mathbb{T})$.

In fact, Statement 1 follows from Theorem 50, Statement 2 from Statement 1 (second equality), 3 from 2 (since the difference $\sigma(T_\varphi)\backslash\sigma_{\text{ess}}(T_\varphi)$ is an open set whose boundary is contained in $\sigma_{\text{ess}}(T_\varphi)$). It remains to check Statement 4. Since $\operatorname{ind} T_{\varphi z^n} = \operatorname{ind} T_{z^n} + \operatorname{ind} T_\varphi = -n + \operatorname{ind} T_\varphi$ and $\operatorname{wind}(\varphi z^n)_r(\mathbb{T}) = n + \operatorname{wind} \varphi_r(\mathbb{T})$ (from the identity $\lim_{r\to 1} \|(\varphi z^n)_r - \varphi_r(z^n)_r\| = 0$ and the fact that $|\varphi_r|$ is uniformly separated from zero it follows that the curves $(\varphi z^n)_r(\mathbb{T})$ and $(z^n)_r(\mathbb{T})\varphi_r(\mathbb{T})$

are homotopic in $\mathbb{C}\setminus\{0\}$) is suffices to prove any of the implications: $0 \notin \sigma(T_\varphi) \Rightarrow \text{wind}\, \varphi_r(\mathbb{T}) = 0$ or $\text{wind}\, \varphi_r(\mathbb{T}) = 0 \Rightarrow 0 \notin \sigma(T_\varphi)$. The first of these implications: if the operator $T_\varphi$ is invertible and the integer $k$, $k \geq 0$, and the function $\psi$, $\psi \in H^\infty$, are chosen such that the norm $\|\varphi - z^{-k}\psi\|_\infty$ is sufficiently small then $0 \notin \sigma(T_z - k_\psi)$ and in particular $0 = \text{ind}\, T_{z^{-k}\psi} = k + \text{ind}\, T_\psi =$ (Lemma 49) $k - \lim_{r\to 1} \text{wind}\, \psi_r(\mathbb{T}) = -\lim_{r\to 1} \text{wind}(z^{-k}\psi)_r(\mathbb{T})$. It is however clear that $\text{wind}\, \varphi_r(\mathbb{T}) = \text{wind}(z^{-k}\psi)_r(\mathbb{T})$. □

**52. Lemma.** *If $f, g \in H^\infty + C$ then*

$$\lim_{r\to 1} \| f_r g_r - (fg)_r \|_\infty = 0,$$

*where $f_r(\zeta) = (\mathscr{H}f)(r\zeta)$, $\zeta \in \mathbb{T}$, $0 < r < 1$.*

*Proof.* Write $f = h_1 + c_1$, $g = h_2 + c_2$ with $h_i \in H^\infty$, $c_i \in C(\mathbb{T})$, $i = 1, 2$. By the multiplicativity of the operator $\mathscr{H}$ on $H^\infty$ the difference in question can be broken up into terms $(h_i)_r(c_j)_r - (h_i c_j)_r$ and $(c_1)_r(c_2)_r - (c_1 c_2)_r$. The last one tends to zero in a trivial manner. In order to estimate the mixed pairs it is sufficient to consider the case $h \in H^\infty$, $c = z^{-n}$, $n > 0$ (as polynomials in $z$ in $z^{-1}$ are dense in $C(\mathbb{T})$ and the case $n \leq 0$ is trivial). Let $p$ denote the $n$ th order Taylor polynomial of $h$. Then

$$\|h_r(z^{-n})_r - (hz^{-n})_r\|_\infty \leq \|(h-p)_r(z^{-n})_r - ((h-p)z^{-n})_r\|_\infty + \|p_r(z^{-n})_r - (pz^{-n})_r\|_\infty$$

where the last term tends to zero as $r \to 1$. The first term contains the function $h - p \stackrel{\text{def}}{=} \sum_{k \geq n+1} a_k z^k$. Computing each member we get

$$(z^k)_r(z^{-n})_r - (z^{k-n})_r = r^k(r^n - r^{-n})z^{k-n}, \quad k \geq n+1;$$

$$(h-p)_r(z^{-n})_r - ((h-p)z^{-n})_r = z^{-n}(r^n - r^{-n})(h-p),$$

whence the desired conclusion. □

**53. The Class $QC$.** By definition $QC$ is the largest self-adjoint subalgebra of $H^\infty + C(\mathbb{T})$, i.e.

$$QC = (H^\infty + C) \cap \overline{(H^\infty + C)}.$$

The elements of $QC$ are called *quasicontinuous* functions. The formulae $\varphi = P_-\varphi + P_+\varphi$, $\tilde{\varphi} = i(P_-\varphi - P_+\varphi + \hat{\varphi}(0))$ give $\{\varphi \in L^\infty(\mathbb{T}) : \tilde{\varphi} \in C(\mathbb{T})\} \subset QC$ so that in particular $QC \neq C(\mathbb{T})$. Moreover from these formulae one readily reads off that

$$QC = VMO \cap L^\infty, \quad VMO \stackrel{\text{def}}{=} \{u + \tilde{v} : u, v \in C(\mathbb{T})\}.$$

In fact, if $\varphi \in H^\infty + C$ then $P_-\varphi \in P_- C(\mathbb{T}) \subset VMO$; analogously $P_-\bar{\varphi} \in VMO$ (if $\varphi \in QC$). Conversely, if $\varphi \in VMO \cap L^\infty$ then $\varphi = u + \tilde{v} = u + iv - 2P_+ v + \text{const} \in L^\infty$ so that $P_+ v \in H^\infty$, i.e. $\varphi \in H^\infty + C$; analogously $\bar{\varphi} \in H^\infty + C$. □

In the theory of Toeplitz operators the unimodular functions in $QC$ are of particular interest because one is lead to them when studying Fredholm operators $T_\varphi$ with symbols in $H^\infty + C$ (if $\varphi \in H^\infty + C$ then by Theorem 50 $0 \notin \sigma_{\text{ess}}(T_\varphi) \Leftrightarrow 0 \notin \sigma_{\text{ess}}(T_{\varphi/\varphi_e}) \Leftrightarrow \varphi/\varphi_e \in QC$; here $\varphi_e$ is an outer function, $|\varphi_e| = |\varphi|$).

Likewise one is lead to functions in $QC$ in the problem of compactness of commutators (cf. 160) and in the problem of describing those operators $T_\varphi$ which are unitary mod $\mathfrak{S}_\infty$, i.e. Toeplitz operators whose images in the Calkin algebra $L(H^2)/\mathfrak{S}_\infty$ are unitary (on the level of $T_\varphi$ itself this means that $T_\varphi T_\varphi^* - I \in \mathfrak{S}_\infty$, $T_\varphi^* T_\varphi - I \in \mathfrak{S}_\infty$).

**54. Theorem.** *Let* $u \in L^\infty$.

1. *The operator* $T_u$ *is unitary* mod $\mathfrak{S}_\infty$ *iff* $|u| = 1$ *a.e. and* $u \in QC$.
2. *The operator* $T_u$ *has the form* $T_u = U + K$, *with* $U$ *unitary and* $K$ *compact, iff* $|u| = 1$ *a.e.,* $u \in QC$ *and* $\lim_{r\to 1} \operatorname{wind} u_r(\mathbb{T}) = 0$.

*Proof*. We use formula (2) and Hartman's Theorem: If $|u| = 1$ a.e., $u \in QC$, then $I - T_u T_u^* = I - T_u T_{\bar u} = H_{\bar u}^* H_{\bar u} \in \mathfrak{S}_\infty$ and analogously $I - T_u^* T_u \in \mathfrak{S}_\infty$. Conversely, if $I - T_u T_u^* \in \mathfrak{S}_\infty$, $I - T_u^* T_u \in \mathfrak{S}_\infty$, then by Corollary 27, $\mathbb{1} - |u|^2 = \mathbb{0}$, i.e. $|u| = 1$ a.e., and then the previous reasoning can be reversed and we get $\bar u \in H^\infty + C$, $u \in H^\infty + C$.

2. If $T_u = U + K$ then $|u| = 1$ a.e., $u \in QC$ by what has just been established, whereas $\lim_{r\to 1} \operatorname{wind} u_r(\mathbb{T}) = -\operatorname{ind} T_u = 0$ (cf. 45, 50, 51).

Conversely, if $|u| = 1$ a.e., $u \in QC$, then $T_u$ is Fredholm (Theorem 50); in fact, it is even invertible, in view of $\operatorname{ind} T_u = -\lim_{r\to 1} \operatorname{wind} u_r(\mathbb{T}) = 0$. Therefore in the polar decomposition $T_u = UR$ the operator $U$ must be unitary (and $R \geq \mathbb{0}$). Hence $k \overset{\text{def}}{=} R - I = (R + I)^{-1}(R^2 - I) = (R + I)^{-1}(T_u^* T_u - I) \in \mathfrak{S}_\infty$, i.e. $T_u = U + K$ and $K \overset{\text{def}}{=} Uk \in \mathfrak{S}_\infty$. □

**55. Some Properties of Quasicontinuous Functions** can be derived on the basis of that portion of the spectral theory of Toeplitz operators that we have already developed. The first of the corollaries to be set forth below is more a curiosity but the other one gives an interesting constructive characterization of the unimodular functions in $QC$.

**56. Corollary.** *If* $u \in H^\infty + C$, $|u| = 1$ *a.e. and if* $\operatorname{dist}(\bar u, H^\infty + C) < 1$ *then* $\bar u \in H^\infty + C$ *(i.e.* $u \in QC$*).*

In fact by Theorem 41 $0 \notin \sigma_{\text{ess}}^l(T_u)$, $0 \notin \sigma_{\text{ess}}^r(T_u)$, i.e. $T_u$ is Fredholm, i.e. $\bar u \in H^\infty + C$ (Theorem 50). □

This corollary also admits a direct proof which uses just the fact that the algebra $H^\infty + C$ is closed: if $v \in H^\infty + C$ and $\|\bar u - v\|_\infty = \|\mathbb{1} - uv\|_\infty < 1$ then $\bar u = v \sum_{n \geq 0} (\mathbb{1} - uv)^n \in H^\infty + C$. □

**57. Corollary (Sarason's Theorem).** *Let* $\varphi \in L^\infty$, $|\varphi| = 1$ *a.e. Then*

$$\varphi \in QC \Leftrightarrow \varphi = z^n e^{i(u + \tilde v)}$$

*with* $n \in \mathbb{Z}$, $u$ *and* $v$ *being continuous real functions on* $\mathbb{T}$.

In the paper Peller-Hruščev [1] this corollary is derived from Theorem 50, 36, Corollary 44 and the standard integral representation of outer functions (Appendix 2). Resisting the temptation to demonstrate the usefulness of Toeplitz operators in applications to function theory, we shall give the more direct proof by Sarason depending only on the structure of the Banach algebra $H^\infty + C$ (and the above integral representation).

*The implication* $\Leftarrow$ is trivial: if $\varphi = z^n \exp i(u + \tilde{v})$ then $\varphi = z^n \exp(v + i\tilde{v}) \cdot \exp(-v + iu) \in H^\infty + C$ and analogously for $\bar{\varphi}$.

*The implication* $\Rightarrow$ uses the invertibility of $\varphi$ in $H^\infty + C$. If the integer $k$, $k \in \mathbb{Z}$, and the function $h$, $h \in H^\infty$, are chosen such that the norm $\|\varphi - z^k h\|_\infty$ is sufficiently small then the function $z^k h$ too is invertible in $H^\infty + C$ so that $h^{-1} \in H^\infty + C$. But then it is clear that the outer and the inner parts $h^e$, $h^i$ are invertible in $H^\infty + C$. This implies that $h^i$ must be a finite Blaschke product. (This is best seen on the basis of Hartman's Theorem: the operator $H_{\bar{h}^i} = P_- \bar{h}^i$ is compact and isometric on the subspace $H^2 \ominus h^i H^2$, i.e. $\dim(H^2 \ominus h^i H^2) = \deg h^i < \infty$. □ Another purely "functional" approach: pick $k \geq 0$, $g \in H^\infty$ such that $\|\bar{h}^i - z^{-k} g\|_\infty = \|z^k - h^i g\|_\infty < 1$: then $\inf\{h^i(\zeta) g(\zeta) : 1 - \delta < |\zeta| < 1\} > \delta$ for small $\delta$, $\delta > 0$, i.e. $h^i$ is a finite Blaschke product. □ From the estimate $\|\mathbb{1} - z^k \bar{\varphi} h\|_\infty < 1$ it follows that the function $z^k \bar{\varphi} h$ has a logarithm in the algebra $H^\infty + C$: $z^k \bar{\varphi} h = \exp(f + g)$, with $g \in H^\infty$, $f \in C(\mathbb{T})$. Thus

$$\varphi = z^k h^i \exp\{\log|h| + i\widetilde{\log|h|} + ic - f - g\},$$

with $c \in \mathbb{R}$. Let us set $v = \log|h| - \operatorname{Re} g$ ($= \operatorname{Re} f$ in view of $|\varphi| = 1$ a.e.), $u_1 = -\operatorname{Im} f + c - \operatorname{Im} \hat{g}(0)$. Then $\tilde{v} = \log|h| - \operatorname{Im} g + \operatorname{Im} \hat{g}(0)$ and $\varphi = z^k h^i \exp i(u_1 + \tilde{v})$. It remains to make the observation that $\operatorname{wind}(z^m h^i) = 0$ for a suitable choice of $m \in \mathbb{Z}$ so that $z^m h^i = \exp iu_2$, $u_2 \in C(\mathbb{T})$, and to put $n = k - m$, $u = u_1 + u_2$. □

**58. Concerning Other Symbol Classes** it is the more difficult to give a spectral analysis the more there are oscillating functions in the classes in question. Examples of a successful analysis of such a kind are rather rare and pertain mainly to operators $T_\varphi$ whose symbols belong to algebras generated by the classes $PC$, $QC$, $H^\infty$ (in various combinations) and their subclasses, the class of "almost periodic" functions (concerning this class cf. below). To give an idea of the problems and methods arising here we shall outline a description of the spectrum of the operators $T_\varphi$ with symbols in the algebra $\bigvee_{L^\infty}\left(\exp\left(\alpha \dfrac{z+\zeta}{z-\zeta}\right) : \alpha \in \mathbb{R}, \zeta \in \mathbb{T}\right)$. First, with the aid of some notions from almost periodic analysis we shall however consider the algebra

$$AP_\zeta \overset{\text{def}}{=} \bigvee\left(\exp\left(\alpha \frac{z+\zeta}{z-\zeta}\right) : \alpha \in \mathbb{R}\right), \quad \zeta \in \mathbb{T},$$

and then we will derive from here the general result using an elementary variant of the "technique of localization". Another case to be discussed below are the operators $T_\varphi$ with symbols in $H^\infty_{\bar{\Theta}}$, where $\Theta$ is an inner function.

The formula $T_{\bar{f}gh} = T_{\bar{f}} T_g T_h$ ($g \in L^\infty$; $\bar{f}$, $h \in H^\infty$), which was mentioned in 36, might indicate that it would suffice to take account of the "oscillation" mentioned in the previous paragraph only "modulo" $H^\infty$, i.e., that it would be possible to pass from the spectral theory for symbols in a given class $\mathscr{F}$ to symbols in

$\overline{H^\infty} \cdot \mathscr{F} \cdot H^\infty$. However, no such universal algorithm exists and there are in fact both successful examples of such an enlargement on the expense of a factor $H^\infty$ (cf. 50, 51) as well as rather unsuccessful ones (89–102). That the spectral analysis of the operators $T_\varphi$ is intimately connected with such an enlargement follows however from the following result, where (as in what follows) $H_M^\infty$ *stands for the least uniformly closed subalgebra of* $L^\infty(\mathbb{T})$ *which contains* $H^\infty$ *and the set* $M$ ($M \subset L^\infty(\mathbb{T})$).

If $X$ is a subalgebra of $L^\infty$ containing the unit and if $\varphi \in X$ then $\sigma_X(\varphi)$ is *the spectrum of* $\varphi$ *in the algebra* $X$.

**59. Douglas's Lemma.** *Let* $\varphi \in L^\infty$.

1. $\sigma_{H^\infty_{\varphi,C}}(\varphi) \subset \sigma^r_{\text{ess}}(T_\varphi)$.
2. $\sigma_{H^\infty_\varphi}(\varphi) \subset \sigma^r(T_\varphi)$.

*Proof*. 1. If $T_\varphi$ is essentially invertible from the right then by Theorem 30 $\varphi^{-1} \in L^\infty$. Let us set $u = \varphi/\varphi_e$ where $\varphi_e$ is an outer function with modulus $|\varphi|$. Then $T_\varphi = T_u T_{\varphi_e}$ and thus $0 \notin \sigma^r_{\text{ess}}(T_u)$. By Theorem 41 $\operatorname{dist}(\bar{u}, H^\infty + C) < 1$, i.e. there exists a function $h$, $h \in H^\infty + C$, such that $\|\bar{u} - h\|_\infty = \|\mathbb{1} - uh\|_\infty < 1$. It follows that $uh$ is invertible in $H^\infty_{u,C} = H^\infty_{\varphi,C}$ and therefore so are also the functions $u$ and $u\varphi_e$.

2. Obvious modification of the proof of Statement 1: replace $\sigma^r_{\text{ess}}$ by $\sigma^r$; by Theorem 41 then $\operatorname{dist}(\bar{u}, H^\infty) < 1$ etc. □

**60. Remark.** It is well-known that $H^\infty_\varphi = H^\infty_{\varphi,C}$ if $\varphi \subset L^\infty \backslash H^\infty$ (i.e. $C(\mathbb{T}) \subset H_\varphi$ in this hypothesis); for a proof based on maximal ideals see 165.60. A special case ($\bar{\varphi}$ an inner function) to be used in the following Sect. 60 will be considered in 90.

**61. Corollary.** *If* $\varphi = \varphi_1 \bar{\varphi}_2$, $\varphi_k$ *being a non-constant inner function* ($k = 1, 2$), *and* $0 \notin \sigma^r_{\text{ess}}(T_\varphi)$, *then* $\bar{\varphi}_1 \in H^\infty_{\bar{\varphi}_2}$. *If the operator* $T_\varphi$ *is Fredholm then* $H^\infty_{\bar{\varphi}_1} = H^\infty_{\bar{\varphi}_2}$.

In fact, if $0 \notin \sigma^r_{\text{ess}}(T_\varphi)$ then by Lemma 59 $\bar{\varphi} \in H^\infty_{\varphi,C}$, i.e. $\bar{\varphi}_1 \in \bar{\varphi}_2 H^\infty_{\varphi,C} \subset H^\infty_{\bar{\varphi}_2,C}$. 60 implies that $H^\infty_{\bar{\varphi}_i,C} = H^\infty_{\bar{\varphi}_i}$. If $T_\varphi$ is Fredholm then by symmetry $\bar{\varphi}_2 \in H^\infty_{\bar{\varphi}_1}$ and thus $H^\infty_{\bar{\varphi}_1} = H^\infty_{\bar{\varphi}_2}$. □

**62. The Class** $AP_\zeta \overset{\text{def}}{=} \bigvee(\chi_{\alpha,\zeta} : \alpha \in \mathbb{R})$, $\chi_{\alpha,\zeta} \overset{\text{def}}{=} \exp\left(\alpha \dfrac{z+\zeta}{z-\zeta}\right)$, $\zeta \in \mathbb{T}$. This space is connected with the classical algebra $AP$ of almost periodic functions,

$$AP \overset{\text{def}}{=} \bigvee\nolimits_{L^\infty(\mathbb{R})}(e^{i\alpha x} : \alpha \in \mathbb{R}),$$

by the obvious relation $AP_\zeta = \{f \circ \omega : f \in AP\}$ where $\omega = -i(z+\zeta)(z-\zeta)^{-1}$ is the standard map of the circumference $\mathbb{T}$ onto the axis $\mathbb{R}$, with pole $\zeta$. For the computation of the spectrum of the operator $T_\varphi$ with an almost periodic symbol $\varphi$ a decisive rôle is played by Bohr's theorem on mean motion.

**63. Bohr's Theorem.** *If* $f \in AP$, $\inf_{\mathbb{R}} |f| > 0$, *then the limit*

$$w_f = \lim_{T \to \infty} \frac{1}{2T}(\arg f(T) - \arg f(-T)) \tag{19}$$

*exists and* $f = e^{iw_f x} e^g$, $g \in AP$.

**64. Theorem.** *Let $\varphi \in AP_\zeta$, $\zeta \in \mathbb{T}$, $\varphi = f \circ \omega$ ($f \in AP$). The following statements are equivalent.*

1. *The operator $T_\varphi$ is invertible.*
2. *The operator $T_\varphi$ is Fredholm.*
3. $\inf_{\mathbb{T}} |\varphi| > 0$, $w_f = 0$.

*Proof*. It is clear that $1 \Rightarrow 2$ and that $2 \Rightarrow \inf_{\mathbb{T}} |\varphi| > 0$. As $\varphi = \chi e^{\psi}$, where $\chi = \chi_{w_f, \zeta}$ and $\psi = g \circ \omega$, and $T_\varphi = T_{e^\psi} T_\chi$ if $w_f \geq 0$, $T_\varphi = T_\chi T_{e^\psi}$ if $w_f < 0$, it is for the remainder of the proof sufficient to show that the operator $T_{e^\psi}$ is invertible, noting that $T_\chi$ is not Fredholm if $w_f \neq 0$ (this is evident).

By the definition of $AP_\zeta$ there exist functions $p, q$ of the type $\sum_{\alpha > 0} a_\alpha \chi_{\alpha, \xi}$ (finite sums) such that $\|\psi - p - \bar{q}\|_\infty < \frac{1}{2}$. Since $p$, $q \in H^\infty$ if follows that $T_{e^\psi} = T_{e^{\bar{q}}} T_{e^{\psi - p - \bar{q}}} T_{e^p}$ must be invertible. □

**65. Remark.** It is clear from what has been said that $T_\varphi$, $\varphi \in AP_\zeta$, is either left or right invertible iff $\inf_{\mathbb{T}} |\varphi| > 0$. For $w_f > 0$ $\dim \operatorname{Ker} T_\varphi^* = \infty$ and for $w_f < 0$ $\dim \operatorname{Ker} T_\varphi = \infty$. The mean motion plays the rôle of a kind of "winding number" at the point $\zeta$ for the function $\varphi = f \circ \omega$,

$$w_f = \lim_{t \to +0} \frac{t}{4} [\arg \varphi(\zeta e^{-it}) - \arg \varphi(\zeta e^{it})].$$

*Notation*: $w_f = \operatorname{wind}_\zeta \varphi$.

**66. Jessen's Proof of Bohr's Theorem.** Set $\delta = \inf_{\mathbb{R}} |f|$ and let $p$ be a trigonometric polynomial, $p = \sum_{\alpha \in \mathbb{R}} a_\alpha e^{i\alpha x}$ (finite sum) approximating $f$: $\|f - p\|_\infty < \delta/2$. Then $\inf_{\mathbb{R}} |p| \geq \delta/2$ and $\|\mathbb{1} - f/p\|_\infty < 1$. As $1/p \in AP$ (this follows, for instance, from the classical theorem of Wiener: $p = \sum a_\alpha e^{i\alpha x}$, $\sum |a_\alpha| < \infty$, $|p| \geq \delta/2 \Rightarrow 1/p = \sum b_\alpha e^{i\alpha x}$, $\sum |b_\alpha| < \infty$, or simply from the infinite differentiability of the function $1/p$) we have $f/p \in AP$ and $\log f/p \in AP$. It is likewise clear that the mean motion of $f/p$ equals zero (because $\log f/p \in L^\infty(\mathbb{R})$). It remains to be shown that the theorem is true for polynomials $p$. It is well-known that for any finite set $\{\alpha\}$ of real numbers there exists an *independent* finite set $B$, $B \subset \mathbb{R}$ (i.e. a set such that $\sum_{\beta \in B} k(\beta)\beta = 0$, $k(\beta) \in \mathbb{Z} \Rightarrow k(\beta) = 0$, $\forall \beta$) such that for any $\alpha$ there holds $\alpha = \sum_{\beta \in B} k_\alpha(\beta)\beta$ with suitable integers $k_\alpha(\beta)$. It is likewise well-known that the "winding" $\{(e^{i\beta x})_{\beta \in B} : x \in \mathbb{R}\}$ is dense in the torus $\prod_{\beta \in B} \mathbb{T}$. (All these results originate from Kronecker; see any book in Harmonic Analysis, e.g. Katznelson [1], Bari [2].

Set $m = \operatorname{card} B$ and consider the vector $\mathbb{B}$ in $\mathbb{R}^m$ with coordinates $\{\beta\}_{\beta \in B}$. Write $k_\alpha \stackrel{\text{def}}{=} \{k_\alpha(\beta)\}_{\beta \in B}$ and $(\cdot, \cdot)$ for the scalar product in $\mathbb{R}^m$. Then

$$p = \sum_{\alpha \in \mathbb{R}} a_\alpha e^{i(k_\alpha, \mathbb{B}x)}.$$

Set $q = \sum_{\alpha \in \mathbb{R}} a_\alpha e^{i(k_\alpha, X)}$, $X = (x_1, \ldots, x_m) \in \mathbb{R}^m$. By Kronecker's "winding" theorem just referred to it follows that $q(\mathbb{R}^m) \subset \operatorname{clos} p(\mathbb{R})$. In particular $\inf_{\mathbb{R}^m} |q| \geq \delta/2$. Let $\arg q$ be any continuous branch of the argument of $q$. As this function is $2\pi$-periodic in each of the variables $x_1, \ldots, x_m$ we have $\arg q(X + 2\pi\delta_s) - \arg q(X) = 2\pi h_s$, $X \in \mathbb{R}^m$, where $h_s$ is a suitable integer, $\delta_s = \{\delta_{is}\}_{1 \leq i \leq m}$ being a member of the standard basis of $\mathbb{R}^m$ ($s = 1, 2, \ldots, m$). It follows that the function $\arg q - (h, X)$ is $2\pi$-periodic ($h = \{h_i\}_{1 \le i \le m}$). And so is trivially $\log|q|$. Therefore $q = e^{i(h,X)}e^{\Psi}$, where $\Psi$ is a $2\pi$-periodic function, and thus $p = e^{i(h,\mathbb{B})x}e^{\psi}$ with $\psi = \Psi(\mathbb{B}x)$ almost periodic (it suffices to approximate $\Psi$ with Fejér sums). □

**67. Gorin's Proof of Bohr's Theorem.** This proof is based on some general facts in Harmonic Analysis. Let $K$ be the compactification of the line $\mathbb{R}$ with the aid of the $AP$-functions. From the Stone-Weierstrass theorem it follows that $AP = C(K)$. One can show that $K$ (in its proper topology) is a connected compact Abelian group.

Let $f \in AP$, $\inf_{\mathbb{R}} |f| = \min_K |f| > 0$. Let us assume that $f(0) = 1$ and set

$$F(x, y) = f(x + y)[f(x)f(y)]^{-1}, \quad x, y \in K.$$

It is clear that $F(\cdot, y)$ belongs to the connected component of the group of invertible elements of $C(K)$ which contains the unit function $\mathbb{1}$. Therefore, as is readily seen, $F(\cdot, y) = \exp \psi(\cdot, y)$ where $\psi(\cdot, y) \in C(K)$. Using the connectedness of the compact space $K$, the relation $F(0, y) \equiv F(x, 0) \equiv 1$ and the uniqueness of the function $\psi(\cdot, y)$ with the property $\psi(0, y) = 0$, one sees that $\psi$ is continuous on $K \times K$ and also that $\psi(0, y) \equiv \psi(x, 0) \equiv 0$. One likewise sees at once that $\psi(x, z) + \psi(x + z, y) = \psi(x + y, z) + \psi(x, y)$. Let us integrate over $y$ (with respect to the Haar measure on $K$): $\psi(x, z) = g(x + z) - g(x) - g(z)$ where $g(x) = -\int_K \psi(x, y)\,dy$, $g \in C(K) = AP$. It follows that $\chi \overset{\text{def}}{=} f \exp(-g)$ is a character of the group $K$, and therefore also of $\mathbb{R}$, i.e. $\chi(x) = e^{iwx}$, $w \in \mathbb{R}$.

Formula (19) is now evident, because $\arg e^g = \operatorname{Im} g \in L^\infty(\mathbb{R})$. □

**68. Corollary.** *If $\varphi \in AP_\zeta$ then the spectrum $\sigma(T_\varphi)$ is connected.*

In fact, in the notation of 67 it is the issue of the connectedness of the set $\varphi(K) \cup \{\lambda \in \mathbb{C} \backslash \varphi(K) : \operatorname{wind}_\zeta(\varphi - \lambda) \ne 0\}$. But the image $\varphi(K)$ is connected, because $K$ is connected, and the mean motion $w_f$ is not changed by small (in the norm $\|\cdot\|_\infty$) perturbations of $f$, so that $\operatorname{wind}_\zeta(\varphi - \lambda) \equiv \text{const}$ on every connected component of the set $\mathbb{C} \backslash \varphi(K)$. □

**69. The Class $AP_\sigma$, $\sigma \subset \mathbb{T}$.** By definition

$$AP_\sigma = \bigvee_{L^\infty}(AP_\zeta : \zeta \in \sigma).$$

Let $\sigma$ be finite. Then to each function in the algebraic sum $\sum_{\zeta \in \sigma} AP_\zeta + C(\mathbb{T})$ there corresponds a function $f_\zeta$ in $AP_\zeta$ such $\lim_{t \to \zeta} [f(t) - f_\zeta(t)] = 0$. By a well-known (and easily provable; cf. Katznelson [1], Levitan [1]) property of almost periodic functions

$$\sup_{\mathbb{T}} |f_\zeta| = \varlimsup_{t \to \zeta} |f_\zeta(t)|, \quad \inf_{\mathbb{T}} |f_\zeta| = \varliminf_{t \to \zeta} |f_\zeta(t)|.$$

Hence $\|f_\zeta\|_\infty \le \|f\|_\infty$, $\zeta \in \sigma$, for all functions $f$ of this kind. In particular the $f_\zeta$ in question is unique. It is natural to regard it as the "principal part" of $f$ at the point $\zeta$.

**70. Lemma.** *Let $\sigma$ be a finite set, $\sigma \subset \mathbb{T}$.*

1. *The map* $f \to f_\zeta$ *is a continuous linear projection from the sum* $\sum_{\zeta\in\sigma} AP_\zeta + C(\mathbb{T})$ *onto* $AP_\zeta$.

2. *The sum* $\sum_{\zeta\in\sigma} AP_\zeta + C(\mathbb{T})$ *is a closed subalgebra of* $L^\infty$; *it coincides with* $AP_\sigma + C(\mathbb{T})$.

3. *If* $f \in AP_\sigma + C(\mathbb{T})$ *and* $\inf_{\mathbb{T}}|f| > 0$ *then*

$$f = f_0 \cdot \left(\prod_{\zeta\in\sigma} f_\zeta\right),$$

*where* $f_0 \in C(\mathbb{T})$, $\min_{\mathbb{T}}|f_0| > 0$, $\inf_{\mathbb{T}}|f_\zeta| > 0$ $(\zeta\in\sigma)$.

*Proof*. Statement 1 has already been established.

If $f \in \sum_{\zeta\in\sigma} AP_\zeta + C(\mathbb{T})$ then $f - \sum_{\zeta\in\sigma} f_\zeta \in C(\mathbb{T})$ so that the fact that the sum is closed follows from Statement 1. It is likewise clear that $f \in \sum_{\zeta\in\sigma} AP_\zeta + C(\mathbb{T})$ iff $f \in C(\mathbb{T}\backslash\sigma)$ and $\forall\zeta\in\sigma\exists g\in AP_\zeta : f - g$ is continuous at $\zeta$. From this description it follows that $AP_\sigma$ is a subalgebra $L^\infty$.

From the property of $AP$-functions mentioned in 69 it follows that $\inf_{\mathbb{T}}|f_\zeta| \geq \inf_{\mathbb{T}}|f| > 0$, $\zeta\in\sigma$, which again implies the continuity of the function $f_0 \overset{\text{def}}{=} f/\prod_{\zeta\in\sigma} f_\zeta$ (for instance, the continuity at $\zeta$, $\zeta\in\sigma$, follows from the formula $f_0 = f/f_\zeta(\prod_{\xi\in\sigma\backslash\{\zeta\}} f_\xi)^{-1} = (1 + o(1))\cdot(\prod_{\xi\in\sigma\backslash\{\zeta\}} f_\xi)^{-1})$. □

**71. Definition.** *Let* $\sigma$ *be a finite set,* $\sigma \subset \mathbb{T}$, $\varphi \in AP_\sigma + C(\mathbb{T})$, $\inf_{\mathbb{T}}|\varphi| > 0$. *We set*

$$\operatorname{wind}_\zeta \varphi = \operatorname{wind}_\zeta \varphi_\zeta,$$

*where* $\varphi$ *is the principal part of* $\varphi$ *at the point* $\zeta$, $\zeta\in\sigma$.

**72. Theorem.** *Let* $\sigma$ *be a finite set,* $\sigma \subset \mathbb{T}$, *and let* $\varphi \in AP_\sigma + C(\mathbb{T})$.

1. $0 \notin \sigma_{\text{ess}}(T_\varphi) \Leftrightarrow \inf_{\mathbb{T}}|\varphi| > 0$; $\operatorname{wind}_\zeta \varphi = 0$, $\zeta\in\sigma$.
2. *If* $0 \notin \sigma_{\text{ess}}(T_\varphi)$ *then* $\operatorname{ind} T_\varphi = -\operatorname{wind} \varphi_0$, *where* $\varphi_0 = \varphi(\prod_{\zeta\in\sigma}\varphi_\zeta)^{-1}$.

**73. Preview.** For the proof of Theorem 72 we need a certain variant of the "technique of localization", that is, a statement that says that whether $T_\varphi$ is *Fredholm* or not depends on the local properties of its symbol $\varphi$ (in contrast to the *invertibility*). There are several approaches to localization theorems, based e.g. on the analysis of the commutators of Toeplitz operators and the $C^*$-algebras generated by such operators or on a study of the maximal ideals of the corresponding subalgebra of $L^\infty(\mathbb{T})$. Here we set forth the direct elementary method due to Sarason (74–76). The main results on localization will then arise as simple consequences (77–82). Theorem 72 will turn out to be a consequence of these consequences (88), and the same holds for some extensions connected with the algebra $AP_{\mathbb{T}} + C(\mathbb{T})$ (87, 88).

**74. The Local Distance.** Let $f, g \in L^\infty(\mathbb{T})$, $\lambda\in\mathbb{T}$. By definition

$$\operatorname{dist}_\lambda(f,g) = \varlimsup_{\zeta\to\lambda} |f(\zeta) - g(\zeta)|;$$

here we intend of course $\lim_{\varepsilon\to 0} \operatorname{ess\,sup}\{|f(\zeta) - g(\zeta)| : |\arg\zeta - \arg\lambda| < \varepsilon\}$. If

$M \subset L^\infty$ then

$$\mathrm{dist}_\lambda(f, M) = \inf\{\mathrm{dist}_\lambda(f, h) : h \in M\}.$$

It is not hard to see that the function $\lambda \mapsto \mathrm{dist}_\lambda(f, M)$ is upper semi-continuous (and thus takes its maximum on $\mathbb{T}$), that for any $f$ and $g$ holds $\max\{\mathrm{dist}_\lambda(f, g) : \lambda \in \mathbb{T}\} = \|f - g\|_\infty$ and that $\mathrm{dist}_\lambda(f, g) = \overline{\lim}_{\zeta \to \lambda, |\zeta| < 1} |(\mathscr{H}f)(\zeta) - (\mathscr{H}g)(\zeta)|$. It is likewise clear that for any $f$, $\lambda$ and $M$, $M$ a linear subspace, $c \cdot \mathbb{1} \in M$, $\forall c \in \mathbb{C}$ holds: $\mathrm{dist}_\lambda(f, M) = \mathrm{dist}_\lambda(f, M + C(\mathbb{T}))$.

The symbol dist (without the $\lambda$) stands for the distance in $L^\infty(\mathbb{T})$.

**75. Douglas-Sarason Lemma**

$$\mathrm{dist}(f, H^\infty + C) = \max_{\lambda \in \mathbb{T}} \mathrm{dist}_\lambda(f, H^\infty).$$

*Proof*. The inequality $\geq$ is evident. In order to prove the opposite inequality fix a number $\varepsilon$, $\varepsilon > 0$, and for each $\lambda$, $\lambda \in \mathbb{T}$, pick a function $h_\lambda$ in $H^\infty$ such that

$$\mathrm{dist}_\lambda(f, h_\lambda) < \varepsilon + \max_{\lambda \in \mathbb{T}} \mathrm{dist}_\lambda(f, H^\infty).$$

Let us denote this last $\max_{\lambda \in \mathbb{T}}$ by $M$ and, using the fact that the function $\zeta \mapsto \mathrm{dist}_\zeta(f, h_\lambda)$ is upper semi-continuous, choose an open arc $\mathscr{I}_\lambda$, $\lambda \in \mathscr{I}_\lambda \subset \mathbb{T}$, such that $\mathrm{dist}_\lambda(f, h_\lambda) < M + 2\varepsilon$ for $\zeta \in \mathscr{I}_\lambda$. Let $\{\mathscr{I}_{\lambda_i}\}_{1 \leq i \leq n}$ be a covering of the circumference $\mathbb{T}$ and let $\{p_i\}_{1 \leq i \leq n}$ be a partition of unity subordinate to this covering, i.e. $p_i \in C(\mathbb{T})$, $p_i \geq 0$, $\mathrm{supp}\, p_i \subset \mathscr{I}_{\lambda_i}$, $\sum_{i=1}^n p_i = \mathbb{1}$.

Set $g = \sum_{i=1}^n p_i h_{\lambda_i}$. Since $H^\infty + C$ is an algebra it follows that $g \in H^\infty + C$. Let $\lambda \in \mathbb{T}$. Then

$$\mathrm{dist}_\lambda(f, g) = \mathrm{dist}_\lambda\left(\sum_{i=1}^n p_i f, \sum_{i=1}^n p_i h_{\lambda_i}\right) \leq \sum_{i=1}^n \mathrm{dist}_\lambda(p_i f, p_i h_{\lambda_i}) = \sum_{\lambda \in \mathscr{I}_{\lambda_i}} \mathrm{dist}_\lambda(p_i f, p_i h_{\lambda_i})$$

$$= \sum_{\lambda \in \mathscr{I}_{\lambda_i}} p_i(\lambda) \mathrm{dist}_\lambda(f, h_{\lambda_i}) < (M + 2\varepsilon) \sum_{\lambda \in \mathscr{I}_{\lambda_i}} p_i(\lambda) \leq M + 2\varepsilon.$$

It follows that $\|f - g\|_\infty < M + 2\varepsilon$ and hence $\mathrm{dist}(f, H^\infty + C) < M + 2\varepsilon$. $\varepsilon$ being arbitrary, this proves $\mathrm{dist}(f, H^\infty + C) \leq M$. □

Notice that the proof remains in force if we replace $H^\infty$ by an arbitrary subset $M$ of $L^\infty$ such that $M + C$ is analgebra.

**76. Simonenko's Theorem.** *Let $\varphi \in L^\infty$ and assume that for each $\lambda$, $\lambda \in \mathbb{T}$, there exists a function $\varphi_\lambda$, $\varphi_\lambda \in L^\infty$, such that the operator $T_{\varphi_\lambda}$ is Fredholm (respectively left or right Fredholm) and $\mathrm{dist}_\lambda(\varphi, \varphi_\lambda) = 0$. Then $T_\varphi$ is Fredholm (respectively left or right Fredholm).*

*Proof*. It is sufficient to establish the property of $T_\varphi$ being onesided Fredholm (say, left Fredholm, for the sake of definiteness). By Theorem 30 $\varepsilon_\lambda \stackrel{\text{def}}{=} \mathrm{ess\,inf}_{\mathbb{T}} |\varphi_\lambda| > 0$ for every $\lambda$, $\lambda \in \mathbb{T}$. If $\mathscr{I}_\lambda$ is an arc containing $\lambda$ on which $\mathrm{dist}_\zeta(\varphi, \varphi_\lambda) < \varepsilon_\lambda/2$ ($\zeta \in \mathscr{I}_\lambda$) then a.e. on $\mathscr{I}_\lambda$ there holds $|\varphi| > \mathrm{ess\,inf}_{\mathbb{T}} |\varphi_\lambda| - |\varphi - \varphi_\lambda| > \varepsilon_\lambda/2$. Choosing a finite covering $\{\mathscr{I}_{\lambda_i}\}_{1 \leq i \leq n}$ we get $\mathrm{ess\,inf}_{\mathbb{T}} |\varphi| > 0$. This yields

$\operatorname{dist}_\lambda(\varphi/|\varphi|, \varphi_\lambda/|\varphi_\lambda|) = \operatorname{dist}_\lambda(\varphi, \varphi_\lambda) = 0$. By Lemma 75 now

$$\operatorname{dist}(\varphi/|\varphi|, H^\infty + C) = \max_{\lambda \in \mathbb{T}} \operatorname{dist}_\lambda(\varphi/|\varphi|, H^\infty),$$

$$\begin{aligned}\operatorname{dist}_\lambda(\varphi/|\varphi|, H^\infty) &\le \operatorname{dist}_\lambda(\varphi/|\varphi|, \varphi_\lambda/|\varphi_\lambda|) + \operatorname{dist}_\lambda(\varphi_\lambda/|\varphi_\lambda|, H^\infty)\\ &= \operatorname{dist}_\lambda(\varphi_\lambda/|\varphi_\lambda|, H^\infty + C) \le \operatorname{dist}(\varphi_\lambda/|\varphi_\lambda|, H^\infty + C).\end{aligned}$$

To conclude the proof we just to have to apply Theorem 41. □

**77. Corollary.** *Let $\varphi \in L^\infty$ and assume that in some neighborhood of each point of $\mathbb{T}$ the function $\varphi$ coincides with a function generating a (left, right) Fredholm operator. Then $T_\varphi$ is (left, right) Fredholm.* □

**78. Corollary.** *Let $\varphi \in L^\infty$ and assume that to each point $\lambda$, $\lambda \in \mathbb{T}$, there corresponds an arc $\mathcal{I}_\lambda$ containing $\lambda$ such that $0 \notin \operatorname{conv}\operatorname{Range}(\varphi|\mathcal{I}_\lambda)$. Then $T_\varphi$ is Fredholm.*

In fact, it suffices to continue $\varphi|\mathcal{I}_\lambda$ outside $\mathcal{I}_\lambda$ as a linear function and to apply Theorem 30 and Corollary 77. □

**79. Corollary (on the Separation of Singularities).** *Let $\varphi, \varphi_k \in L^\infty$ $(k = 1, \ldots, n)$, $\varphi = \prod_{k=1}^n \varphi_k$ and $\operatorname{ess\,inf}_{\mathbb{T}} |\varphi| > 0$. If all the functions $\varphi_k$, except at most one, are continuous, at every point of the circumference $\mathbb{T}$, then*

$$T_\varphi \text{ is Fredholm} \Leftrightarrow \text{all the } T_{\varphi_k},\ 1 \le k \le n, \text{ are Fredholm.}$$

In fact, the implication "⇐" follows from Theorem 76 and the identity $\min_k \operatorname{dist}_\lambda(\varphi, c_\lambda \varphi_k) = 0$, $\lambda \in \mathbb{T}$ $(\exists c_\lambda \neq 0)$ and the converse ("⇒") from the same theorem in conjunction with the identities $\operatorname{dist}_\lambda(\varphi_k, \varphi_k(\lambda)\mathbb{1}) = 0$ (at continuity points of $\varphi_k$), $\operatorname{dist}_\lambda\left(\varphi_k, \dfrac{1}{c_\lambda}\varphi\right) = 0$ (at other points). □

**80. Corollary.** *Let $\varphi \in L^\infty$, $\operatorname{ess\,inf}_{\mathbb{T}} |\varphi| > 0$ and assume that $\varphi$ is* locally sectorial, *i.e. that every point of the circle $\mathbb{T}$ has a neighborhood $\mathcal{O}$ such that the image* $\operatorname{Range}(\varphi|\mathcal{O})$ *lies in an angle less than $\pi$ with the vertex at the origin* 0. *Then* $0 \notin \sigma_{\mathrm{ess}}(T_\varphi)$.

This is a restatement of Corollary 78. □

**81. Remark.** It is pertinent to make a small remark on the rôle of the local "smoothness" condition in the description of Fredholm Toeplitz operators, notably in connection with Corollary 80. In terms of the classical (uniform) oscillation of the function $\varphi$ this statement provides the best result; if we define the oscillation $\omega_\zeta$ of the function $\alpha$ at the point $\zeta$, $\zeta \in \mathbb{T}$, by the formula

$$\omega_\zeta(\alpha) = \lim_{\varepsilon \to 0} \operatorname{ess\,sup}\{|\alpha(t) - \alpha(t')| : |t - \zeta| < \varepsilon, |t' - \zeta| < \varepsilon\},$$

then Corollary 80 can be read off as:

$$\operatorname*{ess\,inf}_{\mathbb{T}} |\varphi| > 0; \quad \omega_\zeta(\arg\varphi) < \pi, \forall \zeta \in \mathbb{T} \Rightarrow 0 \notin \sigma_{\mathrm{ess}}(T_\varphi)$$

(here we mean $\arg\varphi$ modulo $2\pi$). It is for instance clear that the condition $\omega_\zeta(\varphi) < \sqrt{3}\lim_{t\to\zeta}|\varphi(t)|$, $\zeta\in\mathbb{T}$, is sufficient for $T_\varphi$ to be Fredholm and so is the already known (47) condition $\varphi\in PC$, $0\notin\varphi_*$ (since $\inf_{\mathbb{T}}|\varphi|>0$; $\omega_\zeta(\arg\varphi)<\pi$, $\zeta\in\mathbb{T}$). In fact, to a locally sectorial function $\varphi$ one can in a natural way associate a *winding number* wind $\varphi$, setting $\operatorname{wind}\varphi = \frac{1}{2\pi}(\alpha(2\pi)-\alpha(0))$ where $\alpha(x)=\alpha(e^{ix})$ is a branch of the argument of $\varphi$ (i.e. $\varphi=|\varphi|e^{i\alpha}$) on $\mathbb{T}\backslash\{1\}$ (such that $\omega_\zeta(\alpha)<\pi$, $\zeta\in\mathbb{T}\backslash\{1\}$). (The existence of such an $\alpha$ and its uniqueness up to a constant term follows from the definition of locally sectorial.) If $\alpha$ is a $2\pi$-periodic function (i.e. $\operatorname{wind}\varphi=0$) then considering the semi-continuous functions $\overline{\alpha}(\lambda)\overset{\text{def}}{=}\overline{\lim}_{\zeta\to\lambda}\alpha(\zeta)$, $\underline{\alpha}(\lambda)\overset{\text{def}}{=}\underline{\lim}_{\zeta\to\lambda}\alpha(\zeta)$, with $\overline{\alpha}(\lambda)-\underline{\alpha}(\lambda)=\omega_\lambda(\alpha)<\pi$ and $\overline{\alpha}-\pi/2<\underline{\alpha}+\pi/2$, one can find a *continuous* function $\beta$ on $\mathbb{T}$ with the property that $\|\alpha-\beta\|_\infty<\pi/2$. It follows that $\operatorname{Range}(\varphi e^{-i\beta})\subset\{\zeta:\operatorname{Re}\zeta>0\}$, i.e., the operator $T_{\varphi e^{-i\beta}}$ is invertible (Corollary 31), but then $T_\varphi$ is invertible too (cf. Theorem 38, or Theorem 84, below). Next, introducing as usual the factor $z^n$, $n=-\operatorname{wind}\varphi$, we get the complete formulation of Corollary 80.

**82. Simonenko's Theorem.** *If $\varphi$ is locally sectorial and* $\operatorname{ess\,inf}_{\mathbb{T}}|\varphi|>0$ *then $T_\varphi$ is Fredholm and* $\operatorname{ind}T_\varphi = -\operatorname{wind}\varphi$. □

**83. Multipliers Preserving the Fredholm Property.** Let us denote by Fred (Inv) the set of all functions $\varphi$, $\varphi\in L^\infty$, corresponding to a Fredholm (invertible) $T_\varphi$. For which functions $\psi$ is it true that multiplication by $\psi$ conserves the set Fred (Inv), that is, for which $\psi$ is it true that

$$\varphi\in\text{Fred}\Rightarrow\varphi\psi\in\text{Fred},$$

$$\varphi\in\text{Inv}\Rightarrow\varphi\psi\in\text{Inv},$$

respectively? This question has already arisen implicitly (e.g., in 47, 81); once more it arises in the study of operators $T_\varphi$ with $\varphi\in AP_\sigma + C(\mathbb{T})$ (cf. 70, 87, 88). Quite generally, it is useful to know which operations on symbols preserve the Fredholm property and the invertibility. For example, from the basic invertibility criterion 38 it follows at once that

$$\begin{aligned}&\varphi\in\text{Inv}\Rightarrow\overline{\varphi}\in\text{Inv},\quad \varphi^{-1}\in\text{Inv},\\ &\varphi\in\text{Inv}\Rightarrow\varphi\circ\omega\in\text{Inv},\end{aligned}\tag{20}$$

If $|\omega|=1$ a.e., $\omega\in H^\infty$ or $\overline{\omega}\in H^\infty$. Contrariwise, multiplication or addition of functions in Inv in general leads outside the set Fred (e.g., if $\varphi=\chi_e+i\chi_{\mathbb{T}\backslash e}$, $0<me<1$ ($\chi_e$ is the characteristic function of the set $e$) then $\varphi\in\text{Inv}$ (Corollary 31) but $\varphi^2\notin\text{Fred}$ (Theorem 33) although $\varphi^3,\varphi^4\in\text{Inv}$; if $\varphi=\chi_e+3\chi_{\mathbb{T}\backslash e}$ then $\varphi\in\text{Inv}$ but $\varphi-2\notin\text{Fred}$ (Theorem 33) although $\inf_{\mathbb{T}}|\varphi-2|>0$). It is likewise clear that

$$\psi\,\text{Fred}\subset\text{Fred}\Leftrightarrow\exists n\in\mathbb{Z}:(\psi z^n)\text{Inv}\subset\text{Inv}.$$

**84. Theorem.** *Let $\psi\in L^\infty$. The following statements are equivalent.*

1. $\psi\,\text{Inv}\subset\text{Inv}$.
2. $\psi\,\text{Inv}=\text{Inv}$.

3. $\operatorname{ess\,inf}_{\mathbb{T}}|\psi| > 0$ *and* $\forall \varepsilon > 0\ \exists u_\varepsilon, v_\varepsilon \in L^\infty \exists c_\varepsilon \in \mathbb{R} : \psi = |\psi| e^{ia}$, $a = u_\varepsilon + \tilde{v}_\varepsilon + c$, $\|u_\varepsilon\|_\infty < \varepsilon$ *($u_\varepsilon$, $v_\varepsilon$ being real functions).*

4. $\operatorname{ess\,inf}_{\mathbb{T}}|\psi| > 0$ *and* $\psi = |\psi| e^{ia}$, $a \in \operatorname{Re} \widetilde{L^\infty} + \operatorname{clos}_{L^\infty} \operatorname{Re} H^\infty$.

**Special cases.** a) *If* $\psi \in H^\infty$ *then* $\psi \operatorname{Inv} \subset \operatorname{Inv} \Leftrightarrow 1/\psi \in H^\infty$.

b) *If* $\psi \in C(\mathbb{T})$ *then* $\psi \operatorname{Fred} \subset \operatorname{Fred} \Leftrightarrow \min_{\mathbb{T}} |\psi| > 0$.

*Proof of the Theorem.* It is clear that $1 \Leftrightarrow 2$ (cf. (20)) and that $3 \Rightarrow 1$ (by virtue of 38). For the proof of the remaining implications we need the following simple proposition, to be verified in 86.

**85. Lemma.** *If* $u_i, v_i \in L^\infty$ *and* $u_1 + \tilde{v}_1 + c_1 = u_2 + \tilde{v}_2 + c_2 + 2\pi k(\cdot)$, *with* $c_i \in \mathbb{R}$, *$k(\cdot)$ being an integer valued function on* $\mathbb{T}$, *and if* $\|u_1 - u_2\|_\infty < \pi$, *then* $k \equiv \text{const}$.

From here it is easy to derive $1 \Rightarrow 3$. Assume that $\psi \operatorname{Inv} \subset \operatorname{Inv}$. Denote by $A_\psi$ the set of all triples $\alpha = (u, v, c)$, with $u, v \in L^\infty$, $\|u\|_\infty < \pi/2$, $c \in \mathbb{R}$ and $\psi = |\psi| e^{ia}$, $a = a_\alpha \overset{\text{def}}{=} u + \tilde{v} + c$, where (to fix the ideas) the constant $c$ is picked among all numbers fulfilling the relation $\psi = |\psi| e^{ia}$ via the normalization $|\hat{u}(0) + c| = \min\{|\hat{u}(0) + c + 2\pi k| : k \in \mathbb{Z}\}$. By Theorem 38 certainly $A_\psi \neq \varnothing$. Let us set $\gamma = \inf\{\|u\|_\infty : \alpha \in A_\psi\}$. If $\gamma > 0$ let us further set $\delta = \pi/2\gamma - 1$ and define the function $\varphi = (\psi/|\psi|)^\delta$ by the formula $\varphi = \exp(i\delta a_\alpha)$, $\alpha \in A_\psi$; that this definition is consistent (i.e., independent of the choice of $\alpha$) follows from Lemma 85. Let $\varepsilon$, $\varepsilon > 0$, be a sufficiently small number and consider $\alpha \in A_\psi$, $\|u\|_\infty < (1 + \varepsilon)\gamma$. Then $\varphi = \exp i\delta(u + \tilde{v} + c)$ and $\|\delta u\|_\infty < (1 + \varepsilon)\delta\gamma = (1 + \varepsilon)(\pi/2 - \gamma) < \pi/2$ so that $\varphi \in \operatorname{Inv}$ (Theorem 38). If now $\varphi\psi = |\psi| \exp i(u' + \tilde{v}' + c')$ and we assume that $\|u'\|_\infty < \pi/2$ then $\|(1 + \delta)u - u'\|_\infty < \pi$ so, by Lemma 85, $u' + \tilde{v}' + c' = (1 + \delta)a_\alpha$, $\alpha \in A_\psi$. Hence $\gamma \le \|(1 + \delta)^{-1}u'\|_\infty < (1 + \delta)^{-1}\pi/2 = \gamma$. This contradiction shows that always $\|u'\|_\infty \ge \pi/2$, i.e. that $\varphi\psi \notin \operatorname{Inv}$ (Theorem 38). Thus $\gamma = 0$, which is precisely Statement 3.

$3 \Rightarrow 4$. If $\alpha, \beta \in A_\psi$ then by Lemma 85 $u_\alpha + \tilde{v}_\alpha + c_\alpha = u_\beta + \tilde{v}_\beta + c_\beta$ and hence $a \overset{\text{def}}{=} \tilde{v}_\alpha + u_\alpha + c_\alpha = \tilde{v}_\alpha + (\tilde{v}_\beta - \tilde{v}_\alpha + c_\beta) + u_\beta$ and in addition $\inf\{\|u_\beta\|_\infty : \beta \in A_\psi\} = 0$ and $\tilde{v}_\beta - \tilde{v}_\alpha + c_\beta \in L^\infty$, $v_\beta - v_\alpha \in L^\infty$, i.e. $\tilde{v}_\beta - \tilde{v}_\alpha + c_\beta \in \operatorname{Re} H^\infty$. Consequently $a \in \operatorname{Re} \widetilde{L^\infty} + \operatorname{clos}_{L^\infty} \operatorname{Re} H^\infty$.

$4 \Rightarrow 3$. If $a \in \operatorname{Re} \widetilde{L^\infty} + \operatorname{clos}_{L^\infty} \operatorname{Re} H^\infty$ and we write $a$ in the form $a = \tilde{v} + \operatorname{Re} h + r$, where $h \in H^\infty$, $\|r\|_\infty < \varepsilon$, then $a = (v - \operatorname{Im} h)^\sim + r + \operatorname{Re} h(0)$, which proves everything. □

In connection with the implication $1 \Rightarrow 3$ let us remark that the condition $\psi^n \in \operatorname{Inv}$, $\forall n \in \mathbb{Z}$, does not at all mean that Statement 3 holds true for $\psi$ (example: $\psi = \psi_{\mathbb{T}^+} + e^{2\pi i/3}\chi_{\mathbb{T}^-}$; then $\psi^n \in \operatorname{Inv}$ by Corollary 31 but the function $\arg\psi = 2\pi i/3 \cdot \chi_{\mathbb{T}^-}$ has discontinuities of the first kind (jumps)). Some other remarks on Theorem 84 will be found in 165.83.

**86. Proof of Lemma 85.** Set $u = u_1 - u_2$, $v = v_1 - v_2$, $c = c_1 - c_2$. Then the operator $T_\varphi$, $\varphi \overset{\text{def}}{=} e^{i\pi k(\cdot)}$, is invertible, by virtue of Theorem 38, because $\pi k = u/2 + \tilde{v}/2 + c/2$ and $\|u/2\|_\infty < \pi/2$. We may assume that $0 \in \operatorname{Range} k$. If the set $\operatorname{Range} k$ contains an odd integer then by Theorem 33 $\sigma(T_\varphi) = [-1, 1]$, which contradicts the invertibility of $T_\varphi$. If not we can apply the same reasoning to the integer valued function $k/2$. Hence $k \equiv 0$. □

**87. Return to the Algebra** $AP_\sigma + C(\mathbb{T})$. For the proof of Theorem 72 we need only a small portion of the version of the local theory set forth – Corollary 79 and the special case b) of Theorem 84 (this special case incidentally is contained in Corollary 79 and follows also immediately from Theorem 76 itself: $\operatorname{dist}_\lambda(\psi\varphi, \psi(\lambda)\varphi) = 0$, $\lambda \in \mathbb{T}$. □) However, we have to add a few things to Theorem 72 in order to get the analogous result for the algebra $AP_\mathbb{T} + C(\mathbb{T})$; cf. 88. To this end we must first define the local "winding number" $\operatorname{wind}_\zeta \varphi$ for all functions $\varphi$ in $AP_\mathbb{T} + C(\mathbb{T})$, not only for functions with a finite number of singularities. Now, it follows from Lemma 70 that the mapping $\varphi \mapsto \varphi_\zeta$, assigning to the function $\varphi$ its "principal part" at the point $\zeta$, extends from $\bigcup_{\operatorname{card}\sigma<\infty} AP_\sigma + C(\mathbb{T})$ to the closure of that set. If $\varphi \in AP_\mathbb{T} + C(\mathbb{T})$ we therefore just have to set $\operatorname{wind}_\zeta \varphi = \operatorname{wind}_\zeta \varphi_\zeta$, $\zeta \in \mathbb{T}$.

**88. Theorem.** *Let* $\varphi \in AP_\mathbb{T} + C(\mathbb{T})$. *Then* $0 \notin \sigma_{\text{ess}}(T_\varphi) \Leftrightarrow \operatorname{ess\,inf}_\mathbb{T} |\varphi| > 0$; $\operatorname{wind}_\zeta \varphi = 0$, $\forall \zeta \in \mathbb{T}$.

*Proof of Theorem 88 and 72.* Being a uniform limit of functions in $AP_\sigma + C(\mathbb{T})$, $\operatorname{card}\sigma < \infty$, the function $\varphi$ is continuous everywhere except on a countable set. It is also clear that for every $\varepsilon > 0$ the set $\sigma \stackrel{\text{def}}{=} \{\zeta : \omega_\zeta(\varphi) > \varepsilon\}$ is finite and that in the hypothesis $\operatorname{ess\,inf}_\mathbb{T} |\varphi| > 0$ the function $\varphi/\varphi_\zeta$ is continuous at the point $\zeta$ (because this is true if $\varphi \in AP_\sigma + C(\mathbb{T})$, $\operatorname{card}\sigma < \infty$). It follows that $\varphi = \varphi_0 \prod_{\zeta\in\sigma} \varphi_\zeta$. Also $\varphi_0$ is for small $\varepsilon$ locally sectorial ($\omega_\lambda(\varphi_0) \le \varepsilon < \operatorname{ess\,inf}_\mathbb{T} |\varphi_0|$, because $|\varphi_0| \ge |\varphi|$ in the hypothesis $\|\varphi\|_\infty = 1$) and continuous at the points of the set $\sigma$. If we decrease $\varepsilon$ and apply 79, 80 and 64, we complete the proof of Theorem 88.

Moreover, as $\operatorname{wind}_\zeta \varphi_\zeta = 0$, $T_{\varphi_\zeta}$ must be an invertible operator (Theorem 64), so that $\operatorname{ind} T_\varphi = \operatorname{ind} T_{\varphi_0} = -\operatorname{wind} \varphi_0$. □

**89. Enigmatic Toeplitz Operators** arise when we permit symbols with a more complicated argument behavior than for the classes $PC$, $AP$, $H^\infty$. Such are the symbols in the algebras $H^\infty_{\bar\Theta}$, $\Theta$ an inner function. Recall that by definition $H^\infty_{\bar\Theta}$ is the least uniformly closed algebra containing $H^\infty$ and the function $\bar\Theta$ (this function gives in fact rise to a new type of argument behavior (in comparison with $H^\infty$), i.e. $H^\infty_{\bar\Theta} = \operatorname{clos} \bigcup_{n\ge0} \bar\Theta^n H^\infty$. (As a matter of fact, the set of inner functions $\vartheta$ such that $\bar\vartheta \in H^\infty_{\bar\Theta}$ generates a closed semigroup which along with any of its members likewise contains all the inner divisors of the latter.) If $\Theta = \text{const}$ then $H^\infty_{\bar\Theta} = H^\infty$ and if $\Theta$ is a finite Blaschke product then $H^\infty_{\bar\Theta} = H^\infty + C$; in any other algebra $H^\infty_{\bar\Theta}$ some of the features of the previous spectral theory of Toeplitz operators will be absent. We will see this by scrutinizing in some greater detail the problems of such a theory in the important (but from the general point of view simplest) case $\Theta = \exp \dfrac{z+\mathbb{1}}{z-\mathbb{1}}$. First let us mention the following simple fact.

**90. Lemma.** *If* $\Theta$ *is an inner function,* $\Theta \not\equiv \text{const}$, *then* $H^\infty + C \subset H^\infty_{\bar\Theta}$.

*Proof*. As $(\bar{\Theta} - \overline{\Theta(0)})(\mathbb{1} - \Theta(0)\bar{\Theta})^{-1} \in H^\infty_{\bar{\Theta}}$ and $\bar{z}(\Theta - \Theta(0))(\mathbb{1} - \overline{\Theta(0)}\Theta)^{-1} \in H^\infty$ we see that $\bar{z} \in H_{\bar{\Theta}}$. □

**91. Symbols of the Form $\Theta_1\bar{\Theta}_2$.** The question of invertibility on Fredholmness of the operator $T_\varphi$ with the simple symbol of type $\varphi = \bar{\Theta}h$, $h \in H^\infty$, reduces (in view of the necessary condition $\operatorname{ess\,inf}_{\mathbb{T}}|\varphi| > 0$, we can remove from it the invertible factor $T_{h^e}$) to the same question for $T_{\bar{\Theta}\Theta_1}$, $\Theta_1$ an inner function. There arise however here considerable difficulties. Let us recall some well-known facts:

a) the operator $T_{\Theta_1\bar{\Theta}_2}$ differs only by unitary factors (plus an unessential term) from the restriction $P_{\Theta_1}|K_{\Theta_2}: K_{\Theta_2} \to K_{\Theta_1}$, where $K_{\Theta_i} = H^2 \ominus \mathbf{\Theta}_i H^2$ is the model space; in an implicit form this statement is contained in Lecture VIII, Sect. 6 (here is a direct proof: let $\mathcal{J}f = \overline{zf}$, $f \in L^2(\mathbb{T})$; then $T_{\Theta_1\bar{\Theta}_2} = \mathcal{J}(P_-\bar{\mathbf{\Theta}}_1\mathbf{\Theta}_2|H^2_-)\mathcal{J}$, also $\mathcal{J}$ is a conjugate linear isometry mapping $H^2$ onto $H^2_-$ and $H^2_-$ onto $H^2$; moreover $\mathbf{\Theta}_1(P_-\bar{\mathbf{\Theta}}_1\mathbf{\Theta}_2|H^2_-) = (\mathbf{\Theta}_1 P_-\bar{\mathbf{\Theta}}_1|\mathbf{\Theta}_2 H^2_-)\mathbf{\Theta}_2 =$ (by definition of the spaces $K_{\Theta_i}$: $\mathbf{\Theta}_2 H^2_- = H^2_- \oplus K_{\Theta_2}$, $\mathbf{\Theta}_1 P_-\bar{\mathbf{\Theta}}_1 = I_{H^2} \oplus P_{\Theta_1}$) $(I_{H^2_-} \oplus P_{\Theta_1}|K_{\Theta_2})\mathbf{\Theta}_2$, the multiplication operators $\mathbf{\Theta}_1$, $\mathbf{\Theta}_2$ being isometric. □);

b) $0 \notin \sigma^l(T_{\Theta_1\bar{\Theta}_2}) \Leftrightarrow P_{\Theta_1}|K_{\Theta_2}$ is an isomorphic imbedding $\Leftrightarrow \operatorname{dist}(\Theta_1\bar{\Theta}_2, H^\infty) = |\Theta_1(\mathbf{T}_{\Theta_2})| < 1$ (here $\mathbf{T}_{\Theta_1} = P_{\Theta_1}\mathbf{z}|K_{\Theta_1}$ is the model operator); (Lecture VIII, Sect. 6);

c) $0 \notin \sigma(T_{\Theta_1\bar{\Theta}_2}) \Leftrightarrow P_{\Theta_1}|K_{\Theta_2}$ is an isomorphism $K_{\Theta_2} \to K_{\Theta_1} \Leftrightarrow K_{\Theta_1} + \mathbf{\Theta}_2 H^2 = H^2$, $K_{\Theta_2} + \mathbf{\Theta}_1 H^2 = H^2 \Leftrightarrow \operatorname{dist}(\Theta_1\bar{\Theta}_2, H^\infty) < 1$, $\mathbf{\Theta}_2 H^2 \cap K_{\Theta_1} = \{\mathbb{0}\}$; (Lecture VIII, Sect. 6); for other (general) conditions for invertibility see the present Appendix;

d) $\sigma^r(T_{\Theta_1\bar{\Theta}_2}) \supset \sigma_{H^\infty_{\Theta_1\bar{\Theta}_2}}(\Theta_1\bar{\Theta}_2)$ (cf. 59, 60);

e) if $0 \notin \sigma^r_{\text{ess}}(T_{\Theta_1\bar{\Theta}_2})$ then $\bar{\Theta}_1 \in H^\infty_{\bar{\Theta}_2}$; if $0 \notin \sigma_{\text{ess}}(T_{\Theta_1\bar{\Theta}_2})$ then $H^\infty_{\bar{\Theta}_1} = H^\infty_{\bar{\Theta}_2}$; (cf. 61).

**92. The Spectral Inclusion Theorem** for the operators $T_{\Theta_1\bar{\Theta}_2}$ of the type in question will be based on Douglas's Lemma (cf. 59) and is in an essential way connected with the values of the function $\Theta_1/\Theta_2$ on that part of the space of maximal ideals of the algebra $H^\infty$ on which $|\Theta_2| = 1$. Indeed, let $\varepsilon > 0$ and set $\Omega_\varepsilon = \{\zeta \in \mathbb{D} : |\Theta_2(\zeta)| > 1 - \varepsilon\}$, $\widehat{\operatorname{Range}}\,\Theta_1/\Theta_2 \overset{\text{def}}{=} \bigcap_{\varepsilon > 0} \operatorname{clos}[\Theta_1/\Theta_2(\Omega_\varepsilon)]$. It is clear that $\lambda \in \widehat{\operatorname{Range}}\,\Theta_1/\Theta_2 \Leftrightarrow \lambda = \lim_n \Theta_1(\zeta_n)/\Theta_2(\zeta_n)$ for some sequence $\{\zeta_n\}_{n \geq 1}$, $|\zeta_n| < 1$, such that $\lim_n |\Theta_2(\zeta_n)| = 1$.

**93. Lemma.** 1. $\widehat{\operatorname{Range}}\,\Theta_1/\Theta_2 \subset \sigma^r_{\text{ess}}(T_{\Theta_1\bar{\Theta}_2})$, $\overline{\widehat{\operatorname{Range}}\,\Theta_2/\Theta_1 \cup \widehat{\operatorname{Range}}\,\Theta_1/\Theta_2} \subset \sigma_{\text{ess}}(T_{\Theta_1\bar{\Theta}_2})$.

2. *If* $\sigma(\Theta_1) \cap \mathbb{T} \neq \sigma(\Theta_2) \cap \mathbb{T}$ *then* $\sigma_{\text{ess}}(T_{\Theta_1\bar{\Theta}_2}) = \operatorname{clos}\mathbb{D}$. *Here* $\sigma(\Theta_i)$ *is the spectrum of the inner function* $\Theta_i$ (cf. Lecture III).

*Proof*. 1. Let $\lambda \in \widehat{\operatorname{Range}}\,\Theta_1/\Theta_2$ and pick $\zeta_n \in \mathbb{D}$ with $\lim_n|\Theta_2(\zeta_n)| = 1$, $\lim_n \Theta_1(\zeta_n)/\Theta_2(\zeta_n) = \lambda$. If $\lambda \in \operatorname{Range}\Theta_1\bar{\Theta}_2$ then by Theorem 30 $\lambda \in \sigma^r_{\text{ess}}(T_{\Theta_1\bar{\Theta}_2})$. If $\lambda \notin \operatorname{Range}\Theta_1\bar{\Theta}_2$ let us define an outer function $h$ by the identity $|h| = |\Theta_1 - \lambda\Theta_2|$. Then $T_h$ is invertible and

$$T_{\Theta_1\bar{\Theta}_2} - \lambda I = T_u T_h,$$

where $u = (\Theta_1 - \lambda\Theta_2)h^{-1}\bar{\Theta}_2$. By Corollary 61 (cf. e) in 91) it is now sufficient to show that $h(\Theta_1 - \lambda\Theta_2)^{-1} \notin H^\infty_{\bar{\Theta}_2}$. Let $g \in H^\infty$ and $n \geq 0$. Then

$$\|h(\Theta_1 - \lambda\Theta_2)^{-1} - \bar{\Theta}_2^n g\|_\infty = \|\Theta_2^n - h^{-1}(\Theta_1 - \lambda\Theta_2)g\|_\infty$$
$$\geq \lim_k |\Theta_2(\zeta_k)^n - h(\zeta_k)^{-1}[\Theta_1(\zeta_k) - \lambda\Theta_2(\zeta_k)]g(\zeta_k)| = 1.$$

Thus $0 \in \sigma_{\text{ess}}^r(T_u)$ and finally $\lambda \in \sigma_{\text{ess}}^r(T_{\Theta_1\bar{\Theta}_2})$.

2. Let $\zeta \in \sigma(\Theta_1)\backslash\sigma(\Theta_2)$, $|\zeta| = 1$. If the inner function $\vartheta \stackrel{\text{def}}{=} (\Theta_1 - \lambda\Theta_2(\zeta))$ $(\mathbb{1} - \overline{\lambda\Theta_2(\zeta)}\Theta_1)^{-1}$ admits analytic continuation across the point $\zeta$ then this is true for $\Theta_1$ too, but this is not the case (Theorem on the Spectrum of a Function, Lecture III). By this theorem $\lim_n \vartheta(\zeta_n) = 0$ for some sequence of points $\{\zeta_n\}_{n\geq 1}$, $|\zeta_n| < 1$, converging to $\zeta$. Hence $\lambda\Theta_2(\zeta) = \lim_n \Theta_1(\zeta_n)$ and $\lambda \in \hat{\text{R}}\text{ange}\,\Theta_1/\Theta_2$. Consequently $\sigma_{\text{ess}}^r(T_{\Theta_1\bar{\Theta}_2}) = \operatorname{clos}\mathbb{D}$.

In case $\mathbb{T} \cap (\sigma(\Theta_2)\backslash\sigma(\Theta_1)) \neq \varnothing$ we get in exactly the same way $\sigma_{\text{ess}}^l(T_{\Theta_1\bar{\Theta}_2}) = \operatorname{clos}\mathbb{D}$. □

**94. The Function $\Theta_2$ is a Blaschke Product.** This special case is the principal one: a) if we approximate the inner function $\Theta_2$ by Blaschke product $B$ and if we know the spectral properties of the operators $T_{\Theta_1\bar{B}}$ we can get information on $T_{\Theta_1\bar{\Theta}_2}$ too; for instance $0 \in \sigma(T_{\Theta_1\bar{B}_n})$, $\|B_n - \Theta_2\|_\infty \to 0 \Rightarrow 0 \in \sigma(T_{\Theta_1\bar{\Theta}_2})$; b) if $\sigma(\Theta_1) = \{\lambda\}$, $|\lambda| = 1$, then *without loss of generality* we can take $\Theta_2$ to be a Blaschke product (because for a meaningful spectral analysis it is necessary that $\sigma(\Theta_2) \cap \mathbb{T} = \{\lambda\}$ (Lemma 93) and then, disregarding the trivial case $\bar{\Theta}_1\Theta_2 = H^\infty$, we have $\Theta_1\bar{\Theta}_2 = \exp\left(-a\dfrac{\lambda+z}{\lambda-z}\right)\cdot\bar{B}$, with $a > 0$, $B$ being a Blaschke product); c) in this case in studying the operator $T_{\Theta_1\bar{B}}$ we can use the geometry of the reproducing kernel of the model space $K_{\Theta_1}$ and in particular if $\Theta_1 = \exp\left(-\dfrac{1+z}{1-z}\right)$ the question is reduced to the classical problems for families of exponentials $\{e^{i\lambda_n x}\}_{n\geq 1}$ in $L^2(0, 1)$.

Thus *everywhere in what follows* we take $\Theta_2 = B$, a Blaschke product, and $\Theta_1 = \Theta$, an inner function. For simplicity's sake the zeros of $B$ are assumed to be *simple*, $B = \prod_{\lambda\in\sigma} b_\lambda$, where $\sigma \subset \mathbb{D}$, $\prod_{\lambda\in\sigma}(1 - |\lambda|) < \infty$, $b_\lambda = |\lambda|\lambda^{-1}(\lambda - z) \times (\mathbb{1} - \bar{\lambda}z)^{-1}$. By the symbol $k_\Theta$ we denote the *reproducing kernel* of the space $K_\Theta = H^2 \ominus \Theta H^2 : k_\Theta(\cdot, \lambda) = P_\Theta(\mathbb{1} - \bar{\lambda}z)^{-1} = (\mathbb{1} - \overline{\Theta(\lambda)}\Theta)(\mathbb{1} - \bar{\lambda}z)^{-1}$, $\lambda \in \mathbb{D}$.

**95. Lemma.** *Let $B = \prod_{\lambda\in\sigma} b_\lambda$, $\sigma \subset \mathbb{D}$, and let $\Theta$ be an inner function, $K_\Theta = H^2 \ominus \Theta H^2$. The following statements are equivalent.*

1. $0 \notin \sigma^l(T_{\Theta\bar{B}})$.
2. $H^2|\sigma = K_\Theta|\sigma$. (21)
3. $\inf\{\|P_\Theta f\| \cdot \|f\|^{-1} : f \in K_B\} > 0$. (22)

*Proof*. Clearly $1 \Rightarrow 3$. In view of Statement a) in 91 $T_{\bar{\Theta}B}H^2 = H^2 \Leftrightarrow P_B K_\Theta = K_B$ and, the restriction relation $f|\sigma = g|\sigma$ being equivalent to the identity $P_B f = P_B g$, it is clear that $1 \Leftrightarrow 2$. □

**96. Corollary.** *If $0 \notin \sigma^l(T_{\Theta\bar{B}})$ then*

a) $\sup\{|\Theta(\lambda)| : \lambda \in \sigma\} < 1$; (23)

b) *the family*

$$R_\sigma \stackrel{\text{def}}{=} \{k_\Theta(\cdot, \lambda) : \lambda \in \sigma\}$$

*is* minimal, *i.e.*

$$\operatorname{dist}(k_\Theta(\cdot, \lambda), \bigvee(k_\Theta(\cdot, \mu) : \mu \in \sigma \setminus \{\lambda\})) > 0, \quad \forall \lambda \in \sigma. \tag{24}$$

In fact, Statement a) follows in a trivial manner from Statement 3 of Lemma 95 (take $f = (\mathbb{1} - \bar{\lambda} z)^{-1}$, $\lambda \in \sigma$) and Statement b) from the same Statement 3 together with the minimality of the family $\{(\mathbb{1} - \bar{\lambda} z)^{-1} : \lambda \in \sigma\}$, cf. Lecture VI, Sect. 2. □

**97. Lemma.** dim Ker $T_{\bar{\Theta} B}$ = codim $\bigvee(R_\sigma)$. *In particular* $\bigvee(R_\sigma) = K_\Theta$ *if* $T_{\bar{\Theta} B}$ *is injective.*

*Proof*. dim Ker $T_{\bar{\Theta} B}$ = codim clos $T_{\Theta \bar{B}} H^2$ = (cf. a) in 91) codim $P_\Theta K_B$ = codim $\bigvee(R_\sigma)$. □

**98. Discussion.** It is clear that Sects. 93, 95–97 are very close indeed to the contents of Sects. 6, 7 of Lecture VIII and Sects. 2, 3 of Lecture XI. In particular, condition (23) played decisive rôle in the study of bases of reproducing kernels. As in the problem of bases in Lectures VIII and IX the (left, right) invertibility problem for the operator $T_{\Theta \bar{B}}$ in the special case $\Theta = \Theta_a \stackrel{\text{def}}{=} \exp\left(-a \frac{\mathbb{1} + z}{\mathbb{1} - z}\right)$, $a > 0$, is reduced to problems on the geometric properties of the exponentials $\mathscr{E}_\tau = \{e^{i\mu x} : \mu \in \tau\}$,

$$\tau = -\overline{\omega(\sigma)}, \quad \omega \stackrel{\text{def}}{=} i(\mathbb{1} + z)(\mathbb{1} - z)^{-1} \tag{25}$$

in the space $L^2(0, a)$ (concerning the details of this *unitary* transfer see Sect 2, Lecture XI). In particular, the identity (21) can be converted into the following statement on interpolation with entire functions, similar to Statement 2′ on p. 262 (line 5 from the bottom):

*for any function f in the Hardy space* $H^2_\Pi$, $\Pi = \{\zeta \in \mathbb{C} : \operatorname{Im} \zeta > 0\}$, *there exists an entire function g of exponential type not exceeding a/2 and square integrable on the axis* $\mathbb{R}$ *such that*

$$f(\mu) = e^{-i(a/2)\mu} g(\mu), \quad \mu \in \tau.$$

The estimate (22) leads to a result likewise mentioned in Lecture XI (Sect. 3) (the description of the sets $\tau$, $\tau \subset \Pi$ for which this statement is true is the problem of Laurent Schwartz):

*there exists a number c, c > 0, such that*

$$\int_0^\infty \left| \sum_{\mu \in \tau} c_\mu e^{i\mu x} \right|^2 dx \le c \int_0^a \left| \sum_{\mu \in \tau} c_\mu e^{i\mu x} \right|^2 dx$$

*for any finite family of numbers* $\{c_\mu\}_{\mu \in \tau}$.

Both these statements are thus equivalent to the left invertibility of the Toeplitz operator $T_{\Theta_a \bar{B}}$, $B = \prod_{\lambda \in \sigma} b_\lambda$ (the sets $\sigma$ and $\tau$ being connected by the conformal mapping (25)).

All this concerns however only the statement of the problem. How useful these reformulations are is illustrated by the Ingham-Kadec Theorem, partially proved in Lecture XI (in its most essential part). Regarding implication $1 \Rightarrow 2$, cf. Levinson [1], Hruščev-Nikol'skiĭ-Pavlov [1].

**99. Ingham-Kadec Theorem.** *Let* $\alpha$, $\delta > 0$ *and consider the class* $\mathscr{B}_\delta$ *of all Blaschke products* $B = \prod_{\lambda\in\sigma} b_\lambda$ *whose zeros are the inverse images of sets of the type* $\tau = \{n + \delta_n + i\alpha : n\in\mathbb{Z}\}$, $\sup_n|\delta_n| \le \delta$, *under the standard conformal mapping* ($\sigma = \omega^{-1}(-\bar\tau)$, (*cf.* (25)). *Then the following statements are equivalent.*

1. *If* $B\in\mathscr{B}_\delta$ *then* $T_{\Theta_{2\pi}\bar B}$ *is invertible.*
2. $\delta < 1/4$.

**100. The Latest Developments** so far (Jan. 82) are connected with auxiliary assumptions on the geometry of the space $K_B = \bigvee((\mathbb{1} - \bar\lambda z)^{-1} : \lambda\in\sigma)$. The principal of the latter is that the family $\{(\mathbb{1} - \bar\lambda z)^{-1} : \lambda\in\sigma\}$ should be not only minimal (this is always true) but in fact constitute a *basis* for $K_B$ (by necessity an unconditional one, cf. Lectures VI, VII). This requirement is, as we know (cf. Lectures VI–VIII), equivalent to the Carleson condition $\sigma\in(C)$. The rôle of this restriction in the language of the operator $T_{\Theta\bar B}$ is explained in Lemma 101 and the following comments.

**101. Lemma.** *Let* $\Theta$ *be an inner function,* $\sigma\subset\mathbb{D}$ *and* $B = \prod_{\lambda\in\sigma} b_\lambda$, *and assume that condition* (23) *is fulfilled.*

1. *If* $\sigma\in(C)$ *and* $0\notin\sigma(T_{\Theta\bar B})$ *then the family* $R_\sigma$ *generates an unconditional basis in the space* $K_\Theta$.

2. *If* $R_\sigma$ *is an unconditional basis in* $K_\Theta$ *then* $\sigma\in(C)$ *and* $0\notin\sigma(T_{\Theta\bar B})$.

This lemma is proved in Lecture VIII. Here we add two further explanatory remarks. Firstly: if we wish to obtain a criterion for the invertibility of the operator $T_{\Theta\bar B}$ we have to keep in mind the *necessary* condition (21). Under the additional assumption $\sigma\in(C)$ (21) becomes stronger and means that free interpolation is possible ($\forall f\in K_\Theta, \forall a_\lambda : |a_\lambda| \le |f(\lambda)|, \lambda\in\sigma \Rightarrow \exists g\in K_\Theta : g(\lambda) = a_\lambda, \lambda\in\sigma$; cf. Lectures VI, VIII). Secondly: likewise condition (24), being necessary for the invertibility of $T_{\Theta\bar B}$, becomes stronger under the assumption $\sigma\in(C)$. In this case the family $R_\sigma$ is minimal iff it is *uniformly minimal*, i.e.

$$\inf_{\lambda\in\sigma} \operatorname{dist}(k_\Theta(\cdot,\lambda)/\|k_\Theta(\cdot,\lambda)\|, \bigvee(k_\Theta(\cdot,\mu) : \mu\in\sigma\backslash\{\lambda\})) > 0,$$

cf. now Lecture VIII, Sect. 6.

In the special case $\Theta = \Theta_a \overset{\text{def}}{=} \exp\left(-a\dfrac{\mathbb{1}+z}{\mathbb{1}-z}\right)$ we have the following result concerning the invertibility of $T_{\Theta\bar B}$, expressed in terms of the *density* with which the zeros of the Blaschke product $B$ accumulate at the point 1 ($\{1\} = \sigma(\Theta_a)$). It is convenient to describe this density in terms pertaining to the halfplane $\Pi : \tau = \{\mu\}$, $\mu = -\overline{\omega(\lambda)}$ $(\lambda\in\sigma)$; $L^\infty = L^\infty(\mathbb{R})$; $\mathrm{BMO} = \{u + \tilde v : u, v\in L^\infty\}$ etc.

**102. Hruščev's Theorem.** *Let* $\Theta = \Theta_a$, $B = \prod_{\lambda\in\sigma} b_\lambda$ *and assume that condition* (23) *is fulfilled* (*i.e. that* $\inf\{|\operatorname{Im}\mu| : \mu\in\tau\} > 0$). *Assume further that* $\sigma\in(C)$. *The following statements are then equivalent.*

1. *The operator $T_{\Theta_a \bar{B}}$ is invertible.*

2. $\operatorname{dist}_{L^\infty}(\alpha_\tau, \widetilde{L^\infty} + \mathbb{R}) < \pi/2$ *where* $\alpha_\tau(x) = 2\int_0^x \sum_{\mu\in\tau} \operatorname{Im}\mu/|\mu - t|^2\,dt - ax$, $x \in \mathbb{R}$.

*If, in addition,* $\inf\{|\operatorname{Re}\mu - \operatorname{Re}\mu'| : \mu \neq \mu'; \mu, \mu' \in \tau\} > 0$ *and*

$$\sup\{|\operatorname{Im}\mu| : \mu\in\tau\} < \infty, \tag{26}$$

*then these statements are equivalent to Statement 3.*

3. *The function* $f \stackrel{\text{def}}{=} 2\pi N_\tau - ax$ *belongs to* BMO *and its harmonic continuation* $\mathscr{H}f$ *to the halfplane* $\Pi$ *has for some* $y > 0$ *the representation*

$$\mathscr{H}f(x+iy) = u(x) + \tilde{v}(x) + c; \quad u, v \in L^\infty, \quad \|v\|_\infty < \pi/2;$$

*here* $N_\tau$ *is the counting function for the projection* $\operatorname{Re}\tau$:

$$N_\tau(x) \stackrel{\text{def}}{=} \begin{cases} \operatorname{card}([0,x]\cap \operatorname{Re}\tau) & x \geq 0 \\ -\operatorname{card}([x,0)\cap\operatorname{Re}\tau), & x < 0. \end{cases}$$

This theorem is proved in the paper Nikol'skiĭ-Pavlov-Hruščev [1]. Condition (26) together with (23) means that we deal with a Blaschke product whose null set lies in a domain of the type

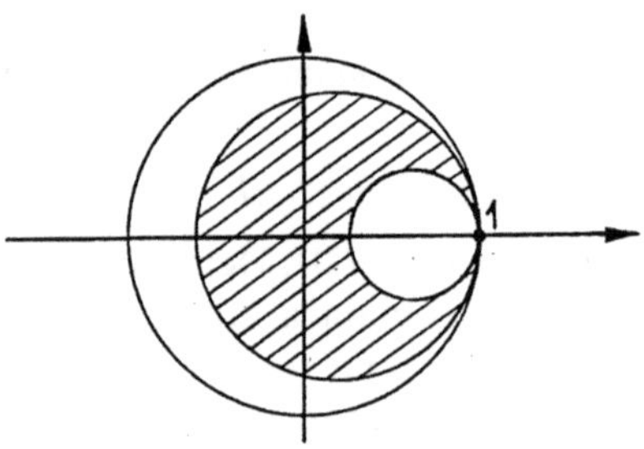


Fig. 7

**103. Algebras Generated by Toeplitz Operators.** Such algebras consist of operators of the type $T = \sum_i \sum_j T_{\varphi_{ij}}$, $\varphi_{ij} \in L^\infty(\mathbb{T})$. It is natural to try to extend to this more general class the methods and results of the theory already set forth, especially as simple combinations of Toeplitz operators play a major rôle in the solution of various boundary problems in Function Theory. The study of these algebras sheds also new light on the structure of the Toeplitz operators themselves, for instance, on those of their properties which are connected with commutation or "approximate commutation" of operators $T_\varphi$ with different $\varphi$. The rule is that such operators do not commute. However, the commutator

$$[T_\varphi, T_\psi] \stackrel{\text{def}}{=} T_\varphi T_\psi - T_\psi T_\varphi$$

may turn out to be "small" in an appropriate sense and, neglecting such small terms, one can construct a theory which retains part of the information on the spectral properties of the Toeplitz operators involved.

**104. Abelianization.** If $X \subset L^\infty(\mathbb{T})$ then $T_X \overset{\text{def}}{=} \{T_x : x \in X\}$ and alg $T_X$ stands for the least (norm) closed operator algebra containing the set $T_X$. We denote by Com $T_X$ the *commutator ideal* of alg $T_X$, i.e. the least two-sided closed ideal in alg $T_X$ containing all commutators $[B, C]$, $B, C \in \text{alg}\, T_X$. It is clear that the factor algebra $\hat{X}$,

$$\hat{X} \overset{\text{def}}{=} \text{alg}\, T_X/\text{Com}\, T_X,$$

is commutative. It is likewise not hard to see that if alg $T_X$ is generated (as an algebra) by a set of operators $T_{X_0}$, $X_0 \subset X$, then Com $T_X$ is generated (as an ideal in alg $T_X$) by the commutators $[T_x, T_y]$, $x, y \in X_0$.

**105. The Plan of the Theory** is now, starting with a given *subalgebra* $X$ of $L^\infty(\mathbb{T})$, (below we consider only subalgebras containing the function $\mathbb{1}$; we will not state this explicitely) to try to take advantage of the maximal ideals of the algebra $\hat{X}$ and, by the same token, to obtain criteria for the invertibility of an operator in alg $T_X$ "modulo" the ideal Com $T_X$. Then one hopes that the space of maximal ideals $\mathcal{M}(\hat{X})$ can be computed effectively, that the ideal Com $T_X$ can be put in relation to the spectral theory in the space $H^2$ and that invertibility in alg $T_X$ can be connected with invertibility in $L(H^2)$. All the desiderata listed occur together rather rarely but the analysis of each of the links of this chain leads in concrete cases to interesting problems and opens up new vistas. It is equally important that a closely allied algebraic approach is possible also to the "localization technique" (cf. 157–163) and further to Toeplitz operators with vector or operator symbols (cf. Douglas [3]) and that this again directly leads to the so-called "operator $K$-theory".

**106. Symbolic Calculus.** The natural imbedding $j$,

$$j : L^\infty \to \text{alg}\, T_{L^\infty}; \quad jf = T_f \quad (f \in L^\infty),$$

is an isometry (cf. 26). It turns out that it admits a left inverse and that among all these inverses there exists a (unique) multiplicative map Sym, called the symbol map. This fact will be established in Lemma 107 and then we verify that Ker Sym $=$ Com $T_{L^\infty}$. For the restriction $\text{Sym}_X = \text{Sym}|\text{alg}\, T_X$, which – in the case when $X$ is a subalgebra of $L^\infty$ – is a left inverse of the restriction $j_X = j|X : X \to \text{alg}\, T_X$, the more precise relation Ker $\text{Sym}_X = \text{Com}\, T_X$ is by no means always true (cf. 109, 122); as a trivial consequence of the definition we have only Ker $\text{Sym}_X =$ alg $T_X \cap$ Ker Sym (and so, by what we just said, Ker $\text{Sym}_X = \text{alg}\, T_X \cap$ Com $T_{L^\infty}$). This circumstance constitutes a first obstacle towards an identification of the algebras $X$ and $\hat{X} = \text{alg}\, T_X/\text{Com}\, T_X$.

The following lemma, apart from being part of this program, contains also a more general fact: an estimate for the distance of an operator of the form $\sum_i \prod_j T_{\varphi_{ij}}$ from the set $\mathfrak{S}^0$,

$$\mathfrak{S}^0 \overset{\text{def}}{=} \{C \in L(H^2) : \lim_n \|Cz^n f\| = 0, \forall f \in H^2\}.$$

Clearly $\mathfrak{S}^0$ is a closed left ideal in the operator algebra $L(H^2)$ containing the ideal of compact operators $\mathfrak{S}_\infty$. Below we likewise show that all commutators of Toeplitz operators lie in $\mathfrak{S}^0$; these include notably the projections $P_\Theta$ onto coinvariant subspaces $K_\Theta = H^2 \ominus \Theta H^2$, $\Theta$ an inner function (the inclusion $P_\Theta \in \mathfrak{S}^0$ is part of the Calculus Lemma, Lecture III, but is also contained in Lemma 107, since $P_\Theta = [T_{\bar{\Theta}}, T_\Theta]$).

**107. Lemma.** *Let $\varphi_{ij}$ be functions in $L^\infty$ (a finite family) and set $T = \sum_i \prod_j \varphi_{ij}$. Then*

1. $\sup\{\lim_n \| Tz^n f \| : f \in H^2, \| f \| \le 1\} = \|\sum_i \prod_j \varphi_{ij}\|_\infty$.
2. $|T| \ge \operatorname{dist}(T, \mathfrak{S}^0) \ge \|\sum_i \prod_j \varphi_{ij}\|_\infty$.
3. $\operatorname{Com} T_{L^\infty} \subset \mathfrak{S}^0$, $T_{L^\infty} \cap \mathfrak{S}^0 = \{\mathbb{O}\}$.
4. *The map* Sym,

$$\operatorname{Sym} : \sum_i \prod_j T_{\varphi_{ij}} \mapsto \sum_i \prod_j \varphi_{ij},$$

*extends by continuity to an algebra homomorphism* $\operatorname{alg} T_{L^\infty} \to L^\infty$. *Moreover, this map induces a homomorphism (a contraction)* $\widehat{\operatorname{Sym}} : T + \operatorname{Com} T_{L^\infty} \mapsto \operatorname{Sym} T$ *of the quotient algebra* $\operatorname{alg} T_{L^\infty}/\operatorname{Com} T_{L^\infty}$ *into* $L^\infty$ *which is a left inverse of the imbedding* $\hat{j} \overset{\text{def}}{=} \pi j$ *($\pi$ being the natural projection* $\operatorname{alg} T_{L^\infty} \to \operatorname{alg} T_{L^\infty}/\operatorname{Com} T_{L^\infty}$): $\widehat{\operatorname{Sym}}(\hat{j}\varphi) = \varphi$, $\forall \varphi \in L^\infty$.

*Proof*. Statement 1 is in essence contained in 26: by formula (13) $\lim_n \| T_\varphi z^n g - z^n \varphi g \| = 0$, $\forall g \in H^2$, $\forall \varphi \in L^\infty$, whence $\lim_n \| \prod_j T_{\varphi_{ij}} z^n f - (\prod_j \varphi_{ij}) z^n f \| = 0$ (for any $i$) and $\lim_n \| Tz^n f \| = \|(\sum_i \prod_j \varphi_{ij}) f \|$. This gives the desired conclusion.

The Statements 2 and 3 follow from 1 and the definition of the ideals $\mathfrak{S}^0$ and $\operatorname{Com} T_{L^\infty}$.

Statement 4 follows from 2 and 3. □

**108. Corollary.** *Let $X$ be a subalgebra of $L^\infty$, and write $j_X = j|X$. Then we have the following commutative diagram:*

$$\begin{array}{ccc} X & \xrightarrow{\;j_X\;} & \operatorname{alg} T_X \\ {\scriptstyle I_X}\downarrow & & \downarrow{\scriptstyle \pi_X} \\ X & \xleftarrow[\widehat{\operatorname{Sym}}_X]{} & \hat{X} = \operatorname{alg} T_X/\operatorname{Com} T_X \end{array} \tag{27}$$

*where* $\widehat{\operatorname{Sym}}_X(T + \operatorname{Com} T_X) \overset{\text{def}}{=} \operatorname{Sym} T$, $\hat{j}_X \overset{\text{def}}{=} \pi_X j_X$ *being an isometric imbedding and* $\widehat{\operatorname{Sym}}_X$ *a contractive homomorphism "onto"; $I_X$ denotes the identity map on $X$.*

In fact, the map $\widehat{\operatorname{Sym}}_X$ was gotten by factoring the restriction $\operatorname{Sym}_X = \operatorname{Sym}|\operatorname{alg} T_X$ by the ideal $\operatorname{Com} T_X$ ($\subset \operatorname{Com} T_{L^\infty} \subset \mathfrak{S}^0$) and is therefore multiplicative and contractive. From the identity $\widehat{\operatorname{Sym}}_X \hat{j}_X = I_X$ it follows that the maps $\hat{j}_X$ and $\widehat{\operatorname{Sym}}_X | \hat{j} X$ are isometric. □

**109. Lemma** *Let $X$ be a subalgebra of $L^\infty$ and consider the maps $\hat{j}_X = \pi_X j_X$ and* $\operatorname{Sym}_X$ *defined by the diagram* (27). *The following statements are equivalent.*

1. $\hat{j}_X X = \operatorname{alg} T_X/\operatorname{Com} T_X$.
2. $\widehat{\operatorname{Sym}}_X$ *is an isometry.*
3. $\operatorname{Ker} \widehat{\operatorname{Sym}}_X = \{\mathbb{O}\}$.

4. $\operatorname{Com} T_X = \operatorname{Ker} \operatorname{Sym}_X$.
5. $\operatorname{alg} T_X = T_X + \operatorname{Com} T_X$.
6. $f, g \in X \Rightarrow T_f T_g - T_{fg} \in \operatorname{Com} T_X$.
7. $\hat{j}_X$ *is a homomorphism.*

*Proof.* $1 \Rightarrow 2$ since $\widehat{\operatorname{Sym}}_X \cdot \hat{j}_X = I_X$ and $\hat{j}_X$ is an isometry.

$2 \Rightarrow 3 \Rightarrow 4$. Trivial.

$4 \Rightarrow 5$. If $T \in \operatorname{alg} T_X$ then $T - T_{\operatorname{Sym} T} \in \operatorname{Com} T_X$ by 4. Note that the *sum* in Statement 5 is *direct*, because $T_X \cap \operatorname{Com} T_X = \{\mathbb{O}\}$ (cf. 107).

$5 \Rightarrow 6$. As $\operatorname{Sym}(T_f T_g - T_{fg}) = \mathbb{O}$ in the expansion $T_f T_g - T_{fg} = T_\varphi + C$ ($C \in \operatorname{Com} T_X$) we must by necessity have $\varphi = \mathbb{O}$.

$6 \Rightarrow 7$. Trivial.

$7 \Rightarrow 1$. Since $\hat{j}_X$ is an isometry and the algebra $\operatorname{alg} T_X$ is generated by the image $jX$, the closed *subalgebra* $\hat{j}_X X$ must coincide with $\operatorname{alg} T_X / \operatorname{Com} T_X$. □

**110. Corollary.** *The maps* $\hat{j}_X$ *and* (*or*) $\widehat{\operatorname{Sym}}_X$ *give rise to an* (*isometric*) *isomorphism between the algebras* $X$ *and* $\hat{X} = \operatorname{alg} T_X / \operatorname{Com} T_X$ *iff one* (*all*) *of the Statements 1–7 of Lemma 109 is* (*are*) *true.* □

**111. Remarks.** Let $\mathfrak{S}$ be a twosided closed ideal in the algebra $\operatorname{alg} T_X$, $X \subset L^\infty$, with $\mathfrak{S} \subset \mathfrak{S}^0$. Then it is clear that the map $\widehat{\operatorname{Sym}}_{\mathfrak{S}} : (T + \mathfrak{S}) \mapsto \operatorname{Sym} T$, $T \in \operatorname{alg} T_X$, is an algebra homomorphism $\operatorname{alg} T_X/\mathfrak{S} \to X$. The results 108–110 remain in force if we there replace the ideal $\operatorname{Com} T_X$ by the arbitrary ideal $\mathfrak{S}$ (the projection $\pi_X$ has to be replaced by $\pi_{\mathfrak{S}} : \operatorname{alg} T_X \to \operatorname{alg} T_X/\mathfrak{S}$; $\hat{j}_X$ by $\hat{j}_{\mathfrak{S}} = \pi_{\mathfrak{S}} j_X$; $\widehat{\operatorname{Sym}}_X$ by $\widehat{\operatorname{Sym}}_{\mathfrak{S}}$). Such analogues of propositions 109 and 110 are of interest only if $\operatorname{Com} T_X \subset \mathfrak{S}$ (since $X$ is commutative).

**112. Closer to the Spectrum.** Thus in order to compute the commutator ideal one needs a description of the commutators $T_f T_g - T_g T_f$ ($f, g \in X$) and in order to identify the algebras $\operatorname{alg} T_X/\operatorname{Com} T_X$ and $X$ of the operators $T_{fg} - T_f T_g$ ($f, g \in X$), the so-called "semi-commutators". Among "spectral theories modulo an (some) ideal $\mathfrak{S}$" it is the case $\mathfrak{S} = \mathfrak{S}_\infty$ that bears the closest relation to the ordinary spectral theory. Therefore in what follows we wish to concentrate on algebras $X$ such that $\operatorname{Com} T_X = \mathfrak{S}_\infty$. The basic results in this case are contained in 121, 123, 126, 129–141.

First we shall however study the question of the connection between the spectra $\sigma(T)$, $\sigma_{\operatorname{alg} T_X}(T)$, $\sigma_{\hat{X}}(T) \stackrel{\text{def}}{=} \sigma_{\hat{X}}(\pi_X T)$, $\sigma_X(\operatorname{Sym} T)$ for operators in the algebra $\operatorname{alg} T_X$, $X \subset L^\infty$. Taking account of the multiplicativity of the maps $\pi_X$ and $\widehat{\operatorname{Sym}}_X$ (cf. 27)) it is not hard to see that always

$$\begin{aligned} &\sigma(T) \cup \sigma_{\hat{X}}(T) \subset \sigma_{\operatorname{alg} T_X}(T), \\ &\sigma_X(\operatorname{Sym} T) \subset \sigma_{\hat{X}}(T). \end{aligned} \tag{28}$$

Some other inclusions will hold true under auxiliary assumptions. For instance, if $X \cong \hat{X}$ (cf. 109 and 110) then

$$\sigma_X(\operatorname{Sym} T) = \sigma_{\hat{X}}(T).$$

A major problem is the identification of the spectra $\sigma(T)$ (the usual spectrum) and

$\sigma_{\mathrm{alg}\,T_X}(T)$; this will be done in 113, 118 in two cases: when $X$ is a *symmetric* algebra (i.e. $f \in X \Rightarrow \bar{f} \in X$) and $T \in \mathrm{alg}\, T_X$, and when $X$ is not necessarily symmetric but $X \supset H^\infty$ or $X \supset \bar{H}^\infty$ and $T \doteq T_\varphi$, $\varphi \in X$. Whenever possible we consider in this connection also the essential spectrum $\sigma_{\mathrm{ess}}(T)$.

**113. Theorem.** *Let $X$ be a symmetric subalgebra of $L^\infty$, $T \in \mathrm{alg}\, T_X$.*

1. $\sigma(T) = \sigma_{\mathrm{alg}\,T_X}(T)$.
2. $\sigma_{\mathrm{ess}}(T) \supset \sigma_X(\mathrm{Sym}\, T)$.

First let us remark that from Statement 1 and the inclusions (28) it follows that $\sigma(T) \supset \sigma_X(\mathrm{Sym}\, T)$ so that the second statement of the theorem constitutes a refinement of the first one. Sometimes the inclusion in Statement 2 admits a converse; for instance if $X \cong \hat{X}$, $\mathrm{Com}\, T_X = \mathfrak{S}_\infty$, then $\sigma_X(\mathrm{Sym}\, T) \supset \sigma_{\mathrm{ess}}(T)$, regardless of whether $X$ is symmetric or not. Illustrations can be found in 123.3, 126.3, 139.

The proof of the theorem will be based on two lemmata on $C^*$-algebras. Let us recall that a *$C^*$-algebra* is a symmetric ($T \in \mathfrak{A} \Rightarrow T^* \in \mathfrak{A}$) and (norm) closed algebra of operators in Hilbert space containing the identity operator $I$. The same appellation is also used for (abstract) Banach algebras with involution which are $*$-isometric to a $C^*$-algebra.

**114. Lemma.** *Let $H$ be a Hilbert space and $\mathfrak{A}$ a $C^*$-algebra of operators in $H$. Then*

1. *$\mathfrak{A}$ is (as a vector space) generated by positive operators.*
2. *If $T \in \mathfrak{A}$ then $\sigma_{\mathfrak{A}}(T) = \sigma(T)$.*
3. *If $\mathfrak{S}$ is a two sided $*$-ideal in $\mathfrak{A}$ then $\mathfrak{A}/\mathfrak{S}$ is a $C^*$-algebra. In particular, this is the case for the Calkin algebra $L(H)/\mathfrak{S}_\infty$.*
4. *If $\mathfrak{B}$ is another $C^*$-algebra and $\alpha : \mathfrak{A} \to \mathfrak{B}$ a (contractive) $*$-homomorphism then the quotient map $\hat{\alpha} : \mathfrak{A}/\mathrm{Ker}\, \alpha \to \mathfrak{B}$ is an isometry (and, in particular, its image $\alpha(\mathfrak{A})$ is closed in $\mathfrak{B}$).*

*Proof*. All four statements of the lemma are classical attributes of the theory of $C^*$-algebras. The Statements 1, 2, 4 are in essence due to Gel'fand and Naĭmark and are rather simple to prove (the proofs are displayed below), whereas 3 constitutes Kaplansky's theorem (in the special case due to Calkin) the proof of which (and in general regarding information on $C^*$-algebras) one can look for in the books Naĭmark [1], Douglas [2], Dixmier [2].

1. If $T \in \mathfrak{A}$ then the self-adjoint operators $\mathrm{Re}\, T = (T + T^*)/2$, $\mathrm{Im}\, T = (T - T^*)/2i$ too belong to $\mathfrak{A}$, along with all continuous functions of these operators (because the latter can be approximated (in norm) by polynomials in the operators). Therefore $x^\pm(\mathrm{Re}\, T) \in \mathfrak{A}$ where $x^\pm \stackrel{\mathrm{def}}{=} (|x| \pm x)/2$. As $x^\pm(\mathrm{Re}\, T) \geq \mathbb{O}$, $x^\pm(\mathrm{Im}\, T) \geq \mathbb{O}$ this proves Statement 1.

2. It is required to prove that the invertibility of $T$, $T \in \mathfrak{A}$, implies that $T^{-1} \in \mathfrak{A}$. Consider the polar decomposition $T = U(T^*T)^{1/2}$, $U$ unitary. Then $0 \notin \sigma(T^*T)$, $T^*T \in \mathfrak{A}$ so that $(T^*T)^{-1/2} \in \mathfrak{A}$ (the function $x^{-1/2}$ is continuous on the spectrum $\sigma(T^*T)$). Hence $U = T(T^*T)^{-1/2} \in \mathfrak{A}$, $T^{-1} = (T^*T)^{-1/2} \in \mathfrak{A}$.

4. It is clear that $\mathrm{Ker}\, \alpha$ is a twosided $*$-ideal in $\mathfrak{A}$ so that $\mathfrak{A}/\mathrm{Ker}\, \alpha$ is a $C^*$-

algebra (Statement 3). It remains to be shown that any injective $*$-homomorphism ($\hat{\alpha}$ in the case at hand) is an isometry. Assume thus that $\operatorname{Ker}\alpha = \{\mathbb{O}\}$. If for some $T$, $T \in \mathfrak{A}$, we have $|T| = 1$, $|\alpha(T)| < 1$ then $|T^*T| = 1$, $|\alpha(T^*T)| < 1$. Let $f$ be a continuous function on the axis $\mathbb{R}$, equal to zero on the spectrum of the (self-adjoint) element $\alpha(T^*T)$, with $f(1) = 1$. Approximating $f$ by polynomials we see that $\alpha(f(T^*T)) = f(\alpha(T^*T))$. But $|f(T^*T)| \geq 1$, $f(\alpha(T^*T)) = \mathbb{O}$. Contradiction. □

**115. Corollary.** *If $\mathfrak{A}$, $\mathfrak{B}$ are $C^*$-algebras and $T \in \mathfrak{A} \subset \mathfrak{B}$ then $\sigma_{\mathfrak{A}}(T) = \sigma_{\mathfrak{B}}(T)$.* □

**116. Lemma.** *Let $X$ be a closed subalgebra of $L^\infty$.*
1. *$T_{L^\infty} \cap \operatorname{alg} T_X = T_X$.*
2. *$X$ is symmetric $\Leftrightarrow$ $\operatorname{alg} T_X$ is a $C^*$-algebra.*

*Proof*. As $\operatorname{Sym}(\operatorname{alg} T_X) = X$ we see that from $T_\varphi \in \operatorname{alg} T_X$ follows $\varphi = \operatorname{Sym} T_\varphi \in X$.

2. The implication $\Rightarrow$ is trivial. The implication $\Leftarrow$ again is an immediate consequence of Statement 1. □

**117. Proof of Theorem 113.** 1. Follows from 116, Statement 2 and 114, Statement 2.

2. Set $\mathfrak{B} = \operatorname{alg} T_X/\mathfrak{S}_\infty$ (this is a $C^*$-algebra since it is the image of $\operatorname{alg} T_X$ under the canonical projection into $L(H^2)/\mathfrak{S}_\infty$; cf. also 114, Statement 4). If $0 \notin \sigma_{\text{ess}}(T)$, $T \in \operatorname{alg} T_X$, then $0 \notin \sigma_{\mathfrak{B}}(T)$ (Corollary 115 for the pair $\mathfrak{A} \overset{\text{def}}{=} L(H^2)/\mathfrak{S}_\infty, \mathfrak{B}$). From the invertibility of $T$ in $\operatorname{alg} T_X/\mathfrak{S}_\infty$ we get the invertibility of the symbol $\operatorname{sym} T$ in $X$ (Lemma 108 and remark 111 with $\mathfrak{S} = \mathfrak{S}_\infty \cap \operatorname{alg} T_X$). □

**118. Douglas's Theorem**. *Let $X$ be a (closed) subalgebra of $L^\infty$.*
1. *If $X \supset H^\infty$ then $\sigma(T_\varphi) = \sigma_{\operatorname{alg} T_X}(T_\varphi) \supset \sigma_X(\varphi)$, $\forall \varphi \in X$.*
2. *If $X \supset H^\infty + C$ then $\sigma_{\text{ess}}(T_\varphi) \supset \sigma_X(\varphi)$, $\forall \varphi \in X$.*

*Proof*. Both inclusions are established in Lemma 59. The identity $\sigma(T_\varphi) = \sigma_{\operatorname{alg} T_X}(T_\varphi)$ is contained in Lemma 120. □

**119. The Inclusions Become Identities** iff $X = H^\infty$ (for the inclusion $\sigma(T_\varphi) \supset \sigma_X(\varphi)$, $\forall \varphi \in X$) or $X = H^\infty + C$ (for $\sigma_{\text{ess}}(T_\varphi) \supset \sigma_X(\varphi)$, $\forall \varphi \in X$). In fact

a) If $X = H^\infty$ the desired identity is contained in 49;

b) if $X \neq H^\infty$, $X \supset H^\infty$ then 60 implies that $z^n \in X$ for all $n \in \mathbb{Z}$; therefore $\sigma_X(z) = \mathbb{T}$, $\sigma(T_z) = \operatorname{clos} \mathbb{D}$.

c) if $X = H^\infty + C$ the desired identity is contained in 50 (cf. likewise 126.3).

d) if $X \neq H^\infty + C, X \supset H^\infty + C$ then the Marshall-Chang Theorem (cf. 124) gives the existence of an *inner* function $\varphi$ with $\bar{\varphi} \in X \backslash (H^\infty + C)$; therefore $\sigma_X(\varphi) = \mathbb{T}$ while $\sigma_{\text{ess}}(T_\varphi) = \operatorname{clos} \mathbb{D}$ (cf. 49). □

**120. Lemma.** *Let $\mathfrak{A}$ be an algebra of operators in the space $H^2$, $T_{H^\infty} \subset \mathfrak{A}$, and let $T_\varphi \in \mathfrak{A}$, $\varphi \in L^\infty$. Then*

$$\sigma_{\mathfrak{A}}(T_\varphi) = \sigma(T_\varphi).$$

*Proof*. If the operator $T_\varphi$ *is invertible* (in $L(H^2)$) then by Theorem 30 $\varphi = u\varphi_e$, $|u| = 1$ a.e., $\varphi_e$ outer, $\varphi_e^{-1} \in H^\infty$. Therefore $T_u = T_\varphi(T_{\varphi_e})^{-1}$ and $T_{\bar{u}}$ are invertible too. 36 gives the existence of an invertible function $\psi$ in $H^\infty$ with $\|\mathbb{1} - u\psi\|_\infty = \|\bar{u} - \psi\|_\infty < 1$. It is clear that $T_{u\psi} = T_u T_\psi \in \mathfrak{A}$ and $|T_{u\psi} - I| < 1$. The invertibility in $\mathfrak{A}$ of the operators $T_{u\psi} = T_u T_\psi$ and $T_\psi$ now entails the same thing for $T_u$ and $T_\varphi = T_u T_{\varphi_e}$. □

**121. The Identity** Com $T_X = \mathfrak{S}_\infty$. **The Rôle of the Semi-Commutators.** As we have already remarked, there holds an immediate relation to the computation of the spectrum only for two commutators: Com $T_X = \{\mathbb{0}\}$ and Com $T_X = \mathfrak{S}_\infty$. The first case is trivial (cf. 165.156). The second leads at once to the identity alg $T_X$/Com $T_X \overset{\text{def}}{=} \hat{X} =$ alg $T_X/\mathfrak{S}_\infty$ and so to the inclusion

$$\sigma_{\hat{X}}(T) \supset \sigma_{\text{ess}}(T), \quad T \in \text{alg}\, T_X, \tag{29}$$

a counterpart of the "spectral inclusion theorems" 113, 118, (28). For a *symmetric algebra* $X$ the condition Com $T_X = \mathfrak{S}_\infty$ allows us to identify (by virtue of 114, 115) the essential spectrum of the operator with its spectrum in the commutative algebra $\hat{X}$:

$$\sigma_{\hat{X}}(T) = \sigma_{\text{ess}}(T). \tag{30}$$

It remains to determine the space of maximal ideals $\mathfrak{M}(\hat{X})$ of $\hat{X}$. Incidentally, from (30) and (28) we get once more Statement 2 of Theorem 113 (in the case Com $T_X = \mathfrak{S}_\infty$). Concerning the space $\mathfrak{M}(\hat{X})$, if we take the algebra $X$ as given, the simplest case which may occur is the case $\tilde{\hat{X}} \cong X$. A sufficient condition for this is stated in Lemma 122 but in our hypothesis (Com $T_X = \mathfrak{S}_\infty$) the requirement $\hat{X} \cong X$ is equivalent (cf. 109, 110) to the following condition on the semi-commutators:

$$T_\varphi T_\psi - T_{\varphi\psi} \in \mathfrak{S}_\infty, \quad \forall \varphi, \psi \in X.$$

On the other hand, it is clear that this last assumptions again guarantees the inclusion Com $T_X \subset \mathfrak{S}_\infty$.

For *non-symmetric algebras* $X$ we have at our disposal an inclusion, opposite to the inclusion (29), only in such assumptions (cf. 118) which *in fact* (cf. 122, 124) reduce to the coincidence of the algebras $\hat{X}$ and $X$. Together with the previous hypothesis Com $T_X = \mathfrak{S}_\infty$ this would imply the remarkable equality

$$\sigma_{\text{ess}}(T_\varphi) = \sigma_X(\varphi), \quad \forall \varphi \in X,$$

were there not an "enhancing" circumstance: the hypotheses of Theorem 118 and the equality Com $T_X = \mathfrak{S}_\infty$ are compatible only in one single case: $X = H^\infty + C$.

The influence of the semi-commutators in the framework of the algebraic approach is – besides the things already set forth – also tied to the formula (2): $T_{\varphi\psi} - T_\varphi T_\psi = H_{\bar{\varphi}}^* H_\psi$, which permits us to "estimate" the semi-commutator, and further the fact that commutators appear as differences of semi-commutators.

Returning to the identity Com $T_X = \mathfrak{S}_\infty$ let us in conclusion remark that even its purely psychological impact is great, for *this is the only case* (besides the trivial one $\operatorname{Com} T_X = \{\mathbb{O}\}$) when the ideal $\operatorname{Com} T_X$ can be identified with something known from other sources. We should however not overestimate this, because if $\operatorname{Com} T_X = \mathfrak{S}_\infty$ then $\operatorname{alg} T_X$ turns out to be rather poor in its supply of operators (it is commutative mod $\mathfrak{S}_\infty$ and Toeplitz operators practically never commute (cf. 165.156); also the meaning cf. for instance, the inclusion $T_\varphi T_\psi - T_{\varphi\psi} \in \mathfrak{S}_\infty$ is that $\psi$ is "almost analytic" or $\varphi$ "almost anti-analytic"; cf. 136, 146). For general algebras $X$ the structure of the ideal $\operatorname{Com} T_X$ as well as the structure of the commutators and the semi-commutators themselves remain largely in obscurity (cf. however 127, 128).

The following lemma contains a condition which is sufficient for the validity of Statements 1–7 of Lemma 109 and thus for the identity $\hat{X} \cong X$. From this condition one can also derive some information pertaining to the ideals $\operatorname{Com} T_X$ (cf. 124–126). After first enumerating some examples (123–126) we then pass to the case $\operatorname{Com} T_X = \mathfrak{S}_\infty$ (129 and so forth).

Slightly modifying a known notion (cf., e.g., Gamelin [1]) we say that a symmetric subalgebra $X$ of $L^\infty(\mathbb{T})$ is *logarithmic* if $\operatorname{Re} X = \operatorname{clos}_{L^\infty}(L^\infty \cap \log |X_A|)$, $X_A \overset{\text{def}}{=} X \cap H^\infty$, i.e. if every real function in $X$ can be approximated arbitrarily well (in the norm $\|\cdot\|_\infty$) with logarithms $\log|g|$ of moduli of functions $g$ in $X_A$. Equivalent formulation: $\forall f \in X$ $(f \geq 0)$ $\forall \varepsilon > 0$ $\exists g \in X_A : \| f - |g| \|_\infty < \varepsilon$. A more restrictive requirement is *logmodularity*: $\operatorname{Re} X = \operatorname{clos} \log |X_A^{-1}|$, $X_A^{-1} = \{f \in X_A,$ $f^{-1} \in X_A\}$.

**122. Lemma.** 1. *Let $M_i \subset H^\infty$ $(i = 1, 2)$ and assume that*

$$X = \operatorname{alg}(M_1, \overline{M_2}) \overset{\text{def}}{=} \bigvee\nolimits_{L^\infty}(\varphi^k \overline{\psi}^n : k, n \geq 0;\ \varphi \in M_1, \psi \in M_2). \tag{31}$$

*Then the Statements 1–7 of Lemma 109 hold true, so that the algebras $X$ and $\hat{X} = \operatorname{alg} T_X / \operatorname{Com} T_X$ are canonically isomorphic ($X \cong \hat{X}$).*

2. *Let $X$ be a logarithmic symmetric subalgebra of $L^\infty$. Then $X = \operatorname{alg}(X_A, \overline{X}_A)$.*

*Proof*. 1. Let us verify Statement 6 of Lemma 109. Since the ideal $\operatorname{Com} T_X$ is closed and moreover the imbedding $j_X$ is continuous and linear it is sufficient to consider the case $f = \varphi_1^k \overline{\psi}_1^l$, $g = \varphi_2^m \overline{\psi}_2^n$:

$$\begin{aligned} T_f T_g - T_{fg} &= T_{\varphi_1^k \overline{\psi}_1^l} T_{\varphi_2^m \overline{\psi}_2^n} - T_{\varphi_1^k \overline{\psi}_1^l \varphi_2^m \overline{\psi}_2^n} = T_{\overline{\psi}_1^l}(T_{\varphi_1^k} T_{\overline{\psi}_2^n} - T_{\varphi_1^k \overline{\psi}_2^n}) T_{\varphi_2^m} \\ &= T_{\overline{\psi}_1^l} [T_{\varphi_1^k}, T_{\overline{\psi}_2^n}] T_{\varphi_2^m}. \end{aligned}$$

Hence $T_{fg} - T_f T_g \in \operatorname{Com} T_X$.

2. In view of Statement 1 of 114 it is sufficient to show that positive functions in $X$ are contained in $\operatorname{alg}(X_A, \overline{X}_A)$. If $f \in X$, $f \geq 0$, then $f^{1/2} \in X$ so that by the hypothesis $\forall \varepsilon > 0$ $\exists g \in X_A : \| f^{1/2} - |g| \|_\infty < \varepsilon$. Hence $\| f - g\overline{g} \|_\infty < \varepsilon(\varepsilon + 2\| f \|_\infty^{1/2})$. □

**123. Tables of Spectra. Symmetric Algebras.** For convenience let us now list the spectral inclusion theorems (which we have already proved) valid in various combinations of the hypotheses. Everywhere below we let $\overline{X} = \{\overline{x} : x \in X\}$, $X$ being a subalgebra of $L^\infty(\mathbb{T})$, $T \in \operatorname{alg} T_X$.

**123.1. Hypothesis:** $X = \bar{X}$.

$$\begin{array}{ccccc} & \subset & \sigma_{\hat{X}}(T) & \subset & \\ \sigma_X(\operatorname{Sym} T) & & & & \sigma_{\operatorname{alg} T_X}(T) = \sigma(T). \\ & \subset & \sigma_{\text{ess}}(T) = \sigma_{\operatorname{alg} T_X/\mathfrak{S}_\infty}(T) & \subset & \end{array}$$

**123.2. Hypothesis:** $X = \bar{X} = \operatorname{alg}(X_A, \bar{X}_A)$.

$$\sigma_X(\operatorname{Sym} T) \subset \sigma_{\hat{X}}(T) \subset \sigma_{\text{ess}}(T) = \sigma_{\operatorname{alg} T_X/\mathfrak{S}_\infty}(T) \subset \sigma_{\operatorname{alg} T_X}(T) = \sigma(T).$$

**123.3 Hypothesis:** $X = \bar{X} = \operatorname{alg}(X_A, \bar{X}_A)$, $\operatorname{Com} T_X \subset \mathfrak{S}_\infty$.

$$\sigma_X(\operatorname{Sym} T) = \sigma_{\text{ess}}(T) = \sigma_{\hat{X}}(T) \subset \sigma_{\operatorname{alg} T_X} = \sigma(T).$$

**123.4 Hypothesis:** $X = \bar{X} = \operatorname{alg}(X_A, \bar{X}_A)$, $\mathfrak{S}_\infty \cap \operatorname{alg} T_X = \{\mathbb{0}\}$.

$$\sigma_X(\operatorname{Sym} T) = \sigma_{\hat{X}}(T) = \sigma_{\text{ess}}(T) = \sigma_{\operatorname{alg} T_X}(T) = \sigma(T).$$

**123.5. Hypothesis:** $X = \bar{X} = \operatorname{alg}(X_A, \bar{X}_A)$, $\operatorname{Com} T_X = \mathfrak{S}_\infty$.

$$\sigma_X(\operatorname{Sym} T) = \sigma_{\text{ess}}(T) \subset \sigma_{\operatorname{alg} T_X}(T) = \sigma(T).$$

Concrete examples to this table are given in 141.

**124. Comments. Douglas Algebras and Sarason Algebras.** How to verify condition (31) in Lemma 121, the single general conditon, which leads to the identities

$$X \cong \operatorname{alg} T_X/\operatorname{Com} T_X, \quad \operatorname{Com} T_X = \operatorname{Ker} \operatorname{Sym}_X, \quad \operatorname{alg} T_X = T_X + \operatorname{Com} T_X$$

(and other ones in 109, 128) and to the spectral inclusion theorems of table 123? For *symmetric algebras X* one recipe is indicated in Lemma 122 itself – we have to show that the algebra is logarithmic. Almost all the algebras which we have met are indeed logarithmic: $C(\mathbb{T})$, $QC$, $AP_\zeta + C(\mathbb{T})$, $AP_\mathbb{T} + C(\mathbb{T})$, $L^\infty(\mathbb{T})$. Besides, for the algebras $C$, $AP_\zeta + C$, $AP_\mathbb{T} + C$ the possibility of a representation of the type (31) is completely trivial, while for the algebra $L^\infty$ it follows immediately from the Douglas-Rudin Theorem (Appendix 2). In the last case it is however simpler to check the logaritmicity: if $f \in L^\infty$, $f \geq \mathbb{0}$, and $\delta > 0$, then there exists an (outer) function $g$ in $H^\infty$ such that $f + \delta = |g|$. □

*The algebra QC is logmodular.* In fact, $QC$ is even a *Dirichlet algebra* in the sense that $\operatorname{Re} QC = \operatorname{clos} \operatorname{Re} QC_A$. (If a symmetric algebra $X$ is a Dirichlet algebra then it is logmodular: $f \geq 0$, $f \in X$, $\delta > 0 \Rightarrow \forall \varepsilon > 0 \exists g \in X_A : \|\log(f + \delta) - \operatorname{Re} g\|_\infty < \varepsilon \Rightarrow \forall \varepsilon > 0\ \exists g \in X_A : \|(f + \delta) - |e^g|\,\|_\infty < \text{const} \cdot \varepsilon$. Let now $\delta \to 0$. □) Indeed, if $f = \bar{f} \in QC$ and $f = u + \tilde{v}$ ($u$, $v \in C(\mathbb{T})$) then we can pick a trigonometric polynomial $p$ such that $\|u - p\|_\infty < \varepsilon$ and set $g = (p + i\tilde{p}) + \tilde{v} - iv$. Then $g \in QC \cap H^\infty = QC_A$ and $\|f - \operatorname{Re} g\|_\infty < \varepsilon$. □

Of course, not all symmetric algebras can be represented in the form (31). For instance, this is not possible for the algebra $PC$ (it is not hard to see that $PC_A = C_A$); moreover $\widehat{PC} \not\supseteq PC$ (cf. 140). However, we shall soon display another large class of symmetric algebras $X$ for which (31) is true (i.e. $X = \operatorname{alg}(X_A, \bar{X}_A)$). First we consider certain algebras which are in general not symmetric, the so-called Douglas algebras. (The only symmetric Douglas algebra is $L^\infty$; that it is a Douglas algebra follows from the Douglas-Rudin Theorem.)

**Definition.** The algebra $\mathscr{D}$, $\mathscr{D} \subset L^\infty(\mathbb{T})$, is called a *Douglas algebra* if there exists a set $\Sigma$ of inner functions such that $\mathscr{D} = H^\infty_{\bar{\Sigma}}$ $(= \operatorname{alg}(H^\infty, \bar{\Sigma}))$.

**First Chang-Marshall Theorem.** *Let $X$ be a (closed) subalgebra of $L^\infty(\mathbb{T})$. The following conditions are equivalent.*

1. *$X \supset H^\infty$.*
2. *$X$ is a Douglas algebra.*

This deep theorem is connected with many problems in function theory, Banach algebras and the theory of Toeplitz operators (for the proof see Chang [1], Marshall [1]; Garnett [2], Sarason [9], [14]; cf. likewise 165.124). For instance, it shows that every intermediate algebra $X$ between $H^\infty$ and $L^\infty$ $(H^\infty \subset X \subset L^\infty)$ admits the representation (31). With every Douglas algebra one can associate two very interesting symmetric algebras.

**Definition.** Let $\mathscr{D}$ be a Douglas algebra. The following two algebras are called *Sarason algebras*:

$Q_{\mathscr{D}} \overset{\text{def}}{=} \mathscr{D} \cap \bar{\mathscr{D}}$,

$C_{\mathscr{D}} \overset{\text{def}}{=}$ the symmetric subalgebra of $L^\infty$ generated by those inner functions which are invertible in $\mathscr{D}$.

If $\mathscr{D} = L^\infty$ we have $Q_{\mathscr{D}} = L^\infty$, $C_{\mathscr{D}} = L^\infty$ (the last identity too is the Douglas-Rudin Theorem [1], cf. Garnett [2] and Appendix 2), if $\mathscr{D} = H^\infty + C$ we similarly have $Q_{\mathscr{D}} = QC$, $C_{\mathscr{D}} = C(\mathbb{T})$ (here is a short proof: if $\Theta$ is an inner function then $I - T_\Theta T_{\bar{\Theta}} = P_\Theta$, the orthogonal projection onto $K_\Theta = H^2 \ominus \Theta H^2$, and by virtue of formula (2) $I - T_\Theta T_{\bar{\Theta}} = H^*_{\bar{\Theta}} H_{\bar{\Theta}}$ so that the hypothesis $\bar{\Theta} \in H^\infty + C$ gives $P_\Theta \in \mathfrak{S}_\infty$ (Hartman's Theorem) and $\deg \Theta < \infty$. □) It follows immediately from the definition that

$$C_{\mathscr{D}} = \operatorname{alg}(C_{\mathscr{D}A}, \bar{C}_{\mathscr{D}A}),$$

i.e. for every Sarason algebra $\mathscr{D}$ Lemma 122 holds true (and with it all of its consequences). It turns out that an analogous statement holds also true for the algebra $Q_{\mathscr{D}}$.

**Second Chang-Marshall Theorem.** *Let $\mathscr{D}$ be a Douglas algebra. The Sarason algebras $C_{\mathscr{D}}$ and $Q_{\mathscr{D}}$ are logmodular.*

The proof of this theorem will be deferred to 165.124. There we also prove that always $\mathscr{D} = H^\infty + C_\mathscr{D}$ (Chang's Theorem).

**125. Corollary.** *The algebras X listed below have the properties* $\hat{X} \cong X$, $\operatorname{alg} T_X = T_X + \operatorname{Com} T_X$ *(as well as other properties from* 109, 128*):*

*$C(\mathbb{T})$, $QC$, $L^\infty(\mathbb{T})$, $AP_\zeta$, $AP_\zeta + C$, $AP_\mathbb{T}$, $AP_\mathbb{T} + C$,*
*the Douglas algebras $\mathscr{D}$,*
*the Sarason algebra $C_\mathscr{D}$, $Q_\mathscr{D}$.* □

**126. Tables of Spectra. Non-symmetric Algebra.** Here we actually utilize Theorem 118 only and so we may restrict ourselves to *Douglas algebras* $\mathscr{D}$. The case $\mathscr{D} = H^\infty$ is trivial (cf. 49).

**126.1. Hypothesis:** $\mathscr{D} \neq H^\infty$, $T \in \operatorname{alg} T_\mathscr{D}$.

$$\begin{matrix} \sigma_\mathscr{D}(\operatorname{Sym} T) = \sigma_\mathscr{D}(T) \\ \sigma(T) \end{matrix} \subsetneqq \sigma_{\operatorname{alg} T_\mathscr{D}}(T).$$

**126.2. Hypothesis:** $\mathscr{D} \neq H^\infty$; $T = T_\varphi$, $\varphi \in \mathscr{D}$.

$$\sigma_\mathscr{D}(\varphi) = \sigma_{\hat{\mathscr{D}}}(T_\varphi) \subset \sigma_{\text{ess}}(T_\varphi) \subset \sigma(T_\varphi) = \sigma_{\operatorname{alg} T_\mathscr{D}}(T_\varphi).$$

**126.3. Hypothesis:** $\mathscr{D} \neq H^\infty$; $\operatorname{Com} T_\mathscr{D} = \mathfrak{S}_\infty$; $T = T_\varphi$, $\varphi \in \mathscr{D}$.

$$\sigma_\mathscr{D}(\varphi) = \sigma_{\text{ess}}(T_\varphi).$$

Table 126.3 (contained in the union of 118, (29) and 125) is mostly given for the completeness of the picture only, becasue in reality only the algebra $\mathscr{D} = H^\infty + C$ can be computed according to this table (cf. 144, 159). For $H^\infty + C$ (after the ideal $\operatorname{Com} T_\mathscr{D}$ being determined) we get an even stronger assertion (cf. 139). Concrete examples for Table 126 can be found in 141.

**127. Corollary.** 1. $\operatorname{Com} T_{L^\infty} = \mathfrak{S}^0 \cap \operatorname{alg} T_{L^\infty}$.

2. *Let $\Sigma$ be a semi-group of inner functions (i.e. $\Theta_1, \Theta_2 \in \Sigma \Rightarrow \Theta_1\Theta_2 \in \Sigma$), $M \subset H^\infty$ and consider $X = \operatorname{alg}(\Sigma, M)$. Then*

$$\operatorname{Com} T_X = \mathfrak{S}^0 \cap \operatorname{alg} T_X = \{C \in \operatorname{alg} T_X : \inf_{\Theta \in \Sigma} |CT_\Theta| = 0\}.$$

*In particular, as X we may take the algebra $C(\mathbb{T})$, $QC$, $L^\infty(T)$, any Douglas algebra $\mathscr{D}$, $Q_\mathscr{D}$, $C_\mathscr{D}$, $AP_\zeta$, $AP_\zeta + C$, $AP_\mathbb{T}$, $AP_\mathbb{T} + C$.*

In fact, the inclusion $\operatorname{Com} T_{L^\infty} \subset \mathfrak{S}^0 \cap \operatorname{alg} T_{L^\infty}$ was mentioned in 107; the opposite inclusion is likewise connected with that lemma: if $C \in \mathfrak{S}^0 \cap \operatorname{alg} T_{L^\infty}$ then $\|\operatorname{Sym} C\|_\infty \leq \operatorname{dist}(C, \mathfrak{S}^0) = 0$ and $C \in \operatorname{Ker} \operatorname{Sym} = \operatorname{Com} T_{L^\infty}$ in view of 125. Concerning the identity $\operatorname{Com} T_X = \mathfrak{S}^0 \cap \operatorname{alg} T_X$ cf. 122, 128.

The equality $\operatorname{Com} T_X = \{\ldots\}$ (the set in the second assertion of the corollary) is established similarly: if $C \in \{\ldots\}$ then $\|\operatorname{Sym} C\|_\infty = \inf_\Theta \|(\operatorname{Sym} C)\Theta\|_\infty = \inf_\Theta \|\operatorname{Sym}(CT_\Theta)\|_\infty \le \inf_\Theta |CT_\Theta| = 0$ and in view of 122 $C \in \operatorname{Com} T_X$. For the proof of the opposite inclusion it suffices to verify the identity $\inf_\Theta |CT_\Theta| = 0$ just for members of the total (in $\operatorname{Com} T_X$) set of operators of the form

$$\left(\prod_1^m T_{\varphi_j}\right)[T_\varphi, T_\psi]\left(\prod_1^n T_{\psi_i}\right),$$

where $\varphi = \bar{\Theta}_1 h_1$, $\psi = \bar{\Theta}_2 h_2$, $\psi_i = \bar{\vartheta}_i g_i$ and $\Theta_1, \Theta_2, \vartheta_i \in \Sigma$, $\varphi_j \in X$, the functions $h_1$, $h_2$, $g_i$ being products of functions in $M$. For such an operator $C$ it suffices to take $\Theta = \Theta_1 \Theta_2 \prod_1^n \vartheta_i$. Then $\varphi\psi(\prod_1^n \psi_i)\Theta \in H^\infty$ and by the formula $T_f T_g = T_{fg}$, $\forall f$, $\forall g \in H^\infty$ (applied repeatedly from the right to the left) we get

$$CT_\Theta = \left(\prod_1^m T_{\varphi_j}\right)(T_{\varphi\psi(\Pi\psi_i)\Theta} - T_{\psi\varphi(\Pi\psi_i)\Theta}) = \mathbb{O}.$$

All the concrete algebras considered in Statement 2 can be represented in this form (for the algebras $Q_{\mathscr{D}}$ this follows from the theorem of Chang quoted just (last paragraph of 124): if $f \in Q_{\mathscr{D}}$ and $f = h + c$, $h \in H^\infty$, $c \in C_{\mathscr{D}}$ is the expansion of $f$ according to Chang's theorem then $h \in Q_{\mathscr{D}} \cap H^\infty = Q_{\mathscr{D}A}$; consequently $Q_{\mathscr{D}} = Q_{\mathscr{D}A} + C_{\mathscr{D}}$, so that for the algebra $Q_{\mathscr{D}}$ as $\Sigma$ we can take the same set as for $C_{\mathscr{D}}$, namely the set of all inner functions in $\mathscr{D}$ which are invertible in that algebra). □

**128. Corollary.** *Let $X$ be a subalgebra of $L^\infty$. Each of the following identities is equivalent to Statements 1–7 of Lemma 109:*

8. $\operatorname{Com} T_X = \operatorname{Com} T_{L^\infty} \cap \operatorname{alg} T_X$.
9. $\operatorname{Com} T_X = \mathfrak{S}^0 \cap \operatorname{alg} T_X$.

In fact, the obvious identity $\operatorname{Ker}\operatorname{Sym}_X = \operatorname{alg} T_X \cap \operatorname{Ker}\operatorname{Sym}$ along with 125, 127 shows that the right members of the identities 8, 9 coincide with $\operatorname{Ker}\operatorname{Sym}_X$. □

**129. The Inclusion $\mathfrak{S}_\infty \subset \operatorname{Com} T_X$.** If an algebra is transitive (that is, has no non-trivial invariant subspaces) and contains at least one rank 1 operator then it contains all compact operators. This observation is good to have in mind in assertions of the type of Theorem 130 and Corollary 132 (Coburn's Theorem) (although we will not make explicit use of it). Another useful statement will be formulated as a lemma.

**Lemma.** *Let $\mathscr{A}$ be an operator algebra, $\mathfrak{S}_\infty \subset \mathscr{A}$. Then $\mathfrak{S}_\infty$ is the smallest bilateral ideal in $\mathscr{A}$ different from $\{\mathbb{O}\}$. In particular $\mathfrak{S}_\infty \subset \operatorname{Com} \mathscr{A}$.*

*Proof.* Let $\mathfrak{S}$ be any twosided ideal, $T \in \mathfrak{S}$, $T \ne \mathbb{O}$. Then $\alpha T \beta \in \mathfrak{S}$ for all operators $\alpha$, $\beta$ of rank 1. The set $\{\alpha T \beta\}$ coincides with the set of all rank 1 operators. Hence $\mathfrak{S}_\infty \subset \mathfrak{S}$. □

**130. Theorem.** *Let $X$ be a (closed) subalgebra of $L^\infty$ and assume that $z \in X$, $X \not\subset H^\infty$. Then $\mathfrak{S}_\infty$ is the smallest closed ideal in $\operatorname{alg} T_X$ different from $\{\mathbb{O}\}$. In particular $\mathfrak{S}_\infty \subset \operatorname{Com} T_X$.*

*Proof.* Let $f \in X$, $p \in \mathscr{P}_A$ (polynomials in $z$). Then

$$[T_z, T_f] = T_z T_f - T_{fz} = (\text{formula (2)}) - H_{\bar z}^* H_f = -(\cdot, P_+ \overline{fz})\mathbb{1},$$
$$T_p[T_z, T_f] = -(\cdot, P_+ \overline{fz})\mathbf{p}.$$

It is clear that the last operator belongs to alg $T_X$. If we now can show that

$$\forall g \in H^2 \ \exists f_n \in X : \lim_n \| P_+ \bar f_n \bar z - g \|_{L^2} = 0, \tag{32}$$

then we get $(\cdot, g)p \in \text{alg}\, T_X$ ($\forall g \in H^2$, $\forall p \in \mathscr{P}_A$) and thus $\mathfrak{S}_\infty \subset \text{alg}\, T_X$. The rest will follow from Lemma 129.

The assertion (32) follows immediately from Lemma 131, reminiscent of Wermer's well-known maximality theorem, cf. Hoffman [1], Gamelin [1], Garnett [2].

**131. Lemma.** *Let $X$ be a subalgebra (recall that always $\mathbb{1} \in X$) of $L^\infty$, $z \in X$. Then*

$$\textit{either } \text{clos}_{L^2} X = H^2 \textit{ or } \text{clos}_{L^2} X = L^2.$$

*Proof of the Lemma.* Set $E = \text{clos}_{L^2} P_- X$.

Assume that $E \neq H^2_-$. Since the operator $f \mapsto P_- zf$ maps $E$ into itself there exists an inner function $\Theta$ such that $E = H^2_- \ominus \bar\Theta H^2_-$ (Beurling-Helson Theorem, Lecture I). This shows that $E = \{\mathbb{0}\}$. In fact, if $f \in X$ then $\Theta f \in H^2$ and so $f = h\bar\vartheta$, with $h \in H^2$, $\vartheta$ being an inner divisor of $\Theta$, relatively prime to the inner part $h^i$. The condition $f^n \in X$, $n \geq 1$, now gives $\vartheta h^n \bar\Theta^n \in H^2$, $n \geq 1$, but this is possible only for $\vartheta = \mathbb{1}$. Thus $f \in H^2$ and $X \subset H^2$, i.e. $\text{clos}_{L^2} X = H^2$.

Assume that $E = H^2_-$. Then it is clear that $\text{clos}_{L^2} X = L^2$. □

**132. Corollary (Coburn's Theorem).** 1. *$\mathfrak{S}_\infty$ is the least closed two-sided ideal in* alg $T_{C(\mathbb{T})}$.

2. Com $T_{C(\mathbb{T})} = \mathfrak{S}_\infty$.

In fact, Statement 1 follows from Lemma 129 and Statement 2. The inclusion Com $T_C \subset \mathfrak{S}_\infty$ is contained in the (trivial) Lemma 133. The inclusion $\mathfrak{S}_\infty \subset$ Com $T_C$ follows from 130. □

*Alternative proof* (by Coburn) of the inclusion $\mathfrak{S}_\infty \subset$ Com $T_C$ (basic for 132). Let us verify that

$$\bigvee\nolimits_{L(H^2)} (T_p [T_{\bar z}, T_z] T_{\bar q} : q, p \in \mathscr{P}_A) = \mathfrak{S}_\infty.$$

We have

$$[T_{\bar z}, T_z] T_{\bar q} f = (P_+ \bar q f, \mathbb{1})\mathbb{1} = (f, q)\mathbb{1}, \quad f \in H^2;$$
$$T_p [T_{\bar z}, T_z] T_{\bar q} f = (f, q)p, \quad f \in H^2,$$

and the desired identity follows. □

**133. Lemma.** *Let $\varphi, \psi \in L^\infty$. If one of the functions $\bar\varphi, \psi$ belongs to $H^\infty + C$ then $T_\varphi T_\psi - T_{\varphi\psi} \in \mathfrak{S}_\infty$. If $\varphi, \psi \in H^\infty + C$ then $[T_\varphi, T_\psi] \in \mathfrak{S}_\infty$.*

This simple fact has already been used in the foregoing (47, 50, 124) but has not been mentioned explicitly.

*Proof*. By formula (2) $T_{\varphi\psi} - T_\varphi T_\psi = H^*_{\bar{\varphi}} H_\psi$ and either $H_{\bar{\varphi}} \in \mathfrak{S}_\infty$ or $H_\psi \in \mathfrak{S}_\infty$ (the trivial part of Hartman's Theorem, Lecture. VIII). □

**134. The Inclusion** Com $T_X \subset \mathfrak{S}_\infty$. As already noted, this inclusion does not occur often and in order to verify it one usually has to invoke the semi-commutators. *Rule*: if the semi-commutators in alg $T_X$ are (in general) not compact then the same is true for the commutators. Only known exception: $X = PC$ (Theorem 140). It remains to study in as great detail as possible the reasons making the semi-commutators compact. This will be done in 135–137, 142–154. Several known results are here obtained as corollaries (138, 141, 159, 160).

**135. Sarason's Theorem.** *Let* $\varphi, \psi \in L^\infty$. *If*

$$\operatorname{dist}_\lambda(\bar{\varphi}, H^\infty) \cdot \operatorname{dist}_\lambda(\psi, H^\infty) \equiv 0, \quad \lambda \in \mathbb{T}, \tag{33}$$

*then* $T_\varphi T_\psi - T_{\varphi\psi} \in \mathfrak{S}_\infty$.

(For the definition of the local distance cf. 74.)

*Proof*. Let $\varepsilon > 0$. Set $\alpha = \{\lambda \in \mathbb{T} : \operatorname{dist}_\lambda(\bar{\varphi}, H^\infty) \geq \varepsilon\}$, $\beta = \{\lambda \in \mathbb{T} : \operatorname{dist}_\lambda(\psi, H^\infty) \geq \varepsilon\}$. By hypothesis $\alpha \cap \beta = \varnothing$ while 74 shows that $\alpha$ and $\beta$ are closed sets. Pick a real continuous function $u$ on $\mathbb{T}$ with $0 \leq u \leq 1$, $u|\alpha \equiv 0$, $u|\beta \equiv 1$ and set $v = \mathbb{1} - u$. All compact operators encountered will be denoted by the letter $K$. In view of 133

$$T_\varphi T_\psi = T_\varphi T_u T_\psi + T_\varphi T_v T_\psi = (T_{\varphi u} T_\psi + K) + (T_\varphi T_{v\psi} + K).$$

By the choice of $u$, $v$ holds $\operatorname{dist}_\lambda(\bar{\varphi}u, H^\infty) < \varepsilon$, $\operatorname{dist}_\lambda(\psi v, H^\infty) < \varepsilon$ $(\forall \lambda \in \mathbb{T})$ and so by Lemma 75 $\operatorname{dist}(\bar{\varphi}u, H^\infty + C) < \varepsilon$, $\operatorname{dist}(\psi v, H^\infty + C) < \varepsilon$. Set $\bar{\varphi}u = h + r$ where $h \in H^\infty + C$, $\|r\|_\infty < \varepsilon$. Then by 133

$$T_{\varphi u} T_\psi - T_{\varphi u\psi} = T_{\bar{h}} T_\psi - T_{\bar{h}\psi} + T_{\bar{r}} T_\psi - T_{\bar{r}\psi} = K + R_1,$$

with $|R_1| \leq 2\varepsilon \|\psi\|_\infty$. Analogously $T_\varphi T_{v\psi} - T_{\varphi v\psi} = K + R_2$, with $|R_2| \leq 2\varepsilon \|\varphi\|_\infty$. Adding up we get

$$T_\varphi T_\psi = T_{\varphi u\psi} + T_{\varphi v\psi} + K + R_1 + R_2 = T_{\varphi\psi} + K + R$$

and $|R| \leq 2\varepsilon(\|\varphi\|_\infty + \|\psi\|_\infty)$. Hence $T_\varphi T_\psi - T_\varphi \in \mathfrak{S}_\infty$. □

**136. Comments.** In the proof of Theorem 135 one can of course use instead of 133 directly formula (2) and the estimate 21 for the essential norm, but this does not simplify the argument. Theorem 135 guarantees the compactness of the semi-commutator in the hypothesis that at each point either $\varphi$ is "asymptotically anti-analytic" or $\psi$ "asymptotically analytic". However, such a localization of analyticity and anti-analyticity does not completely account for the phenomenon of the compactness of the semi-commutators $T_\varphi T_\psi - T_{\varphi\psi}$ (an example showing that condition (33) is not necessary was displayed by Sarason [10]). A partial converse of Theorem 135 (and by the way Lemma 133 too) *in a subclass* of functions $\varphi$ (but not individually) nevertheless is true, cf. 144, 159, 142, 146 where we study in some greater detail compactness (and other forms of "smoothness" as well) for semi-commutators and commutators.

| Algebra $X$ | Com $T_X$ | alg $T_X$/Com $T_X$ | $A \in$ alg $T_X$ | $\sigma_{ess}(A)$, $A \in$ alg $T_X$ | $\sigma_X(\mathrm{Sym}\,A)$ |
|---|---|---|---|---|---|
| Symmetric algebras | | | | | |
| $C(\mathbb{T})$ | $\mathfrak{S}_\infty$ 138, 130, 132, 127 | $C(\mathbb{T})$ 125 | $A = T_{\mathrm{sym}\,A} + K$, $K \in \mathfrak{S}_\infty$. 125, 138 | $(\mathrm{Sym}\,A)(\mathbb{T})$ 139, 47 | $(\mathrm{Sym}\,A)(\mathbb{T})$ |
| $QC$ | $\mathfrak{S}_\infty$ 138, 130, 127 | $QC$ 125 | $A = T_{\mathrm{sym}\,A} + K$, $K \in \mathfrak{S}_\infty$. 125, 138 | Range$(\mathrm{Sym}\,A)$ 139 | Range$(\mathrm{Sym}\,A)$ |
| $PC$ | $\mathfrak{S}_\infty$ 140, 130 | $C(\mathbb{T}_*)$ $\mathbb{T}_* = 165.140$ | ? | $(\mathrm{Sym}\,A)_*(\mathbb{T})$ $= (\mathrm{Sym}\,A)(\mathbb{T}_*)$. 121, 140, 165.140.1, 47 | $(\mathrm{Sym}\,A)(\mathbb{T})$ |
| $L^\infty(\mathbb{T})$ | $\mathfrak{S}^0 \cap$ alg $T_{L^\infty}$ 127, 130 | $L^\infty(\mathbb{T})$ 125 | $A = T_{\mathrm{sym}\,A} + C$, $C \in \mathfrak{S}^0$. 125, 127 | ? 113, 30, 38, 44, 123 | Range$(\mathrm{Sym}\,A)$ |
| $AP_\zeta$, $AP_\zeta + C(\mathbb{T})$ | $\mathfrak{S}^0 \cap$ alg $T_X$ 127, 130, 165.65 165.129 | $AP_\zeta$, $AP_\zeta + C(\mathbb{T})$ 125 | $A = T_{\mathrm{sym}\,A} + C$, $C \in \mathfrak{S}^0$. 125, 127 | $\sigma_{ess}(A) = \sigma(A)$ $\sigma_{ess}(T_\varphi) = \mathrm{Range}\,\varphi$ $\cup \{\lambda : \mathrm{wind}_\zeta(\varphi - \lambda) \neq 0\}$ 64, 72, 123.4, 165.65 | Range$(\mathrm{Sym}\,A)$ |
| $AP_\mathbb{T}$ $AP_\mathbb{T} + C(\mathbb{T})$ | $\mathfrak{S}^0 \cap$ alg $T_X$ 127, 130, 165.65, 165.129 | $AP_\mathbb{T}$ $AP_\mathbb{T} + C(\mathbb{T})$ 125 | $A = T_{\mathrm{sym}\,A} + C$, $C \in \mathfrak{S}^0$. 125, 127 | $\sigma_{ess}(A) = \sigma(A)$ $\sigma_{ess}(T_\varphi) = \mathrm{Range}\,\varphi$ $\cup \{\lambda : \exists\zeta \in \mathbb{T} : \mathrm{wind}_\zeta(\varphi - \lambda) \neq 0\}$ 88, 123.4, 165.65 | Range$(\mathrm{Sym}\,A)$ |

| | | | | | |
|---|---|---|---|---|---|
| $C_{\mathscr{D}}, Q_{\mathscr{D}}$ | $\breve{\mathfrak{S}}^0 \cap \operatorname{alg} T_X$<br>127, 130 | $C_{\mathscr{D}}, Q_{\mathscr{D}}$<br>125 | $A = T_{\operatorname{sym} A} + C$,<br>$C \in \mathfrak{S}^0$.<br>125, 127 | ?<br>113, 30, 123 | Range(Sym $A$) |
| | | | Douglas algebras | | |
| $H^\infty$ | $\{\mathbb{0}\}$ | $H^\infty$ | $A = T_{\operatorname{sym} A}$ | Range(Sym $A$)<br>$\cup \{\lambda : \deg(\operatorname{Sym} A - \lambda)^i = \infty\}$.<br>49 | clos(Sym $A$)($\mathbb{D}$) |
| $H^\infty + C(\mathbb{T})$ | $\mathfrak{S}_\infty$<br>138, 130, 127 | $H^\infty + C(\mathbb{T})$<br>125 | $A = T_{\operatorname{sym} A} + K$,<br>$K \in \mathfrak{S}_\infty$<br>125, 138 | $\sigma_X(\operatorname{Sym} A)$<br>139, 50 | Range(Sym $A$)<br>$\cup \left\{\lambda : \frac{\operatorname{sym} A - \lambda}{(\operatorname{sym} A - \lambda)_e} \notin QC\right\}$<br>$= \left\{\lambda : \lim_{\lvert\zeta\rvert \to 1} \lvert\mathscr{H}(\operatorname{Sym} A)(\zeta) - \lambda\rvert = 0\right\}$<br>$= \{\lambda : (\operatorname{Sym} A - \lambda)$<br>$\neq h \cdot c, \forall h^{\pm 1} \in H^\infty, \forall c^{\pm 1} \in C\}$<br>50, 165.50.1, 57 |
| $\mathscr{D}$<br>(Douglas algebra<br>$\mathscr{D} \neq H^\infty$) | $\mathfrak{S}^0 \cap \operatorname{alg} T_{\mathscr{D}}$<br>127, 130 | $\mathscr{D}$<br>125 | $A = T_{\operatorname{sym} A} + C$,<br>$C \in \mathfrak{S}^0$<br>125, 127 | ?<br>59, 120, 126, 93<br>165.91.1 | ?<br>93, 95, 165.95.1 |

**137. Corollary.** *If at each point of T either $\varphi$ or $\psi$ is continuous then* $T_\varphi T_\psi - T_{\varphi\psi}$, $T_\psi T_\varphi - T_{\varphi\psi} \in \mathfrak{S}_\infty$. □

**138. Corollary.** Com $T_{C(\mathbb{T})} = \mathfrak{S}_\infty$, Com $T_{QC} = \mathfrak{S}_\infty$, Com $T_{H^\infty + C} = \mathfrak{S}_\infty$. Cf. 132, 133. □

**139. Corollary.** *Let $X = C(\mathbb{T})$, $QC$, $H^\infty + C(\mathbb{T})$ and assume that $T \in \operatorname{alg} T_X$. Then*

$$T \text{ Fredholm} \Leftrightarrow \text{The symbol Sym } T \text{ is invertible in } X.$$

For the algebras $C$ and $QC$ this is contained in 123.5 + 125 + 138 and for $H^\infty + C$ in 126.3 + 138 + 125. □

**140. Gohberg-Krupnik Theorem.** Com $T_{PC} = \mathfrak{S}_\infty$, *but the algebra* alg $T_{PC}/\mathfrak{S}_\infty$ *is not canonically isomorphic to PC.*

*Proof*. The inclusion $\mathfrak{S}_\infty \subset$ Com $T_{PC}$ is contained in 132. For the opposite inclusion it suffices to verify that $[T_\varphi, T_\psi] \in \mathfrak{S}_\infty$ for a suitable total (in $PC$) set of symbols $\varphi, \psi$. For this we can take the set of functions in $PC$ each having just one discontinuity (jump). Let thus $\varphi$ and $\psi$ be such functions. If their points of discontinuity do not coincide then $[T_\varphi, T_\psi] \in \mathfrak{S}_\infty$ by virtue of 137. Otherwise there exist a number $\alpha$ and a function $c$, $c \in C(\mathbb{T})$, such that $\varphi = \alpha\psi + c$. Then

$$[T_\varphi, T_\psi] = [T_{\alpha\psi}, T_\psi] + [T_c, T_\psi] = [T_c, T_\psi] \in \mathfrak{S}_\infty$$

in view of 133. Thus Com $T_{PC} \subset \mathfrak{S}_\infty$.

However $\widehat{PC} \not\cong PC$, because (cf. 109, 110) semi-commutators with symbols in $PC$ are not by necessity compact. Example: $\varphi = \chi_{\mathbb{T}_+} - \chi_{\mathbb{T}_-}$; $T_\varphi^2 - I \notin \mathfrak{S}_\infty$ since $\sigma(T_\varphi^2) = \sigma(T_\varphi)^2 = [0, 1]$ (Theorem 33). □

**141. Summary (table). Concrete Algebras.** The table (pp. 352–353) lists various facts for Toeplitz operators $T_\varphi$ whose symbols lie in a given function algebra $X$ and further gives information on arbitrary operators $A$ in alg $T_X$ and also on the algebras alg $T_X$ themselves. For instance, it is clear from what already has been said that if $A \in \operatorname{alg} T_X$, $A$ Fredholm, then its symbol Sym $A$ is invertible in the least symmetric function algebra containing $X$ (and, of course, invertible in $L^\infty$); we have also noted that if $X \supset H^\infty$ then $\sigma_{\operatorname{alg} T_X}(T_\varphi) \subset \sigma(T_\varphi)$, $\varphi \in X$, etc. For a *continuation* of the table cf. 165.141.

The sign "?" indicates that we lack a good description of the object in question. The natural numbers in the boxes of the table refer to the sections in the text where the information mentioned in the table (as well as other facts) can be located. The phrase "$A \in \operatorname{alg} T_X$" has to be read as "if $A \in \operatorname{alg} T_X$ then $A = \ldots$". The converse is true without any reserve in all cases for which Com $T_X = \mathfrak{S}_\infty$; in the remaining cases one has to require that $C \in \operatorname{alg} T_X$.

**142. The Compactness of Semi-Commutators and Commutators** of Toeplitz operators has, besides its usefulness in the "algebraic schemes" of spectral theory as described above, yet another important function – it contains information on the symbols of the operators in question. Supplementing what was said in 121,

129, 134, 136 we now give some results in this "individual" direction. We will also obtain "algebraic" corollaries.

Let us begin with the simplest of the symmetric and analytic cases.

*Notation*: $[T_f, T_g) \stackrel{\text{def}}{=} T_f T_g - T_{fg}$; $[T_M, T_N) = \{[T_f, T_g) : f \in M, g \in N\}$.

**143. Lemma.** 1. *Let* $f \in L^\infty$. *Then* $[T_{\bar{f}}, T_f) \in \mathfrak{S}_\infty \Leftrightarrow f \in H^\infty + C$.
2. *Let* $X \subset L^\infty$, $X = \{\bar{f} : f \in X\}$. *Then* $[T_X, T_X) \in \mathfrak{S}_\infty \Leftrightarrow X \subset QC$.

*Proof*. 1. Formula (2) and Hartman's Theorem. 2. Follows from Statement 1 and Lemma 133. □

**144. Sarason's Lemma.** 1. *Let* $M, N \subset L^\infty$. *Then* $[T_M, T_N) \subset \mathfrak{S}_\infty \Rightarrow [T_{\bigvee(\overline{H^\infty} \cdot M)}, T_{\bigvee(N \cdot H^\infty)}) \subset \mathfrak{S}_\infty$.

2. *Let* $g \in L^\infty$. *The following statements are equivalent.*

a) $g \in H^\infty + C$.

b) $[T_\Theta, T_g) \in \mathfrak{S}_\infty$ *for any inner function* $\Theta$ *(notice that* $[T_\varphi, T_g) = [T_\varphi, T_g]$, $\forall \varphi \in H^\infty$).

c) $[T_{L^\infty}, T_g) \subset \mathfrak{S}_\infty$.

*Proof*. 1. This follows from the formula $T_{\bar{\varphi}}[T_f, T_g) T_\psi = [T_{\bar{\varphi} f}, T_{g\psi})$ valid for $\varphi$, $\psi \in H^\infty$.

2. In view of 133 a ⇒ b.

b ⇒ c. It follows from Statement 1 that $[T_{\bigvee(\overline{H^\infty} \cdot \Theta : \Theta \text{ inner})}, T_g) \subset \mathfrak{S}_\infty$ and from the Douglas-Rudin Theorem (Appendix 2) that $\bigvee(\overline{H^\infty} \Theta : \Theta \text{ inner}) = L^\infty$.

c ⇒ a. Take $f = g$ in Statement 1 of Lemma 143. □

**145. Corollary.** *If* $X$ *is an algebra,* $X \subset L^\infty$, $[T_X, T_X) \subset \mathfrak{S}_\infty$ *then* $H^\infty_{\bar{X}} \cap H^\infty_X \subset H^\infty + C$.

In fact, by 144, Statement 1, $[T_{\overline{H^\infty_{\bar{X}}}}, T_{H^\infty_X}) \subset \mathfrak{S}_\infty$ so if $h \in H^\infty_{\bar{X}} \cap H^\infty_X$ then $[T_{\bar{h}}, T_h) \subset \mathfrak{S}_\infty$. Thus by 143, Statement 1 we get $h \in H^\infty + C$. □

**146. The Compactness Criterion for the Semi-Commutator** $[T_\varphi, T_\psi)$ which we will now announce attaches a precise meaning to all the preceding allusions to the "almost analyticity" of $\bar{\varphi}$ or $\psi$ in the condition $[T_\varphi, T_\psi) \in \mathfrak{S}_\infty$. This criterion as well as Corollary 145 can be stated in terms of Douglas algebras.

**First AChSV Theorem.** *Let* $\varphi, \psi \in L^\infty$. *A necessary and sufficient condition for the operator* $[T_\varphi, T_\psi)$ *to be compact is that*

$$H^\infty_{\bar{\varphi}} \cap H^\infty_\psi \subset H^\infty + C(\mathbb{T}). \tag{34}$$

The sufficiency is proved in the paper Axler-Chang-Sarason [1] and the necessity in Vol'berg [1]. Both proofs make use of the space of maximal ideals of the algebras $H^\infty$, $H^\infty_{\bar{\varphi}}$, $H^\infty_\psi$ but admit also an elementary version omitting these details. Passing to one such elementary proof let us remark that some of the particulars might be of interest also for the spectral theory of the model operators $\mathbf{T}_\Theta$. Besides the basic AChSV Theorem we likewise prove a related simpler result due to Axler-Chang-Sarason which might be termed a "children's variant" of this theorem (cf. 155).

*The plan of the proof* consists of *first* (148–152) reducing the general case of the AChSV Theorem to the special case when the functions $\varphi$ and $\bar{\psi}$ are inner. In this case the proof is completed by invoking new interesting particulars and reduces to the following statement.

**Second AChSV Theorem.** *Let $\Theta$, $\vartheta$ be relatively prime inner functions. The following statements are equivalent.*

1. $[T_\Theta, T_{\bar{\vartheta}}) \in \mathfrak{S}_\infty$.
2. $H^\infty_{\bar{\Theta}} \cap H^\infty_{\bar{\vartheta}} \subset H^\infty + C$.
3. $\lim_{|\zeta|\to 1} \max\{|\Theta(\zeta)|, |\vartheta(\zeta)|\} = 1$.
4. $\forall \varepsilon > 0 \exists u, v \in H^\infty + C : u\Theta + v\vartheta = \vartheta, \|u\|_\infty < 1 + \varepsilon$.
5. $H_{\bar{\Theta}}|K_\vartheta \in \mathfrak{S}_\infty$ *where* $K_\vartheta = H^2 \ominus \vartheta H^2$.
6. $P_\Theta P_\vartheta \in \mathfrak{S}_\infty$ *where* $P_\Theta$ *and* $P_\vartheta$ *are the orthogonal projections onto* $K_\Theta$ *and* $K_\vartheta$ *respectively.*
7. $T_{\bar{\Theta}}|K_\vartheta = U + K$ *where* $U$ *is a unitary and* $K$ *a compact operator in the space* $K_\vartheta$ ($\stackrel{\text{def}}{=} H^2 \ominus \vartheta H^2$).
8. $\Theta(\mathbf{T}_\vartheta) = U + K$ *with* $U$ *unitary and* $K$ *compact in* $K_\vartheta, \mathbf{T}_\vartheta = P_\vartheta S|K_\vartheta$ *being the model operator.*

Statement 6 can be interpreted as the asymptotic orthogonality of the model spaces $K_\Theta$ and $K_\vartheta$. Statement 3 is very convenient for practical verifications: for instance, if the spectra of *the inner functions $\Theta$ and $\vartheta$ do not intersect* ($\sigma(\Theta) \cap \sigma(\vartheta) = \varnothing$) then it is clear that 3 holds true and along with it all the remaining statements. In this simple case these assertions can however also be verified directly. From this we get the following *strange corollary*: $H^\infty + C = \bigcap \mathscr{D}$ where the intersection is taken over all Douglas algebras $\mathscr{D}$ with $H^\infty + C \subset \mathscr{D}$, $H^\infty + C \neq \mathscr{D}$ (it suffices indeed to take two algebras $H^\infty_{\bar{\Theta}}$ and $H^\infty_{\bar{\vartheta}}$ with $\sigma(\Theta \cap \sigma(\vartheta) = \varnothing$, $\deg \Theta = \deg \vartheta = \infty$). (This example shows that the product $H^*_{\bar{\Theta}} H_{\bar{\vartheta}}$ may be compact (or even considerably more "smooth") without any of the factors $H^*_{\bar{\Theta}}$ and $H_{\bar{\vartheta}}$ having this property.)

Returning to the plan of the proof we observe that the conditions of the Second Theorem naturally fall into the groups – conditions on the operators (1, 5–8) and conditions on the functions (2, 3, 4). After the reduction just referred to (First Theorem via the Second Theorem) we will establish the equivalence of 1, 5–8 (cf. 153) and *then* relate them to Statements 2–4 (cf. 154). One of the implications needed here (3 $\Rightarrow$ 4) will be based on an estimate for the solving functions of the Corona Theorem mentioned in Appendix 3, Sect. 14. This estimate is due to Jones [2] (cf. likewise Uchiyama [1]) and differs somewhat from the main estimate of Appendix 3: *let $f_1, f_2 \in H^\infty$, $\delta \leq \max(|f_1(\zeta)|, |f_2(\zeta)|) \leq 1$ for $\zeta \in \mathbb{D}$; then there exist functions $g_1, g_2$ in $H^\infty$ such that $f_1 g_1 + f_2 g_2 = \mathbb{1}$ and $\|g_i\|_\infty \leq k(\delta)$, $\lim_{\delta\to 1} k(\delta) = 1$.*

Let us pass to the realization of our plane. The basic observation in the first part of the proof (cf. 148) is that the compactness of the semi-commutator $[T_\varphi, T_\psi)$ always *extends to the algebra* generated by the symbols $\varphi, \psi$. First let us however mention a consequence of the AChSV Theorem which allows us to reverse Corollary 145 and to state it in the following form.

**147. Corollary.** *Let $X$ be an algebra, $X \subset L^\infty$. Then*

$$[T_X, T_X) \subset \mathfrak{S}_\infty \Leftrightarrow H^\infty_{\bar{X}} \cap H^\infty_X \subset H^\infty + C. \ \square$$

**148. Vol'berg's Lemma.** *Let $f \in L^\infty$ and set*

$$A_f \stackrel{\text{def}}{=} \{g \in L^\infty : [T_f, T_g) \in \mathfrak{S}_\infty\}.$$

*Then $A_f$ is a (closed) subalgebra of $L^\infty$ contained in $H^\infty + C$.*

*Proof.* The inclusion $H^\infty + C \subset A_f$ follows from 133. It is clear that $A_f$ is closed and linear and to see that it is an algebra it is sufficient to prove the implication

$$g \in A_f \Rightarrow g^2 \in A_f.$$

Without loss of generality we may assume that the operators $T_g$ and $T_{g^2}$ are invertible (if not consider $\lambda \mathbb{1} + g$ for large values of $|\lambda|$ and apply 38, Statement 4). Let $g_e$ be an outer function, $|g_e| = |g|$ a.e., and set $u \stackrel{\text{def}}{=} g g_e^{-1}$. Then $T_f T_g - T_{fg} = (T_f T_u - T_{fu}) T_{g_e}$, $u \in A_f$, and it suffices to show that $u^2 \in A_f$.

The operators $T_u$ and $T_{u^2}$ are invertible so by Theorem 36, Statement 5, there exists an outer function $h_2$ ($h_2, h_2^{-1} \in H^2$) such that $u^2 = \bar{h}_2 h_2^{-1}$, $|h_2|^2 \in (\text{HS})$. Set $h = h_2^{1/2}$. A direct computation shows that on the set of polynomials $\mathscr{P}_A$ there holds the identity

$$\begin{aligned} T_f T_{u^2} - T_{fu^2} &= (T_f T_{\bar{h}^2} - T_{f\bar{h}^2} + T_{\bar{h}} T_f T_{\bar{h}} - T_{\bar{h}} T_f T_{\bar{h}}) T_{1/h} T_{1/h} \\ &= T_{\bar{h}}(T_f T_{\bar{h}} - T_{f\bar{h}}) T_{1/h} T_{1/h} + (T_f T_{\bar{h}} - T_{\bar{h}f}) T_{\bar{h}} T_{1/h} T_{1/h} \\ &= T_{\bar{h}}(T_f T_u - T_{fu}) T_{1/h} + (T_f T_u - T_{fu}) T_h T_u T_{1/h} \stackrel{\text{def}}{=} K_1 + K_2. \end{aligned}$$

(In general $h$ and $h^{-1}$ are not bounded but on the dense set of polynomials the operators $T_h = P_+ h$ and $T_{1/h} = P_+ h^{-1}$ as well as their fourfold products make sense.) The desired containments $K_i \in \mathfrak{S}_\infty$, $i = 1, 2$, now follow from 149. $\square$

**149. Lemma** *Let $h$ be an outer function, $h$, $1/h \in H^2$, $|h|^2 \in (HS)$, and let $\varphi_{ij} \in L^\infty$ (a finite set), putting $T \stackrel{\text{def}}{=} \sum_i \prod_j T_{\varphi_{ij}}$, $\mathscr{T} \stackrel{\text{def}}{=} \mathbf{h} \mathbf{T} \mathbf{h}^{-1}$ (on the dense set $\mathscr{P}_A$ of all polynomials). Then $T$ extends by continuity to a linear map $H^2 \to H^2$. If $T \in \mathfrak{S}_\infty$ then $\mathscr{T} \in \mathscr{S}_\infty$.*

*Proof.* By the Helson-Szegö Theorem (Lecture VII, Sect. 6) the operator $P_+^h \stackrel{\text{def}}{=} h P_+ h^{-1}$ is continuous on $H^2$ and along with it the operators $\mathscr{T}_{\varphi_{ij}} = P_+^h \varphi_{ij}$, $\mathscr{T} = \sum_i \prod_j T_{\varphi_{ij}}$.

As $|h|^2$ satisfies the Helson-Szegö condition ($HS$) so will the function $|h|^{2(1+\varepsilon)}$ for sufficiently small $\varepsilon$, $\varepsilon > 0$. By what we have shown the operator $h^{1+\varepsilon} T h^{-(1+\varepsilon)}$ is then continuous in $H^2$ and therefore the operator $A \stackrel{\text{def}}{=} T P_+$ has to be continuous in the space $L^2(\mathbb{T}, |h|^{2(1+\varepsilon)})$. So if $A$ is compact in $L^2(\mathbb{T})$ then by 150 it will be compact in $L^2(\mathbb{T}, |h|^2\, dm)$. Hence $\mathscr{T} \in \mathfrak{S}_\infty$. $\square$

**150. Lemma.** *Let $(\mathscr{X}, \mu)$ be a measure space and $w_k$ ($k = 0, 1$) positive (measurable) functions, $w_\alpha \stackrel{\text{def}}{=} w_0^{1-\alpha} w_1^\alpha$, $0 \le \alpha \le 1$. Suppose that operator $A$ acts continuously on both $L^2(w_k) = L^2(w_k\, d\mu)$, $k = 0, 1$. Then:*

1. *The operator $A$ is continuous on $L^2(w_\alpha)$ and*

$$|A|_{L^2(w_\alpha)} \leq |A|^{1-\alpha}_{L^2(w_0)} |A|^{\alpha}_{L^2(w_1)}.$$

2. *Assume that there exists a sequence $\{\pi_n\}_{n\geq 1}$ of operators of finite rank which in both of the spaces $L^2(w_k\,d\mu)$, $k = 0, 1$, converges (strongly) to the identity operators: $\lim_n \pi_n x = x$, $\forall x$. Then $A$ is compact in all the spaces $L^2(w_\alpha\,d\mu)$, $0 < \alpha < 1$, provided it is compact in one of the spaces $L^2(w_k\,d\mu)$, $k = 0, 1$.*

3. *If $\mathscr{X} = \mathbb{T}$, $\mu = m$ and if $w_k \in (HS)$, $k = 0, 1$, then as the operators $\pi_n$ in Statement 2 we can take the Fourier projections: $\pi_n f \overset{\text{def}}{=} \sum_{j=-n}^{n} \hat{f}(j) z^j$. In particular, the property of compactness is susceptible to interpolation in the scale $L^2(w_\alpha\,d\mu)$, $0 \leq \alpha \leq 1$.*

*Proof.* 1. This is a part of the well-known theorem of Stein-Weiss, cf., for instance, Bergh-Löfström [1], theorem 5.4.1. For the sake of completeness let us give an elementary and elegant proof following Nieto [1]; see also Halmos [4] for a slightly different approach.

First of all some trivial reductions.

If our assertion is already proved for $\alpha = \frac{1}{2}$ so it is proved (by induction) for every $\alpha$ of the form $\alpha = m2^{-n}$, $0 \leq m \leq 2^n$, $n \geq 1$. Then we can pass to the limit in the inequality $(\int |Af|^2 w_\alpha)^{1/2} \leq \|f\|_{L^2(w_\alpha)} |A|^{1-\alpha}_{L^2(w_0)} |A|^{\alpha}_{L^2(w_1)}$ for a fixed $f$ and for $\alpha = m2^{-n}$ tending to an arbitrary $\alpha$, $0 \leq \alpha \leq 1$.

So in what follows we can assume $\alpha = \frac{1}{2}$. Without loss of generality we can also assume that $w_0 > 0$, $w_1 > 0$ everywhere. Let $dv = (w_0 w_1)^{1/2}\,d\mu$. Then $L^2(w_0\,d\mu) = L^2((w_0/w_1)^{1/2}\,dv)$, $L^2(w_1\,d\mu) = L^2((w_1/w_0)^{1/2}\,dv)$, $L^2(w_0^{1/2} w_1^{1/2}\,d\mu) = L^2(dv)$ that is our problem is reduced to the case $w_0 w_1 \equiv 1$.

The next reduction: the case of bounded weights $w_0$, $w_1$ (i.e. the case $\operatorname{ess\,inf} w_0 > 0$, $\operatorname{ess\,sup} w_0 < \infty$ taking into account the identity $w_0 w_1 \equiv 1$). Let $w$ be a strictly positive function and let $e_\varepsilon = \{x : \varepsilon < w(x) < \varepsilon^{-1}\}$, $\varepsilon > 0$. Defining the operator $A_\varepsilon$ by $A_\varepsilon f = \chi_{e_\varepsilon} Af$ we see that $A_\varepsilon$ acts on $L^2(\chi_{e_\varepsilon} w\,d\mu)$ if $A$ does on $L^2(w\,d\mu)$. Moreover $|A_\varepsilon| \leq |A|$ and $\lim_{\varepsilon\to 0} |A_\varepsilon| = |A|$ (because $\bigcup_{\varepsilon>0} L^2(\chi_{e_\varepsilon} w\,d\mu)$ is dense in $L^2(w\,d\mu)$). Now it is clear that the inequalities for $|A_\varepsilon|_{L^2(\chi_{e_\varepsilon} w_\alpha d\mu)}$ imply the desired one for $|A|_{L^2(w_\alpha d\mu)}$.

Now, let $w$, $w^{-1} \in L^\infty(\mu)$ and let the operator $A$ be continuous on both the spaces $L^2(w\,d\mu) \overset{\text{def}}{=} X$ and $L^2(w^{-1}\,d\mu) = X^*$ (the star stands for the dual with respect to the form $(f, g) = \int f\bar{g}\,d\mu$). The rest of the proof does not depend on the specific nature of $X$. Indeed, we can invoke the following general result.

**Sublemma** (Lions-Peetre [1]). *Let $X$ be a reflexive Banach space continuously and densely imbedded into a Hilbert space $H$; the functionals $x \to (x, y)_H$, being continuous on $X$, provide also the second imbedding $H \subset X^*$. Let $A$ be a linear operator acting continuously both on $X$ and $X^*$. Then $A$ is continuous on $H$ and $|A|_H \leq (|A|_X |A|_{X^*})^{1/2}$.*

*Proof of the Sublemma* (Nieto [1]). Let $j$ be the canonical map of $X$ onto $X^{**}$. Then the operators $B \overset{\text{def}}{=} j^{-1}(A|X^*)^* j$ and $C = B \cdot (A|X)$ are continuous on $X$ and $|C|_X \leq |B|_X \cdot |A|_X = |A|_{X^*} \cdot |A|_X$. It is easy to see that $(x, y)_H = (Ax, Ay)_H = (x, Cy)_H$ for $x$, $y \in X$. The last equalities lead to $\|Cx\|_H^2 \leq \|C^2 x\|_H$ for $x \in X$, $\|x\|_H = 1$. Further, for the same $x$,

$$\|Cx\|_H^{2^n} \le \|C^{2^n}x\|_H,\ \|Cx\|_H \le \overline{\lim_n} \|C^{2^n}x\|_H^{2^{-n}} \le \overline{\lim_n}(\text{const}\cdot\|C^{2^n}x\|_X)^{2^{-n}}$$
$$\le \overline{\lim_n}(|C^{2^n}|_X\|x\|_X)^{2^{-n}} \le |C|_X.$$

Thus $|C|_H \le |C|_X$ and together with the previous estimate this gives us

$$\|Ax\|_H^2 = (Cx, x)_H \le \|Cx\|_H \le |C|_H \le |A|_{X^*}\cdot|A|_X$$

for every $x \in X$, $\|x\|_H = 1$. □

2. It follows from 1 that

$$|(I - \pi_n)A|_{L^2(w_\alpha)}$$
$$\le \text{const}\,|(I - \pi_n)A|_{L^2(w_0)}^{1-\alpha}|(I - \pi_n)A|_{L^2(w_1)}^{\alpha},\ 0 < \alpha < 1,$$

and from the hypothesis of the statement that the right hand side of the last inequality tends to zero. Therefore $A$ is compact in $L^2(\mathbb{T}, w_\alpha\, dm)$, $0 < \alpha < 1$.

3. Indeed the convergence of the Fourier series is in the said assumptions part of the Helson-Szegö Theorem (Lecture VIII, Sect. 6). □

**151. Corollary.** *Let* $f, g \in L^\infty$. *Then*

1. $g \in A_f \Leftrightarrow \bar f \in A_{\bar g}$.
2. *If* $g \in A_f$, $\bar h \in H_{\bar f}^\infty$, *then* $[T_h, T_g) \in \mathfrak{S}_\infty$.
3. $H_{\bar f}^\infty \cap A_f$; *in particular, if* $g \in A_f$ *then* $H_{\bar f}^\infty \cap H_g^\infty \subset H^\infty + C$.

In fact, Statement 1 follows from $[T_f, T_g)^* = [T_{\bar g}, T_{\bar f})$ and 2 follows from 1: $g \in A_f \Rightarrow \bar f \in A_{\bar g} \Rightarrow$ (Lemma 148) $H_{\bar f}^\infty \subset A_{\bar g} \Rightarrow \bar h \in A_{\bar g} \Rightarrow g \in A_h$, as requested.

Statement 3: if $h \in A_f$, $h \in H_{\bar f}^\infty$, then by 2 $[T_{\bar h}, T_h) \in \mathfrak{S}_\infty$ which means that $h \in H^\infty + C$ (cf. 143). By Lemma 148 the condition $g \in A_f$ now entails that $H_g^\infty \subset A_f$. □

**152. Derivation of The First AChSV Theorem from the Second One.** The preceding corollary (Statement 3) shows that from the Second AChSV Theorem one only needs to derive the sufficiency of the First Theorem. Assume that condition (34) is fulfilled and let $\Theta$ and $\vartheta$ be inner functions, $\bar\Theta \in H_{\bar\varphi}^\infty$, $\bar\vartheta \in H_{\bar\psi}^\infty$. It follows from (34) that $H_{\bar\Theta}^\infty \cap H_{\bar\vartheta}^\infty \subset H^\infty + C$ and then by the Second AChSV Theorem $[T_\Theta, T_{\bar\vartheta}) \in \mathfrak{S}_\infty$ so that $[T_{\Theta\bar h_1}, T_{\bar\vartheta h_2}) \in \mathfrak{S}_\infty$, $\forall h_i \in H^\infty$ (cf. 144, Statement 1). From the First Marshall-Chang Theorem (cf. 124) it follows that one can choose the functions $\Theta$, $\vartheta$, $h_1$, $h_2$ such that $\|\psi - \bar\Theta h_2\|_\infty < \varepsilon$, $\|\varphi - \bar\vartheta h_1\|_\infty < \varepsilon$ ($\varepsilon$ being any positive number prescribed in advance). Hence $[T_\varphi, T_\psi) \in \mathfrak{S}_\infty$. □

**153. Lemma.** *The Statements* 1, 5, 6, 7, 8 *of the Second AChSV Theorem are equivalent.*

*Proof.* $1 \Leftrightarrow 5 \Leftrightarrow 6$. It is clear that $[T_\Theta, T_{\bar\vartheta})|\vartheta H^2 = \mathbb{O}$, $[T_\Theta, T_{\bar\vartheta})|K_\vartheta = -T_{\Theta\bar\vartheta}|K_\vartheta$. The compactness of the last operator is equivalent to the compactness of the operator $\mathcal{J}T_{\Theta\bar\vartheta}|K_\vartheta = P_-\bar\Theta\mathcal{J}\vartheta|K_\vartheta$, where $\mathcal{J}f = \overline{zf}$, $\forall f \in L^2$, or – since $\bar\vartheta$ is unimodular – the compactness of the operators $P_-\bar\Theta|J\bar\vartheta K_\vartheta$ or $P_\Theta|\mathcal{J}\bar\vartheta K_\vartheta$. It just remains to make the observation that $\mathcal{J}\bar\vartheta K_\vartheta = K_\vartheta$ (Lemma on Lat $S^*$, Lecture III). Thus $1 \Leftrightarrow 5$, $1 \Leftrightarrow 6$.

$6 \Rightarrow 7$. By the obvious relation $T_{\bar\Theta} + H_{\bar\Theta} = \bar\Theta$ we have $\Theta T_{\bar\Theta} + P_\Theta P_\vartheta = I$ in the subspace $K_{\bar{\vartheta}}$. Hence $(T_{\bar{\Theta}}|K_\vartheta)^*(T_{\bar{\Theta}}|K_\vartheta) = I + C$, $C \in \mathfrak{S}_\infty$. The functions $\Theta$ and $\vartheta$ being relatively prime we see that the operator $T_{\bar{\Theta}}|K_\vartheta$ has a trivial kernel, but then (as in Theorem 54, Statement 2) $T_{\bar{\Theta}}|K_\vartheta = U + K$, $U$ unitary, $K \in \mathfrak{S}_\infty$.

$7 \Leftrightarrow 8$. Trivial because of $(T_{\bar{\Theta}}|K_\vartheta)^* = P_+\Theta|K_\vartheta = \Theta(\mathbf{T}_\vartheta)$.

$8 \Rightarrow 6$. Assume that the operator $\Theta T_{\bar{\Theta}} P_\vartheta \equiv (I - P_\Theta)P_\vartheta$ is the sum of an isometric operator (on $K_\vartheta$) and a compact one. Then

$$((I - P_\Theta)P_\vartheta)^*((I - P_\Theta)P_\vartheta) \equiv P_\vartheta - P_\vartheta P_\Theta P_\vartheta = P_\vartheta + C,$$

$C \in S_\infty$, and therefore $P_\vartheta P_\Theta P_\vartheta \in \mathfrak{S}_\infty$, i.e. $P_\Theta P_\vartheta \in \mathfrak{S}_\infty$. □

**154. Proof of the Second AChSV Theorem.** We have already shown that the Statements 1, 5–8 are equivalent and that $1 \Rightarrow 2$ (Corollary 151, Statement 3). Let us now check that $2 \Rightarrow 3 \Rightarrow 4 \Rightarrow 5$.

$2 \Rightarrow 3$. If 3 does not hold true there exists a sequence $\{\zeta_n\}_{n\geq 1}$, $|\zeta_n| < 1$, such that $\lim_n |\zeta_n| = 1$, $\lim_n \Theta(\zeta_n) = \alpha$, $\lim_n \vartheta(\zeta_n) = \beta$ and $|\alpha| < 1$, $|\beta| < 1$. If necessary, passing to a subsequence, we can assume that the sequence $\{\zeta_n\}_{n\geq 1}$ satisfies the Carleson condition $(C)$ (Kabaĭla-Newman Lemma, Lecture VII). Let $B$ be the Blaschke product built with the sequence $\{\zeta_n\}_{n\geq 1}$ and set $\varphi \stackrel{\text{def}}{=} (\Theta - \alpha\mathbb{1})(\mathbb{1} - \bar{\alpha}\Theta)^{-1}$, $\psi \stackrel{\text{def}}{=} (\vartheta - \beta\mathbb{1})(\mathbb{1} - \bar{\beta}\vartheta)^{-1}$. Then $\lim_n \varphi(\zeta_n) = \lim_n \psi(\zeta_n) = 0$ so that $\varphi(\mathbf{T}_B) \in \mathfrak{S}_\infty$, $\psi(\mathbf{T}_B) \in \mathfrak{S}_\infty$ (recall that (cf. Corollary to Statement 3, Lecture VI) $\mathbf{T}_B$ is similar to a diagonal operator in $l^2$ with the numbers $\zeta_n$ in the diagonal). By the Hartman-Sarason Theorem (Lecture VIII, Sect. 3) $\bar{B}\varphi \in H^\infty + C$, $\bar{B}\psi \in H^\infty + C$. Hence $\bar{B} \in H^\infty_{\bar{\varphi}}$, $\bar{B} \in H^\infty_{\bar{\psi}}$ (the functions $\varphi$ and $\psi$ are not constant). By Statement 2 we now get $\bar{B} \in H^\infty + C$. Contradiction.

$3 \Rightarrow 4$. Here we will utilize Jones's Theorem (cf. 146) but not in the disk $\mathbb{D}$ but in an annulus $A_\delta \stackrel{\text{def}}{=} \{\zeta : 1 - \delta < |\zeta| < 1\}$. To prove the variant we need let us remark that the original form of the theorem was conformally invariant and therefore it holds not only in the disk but also in a strip (for example, the strip $\Pi \equiv \{\xi : 0 < \operatorname{Im} \xi < 1\}$). If now the functions $f_1, f_2$ in the Corona Theorem for the strip $\Pi$ are $2\pi$-periodic then the solutions $g_1, g_2$ can be choosen to be $2\pi$-periodic too (it suffices to take a mean of arbitrary solutions considering a limit point of the sequence $(2n + 1)^{-1} \sum_{|k|\leq n} g_i(z + 2\pi k)$, $i = 1, 2$).

Let $0 < \varepsilon < 1$. Pick a small number $\delta$, $\delta > 0$, and apply Statement 3 and Jones's Theorem with the pair $\Theta|A_\delta$, $\vartheta|A_\delta$. We then find functions $u$, $v$ (analytic and bounded in $A_\delta$) such that $\Theta u + \vartheta v = 1$ and $\sup_{A_\delta} |u| < 1 + \varepsilon$. It remains to make the observation that the boundary values of $u$ and $v$ belong to $H^\infty + C$, as the Fourier coefficients with negative indices decay exponentially.

$4 \Rightarrow 5$. As $u, v \in H^\infty + C$ we have by 133 $T_{\bar{u}} T_{\bar{\Theta}} + T_{\bar{v}} T_{\bar{\vartheta}} = I + C_0$, $C_0 \in \mathfrak{S}_\infty$. *On the subspace* $K_\vartheta$ this gives $T_{\bar{u}} T_{\bar{\Theta}} = I + C_0$. As $C_0 \in \mathfrak{S}_\infty$ and $\|u\|_\infty < 1 + \varepsilon$, we have on a suitable subspace $E_\varepsilon$, $E_\varepsilon \subset K_\vartheta$, $\dim K_\vartheta \ominus E_\varepsilon < \infty$, the estimate $\|T_{\bar{\Theta}} x\| \geq (1 + \varepsilon)^{-1} \cdot \|x\|$, $x \in E_\varepsilon$. But $\|T_{\bar{\Theta}} x\|^2 + \|H_{\bar{\Theta}} x\|^2 = \|x\|^2$, $\forall x$, so that $\|H_{\bar{\Theta}} x\| \leq 2\sqrt{\varepsilon}\|x\|$, $x \in E_\varepsilon$. Hence $H_{\bar{\Theta}}|K_\vartheta \in \mathfrak{S}_\infty$. □

**155. Children's Version of the AChSV Theorem.** *Let* $\varphi, \psi \in L^\infty$; *denote by* $\mathscr{K}$ *the ideal of all operators of finite rank and by* $\mathscr{R}$ *the set of rational functions with no poles in* $\mathbb{T}$. *The following statements are equivalent.*

1. $[T_\varphi, T_\psi) \in \mathscr{K}$.
2. *Either* $P_-\bar{\varphi} \in \mathscr{R}$ *or* $P_-\psi \in \mathscr{R}$.
3. *Either* $\operatorname{clos}(\bar{\varphi}H^\infty) \subset H^\infty + \mathscr{R}$ *or* $\operatorname{clos}(\psi H^\infty) \subset H^\infty + \mathscr{R}$. *Moreover*

$$\operatorname{rank}[T_\varphi, T_\psi) = \operatorname{rank} H_{\bar{\varphi}}^* H_\psi = \min(\operatorname{rank} H_{\bar{\varphi}}, \operatorname{rank} H_\psi) = \min(\deg P_-\bar{\varphi}, \deg P_-\psi).$$

*Proof*. It is clear that $2 \Leftrightarrow 3$. It follows from Kronecker's Theorem (Lecture VIII) and formula (2) ($[T_\varphi, T_\psi) = -H_{\bar{\varphi}}^* H_\psi$) that $2 \Rightarrow 1$.

Let us verify the implication $1 \Rightarrow 2$, along with the formula for the rank of the semi-commutator. It is clear that $\operatorname{rank}[T_\varphi, T_\psi) \le \min(\operatorname{rank} H_{\bar{\varphi}}, \operatorname{rank} H_\psi)$ and therefore it is sufficient to consider the case $\operatorname{rank}[T_\varphi, T_\psi) < \infty$.

Set $E = [T_\varphi, T_\psi)H^2$, $\dim E < \infty$. Then $[T_\varphi, T_\psi)f = -P_+\varphi P_-\psi f \in E$, $\forall f \in H^2$. Also $\operatorname{clos}\{P_-\psi f : f \in H^2\} \equiv \operatorname{clos} H_\psi(H^2)$ is a $P_-\mathbf{z}$-invariant subspace of $H_-^2$. Consequently (Beurling-Helson Theorem, Lecture I), it has the form $\bar{\mathbf{z}}K_\Theta$, $\Theta$ an inner function, $K_\Theta = H^2 \ominus \Theta H^2$, and

$$\deg \vartheta = \deg P_-\psi.$$

(As usual $\deg P_-\psi \stackrel{\text{def}}{=}$ if $P_-\psi \notin \mathscr{R}$.) Passing to the complex conjugate we find $P_-\bar{\varphi}K_\Theta \subset \bar{\mathbf{z}}\bar{E}$. This means that the subspace $K_\Theta$ has finite dimension ($\le \dim \bar{E}$) modulo the subspace $\operatorname{Ker} H_{\bar{\varphi}}$. By Lemma 5 there exists an inner function $\Theta$ such that $\operatorname{Ker} H_{\bar{\varphi}} = \vartheta H^2$,

$$\deg \vartheta = \deg P_-\bar{\varphi}.$$

Assume now that $\deg \Theta > \dim E$. Then there exists a number $\lambda$, $|\lambda| < 1$, such that the set $\sigma \stackrel{\text{def}}{=} \{\zeta : \Theta(\zeta) = \lambda\}$ contains more than $\dim E$ points (or even infinitely many points if $\deg \Theta = \infty$; cf., for instance, Frostman's Theorem, Lecture II, Sect. 5). It follows that a family of reproducing kernels $\{(\mathbb{1} - \overline{\Theta(\zeta)}\Theta)\cdot(\mathbb{1} - \bar{\zeta}z)^{-1} : \zeta \in \sigma'\}$, $\sigma' \subset \sigma$, $\operatorname{card}\sigma' = \dim E + 1$, is linearly dependent over $\vartheta H^2$, i.e. that there exists a finite family of numbers $\{c_\zeta\}$, $c_\zeta \ne 0$ such that $\sum_{\zeta \in \sigma'} c_\zeta(\mathbb{1} - \bar{\lambda}\Theta)(\mathbb{1} - \bar{\zeta}z)^{-1} \equiv \vartheta h$, with $h \in H^2$. The function $\mathbb{1} - \bar{\lambda}\Theta$ is outer while $r = \sum_{\zeta \in \sigma'} c_\zeta(\mathbb{1} - \bar{\zeta}z)^{-1}$ is rational and $\not\equiv 0$. Hence $\deg \vartheta \le \deg r \le \dim E$. □

**156. Corollary.** *Let* $\varphi, \psi \in L^\infty$. *Then*

1. $T_\varphi T_\psi$ *is a Toeplitz operator* $\Leftrightarrow T_\varphi T_\psi = T_{\varphi\psi} \Leftrightarrow \deg P_-\varphi = 0$ *or* $\deg P_-\psi = 0 \Leftrightarrow \bar{\varphi} \in H^\infty$ *or* $\psi \in H^\infty$.
2. $T_\varphi T_\psi = \mathbb{O} \Leftrightarrow \varphi = \mathbb{O}$ *or* $\psi = \mathbb{O}$.
3. $T_\varphi^{-1}$ *is a Toeplitz operator* $\Leftrightarrow \varphi^{-1} \in L^\infty$, $T_\varphi^{-1} = T_{\varphi^{-1}} \Leftrightarrow \varphi^{\pm 1} \in H^\infty$ *or* $\bar{\varphi}^{\pm 1} \in H^\infty$. □

**157. Commutants and Centers.** The *commutant* of a set of operators $M$ is $M' \stackrel{\text{def}}{=} \{A : [A, B] = 0, \forall B \in M\}$ and the *center* of an algebra of operators $\mathfrak{A}$ is $Z(\mathfrak{A}) \stackrel{\text{def}}{=} \mathfrak{A}' \cap \mathfrak{A}$. It is clear that $Z(\mathfrak{A})$ is always a commutative algebra and that if $\mathfrak{A}$ is a $C^*$-algebra then so is the center $Z(\mathfrak{A})$. Not only the *center of the algebra* $\operatorname{alg} T_{L^\infty}$ but also the *commutant of the set* $T_{L^\infty}$ are trivial: if $A \in L(H^2)$ and $T_z A = AT_z$ then $A = T_\varphi$, $\varphi \in H^\infty$ and if $T_{\bar{z}}A = AT_{\bar{z}}$ then $A = T_{\bar{\psi}}$, $\psi \in H^\infty$ so that if $A \in T_{L^\infty}'$ then $A = \lambda I$, $\lambda \in \mathbb{C}$. □

The situation changes drastically if we pass to the Calkin algebra $L(H^2)/\mathfrak{S}_\infty$.

The center of the algebra alg $T_{L^\infty}/\mathfrak{S}_\infty$, to be described in 160, plays an essential rôle in the algebraic variant of the technique of localization (162, 163). Let us remark that some things on operators $A$ commuting mod $\mathfrak{S}_\infty$ with an arbitrary Toeplitz operator (and thus with any operator in alg $T_{L^\infty}$ too) are already familiar to us: for instance, operators of the form $A = T_\varphi + K$ with $\varphi \in QC$, $K \in \mathfrak{S}_\infty$ belong here (cf. 133). It turns out (cf. 160) that there holds the identity $T_{QC} + \mathfrak{S}_\infty = (\text{alg } T_{L^\infty}/\mathfrak{S}_\infty)'$.

**158. Lemma.** *Let $A \in \text{alg } T_{L^\infty}$. Then*

$$A - T_{\text{Sym } A} \in \mathfrak{S}_\infty \Leftrightarrow T_\Theta^* A T_\Theta - A \in \mathfrak{S}_\infty \textit{ for any inner } \Theta.$$

One of these statements can be strengthened: if $X$ is an algebra of the type considered in 127, Statement 2, then it follows from $A \in \text{alg } T_X$ and $T_\Theta^* A T_\Theta - A \in \mathfrak{S}_\infty$, $\forall A \in \Sigma$, that $A - T_{\text{Sym } A} \in \mathfrak{S}_\infty$.

*Proof.* $\Rightarrow$. $T_\Theta^* A T_\Theta - A = T_\Theta^*(A - T_{\text{Sym } A}) T_\Theta - (A - T_{\text{Sym } A}) \in \mathfrak{S}_\infty$.

$\Leftarrow$. In view of 125, $A = T_{\text{Sym } A} + C$, $C \in \text{Com } T_{L^\infty}$. In view of 127, $\inf_\Theta |C T_\Theta| = 0$. Moreover $T_\Theta^* C T_\Theta - C = T_\Theta^* A T_\Theta - A \in \mathfrak{S}_\infty$, whence $C \in \mathfrak{S}_\infty$. □

**159. Lemma.** *Let $A \in \text{alg } T_{L^\infty}$. The following statements are equivalent.*

1. $[T_{H^\infty + C}, A] \in \mathfrak{S}_\infty$.
2. $[T_\Theta, A] \in \mathfrak{S}_\infty$ *for any inner* $\Theta$.
3. $A = T_\varphi + K$; $\varphi \in H^\infty + C$, $K \in \mathfrak{S}_\infty$.

*Proof.* 1 $\Rightarrow$ 2. Trivial.

2 $\Rightarrow$ 3. $T_\Theta^* A T_\Theta - A = T_\Theta^*[A, T_\Theta] \in \mathfrak{S}_\infty$ for every inner function $\Theta$. It follows now from 158 that $A = T_{\text{Sym } A} + K$, $K \in \mathfrak{S}_\infty$. Again 144 and Statement 2 give $\text{Sym } A \in H^\infty + C$.

3 $\Rightarrow$ 1. Contained in 133. □

**160. Douglas's Theorem.** *Let $A \in \text{alg } T_{L^\infty}$. The following statements are equivalent.*

1. $A \in Z(\text{alg } T_{L^\infty}/\mathfrak{S}_\infty)$.
2. $[\text{alg } T_{L^\infty}, A] \in \mathfrak{S}_\infty$.
3. $[T_\Theta, A]$, $[T_{\bar\Theta}, A] \in \mathfrak{S}_\infty$ *for all inner* $\Theta$.
4. $A = T_\varphi + K$; $\varphi \in QC$, $K \in \mathfrak{S}_\infty$. □

**161. Comments.** Thus the center of the algebra alg $T_{L^\infty}/\mathfrak{S}_\infty$ is canonically $*$-isomorphic to $QC$. In fact, Lemma 159 *holds true* in a stronger variant – *without the assumption* $A \in \text{alg } T_{L^\infty}$; also the corresponding analogue of Lemma 158 is valid: if $A \in L(H^2)$ then

$$A \in T_{L^\infty} + \mathfrak{S}_\infty \Leftrightarrow T_\Theta^* A T_\Theta - A \in \mathfrak{S}_\infty \text{ all inner } \Theta.$$

Both these results are due to Davidson [1] and will be treated in 165. 161.

The following two Sections give a short survey of the algebraic approach to localization of Toeplitz operators.

**162. Localization over the Center.** First we state Douglas's general localization theorem [2] for $C^*$-algebras and then we apply it to Toeplitz operators mod $\mathfrak{S}_\infty$.

Let $\mathfrak{A}$ be a $C^*$-algebra, $Z$ its center, $X$ a $C^*$-algebra contained in $Z$. Let

$\mathfrak{M} = \mathfrak{M}(X)$ be the maximal ideal space of $X$; by the Stone-Weierstrass theorem $X \cong C(\mathfrak{M})$ (the isomorphism being realized by the Gel'fand transformation $x \mapsto \hat{x}$). If $\mu \in \mathfrak{M}$ we denote by $\mathcal{J}_\mu$ the two-sided ideal in $\mathfrak{A}$ generated by the maximal ideal $\{x \in X : \hat{x}(\mu) = 0\}$.

**Douglas's Localization Theorem.** *Let $\mathfrak{A}$, $X$, $\mathfrak{M}$, $\mathcal{J}_\mu$ have the above significance and let*

$$\pi_\mu : \mathfrak{A} \to \mathfrak{A}/\mathcal{J}_\mu \overset{\text{def}}{=} \mathfrak{A}_\mu, \ \mu \in \mathcal{M},$$

*be the natural projections. Then the map $\Phi$,*

$$\Phi \overset{\text{def}}{=} \sum_{\mu \in \mathfrak{M}} \oplus \pi_\mu,$$

*is an injective (and therefore isometric) $*$-homomorphism of $\mathfrak{A}$ into the $C^*$-algebra $\sum_{\mu \in \mathfrak{M}} \oplus \mathfrak{A}_\mu$ (equipped with the* sup-*norm). The element $a$, $a \in \mathfrak{A}$, is invertible in $\mathfrak{A}$ iff its projections $\pi_\mu$ are invertible in $\mathfrak{A}_\mu$, $\forall \mu \in \mathfrak{M}$.*

Let us apply this theorem to the algebra $\mathfrak{A} = \text{alg}\, T_{L^\infty}/\mathfrak{S}_\infty$. 160 gives then $Z(\mathfrak{A}) \cong QC$. The symmetric subalgebra of $QC$ with the simplest possible maximal ideal space apparently is $C(\mathbb{T})$; $\mathcal{M} = \mathbb{T}$. Let us therefore take $X = C(\mathbb{T})$ in Douglas's Theorem and, remembering the origin of the algebra $\mathfrak{A}$, extend the projections $\pi_\mu$, $\mu \in \mathbb{T}$, to projection $\pi_\mu^*$,

$$\pi_\mu^* : \text{alg}\, T_{L^\infty} \to \text{alg}\, T_{L^\infty}/\mathfrak{S}_\infty \overset{\pi_\mu}{\to} \mathfrak{A}_\mu.$$

Since the invertibility in $\text{alg}\, T_{L^\infty}/\mathfrak{S}_\infty$ is equivalent to being Fredholm, an application of the theorem yields the following result.

**163. Corollary.** *The operator $A$, $A \in \text{alg}\, T_{L^\infty}$, is Fredholm iff all the "local operators" $\pi_\mu^* A$, $\mu \in \mathbb{T}$, are invertible.*

One can prove (cf. Douglas [2]) that all the "fibers" $(\text{alg}\, T_{L^\infty}/\mathfrak{S}_\infty)_\mu$ are canonically $*$-isomorphic and that the local operator $\pi_\mu^*(\sum_i \prod_j T_{\varphi_{ij}})$, $\mu \in \mathbb{T}$, depends only on the values of $\hat{\varphi}_{ij}$ on the corresponding fiber $F_\mu$ of the maximal ideal space $\mathcal{M}(H^\infty)$. More precisely:

$$\pi_\mu^*(\mathbb{T}_\varphi) = \mathbb{O} \Leftrightarrow \hat{\varphi} | F_\mu \equiv 0 \Leftrightarrow \lim_{\zeta \to \mu} |\varphi(\zeta)| = 0.$$

From these considerations we can derive anew the theorems on the local nature of Fredholmness (76–82), Sarason's Theorem for semi-commutators (135) as well as a few more results. Localization is also possible over some other subalgebras of $QC$, cf. 165.162.

**164. Concluding Remarks.** Here are collected: A) bibliographical references to Sects. 1–163 and short comments on them; B) brief description of parts of the theory not touched upon in the main text.

Part A (165.1–165.163) contains references to sources used in writing the corresponding sections of the text and likewise mention of results supplementing the text.

These results are as a rule stated without proof. *Exception*: proof of the Second Chang-Marshall Theorem and Chang's Theorem (165.124). Omission of a number indicates the absence of comments.

Part B (162–172) contains *short expositions* of the following portions of the theory of Hankel and Toeplitz operators: 1) the connection with Wiener-Hopf equations, dual equations and the boundary value problems of function theory; 2) vectorial Hankel and Toeplitz operators, with an indication of their connection with the Corona Theorem of Carleson and with a *proof of a theorem* concerning this connection (169); 3) other extensions of the theory of Hankel and Toeplitz operators ("abstract" theory, multivariate theory); 4) Toeplitz operators up to similarity (the work of Clark); 5) projection methods for inversion and Szegö determinants; asymptotic distribution of spectra.

**165. 1–6. General Information.** Books and surveys close to this Appendix Gohberg-Krupnik [1], Douglas [2], [3], Sarason [9], Gohberg-Kreĭn [1], Butz [1]. A few more sources each containing notably an extensive bibliography: Ahieser [2], Kreĭn-Nudel'man [1] (in connection with interpolation and moment problems); Iohvidov [1] (the algebraic aspect); Power [1], Kahane [2], Malyšev [1], Peller-Hruščev [1] (Hankel and Toeplitz operators, application to best approximation and probability theory); Kreĭn [1], Mushelišvili [1], Gahov [1], Litvinčuk [1], Prössdorf [1] (connection with integral and other equations, boundary problems), Gohberg-Fel'dman [1] (projection methods), Grenander-Szegö [1], Kac [1], [2] (asymptotics spectral, various applications), Halmos [2]. Also: Böttcher-Silbermann [1], [3], Havin-Hruščev-Nikol'skiĭ [1].

Concerning the evolution of the rôle of Hankel and Toeplitz operators (matrices) in mathematical analysis we will limit ourselves to some cursory remarks. Both types of operators arise there in connection with extremal problems, interpolation problems and questions of orthogonal polynomials, further in the spectral analysis of Jacobi matrices (Jacobi, Hermite, Stieltjes, Frobenius, Carathéodory, Weyl, Szegö, ...). In particular, there is an immediate relation between the material of the present Appendix and the work of F. Riesz and Szegö on quadratic forms of the type $\int_{\mathbb{T}} |\sum a_k z^k|^2 \, d\mu$ (spectral distribution; extremal problems, in the final analysis reducing themselves to the Helson-Szegö Theorem, Lecture VIII). Another, apparently unrelated line of development comes from the problems of Riemann and Hilbert and the solution of singular integral equations (Hilbert, Plemelj, Carleman, Hopf, Wiener, ...). In the work of Hankel and Toeplitz themselves the operators in question appear only quite episodically... (e.g. Toeplitz studied the spectral theory of the multiplication operators $f \mapsto \varphi f$, which differ in an essential way from the operators $P_+ \varphi$, cf. Toeplitz [1]; however in [1] it is also proved that $P_+ \varphi$ is continuous $\Leftrightarrow \varphi \in L^\infty$).

**165.3. Widom's Theorem** is proved in Widom [1]; the connection with the Carleson Imbedding Theorem is pointed out in Power [1]. Another result of Widom's: the compactness of the Hankel operator $H[\nu]$ is equivalent to the condition $\nu((\alpha, 1)) + \nu((-1, \alpha)) = o((1-\alpha))$ as $\alpha \to 1$ (vanishing Carleson measure). The Lebesgue measure on the interval (0, 1) (i.e. $v = \chi_{(0,1)}\,dx$) gives rise to the classical *Hilbert matrix* $H[v] = \left\{\frac{1}{n+k+1}\right\}_{n,k\geq 0}$, the study of which has played a prominent rôle in the development of the entire theory.

**165.6. The Important Formula** (2) is due to Sarason [10].

**165.10.** Theorem 10 is proved in Adamyan-Arov-Kreĭn [5]. The case of "self-adjoint" Hankel operators (i.e. operators such that the matrix $\{(H_\varphi z^n, \bar{z}^{k+1})\}_{n,k\geq 0}$ is Hermitian symmetric or, equivalently, $\operatorname{Im}\hat{\varphi}(-n) = 0$, $n \geq 1$) was earlier studied by Clark [2]. Subsequently Clark's method, really originating from Hartman [1], was developed by Butz [1], [2] who anew got a result equivalent to Theorem 10; other results by AAK (cf. below) are not covered by that method. The method discussed there in its broad outlines amounts to the following: a) show that the $s_n$-functions of the operator $H_\varphi$ with $\varphi \in \mathscr{P} + H^\infty$ ($\mathscr{P}$ = the set of all polynomials in $z$ and $\bar{z}$) do not have less than $n$ and not more than $n + k$ zeros in the disk $\mathbb{D}$ (where $k$ is the multiplicity of the $s$-number $s: \cdots < s = s_n = \cdots = s_{n+k} < \cdots$); b) approximate an arbitrary Hankel operator $H_\varphi$ with operators of finite rank $H_{\varphi_n}$ ($\varphi_n \in \mathscr{P} + H^\infty$) so that the $s$-numbers and $s$-functions tend to the appropriate limits; c) derive from a) and b) a statement analogous to a) but now with general symbols $\varphi$, $\varphi \in L^\infty$; d) derive from c) Corollary 17, which is equivalent to Theorem 10. For other particulars connected with Theorem 10 see 165.18.

**165.11. The Remark on the Projection** onto the subspace Hank made in the text is due to Kislyakov and is taken over from Peller's paper [3] (it is essentially the issue of the subspace $\mathscr{S}_\infty \cap$ Hank not being complemented in $\mathfrak{S}_\infty$). It is interesting to note that in other symmetric normed ideals the situation may be different: if $1 < p < \infty$ the natural projection (for the matrices: $\{\alpha_{n,k}\}_{n,k\geq 0} \mapsto \{\beta_{n+k}\}_{n,k\geq 0}$, $\beta_m \overset{\text{def}}{=} (m+1)^{-1}\sum_{n+k=m}\alpha_{n,k}$) is continuous in $\mathfrak{S}_p$ (Peller [3]), for $p = 1$ it is however discontinuous (Peller [3]) but the space $\mathfrak{S}_1 \cap$ Hank is complemented in $\mathfrak{S}_1$, a continuous projection being given by the formula $\beta_m = k_m^{-1}\sum_{n+k=m}(n+1)^\varepsilon(k+1)^\varepsilon\alpha_{n,k}$, where $k_m = \sum_{n+k=m}(n+1)^\varepsilon(k+1)^\varepsilon$, $\varepsilon > 0$ (Aleksandrov, Peller, cf. Peller [3]). See also Peller [7], [8] for some additional properties of the natural averaging projection and Kislyakov [1] for some properties of other projections onto the set of Hankel matrices.

**165.12, 13. The Identity for the Moduli,** $|Af| = |sf|$ a.e. for the successive $s$-functions, holds not only for Hankel operators but for all operators $A: H^2 \to H^2_-$ such that the matrix $\alpha_{n,k} \overset{\text{def}}{=} (Az^n, \bar{z}^{k+1})$ is symmetric: $\alpha_{n,k} = \alpha_{k,n}$ (Butz [2]). For Hankel operators this identity can be found in Adamjan-Arov-Kreĭn [5].

**165.15 The Interpretation Offered** here of the original proof (cf. Adamyan-Arov-Kreĭn [5]) is in essence due to Peller and Hruščev [1].

**165.16. The Case $s_n(H_\varphi) = s_\infty(H_\varphi)$ of Theorem 10.** We have not succeeded in presenting the proof in greater clarity than actually is done in the text. We follow here the original version (Adamyan-Arov-Kreĭn [5]) but the computations are more detailed. The idea of a one dimensional perturbation of the operator $H_\varphi^* H_\varphi$ and the identities necessary for a successful application of the method (cf. the formulae (8)–(11)) are apparently taken over from the technique of unitary extensions in the indefinite metric, which the authors of AAK [5] mention as the point of departure of their work (cf. likewise AAK [3], [1]).

**165.17.** Clark [2], Butz [2]; Adamyan-Arov-Kreĭn [5].

**165.18–19. More General Problems and Description of All Solutions.** We are referring to the problems discussed in Sects. 18–19. In fact, the papers of Adamjan-Arov-Kreĭn [3–5] contain the more general question concerning the description of the set

$$K(\varphi,\rho,n) \stackrel{\text{def}}{=} \{f : f \in \varphi + \mathscr{R}_n + H^\infty,\ \|f\|_\infty \le \rho\},$$

where $\varphi$ is a given function, $\rho$ a given number, $n \ge 0$. This set is non-empty iff $s_n(H_\varphi) \le \rho$ (Theorem 10). The condition when $K(\varphi,\rho,n)$ consists of just one function, mentioned in Sect. 19, can for $n = 0$ be stated in other terms: $\mathbb{1} \notin |\rho^2 I - H_\varphi^* H_\varphi|^{1/2} H^2$, cf. AAK [3]. It is not always clear how to verify for a given function $\varphi$ conditions of this type (the present one or the condition in Sect. 19); nor are the known methods for finding the unique element in $K(\varphi,\rho,n)$ (cf. AAK [3], [5]) particularly constructive. Unfortunately this last remark also extends to the methods for finding $\rho$-Schmidt functions ($\rho = s_n(H_\varphi)$). On the other hand if the set $K(\varphi,\rho,n)$ *contains more than one function* then it admits a complete description (AAK [3], [5]). For example, for $n = 0$, the following parametric *formula of Adamyan-Arov-Kreĭn* holds true: if $K(\varphi,\rho,0) \ne \{f\}, f \in L^\infty$ then there exists an outer function $\psi$ such that

$$\rho \cdot \psi/\bar{\psi} \in K(\varphi,\rho,0),$$

$$K(\varphi,\rho,0) = \left\{\rho\frac{\psi}{\bar{\psi}}\cdot\frac{\mathbb{1}-\bar{\chi}}{\mathbb{1}-\chi}\cdot\frac{\mathscr{E}-\bar{\chi}}{1-\mathscr{E}\chi} : \mathscr{E} \in H^\infty,\ \|\mathscr{E}\|_\infty \le 1\right\},$$

where the function $\chi$ ($\chi \in H^2$) is defined by the formula

$$\frac{\mathbb{1}+\chi}{\mathbb{1}-\chi} = \int_{\mathbb{T}} \frac{\zeta + z}{\zeta - z} |\psi(\zeta)|^2\, dm(\zeta).$$

One has similar formulae for other values of $n$ (cf. AAK [5]). Moreover the so-called "canonical functions" in $K(\varphi,\rho,n)$ (canonical, because of a connection with a certain natural operator condition), with $\mathscr{E} \equiv \text{const}, |\mathscr{E}| = 1$, correspond exactly to the points $\alpha = \varepsilon/\rho r - 1/r \cdot (a, R\mathbb{1})$ of the circle (8) in Sect. 16.

In the special case $n = 0$, $\varphi = g\bar{B}$ ($B$, a Blaschke product; $g \in H^\infty$) the formula just mentioned gives all solutions of the interpolation problem $h - g \in BH^\infty$, $h \in H^\infty$, and is due to Nevanlinna [2]. Regarding this case cf. further Adamyan-Arov-Kreĭn [4]. Nevanlinna's proof can also be found in Garnett's book [2]; there is moreover given a deduction of the Adamyan-Arov-Kreĭn formula for $K(\varphi,\rho,0)$ based on function theoretic ideas. In particular, an analysis is given of one of the reasons leading to non-uniqueness ($K(\varphi,\rho,0) \ne \{f\}$ if $\rho = s_0(H_\varphi) =$

$|H_\varphi|$), viz. the impossibility of reconstructing in a unique way a function $h$, $h \in H^1$, from its argument. Yet another proof of Nevanlinna's formula is due to Stray [1].

A new approach to the Adamyan-Arov-Kreĭn formulae is given in their paper [1]. This method is in essence based on Kreĭn's theory [2] of extensions of symmetric operators and makes use of the notion of "1-step extension". In [1] AAK set out with the problem of the description of $K(\varphi, \rho, 0)$ (from the start on in the case of Hankel operators with operator symbols in spaces of vector valued functions) and treat it in the following manner. Let $A$ $(= H_\varphi)$ be a Hankel operator, $\rho \geq |A|$, and let $A_1 : H^2 \to \mathbf{z}H^2_-$ be a "Hankel operator" whose matrix is gotten from the one for $A$ by adding one row: $P_1 \mathbf{z} A_1 = A_1 \mathbf{z}$ where $P_1 \overset{\text{def}}{=} \mathbf{z}P_-\overline{\mathbf{z}}$ is the orthogonal projection onto $\mathbf{z}H^2_-$, $P_- A_1 = A$. If we write down all such operators $A_1$ with $|A_1| \leq \rho$ we will get a description of the set $K(\varphi, \rho, 0)$. In fact, starting with the operator $A_1$ we can construct another one $A_2$, a "1-step extension" of $A_1$, $A_2 : H^2 \to \mathbf{z}^2 H^2_-$,

$$P_2 \mathbf{z} A_2 = A_2 \mathbf{z}, \quad P_2 \overset{\text{def}}{=} \mathbf{z}^2 P_- \overline{\mathbf{z}}^2, \quad P_1 A_2 = A_1, \quad |A_2| \leq \rho,$$

etc. It is clear that the operators $A_n$ converge strongly to some operator $B$, $B: H^2 \to L^2(\mathbb{T})$, with $\mathbf{z}B = B\mathbf{z}$, $|B| \leq \rho$, $P_- B = A$. It is likewise perfectly clear that *all* operators $B$ with these properties can be obtained in this way, that they are multiplication operators, $Bx = fx$, $x \in H^2$, and that the corresponding symbols $f$ fill up the set $K(\varphi, \rho, 0)$. (By the same token, in AAK [1] there is given a proof of Nehari's Theorem as well as of its vectorial analogues (Sarason [6], Page [1], [2]; regarding these theorems cf. Lecture VIII).) Now the problem is reduced to finding the 1-step continuations (and to a subsequent unification of them in a single formula). This is done in the spirit of Kreĭn's theory just referred to. In the scalar case (which we – but not AAK – always are concerned with) it is easy to describe all extensions, because from the equations for $A_1$ it follows that $A_1 x = Ax + (x, \alpha + zA^*\overline{z})$, $x \in H^2$, where $\alpha = (\mathbb{1}, A_1 \mathbb{1})$, and the condition $|A_1| \leq \rho$ gets reduced to an inequality for the free parameter $\alpha$:

$$(R(\alpha + zA^*\overline{z}), \alpha + zA^*\overline{z}) \leq 1,$$

where $R = (\rho^2 I - A^*A)^{-1}$ (here we assume for simplicity that $\rho > |A|$; the general case can be reduced to this one). This inequality has already been solved in Sect. 16 (cf. the formulae (8)–(11)). □

The problem of 1-step extension is a special case of the following problem. Given operators $T_{12} : H_1 \to K_2$, $T_{21} : H_2 \to K_1$, $T_{22} : H_2 \to K_2$; find a description of all operators $X : H_1 \to K_1$ with $|T_X| \leq \rho$,

$$T_X \overset{\text{def}}{=} \begin{pmatrix} X & T_{21} \\ T_{12} & T_{22} \end{pmatrix}.$$

In fact, in AAK [1] it is proved that if $H_1 = H_2$, $K_1 = K_2$, $T_{12} = T_{21}$, there exists such an $X$ iff $\rho \geq \max(|T_0 P_{H_2}|, |P_{K_2} T_0|)$, and further there is given a description of all these $X$. Parrott [1] remarked that in the existence part a similar result also in the general case at once follows from Kreĭn's theory (part of this theory, sufficient for the present purposes, can be found e.g. in the textbook by Riesz and Sz.-Nagy [1], Sect. 125). Arsene and Gheondea [1] generalized the AAK formula [1] and found that all the unknown operators $X$ for $\rho = 1$ are given by the formula

$$X = -\Gamma_2 T_{22}^* \Gamma_1 + D_{\Gamma_2^*} \Gamma D_{\Gamma_1},$$

where the $\Gamma_i$ are contractions defined by the formulae $T_{21} = \Gamma_2 D_{T_{22}}$, $T_{12} = D_{T_{22}^*}\Gamma_1$ and $\Gamma$ an arbitrary contractive operator mapping Range $D_{\Gamma_1}$ into Range $D_{\Gamma_2^*}$, $D_\alpha \stackrel{\text{def}}{=} (I - \alpha^*\alpha)^{-1}$ being the defect operators.

More information and an extensive bibliography can be found in the works quoted in this Section. See also Davis [1].

**165.21. Spectral Characteristics of Hankel Operators.** The formula for the essential norm $|H_\varphi|_{\text{ess}} = \lim_n |H_{\varphi z^n}| = \text{dist}(\varphi, H^\infty + C)$ appeared in Adamyan-Arov-Kreĭn [2] and simultaneously (for the special case of self-adjoint Hankel operators) in Clark [2]. Here is a *direct proof of this formula*, not using Theorem 10 and essentially copied on the proof of Adamyan-Arov-Kreĭn [2] (cf. likewise Sarason [1]): if $K \in \mathfrak{S}_\infty$ then $|H_\varphi + K| \geq |H_\varphi S^n + KS^n| \geq |H_\varphi S^n| - |KS^n| = \text{dist}(\varphi, \bar{z}^n H^\infty) - |KS^n|$ while $\lim_n |KS^n| = 0$. Hence $|H_\varphi|_{\text{ess}} \geq \lim_n \text{dist}(\varphi, \bar{z}^n H^\infty)$. But the formula $H_\varphi S^n = P_- z^n H_\varphi$ shows that the operator $\bar{z}^n H_\varphi S^n$ differs from $H_\varphi$ by a finite rank operator, i.e. $|H_\varphi S^n| = \text{dist}(\varphi, \bar{z}^n H) \geq |H_\varphi|_{\text{ess}}$. Hence $|H_\varphi|_{\text{ess}} \leq \lim_n \text{dist}(\varphi, \bar{z}^n H^\infty)$. □

From 21 and 75 it is easy to see that $|H_\varphi|_{\text{ess}} = \frac{1}{2}\max_{\zeta \in \mathbb{T}} \omega_\zeta(\varphi)$ for $\varphi \in PC$, $\omega_\zeta(\varphi)$ being the oscillation of the function $\varphi$ at the point $\zeta$ (cf. 81); in a special case this formula is due to Clark [2]; a more general formula (for $\varphi \in PQC$) is given in 165.50.6.

The distance $\text{dist}(\varphi, H^\infty + C)$ *is always attained* but the extremal function is unique only in the trivial case $\varphi \in H^\infty + C$ (Axler-Berg-Jewell-Shields [1]).

For the operators $\Gamma_\varphi \stackrel{\text{def}}{=} \mathscr{I} H_\varphi$, $\mathscr{I}f \stackrel{\text{def}}{=} \bar{z}f(\bar{z})$ (likewise termed Hankel operators) acting in the space $H^2$ there are several papers devoted to the *problem of the computation of the spectrum, the essential spectrum* etc. This problem is "almost imbedded" in the theory of the algebras alg $T_X$ generated by Toeplitz operators, because the product of any two Hankel operators belongs to alg $T_{L^\infty}$: $\Gamma_\varphi \Gamma_\psi = \Gamma_{\varphi^t}^* \Gamma_\psi = H_{\varphi^t}^* H_\psi = $ (formula (2)) $= T_{\bar{\varphi}} t_\psi - T_{\bar{\varphi}^t} T_\psi$ with $\varphi^t = \overline{\varphi(\bar{z})}$. In particular, if $\Gamma_\varphi \geq \mathbb{0}$ (e.g. if $\Gamma_\varphi = H[v]$, $v \geq 0$, cf. 2, 3) then $\Gamma_\varphi \in \text{alg}\, T_{L^\infty}$. Axler has made the observation (cf. Power [1]) that not every Hankel operator is in alg $T_{L^\infty}$, because of the inclusion $A \in \text{Alg}\, T_{L^\infty}$ *by necessity* entails that $[T_\psi, A] \in \mathfrak{S}_\infty$, $\forall \psi \in QC$ (cf. 133, 160). For the operator $A = \Gamma_\varphi$ this condition becomes $\varphi \cdot (\psi(\bar{z}) - \psi(z)) \in H^\infty + C$, $\forall \psi \in QC$. It is plain that the last relation is far from true for all symbols $\varphi$, $\varphi \in L^\infty$. □ Cf. further 165.161. However, if $\Gamma_\varphi \in \text{alg}\, T_{L^\infty}$ then be necessity $\Gamma_\varphi \in \text{Com}\, T_{L^\infty}$, in view of 127, because $\inf_\Theta |\Gamma_\varphi T_\Theta| = \inf_\Theta \inf_{h \in H^\infty} |(\Gamma_\varphi - \Gamma_{\bar{\Theta} h}) T_\Theta| = 0$ (by the Douglas-Rudin Theorem). Power described the essential spectrum $\sigma_{\text{ess}}(\Gamma_\varphi)$ for symbols $\varphi$ in $PC$ and some other algebras; cf. Power [1], [3], where there is likewise reference to some other works on the same subject (Carleman, Magnus, Clark). The nature of the results: $\sigma_{\text{ess}}(\Gamma_\varphi)$ is the union of segments passing through the point 0 (for $\varphi \in PC$ that is a countable union of segments whose length depends on the size of the jumps).

*Two Examples*: 1) $\Gamma_\varphi$ is the operator corresponding to the Hilbert matrix $\{(n+k+1)^{-1}\}_{n,k\geq 0}$, i.e. $\varphi = i(x-\pi)$, $x = \arg z \in (0, 2\pi)$. Then the formula mentioned above gives $\Gamma_\varphi^2 = T_{(x-\pi)^2} - T_{(x-\pi)}^2$ so by 141 and 165.140.1 $\sigma_{\text{ess}}(\Gamma_\varphi^2) = (\operatorname{Sym}\Gamma_\varphi^2)_*(\mathbb{T}) = ([(x-\pi)^2]_* - [(x-\pi)_*]^2)(\mathbb{T}) = [0, \pi^2]$; as $\Gamma_\varphi \geq 0$ it follows that $\sigma_{\text{ess}}(\Gamma_\varphi) = [0, \pi]$ which together with $|\Gamma_\varphi| \leq \|\varphi\|_\infty = \pi$ gives $\sigma(\Gamma_\varphi) = [0, \pi]$. □

2) The operator $H[v]$, $dv = g\,dx$ where $g$ is continuous and in an appropriate sense slowly oscillating on the segment $(-1, 1)$; then $\sigma_{\text{ess}}(H[v]) = \bigcup [0, \lambda\pi]$, union over all numbers $\lambda$ that are limit values of $g$ at the points $\pm 1$. This follows from a localization over the center of the $C^*$-algebra generated by Hankel operators (in the spirit of 162, 163). Cf. Power [1], [3].

In fact, the inclusion $0 \in \sigma(\Gamma_\varphi)$, $\forall \varphi \in L^\infty$, is trivial: $\lim_n \|\Gamma_\varphi z^n\|_\infty = 0$; this shows also that always $\Gamma_\varphi H^2 \neq H^2$. Moreover, as we have already pointed out, the operators $\Gamma_\varphi$ are by necessity "small" on suitable "large" subspaces (of the form $\Theta H^2$): $\inf_\Theta |\Gamma_\varphi T_\Theta| = 0$. However, on the complementary subspaces $K_\Theta$ the Hankel operators might be "big", e.g. Clark [3] proved that the operator $H_{\bar{\Theta} f}$ ($\Theta$ inner, $f \in H^\infty$, $\operatorname{GCD}(\Theta, f^i) = \mathbb{1}$) has a closed range iff $\inf_{\mathbb{D}}(|\Theta| + |f|) > 0$. (This is an immediate consequence of the formula $H_{\bar{\Theta} f} = \bar{\Theta} f(\mathbf{T}_\Theta) P_\Theta$ and the Spectral Mapping Theorem, Lecture III); Fuhrmann [2] has found a criterion for the range to be closed in the case of a sum $H_{\bar{\Theta}_1 f_1} + H_{\bar{\Theta}_2 f_2}$ of operators with closed range, the $\Theta_1$ and $\Theta_2$ being relatively prime ($\Leftrightarrow \inf_{\mathbb{D}}(|\Theta_1| + |\Theta_2|) > 0$). This fact also follows from what we proved in Lectures III, VIII, IX: if the operators $H_{\bar{\Theta}_i f_i}$ are left invertible then the $f_i(\mathbf{T}_{\Theta_i})$ are left invertible and then $f_i(\mathbf{T}_{\Theta_i}) K_{\Theta_i} = K_{\Theta_i}$ so that

$$\sum_i H_{\bar{\Theta}_i f_i} H^2 = \sum_i \bar{\Theta}_i K_{\Theta_i} = \sum_i \mathcal{J} K_{\Theta_i} = \mathcal{J}\left(\sum_i K_{\Theta_i}\right),$$

where $\mathcal{J} f = \overline{zf}$; the last sum is closed iff the family $\{\Theta_i\}_{i=1,2}$ satisfies the Carleson-Vasyunin Condition ($CV$) in Lecture IX, i.e. in the case at hand the condition $\inf_{\mathbb{D}}(|\Theta_1| + |\Theta_2|) > 0$. □

**165.22. Concerning Hartman's Theorem** cf. Lecture VIII, Sect. 7. As Coburn has remarked (cf. Sarason [9]) the fact that $H^\infty + C$ is closed follows in a trivial manner from Hartman's Theorem (if the latter has already been proved independently as, for instance, in AAK [2]). This theorem gives further (cf. Power [1]) a *compactness criterion for the commutator* $[P_+, \varphi] = P_+\varphi - \varphi P_+$ in $L^2$: $[P_+, \varphi] \in \mathfrak{S}_\infty \Leftrightarrow P_+\varphi P_- \in \mathfrak{S}_\infty$, $P_-\varphi P_+ \in \mathfrak{S}_\infty \Leftrightarrow \bar{\varphi} \in H^\infty + C$, $\varphi \in H^\infty + C \Leftrightarrow \varphi \in QC$. □

**165.23. Estimates for the *s*-Numbers.** Concerning Theorem 23 cf. Peller [3], [4] (vectorial case) where auxiliary references to the literature may be found, as well as Peller [6] where the case $0 < p < 1$ is considered. The asymptotics of the $s$-numbers (eigenvalues) for some operators $H[v]$ has been considered by Widom [1], Parfenov [1]; cf. also the survey by Birman-Solomyak [1].

**165.24–26, 28, 31–32, 34. The First Part of the Theory** connecting the properties of the operators $T_\varphi$ with the properties of the multiplication operators $\varphi$ follows in essence the paper Brown-Halmos [1]; cf. further Halmos [2]. The equation $S^*TS = T$ has subsequently served as a basis for generalizations, cf. 168. The term "sectorial" is taken over from Douglas-Widom [1].

**165.27.** Hartman-Wintner [1], Gohberg [3], Brown-Halmos [1], Douglas [3].

**165.29.** For these topics see Kato [1].

**165.30.** Statement 2 = Brown-Halmos [1], Statement 1 = Hartman-Wintner [2], Douglas [3]. For an alternative proof cf. 38.

**165.33, 34.** Hartman-Wintner [1], Brown-Halmos [1]. The point spectrum of the Toeplitz operator $T_\varphi$ is usually connected with the winding number wind $(\varphi - \lambda)$. Some works are devoted to a search for necessary and (or) sufficient conditions for Ker $T_\varphi = \{\mathbb{O}\}$; for a survey (and new results) see Clark [4].

**165.35–38.** This series of conditions is in essence due to Widom [1] who used ideas from Helson-Szegö [1]. The presentation of the text is close to Devinatz [1]; that author shows all theorems at once for $H^2$ spaces generated by an arbitrary Dirichlet algebra. For yet another criterion see 165.64. The connection with the norm $\|\cdot\|_{BMO}$ is discussed in 165.53.3.

Let us make *some complements* to Theorem 36:

a) if $|\varphi| = 1$ a.e., $0 \notin \sigma(T_\varphi)$ and $\|\varphi - h\|_\infty < 1$ $(h \in H^\infty)$ then $h$ is outer (because $\|1 - \bar{\varphi}h\|_\infty < 1 \Rightarrow |I - T_\varphi^* T_h| < 1 \Rightarrow 0 \notin \sigma(T_h)$. $\square$

b) if $h$ is outer, $h \in H^2$, $\mathrm{dist}(\bar{h}/h, H^\infty) < 1$ then $h^{-1} \in H^2$. (Consequence of the Helson-Szegö Theorem, Lecture VIII, Sect. 6.)

c) uniqueness of the representation $\varphi = \bar{h}h^{-1}$: if $h_i^{\pm 1} \in H^2$, $i = 1, 2$, and $\bar{h}_1 h_1^{-1} = \bar{h}_2 h_2^{-1}$, then $h_1 = \lambda h_2$, $\lambda = \text{const}$ (because $\bar{h}_1 \bar{h}_2^{-1} = h_1 h_2^{-1} \in \bar{H}^2 \cap H^2$. $\square$).

d) if $\varphi = e^{i(u+\tilde{v})}$, $u \in C(\mathbb{T})$, $v \in L^\infty$, then the operator $T_\varphi$ is invertible (because $u = (u - p) + \tilde{\tilde{p}} + c$, where $p$ is a trigonometric polynomial, $\|u - p\|_\infty < \varepsilon$, $p \in C(\mathbb{T})$, $c = \text{const}$. $\square$).

e) if $\varphi = e^{i(u+\tilde{v})}$, $u \in C(\mathbb{T})$, $v \in L^\infty$, then $\varphi = \bar{h}h^{-1}$, with $h^{\pm 1} \in \bigcap_{p<\infty} H^p$ (because $\varphi^n = \bar{h}_n h_n^{-1}$, $h_n^{\pm} \in H^2$ (in view of d) and and Theorem 36) and $h = \lambda h_n^{1/n}$ by c). $\square$).

**165.41.** Statement 1 = Widom [1], Devinatz [1] (implicit), Rabindranathan [1]. Statement 2 = Douglas-Sarason [1].

**165.42.** Concerning the algorithm mentioned in 42 for the decomposition of the function as a sum $u + v$ see 165.53.3.

**165.43.** Coburn [1]. For multivariate and vectorial Toeplitz operators this property is not any longer true.

**165.47.** Widom [1] in the case that $\arg \varphi$ has only a finite number of jumps, Devinatz [1] in general. Employing a different language (cf. below) a much more general result was established by Simonenko [1]. The spectral theory of operators $T_\varphi$ with symbols $\varphi$ whose Fourier series are absolutely convergent, $\sum_{n\in\mathbb{Z}} |\hat{\varphi}(n)| < \infty$, was erected earlier by Kreĭn [1].

Until quite recently one has witnessed three formally independent lines of development of the spectral theory established here: boundary problems, Wiener-Hopf equations, pure Toeplitz operators. In 1948 Rapoport [1] unified the spectral theory of Wiener-Hopf operators and singular integral operators of the form $\varphi P_+ + \psi P_-$ (in the space $L^2$), to which the classical Riemann-Hilbert problem etc. can be reduced (cf. 167). There the rôle of the winding number wind in the spectral theory of Wiener-Hopf operators was discovered. In the paper [2] by the same author there is considered, apparently for the first time, the question of the invertibility (search of the spectrum) for "Toeplitz operators" proper (discrete analogues of the Wiener-Hopf operators). The theorem of the unitary equivalence of the two theories – Toeplitz operators and Wiener-Hopf operators – was established only in 1965 by Rosenblum [2], while the simplest and most natural way of identifying the two subjects was found in 1967 by Devinatz [2] (for this refer to 166). A major contribution to the development of the spectral theory of integral operators was provided by the Georgian school of mathematics (Mushelišvili, Vekua, Gahov, Hvedelidze etc.), especially in the initial phase and what subsequent generalization goes (weighted spaces, $L^p$ spaces etc.); for the history of the question see the books mentioned in 165.1. The disconnection between the essentially identical directions referred to at the beginning of this paragraph has resulted in a multiple repetition in the literature of many of the discoveries of the theory.

Theorem 47 for continuous symbols $\varphi$ and within the framework of singular integral equations was in essential parts proved by Mihlin [1] and Gohberg [1]. Simonenko (still in the language of singular integral equations) in his paper [1], which anticipated the discovery of the localization principle, established a theorem which is considerably more general than 47 (and in fact contains Statements 78, 80; from the localization principle 76 and the elementary Theorem 30, Statement 2, 47 immediately follows).

Gohberg [2] remarked that for continuous symbols $\varphi$ holds

$$\boldsymbol{\varphi} = P_+\boldsymbol{\varphi}P_+ + P_-\boldsymbol{\varphi}P_- + P_-\boldsymbol{\varphi}P_+ + P_+\boldsymbol{\varphi}P_- = P_+\boldsymbol{\varphi}P_+ + P_-\boldsymbol{\varphi}P_- + K,\ K\in\mathfrak{S}_\infty, \tag{35}$$

and again that from this one *in a trivial way obtains* a criterion "$\min_{\mathbb{T}} |\varphi| > 0$" for the operator $T_\varphi$ to be Fredholm. Subsequently, extending this commutator approach, Gohberg and Krupnik [2] (cf. 165.140.1) proved an analogue of Theorem 47 for operators of the form $T = \sum_i \prod_j T_{\varphi_{ij}} \in PC$. Sarason [11] then did the same thing for $\varphi_{ij} \in PQC \overset{\text{def}}{=} \operatorname{alg}(PC, QC)$ (regarding this last algebra cf. 165.50). In the same paper Sarason also gives a description of the set $\varphi_*(\mathbb{T})$ in the language of the harmonic continuation $\mathscr{H}\varphi$ of $\varphi$ (cf. 165.50).

**165.48. The Class** $H^\infty + C(\mathbb{T})$ was introduced by Sarason [6] and was in connection with Toeplitz operators considered by Douglas [4]. That $H^\infty + C$ is closed was established by Sarason [6] with the aid of a chain of dualities: $(C(\mathbb{T})/C_A)^* = H_0^1$, $(H_0^1)^* = L^\infty/H^\infty$ (natural isomorphisms), showing that the imbedding $j: C/C_A \to L^\infty/H^\infty$ is isometric and therefore the pre-image $\pi^{-1}j(C/C_A) = H^\infty + C$, $\pi: L^\infty \to L^\infty/H^\infty$, closed in $L^\infty$. □ In Lecture VIII we used instead of the duality a remark by Zal'cman ($\operatorname{dist}(f, C_A) = \operatorname{dist}(f, H^\infty)$, $\forall f \in C(\mathbb{T})$). Other proofs for the fact that $H^\infty + C$ is closed can be found in Adamyan-Arov-Kreĭn [2], Rudin [3]. Sarason [9] has remarked that from $f \in L^\infty$, $|\mathscr{H}f| \in C(\operatorname{clos}\mathbb{D})$ it follows that $f \in H^\infty + C$. An interesting property of the algebra $H^\infty + C$ was discovered by Axler [1].

**165.48.1. Axler's Theorem.** *Let $f \in L^\infty(T)$. There exist a Blaschke product $B$ and a function $g$, $g \in H^\infty + C$, such that $f = g/B$.*

*Proof.* By the Douglas-Rudin Theorem (Appendix 2) there exist Blaschke products $B_n$ and functions $g_n$, $g_n \in H^\infty$, such that $\lim_n \|f - g_n/B_n\|_\infty = 0$. Write $B_n = B_n' \cdot B_n''$ where $B_n'$ is a finite product, $|B_n''(0)| > 1 - 2^{-n}$, and set $B = \prod_{n\geq 1} B_n''$. Then $\lim_n \|Bf - Bg_n/B_n\|_\infty = 0$ and $Bg_n/B_n = (B/B_n'')\cdot g_n \cdot 1/B_n' \in H^\infty + C$, because $1/B_n' \in C$. Consequently $Bf \in H^\infty + C$. □

Later Wolff [1], answering a question by Sarason, established a in a sense more refined result (cf. 165.53); it follows in particular from Wolff's theorem that if $\varphi \in H^\infty + C$, $\operatorname{ess\,inf}_{\mathbb{T}} |\varphi| > 0$, then $\varphi = h \cdot c$ with $h \in H^\infty$, $c \in C(\mathbb{T})$.

**165.49.** Wintner [1]; cf. also Halmos [2].

**165.50.** Douglas [4], [5]; these papers contain also the propositions 51, 52. The most useful formula (18) was found by Peller and Hruščev [1]. Wolff's Theorem, mentioned six lines above, leads (independently of Theorem 50) to the following result.

**165.50.1. Corollary.**

$$\varphi \in H^\infty + C,\ 0 \notin \sigma_{\mathrm{ess}}(T_\varphi) \Leftrightarrow \varphi = f \cdot g;\ f^{\pm 1} \in H^\infty,\ g^{\pm 1} \in C(\mathbb{T});\ \operatorname{ind} T_\varphi = -\mathit{wind}\, g.$$

In fact, if $0 \notin \sigma_{\mathrm{ess}}(T_\varphi)$ then $\operatorname{ess}_T \inf|\varphi| > 0$ and $\varphi = h \cdot c$ ($h \in H^\infty$, $c \in C(\mathbb{T})$); it is clear that $\operatorname{ess\,inf}_{\mathbb{T}} |h| > 0$ and $(h^e)^{-1} \in H^\infty$ (the outer factor). This together with the formula $T_\varphi = T_c T_{h^i} T_{h^e}$ shows that $T_{h^i}$ is Fredholm, that is, $\deg h^i < \infty$. It remains to set $f = h^e$, $g = h^i c$.

The converse statement is trivial. □

Let us state yet another form of the criterion for operators $T_\varphi$, $\varphi \in \mathrm{H}^\infty + \mathrm{C}$, to be Fredholm: $\operatorname{ess\,inf}_{\mathbb{T}} |\varphi| > 0$, $\varphi/\varphi^e = z^n e^{i(u+v)}$ with $n \in \mathbb{Z}$, $u, v \in C(\mathbb{T})$ (follows from 50 and 57).

**The Rôle of Harmonic Continuation** in the Fredholm problem for Toeplitz operators has lead to an independent problem in Douglas [3] and Sarason [11]. It follows from 33 and elementary considerations that an operator $T_\varphi$ with real symbol $\varphi$ is Fredholm (or invertible) iff $\inf\{|(\mathscr{H}\varphi)(\zeta)| : 1 - \delta < |\zeta| < 1\} > 0$ with a suitable $\delta$, $\delta > 0$. To the same condition reduces also the Fredholmness of $T_\varphi$ for $\varphi \in H^\infty + C$ (Theorem 50). The problem in question consists now of deciding for which class of symbols $\varphi$ the Fredholmness (respectively invertibility) of $T_\varphi$ is connected with the modulus $|\mathscr{H}\varphi|$ of the harmonic continuation. To what has already been said on this subject we may add the following.

**165.50.2.** *If* $\varphi \in H^\infty + C$ *and* $\inf_{\mathbb{D}} |\mathscr{H}\varphi| > 0$ *then* $0 \notin \sigma(T_\varphi)$.

In fact, if this is so then $T_\varphi$ is Fredholm an $\operatorname{ind} T_\varphi = -\lim_{r\to 1} \operatorname{wind} \varphi_r(\mathbb{T})$ (50 and 51); since the curves $\varphi_r$, $0 \le r \le 1$, are homotopic (in $\mathbb{C}\backslash\{0\}$) the winding number does not depend on $r$ and $\operatorname{ind} T_\varphi = -\operatorname{wind} \varphi_0(\mathbb{T}) = 0$. Hence $0 \notin \sigma(T_\varphi)$. □

**165.50.3.** *Let* $\varphi \in L^\infty$ *and* $0 \in \operatorname{int}(\operatorname{conv} \operatorname{Range} \varphi)$. *Then there exists a real function* $\psi$, $\psi \ge 1$, $\psi \in L^\infty$, *such that* $\int_{\mathbb{T}} \varphi\psi \, dm = 0$. *If* $\varphi \in C$ *(or* $\varphi \in \mathscr{P}$*) then* $\psi$ *can be taken in this class too.*

This can be proved by elementary considerations with convex hulls. □

As the following example shows, the invertibility of $T_\varphi$ can be connected with the condition $\inf_{\mathbb{D}} |\mathscr{H}\varphi| > 0$ perhaps only under special restrictions on the modulus of $\varphi$ (for instance the question is open for unimodular $\varphi$, in which terms the invertibility criterion 37, 38 is stated).

**165.50.4.** *Douglas's example* [3]. There exists an invertible operator $T_\varphi$ with $\inf\{|\mathscr{H}\varphi(\zeta)| : 1 - \delta < |\zeta| < 1\} = 0, \forall\delta > 0$.

Let $p$ be a trigonometric polynomial with $T_p$ invertible but $0 \in \operatorname{int}(\operatorname{conv} \operatorname{Range} p)$. Let $\psi$ be the polynomial gotten from $p$ by 165.50.3; then $0 \notin \sigma(T_{p\psi})$ but $\mathscr{H}(p\psi)(0) = 0$. Let further $B$ be an infinite Blaschke product and set $\varphi = (p\psi) \circ B$. Then $0 \notin \sigma(T_\varphi)$ (cf. 83) but the function $\mathscr{H}\varphi = (\mathscr{H}p\psi) \circ B$ vanishes at all points $\zeta$ where $B(\zeta) = 0$. □

Recently Wolff [2] constructed an operator $T_\varphi$ which is not invertible, although $\inf_{\mathbb{D}} |H_\varphi| > 0$. Here we prove instead

**165.50.5. Tolokonnikov's Theorem** [2]. *Let* $\varphi \in L^\infty$ *and assume that* $\frac{23}{24} < \delta \overset{\text{def}}{=} \inf_{\mathbb{D}} |\mathscr{H}\varphi|$, $|\varphi| \le 1$. *Then* $0 \notin \sigma(T_\varphi)$ *and* $|T_\varphi^{-1}| \le (24\delta - 23)^{-1/2}$.

*Proof.* The theorem follows from the identity $\|\varphi f\|^2 = \|T_\varphi f\|^2 + \|H_\varphi f\|^2$, $f \in H^2$, and the estimate $|H_\varphi| \le 2\sqrt{2e(1-\delta)}$; the latter is established in the style of Appendix 3. By Nehari's Theorem $|H_\varphi| = \sup\{|\int_{\mathbb{T}} F\varphi \, dm| : \|F\|_1 \le 1, F \in H_0^1\}$, whereas by formula (2) in Appendix 3 (for simplicity we write $\varphi$ in place of $\mathscr{H}\varphi$)

$$\left|\int_{\mathbb{T}} F\varphi \, dm\right| = \left|\frac{1}{2\pi} \iint_{\mathbb{D}} |\Delta(F\varphi)| \log 1/|z| \, dx\, dy\right|$$

$$= \left|\frac{2}{\pi} \iint_{\mathbb{D}} |F'\bar{\partial}\varphi| \log 1/|z| \, dx\, dy\right| \le$$

$$\le \frac{2}{\pi} \left(\iint_{\mathbb{D}} |F'|^2/|\Phi|^2 \log 1/|z| \, dx\, dy\right)^{1/2} \cdot$$

$$\cdot \left(\iint_{\mathbb{D}} |\bar{\partial}\varphi|^2 |\Phi|^2 \log 1/|z| \, dx\, dy\right)^{1/2}$$

where $\Phi$ is an outer function, $|F| = |\Phi|^2$. The integrals are estimated as in Appendix 3. The first of them is the integral $I_1'$ from Sect. 5 of that Appendix and therefore does not exceed $(2\pi)^{1/2}$; the second one is similar to the integral $I_2$ in Sect. 6 of Appendix 3: $\Delta(|\varphi|^2) = 4(|\partial\varphi|^2 + |\bar{\partial}\varphi|^2)$,

$$\iint_{\mathbb{D}} |\bar{\partial}\varphi|^2 |\Phi|^2 \log 1/|z|\, dx\, dy$$

$$\leq \frac{1}{4} \iint_{\mathbb{D}} |\Phi|^2 \Delta(|\varphi|^2) \log 1/|z|$$

$$\leq \text{(Lemma 6, Appendix 3, with } a = |\varphi|^2 - \delta^2) \; \frac{\pi e}{2}(1 - \delta^2).$$

Hence $|H_\varphi| \leq 2(3 + \sqrt{6})^{1/2}(1 - \delta^2)^{1/2} \leq 2\sqrt{e(1 - \delta^2)}$. □

**165.50.6. The Fredholm Criterion in Terms of $|\mathscr{H}\varphi|$ extends to the Class** $PQC \overset{\text{def}}{=} \text{alg}(PC, QC)$, Sarason [11]. On the other hand in the same paper it is shown that such a criterion also can be expressed in terms similar to 47 (by means an "averaging of the jumps" of the symbol $\varphi$). Let $\varphi \in PQC$, $\lambda = e^{i\theta} \in \mathbb{T}$. Set

$$\gamma_\lambda(\varphi) = \overline{\lim_{\delta \to 1}} \left| \frac{1}{\delta} \int_\theta^{\theta+\delta} \varphi(e^{it})\, dt - \frac{1}{\delta} \int_{\theta-\delta}^{\theta} \varphi(e^{it})\, dt \right|,$$

$$\beta_\lambda(\varphi) = \underline{\lim_{\delta \to 0}} \operatorname{dist}\left(0, \left[\frac{1}{\delta} \int_\theta^{\theta+\delta} \varphi(e^{it})\, dt, \frac{1}{\delta} \int_{\theta-\delta}^{\theta} \varphi(e^{it})\, dt\right]\right).$$

**Sarason's Theorem.** *Let $\varphi \in PQC$, $\operatorname{ess\,inf}_{\mathbb{T}} |\varphi| > 0$. The following statements are equivalent.*

1. $0 \notin \sigma_{\text{ess}}(T_\varphi)$.
2. $\beta_\lambda(\varphi) > 0$, $\forall \lambda \in \mathbb{T}$.
3. $\gamma_\lambda(\varphi/|\varphi|) < 2$, $\forall \lambda \in \mathbb{T}$.
4. $\inf\{|\mathscr{H}\varphi(\zeta)| : 1 - \delta < |\zeta| < 1\} > 0$ *for some* $\delta > 0$.

The proof depends on the identity $QC = VMO \cap L^\infty$ (cf. 53) and some estimates for the mean oscillation by means of which one derives for $\varphi \in PQC$ the identity (cf. 75, 165.21)

$$\operatorname{dist}(\varphi, H^\infty + C) = \max_{\lambda \in \mathbb{T}} \frac{1}{2} \gamma_\lambda(\varphi).$$

The implications $2 \Leftrightarrow 3$ are elementary while the proof of $1 \Leftrightarrow 2$ utilizes 41 and the preceding formula for $\operatorname{dist}(\varphi, H^\infty + C)$. The implication $2 \Leftrightarrow 4$ reflects in an essential way the "rôle of the harmonic continuation"; in particular, it gives another proof of Theorem 47. In the case of "pure jumps" (i.e. when the left and right limits $S_l$ and $S_r$ exist at the point $\lambda$) the statement $2 \Leftrightarrow 4$ is geometrically transparent, because it is clear that the limit values of the function $\mathscr{H}\varphi(\zeta)$ along the straight line $\{\zeta : \arg(\lambda - \zeta) = \alpha\}$, $|\alpha| < \pi/2$, cover the whole interval $(S_l, S_r)$ as the angle $\alpha$ runs through all the permissible values. A refinement of this reasoning (Sarason [11]) reveals that $\beta_\lambda(\varphi) = 0 \Leftrightarrow 0$ is a limit value of $\mathscr{H}\varphi$ at the point $\lambda$; here $\varphi \in PQC$, $\operatorname{ess\,inf}_{\mathbb{T}} |\varphi| > 0$.

These reasonings might give the impression that Statement 4 in Sarason's Theorem *always* forces $\varphi$ to be locally sectorial (cf. 80) and thus $T_\varphi$ to be Fredholm but this is not the case (private communication by A. B. Aleksandrov). Let us notice that the converse is trivial: if $\operatorname{ess\,inf}_{\mathbb{T}}|\varphi| > 0$, $\varphi$ locally sectorial, then Statement 4 holds true. It follows from this and Theorem 38, Statement 4, that the symbol of an invertible Toeplitz operator $T_\varphi$ can be represented in the form $\varphi = h\varphi_1$ where $h^{\pm 1} \in H^\infty$ and $\inf\{|\mathscr{H}\varphi_1(\zeta)| : 1 - \delta < |\zeta| < 1\} > 0$ for some $\delta$, $\delta > 0$.

**165.51.** Douglas [4], [5]. In connection with Statements 2 and 3 let us point out that *the spectrum $\sigma(T_\varphi)$ is indeed connected for any symbol $\varphi$, $\varphi \in L^\infty$* (Widom [3]); cf. also Douglas [2] where it is shown that even the essential spectrum $\sigma_{ess}(T_\varphi)$ is connected. The formula $\operatorname{ind} T_\varphi = -\lim_{r \to 1-0} \operatorname{wind} \varphi_r(\mathbb{T})$ holds also for $\varphi \in PQC$ (Sarason [11]).

**165.52.** Douglas [4], [2]. If $f \in QC$, $g \in L^\infty$, likewise $\lim_{r \to 1} \|f_r g_r - (fg)_r\|_\infty = 0$ is valid (Sarason [11]). However, the harmonic continuation $\mathscr{H}$ cannot be asymptotically multiplicative in any (closed) algebra larger than $H^\infty + C$.

**165.53. The Class $QC$.** The basic material can be found in the papers Douglas [4], [3], [6], Sarason [12], [9]. The identity $QC = VMO \cap L^\infty$ was proved by Sarason [12]. On the rôle of this class in the theory of localization of Toeplitz operators (in the spirit of 159–163) cf. Douglas [3], [6].

Formula (35) is valid also for $\varphi \in QC$, which at once yields a Fredholm criterion.

**165.53.1. Corollary.** *If $\varphi \in QC$ then*

$$0 \notin \sigma_{ess}(T_\varphi) \Leftrightarrow \operatorname{ess\,inf}_{\mathbb{T}} |\varphi| > 0. \ \square$$

This result follows of course also from 50, 53. From the relation $QC = VMO \cap L^\infty$ and the description of the class $VMO$ in terms of the harmonic continuation $\mathscr{H}$ (165.53.3) it again follows that a *unimodular* function $u$ is in $QC$ iff $|\mathscr{H}u| \in C(\operatorname{clos} \mathbb{D})$; cf. Sarason [9].

Important information on the class $QC$ was provided by Wolff [1].

**Wolff's Theorem.** 1. *If $f \in L^\infty$ then there exist functions $g$, $h$ such that $f = g/h$, $g \in QC$, $h$ an outer function in $QC_A = QC \cap H^\infty$.*

2. *If $u \in H^\infty + C$, $|u| = 1$ a.e., then $u = \Theta v$ for a suitable inner function $\Theta$ and with $v \in QC$.* (In fact, $\Theta$ can be taken to be a Blaschke product.)

**165.53.2. Complement.** *Statement 2 in Wolff's Theorem is equivalent to the following*:

$$f \in H^\infty + C,\ \operatorname{ess}_{\mathbb{T}} \inf |f| > 0 \Rightarrow f = f_0 f_1 \text{ where } f_0 \in H^\infty,\ f_1 \in C(\mathbb{T}). \tag{36}$$

In fact, it follows from Wolff's Theorem that the function $u \stackrel{\text{def}}{=} f/f_e$ ($u \in H^\infty + C$) has the representation $u = \Theta \cdot v$, $v \in QC$. From 57 we get $v = z^n e^{i(u + \tilde{v})} = z^n e^{v + i\tilde{v}} e^{-v + iu}$. Thus we may take $f_0 = f_e e^{v + i\tilde{v}}$, $f_1 = z^n e^{-v + iu}$.

Conversely, assume that $u \in H^\infty + C$, $|u| = 1$ a.e. By (36) $u = f_0 f_1$ where clearly $\operatorname{ess\,inf}_{\mathbb{T}} |f_0| > 0$, $\min_{\mathbb{T}} |f_1| > 0$. Hence $u = \Theta \cdot v$ where $\Theta = f_0^i$ (the inner part of $f_0$) and $v = f_0^e f_1$. Also $v \in H^\infty + C$ and $\bar{v} = v^{-1} = \dfrac{1}{f_0^e} \cdot \dfrac{1}{f_1} \in H^\infty + C$, i.e. $v \in QC$. $\square$

**165.53.3. The Classes** $VMO$ **and** $BMO$ have appeared in this Appendix mostly only episodically (and mainly implicitly) although they play a prominent rôle in contemporary Function Theory, cf. Garnett [2], Koosis [3], Sarason [9]. As definition one may take any of the following equivalent descriptions of $VMO$ and $BMO$; in order to shorten the notation we put in these formulae $f_I = (mI)^{-1}\int_I f\,dm$, $I$ any interval (arc) on the circumference $\mathbb{T}$; $f_\zeta = (\mathscr{H}f)(\zeta)$, $\zeta \in \mathbb{D}$.

$$\begin{aligned} BMO &= \left\{ f \in L^1(\mathbb{T}) : \sup_I |f - f_I|_I < \infty \right\} \\ &= \left\{ f \in L^1(\mathbb{T}) : \sup_\zeta |f - f_\zeta|_\zeta < \infty \right\} \\ &= \{ u + \tilde{v} : u, v \in L^\infty(\mathbb{T}) \}, \\ VMO &= \left\{ f \in L^\infty(\mathbb{T}) : \lim_{\varepsilon \to 0} \sup_{I, mI < \varepsilon} |f - f_I|_I = 0 \right\} \\ &= \left\{ f \in L^1(\mathbb{T}) : \lim_{\varepsilon \to 0} \sup_{1-\varepsilon < |\zeta| < 1} |f - f_\zeta|_\zeta = 0 \right\} \\ &= \{ u + \tilde{v} : u, v \in C(\mathbb{T}) \}. \end{aligned}$$

The norms, which in the natural way correspond to these descriptions of $BMO$ and $VMO$, are all equivalent. One of these norms, $\|f\|_{BMO} \sim \inf\{\|u\|_\infty + \|v\|_\infty : f = u + \tilde{v}\}$, appears in fact in the basic invertibility criterion 36, 38: for a suitable branch of the argument $\arg\varphi$ the invertibility of $T_\varphi$ ($\operatorname{ess\,inf}_{\mathbb{T}} |\varphi| > 0$) is equivalent to the condition $\operatorname{dist}_{BMO}(\arg\varphi, \widetilde{L}^\infty + \mathbb{R}) < \pi/2$. Indeed, here $\operatorname{dist}_{BMO}$ coincides with $\operatorname{dist}_{L^\infty}$. Jones and Garnett (cf. Garnett [2]) have shown that if $f$ is a real function then

$$\operatorname{dist}_{BMO}(f, L^\infty) \overset{\text{def}}{=} \inf\{\|v\|_\infty : f = u + \tilde{v},\, u \in L^\infty\} = \frac{\pi}{2}\varepsilon(f)$$

where $\varepsilon(f)^{-1} = \sup\{c : \sup_I[(mI)^{-1}m\{x \in I : |f - f_I| > \lambda\}] \le e^{-c}$ for $\lambda > \lambda(c)\}$ is the least constant in the John-Nirenberg inequality; moreover $\varepsilon(f) = e \cdot \overline{\lim_{p\to\infty}} \frac{1}{p} \cdot [(|f - f_I|^p)_I]^{1/p}$.

It is clear that in the language of $BMO$ and $VMO$ the theorems of Nehari and Hartman admit the following formulation: let $\varphi \in L^2$ and set $H_\varphi x = P_-\varphi x$, $x \in \mathscr{P}_A$; a necessary and sufficient condition for $H_\varphi$ to be bounded (compact) is that $P_-\varphi \in BMO$ (respectively $P_-\varphi \in VMO$).

**165.54.** Douglas [3]. The connection between the "unitarity mod $\mathfrak{S}_\infty$" for $T$ and the possibility of representing the operator in the form $T = U + K$, $K \in S_\infty$, becomes so simple only because of the fact that the spectrum of a unitary operator is "thin". The corresponding connection between "normality mod $\mathfrak{S}_\infty$"

($[T, T^*] \in \mathfrak{S}_\infty$) and the possibility of representing the operator in the form $T = N + K$ ($N$ normal, $K$ compact) is considerably more complicated and also more interesting, cf. the books Fillmore [1], Douglas [7].

**165.56. Peller-Hruščev** [1]. It is equally easy to establish the following result of Sarason's, which will be used in 165.124.9.

**165.56.1. Lemma.** *If* $|u| = 1$ *a.e.,* $\operatorname{dist}(u, H^\infty) < 1$, *but* $\operatorname{dist}(u, zH^\infty) = 1$ *then* $\bar{u} \in H_u^\infty$.

*Proof.* It follows from 41 that $T_U$ is left invertible; by the hypothesis of the lemma and using Statement d); Lecture VIII, Sect. 6, we see that $T_U$ is then invertible too. By 36 (or once more by 41) we get $\operatorname{dist}(\bar{u}, H^\infty) < 1$ and finish as in 56: $\|\bar{u} - v\|_\infty = \|1 - uv\|_\infty < 1$ $(v \in H^\infty) \Rightarrow \bar{u} = v \sum_{n \geq 0} (1 - uv)^n \in H_u^\infty$. □

**165.57.** Sarason [12]. The number $n$ in Sarason's representation is of course $-\operatorname{ind} T_\varphi$.

**165.58. The Algebras** $H_M^\infty$ were in connection with the spectral theory of Toeplitz operators first considered by Douglas [5]. For these algebras cf. 165.124.

**165.59.** Douglas [5], [2], Lee-Sarason [1] (Statement 2*c* with $\sigma^r(T_\varphi)$ replaced by $\sigma(T_\varphi)$); Sarason [9] (the general case). On the computation of the spectrum $\sigma_{H_M^\infty}(\varphi)$ cf. 93, 165.93. In the more general context of the algebras alg $T_X$ a "spectral inclusion theorem" is discussed in 112, 118 and elsewhere. Let us also notice that the inclusion $\sigma_{H_\varphi}(\varphi) \subset \sigma(T_\varphi)$ trivially follows from 38, Statement 4: if $0 \notin \sigma(T_\varphi)$ then $\varphi = h\psi$ with $h^{\pm 1} \in H^\infty$ and $\psi$ sectorial with $\operatorname{ess\,inf}_{\mathbb{T}} |\psi| > 0$; the function $1/z$ being the uniform limit of polynomials on the set Range $\psi$ this gives $1/\psi \in \bigvee(\psi^n : n \geq 0)$. □

**165.60. The Proof of the Theorem on the Minimality** of the algebra $H^\infty + C$, as formulated in Remark 60, is due to Hoffman and Singer (cf. Hoffman [1], chap. VI; Garnett [2], chap. IX, Sect. 1). It depends on two results on uniform algebras: a) the algebra $H^\infty$ is logmodular on the maximal ideal space $\mathfrak{M}$ of the algebra $L^\infty(\mathbb{T})$ (trivial); b) every multiplicative functional ($\gamma$ say) of a logmodular algebra has a unique representing measure, i.e. a measure $\mu$, $\mu \geq \mathbb{0}$, such that $\gamma(f) = \int f\, d\mu$, $\forall f$ (Hoffman's Theorem; cf. Gamelin [1], Theorem 4.2, chap. 2).

*One has to prove* that if $\mathscr{D}$ is a subalgebra of $L^\infty$, $H^\infty \subset \mathscr{D}$ and $H^\infty \neq \mathscr{D}$, then $H^\infty + C \subset \mathscr{D}$. To this end it suffices to show that $\bar{z} \in \mathscr{D}$, i.e., that $z$ is an invertible element in $\mathscr{D}$.

Assume that there exists in $\mathscr{D}$ a multiplicative functional $\gamma$ with $\gamma(z) = 0$. Set $\beta = \gamma | H^\infty$. Then $\beta(zH^\infty) = 0$ and for all $f$, $f \in H^\infty$,

$$\beta(f) = \beta(f(0)\mathbb{1} + (f - f(0)\mathbb{1})) = f(0)\beta(\mathbb{1}) = f(0) = \int f\, dm,$$

so that $m$ is a representing measure for $\beta$. But then by Hoffman's Theorem it represents also the functional $\gamma$. Therefore for all $f$, $f \in \mathscr{D}$, and all $n$, $n \geq 1$,

$$\int z^n f\,dm = \gamma(z^n f) = \gamma(z^n)\gamma(f) = 0,$$

i.e. $f \in H^\infty$. This contradiction shows that $z$ is invertible in $\mathscr{D}$. □

In the case $(\mathscr{D} \cap C(\mathbb{T}))\backslash H^\infty \neq \varnothing$ there is a direct elementary proof due to P. Cohen, cf. Hoffman [1].

**165.61.** Douglas-Sarason [2] (with $\sigma^r_{\text{ess}}(T_\varphi)$ replaced by $\sigma(T_\varphi)$).

**165.62, 63, 66.** Regarding almost periodic functions see Katznelson [1], Levitan [1].

**165.64.** Gohberg-Fel'dman [1], [2], Coburn-Douglas [1]. Of course, the relation $w_f = 0\,(f \in AP)$ does not entail that the function $\varphi = f \circ \omega$ is continuous at the point $\zeta$. The proof of 64 is essentially the only place in Appendix 4 where we employ the elements of the original Wiener-Hopf technique: if $\varphi = \varphi_- \varphi_0 \varphi_+$, $\varphi_+$, $\bar{\varphi}_- \in H^\infty$, $\varphi_+^{-1}$, $\bar{\varphi}_-^{-1} \in H^\infty$ and $\|\mathbb{1} - \varphi_0\|_\infty < 1$ then $0 \notin \sigma(T_\varphi)$. Of course, it was development of this approach that lead straight to the Theorems 36–41, using precisely the analytic and anti-analytic "contributions" of the argument $\arg\varphi$ of the symbol of an invertible Toeplitz operator. Moreover Widom [2] discovered the final form of the "Wiener-Hopf method": $0 \notin \sigma(T_\varphi) \Leftrightarrow \exists \varphi_i \in H^2$ such that $\varphi = \varphi_1 \bar{\varphi}_2$, $\varphi_i^{-1} \in H^2$ and the operator $f \mapsto (1/\varphi_1)P_+(1/\bar{\varphi}_2)f$ has a continuous extension to $H^2$.

**165.65.** The number $\text{wind}_\zeta\,\varphi$, $\varphi \in AP_\zeta$, may be identified with the "generalized index" of the operator $T_\varphi$. To this end let us consider a representation $\rho$ of the algebra $\text{alg}\,T_{AP_\zeta}$ in a suitable hyperfinite factor $\mathfrak{A}$ of type $I$; for the operators in $\mathfrak{A}$ one has the notion of "generalized Fredholmness" (= invertibility of their natural projections in the quotient algebra $\mathfrak{A}/I$ by the (unique) closed twosided ideal $I$ in $\mathfrak{A}$; this ideal plays the same rôle as the ideal $\mathfrak{S}_\infty$ in the usual Fredholm theory), of the relative dimension of the kernels $\text{Ker}\,T$, $\text{Ker}\,T^*$ $(T \in \mathfrak{A})$ and of the generalized index $\text{ind}_A T$. One can show that $\text{ind}_A\,\rho A = -\text{wind}_\zeta(\text{Sym}\,A)$ for any $A$, $A \in \text{alg}\,T_{AP_\zeta}$. For this cf. Coburn-Douglas-Schaeffer-Singer [1], Douglas [3]. Besides, the above ideal $I$ hardly admits a description in terms of the operators in $\text{alg}\,T_{AP_\zeta}$.

**165.67.** Gorin [1].

**165.68.** Douglas [3].

**165.69–72, 87, 88.** Gohberg-Semencul [1], Gohberg-Krupnik [1] (here one finds references also to auxiliary literature), Douglas [3]. The presentation of the text is based on the last two sources. (Incidentally, Gohberg (and co-workers) use a somewhat different localization technique, cf. 165.79.1.) To Theorem 72 we may add the following complement: an operator $T_\varphi$, $\varphi \in AP_\sigma + C$, is left (right) invertible iff $\text{ess inf}_{\mathbb{T}}|\varphi| > 0$, $\text{wind}_\zeta\,\varphi \geq 0$, $\forall \zeta \in \sigma$ (respectively $\text{wind}_\zeta\,\varphi \leq 0$); cf. Gohberg-Krupnik [1].

**165.69.1. An Interesting Problem** (in connection with this theory) is stated in

Douglas [3]: for which symmetric algebras $X_M \stackrel{\text{def}}{=} \bigvee(M^n\overline{M}^k : n, k \geq 0)$, generated by a set $M$ of inner functions, do we have an analogue of the results 64, 72, 88? In 64, 72, 88 one takes for $M$ a set of elementary exponentials $\exp\left\{\alpha\frac{z+\zeta}{z-\zeta}\right\}$, $\alpha > 0$, $\zeta \in \mathbb{T}$. The question is reduced to the "independence" of algebras $X_\Theta$ corresponding to different generators $\Theta$, $\Theta \in M$. In particular, if the analogue just discussed is valid for $X_M$ and if $\Theta_1$, $\Theta_2$ are relatively prime inner function in $M$ then the operator $T_{\Theta_1\overline{\Theta}_2}$ is by necessity non-invertible (concerning this subject cf. 91–102). For singular inner functions $\Theta_1$, $\Theta_2$ whose representing measures have finite supports the latter assertion certainly holds. However, there exist relatively prime $\Theta_1, \Theta_2$ *with pure point measures* such that $\|\Theta_1 - \Theta_2\|_\infty < 1$ (and even such with the norm $\|\Theta_1 - \Theta_2\|_\infty$ arbitrarily small), cf. for instance Stephenson [1]. In this case $0 \notin \sigma(T_{\Theta_1\overline{\Theta}_2})$ and the analogue of the $AP_{\mathbb{T}}$-theory does not hold true for the algebra $X_{\Theta_1,\Theta_2}$. Within the framework of this circle of ideas it might also be of interest to describe all inner functions contained in the algebra $X_M$; to these belong for instance the functions $(\lambda - \Theta)(\mathbb{1} - \bar{\lambda}\Theta)^{-1}$, $|\lambda| < 1$, $\Theta \in M$, and their finite products; in general $f \circ \Theta \in X_M$ if $f \in C_A$.

**165.92.2.** Other problems of the generalized $AP$-theory of Toeplitz operators are connected with the Wiener-Hopf equations, cf. 166.

**165.69.3.** For more extensive algebras of symbols than $AP_\zeta + C$, $AP_{\mathbb{T}} + C$, in particular admitting other types of discontinuities, Fredholm criteria and index formulae can be found in the following papers: 1) Sarason [13], the algebra $SAP \stackrel{\text{def}}{=} \text{alg}(AP_\zeta, PC_\zeta)$, $PC_\zeta = PC \cap C(\mathbb{T}\backslash\{\zeta\})$; one defines the left and right winding numbers $\text{wind}_\zeta^\pm$ as well as the left and right "mean values" $M_\pm$ $\left(M_\pm(f) = \lim_{T\to\infty}\frac{1}{T}\int_0^T f(\pm t)\,dt\right.$ after a transition to the axis $\mathbb{R}$). Criterion for $T_\varphi$, $\varphi \in SAP$, to be Fredholm: $\text{ess inf}_{\mathbb{T}}|\varphi| > 0$, $\text{wind}_\zeta^\pm(\varphi) = 0$, $\min_{k\in\mathbb{Z}}|M_+(\arg\varphi) - M_-(\arg\varphi) + k\pi| < \pi/2$. 2) Power [2], the algebras $\text{alg}(AP_\zeta, SO_\zeta)$, $\text{alg}(SAP, SO_\zeta)$, $\text{alg}(MP_\zeta, SO_\zeta)$. Here $SO_\zeta = SO \circ \omega$ is the transplantation from the axis $\mathbb{R}$ onto the circle $\mathbb{T}$ of the class of *slowly oscillating* functions ($f \in SO \Leftrightarrow f \in C(\mathbb{R})$ and $\lim_{x\to\infty}\omega_{I_n}(f) = 0$, with $I_x = [-2x, -x] \cup [x, 2x]$, $\omega_I(f)$ being the oscillation of $f$ on the set $I$), $MP = \{f \in C(\mathbb{R}) : f(x) = f(2x), x \in \mathbb{R}\}$ the class of multiplicatively periodic functions. For the algebras $\text{alg}(SAP, SO_\zeta)$ and $\text{alg}(MP_\zeta, SO_\zeta)$ there is found a sufficient condition for an operator $T_\varphi$ (with the symbol in the algebra) to be Fredholm, for $\text{alg}(AP_\zeta, SO_\zeta)$ a necessary and sufficient one. The latter amounts to $\text{ess inf}_{\mathbb{T}}|\varphi| > 0$ and $\lim_{|t|\to\infty} t^{-1}(\arg f(2t) - \arg f(t)) = 0$ if $\varphi = f \circ \omega$ ($f$ a function on $\mathbb{R}$).

In Power [2] also some other algebras are considered; the method is based on Douglas's Localization Theorem 162 and on the "asymptotic independence" of the function algebras under study (the simplest instance of a similar independence can be located in 70 and 87). Let us further remark that $SO_\zeta \subset QC$, that for the algebra $\text{alg}(AP_\zeta, QC)$ only partial results are available (Douglas [6]) and that the

Toeplitz operators with symbols in $\operatorname{alg}(AP_\zeta, H^\infty) = \operatorname{alg}\left(\exp\left(\frac{\zeta + z}{\zeta - z}\right), H^\infty\right)$ are indeed very numerous (cf. 89–102 and the corresponding remarks).

Cf. also 165.125.

**165.73. The Method of Localization** of Toeplitz operators is one of the most spectacular pages of the entire theory. The credit for the discovery of the method goes to Simonenko [2], [3], [4], who also found numerous applications of it. Incidentally, the utilization of the idea of a local description of (commutative!) Banach algebras (including functions of elements of the algebra; in particular, the local nature of invertibility) may be traced back to Wiener [1] and as an independent general method it is proposed in the work of Šilov [1], [2] and Kreĭn [3].

*Simonenko's Theory* is stated in the language of "operators of local type" ($OLT$). An operator $T: L^p(K) \to L^p(K)$, $1 \le p < \infty$, $K$ a compact space, is called an $OLT$ if

$$\chi_{\Delta_1} T \chi_{\Delta_2} \in \mathfrak{S}_\infty$$

for arbitrary $\Delta_i \in K$, $\Delta_i = \operatorname{clos} \Delta_i$, $\Delta_1 \cap \Delta_2 = \varnothing$. The $OLT$ form an algebra, closed with respect to $|\cdot|_{\text{ess}}$ and self-adjoint in a natural sense. If $T \in OLT$ and if $T$ is invertible (Fredholm) then $T^{-1} \in OLT$ (the left and right regularizers of $T$ are in $OLT$). In a natural way there arises now a notion of equivalence of two operators $T_1$ and $T_2$ at the point $x$, $x \in K$: $T_1 \stackrel{x}{\sim} T_2$ if

$$\inf\{|(T_1 - T_2)\chi\mathcal{O}|_{\text{ess}} + |\chi\mathcal{O}(T_1 - T_2)|_{\text{ess}} : \mathcal{O} \text{ open}, x \in \mathcal{O}\} = 0.$$

If $T_i \in OLT$ $(i = 1, 2)$ then it suffices to take in this definition just one term (any of the terms). Further, the operator $T$ is called Fredholm at the point $x$, $x \in K$, if there exist operators $R_1, R_2$ such that $R_1 T \stackrel{x}{\sim} I$, $TR_2 \stackrel{x}{\sim} I$, and locally Fredholm if it is Fredholm at every point.

**Theorem.** *If $T \in OLT$ and locally Fredholm then it is Fredholm.*

**Corollary.** *If $T \in OLT$ and $\forall x \in K$ $\exists T_x$, a Fredholm operator, $T \stackrel{x}{\sim} T_x$, then $T$ is Fredholm.*

**Theorem.** *If the family of operators $\{T_x\}_{x \in K}$ is locally continuous in the sense that*

$$\inf_{\substack{\mathcal{O} \\ x \in \mathcal{O}}} \sup_{y, y \in \mathcal{O}} |(T_y - T_x)\chi_{\mathcal{O}}|_{\text{ess}} = 0, \forall x \in K,$$

*and if $T_x \in OLT$, $\forall x \in K$, then there exists an operator $T$, $T \in OLT$, such that $T_x \stackrel{x}{\sim} T$, $\forall x \in K$, and this operator is unique* mod $\mathfrak{S}_\infty$.

Simonenko [2–5] gives a variety of examples of operators in $OLT$ and applies to them the theory developed (operators of the Riemann-Hilbert boundary value problem, singular integral operators, pseudo differential operators, operators with Calderón-Zygmund type kernels etc.). In particular, our Theorems 47, 76–78, 80, 82 are contained in [2–5]. In fact, 47, 80, 82 were established already in Simonenko [1].

*Sarason's Method*, as contained in 74, 75, is set out in Sarason [10]. Yet another variant of the method of localization, relying on a notion equivalent to the one of the local distance, is outlined in Gohberg-Krupnik [3], [1].

*Concerning Douglas's Method* (localization in the fibers of the maximal ideal space of the algebras in question) cf. 165.162.

**165.75.** Douglas-Sarason [1] (utilizing the following "local" description of the algebra $H^\infty + C: f \in H^\infty + C \Leftrightarrow \hat{f} | \mathfrak{M}_\lambda \in \widehat{H^\infty} | \mathfrak{M}_\lambda$, $\forall \lambda \in \mathbb{T}$; here $\hat{f}$ is the Gel'fand transform of the element $f$, $f \in L^\infty$, $\mathfrak{M}_\lambda$ the fiber of the maximal ideal space $\mathfrak{M}(L^\infty)$ over the point $\lambda$. $\lambda \in \mathbb{T}$). The elegant treatment of the text is due to Sarason [10]. Sarason also found an explicit formula for $\text{dist}_\lambda(\varphi, H^\infty + C)$ for $\varphi \in PQC$, cf. 165.50.6, 165.21.

**165.76–78.** Simonenko [4], [5], Douglas-Sarason [1], Gohberg-Krupnik [1].

**165.79.** Gohberg-Krupnik [1]. In that book the result 79 is not derived on a local basis but from the following elementary proposition, implicit in [1] and in essence related to the "commutator approach" to the Fredholm problem.

**165.79.1. Lemma.** *Let* $[T_i, T_j] \in \mathfrak{S}_\infty$; $1 \le i, j \le n$. *Then the operator* $T = \prod_{i=1}^n T_i$ *is Fredholm iff all the operators* $T_i$, $1 \le i \le n$, *are Fredholm.*

*Proof.* If the $T_i$ are Fredholm so will be $T$. If however $T$ is Fredholm then clearly the operator $T_j(\prod_{i \ne j} T_i) = T + K_j$, $K_j \in \mathfrak{S}_\infty$, too is Fredholm. Therefore the image of $T_j$ contains the image of the (Fredholm) operator $T + K_j$ and as a consequence is a closed subspace of finite codimension. An analogous conclusion holds then with $T_j^*$. □

**165.80.** Simonenko [1], [6], Douglas-Widom [1] (the term "locally sectorial" is due to these authors). Gohberg-Fel'dman [1] employ the following equivalent form of local sectoriality, borrowed from Simonenko's papers: $\exists p, r$ ($p$ a trigonometric polynomial, $r \in L^\infty$, $\|r\|_\infty < 1$), such that $\varphi = p(1 + r)$; then $\text{wind}\, \varphi \overset{\text{def}}{=} \text{wind}\, p$. The equivalence mentioned is in essence established in Douglas-Widom [1].

**165.81. The Rôle of Local Sectoriality** in the Fredholm problem was discussed already in 165.50. It remains unclear for which classes of symbols the property of $T_\varphi$ being Fredholm is equivalent to $\varphi$ being locally sectorial. For example, one has this equivalence for $\varphi \in PC$, $\text{ess inf}_{\mathbb{T}} |\varphi| > 0$, and further for functions taking two values (cf. 33). For functions taking three values this is not any longer so.

*Example.* Let $\varepsilon$ be a positive function on $(0, \pi/2)$, $\lim_{x \to 0} \varepsilon(x) = 0$, $\Omega = \{x + iy : 0 < x < \pi/2, |y - \sin(1/x)| < \varepsilon(x)\}$, $\omega$ a conformal map of the disk onto $\Omega$. Let $u$ be a real function on $\mathbb{T}$ such that

$$\alpha \stackrel{\text{def}}{=} \operatorname{Im}\omega + u = \begin{cases} \pi/2, \text{ if } \operatorname{Im}\omega(\zeta) > 1/2 \\ -\pi/2, \text{ if } \operatorname{Im}\omega(\zeta) < -1/2. \\ 0, \text{ if } |\operatorname{Im}\omega(\zeta)| \le 1/2 \end{cases}$$

Then $\|u\|_\infty \le \pi/2 - 1/2$ and by Theorem 36 $T_\varphi$, $\varphi \stackrel{\text{def}}{=} e^{i\alpha}$, is invertible, although $\varphi$ is not locally sectorial. □

**165.82.** Simonenko [1].

**165.83.** The problem of describing the multipliers which preserve the Fredholm property (the property of being invertible) apparently was first mentioned in Frolov's paper [1]; cf. also Gohberg-Krupnik [1]. In particular, it has been proved that a piecewise continuous function $\psi$ is a multiplier preserving Inv iff $\operatorname{ess\,inf}_{\mathbb{T}} |\psi| > 0$, $\psi/|\psi| \in C$ (this result – also for all $\psi \in PC$ – follows from 84, Statement 3, and likewise immediately from 47).

The existence of a function $\varphi$, $\varphi \in \operatorname{Inv}$, with $\varphi^2 \notin \operatorname{Fred}$, is noted in Widom [2]. A certain class of additive perturbations (with integral, but not Toeplitz operators) preserving the Fredholm property of the operator $T_\varphi$, $\varphi \in PC$, is described in Krupnik-Špigel' [1].

**165.84, 85.** The implication $1 \Rightarrow 3$ is contained in the results of Krupnik [1]' while the complete formulation of Theorem 84 (in a version slightly different from 84) is due Spitkovskiĭ [1]. In particular, Statement 4 is in that paper replaced by

$$4'.\ \psi \in \operatorname{clos}_{(L^\infty)^{-1}}(H^\infty)^{-1} \cdot (\overline{H^\infty})^{-1},$$

where $(H^\infty)^{-1}$ and $(L^\infty)^{-1}$ are the sets of all invertible functions in the algebras $H^\infty$ and $L^\infty$ respectively. The proof of Theorem 84 given in the text seems to be new. It follows immediately from the definition (and from 70) that the property of "being a multiplier on the set Fred" is a local one (in a sense analogous to 76).

Spitkovskiĭ [2] proved also that the supply of Fredholm multipliers for Toeplitz operators is the same in all spaces $L^p(\mathbb{T})$, $1 < p < \infty$, and that $\operatorname{ind} T_{\varphi\psi} = \operatorname{ind} T_\varphi + \operatorname{ind} T_\psi$ if $\varphi \in \operatorname{Fred}$ and $\psi$ is a multiplier. In the same paper the analogous problem for matrix Toeplitz operators too is considered (cf. likewise the literature mentioned in 168).

Let us further mention that it follows immediately from 84 (e.g. with the aid of 165.50.1) that all invertible functions in $H^\infty + C$ are multipliers preserving Fredholmness, and so are all functions in $(H^\infty + C)^{-1} \cdot (\overline{H^\infty} + C)^{-1}$. The question of describing the set $\operatorname{clos}_{L^\infty} \operatorname{Re} H^\infty$ (Spitkovskiĭ turns the attention to this circumstance) remains open. This again is equivalent to describing the closure $\operatorname{clos}_{BMO} \widetilde{L}^\infty$ $(\equiv \widetilde{L^\infty} + \operatorname{clos}_{L^\infty} \operatorname{Re} H^\infty)$. One of the possible conjectures here might be $\operatorname{clos}_{L^\infty} \operatorname{Re} H^\infty = \operatorname{Re} H^\infty + \operatorname{Re} C$, but this is not the case. Indeed, the set $\operatorname{Re}(H^\infty + C)$ is even not closed, since otherwise the algebra $H^\infty + C$ would be symmetric (Bernard [1]). In connection with the description of $\operatorname{clos}_{BMO} L^\infty = (\operatorname{clos}_{BMO} L^\infty)^\sim$ let us once more refer to the formula for $\operatorname{dist}_{BMO}(f, L^\infty)$ in 165.53.3.

**165.87, 88.** Cf. 165.69. The formulation of 88 is taken over from Douglas [3].

**165.90.** Cf. 165.60.

**165.91.** The study of operators of the type $T_{\Theta_1\bar{\Theta}_2}$ was initiated in Lee-Sarason [1], Douglas-Sarason [2]. These authors did not connect $T_{\Theta_1\bar{\Theta}_2}$ with model operators and invoked the metric properties of the restrictions $P_{\Theta_1}|K_{\Theta_2}$ only in an implicit form; the connection of the latter with problems close to 91–102 was noted in Axler-Chang-Sarason [1]. A comprehensive systematic study of all these topics mentioned was begun in Lecture VIII and in the papers Nikol'skiĭ [11], Hruščev-Nikol'skiĭ-Pavlov [1].

Concerning statements d) and e) cf. 165.59–61.

If we are interested in the question of getting Fredholm criteria for the Toeplitz operator $T_\varphi$ then, as the following lemma indicates, everything can be reduced to operators of the form $T_{\Theta_1\bar{\Theta}_2}$. It is plain that in this context we may restrict ourselves to symbols $\varphi$ with $\operatorname{ess\,inf}_{\mathbb{T}}|\varphi| > 0$.

**165.91.1. Lemma.** *Let $\varphi \in L^\infty$, $\operatorname{ess\,inf}_{\mathbb{T}}|\varphi| > 0$. Then there exist inner functions $\Theta_1$ and $\Theta_2$, an outer function $h$ and a continuous one $g$ such that*

$$\varphi = \Theta_1\bar{\Theta}_2 gh \tag{37}$$

*and moreover $0 \notin \sigma_{\text{ess}}(T_\varphi) \Leftrightarrow 0 \notin \sigma_{\text{ess}}(T_{\Theta_1\bar{\Theta}_2})$.*

*Proof.* Write $\varphi = \varphi_e u$, $\varphi_e$ outer, $|u| = 1$ a.e. By 165.48.1 $u = f\bar{B}$ ($f \in H^\infty + C$, $B$ a Blaschke product) and by 165.53.2: $f = f_0 g$ ($f_0 \in H^\infty$, $g \in C(\mathbb{T})$). Setting $h = \varphi_e \cdot f_0^e$, $\Theta_1 = f_0^i$, $\Theta_2 = B$, we get the desired representation of $\varphi$. That $T_\varphi = T_{\Theta_1\bar{\Theta}_2 g}T_h$ is Fredholm is now equivalent to $T_{\Theta_1\bar{\Theta}_2 g}$ being Fredholm and further, in view of the inclusion $T_{\Theta_1\bar{\Theta}_2 g} - T_{\Theta_1\bar{\Theta}_2}T_g \in \mathfrak{S}_\infty$, to $T_{\Theta_1\bar{\Theta}_2}T_g$ being Fredholm. As $\min_{\mathbb{T}}|g| > 0$ we have $0 \notin \sigma_{\text{ess}}(T_g)$ so that $0 \notin \sigma_{\text{ess}}(T_\varphi) \Leftrightarrow 0 \notin \sigma_{\text{ess}}(T_{\Theta_1\bar{\Theta}_2})$. □

Concerning representations of the type (33) cf. also Guillori-Sarason [1].

**165.92.** For a general Douglas algebra $H^\infty_{\bar{\Sigma}}$($\Sigma$, a suitable family of functions) the maximal ideal space is found in an analogous way as in 93 (Douglas [5], [2]):

$$\begin{aligned}\mathfrak{M}(H^\infty_{\bar{\Sigma}})\,&\{\gamma \in \mathfrak{M}(H^\infty) : |\hat{\Theta}(\gamma)| = 1,\ \forall\Theta \in \Sigma\}\\ &= \bigcap_{\Theta\in\Sigma}\bigcap_{\varepsilon>0} \operatorname{clos}_{\mathfrak{M}}\{\zeta \in \mathbb{D} : |\Theta(\zeta)| > 1 - \varepsilon\};\end{aligned}$$

here $\hat{\Theta}$ denotes the Gel'fand transform of the function $\Theta$, $\operatorname{clos}_{\mathfrak{M}}$ standing for the closure of the set in question in the maximal ideal space $\mathfrak{M}(H^\infty)$. In particular, a function $\varphi$, $\varphi \in H^\infty_{\bar{\Sigma}}$, is invertible in $H_{\bar{\Sigma}}$ iff $\exists\Theta \in \Sigma\ \exists\varepsilon > 0$ such that $|\Theta(\zeta)| \geq 1 - \varepsilon$, $\zeta \in \mathbb{D} \Rightarrow |\varphi(\zeta)| \geq \varepsilon$.

**165.93.** Statement 1 is of course contained in the union of 91 d) and formula 165.92. The elementary proof of the text has general features in common with a reasoning by Lee-Sarason [1] showing that $0 \in \widehat{\operatorname{Range}}\,\Theta_1/\Theta_2 \Rightarrow 0 \in \sigma^r(T_{\Theta_1\bar{\Theta}_2})$.

Statement 2 was proved (in a somewhat different way) in the same paper Lee-Sarason [1].

**165.95–102.** A detailed discussion is given in 98, 101. Cf. further Nikol'skiĭ

[11], Hruščev-Nikol'skiĭ-Pavlov [1]. Statements 96 b) and 97 are in essence contained in Lee-Sarason [1].

Below we give some complementary information on the spectral properties of the operators $T_{\Theta_1\bar{\Theta}_2}$.

**165.95.1.** In view of 91 d) a necessary and sufficient condition for the invertibility of the operator $T_{\Theta_1\bar{\Theta}_2}$ is the invertibility of the function $\Theta_i$ in the algebra $H^\infty_{\Sigma_j}$ $(i, j = 1, 2)$. Therefore (and also in connection with Chang's Theorem, cf. 124) it is useful to have a criterion for the invertibility of the inner function $\vartheta$ in the algebra $H^\infty_\Sigma$. For the algebras

$$\mathscr{D}_1 \stackrel{\text{def}}{=} H^\infty_{\bar{\Theta}_1}, \quad \Theta_1 \stackrel{\text{def}}{=} \exp\left(\frac{z+\mathbb{1}}{z-\mathbb{1}}\right),$$

which were considered in 98–102 and in Lecture VIII, XI, such a criterion was found by Douglas-Sarason [2]: *the following statements are equivalent*:

a) the inner function $\vartheta$ is invertible in the algebra $\mathscr{D}_1$,

b) $\sigma(\vartheta) \cap \mathbb{T} = \{1\}$ and $|\vartheta(\zeta)| \geq \delta$ in some crescent $\{\zeta : |\Theta(\zeta)| \geq 1 - \delta\}$, $\delta > 0$;

c) $\sigma(\vartheta) \cap \mathbb{T} = \{1\}$ and $\sup_{x \in \mathbb{R}} |\Theta'(x)| = 2 \sup_{\zeta \in \mathbb{T}} |\vartheta'(\zeta)| |1 - \zeta|^{-2} < \infty$, where $\Theta(x) = \vartheta\left(\frac{x-i}{x+i}\right)$, $x \in \mathbb{R}$.

Analogous criteria for many other Douglas algebras are located in Chang's paper [3].

**165.95.2. Kober's Theorem.** *Let BUC be the set of bounded uniformly continuous functions on* $\mathbb{R}$, $\omega|\zeta| \stackrel{\text{def}}{=} i(1+\zeta)(1-\zeta)^{-1}$, $\zeta \in \mathbb{T}$ (cf. formula (25)). *Then*

$$BUC \circ \omega \subset \mathscr{D}_1.$$

*Proof.* Set $k(t) = 2\pi^{-1}(t^{-1} \sin 2^{-1}t)^2$, $t \in \mathbb{R}$, and let $f \in BUC$, $g(\zeta) \stackrel{\text{def}}{=} a \int_{\mathbb{R}} k(a(t - \zeta)) f(t)\, dt$, $\zeta \in \mathbb{C}$. As $k$ is an entire function of exponential type $a$, $g$ too is an entire function of exponential type $a$, and furthermore, as $\int_{\mathbb{R}} k = 1$ and $g(x) = \int_{\mathbb{R}} k(t) \cdot f(x + t/a)\, dt$, $x \in \mathbb{R}$, we find

$$|g(x) - f(x)| = \left|\int_{\mathbb{R}} k(t)(f(x + t/a) - f(x))\, dt\right|$$

$$\leq \int_{\mathbb{R}} |k(t)| \cdot \sup_x |f(x + t/a) - f(x)|\, dt.$$

Thus, by Lebesgue's theorem, $\|g - f\|_\infty < \varepsilon$ for sufficiently large $a$, $a > 0$. On the other hand, if $b > a$ then the function $g \cdot e^{ibz}$ will be bounded in the halfplane $\operatorname{Im} z > 0$ (Phragmén-Lindelöf Theorem, Lecture I, Sect. 7). Hence $g \in \mathscr{D}_1 \circ \omega^{-1}$ and also $f \in \mathscr{D}_1 \circ \omega^{-1}$ (because of the arbitrariness of $\varepsilon$, $\varepsilon > 0$). □

**165.95.3. Corollary.** $\mathscr{D}_1 = H^\infty + BUC \circ \omega$.

In fact, by Chang's Theorem (cf. 124, 165.124.6) $\mathscr{D}_1 = H^\infty + C_{\mathscr{D}_1}$. Also $C_{\mathscr{D}_1}$ (as symmetric algebra) is generated by those inner functions which are invertible in $\mathscr{D}_1$. From 165.95.1 c) it therefore follows that $C_{\mathscr{D}_1} \subset BUC \circ \omega$. The converse is obvious because of 165.95.2. □

**165.95.4.** The interesting results by Koosis [2], Lax [1], Dym-Shapiro [1], Vol'berg-Hruščev [1] on "inner compactness" (i.e. concerning pairs $\vartheta$, $\Theta$ such that $T_{\bar{\Theta}}h|K_\vartheta \in \mathfrak{S}_\infty$, $\forall h > 0$; here $\Theta$ is a purely singular inner function) have some pertinence to the operators $T_{\vartheta\bar{\Theta}}$. Cf. further Lee-Sarason [1]. Other problems are mentioned in 165.69.1.

**165.95.5.** Among issues concerning the operators $T_{\vartheta\bar{\Theta}}$ which are not yet settled let us mention the following two: a) it is not clear whether the following conditions, which are necessary for the invertibility of $T_{\bar{B}\Theta}$, $B$ a Blaschke product, also are sufficient: i) $\bigvee R_\sigma = K_\Theta$ (cf. 97); ii) the minimality (24) of the family $R_\sigma$; iii) $H^\infty_{\bar{B}} = H^\infty_{\bar{\Theta}}$ (cf. 91d). The case $\Theta = \exp(z + \mathbb{1})\cdot(z - \mathbb{1})^{-1}$ is discussed in Lee-Sarason [1]; in particular there is constructed an example displaying that i & ii $\not\Rightarrow$ iii. b) it is not clear if Theorem 102 holds without the assumption $\sigma \in (C)$ (and if not, what the substitute might be).

**165.103. The Methods of Banach Algebras** are used actively in the spectral theory of Toeplitz operators, beginning with Kreĭn's pioneering work [1], where the Fredholmness of Wiener-Hopf operators is connected with the invertibility of the symbol in the appropriate algebra (the Wiener algebra). Even earlier Mihlin [1] had studied the algebra generated by singular integrals with continuous symbols as well as their quotients by the ideal $\mathfrak{S}_\infty$, founded a symbolic calculus and exploited it in the research of Fredholm criteria (From this work there has grown out an impressive edifice which certainly cannot be described in a few words, cf., for example, Mihlin [2] and the collection PDO [1].). Among the very first paper in this direction let us mention Gohberg [3] which in its methodological arrangement comes rather close to the theory presented here and also anticipates several of its basic constructions.

The algebra alg $T_X$ as an independent object of study first appeared in the work of Coburn [2], [3] but the beginning of its utilization for the study of the operators in alg $T_X$ is linked with the papers by Douglas [8], [5] and Gohberg-Krupnik [2]. Subsequently this "algebraic approach" to the spectral theory of Toeplitz operators was generally acknowledged and now clearly occupies a dominating position. Most completely (including a complete bibliography) this approach is reflected in the beautiful books by Douglas [2], [3]. In Sect. 104–164 we attempt to complete this theory by some new strokes and to make more clear various details of the method. The exposition is *not* written for specialists on operator algebra.

**165.106.** The notion of symbol of a singular integral operator (the very word) was introduced by Mihlin [1].

**165.107.** Regarding Statements 1, 2 see 26, 27; Statement 3 is strengthened in 127 while Statement 4 is contained in the books Douglas [2], [3].

**165.108–110.** Some concrete cases ($X = L^\infty$, $X = C$, ...) might have been known earlier but perhaps not in exactly the same form.

**165.112.** Several concrete "spectral inclusion theorems" have already been discussed but within its natural limits the question appears only in the present algebraic setting. It is however unclear for which function algebras $X$ the invertibility (or just the Fredholm property) of an operator $T_\varphi$, $\varphi \in X$, by necessity entails $\varphi^{-1} \in X$. In Sect. 59, 112–120 some sufficient conditions for this are stated. Let us however observe that *in the smallest algebra containing* $\varphi$, *viz.* $X_\varphi = \bigvee(\varphi^n : n \geq 0)$, *this invertibility need not hold true*. In fact, it suffices to take for $\varphi$ an outer function in $H^\infty$ such that the image $\varphi(\mathbb{D})$ lies in $\{\zeta \in \mathbb{C} : |\zeta| \geq \varepsilon\}$ but contains an annulus $A = \{\zeta \in \mathbb{C} : \varepsilon_1 \leq |\zeta| \leq \varepsilon_2\}$. By the maximum modulus principle $\varphi^{-1} \in X_\varphi$. □

**165.113–115.** References to $C^*$-algebras can be found in the text. Theorem 113, Statement 2 goes back to Mihlin [1], Kreĭn [1], Gohberg [3], Douglas [2].

**165.118.** Cf. 165.59.

**165.120.** Douglas [2].

**165.121.** A "naive" utilization of semi-commutators in the study of the spectra of Toeplitz operators (and operators in alg $T_{L^\infty}$) appears already in the very first papers in this direction (cf. the bibliography of the Sect. 165.103, 165.47). The scheme of reasoning: if $T_f T_g - T_{fg} \in \mathfrak{S}_\infty$ and if one of the operators $T_f$, $T_g$ is Fredholm, then the Fredholmness of the other one must be equivalent to $T_{fg}$ being Fredholm. From this follows, for instance, once more that if $f \in QC$, $\operatorname{ess\,inf}_{\mathbb{T}} |f| > 0$ then $0 \notin \sigma_{\mathrm{ess}}(T_f)$ $(T_f T_{1/f} - I \in \mathfrak{S}_\infty)$.

Cf. further the subsequent branching off of the problem "Com $T_X = \mathfrak{S}_\infty$" in 129, 134; likewise cf. 165.129, 165.125.

**165.122.** The rôle of the representations $X = \operatorname{alg}(M_1, \overline{M_2})$ in the problem of inversion of operators (with concrete algebras at hand) was noticed already by Kreĭn [1], Gohberg-Kreĭn [5] and has its origin in the Wiener-Hopf factorization method. In other concrete cases ($X = L^\infty, AP_\zeta, H^\infty_\Sigma$) these representations are used (in the spirit of 122) by Douglas [2], [3], [5]. Generalizing the factorization method, Budyanu-Gohberg [1] erected the theory of "decomposing Banach algebras"; cf. further Gohberg-Fel'dman [1].

**165.124.** It has recently become known that the algebras $Q_{\mathscr{D}}$, $C_{\mathscr{D}}$ can be Dirichlet algebras for various Douglas algebras $\mathscr{D}$ different from $H^\infty + C$ (Vol'berg-Tolokonnikov [1]). The statement in the text (concerning the algebra $QC$) is contained in Wolff [1] and is there derived from the formula $QC = QC_A + C(\mathbb{T})$, $QC_A \stackrel{\text{def}}{=} QC \cap H^\infty$.

The Douglas algebras were defined by Douglas [5] and exploited by him within the framework of his algebraic approach to the theory of Toeplitz operators. The Sarason algebras $Q_{\mathscr{D}}$ and $C_{\mathscr{D}}$ were introduced by Sarason [14], who also stated Chang's Theorem ($\mathscr{D} = H^\infty + C_{\mathscr{D}}$) as a conjecture. The First Chang-Marshall Theorem was proved in the papers Chang [1], Marshall [1]; it is set forth also in the book Garnett [2]. A beautiful introduction to this subject is given in Sarason's surveys [9], [14] which contain all details of the proof, except for a special Blaschke product, the construction of which is based on the laborious construction of contours in the original proof of Carleson's Corona Theorem. The Second Chang-Marshall Theorem is by comparison essentially simpler. It is established in the paper [1] of these authors (Lemma 3.1). Let us remark that, besides the results mentioned in the text and to be discussed below, the theory of Douglas algebras contains many other interesting theorems for which we refer to the literature quoted in this Section.

Let us pass to the proof of the Second Chang-Marshall Theorem, Chang's Theorem and some allied results. The presentation differs from the original one (which utilized maximal ideals).

**165.124.1. Lemma.** *A symmetric algebra $X$, $X\in L^\infty(\mathbb{T})$, is logarithmic if*

$$X\subset \operatorname{clos}_{L^\infty}(ug: g\in X_A;\ u\in L^\infty,\ |u|=1\ a.e.). \tag{38}$$

*Proof.* If $f\in X$, $f\ge 0$, then

$$\inf_{g\in X_A}\|f-|g|\|_\infty\le\inf\{\|f-ug\|_\infty: g\in X_A;\ u\in L^\infty,\ |u|=1\text{ a.e.}\}=0. \square$$

**165.124.2. Lemma.** *If $X$ is a symmetric algebra with the property* (38) *and* if division is possible in the algebra $X_A$ (*i.e.* $g\in X_A$, $\operatorname{ess\,inf}_{\mathbb{T}}|g|>0\Rightarrow g^e\in X_A$) *then $X$ is logmodular.*

*Proof.* Let $f\in X$, $f\ge\delta>0$, and let $g$, $g\in X_A$, be a function such that $\|f-|g|\|_\infty<\delta/2$. Then $g=g^ig^e$ with the outer part invertible in $H^\infty$. By the hypothesis $g^e\in X_A\subset X$ so that $(g^e)^{-1}\in X$ and consequently $(g^e)^{-1}\in X_A$. $\square$

**165.124.3. Lemma.** *Let $\mathscr{D}$ be a Douglas algebra and $f\in Q_{\mathscr{D}}$*, $\operatorname{ess\,inf}_{\mathbb{T}}|f|>0$. *Then $f^e\in Q_{\mathscr{D}A}$* ($\overset{\text{def}}{=}Q_{\mathscr{D}}\cap H^\infty$).

*Proof.* Let $f=uf^e$, $|u|=1$ a.e. As $(f^e)^{-1}\in H^\infty$ we get

$$u=f(f^e)^{-1}\in\mathscr{D},\ u^{-1}=f^ef^{-1}\in\mathscr{D},$$

i.e. $u\in\mathscr{D}\cap\bar{\mathscr{D}}=Q_{\mathscr{D}}$. Hence $f^e=\bar{u}f\in Q_{\mathscr{D}}\cap H^\infty$. $\square$

**165.124.4. Lemma.** *Division is possible in $C_{\mathscr{D}A}$.*

*Proof.* Let $g\in C_{\mathscr{D}A}$, $\operatorname{ess\,inf}_{\mathbb{T}}|g|>0$, $g=g^ig^e$ being the outer-inner factorization of $g$. In view 165.124.3 $g^e\in Q_{\mathscr{D}}$ so that $(g^e)^{-1}\in Q\mathscr{D}$. Hence $g^i=g(g^e)^{-1}\in Q_{\mathscr{D}}$ and thus $\overline{g^i}\in D_{\mathscr{D}}$, which together yields $g^i\in C_{\mathscr{D}}$. Consequently $g^e=\overline{g^i}\cdot g\in C_{\mathscr{D}}$. $\square$

**165.124.5.** *Proof of the Second Chang-Marshall Theorem.* The algebra $Q_{\mathscr{D}}$ is in fact "superlogarithmic": $|Q_{\mathscr{D}}^{-1}|=|Q_{\mathscr{D}A}^{-1}|$ (cf. 165.124.3).

The algebra $C_{\mathscr{D}}$ is logmodular in view of 165.124.4 and 165.124.2 (property (38) holds true *by the definition of $C_{\mathscr{D}}$*). $\square$

Let us remark that instead of "superlogarithmic" one usually says "factorizable", because this property (for an arbitrary symmetric algebra $X$) is equivalent to the possibility of a representing a positive function $f$, $f\in X$, in the form $f=|g|^2$, $g\in X_A$. Such a factorization is essential in the theory of Wiener-Hopf equations. Cf., for instance, Budyanu-Gohberg [1] and the extensive bibliography listed there. It is interesting to remark that the very simplest factorization theorem for the algebra $\mathscr{P}$ of trigonometric polynomials was proved by Fejér and

Riesz (and even by Stieltjes, for this refer to Grenander-Szegö [1]) in connection with indeed the same problems (moment problems, orthogonal polynomials, Hankel operators...).

**165.124.6. Chang's Theorem.** $\mathscr{D} = H^\infty + C_{\mathscr{D}}$.

*Proof.* Follows the scheme of the paper Chang-Marshall [1] and is contained in the following three lemmata, of which the first two imitate the proof of the closedness of the sum $H^\infty + C$ as given in Lecture VIII. □

**165.124.7. Lemma.** *If* $f \in C_{\mathscr{D}}$ *then* $\operatorname{dist}(f, C_{\mathscr{D}A}) = \operatorname{dist}(f, H^\infty)$.

*Proof.* We have to show that $\operatorname{dist}(f, C_{\mathscr{D}A}) \le \operatorname{dist}(f, H^\infty)$. Without loss of generality we may assume that $d \stackrel{\text{def}}{=} \operatorname{dist}(f, H^\infty) > 0$. Set $F = (1 - 2\varepsilon) d^{-1} f$, $\varepsilon > 0$. Then $\operatorname{dist}(F, H^\infty) = 1 - 2\varepsilon$. As $F \in C_{\mathscr{D}}$ there exist an inner function $\Theta$, $\Theta \in C_{\mathscr{D}}$, and a function $g$ in $H^\infty \cap C_{\mathscr{D}}$ such that $\|F - \bar{\Theta} g\|_\infty < \varepsilon$. Hence $\operatorname{dist}(\bar{\Theta} g, H^\infty) < 1 - \varepsilon$. □

**165.124.8. Sublemma** (cf. Chang [2]). *Let* $g \in H^\infty$, $\Theta \in C_{\mathscr{D}}$ *(an inner function) and* $\operatorname{dist}(g\bar{\Theta}, H^\infty) < 1$. *Let further u be a unimodular function in* $g\bar{\Theta} + H^\infty$ *with* $\operatorname{dist}(\bar{z}u, H^\infty) = 1$ (the existence of such a $u$ is guaranteed by the *AAK* Theorem, Lecture VIII). *Then* $u \in C_{\mathscr{D}}$.

Continuing the proof of the lemma and recalling that $\bar{\Theta} g \in C_{\mathscr{D}}$ we find that $\operatorname{dist}(\bar{\Theta} g,\ H^\infty \cap C_{\mathscr{D}}) \le \|u\|_\infty = 1$ and therefore $\operatorname{dist}(F,\ H^\infty \cap C_{\mathscr{D}}) < 1 + \varepsilon$, $\operatorname{dist}(f,\ H^\infty \cap C_{\mathscr{D}}) < (1 + \varepsilon)(1 - 2\varepsilon)^{-1} d$. As $\varepsilon$, $\varepsilon > 0$, was arbitrary, everything is proved. □

**165.124.9.** *Proof of the sublemma.* From the conditions $\operatorname{dist}(u, H^\infty) < 1$, $\operatorname{dist}(\bar{z}u, H^\infty) = 1$ and by 165.56.1 it follows that $\bar{u} \in H_u^\infty$. As $u \in g\bar{\Theta} + H^\infty$ we get $H_u^\infty \subset H_{g\bar{\Theta}}^\infty \subset \mathscr{D}$. Hence $\bar{u} \in \mathscr{D}$ and (because $\bar{\Theta} \in C_{\mathscr{D}}$) $\overline{u\Theta} \in \mathscr{D}$. Thus $u\Theta$ is a unimodular function in $H^\infty$ and $\overline{u\Theta} \in \mathscr{D}$. This means that $u\Theta \in C_{\mathscr{D}}$ and then also $u \in C_{\mathscr{D}}$. □

**165.124.10. Lemma.** *The set* $H^\infty + C_{\mathscr{D}}$ *is closed.*

*Proof.* It follows from 165.124.7 that the natural imbedding $C_{\mathscr{D}}/H^\infty \cap C_{\mathscr{D}} \to L^\infty/H^\infty$ $(f + (H^\infty \cap C_{\mathscr{D}}) \mapsto f + H^\infty)$ is an isometry. □

**165.124.11. Lemma.** *The set* $H^\infty + C_{\mathscr{D}}$ *is dense in* $\mathscr{D}$.

*The proof* is based on the first Chang-Marshall Theorem according to which the set of functions of the form $\bar{\Theta} g$, $\Theta$ inner, $\Theta \in C_{\mathscr{D}}$, $g \in H^\infty$, is dense in $\mathscr{D}$. But in view of 165.124.8 this set is contained in $H^\infty + C_{\mathscr{D}}$. □

**165.124.12. Corollary** (cf. Chang [2]). $Q_{\mathscr{D}} = VMO_{\mathscr{D}} \cap L^\infty$ *where* $VMO_{\mathscr{D}} \stackrel{\text{def}}{=} C_{\mathscr{D}} + \tilde{C}_{\mathscr{D}}$.

The deduction of this formula from Chang's Theorem *repeats word by word* the proof in the special case $\mathscr{D} = H^\infty + C(\mathbb{T})$ (cf. 53), provided we in the argument in question make the obvious substitutions $C \to C_{\mathscr{D}}$, $VMO \to VMO_{\mathscr{D}}$. □

Let us remark that the class $VMO_{\mathscr{D}}$ (as well as the corresponding class $BMO_{\mathscr{D}}$)

admits an interesting local description in the spirit of 165.53.3 (Chang [2], cf. also Garnett [2]).

**165.125.** Part of this result is perhaps known; cf. Douglas [2], [3]. In particular, Douglas proves the canonical isomorphism alg $T_{L^\infty}$/Com $T_{L^\infty} \cong L^\infty$ in various ways. In the book Douglas [2] there is developed a method which has served as a basis for the presentation in 103–129; the lectures Douglas [3] contain also a reasoning based on a theorem by Bunce [1]: if $\mathscr{T}$ is a set of commuting subnormal operators and alg* $\mathscr{T} \stackrel{\text{def}}{=} \text{alg}(\mathscr{T}, \mathscr{T}^*)$ then alg* $\mathscr{T}$/Com alg* $\mathscr{T}$ is canonically isomorphic to $C(\mathfrak{M})$, where $\mathfrak{M}$ is the "joint spectrum" of $\mathscr{T}$ (in a suitable sense). Setting $\mathscr{T} = T_{H^\infty}$ and applying the Douglas-Rudin Theorem we get alg $T_{L^\infty}$/ Com $T_{L^\infty} \cong L^\infty$.

Power [2] showed that the addition of slowly oscillating functions (165.69.3) does not change results of the type 125: if $\hat{X} \cong X$ then $\text{alg}(X, SO_\zeta)^\wedge \cong \text{alg}(X, SO_\zeta)$. For the algebra $SAP_\zeta$ this assertion is contained in Sarason [13].

**165.127.** The inclusion Com $T_{L^\infty} \subset \{C \in \text{alg}\, T_{L^\infty} : \inf_\Theta |CT_\Theta| = 0\}$ is in Douglas [6], [2]. In the proof of 127 the same method is used. Let us further mention Power's formula [3] for Com $T_X$, proved for symmetric algebras $X$ of the form $\text{alg}(\bar{\Sigma}, M)$, $M \subset H^\infty$, in the hypothesis $X = \{f(\bar{z}) : f \in X\}$ : Com $T_X = \text{alg}_{\text{even}}(J\, \text{Hank}_X)$; here, by definition, $\text{alg}_{\text{even}}$ is the closure of the linear hull of all even products $\prod_{1 \le k \le 2n} JH_{\varphi_k}$, $\varphi_k \in X$; $Jf \stackrel{\text{def}}{=} \bar{z}f(\bar{z})$. In the very last moment arrived also the pre-print Barria-Halmos [1], containing Identity 1 in Corollary 127.

**165.129.** These facts are well-known; cf. for example Douglas [2]. A case, in some sense in opposition to the inclusion $\mathfrak{S}_\infty \subset$ Com $T_X$ considered here, is the relation $\mathfrak{S}_\infty \cap \text{alg}\, T_X = \{\mathbb{0}\}$. For a $C^*$-algebra $\mathfrak{A}$ (in particular, for $\mathfrak{A} = \text{alg}\, T_X$, $X = \bar{X}$) the identity $S_\infty \cap \mathfrak{A} = \{\mathbb{0}\}$ is equivalent to the coincidence of the norms $|\cdot|$ and $|\cdot|_{\text{ess}}$ on $A$ (if $\mathfrak{S}_\infty \cap \mathfrak{A} = \{\mathbb{0}\}$ and $T \in \mathfrak{A}$ then the operator $T^*T$ has no eigenvalues of finite multiplicity so that $|T|^2 = |T^*T| = |T^*T|_{\text{ess}} = |T|^2_{\text{ess}}$; the converse is obvious. □).

**165.132.** Coburn [2]. For a generalization to algebras of operators acting in $L^2$-spaces on locally compact groups, cf. Coburn-Douglas-Schaeffer-Singer [1].

**165.133,135.** Sarason [10].

**165.137.** Gohberg-Krupnik [2]. There it is shown that if $\varphi, \psi \in PC$ and have a common discontinuity point, then $T_\varphi T_\psi - T_{\varphi\psi} \notin \mathfrak{S}_\infty$.

**165.138.** Coburn [2], Sarason [10].

**165.139.** Douglas [2], [3]. For the operators $T = T_\varphi$ cf. 165.47, 165.50.

**165.140.** Gohberg-Krupnik [2]. Here it is likewise established (with the aid of 47) that $\widehat{PC} \stackrel{\text{def}}{=} \text{alg}\, T_{PC}/\text{Com}\, T_{PC} = C(\mathbb{T}_*)$, where the compact space $\mathbb{T}_*$ is the cylinder $\mathbb{T} \times [0,1]$ in the topology generated by the sets $(a,b) \times [0,1]$, $\{(a,b) \times [0,1]\} \cup \{a\} \times (t,1]$, $\{(a,b) \times [0,1]\} \cup \{b\} \times [0,t)$, where $a$, $b \in \mathbb{T}$,

$t \in (0,1)$. This topology on $\mathbb{T}_*$ is defined by the functions $\varphi^\#(e^{it},\lambda) \stackrel{\text{def}}{=} \lambda\varphi(e^{it-0}) + (1-\lambda)\varphi(e^{it+0})$, $\varphi \in PC$, and the algebra isomorphism described above is generated effectively by the map $T_\varphi + \mathfrak{S}_\infty \mapsto \varphi^\#$. Of course this gives anew 47 and moreover

**165.140.1. Corollary.** *Let $\varphi_{ij} \in PC$. The operator $T = \sum_i \prod_j T_{\varphi_{ij}}$ is Fredholm iff* $\mathrm{Sym}_\# T \stackrel{\text{def}}{=} \sum_i \prod_j \varphi_{ij}^\# \neq 0$ *on* $\mathbb{T}_*$. □

Analogous results for the algebra $PQC$ (165.50.6) were obtained by Sarason [1].

**165.141.** Continuation of Table 141. See p. 391.

**165.144.** Sarason [10].

**165.146.** The Second AChSV Theorem is also contained in the union of the papers Axler-Chang-Sarason [1], Vol'berg [1]. The plan of the proof presented here follows basically the first of these papers. The idea to use Jones's result comes from Vol'berg. Criteria for the compactness of the imbedding $K_\Theta \to L^2(wdm)$, which are close in spirit to the AChSV Theorem, and criteria for the equivalence of the norms $\|\cdot\|_{L^2}$, $\|\cdot\|_{L^2(wdm)}$ on $K_\Theta$ can be found in Vol'berg [2].

**165.148–151.** In Vol'berg's paper [1] all this is done differently; the present approach is the outcome on his active participation. The proof of 148 trivializes if one knows in advance that the set $A_f$ is generated by functions of the form $g_1\bar{g}_2$, $g_i \in H^\infty$. For special cases of 149 see Gohberg-Krupnik [1].

**165.152–154.** Here we follow mainly the paper Axler-Chang-Sarason [1] (concerning the implication $3 \Rightarrow 4$ cf. 165.146). This paper contains many other interesting particulars; notably there is a direct proof of the implication $7 \Rightarrow 4$, based on Sarason's Commutant Theorem, Lecture VIII, Sect. 1, as well as a derivation of $3 \Rightarrow 2$ from the First Chang-Marshall Theorem. The implication $1 \Rightarrow 3$ can be obtained on the basis of an analysis of the action of the semi-commutator $[T_\Theta, T_{\bar{g}})$ on the rational fractions $(\mathbb{1} - \bar{\lambda}z)^{-1}$, $\lambda \in \mathbb{D}$, cf. Vol'berg [1].

**165.155.** Axler-Chang-Sarason [1] (without the formula for rank); for the proof the authors use a part of the main AChSV Theorem.

**165.156.** Brown-Halmos [1], Halmos [2]. In the latter paper it is proven that $[T_\varphi, T_\psi] = \mathbb{0}$ iff either $\varphi, \psi \in H^\infty$ or $\bar{\varphi}, \bar{\psi} \in H^\infty$ or $\varphi, \psi$ and $\mathbb{1}$ are linearly dependent.

**165.157.** In 157–163 we outline Douglas's approach to the "localization technique" for the algebra alg $T_{L^\infty}$; cf. Douglas [2], [3], [6].

**165.158–160.** Douglas [3], [6] (with different proofs). Let us remark that in general one cannot restrict oneself in 158 to just a portion of the inner functions (with preservation of the equivalence). For instance, one does not know the structure of the essential commutant of the shift operator $\{A : [A, S] \in \mathfrak{S}_\infty\} = \{A : T_{z^n}^* A T_{z^n} - A \in \mathfrak{S}_\infty, n \geq 0\}$, cf. Barria-Halmos [1].

**165.161.** Some remarks by Davidson [1] are utilized in 158–159. Let us

| Algebra $X$ | Com $T_X$ | alg $T_X$/Com $T_X$ | $\sigma_{ess}(T_\varphi)$, $\varphi \in X$ | $\sigma_X(\varphi)$ |
|---|---|---|---|---|
| $PQC$ 165.50.6 | $\mathfrak{S}_\infty$ 133, 140 | $[\operatorname{alg} \mathscr{H}(PQC)]/C_0$ $C_0 = \{f \in C(\operatorname{clos} \mathbb{D}) : f \mid \mathbb{T} \equiv 0\}$; Sarason [11] | $\left\{\lambda \in \mathbb{C} : \lim\limits_{\lvert\zeta\rvert \to 1} \lvert \mathscr{H}_\varphi(\zeta) - \lambda \rvert = 0\right\}$ 165.50.6 | Range $\varphi$ |
| $SAP_\zeta$ 165.69.3 | ? | $SAP_\zeta$ Sarason [13] | Range $\varphi \cup \{\lambda : \operatorname{wind}_\zeta^\pm(\varphi - \lambda) \neq 0\}$ $\cup \left\{\lambda : \min\limits_{k \in \mathbb{Z}} \lvert M_+(\arg(\varphi - \lambda)) - M_-(\arg(\varphi - \lambda)) + k\pi \rvert \geq \pi/2\right\}$, 169.69.2 | Range $\varphi$ |
| $\operatorname{alg}(SO_\zeta, AP_\zeta)$ 169.69.3 | ? | $\operatorname{alg}(SO_\zeta, AP_\zeta)$ 165.125 | Range $\varphi$ $\cup \left\{\lambda : \lim\limits_{\lvert t\rvert \to \infty} t^{-1}(\arg f_\lambda(2t) - \arg f_\lambda(t)) \neq 0\right\}$, $f_\lambda \circ \omega_\zeta \stackrel{\text{def}}{=} \varphi - \lambda$ 165.69.3 | Range $\varphi$ |
| $\mathscr{D} = \operatorname{alg}(H^\infty, e^{\frac{1+z}{1-z}})$ $= H^\infty + BUC \circ \omega$, 165.95.3 | $\mathfrak{S}^0 \cap \operatorname{alg} T_{\mathscr{D}}$ 127, 130 | $\mathscr{D}$ 125 | ? 59, 120 | 165.92; for the case $\varphi \in H^\infty$ – 165.95.1 |

mention yet another result of Power's [3]: the center of the $C^*$-algebra $\mathrm{alg}^*(T_{L^\infty}, \mathscr{I}\,\mathrm{Hank})/\mathfrak{S}_\infty$ consists exactly of the cosets $T_\varphi + \mathfrak{S}_\infty$ where $\varphi \in SQC \overset{\mathrm{def}}{=} \{\varphi \in QC : \varphi(z) = \varphi(\bar{z})\}$; here $\mathscr{I}\,\mathrm{Hank} = \{\mathscr{I}H_\varphi : \varphi \in L^\infty\}$, $\mathscr{I}f = \bar{z}f(\bar{z})$. A simple proof of Power's theorem was found by Hoffman [1]. The idea is to "continue" the operator $A$ belonging to the essential commutant of $\mathrm{alg}^*(T_{L^\infty}, \mathscr{I}\,\mathrm{Hank})$ to an operator $B \overset{\mathrm{def}}{=} UAP_+U + AP_+$ in $L^2$; here $Uf = f(\bar{z})$. It is not hard to see that $B\boldsymbol{\psi} - \boldsymbol{\psi}B \in \mathfrak{S}_\infty$ for all $\psi$, $\psi \in L^\infty$, and to derive from this the representation $B = \boldsymbol{\varphi} + K$, $K \in \mathfrak{S}_\infty$ (here one uses the theorem of Johnson and Parrott on the essential commutant of the algebra $\{\boldsymbol{\psi} : \psi \in L^\infty\}$ in $L^2(\mathbb{T})$). Returning to the formula for the operator $B$ we get first $A - T_\varphi \in \mathfrak{S}_\infty$ and then $\varphi \in SQC$ (without any appeal to 159, 160). □

**165.162,163.** For the details see Douglas [2], [3], [6], Power [3], [4].

**166. The Connection with the Wiener-Hopf Equations.** The Wiener-Hopf operator with kernel $k$ is defined for functions $f$ given on the halfaxis $\mathbb{R}_+ = (0, \infty)$ by the formula

$$W_k f(x) = \int_{\mathbb{R}_+} k(x - t) f(t)\, dt,\ x \in \mathbb{R}_+,$$

a Wiener-Hopf equation being an equation of the type $\lambda f - W_k f = g$. Such equations arise in various problems of physics and mathematics (cf. the references in 165.1). In all cases when the Fourier transform is applicable the $WH$-operators reduce to Toeplitz operators. Let us for instance, following Devinatz [2], consider the case $W_k : L^2(\mathbb{R}_+) \to L^2(\mathbb{R}_+)$.

Let $\hat{k}$ be a bounded function on $\mathbb{R}$, $P^+$ the orthogonal projection in $L^2(\mathbb{R})$ onto the subspace $H^2_\Pi$ (Lecture XI). Then the operator $T$, $Tf \overset{\mathrm{def}}{=} P^+\hat{k}f$, $f \in H^2$, is unitarily equivalent to the Toeplitz operator $T_\varphi = U^{-1}TU$, where $Uf$ $\pi^{-1/2}(x + i)^{-1} f \circ \omega^{-1}$ is the unitary map from $H^2$ onto $H^2_\Pi$ of Lecture XI, $\varphi = \hat{k} \circ \omega$, $\omega(\zeta) = i(1 + \zeta)(1 - \zeta)^{-1}$. On the other hand, the operator $T$ reduces to a $WH$ operator after a Fourier transformation: $\mathscr{F}T\mathscr{F}^{-1}g = \mathscr{F}P^+\mathscr{F}^{-1}(\mathscr{F}\hat{k}\mathscr{F}^{-1}g) = \mathscr{F}P^+\mathscr{F}^{-1}((\mathscr{F}\hat{k}) * g) = \chi_{\mathbb{R}_+}(k * g)$ (use the Paley-Wiener Theorem, Lecture XI; the transform $k \overset{\mathrm{def}}{=} \mathscr{F}\hat{k}$ and convolution are taken in distribution sense; $\chi_{\mathbb{R}_+}$ is multiplication by the characteristic function $\chi_{\mathbb{R}_+}$, i.e. the orthogonal projection in $L^2(\mathbb{R})$ onto $L^2(\mathbb{R}_+)$). Thus $W_k = (\mathscr{F}U)T_\varphi(U^{-1}\mathscr{F}^{-1})$ and, by the same token, the bounded $WH$-operators are precisely those which correspond to symbols $\varphi$ in $L^\infty(\mathbb{T})$. □

As already has been pointed out, the theory of $WH$-operators (and, in particular, the study of $WH$-equations what normal solvability concerns) for a long time developed independently of the theory of Toeplitz operators, cf. the bibliography in 165.47. There are special problems of great interest for $WH$-operators $W_k$ in the case when the distribution $k$ is a finite measure on $\mathbb{R}$. In the papers Gohberg-Krupnik [4], Coburn-Douglas [1] there are given Fredholm and invertibility criteria for operators $W_k$ for measures without the continuous singular part $k_s$ (i.e. $k = \sum_{s \geq 1} \alpha_s \delta_{t_s} + k_a\, dt$); after the transformation of the previous paragraph the problem is reduced to an operator $T_\varphi$, $\varphi \in AP_1 + C(\mathbb{T})$, and thus the answer reads: $W_k$ is Fredholm iff $\inf_{\mathbb{R}}|\mathscr{F}k| > 0$, $w \overset{\text{def}}{=} w_{\mathscr{F}k} = 0$ (mean motion); $\operatorname{ind} W_k = -\operatorname{wind}(\mathscr{F}k/\mathscr{F}(k - k_a))$. If $k_s \neq \mathbb{O}$ but $k$ is invertible in the convolution algebra $M(\mathbb{R})$ a similar result is given in Douglas-Taylor [1].

**167. Boundary Problem, Singular Equations, Dual Equations.** The classical boundary problem in Function Theory (the Riemann-Hilbert problem) is to find the conditions for the solvability of the equation

$$f_+ = gf_- + h,$$

where $g$ and $h$ are given functions on the circumference $\mathbb{T}$ while $f_\pm$ are unknown functions in the classes $H^2_\pm$ respectively. Usually this problem is declared solved if one has found for the problem operator $RH$,

$$(RH)f \overset{\text{def}}{=} P_+f - gP_-f,$$

the conditions for it being Fredholm, the index, the dimension of the kernel and the cokernel, a formula for the inverse operator (in the case of an invertible operator $RH$). We will limit ourselves only to the Fredholm criterion in the space $L^2(\mathbb{T})$ but first let us point out that the same question arises likewise in the solution of the singular integral equation

$$cf + d(SI)f = g,$$

where $c, d, g$ are given, with $(SI)f = (2\pi i)^{-1}\int_{\mathbb{T}}(\zeta - z)^{-1}f(\zeta)\,d\zeta$ (the integral in the sense of principal values); in fact, $SI = \frac{1}{2}(P_+ - P_-)$ so the corresponding operator takes the form

$$T = \mathbf{a}P_+ + \mathbf{b}P_-. \tag{39}$$

By virtue of 166 one can pass to a unitarily equivalent operator in $L^2(\mathbb{R})$:

$$\boldsymbol{\alpha}P^+ + \boldsymbol{\beta}P^-,$$

and, after taking Fourier transforms, to the operator

$$Qf = \hat{\alpha} * (\chi_{\mathbb{R}_+}f) + \hat{\beta} * (\chi_{\mathbb{R}_-}f),$$

corresponding to the equation $Qf = g$, called the "transpose of the dual equation". The "dual equation" itself takes the form

$$Q^*f = g,\ Q^*f = \chi_{\mathbb{R}_+}(\hat{\alpha}_1 * f) + \chi_{\mathbb{R}_-}(\hat{\beta}_1 * f),$$

where $\alpha_1(x) = \overline{\alpha(x)}$, $\beta_1(x) = \overline{\beta(x)}$. It is clear that it is sufficient to have a Fredholm criterion (invertibility criterion) for the operator (39) in order to be able to solve the corresponding problem for all the other operators mentioned.

*Solution.* 1. The operator $T$ is bounded iff $a, b \in L^\infty(\mathbb{T})$ (follows from 26).

2. If $T$ is Fredholm then $\operatorname{ess\,inf}_{\mathbb{T}}|a| > 0$, $\operatorname{ess\,inf}_{\mathbb{T}}|b| > 0$ (follows from (30) or directly by an obvious modification of formula (13) applied to $T^*$).

3. If Condition 2 is fulfilled then $T = \mathbf{b}\left(\frac{\mathbf{a}}{\mathbf{b}}P_+ + P_-\right) = \mathbf{b}(\boldsymbol{\varphi}P_+ + P_-) = \mathbf{b}(P_+\boldsymbol{\varphi}P_+ + P_-)(I + P_-\boldsymbol{\varphi}P_+)$, where the operators $\mathbf{b}$ and $I + P_-\boldsymbol{\varphi}P_+$ are invertible in $L^2(\mathbb{T})$ ($(I + P_-\boldsymbol{\varphi}P_+)^{-1} = I - P_-\boldsymbol{\varphi}P_+$).

**Corollary.** *The operator* (39) *is Fredholm iff*

$$\operatorname*{ess\,inf}_{\mathbb{T}}|a| > 0,\ \operatorname*{ess\,inf}_{\mathbb{T}}|b| > 0,\ 0 \notin \sigma_{\text{ess}}(T_\varphi);$$

*moreover then* ind $T = \text{ind}\, T_\varphi$; $\varphi = a/b$. □

The boundary problems, the singular and allied equations have a rich independent history and their own set of problems for which we refer to the literature mentioned in 165.1, 165.47. An algebraic approach to these problems (in the spirit of 162, 163) is outlined in Douglas [6], Power [4]. A spectral theory of singular integral operators in the spirit of "Pincus' canonical model" is developed in the book by Clancey [1].

**168. Vectorial, Multivariate and "Abstract" Operators of Toeplitz and Hankel Type.** As in 166, 167 this circle of ideals is extremely wide and we can only in passing touch upon some themata. One gets *vectorial* Hankel and Toeplitz operators upon replacing the space $L^2(\mathbb{T})$ by $L^2(E, \mathbb{T}) = L^2(E)$ of $E$-valued $L^2$ functions on the circumference $\mathbb{T}$. Omitting the stage when one considers matrices of such operators with respect to the standard basis of the subspaces $\mathbf{z}^n E$, $n \in \mathbb{Z}$ (regarding this and the defining relations $S_E^* T S_E = T$, $S_E^* H = H S_E$, cf., for example, Bonsall-Power [1] and the references listed there), let us notice that the formulae

$$T_\varphi f = P_+ \varphi f,\ H_\varphi f = P_- \varphi f,\ f \in H^2(E),$$

remain in force but $\varphi$ now has to be interpreted as an operator valued function, $\varphi \in L^\infty(L(E))$, $L(E)$ being the space of all operators from $E$ into $E$ and $P_\pm$ the orthogonal projections onto $H^2(E)$, $H^2_-(E)$ respectively. For the analogues of the theorems of Nehari and Hartman for the operators $H_\varphi$ see Lecture VIII, Sect. 7. An analogue of Theorem 23 is proved in Peller [4].

The spectral theory of vectorial Toeplitz operators $T_\varphi$ develops parallel to the "scalar" theory. Among exhausting treatments we mention Gohberg-Kreĭn [5], Simonenko [6], [5], Gohberg-Fel'dman [1], Douglas [3] and a large group of papers (mainly by American mathematicians) coming from the spectral theory of multivariate stationary processes; for this refer to Masani-Wiener [1], [2], Rozanov [2], Rosenblum [1], [3] and the references listed there. The fundamental method of the theory is the factorization of operator functions (symbols) of the type $\varphi = \varphi_- D \varphi_+$, where $\varphi_\pm \in H^2_\pm(L(E))$ and $D$ is a diagonal matrix, defining the so-called "partial indices" of the operator. In the case of a continuous symbol $\varphi$ one has for $\dim E < \infty$ a "determinant criterion": $0 \notin \sigma_{\text{ess}}(T_\varphi) \Leftrightarrow \det \varphi(\zeta) \neq 0$, ind $T_\varphi = -\text{wind}(\det \varphi)$ (Gohberg-Kreĭn [5]). However Coburn's Lemma 43 is of course not any longer valid for vectorial Toeplitz operators. Regarding other more sophisticated questions see the papers mentioned in the beginning of this paragraph and furthermore the works of Spitkovskiĭ (especially the paper [3] which has an extensive bibliography).

Various "abstract" variants of Hankel and Toeplitz operators have been considered by several authors. Regarding Hankel operators (among the latter we likewise count the Hankel operators on the unit sphere of $\mathbb{C}^n$, $n > 1$) see Butz [1] and regarding Toeplitz operators see Gohberg-Feld'man [1], Gohberg-Krupnik [1], Devinatz-Shinbrot [1]. In particular, the last paper has the following generalization of Theorem 36–38. Let $H$ be a Hilbert space, $H_+$ a subspace of $H$, $P_+ = P_{H_+}$; 1) if $U$ is unitary in $H$ then the following statements are equivalent: a) the operator $P_+U|H_+$ is invertible, b) $|(I-P_+)UP_+| < 1$, $|P_+U(I-P_+)| < 1$, c) there exists an invertible operator $G$ in $H$ such that $GH_+ = H_+$ and $|U-G| < 1$; 2) let $\Phi$ be an invertible opertor in $H$. Then there exists a unitary operator $U$ in $H$ and an invertible one $B$ such that $\Phi = UB$ and $BH_+ = H_+$, the operator $P_+\Phi|H_+$ is invertible iff the operator $P_+U|H_+$ is invertible.

Regarding multivariate Toeplitz operators (i.e. Toeplitz operators defined on spheres and tori in $\mathbb{C}^n$, $n \geq 1$) see Douglas [3], Krupnik-Fel'dman [1] and the references found there.

Vectorial Hankel and Toeplitz operators are intimately connected with the Carleson Corona Theorem (cf. Appendix 3). In particular, one has the following theorem due to Sz.-Nagy and Foiaş [10] and Schubert [1] (it was preceded by a less complete result by Arveson [1]).

**169. Theorem.** *Let $f_i \in H^\infty$, $1 \leq i \leq n$, $\delta > 0$. The following statements are equivalent.*

1. *There exist functions $g_i$, $g_i \in H^\infty$, $1 \leq i \leq n$, such that*

$$\sum_i f_i g_i = \mathbb{1}, \ \sup_{\mathbb{D}} \sum_i |g_i|^2 \leq \delta^{-2}.$$

2. $\sum_i \| T_{\bar{f}_i} x \|^2 \geq \delta^2 \| x \|^2$, $x \in H^2$.

*If $f = \{f_i\}_{i=1}^n$ is a left inner function (i.e. $|f| \stackrel{\text{def}}{=} (\sum_i |f_i|^2)^{1/2} = 1$ a.e. on $\mathbb{T}$) then Statements 1 and 2 are equivalent to*

3. $\operatorname{dist}(\bar{f}, H_n^\infty) \leq (1-\delta^2)^{1/2}$.

Here $H_n^\infty = H^\infty(L(\mathbb{C}, \mathbb{C}^n))$ is the space of vector valued functions $g = \{g_i\}_{i=1}^n$, $g_i \in H^\infty$, in the norm $\| g \|_\infty = \sup\{|g(\zeta)| : \zeta \in \mathbb{D}\}$. By Page's Theorem (Lecture VIII, Sect. 4) $\operatorname{dist}(\bar{f}, H_n^\infty) = |H_{\bar{f}}|$, where $H_{\bar{f}} x \stackrel{\text{def}}{=} P_- \bar{f} x$ is considered as an operator from $H^2$ into $H^2(\mathbb{C}^n)$.

*Proof.* Since Statement 2 entails the lower estimate $\| T_{\bar{f}} \| \geq \delta \| x \|$, $x \in H^2$, and as furthermore

$$\| T_{\bar{f}} \|_{H^2(\mathbb{C}^n)}^2 + \| H_{\bar{f}} \|_{H^2(\mathbb{C}^n)}^2 = \| fx \|_{H^2(\mathbb{C}^n)}^2,$$

it is clear that $2 \Leftrightarrow 3$ for $|f| = 1$ a.e. on $\mathbb{T}$.

$1 \Rightarrow 2$.
$$\begin{aligned}\sum_i \| T_{\bar{f}_i x} \|^2 &= \sum_i (\| f_i x \|^2 - \| P_- \bar{f}_i x \|^2) \\ &= \sum_i (\| f_i x \|^2 - \| P_-(\bar{f}_i - \delta^2 g_i) x \|^2) \\ &\geq \sum_i (\| f_i x \|^2 - \| (\bar{f}_i - \delta^2 g_i) x \|^2) \\ &= \left(\text{use the identity } \sum_i f_i g_i = 1\right) \int_{\mathbb{T}} \left(2\delta^2 - \delta^4 \sum_i |g_i|^2\right) |x|^2 \\ &\geq \delta^2 \| x \|^2.\end{aligned}$$

$2 \Rightarrow 1$. Here we use the Commutant Theorem from Lecture VIII, Sect. 4. Let $T_f$ be the Toeplitz operator acting from $H^2(\mathbb{C}^n)$ into $H^2$, corresponding to the (analytic) function $f$ ($T_f x = \sum f_i x_i$, $x \in H^2(\mathbb{C}^n)$) and set $K_* = T_f^* H^2$; $K_*$ is a closed subspace of $H^2(\mathbb{C}^n)$. Let us also put $K = K_* \times H^2 \subset H^2(\mathbb{C}^{n+1})$ and let $A$ be the operator in $K$ defined by the formula $A(x, y) = (\mathbb{0}, T_f^{*-1}x)$, $(x, y) \in K_* \times H^2 = K$. It is clear that $S_{n+1}^* K \subset K$ ($S_{n+1}^* \stackrel{\text{def}}{=} S_{\mathbb{C}^{n+1}}^*$ being the backward shift in $H^2(\mathbb{C}^{n+1})$) and that $S_{n+1}^* A = A S_{n+1}^*$.

By the Commutant Theorem just referred to we get a function $G$, $G \in H^\infty(L(\mathbb{C}^{n+1}))$ such that $A = P_k G^* | K$, $GK^\perp \subset K^\perp$ and $\|G\|_\infty = |A| \le \delta^{-1}$. Since $G$ is analytic then $A = P_+ G^* K$ and it follows from the definition of $A$ that $P_+ G^*(T_f^* y, \mathbb{0}) = (\mathbb{0}, y)$, $y \in H^2$. If $j$ denotes the natural imbedding of $H^2$ into $\{\mathbb{0}\} \times H^2 \subset H^2(\mathbb{C}^n) \times H^2$ this gives $T_f P_{H^2(\mathbb{C}^n)} G = I_{H^2}$, i.e. $\sum_i f_i g_i = \mathbb{1}$, $\|g\|_\infty = \|P_{H^2(\mathbb{C}^n)} G\|_\infty \le \delta^{-1}$. $\square$

**170. Self-adjoint Toeplitz Operators** have been studied in a series of papers by Hartman-Wintner [2], Ismagilov [1], Rosenblum [1], [4], [5] (and in some other works quoted there). The main results may be described as follows (we consider only continuous operators $T_\varphi$, $\varphi$ a real function; in the above papers there is the more general case of a locally semi-bounded symbol).

a) The spectral measure of $T_\varphi$ is absolutely continuous with respect to Lebesgue measure, provided $\varphi \ne$ const (Rosenblum [5]).

b) Let $E_m$ be the set of those $\lambda$, $\lambda \in [\operatorname{ess\,inf} \varphi, \operatorname{ess\,sup} \varphi]$, for which the set $\{\zeta \in \mathbb{T} : \varphi(\zeta) < \lambda\}$ consists of exactly $m$ intervals mod 0 ($m = 1, 2, \ldots$), $E_\infty \stackrel{\text{def}}{=} [\operatorname{ess\,inf} \varphi, \operatorname{ess\,sup} \varphi] \setminus \bigcup_{m \ge 1} E_m$. Set moreover $A_m f = xf$ in $L^2(E_m)$ and $B_m = \bigoplus_{i=1}^m A_m$. Then $T_\varphi$ is unitarily equivalent to the operator $B_\infty \oplus \sum_{m \ge 1} \oplus B_m$ (Ismagilov [1]).

c) Rosenblum has given an explicit expression for the unitary correspondence in b) (in [4] for Toeplitz operators, in [6] for singular integral operators of the form $f \mapsto af + (\pi i)^{-1} \cdot b(SI)\bar{b}f$).

**171. The Work of Clark** [5]–[8] constitutes a continuation of the theory of the previous Section and is devoted to finding "canonical forms" for Toeplitz operators up to similarity. Roughly speaking, it is shown that a Toeplitz operator $T_\varphi$ with a rational symbol $\varphi$ is linearly similar to an orthogonal sum $(\sum_i \oplus T_{\varphi_i}) \oplus (\sum \oplus T_{\psi_i}) \oplus V$ of analytic ($T_{\varphi_i}$) and anti-analytic ($T_{\psi_i}$) Toeplitz operators and a normal operator $V$; also these operators admit an effective description. Let us as an illustration quote the main theorem in [5]. Let $\varphi$ be a rational function (no poles on $\mathbb{T}$) and let $\sigma_\varphi$ be a compact set in $\mathbb{C}$, the boundary of which $\partial\sigma_\varphi$ is a simple closed curve. Assume that $\varphi(\mathbb{T}) \subset \partial\sigma_\varphi$ and that $\partial\sigma_\varphi$ is analytic near $\varphi(\mathbb{T})$. Set $v \stackrel{\text{def}}{=} \operatorname{wind}(\varphi - \lambda) \ge 0$, $\lambda \in \operatorname{int} \sigma_\varphi$. Let $\omega$ be a conformal mapping of $\mathbb{D}$ onto $\operatorname{int} \sigma_\varphi$ and let us agree to say that $e^{i\theta}$ is a point of return for $\varphi$ if the function $t \mapsto \arg \omega^{-1}(\varphi(e^{it}))$ decreases at the point $\theta$; let $\Sigma$ be the set of all points of return.

**Theorem.** *If $\varphi(\Sigma) \ne \partial\sigma_\varphi$ then $T_\varphi$ is linearly similar to the operator $V \oplus (\sum_1^v \oplus T_\omega)$ where $V$ is normal, $\sigma(V) \subset \varphi(\Sigma)$. Moreover the multiplicity function (the local multiplicity of the spectrum) of $V$ at a point $\lambda$, $\lambda \in \varphi(\Sigma)$, equals* $\operatorname{card}\{e^{i\theta} : \varphi(e^{i\theta}) = \lambda, e^{i\theta} \in \Sigma\}$.

In particular, if $\varphi = B_1\bar{B}_2$, the $B_i$ being finite Blaschke products with $\deg B_2 \leq \deg B_1$, then the representation $V \oplus (\sum_1^\nu \oplus T_\omega)$ is the Wold-Kolmogorov expansion (Lecture I) of an isometric operator similar to $T_\varphi$. If $\varphi$ is real then the word "linearly similar" may be replaced by "unitarily equivalent" and this reduces to a special case of 170. Let us further mention a result by Douglas ([3], p. 33) on the unitary equivalence mod $\mathfrak{S}_\infty$ of certain Toeplitz operators with normal operators.

**172. Projection Methods for Inversion, Szegö Determinants.** As in 166–168 the number of investigations on this question is extraordinarily large, because it is a question of importance also from the point of view of applications, namely the issue of the constructive inversion of an operator $T_\varphi$ once one already knows that $0 \notin \sigma(T_\varphi)$. Usually one employs the projection method, i.e. one picks a suitable increasing chain of $S^*$-invariant subspaces $\{K(\lambda)\}_{\lambda \in \Lambda}$ which exhaust the space $H^2$, $H^2 = \bigvee(K(\lambda) : \lambda \in \Lambda)$, and considers the "sections" of $T_\varphi$: $T_{\varphi,\lambda} \overset{\text{def}}{=} P_{K(\lambda)} T_\varphi | K(\lambda)$. If for all sufficiently large $\lambda$ the operator $T_{\varphi,\lambda}$ is invertible and if $T_\varphi^{-1} = \lim_{\lambda\to\infty} T_{\varphi,\lambda}^{-1}$ then one says that the projection method based on the chain $\{K(\lambda)\}_{\lambda\in\Lambda}$ is applicable to the operator $T_\varphi$. In the applications ususally $\Lambda = \mathbb{N}$, $K(n) = K_{z^n}$, so that the operator $T_{\varphi,n}$ corresponds to the $n \times n$ "section" of the matrix of $T_\varphi$, or else $\Lambda = \mathbb{R}_+$, $K(\lambda) = \Theta^\lambda$, $\Theta = \exp\left(\frac{z+\mathbb{1}}{z-\mathbb{1}}\right)$, so that, after the transition 166 to the corresponding Wiener-Hopf operator, $T_{\varphi,\lambda}$ here also reduces to a "finite section" $W_{k,\lambda}$ of the corresponding operator $W_k$,

$$W_{k,\lambda} f(x) = \int_0^\lambda f(x-t)k(t)\,dt,\ x \in [0,\lambda].$$

Notice that the invertibility of the operators $T_{\varphi,n}$ is equivalent to the solvability of the equations $T_{\varphi,n}x = \mathbb{1}$, $T_{\varphi,n}y = z^{n-1}$; cf. Iohvidov [1]. There is a vast literature devoted to the question of the applicability of the projection method, starting with Szegö's work on the limit distributions of spectra of finite sections, where the so-called Szegö determinants $D_n \overset{\text{def}}{=} \det(P_{z^n} T_\varphi | K_{z^n})$ arise. Their behavior plays a major rôle in the projection methods; in known (but rather stringent) assumptions on the positive function $\varphi$ it happens that

$$\lim_n D_n \varphi_e(0)^{-n-1} = \exp\left(\pi^{-1} \iint_{\mathbb{D}} |\varphi_e'/\varphi_e|^2\,dx\,dy\right),$$

where $\varphi_e$ is an outer function, $|\varphi_e| = \varphi$ a.e. on $\mathbb{T}$. In the case of the operators $W_{k,\lambda}$ these determinants have to be replaced by corresponding Fredholm determinants. For the asymptotics of $D_n$ for more general symbols $\varphi$ read the excellent exposition in Böttcher-Silbermann [1]. These results are supplemented by the theorems on the "limit distribution of the spectra": if $\{\lambda_i^{(n)}\}_1^n$ are the eigenvalues of $T_{\varphi,n}$, $\varphi = \bar{\varphi}$, and $F$ a continuous function then (in appropriate assumptions on $\varphi$)

$$\lim_n \frac{1}{n} \sum_{i=1}^n F(\lambda_i^{(n)}) = \int_{\mathbb{T}} F(\varphi(\zeta))\,dm(\zeta).$$

For details see the books Grenander-Szegö [1], Böttcher-Silbermann [3].

Let us mention some more literature on the projection methods and Szegö determinants: Kreĭn [4], Gohberg-Fel'dman [1], Devinatz [2], Hirshman [1] (a survey containing a comparison of various inversion methods), Widom [4], [5], [6], Basor [1], Basor-Helton [1], Böttcher-Silbermann [1], [2], Roch-Silbermann [1], Prössdorf-Silbermann [1], Silbermann [1], [2]. The Szegö determinants lend themselves to physical and probabilistic interpretation, cf. Kac [1], [2].

In conclusion let us remark that it remains not clear whether the projection methods are applicable to an arbitrary invertible Toeplitz operator. Let us however state one sufficient condition (Devinatz [2]). Let $T_\varphi$ be an invertible Toeplitz operator and let $\varphi = |\psi|^2 h/\bar{h}$ be the representation of $\varphi$ in 36, Statement 5, $\psi$ an outer function, $|\varphi| = |\psi|^2$ a.e. on $\mathbb{T}$; if moreover either $\psi h/\overline{\psi h}$ or $\psi\bar{h}/\bar{\psi}h$ is in the class $\mathscr{D}_1 = \mathrm{alg}\left(H^\infty, \exp\frac{\mathbb{1}+z}{\mathbb{1}-z}\right)$ then the projection method $\{T_{\varphi,\lambda}\}_{\lambda>0}$ is applicable to $T_\varphi$. In particular, this is the case if $f \in BUC \circ \omega$.

# Appendix 5. Hankel Operators of Schatten-von Neumann Class and Their Application to Stationary Processes and Best Approximations

*by S. V. Hruščev and V. V. Peller*

**0.1. Contents.** In this Appendix there is given a description of Hankel operators of Schatten-von Neumann class $\mathfrak{S}_p$, some new applications of Hankel operators to the spectral theory of stationary Gaussian processes and further a survey of the theory of the best approximation by analytic functions in the uniform metric. In our approach to the subject this theory unifies Hankel operators and stationary processes.

Chap. I collects some auxiliary material on Hankel operators not contained in Appendix 4. In particular we prove Peller's Theorem on Hankel operators of class $\mathfrak{S}_p$ and give a useful new formula connecting the Hankel operators $H_u$ and $H_{\bar{u}}$ with $u$ unimodular.

Chap. II provides a compact presentation of the theory of stationary Gaussian processes based o the technique of Fock spaces. This approach, since a long time ago standard in papers devoted to the probability aspects of quantum theory, replaces the cumbersome apparatus of Hermite polynomials in several variables, as developed by Wiener, and the equivalent machinery of the stochastic integrals of Itô.

Chap. III on the one hand connects Hankel operators with stationary Gaussian processes, on the other hand has an interest of its own. Here we set forth three new (and, as we believe, simple) methods of proving theorems on best approximation by analytic functions in the uniform metric, covering all previously known results, as well as many new ones.

The first method refers to function spaces on the unit circle described in terms of best rational approximation. Important classes of functions of this type were introduced by Gončar [1]. The second method develops beautiful ideas due to Adamjan-Arov-Kreĭn [2]. It covers a wide class of separable Banach algebras in which the Riesz projection is a bounded operator. Finally, the third method allows us to study spaces not contained in the range of action of the first two ones. Among them are found locally convex function spaces such as the Carleman classes, and also some non-separable Banach spaces. The chapter is terminated by some new applications of theorems on best approximation to various problems of Analysis.

In Chap. IV the spectral densities of stationary Gaussian processes, subject to various mixing conditions, are investigated. We give new proofs for many known results and find also a few new ones. The proofs of all the theorems in Chap. IV are founded on one and the same method.

This Appendix does not assume from the reader any particular knowledge of probability theory and all notions necessary on the theory of functions (Besov, Hölder, Sobolev classes), outside the scope of the customary university curriculum and not included in the main body of the book, are collected in a "Review of Function Classes", located at the end of the Appendix. The notation of the main text of the book is employed without special explanation.

**0.2. Let us list the Main Divisions** of this Appendix.



## Chapter I. Hankel Operators and Symmetric Normed Ideals

This Chapter consists of two sections. In Sect. 1.1 we describe the Hankel operators of Schatten-von Neumann class $\mathfrak{S}_p$ in terms of their symbols. We give also some applications of this result.

In Sect. 1.2 we establish a connection between the Hankel operators $H_u$ and $H_{\bar{u}}$ for $u$ unimodular; $u$ is termed unimodular if $|u| \equiv 1$ a.e. Such a connection was first noted in a paper by Solev [1] in the case $H_u \in \mathfrak{S}_2$. We derive a simple algebraic formula expressing the connection between such operators, with the aid of which we get more general results.

**1.1. Hankel Operators of Class $\mathfrak{S}_p$.** The main object of this Section is to obtain a criterion for a Hankel operator to be in the Schatten-von Neumann class $\mathfrak{S}_p$ (cf. Sect. 7, 8, 23 of Appendix 4) in terms of its symbol. It will be proved that a necessary and sufficient condition for this is that the symbol lies in the Besov class

$B_p^{1/p}$ (for Besov classes see the Review of Function Classes p. 453–454). As an application we prove the boundedness in $\mathfrak{S}_p$, $1 < p < \infty$, of the natural projection onto the space of Hankel operators and further we give results on rational approximation in the BMO norm.

**Theorem 1.1** (Peller [3], [5]). *Let* $1 \le p < \infty$, $\varphi \in L^\infty$. *Then*

$$H_\varphi \in \mathfrak{S}_p \Leftrightarrow P_- \varphi \in B_p^{1/p}.$$

Actually the assertion of Theorem 1.1 is also valid for $0 < p < 1$. The implication $P_-\varphi \in B_p^{1/p} \Rightarrow H_\varphi \in \mathfrak{S}_p$ is proved in Peller [3]. The converse was obtained recently by Peller [6] and Semmes [1] using two quite different approaches.

*Remark.* In Chap. III we will show that for every $g$ in $P_- B_p^{1/p}$ there exists a function $f$ in $P_+ B_p^{1/p}$ such that $f + g \in L^\infty$ (cf. Corollary 36). Therefore Theorem 1 can also be formulated as follows: Let $\varphi \in L^\infty$. Then $H_\varphi \in \mathfrak{S}_p$ iff $\varphi \in H^\infty + B_p^{1/p}$.

Instead of the Hankel operators $H_\varphi$ we can consider the operators $\Gamma_f$ in $l^2$ defined by the matrices $\{\hat{f}(n+k)\}_{n,k \ge 0}$, i.e.

$$(\Gamma_f e_n, e_k) = \hat{f}(n+k),$$

where $\{e_n\}_{n \ge 0}$ is the standard orthonormal basis in $l^2$ ($f$ is here an analytic function in $\mathbb{D}$). By Nehari's Theorem we see that $\Gamma_f$ is bounded iff $f \in P_+ L^\infty$, with $|\Gamma_f| = \inf\{\|f\|_{L^\infty} : P_+ f = f\}$.

Obviously Theorem 1.1 can be stated in the following manner.

**Theorem 1.1′.** *Let $f$ be analytic in* $\mathbb{D}$, $1 \le p < \infty$. *Then*

$$\Gamma_f \in \mathfrak{S}_p \Leftrightarrow f \in B_p^{1/p}.$$

This theorem will first be proved for $p = 1$. If $1 < p < \infty$ we will use methods of interpolation of linear operators.

Before passing to the proof we state some results on the classes $\mathfrak{S}_p$, the proof of which can be found in the book Gohberg-Kreĭn [1].

If $T$ is a nuclear operator (i.e. $T \in \mathfrak{S}_1$) in the Hilbert space $\mathscr{H}$ then its *trace* is given by the formula

$$\operatorname{trace} T = \sum_{n \ge 0} (Te_n, e_n),$$

vhere $\{e_n\}_{n \ge n}$ is an (any) orthonormal basis in $\mathscr{H}$. The value of the functional ·ace is thus independent of the choice of the basis and we have $|\operatorname{trace} T|$ $\|T\|_{\mathfrak{S}_1}$. Using the Hermitian duality

$$\langle T, R \rangle = \operatorname{trace}(TR^*)$$

dual spaces $(\mathfrak{S}_p)^*$ can be described as follows (with equality of norms):

$$(\mathfrak{S}_p)^* = \mathfrak{S}_{p'}, \quad 1 < p < \infty,$$

$$(\mathfrak{S}_1)^* = L(\mathscr{H}),$$

$$(\mathfrak{S}_\infty)^* = \mathfrak{S}_1.$$

If the matrices of the operators $T$ and $R$ in an orthonormal basis are $\{t_{mk}\}_{m,k\geq 0}$ and $\{\tau_{mk}\}_{m,k\geq 0}$ then

$$\operatorname{trace} TR^* = \sum_{m,k\geq 0} t_{mk}\overline{\tau_{mk}},$$

at least if the series is absolutely convergent.

*Proof of Theorem* 1.1′. *The case* $p=1$. Let $f\in B_1^1$. Let us show that the operator $\Gamma_f$ is nuclear. As

$$f=\sum_{n\geq 0} f*W_n \quad \text{and} \quad \sum_{n\geq 0} 2^n\|f*W_n\|_{L^1}<\infty,$$

(cf. Review of Function Classes for the definition of the kernels $W_n$), it is easy to see that it suffices to prove the inequality

$$\|\Gamma_\varphi\|_{\mathfrak{S}_1}\leq (m+1)\|\varphi\|_{L^1}$$

for any polynomial $\varphi$ of degree $m$.

To prove the latter inequality let us define elements $x_\zeta$ and $y_\zeta$, $\zeta\in\mathbb{T}$, of the space $l^2$ by the formula

$$x_\zeta(n)=\begin{cases}\zeta^n, 0\leq n\leq m\\ 0,\ n>m\end{cases};\quad y_\zeta(k)=\begin{cases}\varphi(\zeta)\bar\zeta^k, 0\leq k\leq m\\ 0,\ k>m\end{cases}.$$

Define an operator $A_\zeta$ in $l^2$ by the formula $A_\zeta x=(x,x_\zeta)y_\zeta$, $x\in l^2$. Then $A_\zeta\in\mathfrak{S}_1$ and $\|A_\zeta\|_{\mathfrak{S}_1}=\|x_\zeta\|_{l^2}\|y_\zeta\|_{l^2}=(m+1)\cdot|\varphi(\zeta)|$. Let us show that

$$\Gamma_\varphi=\int_{\mathbb{T}} A_\zeta\, dm(\zeta)$$

(the function $\zeta\mapsto A_\zeta$ is continuous so the integral may be taken as a limit of Riemann sums). We have

$$(\Gamma_\varphi e_n, e_k)=\hat\varphi(n+k)=\int_{\mathbb{T}}\varphi(\zeta)\bar\zeta^{n+k}\,dm(\zeta),$$

$$(A_\zeta e_n, e_k)=\varphi(\zeta)\bar\zeta^k\bar\zeta^n.$$

Consequently $\Gamma_\varphi=\int_{\mathbb{T}} A_\zeta\, dm(\zeta)$ and

$$\|\Gamma_\varphi\|_{\mathfrak{S}_1}\leq\int_{\mathbb{T}}\|A_\zeta\|_{\mathfrak{S}_1}\,dm(\zeta)\leq (m+1)\int_{\mathbb{T}}|\varphi(\zeta)|\,dm(\zeta).$$

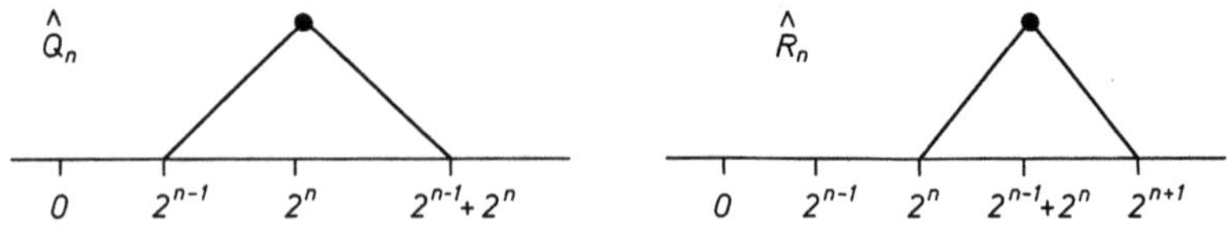


Fig. 8

Assume now that $\Gamma_f \in \mathfrak{S}_1$. Let us show that

$$\sum_{n\geq 0} 2^n \| f * W_n \|_{L^1} < \infty.$$

We define functions $Q_n$ and $R_n$ by the formulae:

$$\hat{Q}_n(k) = \begin{cases} 0, k \leq 2^{n-1} \\ 1 - |k - 2^n|/2^{n-1}, 2^{n-1} \leq k \leq 2^n + 2^{n-1}, \\ 0, k \geq 2^n + 2^{n-1} \end{cases}$$

$$\hat{R}_n(k) = \begin{cases} 0, k \leq 2^n \\ 1 - |k - 2^n - 2^{n-1}|/2^{n-1}, 2^n \leq k \leq 2^{n+1}. \\ 0, k \geq 2^{n+1} \end{cases}$$

Then $W_n = Q_n + \frac{1}{2} R_n$. Let us show that

$$\sum_{n\geq 0} 2^{2n+1} \| f * Q_{2n+1} \|_{L^1} < \infty.$$

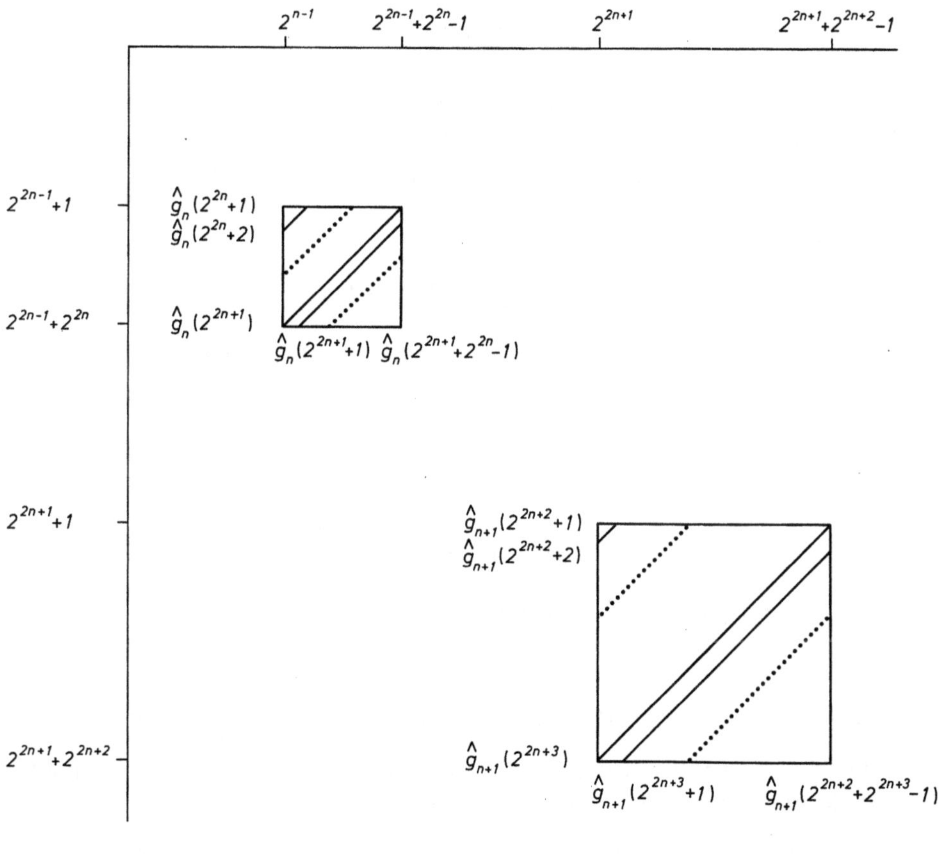


Fig. 9

The inequalities $\sum_{n\geq 0} 2^{2n+1}\|f * R_{2n+1}\|_{L^1} < \infty$, $\sum_{n\geq 1} 2^{2n}\|f * Q_{2n}\|_{L^1} < \infty$, $\sum_{n\geq 1} 2^{2n}\|f * R_{2n}\|_{L^1} < \infty$ can then be verified in exactly the same way.

For the proof of the inequality under consideration we pick an operator $B$ in $L(l^2)$ such that $|B| \leq 1$ and $\langle \Gamma_f, B\rangle = \sum_{n\geq 1} 2^{2n}\cdot \|f * Q_{2n+1}\|_{L^1}$.

Consider in the plane the square $S_n = [2^{2n-1}, 2^{2n-1} + 2^{2n} - 1] \times [2^{2n-1} + 1, 2^{2n-1} + 2^{2n}]$.

Let $\{g_n\}_{n\geq 1}$ be a sequence of functions in $L^\infty$ such that $\|g_n\|_{L^\infty} \leq 1$. We define the matrix $\{b_{mk}\}_{m,k\geq 0}$ of $B$ by the identity

$$b_{mk} = \begin{cases} \hat{g}_n(m+k), & (m,k)\in S_n,\ n\geq 1 \\ 0, & (m,k)\notin \bigcup\limits_{n\geq 1} S_n \end{cases}.$$

Let us show that $|B| \leq 1$. Set

$$\mathscr{H}_n = \operatorname{span}\{e_m : 2^{2n-1} \leq m \leq 2^{2n-1} + 2^{2n} - 1\},$$

$$\mathscr{H}_n' = \operatorname{span}\{e_m : 2^{2n-1} + 1 \leq m \leq 2^{2n-1} + 2^{2n}\}.$$

Then we have the identity

$$B = \sum_{n\geq 1} P_n'\Gamma_{g_n}P_n,$$

where $P_n$ and $P_n'$ are the orthogonal projections onto $\mathscr{H}_n$ and $\mathscr{H}_n'$ respectively. The spaces $\{\mathscr{H}_n\}_{n\geq 1}$ and $\{\mathscr{H}_n'\}_{n\geq 1}$ being pairwise orthogonal we see that

$$|B| = \sup_n |P_n'\Gamma_{g_n}P_n| \leq \sup_n |\Gamma_{g_n}| \leq \sup_n \|g_n\|_{L^\infty} \leq 1.$$

Let us show that

$$\langle \Gamma_f, B\rangle = \sum_{n\geq 0} 2^{2n}(Q_{2n+1} * f, g_n),$$

where as usual $(\varphi, \psi) = \int_{\mathbb{T}} \varphi(\zeta)\overline{\psi(\zeta)}\,dm(\zeta)$, $\varphi\in L^1$, $\psi\in L^\infty$. We have:

$$\begin{aligned}\langle \Gamma_f, B\rangle &= \sum_{n\geq 1} \langle \Gamma_f, P_n'\Gamma_{g_n}P_n\rangle \\ &= \sum_{n\geq 1} \sum_{k=2^{2n}}^{2^{2n}+2^{2n+1}} (2^{2n} - |k - 2^{2n+1}|)\hat{f}(k)\overline{\hat{g}_n(k)} \\ &= \sum_{n\geq 1} 2^{2n}(Q_{2n+1} * f, g_n).\end{aligned}$$

Pick now the sequence $\{g_n\}_{n\geq 1}$ such that $(Q_{2n+1} * f, g_n) = \|Q_{2n+1} * f\|_{L^1}$. Then

$$\langle \Gamma_f, B\rangle = \sum_{n\geq 1} \|Q_{2n+1} * f\|_{L^1}.$$

Consequently $\sum_{n\geq 1} 2^{2n+1}\|Q_{2n+1} * f\|_{L^1} = 2\langle \Gamma_f, B\rangle \leq 2\|\Gamma_f\|_{\mathfrak{S}_1} < \infty$. □

For the proof of the theorem for $1 < p < \infty$ we require the following variant of the Marcinkiewicz interpolation theorem.

**Theorem** (Marcinkiewicz). *Let $(X, \mu)$ be any measure space, $\mathscr{V}$ an operator from $\mathfrak{S}_\infty$ into $L^\infty(X, \mu)$ which is also of weak type $(1, 1)$, i.e.*

$$\mu\{x \in X : |(\mathscr{V}T)(x)| > \delta\} \le \text{const}\, \|T\|_{\mathfrak{S}_1}/\delta, \quad T \in \mathfrak{S}_1, \quad \delta > 0.$$

*Then $\mathscr{V}$ is a bounded operator from $\mathfrak{S}_p$ into $L^p(X, \mu)$ for $1 < p \le \infty$.*

This result follows easily from the classical Marcinkiewicz theorem (cf. Zygmund [1], Chap. 1, Sect. 4). In fact, let $T \in \mathfrak{S}_p$ have the Schmidt expansion $\sum_{k\ge 0} \lambda_k(\cdot, f_k)g_k$. Define an operator $\tilde{\mathscr{V}}$ in the sequence space $c_0 \stackrel{\text{def}}{=} \{\{x_k\}_{k\ge 0} : \lim_{k\to\infty} x_k = 0\}$ by the formula

$$\tilde{\mathscr{V}}\{x_k\}_{k\ge 0} = \mathscr{V}\left(\sum_{k\ge 0} x_k(\cdot, f_k)g_k\right).$$

Then $\tilde{\mathscr{V}}$ is of weak type (1, 1) and is continuous from $c_0$ into $L^\infty(X, \mu)$. Consequently by the Marcinkiewicz theorm it is continuous from $l^p$ into $L^p(X, \mu)$. This means that $\mathscr{V}T = \tilde{\mathscr{V}}\{\lambda_n\}_{n\ge 0} \in L^p(X, \mu)$.

Let us now proceed to the proof of Theorem 1.1′ for $p > 1$.

With the operator $T$ in $l^2$ we associate a function $\mathscr{W}T$ analytic in $\mathbb{D}$ gotten in the following manner:

$$\mathscr{W}T = \sum_{n\ge 0} t_n z^n, \quad t_n \stackrel{\text{def}}{=} \sum_{j=0}^{n} (Te_j, e_{n-j}).$$

**Lemma 1.2.** *If $1 < p < \infty$ then $W$ is a bounded operator from $\mathfrak{S}_p$ into $B_p^{-1/p'}$.*

*Proof.* Let us define a measure $\mu$ on the Borel sets of $\mathbb{D}$ by the formula

$$\mu(E) \stackrel{\text{def}}{=} \iint_E \frac{1}{(1-|z|)^2}\, dx\, dy.$$

Consider the following transformation given on the set of all operators in $l^2$:

$$\mathscr{V}T = (1 - |z|^2)(\mathscr{W}T)'.$$

Let us show that $\mathscr{V}$ is continuous from $\mathfrak{S}_p$ into $L^p(\mathbb{D}, \mu)$ for $1 < p < \infty$. From this it will follow that

$$\iint_D |(\mathscr{W}T)'|^p (1 - |z|)^{2p-2}\, dx\, dy < \infty$$

for $T \in \mathfrak{S}_p$, $1 < p < \infty$, and thus $\mathscr{W}T \in B_p^{-1/p'}$ (see Review of Function Classes).

We begin by the continuity of $\mathscr{V} : \mathfrak{S}_\infty \to L_\infty(\mathbb{D}, \mu)$. This is the same as to say that $\mathscr{W}T \in B_\infty^{-1}$ for $\mathbb{T} \in \mathfrak{S}_\infty$. Let $\varphi$ be a polynomial. Then we have

$$\langle T, \Gamma_\varphi\rangle = \sum_{n\ge 0} \hat{\varphi}(n) \sum_{j=0}^{n} (Te_j, e_{n-j}) = \sum_{n\ge 0} \hat{\varphi}(n)\widehat{\mathscr{W}T}(n) = (\varphi, \mathscr{W}T).$$

Consequently $|(\varphi, \mathscr{W}T)| \le \|T\|_{\mathfrak{S}_\infty} \|\Gamma_\varphi\|_{\mathfrak{S}_1} \le \text{const}\, \|T\|_{\mathfrak{S}_\infty} \|\varphi\|_{B_1^1}$ by the part of Theorem 1.1′ which we have already established, viz. the case $p = 1$. From this it follows that $\mathscr{W}T$ gives rise to a continuous linear functional on $P_+B_1^1$ so that $\mathscr{W}T \in B_\infty^{-1}$ (cf. Review of Function Classes).

Next we show that $\mathscr{V}$ is of weak type (1, 1), i.e. that

$$\mu\{\zeta\in\mathbb{D}:|(\mathscr{V}T)(\zeta)|>\delta\}\le \text{const}\,\|T\|_{\mathfrak{S}_1}/\delta,\ \delta>0.$$

Let $T\in\mathfrak{S}_1$. Then $\mathscr{W}T\in H^1$. In fact, it suffices to establish this for operators $T=(\cdot,x)y$ where $x=\{x_n\}_{n\ge 0}\in l^2$, $y=\{y_n\}_{n\ge 0}\in l^2$. Consider $f=\sum_{n\ge 0}\bar{x}_n z^n$, $g=\sum_{n\ge 0}y_n z^n\in H^2$. Then $\widehat{\mathscr{W}T}(n)=\widehat{fg}(n)$ and consequently $\mathscr{W}T=fg\in H^1$. (If we use the fact that every $H^1$ function is the product of two $H^2$-functions it is seen that $\mathscr{W}\mathfrak{S}_1=H^1$.)

Let us now invoke the Littlewood-Paley theorem (cf. Zygmund [1], Chap. XIV, Sect. 3):

$$\varphi\in H^1\Rightarrow\int_{-\pi}^{\pi}\left(\int_0^1|\varphi'(re^{i\theta}|^2(1-r)\,dr\right)^{1/2}d\theta<\infty.$$

Thus, in order to prove that $\mathscr{V}$ has weak type (1, 1), it suffices to show the following: let $\psi$ be a measurable function on $[0,1]$ and set $E_\delta=\{x\in[0,1]:|\psi(x)|(1-x)^2>\delta\}$; then

$$\int_{E_\delta}\frac{dx}{(1-x)^2}\le \text{const}\left(\int_0^1|\psi(x)|^2(1-x)\,dx\right)^{1/2}\Big/\delta. \tag{1}$$

It follows indeed from (1) that $\mathscr{V}$ has weak type (1, 1), because

$$\mu\{\zeta\in\mathbb{D}:(1-|\zeta|)^2|(\mathscr{W}T)'(\zeta)|>\delta\}$$
$$\le \text{const}\int_{-\pi}^{\pi}\left(\int_0^1|(\mathscr{W}T)'(re^{i\theta})|^2(1-r)\,dr\right)^{1/2}d\theta/\delta.$$

Let us prove (1). We have

$$\int_0^1|\psi(x)|^2(1-x)\,dx\ge\int_{E_\delta}|\psi(x)|^2(1-x)\,dx$$
$$\ge\delta^2\int_{E_\delta}(1-x)^{-3}\,dx.$$

Therefore in order to prove (1) it suffices to prove that

$$\int_E dx/(1-x)^{-2}\le \text{const}\left(\int_E dx/(1-x)^{-3}\right)^{1/2} \tag{2}$$

for arbitrary subsets $E$ of $[0,1)$. To this end we map the interval $[0,1)$ onto the halfaxis $[1,\infty)$ by the transformation $s(x)=\dfrac{1}{1-x}$. It is then clear that (2) is equivalent to the inequality

$$\int_F ds\le \text{const}\left(\int_F s\,ds\right)^{1/2}$$

valid for any measurable subset $F$ of the set $[1,\infty)$. But it is easy to see that

$\int_F s\,ds \le \int_1^{a+1} s\,ds = \frac{1}{2}((a+1)^2 - 1)$ where $a = \int_F ds$ is the Lebesgue measure of $F$. Consequently

$$\left(\int_F s\,ds\right)^{1/2} \ge (a^2/2 + a)^{1/2} \ge a/\sqrt{2} = 1/\sqrt{2}\int_F ds.$$

Thus $\mathscr{V}$ is of weak type (1, 1) and so by Marcinkiewicz's theorem it is a bounded operator from $\mathfrak{S}_p$ into $L^p(\mathbb{D}, \mu)$ for $1 < p \le \infty$. As we have already remarked this implies that $\mathscr{W}$ is bounded from $\mathfrak{S}_p$ into $B_{p'}^{-1/p'}$. □

*Proof of Theorem* 1.1′. *The case* $1 < p < \infty$. Assume that $\Gamma_f \in \mathfrak{S}_p$, $1 < p < \infty$. Let us show that $f \in B_p^{1/p}$. By Lemma 1.2 $\mathscr{W}\Gamma_f \in B_p^{-1/p'}$, which means that

$$\iint_{\mathbb{D}} \left|\sum_{n\ge 0}(n+1)\hat{f}(n)z^n\right|^p (1-|z|)^{p-2}\,dx\,dy < \infty.$$

But this is equivalent to $\iint_{\mathbb{D}} |f'|^p(1-|z|)^{p-2}\,dx\,dy < \infty$. Hence $f \in B_p^{1/p}$.

Next assume that $f \in B_p^{1/p}$, $1 < p < \infty$. Let us show that $\Gamma_f \in \mathfrak{S}_p$. If $T$ is an operator in $l^2$ all of whose matrix elements except for a finite number vanish then

$$\langle \Gamma_f, T\rangle = \sum_{n\ge 0}\hat{f}(n)\sum_{j=0}^{n}\overline{(Te_j, e_{n-j})} = (f, \mathscr{W}T).$$

Consequently, by Lemma 1.2,

$$\begin{aligned}|\langle \Gamma_f, T\rangle| &\le \text{const}\,\|f\|_{B_p^{1/p}}\|\mathscr{W}T\|_{B_{p'}^{-1/p}}\\ &\le \text{const}\,\|f\|_{B_p^{1/p}}\|T\|_{\mathfrak{S}_{p'}}.\end{aligned}$$

This shows that $\Gamma_f \in \mathfrak{S}_p$. □

*Remark.* Not long ago two other proofs in the case $1 < p < \infty$ of Theorem 1.1′ appeared; cf. Peetre-Svensson [1], Rochberg [1]. There the interpolation is performed in a different manner.

The results obtained allow us to prove the boundedness of the natural projection (averaging) onto the set of Hankel matrices. This operator is defined on the set of all operators in $l^2$ by the formula

$$((\mathscr{P}A)e_n, e_m) = \frac{1}{n+m+1}\sum_{j=0}^{m+n}(Ae_j, e_{n+m-j}).$$

It is clear that $\mathscr{P}$ projects onto the set of Hankel matrices.

**Theorem 1.3** (Peller [3], [5]). *If* $1 < p < \infty$ *then* $\mathscr{P}$ *is bounded on* $\mathfrak{S}_p$.

*Proof.* Let $T \in \mathfrak{S}_p$, $1 < p < \infty$. Then $\mathscr{P}T = \Gamma_f$ where the function $f$ is defined by the identity

$$f(n) = \frac{1}{n+1}\sum_{j=0}^{n}(Te_j, e_{n-j}).$$

As by virtue of Lemma 1.2 $\iint_{\mathbb{D}} |\mathscr{W}T|^p(1-|z|)^{p-2}\,dx\,dy < \infty$ this gives

$$\iint_{\mathbb{D}} |f'|^p(1-|z|)^{p-2}\,dx\,dy < \infty,$$

i.e. $f \in B_p^{1/p}$. By Theorem 1.1′ $\mathscr{P}T = \Gamma_f \in \mathfrak{S}_p$. □

The following application of Theorem 1.1′ is concerned with rational approximation. It gives us the possibility to describe in local terms function classes defined in terms of the rate of rational approximation.

Let us consider the set $\mathscr{R}_n$ of rational functions with the poles outside $\mathbb{T}$, the sum of the multiplicities of the latter (including the point $\infty$) not exceeding $n$. If $f \in BMO$ then $r_n(f) \overset{\text{def}}{=} \operatorname{dist}_{BMO}(f, \mathscr{R}_n)$ (cf. Appendix 4, Sect. 165.53.3 concerning the space $BMO$). Among many equivalent norms in $BMO$ we pick the following one:

$$\|f\|_{BMO} \overset{\text{def}}{=} \inf_{g \in BMO_A} \|f-g\|_{L^\infty} + \inf_{g \in BMO_A} \|\bar{f}-g\|_{L^\infty} + |\hat{f}(0)|,$$

where $BMO_A \overset{\text{def}}{=} BMO \cap H^2 = P_+L^\infty = P_+BMO$.

The following theorem allows us to characterize in local terms the functions $f$ such that $\sum_{n\ge 0}(r_n(f))^p < \infty$.

**Theorem 1.4** (Peller [3]). *Let $f \in BMO$. Then*

$$\sum_{n\ge 0}(r_n(f))^p < \infty,\ 1 \le p < \infty,$$

*iff $f \in B_p^{1/p}$.*

*Proof.* In view of $P_+BMO \subset BMO$, $P_+B_p^{1/p} \subset B_p^{1/p}$, $P_+\mathscr{R}_n \subset \mathscr{R}_n$ it clearly suffices to consider the case $f \in BMO_A$.

The result now follows at once from Theorem 1.1′ if we further invoke the following three propositions:

1) Nehari's Theorem (cf. Lecture VIII, Sect. 4 and Appendix 4, Sect. 165.53.3): $|\Gamma_\varphi| = \|\varphi\|_{BMO}$, $\varphi \in BMO_A$.

2) The Adamyan-Arov-Kreĭn Theorem (cf. Appendix 4, Sect. 10): $s_n(\Gamma_\varphi) = \inf\{|\Gamma_\varphi - \Gamma_\psi| : \operatorname{rank}\Gamma_\psi \le n\}$.

3) Kronecker's Lemma (cf. Lecture VIII, Sect. 3): $\operatorname{rank}\Gamma_\psi \le n \Leftrightarrow \psi \in \mathscr{R}_n$. □

**1.2. The Connection between the Hankel Operators $H_u$ and $H_{\bar{u}}$.** In this Sect. we set forth a method with the aid of which one can (under suitable restrictions on the unimodular function $u$) draw information on the Hankel operator $H_{\bar{u}}$ from the Hankel operator $H_u$. It is founded on the following simple but most useful formula connecting the operators $H_u^*H_u$ and $H_{\bar{u}}^*H_{\bar{u}}$, with $u$ unimodular:

$$H_{\bar{u}}^*H_{\bar{u}}T_u = T_uH_u^*H_u. \tag{3}$$

It is easy to give a proof of (3) if we take account the formula

$$T_{\varphi\psi} - T_\varphi T_\psi = H_{\bar{\varphi}}^*H_\psi,\ \varphi, \psi \in L^\infty(\mathbb{T})$$

(cf. Appendix 4, Sect. 6). In fact

$$H_{\bar{u}}^*H_{\bar{u}}T_u = (I - T_uT_{\bar{u}})T_u = T_u(I - T_{\bar{u}}T_u) = T_uH_u^*H_u. \ \square$$

The proof of the main result of this Section will be based on a lemma in Halmos [2], Problem 152.

Recall that every operator $T$ in a Hilbert space $\mathscr{K}$ has a *polar decomposition* $T = U(T^*T)^{1/2}$ where $\operatorname{Ker} U = \operatorname{Ker} T$ whereas the operator $U|\mathscr{K} \ominus \operatorname{Ker} U$ is isometric on $\operatorname{clos} T\mathscr{K}$. The operator $U$ is termed the partially isometric part of $T$.

**Lemma 1.5.** *Let $A$ and $B$ be self-adjoint operators in a Hilbert space and $T$ an operator such that $AT = TB$. Then $AU = UB$ where $U$ is the partially isometric part of $T$.*

*Proof.* It follows from $AT = TB$ that $T^*A = BT^*$ and therefore $T^*TB = T^*AT = BT^*T$. Consequently, $B$ commutes with every function of the self-adjoint operator $T^*T$ and so, in particular, with $(T^*T)^{1/2}$. Consequently

$$AU(T^*T)^{1/2} = U(T^*T)^{1/2}B = UB(T^*T)^{1/2}.$$

If we multiply this identity from the right by $(T^*T)^{1/2}$ we get $AUT^*T = UBT^*T$. This shows that the operators $AU$ and $UB$ coincide on $\operatorname{clos}(T^*TK)$. Let us check that they coincide also on the orthogonal complement of the latter subspace. As $\operatorname{Ker} U = \operatorname{Ker} T$ and as $B$ leaves invariant the kernel of $T$ (this follows from $AT = TB$) we see that $AUx = \mathbb{O} = VBx$ for $x \in \operatorname{Ker} U$. □

If we apply the Lemma to formula (3) we get

$$H_{\bar{u}}^* H_{\bar{u}} U = U H_u^* H_u, \tag{4}$$

where $U$ now is the partially isometric part of the Toeplitz operator $T_u$.

**Theorem 1.6.** *Let $u$ be a unimodular function such that* $\operatorname{clos} T_u H^2 = H^2$ *and set* $E = \{f \in H^2 : H_u^* H_u f = f\}$. *Then $H_{\bar{u}}^* H_{\bar{u}}$ is unitarily equivalent to $H_u^* H_u | H^2 \ominus E$.*

*Proof.* Let us remark that $E = \operatorname{Ker} T_u = \operatorname{Ker} U$ and that $U$ maps the space $H^2 \ominus E$ isometrically onto $H^2$. Formula (4) now gives

$$H_{\bar{u}}^* H_{\bar{u}} = U H_u^* H_u U^* = U(H_u^* H_u | H^2 \ominus E) U^*. \ \square$$

Notice that $\operatorname{clos} T_u H^2 \neq H^2$ iff there exist an outer function $h$ in $H^2$ and an inner function $\Theta$ such that $u = z\Theta h/\bar{h}$.

**Corollary 1.7.** *Let $u$ be a unimodular function such that* $\operatorname{Ker} T_u = \operatorname{Ker} T_u^* = \{\mathbb{O}\}$. *Then $H_{\bar{u}}^* H_{\bar{u}}$ and $H_u^* H_u$ are unitarily equivalent.* □

**Corollary 1.8.** *Let $u$ be a unimodular function in $H^\infty + C(\mathbb{T})$ such that* $\operatorname{clos} T_u H^2 = H^2$. *Then*

$$s_k(H_{\bar{u}}) = s_{k+n}(H_u),\ k \in \mathbb{Z}_+,$$

*where* $n = \dim \operatorname{Ker} T_u = \dim\{f \in H^2 : H_u^* H_u f = f\}$.

*Proof.* This statement follows immediately from Theorem 1.6 and the fact that $|H_u^* H_u | H^2 \ominus E| < 1$. □

**Corollary 1.9.** *Let $u$ be a unimodular function such that* $\operatorname{clos} T_u H^2 = H^2$. *Assume that $H_u \in \mathfrak{S}_p$, $1 \le p < \infty$. Then $H_{\bar{u}} \in \mathfrak{S}_p$ and $\|H_{\bar{u}}\|_{\mathfrak{S}_p} \le \|H_u\|_{\mathfrak{S}_p}$.* □

## Chapter II. From Probability Theory to Hankel Operators

The known methods of reducing probabilistic problems on stationary Gaussian processes to problems of spectral theory of functions use in one way or another orthogonal expansions in Hermite polynomials in infinitely many variables. This method was first proposed by Wiener [1] (for the history of the question see Masani [1]). Subsequently Segal [1] made the observation that such expansions are intimately connected with the Fock spaces as used in quantum theory. The use of Fock spaces indeed turned out to be most fruitful in quantum theory as well as in probability. In this Chapter we discuss the mixing properties of stationary Gaussian processes and, employing the language of Fock spaces, manage to reduce the problems arising here to problems on Hankel operators.

More detailed treatments of Fock spaces can be found in Simon [1], Nelson [1]. Regarding Gaussian processes and the probabilistic terminology utilized see Rozanov [2], Yaglom [1], Lamperti [1], Dym-McKean [1], Ibragimov-Rozanov [1], Reed [1].

**2.1. Stationary Gaussian Processes.** Let $(\Omega, \Sigma, P)$ be a probability space, i.e. $\Omega$ is a space of elementary events, $\Sigma$ a $\sigma$-algebra of subsets of $\Omega$, $P$ a probability measure on $\Sigma$. A measurable function $f: \Omega \to \mathbb{R}$ is called a *Gaussian random variable* if it has a Gaussian distribution:

$$P\{\omega : f(\omega) < x\} = \frac{1}{\sqrt{2\pi\sigma_f^2}} \int_{-\infty}^{\infty} e^{-t^2/2\sigma_f^2}\, dt.$$

(We consider Gaussian variables with zero expectation only.) Here $\sigma_f^2 = \mathbb{E}|f|^2 \overset{\text{def}}{=} \int_\Omega |f|^2\, dP$ is the *dispersion* of the random variable $f$. It will be convenient to regard a function which too vanishes identically as a Gaussian random variable. A family $\{X_\alpha\}_{\alpha \in A}$ of random variables on $\Omega$ is termed Gaussian if its linear hull consists of Gaussian variables.

If $X_1, \dots, X_n$ is a Gaussian family and $\gamma = (\gamma_1, \dots, \gamma_n) \in \mathbb{R}^n$ then $Y = \sum_{k=1}^n \gamma_k X_k$ is a Gaussian random variable. Its dispersion equals

$$\mathbb{E}Y^2 = \sum_{i,j=1}^{n} \gamma_i \gamma_j \mathbb{E}(X_i X_j) = (Q\gamma, \gamma).$$

The matrix $Q$ defined by the leftmost part of this identity apparently is symmetric and positive definite. It is easy to see that $Q$ is invertible iff the functions $X_1, \dots, X_n$ are linearly independent. If this is the case then the joint distribution $\mu_X$ of the family $X_1, \dots, X_n$, by definition the image of $P$ under the map $\omega \mapsto (X_1(\omega), \dots, X_n(\omega))$, is absolutely continuous with respect to Lebesgue measure in $\mathbb{R}^n$. With the aid of the Fourier inversion formula it is then easy to find the density $P_X$ of $\mu_X$:

$$P_X(x) = (2\pi)^{-n/2}(\det Q)^{-1/2} \cdot \exp\left\{-\frac{1}{2}(Q^{-1}x, x)\right\}. \tag{1}$$

From this one readily reads off that two random variables $X_1$ and $X_2$, members of the same Gaussian family, are independent iff they are orthogonal: $\mathbb{E}X_1X_2 = 0$.

A closed subspace $\mathbb{G}$ of the real space $L^2_{\mathbb{R}}(\Omega)$ is called *Gaussian* if it consists of Gaussian variables. The closure in $L^2_{\mathbb{R}}(\Omega)$ of the linear hull (over the reals) of a Gaussian family is a Gaussian subspace, as is shown by the following simple lemma.

**Lemma 2.1.** *If $\{X_n\}_{n\geq 0}$ is a sequence of Gaussian random variables and $X = \lim_n X_n$ a.e. on $\Omega$ then $X$ is a Gaussian variable too.*

*Proof.* By Lebesgue's theorem

$$\mathbb{E}(e^{itX}) = \lim_n \mathbb{E}(e^{itX_n}) = \exp\left\{-\frac{t^2}{2}\lim_n \sigma_n^2\right\}.$$

The result now follows from the Fourier inversion formula. □

Let us fix a Gaussian subspace $\mathbb{G}$ in $L^2_{\mathbb{R}}(\Omega)$. We may assume that $\Sigma$ is the least $\sigma$-algebra with respect to which all functions in $\mathbb{G}$ are measurable. Let $\mathbb{A}$ and $\mathbb{B}$ be two closed subspaces of $\mathbb{G}$ and let $\mathscr{A}$ and $\mathscr{B}$ be the least $\sigma$-algebras with respect to which all functions in $\mathbb{A}$, respectively in $\mathbb{B}$ are measurable. Recall that the subsapces $\mathbb{A}$ and $\mathbb{B}$ are called independent if the $\sigma$-algebras $\mathscr{A}$ and $\mathscr{B}$ are independent. It follows of course from (1) that the independence of $\mathbb{A}$ and $\mathbb{B}$ is equivalent to orthogonality. The following identity:

$$n!\prod_{i=1}^n z_i = \sum_{r=1}^n (-1)^{n-r} \sum_{i_1<\cdots<i_r} (z_{i_1} + \cdots + z_{i_r})^n, \tag{2}$$

valid in any commutative ring with unit, is easily established by induction over $n$.

**Theorem 2.2.** $L^2_{\mathbb{R}}(\Omega) = \operatorname{span}\{X^k : X \in \mathbb{G}, k \in \mathbb{Z}_+\}$.

*Proof.* Consider a function $g$ which is orthogonal to all powers $X^k$, $X \in \mathbb{G}$, $k \in \mathbb{Z}_+$. Formula (2) then shows that $g$ is orthogonal to every function $f(X_1, \ldots, X_n)$, $f$ a polynomial in $n$ variables. As the set of polynomials in $n$ variables is dense in the space of all functions in $\mathbb{R}^n$ which are square summable with respect to the weight $\exp\{-\frac{1}{2}\|x\|^2\}$, this gives $\int_E g\,dP = 0$ for every subset $E$ of $\Omega$ which is measurable with respect to a finite set of elements in $\mathbb{G}$. These subsets generate the whole $\sigma$-algebra $\Sigma$ and therefore $g = \mathbb{0}$. □

**Definition.** Let $\{X_t\}_{t\in\mathscr{T}}$ be a Gaussian family. If $\mathscr{T} = \mathbb{Z}$ or $\mathscr{T} = \mathbb{R}$ we say that it is a *Gaussian process*. The set $\mathscr{T}$ will be interpreted as time. We set $\mathbb{G} = \operatorname{span}\{X_t : t \in \mathscr{T}\}$. In order to avoid pathology we assume, in the case $\mathscr{T} = \mathbb{R}$, that the map $t \mapsto X_t$ is continuous from $\mathbb{R}$ into $\mathbb{G}$.

**Definition.** A Gaussian process $\{X_t\}_{t\in\mathcal{T}}$ is called *stationary*

$$\mathbb{E}X_tX_s = Q(t-s),\ t,\ s\in\mathcal{T}.$$

The function $Q$ is called the *correlation function* of the process.

It is clear that $Q$ is positive definite, i.e.

$$\sum_{i,j=1}^{n} Q(t_i - t_j)\gamma_i\bar{\gamma}_j \geq 0$$

for arbitrary families $(t_1,\ldots,t_n)\in\mathcal{T}^n, (\gamma_1,\ldots,\gamma_n)\in\mathbb{C}^n$. In the case $\mathcal{T} = \mathbb{R}$, thanks to the continuity of the mapping $t\mapsto X_t$, $Q$ will be a continuous function. By the theorem of Riesz-Herglotz ($\mathcal{T} = \mathbb{Z}$) (cf. Riesz and Sz.-Nagy [1], Sect. 5.3) and Bochner ($\mathcal{T} = \mathbb{R}$) (cf. Ahieser-Glazman [1], Sect. 70) there exists a unique positive measure $\Delta$ on $\mathbb{T}$ (respectively on $\mathbb{R}$) such that

$$Q(n) = \int_{\mathbb{T}} \zeta^n \, d\Delta(\zeta) \quad \left(Q(t) = \int_{\mathbb{R}} e^{itx}\, d\Delta(x)\right).$$

The measure $\Delta$ is called the *spectral measure* of the process $\{X_t\}_{t\in\mathcal{T}}$. As $Q(t) = Q(-t)$, $t\in\mathcal{T}$, $\Delta$ is invariant with respect to the map $\zeta\mapsto\bar{\zeta}$ (respectively $x\mapsto -x$). The map $\Phi$, defined by the formula

$$\Phi\left(\sum_{k=1}^{n} c_k X_{t_k}\right) = \sum_{k=1}^{n} c_k e^{it}k^x, \tag{3}$$

extends to an isometry of the complexification $\mathbb{G}_c = \mathbb{G} + i\mathbb{G}$ of $\mathbb{G}$ onto $L^2(\mathbb{T},\Delta)(L^2(\mathbb{R},\Delta))$ and is called the *spectral representation of the process*. It is plain that for each finite positive measure $\Delta$ with the above symmetry property there exists a stationary Gaussian process with spectral measure $\Delta$. For the proof it suffices to take any Gaussian space $\mathbb{G}$ and to map it isometrically onto the subspace of $L^2(\mathbb{T},\Delta)$ $(L^2(\mathbb{R},\Delta))$ consisting of the functions $f$ with $f(\bar{\zeta}) = \overline{f(\zeta)}$, $\zeta\in\mathbb{T}$ $(f(-t) = \overline{f(t)},\ t\in\mathbb{R})$; $X_t$ is then defined as the pre-image of the function $x\mapsto e^{itx}$.

In the following we consider the complex Hilbert space $L^2(\Omega)$. By a complex Gaussian subspace in $L^2(\Omega)$ we mean the complexification of a real Gaussian subspace of $L^2_{\mathbb{R}}(\Omega)$. Furthermore, we usually say simply Gaussian space instead of complex Gaussian space.

**2.2. The Fock Space.** Let $\mathcal{H}$ be a Hilbert space. For every natural number $n$ we consider the *n-fold symmetric tensor product*

$$\mathcal{H}_n \overset{\text{def}}{=} \mathcal{H} \circledS \ldots \circledS \mathcal{H} \subset \mathcal{H}\otimes\cdots\otimes\mathcal{H} \overset{\text{def}}{=} \bigotimes^n \mathcal{H}.$$

The space $\mathcal{H}_n$ will be regarded as a closed subspace of the Hilbert tensor product. The orthogonal projection $\mathrm{Sym}: \bigotimes^n\mathcal{H}\to\mathcal{H}_n$ is defined on decomposable tensors by the formula

$$\mathrm{Sym}(f_1\otimes\cdots\otimes f_n) = \frac{1}{n!}\sum_{\sigma\in S_n} f_{\sigma_1}\otimes\cdots\otimes f_{\sigma_n}.$$

Here $S_n$ is the group of permutations of the set $\{1,2,\ldots,n\}$. The scalar product on $\bigotimes^n \mathscr{H}$ takes the form

$$(g_1 \otimes \cdots \otimes g_n, f_1 \otimes \cdots \otimes f_n)_\otimes = (g_1, f_1) \ldots (g_n, f_n).$$

It is useful to notice that

$$(\operatorname{Sym} g_1 \otimes \cdots \otimes g_n, \operatorname{Sym} f_1 \otimes \cdots \otimes f_n)_\otimes = \frac{1}{n!} \sum_{\sigma \in S_n} (g_{\sigma_1}, f) \ldots (g_{\sigma_n}, f_n). \tag{4}$$

By technical reasons it is convenient to introduce in $\mathscr{H}_n$ the new scalar product

$$(\cdot,\cdot)_{\circledS} \stackrel{\text{def}}{=} n!(\cdot,\cdot)_\otimes.$$

If $\mathscr{H} = L^2(\mathbb{T},\Delta)$ then $\mathscr{H}_n$ can be canonically realized as the subspace of $L^2(\mathbb{T}^n, \bigotimes^n \Delta)$ consisting of functions symmetric with respect to permutation of the coordinates. Then

$$(f,g)_{\circledS} = n! \underbrace{\int \ldots \int}_{n} f(x_1,\ldots,x_n)\overline{g(x_1,\ldots,x_n)}\, d\Delta(x_1)\ldots d\Delta(x_n).$$

We set $\mathscr{H}_0 \stackrel{\text{def}}{=} \mathbb{C}$.

**Definition.** The *Fock space* over $\mathscr{H}$ is the direct sum of Hilbert spaces

$$\mathfrak{F}(\mathscr{H}) = \bigoplus_{n=0}^{\infty} \mathscr{H}_n.$$

The space $\mathfrak{F}(\mathscr{H})$ is likewise referred to as the Boson Fock space.

A vector $h$ in $\mathfrak{F}(\mathscr{H})$ is termed a finite-particle vector if all the projections, but for a finite number, of $h$ onto the subspaces $\mathscr{H}_n$ equal zero. Consider an arbitrary linear operator $T: \mathscr{H} \to \mathscr{K}$. Then the identity

$$\mathfrak{F}(T)|\mathscr{H}_n \stackrel{\text{def}}{=} \underbrace{T \otimes \cdots \otimes T}_{n},\ n \geq 1,\ \mathfrak{F}(T)|\mathscr{H}_0 \stackrel{\text{def}}{=} I,$$

defines a linear operator $\mathfrak{F}(T): \mathfrak{F}(\mathscr{H}) \to \mathfrak{F}(\mathscr{K})$ at least on finite-particle vectors. It is easy to see that $\mathfrak{F}(T)$ is unitary if $T$ is unitary, isometric if $T$ is isometric, a projection if $T$ is a projection. It is further clear that $\mathfrak{F}(ST) = \mathfrak{F}(S)\mathfrak{F}(T)$.

**Lemma 2.3** (Halmos [3]). *Let $\mathscr{H}$ and $\mathscr{K}$ be Hilbert spaces and consider a linear operator $A: \mathscr{H} \to \mathscr{K}$ which is a contraction, that is, $|A| \leq 1$. Then there exists a unitary operator $U$ in $\mathscr{H} \oplus \mathscr{K}$ such that $A = \mathscr{P}U\mathscr{J}$. Here $\mathscr{J}$ is the canonical imbedding of H into $\mathscr{H} \oplus \mathscr{K}$ and $\mathscr{P}$ the projection of $\mathscr{H} \oplus \mathscr{K}$ onto $\mathscr{K}$.*

*Proof.* $U$ is given by the block matrix:

$$U = \begin{pmatrix} (I - A^*A)^{1/2} & A^* \\ A & -(I - AA^*)^{1/2} \end{pmatrix}. \square$$

**Corollary 2.4.** *A is a contraction iff its Fock power $\mathfrak{F}(A)$ is a contraction.* $\square$

Let us now indicate how to introduce a Fock space structure in the Hilbert space $L^2(\Omega)$. For every natural number $n$ let $G^n$ temporarily be the space of homogeneous polynomials of degree $n$ in the elements of a Gaussian subspace $\mathbb{G}$ of $L^2(\Omega)$. Set $\mathbb{G}^0 \stackrel{\text{def}}{=} \mathbb{C}$.

**Lemma 2.5.** *The subspaces* $G^n$, $n \in \mathbb{Z}_+$, *are linearly independent.*

*Proof.* Let $X_1, \ldots, X_m$ be Gaussian variables in $\mathbb{G}$ and consider homogeneous polynomials $\{g_n\}$ of degree $n$, $n \in \{0, 1, \ldots, N\}$ in $m$ variables such that $\sum_{n=0}^{N} g_n(X_1, \ldots, X_m) \equiv 0$ a.e. on $\Omega$. Consider the image of the probability measure under the map $\omega \mapsto (X_1(\omega), \ldots, X_m(\omega))$. It is not hard to see (cf. (1)) that it is concentrated on a certain linear subspace $L$ of $\mathbb{C}^m$ and absolutely continous with respect to Lebesgue measure on $L$. Therefore $\sum_{n=0}^{N} g_n(x_1, \ldots, x_m) \equiv 0$ for $(x_1, \ldots, x_m) \in L$. But the polynomials $g_n$ have different degrees of homogeneity and, consequently, $g_n|L \equiv 0$, $n \in \{0, \ldots, N\}$ so that by necessity $g_n(X_1, \ldots, X_m) \equiv 0$. $\square$

Denote by : : the linear operator which orthogonalizes the subspaces $G^n$ according to the Gram-Schmidt procedure. In other words, if $f \in G^n$, then $:f: = f - \mathscr{P}_n f$ where $\mathscr{P}_n$ denotes orthogonal projection onto $\operatorname{span}\{G^k : 0 \leq k \leq n-1\}$. The mapping $f \mapsto :f:$ is called the *Wick transform*. It is defined on the set of all polynomials in elements of the space $\mathbb{G}$, that is, on a dense subset of $L^2(\Omega)$ (cf. Theorem 2.2). Let us set $\Gamma(\mathbb{G})_n \stackrel{\text{def}}{=} :\mathbb{G}^n:$, $n \in \mathbb{Z}_+$. Then

$$L^2(\Omega) = \bigoplus_{n=0}^{\infty} \Gamma(\mathbb{G})_n. \tag{5}$$

**Lemma 2.6.** *Let $f$ be a real function in $\mathbb{G}$. There exist real numbers $a_0, a_1, \ldots, a_{n-1}$ such that*

$$:f^n: = f^n + a_{n-1} f^{n-1} + \cdots + a_0.$$

*Proof.* Choose the numbers $a_0, a_1, \ldots, a_{n-1}$ in such a way that the function $f_n \stackrel{\text{def}}{=} f^n + a_{n-1} f^{n-1} + \cdots + a_0$ is orthogonal to all the powers $f^0, \ldots, f^{n-1}$. Set $\mathbb{G}(f) = \mathbb{G} \ominus \{\lambda f : \lambda \in \mathbb{C}\}$. Since in $\mathbb{G}$ orthogonality is equivalent to independence we have $\mathbb{E}\varphi(f)\psi(h) = \mathbb{E}\varphi(f)\mathbb{E}\psi(h)$ for arbitrary polynomials $\varphi$ and $\psi$ and every function $h$ in $\mathbb{G}(f)$.

Let us check that $f_n$ is orthogonal to the subspaces $\mathbb{G}^k$ with $k \leq n-1$. By virtue of the polarization formula (2) it suffices to show that $\mathbb{E}f_n(\lambda f + g)^k = 0$ for $g \in \mathbb{G}(f)$; $\lambda \in \mathbb{C}$, $k \leq n-1$. We have

$$\mathbb{E}f_n(\lambda f + g)^k = \sum_{j=0}^{k} \binom{k}{j} \lambda^j \mathbb{E}(f_n f^j g^{k-j}) = \sum_{j=0}^{k} \binom{k}{j} \lambda^j \mathbb{E}f_n f^j \mathbb{E}g^{k-j} = 0. \ \square$$

*Remark.* Let $H_n(x) = x^n + a_{n-1}^{(n)} x^{n-1} + \cdots + a_0^{(n)}$, $n \in \mathbb{Z}_+$, be the system of orthogonal polynomials in $L^2(\mathbb{R}, e^{-x^2} dx)$, that is, the system of Hermite polynomials. Then, if $f$ is a real function in $\mathbb{G}$ and $\mathbb{E}f^2 = 1$, it is easy to see that $:f^n: = H_n(f)$.

**Theorem 2.7.** *For every natural number n the map*

$$:f_1 \ldots f_n: \mapsto \mathrm{Sym}(f_1 \otimes \cdots \otimes f_n)$$

*extends in a unique way to an isometry of* $\Gamma(\mathbb{G})_n$ *onto* $\mathbb{G}_n$.

*Proof.* In view of (4) it suffices to prove the following identity:

$$\mathbb{E}(:g_1 \ldots g_n: :f_1 \ldots f_n:) = \sum_{\sigma \in S_n} \mathbb{E}(g_{\sigma_1} f_1) \ldots \mathbb{E}(g_{\sigma_n} f_n). \tag{6}$$

The polarization formula (2) shows that it suffices to consider the case of real functions $g_1 = \cdots = g_n = g$, $f_1 = \cdots = f_n = f$. Otherwise put, we have to verify that

$$\mathbb{E}(:g^n: :f^n:) = n!(\mathbb{E}fg)^n,$$

where $f$ and $g$ are real functions in the unit sphere of $\mathbb{G}$. It is clear that $g = (\mathbb{E}gf)f + h$ with $h \perp f$. By the definition of the Wick transform we have: $\mathbb{E}(:g^n: :f^n:) = (\mathbb{E}gf)^n \cdot \mathbb{E}(:f^n: :f^n:)$. It remains to remark that the identity $\mathbb{E}(:f^n: :f^n:) = 1/\sqrt{2\pi} \cdot \int H_n(x)^2 e^{-x^2} dx = n!$ is a result of the usual normalization of the Hermite polynomials. □

This theorem together with formula (5) allows us to identify $L^2(\Omega)$ with the Fock space $\mathfrak{F}(\mathbb{G})$. If $\mathbb{A}$ is a Gaussian subspace of $\mathbb{G}$ and $\mathscr{A}$ the $\sigma$-algebra of the sets which are measurable with respect to $\mathbb{A}$ then, as is shown by Theorem 2.7, the space $\mathfrak{F}(\mathbb{A})$ corresponds in this identification to the subspace $L^2(\Omega, \mathscr{A}, P)$ consisting of the $\mathscr{A}$-measurable functions.

**Corollary 2.8.** *Let* $\{e_n\}_{n \geq 0}$ *be a complete orthonormal system in* $\mathbb{G}$. *Then the functions*

$$:e_{i_1}^{n_1} \ldots e_{i_k}^{n_k}:/\sqrt{n_1! \ldots n_k!}, \quad n_1 + \cdots + n_k = n, \quad i_1 < \cdots < i_k,$$

*form a complete orthonormal system for* $\Gamma(\mathbb{G})_n$.

*Proof.* This follows from (6). □

Let $\{X_t\}_{t \in \mathscr{T}}$ be a stationary Gaussian process and $\mathbb{G}$ the corresponding Gaussian subspace of $L^2(\Omega)$. Using the spectral representation of the process (cf. (3)) it is easy to see that the map $V^t x_s = x_{t+s}$ extends uniquely to a unitary operator in $\mathbb{G}$. The strongly continuous group $\{V^t\}_{t \in \mathscr{T}}$ of unitary operators in $\mathbb{G}$ is associated with a group $\{\tau^t\}_{t \in \mathscr{T}}$ of measure preserving transformations of the probability space $(\Omega, \Sigma, P)$, i.e., $(V^t f)(\omega) = f(\tau^t \omega)$. This can be seen if we realize $\Omega$ as the set of maps $\omega : s \mapsto \omega_s$ from $\mathscr{T}$ into $\mathbb{R}$. Then $X_s(\omega) = \omega_s$ and $(\tau^t \omega)_s = \omega_{s+t}$ (cf. Lamperti [1]). The condition of stationarity shows that $\tau^t$ is indeed measure preserving. It is now clear that $V^t$ can be extended to the whole of $L^2(\Omega)$ as a unitary operator $U^t : U^t f(\omega) = f(\tau^t \omega)$, $f \in L^2(\Omega)$. The operator $U^t$ maps $G^n$ into itself so that, since it is unitary, $\Gamma(\mathbb{G})_n$ will be an invariant subspace. It is now easy to see that $U^t = \mathfrak{F}(V^t)$. We are then in a position to write down the spectral representation of the unitary group $\{U^t\}_{t\in\mathcal{T}}$. To fix the ideas consider the case $\mathcal{T}=\mathbb{Z}$ (the case $\mathcal{T}=\mathbb{R}$ can be taken care of in an analogous manner). Let $\mathcal{M}_z$ be the operator of multiplication by $z$ in $L^2(\mathbb{T},\Delta)$. Then

$$U^n = \mathfrak{F}(\Phi)^*\mathfrak{F}(\mathcal{M}_{z^n})\mathfrak{F}(\Phi), \quad n\in\mathbb{Z}, \tag{7}$$

where the unitary map $\Phi$ is defined by formula (3). The spectral properties of the unitary group $\{U^t\}_{t\in\mathcal{T}}$ were first studied using the technique of Fock spaces by Veršik [1], [2].

**2.3. Mixing Properties and Regularity Conditions for Stationary Gaussian Processes.** In this Section we consider various regularity conditions for stationary processes (that is, different kind of conditions for the weak dependence of the future and the past), these conditions being related to mixing properties of the automorphism group $\{\tau^t\}_{t\in\mathcal{T}}$. The problem of classifying in spectral terms stationary Gaussian processes by their regularity properties will be reduced to corresponding problems for Hankel operators.

We will require the following notation. Let $\{X_t\}_{t\in\mathcal{T}}$ be a stationary Gaussian process. We then set

$$\mathbb{G}_t = \operatorname{span}\{X_s : s<t\}, \quad \mathbb{G}^t = \operatorname{span}\{X_s : s\geq t\}.$$

We denote by $\Sigma_t$ the $\sigma$-algebra of subsets measurable with respect to $\mathbb{G}_t$ and by $\Sigma^t$ the corresponding object with $\mathbb{G}^t$.

Let $\mathcal{A}$ be a $\sigma$-subalgebra of $\Sigma$. Then the operator of conditional expectation $\mathbb{E}_{\mathcal{A}}$ is defined as follows:

$$\int_A f\,dP = \int_A \mathbb{E}_{\mathcal{A}} f\,dP, \quad f\in L^2(\Omega), \quad A\in\mathcal{A},$$

the function $\mathbb{E}_{\mathcal{A}} f$ being measurable with respect to $\mathcal{A}$.

It is clear that $\mathbb{E}_{\mathcal{A}}$ is the orthogonal projection in $L^2(\Omega)$ onto $L^2(\Omega,\mathcal{A},P)$.

If $\mathcal{K}$ is a subspace of the Hilbert space $\mathcal{H}$ then the orthogonal projection of $\mathcal{H}$ onto $\mathcal{K}$ will be denoted by $\mathcal{P}_K$.

**Lemma 2.9.** *Let $\mathbb{A}$ be a Gaussian subspace of $\mathbb{G}$ and denote by $\mathcal{A}$ the $\sigma$-algebra of sets measurable with respect to $\mathbb{A}$. Then $\mathfrak{F}(\mathcal{P}_{\mathbb{A}})=\mathbb{E}_{\mathcal{A}}$.*

*Proof.* It is clear that $\mathfrak{F}(\mathbb{A}) = L^2(\Omega,\mathcal{A},P)$ and $\mathfrak{F}(\mathcal{P}_{\mathbb{A}}) = \mathcal{P}_{\mathfrak{F}(\mathbb{A})}$. □

**Ergodicity.** Recall that a group of measure preserving transformations $\{\tau^t\}_{t\in\mathcal{T}}$ in a measure space $(\Omega,\Sigma,P)$ is said to be *ergodic* if $A\in\Sigma$ and $\tau^{-t}A=A$ (for every element $t$ of $\mathcal{T}$) entails that $P(A)=0$ or $P(A)=1$. Let us consider the group of transformations $\{\tau^t\}_{t\in\mathcal{T}}$ introduced in Sect. 2.2. Let $\Delta$ be the spectral measure of the process $\{X_t\}_{t\in\mathcal{T}}$.

**Theorem 2.10** (Maruyama [1]). *The group $\{\tau^t\}_{t\in\mathcal{T}}$ is ergodic iff $\Delta$ is continuous.*

*The proof* will, for the sake of definiteness, be carried out in the case $\mathcal{T}=\mathbb{Z}$. If $\zeta_0\in\mathbb{T}$ and $\Delta(\{\zeta_0\})>0$ then the function $f=\Phi^{-1}(\chi_{\{\zeta_0\}})$ is a non-zero element of $\mathbb{G}$. Let us set $U \stackrel{\text{def}}{=} U^1$. It is clear that $Uf = \zeta_0 f$ and, consequently, $U|f| = |f|$. Therefore the set $A = \{\omega \in \Omega : |f(\omega)| > 1\}$ is invariant and has intermediate measure.

Next let $\Delta$ be continuous and pick a function such that $Uf = f$. By formula (7) there exists a positive integer $n$ and a non-zero function $h$, $h \in \underbrace{L^2(\mathbb{T}, \Delta) ⓢ \dots ⓢ L^2(\mathbb{T}, \Delta)}_{n}$, such that $\underbrace{(\mathscr{M}_z \otimes \cdots \otimes \mathscr{M}_z)}_{n} h = h$. Then

$$\int_{\mathbb{T}^n} (\zeta_1 \dots \zeta_n)^m h \cdot \bar{h}\, d^n\Delta = \int_{\mathbb{T}^n} |h|^2\, d^n\Delta, \quad m \in \mathbb{Z}, \tag{8}$$

as apparently $((\mathscr{M}_z \otimes \cdots \otimes \mathscr{M}_z)^m h, h) = (h, h)$. Consider now the diagonal map $\mathscr{D}_n : (\zeta_1, \dots, \zeta_n) \mapsto \zeta_1 \cdot \dots \cdot \zeta_n$ which sends the measure $\bigotimes^n \Delta$ onto the convolution power of the measure $\Delta$ on the circumference $\mathbb{T}$. From (8) it follows then that the image of the measure $|h|^2 d^m\Delta$ with respect to $\mathscr{D}_n$ equals $\|h\|_{\otimes}^2 \delta_1$ and, as the measure $|h|^2 d^m\Delta$ is absolutely continuous with respect to $d^n\Delta$, $\|h\|_{\otimes}^2 \delta_1$ is then also absolutely continuous with respect to $\Delta^{*n}$. But the convolution of continuous measures is continuous. Hence $h = \mathbb{O}$. □

**Mixing.** A more stringent regularity condition on the group $\{\tau^t\}_{t \in \mathscr{T}}$ is the condition of mixing. We say that $\{\tau^t\}_{t \in \mathscr{T}}$ has the *mixing* property if for arbitrary $A$ and $B$ in $\Sigma$ we have the formula

$$\lim_{t \to \infty} P(A \cap \tau^{-t} B) = P(A) \cdot P(B).$$

In spectral language this mixing condition takes the form: $\lim_{t \to \infty} (U^t f, g) = 0$ for all functions $f$ and $g$ in $L^2(\Omega)$ which are orthogonal to the constants. The following very well-known result, which we for definiteness state for $\mathscr{T} = \mathbb{Z}$, describes mixing for the group $\{\tau^n\}_{n \in \mathbb{Z}}$ in spectral terms.

**Theorem 2.11.** *The group $\{\tau^n\}_{n \in \mathbb{Z}}$ has the mixing property iff* $\lim_{|n| \to \infty} \hat{\Delta}(n) = 0$.

We omit its proof, because it is so similar to the proof of Maruyama's theorem. In ergodic theory it was with the aid of this theorem that one first got an example of mixing with a simple continuous singular spectrum (cf. Girsanov [1]).

**Non-determinstic processes.** A process $\{X_t\}_{t \in \mathscr{T}}$ is called *deterministic* if $\Sigma_t = \Sigma$ for every $t$, $t \in \mathscr{T}$, and *non-deterministic* in the opposite case.

It is easy to see that a process is deterministic iff $\mathbb{G}_t = \mathbb{G}$ for each $t$ in $\mathscr{T}$. Let $\Delta_a$ be the absolutely continuous (with respect to Lebesgue measure) part of the spectral measure $\Delta$ of the process $\{X_t\}_{t \in \mathscr{T}}$.

**Theorem 2.12.** *A stationary Gaussian process $\{X_t\}_{t \in \mathscr{T}}$ is non-deterministic iff*

$$\int_{\mathbb{T}} \log(d\Delta_a/dm)\, dm > -\infty \; (\mathscr{T} = \mathbb{Z}),$$

$$\int_{\mathbb{R}} \log(d\Delta_a/dx)\, dx/(1 + x^2) > -\infty \; (\mathscr{T} = \mathbb{R}) \tag{9}$$

If $\mathcal{T} = \mathbb{Z}$ this theorem follows in a trivial manner from the Szegö-Kolmogorov Theorem (cf. Lecture I, Sect. 7) while in the case $\mathcal{T} = \mathbb{R}$ one has to invoke a theorem of Kreĭn's [5], the continuous analogue of the Szegö-Kolmogorov Theorem.

It is well-known that condition (9) is equivalent to the condition that $d\Delta = |h|^2\,dm + d\Delta_s$ (respectively $d\Delta = |h|^2\,dx + d\Delta_s$) where $h$ is an outer function in the Hardy class $H^2$ on the disk (respectively in the halfplane) (cf. Appendix 2).

**Uniform mixing with respect to the past.** We intend the following property of the group $\{\tau^t\}_{t\in\mathcal{T}}$: for any set $B$ in $\Sigma$

$$\lim_{t\to\infty}\sup_{A\in\Sigma_0}|P(A\cap\tau^{-t}B)-P(A)P(B)|=0.$$

It is easy to see that this condition is equivalent to the following one

$$\lim_{t\to-\infty}\sup_{A\in\Sigma_t}|P(A\cap B)-P(A)P(B)|=0. \tag{10}$$

Processes with this property will be referred to as *regular*. Set now $\xi_A \overset{\text{def}}{=} \chi_A - P(A)$ for $A$ in $\Sigma$.

**Theorem 2.13.** *The following conditions are equivalent*:

1) *the process* $\{X_t\}_{t\in\mathcal{T}}$ *is regular*;
2) $\bigcap_{t\in T}\Sigma_t = \{\varnothing,\Omega\}$;
3) $\bigcap_{t\in T}\mathbb{G}_t = \{\mathbb{O}\}$;
4) $d\Delta = |h|^2\,dm$ *where h is an outer function in* $H^2$.

*Proof.* $1\Rightarrow 2$. If $B\in\bigcap_{t\in\mathcal{T}}\Sigma_t$ then putting $A = B$ in (10) we get $P(B) = P(B)\cdot P(B)$.

$2\Rightarrow 3$. If $f\in\bigcap_{t\in\mathcal{T}}\mathbb{G}_t$ and $f\neq\mathbb{O}$ then $\{\omega\in\Omega : |f(\omega)|>1\}\in\bigcap_{t\in\mathcal{T}}\Sigma_t$.

$3)\Rightarrow 1)$. Statement 3) is equivalent to the condition that the projections $\mathcal{P}_{\mathbb{G}_t}$ tend strongly to zero as $t\to-\infty$. As $F(\mathcal{P}_{\mathbb{G}_t}) = E_{\Sigma_t}$ this shows that $\lim_{t\to-\infty}\|\mathbb{E}_{\Sigma_t}f\|_{L^2} = 0$ for $f\in L^2(\Omega)\ominus\mathbb{C}$ (it is sufficient to check this for finite-particle vectors $f$). Let now $B\in\Sigma$ and $A\in\Sigma_t$. Then

$$|\mathbb{E}\xi_A\xi_B| = |\mathbb{E}\xi_A\cdot\mathbb{E}_{\Sigma_t}\xi_B| \le \|\xi_A\|_{L^2}^{1/2}\cdot\|\mathbb{E}_{\Sigma_t}\xi_B\|_{L^2}^{1/2}$$

and observe that $\mathbb{E}\xi_A\xi_B = P(A\cap B) - P(A)\cdot P(B)$.

$3)\Leftrightarrow 4)$. This follows in a trivial way from the Szegö-Kolmogorov Theorem ($\mathcal{T} = \mathbb{Z}$, cf. Lecture I, Sect. 7) and the theorem of Kreĭn [5] ($\mathcal{T} = \mathbb{R}$). □

This result is well-known. Cf., for example, Ibragimov-Rozanov [1].

**Uniform mixing of the past and the future.** Now we intend the following property of the group $\{\tau^t\}_{t\in\mathcal{T}}$:

$$\begin{aligned}&\sup_{A\in\Sigma_0,B\in\Sigma^0}|P(A\cap\tau^{-t}B)-P(A)\cdot P(B)|\\ &=\sup_{A\in\Sigma_0,B\in\Sigma^t}|P(A\cap B)-P(A)\cdot P(B)|\to 0,\quad t\to\infty.\end{aligned}\tag{11}$$

Let $\mathbb{A}$ and $\mathbb{B}$ be any two subspaces of a Hilbert space. The cosine of the angle between them is defined by the formula

$$\cos(\mathbb{A}, \mathbb{B}) \stackrel{\text{def}}{=} \sup\{|(a,b)| : \|a\| = \|b\| = 1, a \in \mathbb{A}, b \in \mathbb{B}\}.$$

It is not hard to show that $\cos(\mathbb{A}, \mathbb{B}) = |\mathscr{P}_{\mathbb{A}}\mathscr{P}_{\mathbb{B}}|$.

**Theorem 2.14** (Kolmogorov-Rozanov [1]). *Let $\mathbb{A}$ and $\mathbb{B}$ be Gaussian subspaces of a Gaussian subspace $\mathbb{G}$ of $L^2(\Omega)$, $\mathscr{A}$ and $\mathscr{B}$ being the corresponding $\sigma$-algebras. Set*

$$k = k(\mathscr{A}, \mathscr{B}) \stackrel{\text{def}}{=} \sup_{A \in \mathscr{A}, B \in \mathscr{B}} |P(A \cap B) - P(A) \cdot P(B)|.$$

*Then*

$$4k \le \cos(\mathbb{A}, \mathbb{B}) = \cos(L^2(\Omega, \mathscr{A}) \ominus \mathbb{C}, L^2(\Omega, \mathscr{B}) \ominus \mathbb{C}) \le \sin 2\pi k.$$

*Proof.* By our conventions we may identify the space $L^2(\Omega, \mathscr{A}) \ominus \mathbb{C}$ with the space $\mathfrak{F}_1(\mathbb{A}) \stackrel{\text{def}}{=} \bigoplus_{n=1}^{\infty} \mathbb{A}_n = \mathfrak{F}(\mathbb{A}) \ominus \mathbb{C}$ and similarly $L^2(\Omega, \mathscr{B}) \ominus \mathbb{C}$ with $\mathfrak{F}_1(\mathbb{B})$. Then the cosine between the subspaces $\mathfrak{F}_1(\mathbb{A})$ and $\mathfrak{F}_1(\mathbb{B})$, which we regard as subspaces of the Fock space $\mathfrak{F}(\mathbb{G})$, equals $|\mathscr{P}_{\mathfrak{F}_1(\mathbb{A})} \cdot \mathscr{P}_{\mathfrak{F}_1(\mathbb{B})}|$. Next notice that

$$\mathscr{P}_{\mathfrak{F}_1(\mathbb{A})}\mathscr{P}_{\mathfrak{F}_1(\mathbb{B})}x = \mathfrak{F}(\mathscr{P}_{\mathbb{A}})\mathfrak{F}(\mathscr{P}_{\mathbb{B}})x = \mathfrak{F}(\mathscr{P}_{\mathbb{A}}\mathscr{P}_{\mathbb{B}})x$$

for $x \in \mathfrak{F}_1(\mathbb{G})$. Hence $|\mathscr{P}_{\mathfrak{F}_1(\mathbb{A})}\mathscr{P}_{\mathfrak{F}_1(\mathbb{B})}| \le |\mathscr{P}_{\mathbb{A}}\mathscr{P}_{\mathbb{B}}|$. The opposite inequality is trivial.

Let us now prove the left inequality. It follows from the identity just proved that

$$k = \sup |\mathbb{E}\xi_A \xi_B| \le \cos(\mathbb{A}, \mathbb{B}) \sup_{A,B} (\mathbb{E}\xi_A)^{1/2} \cdot (\mathbb{E}\xi_B)^{1/2}$$

$$= \frac{1}{4}\cos(\mathbb{A}, \mathbb{B}).$$

The right inequality may be established by direct computation. Let $a$ and $b$ be unit vectors in $\mathbb{A}$ and $\mathbb{B}$ respectively, with $|\mathbb{E}ab| \ge \cos(\mathbb{A}, \mathbb{B}) - \varepsilon$, $\varepsilon > 0$. Set

$$A = \{\omega : a(\omega) > 0\}, \quad B = \{\omega : b(\omega) > 0\}.$$

An elementary computation reveals that $P(A \cap B) = \frac{1}{4} + \frac{1}{2\pi} \arcsin(|\mathbb{E}ab|)$, $P(A) = P(B) = \frac{1}{2}$. Hence

$$P(A \cap B) - P(A) \cdot P(B) = \frac{1}{2\pi} \arcsin(\mathbb{E}ab). \square$$

*Remark.* $\cos(\mathbb{A}, \mathbb{B}) > 0$ iff $k(\mathbb{A}, \mathbb{B}) > 0$ and $\cos(\mathbb{A}, \mathbb{B}) < 1$ iff $k(\mathbb{A}, \mathbb{B}) < \frac{1}{4}$.

Set now

$$\rho(t) = \cos(\mathbb{G}_0, \mathbb{G}^t), \quad t \in \mathscr{T}.$$

Theorem 2.14 allows us to conclude that condition (11) is equivalent to $\lim_{t \to \infty} \rho(t) = 0$. Processes with this property will be called *completely regular*.

The following lemma allows us to accomplish the passage to Hankel operators.

**Lemma 2.15.** *Let $\{X_t\}_{t\in\mathscr{T}}$ be a regular process with the spectral density $|h^2|$, $h$ an outer function in $H^2$. Then*

$$\mathscr{P}_{G_0}\mathscr{P}_{G^t} = V_1 H_{\bar{h}/h\cdot e_t} P_+ V_2, \tag{12}$$

*where $e_t \stackrel{\text{def}}{=} z^t$ ($\mathscr{T} = \mathbb{Z}$), $e_t(x) \stackrel{\text{def}}{=} e^{itx}$ ($\mathscr{T} = \mathbb{R}$), $V_1$ being a unitary operator from $L^2$ onto $\mathbb{G}$, $V_1 f = \Phi^*(1/\bar{h}\cdot f)$, $V_2$ one from $\mathbb{G}$ onto $L^2$, $V_2 f = \bar{e}_t h\Phi(f)$.*

*Proof.* Just to shorten the notation take $\mathscr{T} = \mathbb{Z}$. The operator of multiplication by the function $1/h$ is an isometry from $L^2$ onto $L^2(|h^2|)$. Let us show that it maps $H^2$ onto $H^2(|h^2|)$. Since $h$ is outer we have $\operatorname{span}\{z^n h : n \geq 0\} = H^2$. Apply now the trivial relation $\frac{1}{h} z^n h = z^n$. In the same way we prove that the isometry $f \mapsto 1/\bar{h}\cdot f$ maps $H^2_-$ onto $H^2_-(|h^2|)$. Taking account of this we get formulae for the orthogonal projection of $L^2(|h^2|)$ onto the subspaces $H^2_-(|h|^2)$ and $z^n H^2_+(|h|^2)$: $f \mapsto 1/\bar{h}\cdot P_-\bar{h}f$, $f \mapsto z^n/h\cdot P_+\bar{z}^n hf$. From this (12) follows immediately. □

This shows that the process $\{X_t\}_{t\in\mathscr{T}}$ is completely regular iff $\lim_{t\to\infty} |H_{\bar{h}/h\cdot e_t}| = 0$. In the case $\mathscr{T} = \mathbb{Z}$ a criterion for the complete regularity in terms of the spectral density will be given in Chap. IV.

**Rapid mixing.** To this class we count processes for which $\rho$ decreases sufficiently fast. For example, in Chap. IV we consider the case of power-like and exponential decrease (as well as some intermediate cases of decrease) of the number $\rho(n)$.

Lemma 2.15 reduces in an obvious way the problem of classification of such processes in spectral terms to the computation of norms of Hankel operators.

Let us now consider a limiting case of rapid mixing.

These are stationary Gaussian processes such that the $\sigma$-algebras $\Sigma_0$ and $\Sigma^t$ are independent for $t \geq T_0$. In Chap. IV we will give a characterization of such processes in terms of the spectral density.

**$\mathfrak{S}_p$-regular processes.** A stationary gaussian process $\{X_t\}_{t\in\mathscr{T}}$ is termed $\mathfrak{S}_p$-*regular* if $\mathbb{E}_{\Sigma_0}\mathbb{E}_{\Sigma^t} \in \mathfrak{S}_p$ for $t > T_0$. The condition of $\mathfrak{S}_p$-regularity is a kind of weak dependence of the far future and the past.

**Lemma 2.16.** *Let $T \in \mathfrak{S}_p$, $1 \leq p \leq \infty$. Then for any natural number $n$*

$$\left\| \bigotimes^n T \right\|_{\mathfrak{S}_p} = \|T\|^n_{\mathfrak{S}_p}.$$

*Proof.* Let $T = \sum_{k=1}^\infty s_k(\cdot, \xi_k)\eta_k$ be the Schmidt expansion for $T$. Then it is easy to see that

$$\bigotimes^n T = \sum s_{i_1}\dots s_{i_n}(\cdot, \xi_{i_1}\otimes\cdots\otimes\xi_{i_n})\eta_{i_1}\otimes\cdots\otimes\eta_{i_n}$$

is the Schmidt expansion for $\bigotimes^n T$. Now just remark that

$$\sum (s_{i_1} \ldots s_{i_n})^p = \left(\sum s_k^p\right)^n. \square$$

**Corollary 2.17.** *If* $T: \mathbb{G} \to \mathbb{G}$ *and* $\|T\|_{\mathfrak{S}_p} < 1$ *then* $\mathfrak{F}(T) \in \mathfrak{S}_p$ *and* $\|\mathfrak{F}(T)\|_{\mathfrak{S}_p} \leq (1 - \|T\|_{\mathfrak{S}_p}^p)^{-1/p}$. $\square$

**Theorem 2.18.** *The stationary Gaussian process* $\{X_t\}_{t \in \mathcal{T}}$ *is* $\mathfrak{S}_p$*-regular iff* $\mathcal{P}_{\mathbb{G}_0}\mathcal{P}_{\mathbb{G}^t} \in \mathfrak{S}_p$ *for* $t > T_0$.

*Proof.* Assume that $\mathcal{P}_{\mathbb{G}_0}\mathcal{P}_{\mathbb{G}^t} \in \mathfrak{S}_p$. By Lemma 2.15 this gives $H_{\bar{h}/h \cdot e_t} \in \mathfrak{S}_p$. Denote by $S^t$ the operator of multiplication by $e_t$, $t > 0$, in $H^2$. Then $\lim_{t \to \infty} (S^t)^* = \mathbb{O}$ in the strong operator topology. Therefore

$$\begin{aligned}\lim_{t \to \infty} \|\mathcal{P}_{\mathbb{G}_0}\mathcal{P}_{\mathbb{G}^t}\|_{\mathfrak{S}_p} &= \lim_{t \to \infty} \|H_{\bar{h}/h \cdot e_t}\|_{\mathfrak{S}_p} \\ &= \lim_{t \to \infty} \|(S^t)^* H_{\bar{h}/h}^*\|_{\mathfrak{S}_p} = 0\end{aligned}$$

and, consequently, $\|\mathcal{P}_{\mathbb{G}_0}\mathcal{P}_{\mathbb{G}^t}\|_{\mathfrak{S}_p} < 1$ for $t > T_1$. Upon applying the preceding corollary we get $E_{\Sigma_0}E_{\Sigma^t} \in \mathfrak{S}_p$ for $t > T_1$. The converse is trivial. $\square$

A criterion for $\mathfrak{S}_p$-regularity in terms of the spectral measure will in the case $\mathcal{T} = \mathbb{Z}$ be deduced in Chap. IV. As we will see there, the requirement of $\mathfrak{S}_\infty$-regularity is in the case $\mathcal{T} = \mathbb{Z}$ equivalent to complete regularity. For continuous time this is not so. In Chap. IV we will likewise consider the case of a rapidly decreasing norm $|\mathcal{P}_{\mathbb{G}_0}\mathcal{P}_{\mathbb{G}^t}|_{\mathfrak{S}_p}$.

Gaussian $\mathfrak{S}_2$-regular processes are also referred to as absolutely regular ones. As $\mathfrak{S}_2$ is a Hilbert space there are many equivalent formulations of the condition of $\mathfrak{S}_2$-regularity. For instance, a Gaussian process is $\mathfrak{S}_2$-regular iff it is information regular (for details see the book Ibragimov-Rozanov [1]).

**Theorem 2.19** (cf. Ibragimov-Rozanov [1], Chap. IV, Sect. 2). *A necessary and sufficient condition for a stationary Gaussian process to be* $\mathfrak{S}_2$*-regular is that*

$$\lim_{t \to \infty} \mathbb{E} \sup_{A \in \Sigma^t} |\mathbb{E}_{\Sigma_0} \xi_A|^2 = 0. \tag{13}$$

*Remark.* It is easy to see that condition (13) is equivalent to the condition

$$\lim_{t \to \infty} \mathbb{E} \sup_{A \in \Sigma^t} |\mathbb{E}_{\Sigma_0} \xi_A|^p = 0 \text{ for } p < \infty.$$

*The proof* of the necessity is based on the following lemma.

**Lemma 2.20.** *Let* $T$ *be an operator in* $L^2(\Omega)$. *Then* $T \in \mathfrak{S}_2$ *iff*

$$\int_\Omega \sup_{\|x\|_2 \leq 1} |(Tx)(\omega)|^2 \, dP(\omega) < \infty.$$

*Moreover, the integral in the left hand side equals* $\|T\|_{\mathfrak{S}_2}^2$.

*Proof of the lemma.* If $T \in \mathfrak{S}_2$ then

$$Tx = \sum_n s_n(x, \xi_n)\eta_n, \quad \sum_n s_n^2 < \infty.$$

Upon applying Cauchy's inequality we get

$$\sup_{\|x\|_2 \le 1} |Tx|^2(\omega) \le \sup_{\|x\|_2 \le 1} \left(\sum_n |(x, \xi_n)|^2\right) \cdot \left(\sum_n s_n^2 |\eta_n(\omega)|^2\right),$$

whence

$$\int_\Omega \sup_{\|x\|_2 \le 1} |Tx|^2 \, dP = \sum_n s_n^2 < \infty.$$

Next assume that the integral in the hypothesis of the lemma converges and let $\mathscr{P}$ be the projection of conditional expectation onto an arbitrary finite subalgebra of the $\sigma$-algebra $\Sigma$. Then $T\mathscr{P} \in \mathfrak{S}_2$ (since $\mathscr{P}$ is of finite rank) and, consequently,

$$(T\mathscr{P}x)(s) = \int k(s,t)x(t)\, dP(t)$$

with $k \in L^2(\Omega \times \Omega)$. Hence

$$\begin{aligned}\int_\Omega |k(s,t)|^2 \, dP(t) &= \sup_{\|x\|_2 \le 1} |(TPx)(s)|^2 \\ &\le \sup_{\|x\|_2 \le 1} |(Tx)(s)|^2.\end{aligned}$$

Therefore $\|TP\|_{\mathfrak{S}_2} \le \int_\Omega \sup_{\|x\|_2 \le 1} |Tx|^2 \, dP$. Passing to the limit through the directed set of all finite subalgebras we get $\|T\|_{\mathfrak{S}_2} < \infty$. □

Assume that $\mathbb{E}_{\Sigma_0}\mathbb{E}_{\Sigma^t} \in \mathfrak{S}_2$. Set $T_t \stackrel{\text{def}}{=} \mathbb{E}_{\Sigma_0}\mathbb{E}_{\Sigma^t}\mathscr{P}_{L^2(\Omega)\ominus\mathbb{C}}$. Then $\lim_{t\to\infty} \|\mathscr{P}_{\mathbb{G}_0}\mathscr{P}_{\mathbb{G}^t}\|_{S_2} = 0$ (cf. the proof of Theorem 2.18). Consequently in view of 2.17 $\lim_{t\to\infty} \|T_t\|_{\mathfrak{S}_2} = 0$. Applying Lemma 2.20 we get

$$\mathbb{E} \sup_{A \in \Sigma^t} |\mathbb{E}_{\Sigma_0}\xi_A|^2 \le \frac{1}{4} \|T_t\|_{\mathfrak{S}_2}^2.$$

In proving the sufficiency of (13) we follow the scheme laid down in the book Dym-McKean [1], pp. 70–71. In view of Theorem 2.18 it is sufficient to check that $\mathscr{P}_{\mathbb{G}_0}\mathscr{P}_{\mathbb{G}^t} \in \mathfrak{S}_2$. Let $L^t$ be an arbitrary finite dimensional subspace of $\mathbb{G}^t$ and let

$$\mathscr{P}_{\mathbb{G}_0}\mathscr{P}_{L^t}x = \sum_{k=1}^n s_k(x, a_k)b_k$$

be the Schmidt expansion of the operator $\mathscr{P}_{\mathbb{G}_0}\mathscr{P}_{L^t}$ (of finite rank). Let us show that $(a_k, b_j) = s_k\delta_{kj}$ ($\delta_{kj} = 0$ for $k \ne j$, $\delta_{kk} = 1$). In fact $(a_k, a_j) = s_j^{-1}(a_k, \mathscr{P}_{\mathbb{G}_0}b_j) = s_j^{-1}(\mathscr{P}_{\mathbb{G}_0}a_k, \mathscr{P}_{\mathbb{G}_0}b_j) = s_j^{-1}s_k s_j(a_k, b_j)$. Denote by $p_{ab}$, $p_a$ and $p_b$ the densities of the joint distributions of the random variables $ab = (a_1, \ldots, a_n, b_1, \ldots, b_n)$, $a =$

$(a_1, \ldots, a_n)$ and $b = (b_1, \ldots, b_n)$ respectively and by $\Sigma(a)$ ($\Sigma(b)$) the $\sigma$-algebra generated by $a$ ($b$). Then

$$\begin{aligned}\mathbb{E} \sup_{A \in \Sigma^t} |\mathbb{E}_{\Sigma_0} \xi_A| &\geq \mathbb{E} \sup_{A \in \Sigma(a)} |\mathbb{E}_{\Sigma(b)} \xi_A| \\ &\geq \int p_b \, db \sup_A \int_A [p_{ab}/p_b - p_a] \, da \\ &= \iint [p_{ab} - p_a p_b]^+ \, da \, db \\ &= \frac{1}{2} \iint |p_{ab} - p_a p_b| \, da \, db \stackrel{\text{def}}{=} K.\end{aligned}$$

For simplicity write $x = \log p_{ab}/p_a p_b$. An elementary computation reveals that for the integral $L = \iint e^{ix} p_a p_b \, da \, db$ holds the following identity:

$$|L|^2 = \prod_{k=1}^n \left| 1 + (2i + 1) \frac{s_k^2}{1 - s_k^2} \right|.$$

Remarking that $2K = \iint |e^x - 1| p_a p_b \, da \, db$ and using the elementary estimate $|e^{ix} - 1| \leq 3|e^x - 1|$ we get

$$\begin{aligned}|L| &\geq 1 - \iint |e^{ix} - 1| p_a p_b \, da \, db \\ &\geq 1 - 3 \iint |e^x - 1| \cdot p_a p_b \, da \, db \\ &\geq 1 - 6K.\end{aligned}$$

Hence

$$(1 - 6K)^{-2} \geq |L|^{-2} \geq \prod_{k=1}^n (1 + s_k^2) \geq 1 + \sum_{k=1}^n s_k^2.$$

Consequently $\|\mathscr{P}_{\mathbb{G}_0} \mathscr{P}_{L^t}\|_{\mathfrak{S}_2} \leq (1 - 6K)^{-1} \leq \text{const}$ provided $K$ is sufficiently small. As $L^t$ was an arbitrary finite dimensional subspace of $\mathbb{G}^t$ this gives $\mathscr{P}_{\mathbb{G}_0} \mathscr{P}_{\mathbb{G}^t} \in \mathfrak{S}_2$. □

A much stronger condition than (13) is the condition

$$\varphi(t) \stackrel{\text{def}}{=} \sup_{\omega \in \Omega} \sup_{A \in \Sigma^t} |E_{\Sigma_0} \xi_A| \to 0, \; t \to \infty. \tag{14}$$

It is clear that it is verified if the $\sigma$-algebras $\Sigma_0$ and $\Sigma^t$ are independent for $t \geq T_0$. It turns out that the converse is likewise true.

**Theorem 2.21** (cf. Ibragimov-Linnik [1], Theorem 17.32). *If a stationary Gaussian process satisfies condition* (14) *then there exists a number* $T_0 > 0$ *such that the* $\sigma$*-algebras* $\Sigma_0$ *and* $\Sigma^t$ *are independent for* $t > T_0$.

*Proof.* Let us verify that

$$\varphi(t) = \sup_{A \in \Sigma^t, B \in \Sigma_0} |P(A \cap B) - P(A)P(B)|/P(B).$$

If $A \in \Sigma^t$, $B \in \Sigma_0$ it is clear that

$$\begin{aligned} |P(A \cap B) - P(A)P(B)| &= |\mathbb{E}(\chi_B \cdot \mathbb{E}_{\Sigma_0} \xi_A)| \le \varphi(t) \|\chi_B\|_{L^1} \\ &= \varphi(t)P(B). \end{aligned}$$

In order to prove the opposite estimate it suffices to take

$$B = \{\omega : (\mathbb{E}_{\Sigma_0} \xi_A)(\omega) > \varphi(t) - \varepsilon\}.$$

Let now $a \in \mathbb{G}_0$ and $b \in \mathbb{G}^t$ be real Gaussian variables such that $\mathbb{E}a^2 = \mathbb{E}b^2 = 1$, $\mathbb{E}ab = \rho > 0$, and $t$ a point, $t \in \mathscr{T}$, such that $\varphi(t) < (2\pi\sqrt{e})^{-1}$. Set $B_k = \{\omega : a(\omega) \ge k/\rho\}$, $A = \{\omega : b(\omega) \in [0, 1]\}$ with $k > 0$. We have

$$\begin{aligned} &P(A \cap B_k) - P(A)P(B_k) \\ &= \int_{k/\rho}^{\infty} \int_0^1 (p_{ab} - p_a p_b)\, db\, da \\ &= \frac{1}{2\pi} \int_{k/\rho}^{\infty} e^{-b^2/2} \int_0^1 \left( \frac{1}{\sqrt{1-\rho^2}} \exp\left( -\frac{1}{2} \cdot \frac{(a - \rho b)^2}{1 - \rho^2} \right) - \exp(-a^2/2) \right) da\, db. \end{aligned}$$

For $b \ge k/\rho$, $0 \le a \le 1$ we then get

$$\begin{aligned} \varphi(t) &\ge \sup_k |P(A \cap B_k) - P(A)P(B_k)|/P(B_k) \\ &\ge \frac{1}{2\pi\sqrt{e}} \sup_k \left| 1 - \frac{1}{\sqrt{1-\rho^2}} \exp\left( -\frac{1}{2} \cdot \frac{(k-1)^2}{1-\rho^2} + \frac{1}{2} \right) \right| \\ &= \frac{1}{2\pi\sqrt{e}}. \end{aligned}$$

Contradiction. □

## Chapter III. Best Approximation by Analytic Functions

In this Chapter we consider two intimately related problems which, as we show below, admit applications to various questions in Analysis and Probability.

**The Problem of Best Approximation by Analytic Functions in the Uniform Norm.** For any function $f$ in $BMO$ (regarding the classes $BMO$ and $VMO$ see Appendix 4, Sect. 16.53.3) there exists a function $f_0$ in $BMO_A \overset{\text{def}}{=} P_+ BMO$ such that $f - f_0 \in L^\infty(\mathbb{T})$ and

$$\|f - f_0\|_\infty = \inf\{\|f - g\|_\infty : g \in BMO_A\}.$$

The function $f_0$ (in general not unique) is called a *best approximation* of $f$.

The problem of best approximation may now be formulated as follows. For which function classes $X$, $X \in BMO$, on $\mathbb{T}$ do we have the implication:

$$f \in X \Rightarrow f_0 \in X?$$

If $f \in VMO$ then by Lecture VIII, Sect. 6 a), b) (cf. likewise Appendix 4, Sect. 165.53.3) the function $f_0$ is unique. In $VMO$ we can therefore define the (non-linear) *operator of best approximation* $\mathscr{A}$, given by the formula $\mathscr{A}f = f_0$.

**The Problem of Restoration of Unimodular Functions.** Let $f \in VMO$. In which auxiliary assumptions on the unimodular function $u$ can we assert that

$$P_{-}u \in X \Rightarrow u \in X?$$

Without entering into details let us indicate the connection between these problems. Let $X \subset VMO$ with $P_{-}X \subset X$. Then, as was established in Lecture VIII, Sect. 6 a), b), $f - \mathscr{A}f = cu$, with $c \in \mathbb{C}$, $u$ a unimodular function. It is clear that $P_{-}u = 1/c \cdot P_{-}f \in X$ so that from the problem of best approximation we come to the problem of restoration of unimodular functions $u$.

If $u$ is a unimodular function such that $T_u H^2 = H^2$ then $-P_{+}\bar{z}u$ is the best approximation of the function $P_{-}\bar{z}u$. In fact, in this hypothesis there exists $f$ in $H^2$ with $T_u f = \mathbb{1}$ such that $T_{\bar{z}u} f = \mathbb{O}$. In view of Appendix 4, Sect. 36 then $\operatorname{dist}_{L^\infty}(\bar{z}u, H^\infty) = 1$. This means precisely that the function $-P_{+}\bar{z}u$ is the best approximation of $P_{-}\bar{z}u$. Thus in this special case the restoration problem reduces to the problem of best approximation.

In this formulation the problem of best approximation was first considered by Shapiro [3], [4], where it was established that the operator of best approximation $\mathscr{A}$ preserves the class of functions analytic in a neighborhood of the circumference $\mathbb{T}$. A major contribution to this theme was given by Adamyan-Arov-Kreĭn [2], who developed a clever method, with the aid of which they were able to solve the best approximation problem in the spaces $Fl^1(w) = \{f \in C(\mathbb{T}) : \sum_{n \in \mathbb{Z}} |\hat{f}(n)| w_{|n|} < \infty\}$ for some classes of sequences $\{w_n\}_{n \geq 0}$. In Carleson-Jacobs [1] the problem of best approximation was solved in the Hölder classes $\Lambda_\alpha$, $\alpha \in \mathbb{R}_{+}$ and furthermore in differentiation invariant Carleman classes $C\{M_n\}$. The survey Kahane [2] is devoted to the problem of best approximation, where the latter is considered likewise for some other norms than the uniform one. Other aspects of the problem are treated in the papers Havinson [1], [2], Macintyre-Rogosinski [1], Rogosinski-Shapiro [1].

The restoration problem for unimodular functions appears in less explicit form in the work of Ibragimov [1], [2] on stationary Gaussian processes and was solved by him for the Hölder classes. Let us notice that from these results it is possible to derive the theorem of best approximation in the classes $\Lambda_\alpha$. For the class of functions with a finite Dirichlet integral the restoration problem was solved in Ibragimov-Solev [1] and, by a different method, in Solev [1].

In Sect. 3.1 we will set forth a method for solving the problems of best approximation and restoration in function spaces admitting a description in terms of rational approximation in the $BMO$ norm. As examples of such spaces we mention the Besov classes $B_p^{1/p}$, $1 \le p < \infty$, the space $VMO$ and the class of rational functions. In particular, we will show that the best rational approximation of a rational function is again a rational function. We likewise give an application of the best approximation problem to the Nevanlinna-Pick interpolation problem.

In Sect. 3.2 we develop an axiomatic approach to the problems of best approximation and restoration. The theorem of best approximation will be proved for a wide class of separable Banach algebras on the circumference $\mathbb{T}$. This approach makes use of an idea in Adamyan-Arov-Kreĭn [2].

In Sect. 3.3 yet another treatment of the same problems will be given. On the basis of the results of Sect. 3.2 it allows us to extend these to some non-separable Banach function spaces. In particular, this method leads to a new proof of the theorem of Carleson-Jacobs [1].

In Sect. 3.4 we consider examples of spaces for which the theorems of restoration and best approximation hold true. In particular, we consider the Hölder classes $\lambda_\alpha$ and $\Lambda_\alpha$; the Besov classes $B_p^\alpha$ with $\alpha > 1/p$; the Sobolev classes $\mathscr{L}_p^s$ with $s > 1/p$; Carleman classes $C\{M_n\}$, including many which are *not* differentiation invariant; the classes $Fl^p(w) = \{f : \sum_{n\in\mathbb{Z}} |\hat{f}(n) w_{|n|}|^p < \infty\}$ in suitable hypotheses on the sequences $w$ as well as some other classes. Some results will be stated without proof. The proofs, along with some other examples of function classes, will be found in Peller-Hruščev [1].

In Sect. 3.5 we describe the arguments of unimodular functions with the aid of the theorems on best approximation established in the previous Sections. It is proved that for a wide class of spaces $X$ the following result holds true. A unimodular function $u$ is in $X$ iff $u = z^n \exp(i\varphi)$ with $n \in \mathbb{Z}$ and $\varphi$ a real function in $X$.

Sect. 3.6 is devoted to applications of the theory of best approximation by analytic functions in the uniform norm. We consider the problem of restoration of measures on the circumference from their Cauchy integrals outside the unit disk, as well as the problem of interpolation by analytic functions of class $B_p^{1/p} \cap C_A$. In particular it is proved that for every complex measure $\mu$ on $\mathbb{T}$ one has the implication

$$\sum_{n\ge1} |\hat{\mu}(-n)|^2/n < \infty \Rightarrow \sum_{n\ge1} |\hat{\mu}(n)|^2/n < \infty.$$

Moreover, for a closed subset $E$, $E \subset \mathbb{T}$, one has the relation $C_A \cap B_2^{1/2}|E = C(E)$ iff $E$ has vanishing logarithmic capacity. Finally, yet another application is that we obtain an analogue of the Fefferman-Stein decomposition for the classes $B_p^{1/p}$. Namely, every function $f$ in $B_p^{1/p}$ may be represented in the form $u + \tilde{v}$ where $u$, $v \in B_p^{1/p} \cap L^\infty$. Some results are stated without proof; for these we have to refer to Peller-Hruščev [1].

**3.1. Function Spaces Described in Terms of Rational Approximation in the BMO Norm.** Recall that in $BMO$ we consider the following norm:

$$\|\varphi\|_{BMO} \stackrel{\text{def}}{=} \inf_{g\in BMO_A} \|\varphi - g\| + \inf_{g\in BMO_A} \|\bar{\varphi} - g\| + |\hat{\varphi}(0)|.$$

Let $n \in \mathbb{Z}_+$. Then $\mathscr{R}_n$ consists of all rational functions with poles outside $\mathbb{T}$, the sum of the multiplicities of which (including the point $\infty$) not exceeding $n$. If $f \in BMO$ then $r_n(f) \stackrel{\text{def}}{=} \text{dist}_{BMO}(f, \mathscr{R}_n)$.

Recall also that a linear sequence space $E$ is called an *ideal* (*Köthe*) *space* if it together with a sequence $\{x_n\}_{n\geq 0}$ also contains any sequence $\{y_n\}_{n\geq 0}$ with $|y_n| \leq |x_n|$.

**Definition.** We say that a function class $X$ admits a description in terms of rational approximation in the $BMO$ norm if there exists an ideal space $E$ such that

$$f \in X \Leftrightarrow \{r_n(f)\}_{n\geq 0} \in E.$$

For the sake of shortness such classes $X$ will be termed *$\mathscr{R}$-spaces.*

It is easy to see that $VMO$ is an $\mathscr{R}$-space. In this case we may take $E = c_0(\mathbb{Z}_+) \stackrel{\text{def}}{=} \{\{x_n\}_{n\geq 0} : \lim_n x_n = 0\}$. Other examples of $\mathscr{R}$-spaces are the Besov spaces $B_p^{1/p}$, $1 \leq p < \infty$; here $E = l^p$ (cf. Theorem 1.4). Also each set $\mathscr{R}_n$ is an $\mathscr{R}$-space (now we take $E$ to be the set of sequences $\{x_m\}_{m\geq 0}$ such that $x_m = 0$ for $m \geq n$). If $E$ is the space of all sequences such that all elements but for a finite number vanish then the corresponding $\mathscr{R}$-space contains precisely all rational functions.

We now list some properties of $\mathscr{R}$-spaces. Let $X$ be an $\mathscr{R}$-space. Then it is clear that $P_+X \subset X$. Furthermore if $f \in H^\infty$ then the operator $T_{\bar{f}}P_+$ maps $X$ into $X$ (this was observed by Nikol'skiĭ (cf. Peller [3])). If $X$ is a Banach $\mathscr{R}$-space (i.e. $E$ carries a norm such that

$$|x_n| \leq |y_n| \Rightarrow \|\{x_n\}_{n\geq 0}\|_E \leq \|\{y_n\}_{n\geq 0}\|_E;$$

$$c_1\|f\|_X \leq \|\{r_n(f)\}_{n\geq 0}\|_E \leq c_2\|f\|_X)$$

then $X$ is conformally invariant (up to a constant), i.e. there exists a constant $K$ such that

$$\inf_{c\in\mathbb{C}} \|\varphi \circ \gamma - c\|_X \leq K \inf_{c\in\mathbb{C}} \|\varphi - c\|_X. \tag{1}$$

for every $\varphi$ in $X$ and every conformal map $\gamma$ of the extended complex plane into itself such that $\gamma(\mathbb{T}) \subset \mathbb{T}$ (this follows from the fact that (1) holds for $X = BMO$ and that $\mathscr{R}_n + \text{const} \subset \mathscr{R}_n$, along with the conformal invariance of $\mathscr{R}_n$).

Consequently, spaces of smooth functions such as $\Lambda_\alpha$, $C^{(n)}$ are *not* $\mathscr{R}$-spaces. Moreover, if $X$ is an $\mathscr{R}$-space such that $X \backslash \bigcup_{n\geq 0} \mathscr{R}_n \neq \varnothing$ then $X \not\subset \text{Lip}\,\omega$ for no modulus of continuity $\omega$ whatsoever (cf. Gončar [1]).

Notice that it follows from the Adamyan-Arov-Kreĭn Theorem (Appendix 4, Sect. 10) and Kronecker's Lemma (cf. Lecture VIII, Sect. 3) that for any function $\varphi$ in *BMO holds*

$$r_n(P_-\varphi) = s_n(H_\varphi),\ n \geq 0. \tag{2}$$

Consequently, if $E$ is an ideal sequence space with $E \subset c_0(\mathbb{Z})$ and

$$\{x_n\}_{n\geq 0} \in E \Rightarrow \{x_{[n/2]}\}_{n\geq 0} \in E$$

($[n/2]$ denotes the integer part of $n/2$) then the function class

$$X = \{\varphi : \{s_n(H_\varphi)\}_{n\geq 0} \in E, \{s_n(H_{\bar{\varphi}})_{n\geq 0} \in E\}$$

is an $\mathscr{R}$-space.

Our next objective is now to show that the problem of best approximation has a solution for $\mathscr{R}$-spaces and to give conditions on a unimodular function which are sufficient for the solution of the restoration problem.

**Theorem 3.1.** *Let $X$ be an $\mathscr{R}$-space and $u$ a unimodular function on $T$ with $P_-u \in X$. If the Toeplitz operator $T_u$ in $H^2$ has a dense range then $P_+u \in X$.*

**Theorem 3.2.** *Consider a linear $\mathscr{R}$-space $X$ such that $\mathbf{z} \cdot X = X$ and let $u$ be a unimodular function on $T$ such that $P_-u \in X$. If $u \in VMO$ then $u \in X$.*

Recall that the (non-linear) operator of best approximation $\mathscr{A}$ is defined in the class $VMO$ as follows:

$$\mathscr{A}f = f_0,\ f_0 \in BMO_A,\ \|f - f_0\|_\infty = \inf_{g \in BMO_A} \|f - g\|_\infty.$$

In view of Lecture VII, Sect. 6 a), b) the function $f_0$ must be unique.

**Theorem 3.3.** *Let $X$ be an $\mathscr{R}$-space. Then the operator of best approximation $\mathscr{A}$ acts in $X$, i.e. $\mathscr{A}X \subset X$. Moreover $r_n(\mathscr{A}f) \leq r_n(f)$, $n \in \mathbb{Z}_+$, and if $P_-f \neq 0$ then $r_n(\mathscr{A}f) \leq r_{n+1}(f)$.*

*Proof of Theorem* 3.1. As by assumption $P_-u \in VMO$ we have $H_u \in \mathfrak{S}_\infty$. Also by Corollary 1.8

$$s_n(H_{\bar{u}}) \leq s_n(H_u),\ n \in \mathbb{Z}_+.$$

Consequently, by the Adamjan-Arov-Kreĭn Theorem (cf. Appendix 4, Sect. 10) and Kronecker's Lemma (cf. Lecture VIII, Sect. 3)

$$r_n(P_+u) = r_n(P_-\bar{u}) \leq r_n(P_-u).$$

Hence $P_+u \in X$. □

*Proof of Theorem* 3.2. In view of Appendix 4, Sect. 50 and 53, $T_u$ is a Fredholm operator so that in particular $T_{z^n u}$ is invertible if $n = \operatorname{ind} T_u$. Since $zX = X$ we have $P_-z^n u \in X$. Consequently, by Theorem 3.1 we may conclude that $P_+z^n u \in X$. Thus $z^n u \in X$ and from this again $u \in X$. □

For the proof of Theorem 3.3 we need the following lemma.

**Lemma 3.4.** *Let $h$ be an outer function in $H^2$, $b$ an inner function and set $u = \bar{b} \cdot \bar{h}/h$. Then $T_u$ has dense range in $H^2$.*

*Proof.* Assume that $f \perp T_u H^2$, $f \neq \mathbb{O}$. Then $(f, \bar{h}/h \cdot \bar{b}g) = 0$ for any $g$ in $H^2$. Write $f = f_i \cdot f_e$ with $f_i$ inner and $f_e$ outer. Set $g = f_i bh$. Then $(f_e, \bar{h}) = 0$ and therefore $f_e(0)h(0) = 0$. But this is impossible since $f_e$ and $h$ both are outer. □

*Proof of Theorem* 3.3. Let $f \in VMO$. By Appendix 4, Sect. 20 the function $f - \mathscr{A}f$ has then a constant modulus $\delta$, say, on $\mathbb{T}$. If $\delta = 0$ the clearly $r_n(\mathscr{A}f) = r_n(f)$, $n \in \mathbb{Z}_+$. If $\delta > 0$ then the unimodular function $u = \delta^{-1}(f - \mathscr{A}f)$

has the form $u = \overline{h}/h \cdot \overline{bz}$ where $h$ is an outer function in $H^2$ and $b$ an inner function (cf. Lecture VIII, Sect. 6, a), b)). By Lemma 3.4 the Toeplitz operator $H_u$ has a dense range in $H^2$. Moreover $\operatorname{Ker} T_u \neq \{\mathbb{0}\}$ as $T_u(hb) = \mathbb{0}$. Consequently, by Corollary 1.8 and formula (2) $r_k(P_+u) = s_k(H_{\bar{u}}) \leq s_{k+1}(H_u) = r_{k+1}(P_-u)$, $k \geq 1$. But this means that $r_k(P_+f - \mathscr{A}f) \leq r_{k+1}(P_-f)$. Fix now a number $n$ in $\mathbb{Z}_+$. Write $r_{n+1}(f) = \|f - g\|_{BMO}$ where $g$ is a rational function in $\mathscr{R}_{n+1}$. We have then $\|f - g\|_{BMO} = \|P_+f - P_+g\|_{BMO} + \|P_-f - P_-g\|_{BMO}$. Let further $k$ be the number of poles of $g$ outside $\mathbb{D}$. Then $P_+g \in \mathscr{R}_k$ and $P_-g \in \mathscr{R}_{n-k+1}$. This yields $r_{k+1}(P_+f) + r_{n-k}(P_-f) \leq r_{n+1}(f)$.

Summing up all the information gathered we get

$$r_n(Af) \leq r_k(P_+f - Af) + r_{n-k}(P_+f) \leq r_{k+1}(P_-f) + r_{n-k}(P_+f) \leq r_{n+1}(f). \square$$

If $X$ is a Banach $\mathscr{R}$-space Theorem 3.1 and 3.3 can be stated with an inclusion of a norm estimate. We give these formulations for the case $X = B_p^{1/p}$, $VMO$.

**Corollary 3.5.** *Let $X = B_p^{1/p}$ or $VMO$. Then the following statements hold true.*

1) *Let $u$ be a unimodular function in $X$ such that* $\operatorname{clos} T_uH^2 = H^2$ *and $P_-u \in X$. Then $u \in X$ and* $\|u\|_X \leq \operatorname{const} \|P_-u\|_X$.

2) *If $u$ is a unimodular function in $VMO$ such that $P_-u \in X$ then $u \in X$.*

3) *If $f \in X$ then $\mathscr{A}f \in X$ and* $\|\mathscr{A}f\|_X \leq \operatorname{const} \|f\|_X$. $\square$

**Corollary 3.6.** *If $f \in B_p^{1/p}$ there exists an analytic function $g$ in $B_p^{1/p}$ such that $f + g \in L^\infty$.* $\square$

**Corollary 3.7.** *Let $u$ be a unimodular function on $\mathbb{T}$ such that the operator $T_u$ is invertible. Then*

$$r_n(P_-u) = r_n(P_+u),\ n \in \mathbb{Z}_+.$$

*Proof.* The statement follows from inequality (2) and Corollary 1.8. $\square$

This result is of interest even if $u = e^{i\alpha}$, $\alpha$ a continuous function.

The following result, which apparently first was formulated in Macintyre-Rogosinski [1], shows that the operator of best approximation $\mathscr{A}$ maps rational functions into rational functions. We give a simple proof of this fact.

**Corollary 3.8.** *Let $f \in \mathscr{R}_n$, $n \in \mathbb{Z}_+$. Then $\mathscr{A}f \in \mathscr{R}_n$. If we also assume that $P_+f \neq \mathbb{0}$ then we may conclude that $\mathscr{A}f \in \mathscr{R}_{n-1}$.*

*Proof.* This follows at once from Theorem 3.3. $\square$

The following application pertains to the Nevalinna-Pick interpolation problem. Let $\{\lambda_1, \ldots, \lambda_N\}$ be a finite subset of the disk $\mathbb{D}$ and $\{\alpha_{nk} : 1 \leq n \leq N, 0 \leq k \leq m_n\}$ a family of complex numbers. Find a function $f$ of minimal norm in $H^\infty$ such that

$$f^{(k)}(\lambda_n) = \alpha_{nk},\ 1 \leq n \leq N, 0 \leq k < m_n. \tag{3}$$

**Corollary 3.9.** *Let $f$ solve* (2) *with minimal norm in $H^\infty$. Then $f = cB$, $c \in \mathbb{C}$, where $B$ is a Blaschke product of class $\mathscr{R}_{m_1+m_2+\cdots+m_N-1}$.*

*Proof.* By the Lagrange interpolation formula there exists a polynomial $Q$ of degree $m_1 + \cdots + m_N$ which solves (3). Let $B_0$ be a Blaschke product having its zeros at the interpolation nodes taken with the multiplicities $m_1, \ldots, m_N$. It is clear that $B_0 \in \mathscr{R}_{m_1+\cdots+m_N}$. Also the general solution of (3) must be of the form $f = Q - B_0 g$, $h \in H^\infty$. Furthermore

$$\inf_{h\in H^\infty} \|Q - B_0 h\|_{L^\infty} = \inf_{h\in H^\infty} \|Q/B_0 - h\|_{L^\infty}. \tag{4}$$

Notice that $Q/B_0 \in \mathscr{R}_{2(m_1+\cdots+m_N)}$. According to Corollary 3.8 $h = \mathscr{A}(Q\bar{B}_0) \in \mathscr{R}_{2(m_1+\cdots+m_N)-1}$. It is now plain that $Q - B_0 h = cB$ with $c \in \mathbb{C}$, $B$ a finite Blaschke product. Let us consider the Toeplitz operator $T_{B\bar{B}_0}$. In view of $\operatorname{dist}(B\bar{B}_0, H^\infty) = 1$ (cf. (4)) it has a non-trivial kernel (cf. Appendix 4, Sect. 36). Let $k$ be the number of zeros of $B$ (with multiplicities). Then by Appendix 4, Sect. 43

$$\dim \operatorname{Ker} T_{B\bar{B}_0} = \operatorname{ind} T_{\bar{B}_0} T_B = \operatorname{ind} T_{\bar{B}_0} + \operatorname{ind} T_B = m_1 + \cdots + m_N - k > 0.$$

Consequently $B \in \mathscr{R}_{m_1+\cdots+m_N-1}$. □

This result was first established by Schur [1].

**3.2. The Operator of Best Approximation in Separable Banach Algebras.** In this Section we develop an axiomatic approach to the best approximation in Banach algebras of functions on the circle $\mathbb{T}$. More precisely, we consider Banach algebras $X$, $X \subset BMO$, subject to the following axioms.

**Axioms.**

(A1) If $f \in X$ then $\bar{f} \in X$ and $P_+ f \in X$.

(A2) $X$ is a Banach algebra with respect to pointwise multiplication.

(A3) The set of trigonometric polynomials is dense in $X$.

(A4) Every multiplicative functional on $X$ coincides with the point evaluation functional at some point $\zeta$, $\zeta \in \mathbb{T} : f \mapsto f(\zeta)$.

*Remark.* As every multiplicative functional on a Banach algebra has norm one it follows from axioms (A3), (A2), (A3) that $X \subset C(\mathbb{T})$, with continuous imbedding. In view of (A4) the maximal ideal space of $X$ coincides with $\mathbb{T}$.

**Theorem 3.10.** *Let $X$ be a space satisfying axioms* (A1)–(A4) *and let $u$ be a unimodular function on $\mathbb{T}$, with $P_- u \in X$. Assume that one of the following conditions is fulfilled*:

1) $u \in VMO$;

2) $\operatorname{clos}(T_u H^2) = H^2$.

*Then $u \in X$.*

**Theorem 3.11.** *Let $X$ be a space satisfying axioms* (A1)–(A4). *Then the operator of best approximation $\mathscr{A}$ acts in $X$, i.e. $\mathscr{A}X \subset X$.*

*Proof of Theorem* 3.10. Let $u$ be in $VMO$. Then by Appendix 4, Sect. 53, 57, 165.35 we know that $u = z^n \bar{h} h^{-1}$ with $n \in \mathbb{Z}$, $h$ an outer function in $H^2$ such that the Toeplitz operator $T_{\bar{h}h^{-1}}$ is invertible. By axioms (A2) and (A3) for every integer

$m$ holds $z^m X \subset X$. Therefore we may assume without loss of generality that $n = -1$. Then the subspace

$$\operatorname{Ker} T_u = \{f \in H^2 : \|H_u f\| = \|f\|\} = \{\lambda h : \lambda \in \mathbb{C}\}$$

clearly must be one dimensional.

Consider now the closed subspace $X_A \overset{\text{def}}{=} P_+ X$ of $X$. According to axioms (A1) and (A2) the Hankel operator $H_u$ is continuous from $X_A$ into $X$ (in fact $H_u f = P_-(P_- u)f$, $f \in X_A$). Therefore the operator $R \overset{\text{def}}{=} H_u^* H_u | X_A$ too must be continuous in $X_A$ (here $H_u^* f = P_+ \bar{u} f$, $f \in H_-^2$). As the operator $H_\varphi$ clearly is of finite rank if $\varphi$ is a trigonometric polynomial, (A3) shows that $R$ is a compact operator in $X_A$.

The dual $X_A^*$ of the space $X_A$ can in a natural way be identified with the space of functions analytic in $\mathbb{D}$ such that the value of the functional $g$, $g \in X_A^*$, on the polynomials is given by the hermitian form

$$(f, g) = \sum_{n \ge 0} \hat{f}(n)\overline{\hat{g}(n)}.$$

As $X_A \subset C_A$ we have the continuous imbeddings

$$X_A \subset H^2 \subset X_A^*.$$

It is clear that

$$H_u^* H_u h = h$$

and that the eigenspace of the operator $H_u^* H_u$ in $H^2$ corresponding to the number 1 is one dimensional (for the proof just remark that the condition $H_u^* H_u f = f$ implies that $f \in \operatorname{Ker} T_u$). The operator $R^*$ in $X_A^*$, the adjoint of $R$, is of course given by the same formula $R^* f = H_u^* H_u f$ at least for $f$ in $H^2$. Consequently, by the Riesz-Schauder theory $R^*$ is compact in $X_A^*$ and the multiplicities of the eigenvalue 1 for the operators $R$ and $R^*$ coincide (cf. Yosida [1], Chap. $X$, sect. 5).

As every eigenvector of $R$ is an eigenvector of $H_u^* H_u$ (in $H^2$) and every eigenvector of $H_u^* H_u$ an eigenvector of $R^*$ it follows that

$$\dim \operatorname{Ker}(I - R) = \dim \operatorname{Ker}(I - H_u^* H_u) = \dim \operatorname{Ker}(I - R^*) = 1$$

Let $f \in X_A$, $f \not\equiv \mathbb{0}$, $Rf = f$. Then $H_u^* H_u f = f$ and consequently $h = \lambda f \in X_A$.

Thus in order to establish the inclusion $u = \overline{zh} \cdot h^{-1} \in X$ it suffices to show that $h$ has no zeros on the circle $\mathbb{T}$. In fact axiom (A1) says that $\bar{h} \in X$ and by axiom (A4) using Gel'fand's theorem (cf. Gamelin [1], Chap. 1, Sect. 2) we then get $h^{-1} \in X$.

If $X \in \lambda_\alpha$ for some positive $\alpha$ then it is easy to complete the proof. Assume that $h(t) = 0$, $t \in \mathbb{T}$. Then $|h(\zeta)| \le \text{const}\, |\zeta - t|^\alpha$. Since $u \in VMO$ we have by Appendix 4, Sect. 165.35 $1/h \in \bigcap_{p<\infty} H^p$. On the other hand

$$\|h^{-1}\|_{H^p}^p \ge c \int_{\mathbb{T}} dm(\zeta)/|\zeta - t|^{p\alpha} = \infty,$$

if $p\alpha > 1$. This contradiction shows that $\inf_{\zeta \in \mathbb{T}} |h(\zeta)| > 0$.

In the general case we have to apply a more subtle argument. Assume that $h$

vanishes on $\mathbb{T}$. To fix the ideas let $h(1) = 0$. We will show that $h(\mathbb{1} - z)^{-1} \in X_A^*$ and that $R^* h(\mathbb{1} - z)^{-1} = h(\mathbb{1} - z)^{-1}$. This clearly contradicts $\dim \operatorname{Ker}(I - R^*) = 1$.

**Lemma 3.12.** *$h(\mathbb{1} - z)^{-1} \in X_A^*$ and $(\varphi, h(\mathbb{1} - z)^{-1}) = (P_+ \bar{h}\varphi)(1)$ for any $\varphi$, $\varphi \in X_A$.*

*Proof.* The axioms (A1), (A2) show that the functional $\varphi \mapsto (P_+ \bar{h}\varphi)(1)$ is continuous. To complete the proof it is thus sufficient to show that $(z^n, h(\mathbb{1} - z)^{-1}) = (P_+ \bar{h} z^n)(1)$ for $n \in \mathbb{Z}_+$. It is however clear that $h(\mathbb{1} - z)^{-1} = \sum_{n \geq 0} z^n (\sum_{k=0}^n \hat{h}(k))$ and that $(P_+ \bar{h} z^n)(1) = \sum_{k=0}^n \overline{\hat{h}(k)}$. □

Let us now check that $R^* h(\mathbb{1} - z)^{-1} = h(\mathbb{1} - z)^{-1}$, i.e., that

$$(H_u^* H_u \varphi, h(\mathbb{1} - z)^{-1}) = (\varphi, h(\mathbb{1} - z)^{-1}),\ \varphi \in X_A.$$

We will establish this formula by expanding the function $(\mathbb{1} - z)^{-1} \cdot h$ in the series $\sum_{n=0}^\infty z^n h$ and then apply the summation method of Cesaro. We have

$$(H_u^* H_u \varphi, z^n h) = (H_u \varphi, u z^n h) = (P_- u\varphi, z^{n-1} \bar{h}) = (u\varphi, P_- z^{n-1} \bar{h})$$
$$= (u\varphi, z^{n-1}\bar{h}) - (u\varphi, P_+ z^{n-1} \bar{h}).$$

Hence

$$(H_u^* H_u \varphi, z^n h) = (\varphi, z^n h) - (u\varphi, P_+ z^{n-1} \bar{h}).$$

Let $K_N(\zeta) \stackrel{\text{def}}{=} \sum_{k=-N}^N (1 - |k|/N)\zeta^k$ be the Fejér kernel and set $K_N^+ \stackrel{\text{def}}{=} P_+ K_N$. Then

$$(H_u^* H_u \varphi, K_N^+ h) = (\varphi, K_N^+ h) - (u\varphi, P_+ K_N \overline{zh}).$$

Let us show that $\lim_{N \to \infty} (u\varphi, P_+ K_N \overline{zh}) = 0$. We have

$$(u\varphi, P_+ K_N \overline{zh}) = (P_+ u\varphi, K_N \overline{zh}) = (u\varphi, K_N \overline{zh}) - (P_- u\varphi, K_N \overline{zh})$$
$$= (zhu\varphi, K_N) - (zhP_- u\varphi, K_N)$$
$$= (\bar{h}\varphi, K_N) - (zhP_- u\varphi, K_N).$$

Notice that $\bar{h}\varphi \in X$ and that $P_- u\varphi \in X$. Hence

$$\lim_{N \to \infty} (\bar{h}\varphi, K_N) = \bar{h}(1)\varphi(1) = 0$$

and

$$\lim_{N \to \infty} (zhP_- u\varphi, K_N) = h(1)(P_- u\varphi)(1) = 0.$$

It remains to be shown that the sequence of functionals $\{K_N^+ h\}$ converges to $h(\mathbb{1} - z)^{-1}$ in the weak topology $\sigma(X_A^*, X_A)$. Let $f \in X_A$. Then

$$(f, K_N^+ h) = (f\bar{h}, K_N^+) = (P_+ f\bar{h}, K_N) \to P_+ f\bar{h}(1),\ N \to \infty.$$

Apply now Lemma 3.12.

Finally assume that assumption 2) of the theorem is fulfilled. As $P_- u \in X \subset C(\mathbb{T})$ then by Theorem 3.1 $u \in VMO$ and we can apply the part of the theorem already established. □

*Proof of Theorem* 3.11. Assume that $f \in X$ with $\mathscr{A}f \neq f$. Then $u = \delta^{-1}(f - \mathscr{A}f)$ is unimodular for a suitable $\delta$, $\delta > 0$, and $P_- u \in X$. As $X \subset VMO$ Theorem

3.3 gives $u \in VMO$. Upon applying Theorem 3.10 we find that $u$ and *a fortiori* $\mathscr{A}f$ belong to $X$. □

*Remark.* In the proof of Theorem 3.10 we have utilized two ideas in the paper Adamyan-Arov-Kreĭn [2]. The first is the idea to apply the Riesz-Schauder theory to the operator $H_u^* H_u$ and the second is the observation that $h(\mathbb{1} - z)^{-1}$ is an eigenfunction of $R^*$, provided $h(1) = 0$, $R^*h = h$. With the aid of these facts it is shown in Adamyan-Arov-Kreĭn [2] that the operator of best approximation preserves the class $\mathscr{F}l^1 = \{f \in C(\mathbb{T}) : \sum_{n \in \mathbb{Z}} |\hat{f}(n)| < \infty\}$ as well as some other weighted classes $Fl^1(w)$. These spaces satisfy axioms (A1)–(A4).

**3.3. How to Cope with Non-Separability and the Absence of a Norm.** Despite the fact that the axioms (A1)–(A4) comprise a rather extended class of function spaces on $\mathbb{T}$ many important instances are not included in the scheme outlined in Sect. 3.2. On the one hand we have non-separable spaces such as $\Lambda_\alpha$, $B_p^\alpha$, on the other hand locally convex non-Banach spaces, for example, the Carleman classes $C\{M_n\}$. In the present Section we add auxiliary information which allows us to extend the best approximation theorem to these cases.

Let us begin by formulating the axioms which the classes considered in this Section have to satisfy:

(B1) $X$ is a vector space of functions defined on the circle, containing the polynomials and such that

$$f \in X \Rightarrow \bar{f}, zf, P_+ f \in X.$$

(B2) There exists a space $Y$ satisfying the axioms (A1)–(A4) such that $X \subset Y$ and $T_{\bar{f}} X_A \subset X_A$ for every $f$ in $Y_A$.

(Recall that $X_A \overset{\text{def}}{=} P_+ X$, $Y_A \overset{\text{def}}{=} P_+ Y$.)

**Lemma 3.13.** *Assume that the space X satisfies the axioms* (B1) *and* (B2). *Let h be an outer function such that* $h, h^{-1} \in H^2$. *Then* $P_- \bar{h}/h \in X$ *iff* $|h|^2 \in X$ *and* $\inf_{t \in \mathbb{T}} |h(t)| > 0$.

*Proof.* Let $P_- \bar{h}/h \in X$. If $h, h^{-1} \in H^2$ then clos $T_{\bar{h}/h} H^2 = H^2$ (cf. Lemma 3.4). Let $Y$ be the space in axiom (B1). Applying Theorem 3.10 with $Y$ and the unimodular function $u = \bar{h} \cdot h^{-1}$ we find $u \in Y$. Moreover, in the course of the proof of Theorem 3.10 we established that $h \in Y$ and $\inf_{t \in \mathbb{T}} |h(t)| > 0$ so that $h^{-1} \in Y$. Let us now show that $|h|^2 \in X$. Using the elementary identity $zP_- \bar{z} g = \overline{P_+ \bar{g}}$ we get $P_+(h/\bar{h}) \in X$. We also have

$$P_+ |h|^2 = P_+(h/\bar{h}) \bar{h}^2 = P_+(\bar{h}^2 P_+(h/\bar{h})) = T_{\bar{h}^2}(P_+ h/\bar{h}),$$

whence by axiom (B2) $P_+ |h|^2 \in X$. The function $|h|^2$ being real we then also have $P_- |h|^2 \in X$.

Conversely let $|h|^2 \in X$ and $\inf_{t \in \mathbb{T}} |h(t)| > 0$. Then

$$P_+(h/\bar{h}) = P_+(1/\bar{h}^2)|h|^2 = T_{\bar{h}^{-2}}(P_+(h/\bar{h})),$$

because $h^{-2} = c|h|^{-2} \exp(-i \log |h|^2) \in Y$, $c \in \mathbb{C}$. In the last inclusion we used that $|h|^2 \in Y$ and that $Y$ satisfies the axioms (A1), (A2), (A4). □

The spaces $\Lambda_\alpha$, $B^\alpha_{p\infty}$ satisfy axioms (B1), (B2); this will be discussed in the following Section.

Let us introduce a further axiom.

(B3) If $f \in X$, $f \geq 0$ and $\inf_{t\in\mathbb{T}} f(t) \geq 0$ then $1/f \in X$.

**Theorem 3.14.** *Assume that X satisfies axioms* (B1)–(B3). *Then the following conditions are fulfilled.*

1) *If u is a unimodular function in VMO such that* $P_- u \in X$ *then* $u \in X$.

2) *If u is a unimodular function such that* $H^2 = \operatorname{clos} T_u H^2$ *and* $P_- u \in X$ *then* $u \in X$.

3) $\mathscr{A} X \subset X$.

*Proof.* Let us first prove Statement 1. As $u \in VMO$ then by Appendix 4, Sect. 53, 57, $u = z^n \bar{h} h^{-1}$, $n \in \mathbb{Z}$, where $h$ is an outer function in $H^2$ such that $1/h \in H^2$ and the Toeplitz operator $T_{\bar{h}h^{-1}}$ is invertible in $H^2$. From this we conclude with the aid of Lemma 3.13 that $|h|^2 \in X$ and $\inf_{t\in\mathbb{T}} |h(t)| > 0$. By axiom (B3) then $|h|^{-2} \in X$ and, as $X \subset C(\mathbb{T})$, we get $\inf_{t\in\mathbb{T}} |h(t)|^{-2} > 0$. Applying now Lemma 3.13 to the function $h^{-1}$ we get $P_- h/\bar{h} = P_- \overline{h^{-1}}/h^{-1} \in X$. Consequently $u = z^n \bar{h} h^{-1} \in X$.

Statements 2 and 3 follow in a trivial manner from Statement 1 and Theorem 3.1 and 3.3. □

The example of the space $X = AC_\delta$ of functions admitting an analytic continuation to the annulus $\{\zeta \in \mathbb{C} : \delta < |\zeta| < 1/\delta\}$, as we will see in Sect. 3.4 (Example 6), shows that axiom (B3) is essential for the validity of Theorem 3.14.

**3.4. Examples and Counter-examples.** This Section is mainly devoted to applications of the abstract theorems of the previous Sections of this Chapter. We display examples of spaces for which the theorems of best approximation and restoration of unimodular functions are valid and we consider also some counter-examples.

Some of the results are communicated without proof. For the proofs and likewise for some further examples we refer to Peller-Hruščev [1].

*Example* 1. The Hölder classes $\Lambda_\alpha$, $0 < \alpha < \infty$, obviously satisfy axioms (A1)–(A4) and therefore we have $\mathscr{A}\lambda_\alpha \subset \lambda_\alpha$. Let us now consider the non-separable Hölder spaces $\Lambda_\alpha$, $0 < \alpha < \infty$. In this case axiom (A3) is not satisfied but axioms (B1)–(B3) hold true for these spaces. Axiom (B1) and (B3) are trivial. In order to verify axiom $(B_2)$ we may take $Y = \Lambda_{\alpha/2}$ and use that the Toeplitz operator $T_{\bar{f}}$, $f \in H^\infty$, are bounded in $P_+\Lambda_\alpha$. This fact, first noticed by Havin [2] and Šamoyan [1], is proved as follows. The space $P_+\Lambda_\alpha$ is dual to the space $P_+ B^{-\alpha}_{11}$:

$$g \in P_+ B^{-\alpha}_{11} \Leftrightarrow \bar{\partial} g = 0, \iint_{\mathbb{D}} |g(z)|(1-|z|)(1-|z|)^{\alpha-1}\, dx\, dy < \infty,$$

the duality being defined by the Hermitian form $(\varphi, g) = \sum_{n\geq 0} \hat{\varphi}(n) \cdot \overline{\hat{g}(n)}$ where $g \in P_+ B^{-\alpha}_{11}$ and $\varphi$ is a polynomial (cf. Review of Function Classes). It is now clear that the operator $T_{\bar{f}}$ is conjugate to the operator of multiplication by the function $f$, acting in the space $B^{-\alpha}_{11}$.

In conclusion let us notice that $C^{(\infty)} = \bigcap_{\alpha<\infty} \lambda_\alpha$ so that $\mathscr{A}C^{(\infty)} \subset C^{(\infty)}$.

*Example* 2. Let us consider the Besov classes $B^\alpha_{pq}$, $1 \le p \le \infty$, $1 \le q \le \infty$, $\alpha \ge 1/p$.

Assume first that $\alpha > 1/p$, $q < \infty$. Axiom (A1) is the fulfilled in view of the description of the spaces in terms of convolution with the kernels $W_k$, cf. the Review of Function Classes. If $\alpha > 1/p$ we have the imbedding $B^\alpha_{pq} \subset \lambda_\beta$ if $\beta < \alpha - 1/p$ (cf. (S.M.) Nikol'skiĭ [1], Chap. 6) and the space $B^\alpha_{pq}$ turns out to be an algebra (cf. Triebel [1], Sect. 2.6.2, Bergh-Löfström [1], Sect. 6.8, Exc. 9). The axioms (A3) and (A4) are established in a trivial manner. In the case $q = 1, \alpha = 1/p$ the space $B^\alpha_{pq}$ is still an algebra (cf. Triebel [1], Bergh-Löfström [1], Sect. 6.8, Exc. 9) and satisfies axioms (A1)–(A4).

Let us now consider the classes $B^\alpha_{p\infty}$ for $\alpha > 1/p$. Axiom (B1) is fulfilled, as the Riesz projection $P_+$ acts in this space. In order to establish (B2) let us remark that $B^\alpha_{p\infty} \subset \lambda_\beta$, $\beta < \alpha - 1/p$, and therefore we can take $Y = \lambda_\beta$. The Toeplitz operator $T_{\bar{f}}$, $f \in H^\infty$, is bounded in $P_+ B^\alpha_{p\infty}$. This fact is established in exactly the same way as in the case of the space $P_+ \Lambda_\alpha$, using the explicit description of the pre-dual of $B^\alpha_{p\infty} = (B^{-\alpha}_{p',1})^*$ along with the description of $P_+ B^{-\alpha}_{p',1}$ in terms of the values in the disk $\mathbb{D}$ (cf. Review of Function Classes).

Consequently, the theorems of restoration of unimodular functions and of best approximation are true for the classes $B^\alpha_{pq}$, $\alpha > 1/p$.

We emphasize that the results obtained give a complete solution of the problem of best approximation for the classes $B^\alpha_p \overset{\text{def}}{=} B^\alpha_{pp}$. The case $\alpha > 1/p$ was just taken care of. The classes $B^{1/p}_p$ admit a description in terms of rational approximation in $BMO$ so the technique of Sect. 3.1 is applicable. If however $\alpha < 1/p$ then $B^\alpha_p \not\subset BMO$ so the problem of best approximation does not even make sense in this case. For the sake of completeness let us give a simple proof of $B^\alpha_p \not\subset BMO$ for $\alpha < 1/p$. Let $r > 1$ and $r^{-1} < \delta = 1/p - \alpha$. If $B^\alpha_p \subset BMO$ we would then in particular have $B^\alpha_p \subset L^r$. By a theorem of Hardy-Littlewood (cf. Zygmund [1], Chap. 12, Sect. 9) the operator of fractional integration $I_\delta$ (cf. Review of Function Classes) maps $L^r$ into the Hölder class $\lambda_{\delta-1/r}$. As $I_\delta B^\alpha_p = B^{\alpha+\delta}_p$ (again cf. Review of Function Classes) this gives

$$B^{1/p}_p = I_\delta B^\alpha_p \subset I_\delta L^r \subset \lambda_{\delta-1/r}.$$

This inclusion apparently contradicts the fact that $B^{1/p}_p$ is conformally invariant (cf. (1)). □

*Example* 3. Let us consider the Sobolev space $\mathscr{L}^s_p$ for $1 < p < \infty, s > 1/p$. The Riesz projection $P_+$ is continuous in $L^s_p$, since it is continuous in $L^p$, $1 < p < \infty$. By the imbedding theorem $\mathscr{L}^s_p \subset \lambda_\beta$, $\beta < s - 1/p$ (cf. (S.M.) Nikol'skiĭ [1], Chap. 9) and for $s > 1/p$ the space $\mathscr{L}^s_p$ is an algebra (cf. Strichartz [1], Triebel [1], Sect. 2.6.2). Thus in these hypotheses on the parameters the spaces satisfy axiom (A1)–(A4). Again if $s < 1/p$ the problem of best approximation does not make sense, as $\mathscr{L}^s_p \not\subset BMO$ (for the proof it is sufficient to remark that $\mathscr{L}^s_p \supset B^{1+s-1/p}_1$). The question of the solution of the problems of restoration and of best approximation in the spaces $\mathscr{L}^{1/p}_p$, $p \ne 2$, remains open.

*Example* 4. Let $Z$ be any of the following Banach spaces: $Z = VMO$, $BMO$, $C_A + \bar{C}_A$, $H^\infty + \overline{H^\infty}$, $H^1 + \overline{H^1}$, $L^1 + \widetilde{L^1}$. Fixing $Z$ we get a whole series of spaces putting

$$Z^{(n)} = \{f \in \mathscr{D}' : f^{(n)} \in Z\}, n = 1, 2, \ldots.$$

**Theorem 3.15.** *If* $Z = VMO, C_A + \bar{C}_A, H^1 + \overline{H^1}$ *and* $n$ *is a positive integer then the space* $Z^{(n)}$ *satisfies the axioms* (A1)–(A4). *The space* $(L^1 + \widetilde{L^1})^{(n)}$ *satisfies the axioms* (A1)–(A4) *for* $n \geq 2$.

*Remark.* The space $(L^1 + \widetilde{L^1})^{(1)}$ is not an algebra, because $(L^1 + L^1)^{(1)} \not\subset H^\infty$.

For $Z = BMO, H^\infty + \overline{H^\infty}$ the space $Z^{(n)}$ is not separable. Therefore in order to solve the problem of best approximation in these classes we need the following theorem.

**Theorem 3.16.** *Let* $n \in \mathbb{Z}_+$, $n \geq 1$, *and take* $Z = BMO$ *or* $Z = H^\infty + \overline{H^\infty}$. *Then the space* $Z^{(n)}$ *satisfies the Axioms* (B1)–(B3).

We omit the proofs of Theorem 3.15 and 3.16. Let us just remark that as $Y$ in Axiom (B2) we may take the space $\{f : \sum_{n \in \mathbb{Z}} |\hat{f}(n)| \log(2 + |n|) < \infty\}$.

Theorem 3.15 and 3.16 show that in the classes $Z^{(n)}$ under consideration the theorems of best approximation by analytic functions and of restoration of unimodular functions indeed hold true. Let us mention an amusing corollary.

**Corollary 3.17.** *Let* $u$ *be a unimodular function on* $\mathbb{T}$ *such that* $u \in VMO$ *or* clos $T_u H^2 = H^2$ *and let* $n \geq 1$. *Then the following statements hold true.*

1) $P_- u \in C^{(n)} \Rightarrow u \in C^{(n)}$.
2) $(P_- u)^{(n)} \in L^\infty \Rightarrow u^{(n)} \in L^\infty$.
3) $(P_- u)^{(n)} \in L^1 \Rightarrow u^{(n)} \in L^1$. □

*Example* 5. Carleman classes. Let $\mathscr{E}$ be a Banach space of functions on $\mathbb{T}$ and $\{M_n\}_{n \geq 0}$ an increasing logarithmically convex sequence, with $M_0 = 1$. The *Carleman class* $C(\mathscr{E}; \{M_n\})$ consists of the infinitely differentiable functions $f$ on $\mathbb{T}$ satisfying the following condition:

$$\|f^{(n)}\|_{\mathscr{E}} \leq C_f Q_f^n \cdot n!\, M_n, \quad n = 0, 1, \ldots$$

In Carleson-Jacobs [1] it is shown that the operator of best approximation preserves differentiation invariant Carleman classes $C(C(\mathbb{T}), \{M_n\})$. The following theorem allows us to draw the same conclusion in considerably less stringent assumptions.

**Theorem 3.18.** *Let* $\{M_n\}_{n \geq 0}$ *be an increasing logarithmically convex sequence of positive numbers. Then the class* $C(L^2, \{M_n\})$ *satisfies the Axioms* (B1)–(B3). *If the sequence* $\{M_n\}_{n \geq 0}$ *also satisfies the condition*

$$M_{n+1}/M_n \leq cQ^{Q^n}$$

*for some positive numbers* $c$, $Q$ *then the class* $C(C(\mathbb{T}), \{M_n\}_{n \geq 0})$ *satisfies the same conditions too.*

As for the space $Y$ in Axiom (B2) we may take the space $\{f : \sum_{n \in \mathbb{Z}} |\hat{f}(n)| \log(|n| + 2) < \infty\}$. For the verification of Axiom (B3) the following result, which we also need in what follows, is needed.

**Lemma 3.19.** *Let $\mathscr{E} = C(\mathbb{T})$ or $\mathscr{E} = L^2$, $g \in C(E, \{M_n\})$, $f$ analytic in a neighborhood of the set $g(\mathbb{T})$. Then $f \circ g \in C(\mathscr{E}, \{M_n\})$.*

*Example* 6. The space $AC_\delta$, $0 < \delta < 1$, is made up by those functions which admit an analytic continuation to the annulus $\{\zeta \in \mathbb{C} : \delta < |\zeta| < \delta^{-1}\}$. The space $AC_\delta$ satisfies Axioms (B1), (B2) (in this case take $Y = \mathscr{F} l^1$). Let us notice that for the space $AC_\delta$ one can somewhat strengthen the conclusion of Lemma 3.13. The following corollary of Lemma 3.13 will be needed in Chap. IV.

**Theorem 3.20.** *Let $h$ be an outer function such that $h$, $h^{-1} \in H^2$. Then $P_-\bar{h} \cdot h^{-1} \in AC_\delta$ iff $|h|^2 \in AC_\delta$.*

*Proof.* This follows readily from Lemma 3.13 if we observe that the assumptions $|h|^2 \in AC_\delta$ and $h^{-1} \in H^2$ entail the condition $\inf_{t \in \mathbb{T}} |h(t)| > 0$. □

The space $AC_\delta$ does not satisfy Axiom (B3) and indeed $\mathscr{A}(AC_\delta) \not\subset AC_\delta$. In fact, let $h = z - \dfrac{1}{1+\delta}$ and set $u = \overline{zh} \cdot h^{-1}$. Then

$$P_- z^2 u = P_- z\left(z - \frac{2}{1+\delta}\right) \cdot \left(z - \frac{2}{1+\delta}\right)^{-1} = 0$$

so that $P_- u \in AC_\delta$. On the other hand

$$P_+ z^2 u = \left(1 - \frac{2}{1+\delta} z\right)\left(z - \frac{2}{1+\delta}\right)^{-1} \notin AC_\delta$$

because $1 < \dfrac{2}{1+\delta} < 1/\delta$. □

*Example* 7. Let $w = \{w_n\}_{n \geq 0}$ be a sequence of positive numbers and let $1 \leq p \leq \infty$. Let us consider the space $\mathscr{F} l^p(w)$ of formal Fourier series

$$f \sim \sum_{n \in \mathbb{Z}} \hat{f}(n) \zeta^n$$

such that $\|f\|_{\mathscr{F} l^p(w)} \overset{\text{def}}{=} (\sum_{n \in \mathbb{Z}} |\hat{f}(n) w_{|n|}|^p)^{1/p} < \infty$, $p < \infty$, $\|f\|_{\mathscr{F} l^\infty(w)} \overset{\text{def}}{=} \sup_n |\hat{f}(n) w_{|n|}|$. Denote by $\mathscr{F} c_0(w)$ the closure of the polynomials in $Fl^\infty(w)$. It is easy to see that in the hypothesis

$$\sup_m \left( \sum_{k+j=m} (w_{|m|}/w_{|k|} \cdot w_{|j|})^{p'} \right)^{1/p'} < \infty \tag{5}$$

$\mathscr{F} l^p(w)$ is an algebra (for details see Nikol'skiĭ [2], Sect. 3.2). If $p = 1$ this condition takes on the simple form $w_{|k+j|} \leq \text{const}\, w_{|k|} \cdot w_{|j|}$. Let us remark that if $\mathscr{F} l^p(w)$ is an algebra then $\inf_n w_n > 0$. In fact $w_0 = \|z^{-n} \cdot z^n\| \leq \|z^{-n}\| \cdot \|z^n\| = w_n^2$. It is clear that if

$$\overline{\lim_{n\to\infty}}\, w_n^{1/n} = 1 \tag{6}$$

then $\mathscr{F}l^p(w) \not\subset AC_\delta$ for any $\delta$, $0 < \delta < 1$.

Summing up what we have just said we get the following result.

**Theorem 3.21.** *The spaces* $\mathscr{F}l^p(w)$, $1 \le p < \infty$, *and* $\mathscr{F}c_0(w)$ *satisfy axioms* (A1)–(A4) *provided the sequence* $\{w_n\}_{n\ge 0}$ *satisfies the conditions* (5) *and* (6).

In the non-separable case $\mathscr{F}l^\infty(w)$ the verificiation of Axiom (B3) causes some difficulty, because it is not clear that the maximal ideal space of $\mathscr{F}l^\infty(w)$ coincides with $\mathbb{T}$. We omit the proof of the following theorem.

**Theorem 3.22.** *Let* $\{w_n\}_{n\ge 0}$ *be a nondecreasing sequence such that* $\sup_n w_{2n}/w_n < \infty$ *and assume that condition* (5) *holds with* $p = 1$. *Then* $\mathscr{F}l^\infty(w)$ *satisfies Axioms* (B1)–(B3).

*Example* 8. The operator of best approximation does not act in the space $C(\mathbb{T})$, i.e. $\mathscr{A}C(\mathbb{T}) \not\subset C(\mathbb{T})$. This was proved by Adamyan-Arov-Kreĭn [2] and an alternative proof was obtained by Carleson-Jacobs [1]. Let us prove it here.

Let $\varphi$ be a real function in $C(\mathbb{T})$ with $\tilde{\varphi} \notin C(\mathbb{T})$. Let us set $u = \bar{z}e^{i\tilde{\varphi}}$. Then $u \notin H^\infty$ (cf. Appendix 4, Sect. 53) and $\operatorname{dist}(u, H^\infty) = 1$. We have $u = f + g$, $g \in H^\infty$, $f \in C(\mathbb{T})$. Then $\operatorname{dist}(f, H^\infty) = 1$ and thus $\mathscr{A}f = g$. However $g \notin C(\mathbb{T})$ as $\tilde{\varphi} \notin C(\mathbb{T})$. □

*Open questions.* Concluding this Section we mention some spaces $X$ for which the problem of best approximation has not been solved.

It seems that the most important and most interesting cases of this problem are the spaces $\mathscr{F}l^\infty(w)$ and $\mathscr{F}c_0(w)$, $w_n = n + 1$, $n \in \mathbb{Z}_+$.

The same question for $C_A + \overline{C_A}$ is a quite old – and a very interesting – open problem (cf. Sarason [9]).

The problem is further open for the classes $\mathscr{L}_p^{1/p}$, $1 < p < \infty$, $p \ne 2$, as well as for the class $\{f = f_1 + f_2 : f_1' \in L^1, \widetilde{f_2'} \in L^1\}$.

**3.5. The Arguments of Unimodular Functions.** In the preceding we have repeatedly used Sarason's Theorem (Appendix 4, Sect. 57) to the effect that a unimodular function $u$ is in $VMO$ iff it can be represented in the form $u = z^n e^{i\varphi}$, $\varphi$ a real function in $VMO$. In Chap. IV we need analogues of this result for other function classes too. Here the spaces $B_p^{1/p}$, $1 < p < \infty$, cause the greatest difficulty, because they contain discontinuous and unbounded functions. For the proof of the analogue of Sarason's Theorem we will use results on best approximation. We know of no other proof for the classes $B_p^{1/p}$, $p \ne 2$. If $p = 2$ there is an alternative proof based on the technique of orthogonal polynomials and the asymptotic formula for Toeplitz determinants (cf. Ibragimov-Solev [1]).

Let us consider a function space $X$ on $\mathbb{T}$ subject to the following axioms.

(C1) $X$ contains the set of polynomials, $X \subset VMO$.

(C2) $f \in X \Rightarrow \bar{f} \in X$, $P_+ f \in X$.

(C3) $\mathscr{A}X \subset X$.

(C4) If $f \in X \cap L^\infty$ and $\varphi$ analytic in a neighborhood of $f(\mathbb{T})$ then $\varphi \circ f \in X$.

*Remark.* Axiom (C4) entails that $X \cap L^\infty$ is an algebra ($4fg = (f+g)^2 - (f-g)^2$).

It is easy to see that all examples of spaces in which the operator $\mathscr{A}$ acts, considered in Sect. 3.4, satisfy Axiom (C1)–(C4), notably the spaces $B_p^{1/p}$ and $C(C(\mathbb{T}), \{M_n\})$, with $\{M_n\}_{n\geq 0}$ satisfying condition (7).

**Theorem 3.23.** *Assume that the space $X$ satisfies Axioms* (C1)–(C4) *and let $u$ be a unimodular function in $X$. Then $u = z^n e^{i\varphi}$ where $n \in \mathbb{Z}$ and $\varphi$ is a real function in $X$.*

*Proof.* As $X \subset VMO$ the Toeplitz operator $T_u$ must be Fredholm (cf. Appendix 4, Sect. 50, 53, 165.53.3). Without loss of generality we may assume that $\operatorname{ind} T_u = 0$, that is, that $T_u$ is an invertible operator. Consequently, $\operatorname{dist}(u, H^\infty) < 1$ and $h = \mathscr{A}u$ is an outer function (cf. Appendix 4, Sect. 165.53). In view of Axiom (C3) $h \in X$. By the foregoing remark $X \cap L^\infty$ is an algebra so that $\bar{u}h \in X$. We have $\|1 - \bar{u}h\|_{L^\infty} = \|u - h\|_{L^\infty} < 1$, which shows that the set $(\bar{u}h)(\mathbb{T})$ is contained in the disk $\{\zeta : |1 - \zeta| < \delta\}$ for some $\delta$, $0 < \delta < 1$. Hence $\bar{u}h = |h| \exp(i\psi)$ with $\psi$ real, $\|\psi\|_{L^\infty} < \pi/2$.

Notice now that $\inf_{\mathbb{T}} |h(t)| > 0$ so that by Axiom (C4) $\log|h| = \frac{1}{2}\log(h \cdot \bar{h}) \in X$ and thus (in view of Axiom (C2)) $\log|h| \in X$. Consequently, utilizing once more Axiom (C4) with the function $\bar{u}h$, $\bar{u}h \in L^\infty \cap X$, we get

$$\log \bar{u}h = \log|h| + i\psi \in X.$$

Thus $\psi \in X$.

Write $u = \exp(i\beta)$ where $\beta$ is a real function in $VMO$ (cf. Appendix 4, Sect. 57). Then $\exp(i(\beta + \psi - \log|h| + c) = 1$ with $c \in \mathbb{R}$. Hence (cf. Appendix 4, Sect. 165.35)

$$\beta = \log|h| - \psi + \text{const} \in X. \square$$

*Remark.* In all the examples encountered in Sect. 3.4 the converse is true too. Namely if $\varphi$ is a real function in $X$ and $n \in \mathbb{Z}$ then $z^n e^{i\varphi} \in X$. This is trivial if $X \subset L^\infty$ but the same is true also for the Besov classes $B_p^{1/p}$, because directly from the definition of $B_p^{1/p}$, $1 < p < \infty$, we readily see that $f \circ \varphi \in B_p^{1/p}$ if $\varphi \in B_p^{1/p}$ and $f$ is a function on $\varphi(\mathbb{T})$ such that $|f(\zeta_1) - f(\zeta_2)| \leq \text{const}\,|\zeta_1 - \zeta_2|$.

**Corollary 3.24.** *Let $u$ be a unimodular function. Then $u \in B_p^{1/p}$, $1 \leq p < \infty$, iff $u$ admits the representation*

$$u = z^n \exp(i\varphi),$$

*where $n \in \mathbb{Z}$ and $\varphi$ is a real function in $B_p^{1/p}$.*

**3.6. Applications: Restoration in Spaces of Measures, Interpolation and the Fefferman-Stein Decomposition in the Classes $B_p^{1/p}$.** The restoration problem in the space $M(\mathbb{T})$ of all finite complex Borel measures on $\mathbb{T}$ may be formulated as follows. Let $X$ be a space of distributions on $\mathbb{T}$. In which hypotheses on $X$ is it true that

$$\mu \in M(\mathbb{T}), \quad P_-\mu \in X \Rightarrow \mu \in X?$$

The well-known theorem of $F$. and $M$. Riesz may be conceived as a restoration theorem. Namely according to that theorem

$$\mu \in M(\mathbb{T}), \quad P_-\mu \in L^1(\mathbb{T}) \Rightarrow \mu \in L^1(\mathbb{T}).$$

Another example is the theorem of Rajchman [1]:

$$\mu \in M(\mathbb{T}), \quad P_-\mu \in \mathscr{F}c_0 \Rightarrow \mu \in \mathscr{F}c_0.$$

Recall that $B_p^{-1/p'}$, $1 < p < \infty$, is the class of distributions on $\mathbb{T}$ such that

$$\sum_{n \in \mathbb{Z}} 2^{-n(p-1)} \| f * W_n \|_{L^p}^p < \infty,$$

and that $f \in P_+ B_p^{-1/p'}$ iff

$$\iint_{\mathbb{D}} |f|^p (1-|z|)^{p-2}\, dx\, dy < \infty$$

(cf. Review of Function Classes). The space $B_p^{-1/p'}$ can in a natural way be identified with the space $(B_{p'}^{1/p'})^*$. If $p = 2$ then $B_2^{-1/2}$ coincides with the space of *distributions of finite energy*,

$$\mathscr{E}(f) = \sum_{n \in \mathbb{Z}, n \neq 0} |\hat{f}(n)|^2/|n| < \infty,$$

**Theorem 3.25.** *Let $1 < p < \infty$ and $\mu \in M(\mathbb{T})$. Then*

$$P_-\mu \in B_p^{-1/p'} \Rightarrow \mu \in B_p^{-1/p'}.$$

*Proof.* For $0 < r < 1$ set $\mu_r = P_r * \mu$, where $P_r$ is the Poisson kernel. Let $g \in P_- B_{p'}^{1/p'}$ be a function such that $\text{const} \cdot \| P_+\mu_r \|_{B_p^{-1/p'}} \le |\int zg\, d\mu_r|$ and $\| g \|_{B_{p'}^{1/p'}} \le 1$. Such a function does exist, as $(B_{p'}^{1/p'})^* = B_p^{-1/p'}$. Let us set $f = \mathscr{A}g$. Then $\| f - g \|_{L^\infty} = \text{dist}_{L^\infty}(g, H^\infty) \le \text{const} \| g \|_{B_{p'}^{1/p'}}$ because $B_{p'}^{1/p'} \subset VMO$. Upon applying the theorem of best approximation for the space $B_{p'}^{1/p'}$ (cf. Cor. 3.5.3) we find that $f \in B_{p'}^{1/p'}$ and $\| f \|_{B_{p'}^{1/p'}} \le \text{const} \| g \|_{B_{p'}^{1/p'}}$. We have

$$\begin{aligned}
\| P_+\mu_r \|_{B_p^{-1/p'}} &\le \text{const} \left| \int zg\, d\mu_r \right| \\
&\le \text{const} \left| \int z(g-f)\, d\mu_r \right| + \text{const} \left| \int zf\, dP_-\mu_r \right| \\
&\le \text{const} \| g - f \|_{L^\infty} \cdot \| \mu \|_{M(\mathbb{T})} + \text{const} \| zf \|_{B_{p'}^{1/p'}} \cdot \| P_-\mu_r \|_{B_p^{-1/p'}} \\
&\le C,
\end{aligned}$$

where $C$ is a positive constant independent of $r$. Consequently $P_+\mu \in B_p^{-1/p'}$. □

For $p = 2$ the theorem admits a very simple formulation.

**Corollary 3.26.** *If $\mu \in M(\mathbb{T})$ then*

$$\sum_{n \ge 1} |\hat{\mu}(-n)|^2/n < \infty \Rightarrow \sum_{n \ge 1} |\hat{\mu}(n)|^2/n < \infty. \ \square$$

*Remark.* From Corollary 3.26 it is easy to deduce the following statement. Let $s \in \mathbb{Z}_+$. Then

$$\sum_{n \geq 1} |\hat{\mu}(-n)|^{2s}/n < \infty \Rightarrow \sum_{n \geq 1} |\hat{\mu}(n)|^{2s}/n < \infty. \tag{11}$$

For the proof it suffices to apply Corollary 3.26 with the measure $\mu * \cdots * \mu$ ($s$ times). There remains a most interesting (open) question: for which values of $s$, $s \geq 1/2$, does the implication (11) hold true?

Theorem 3.25 has applications to the theory of interpolation of analytic functions.

**Definition.** A closed subset $E$ of the unit circle $\mathbb{T}$ is called an *interpolation set* for $C_A \cap B_p^{1/p}$ if $C_A \cap B_p^{1/p}|E = C(E)$.

**Definition.** Let $E$ be a closed subset of $E$. Then

$$\operatorname{cap}_p^{1/p}(E) \stackrel{\text{def}}{=} \inf\{\|\varphi\|_{B_p^{1/p}} : \varphi \in C(\mathbb{T}) \cap B_p^{1/p}, |\varphi|\,|E \geq 1\}.$$

Note that for $p = 2$ this *capacity* is equivalent to the usual logarithmic capacity.

**Theorem 3.27.** *Let* $1 < p < \infty$ *and consider a closed subset* $E$ *of* $\mathbb{T}$. *Then* $E$ *is an interpolation set for* $C_A \cap B_p^{1/p}$ *iff* $\operatorname{cap}_p^{1/p}(E) = 0$.

For the proof of this theorem we refer to Peller-Hruščev [1]. It follows from the results of Sjödin [1] that if $C(\mathbb{T}) \cap B_p^{1/p}|E = C(E)$ then $\operatorname{cap}_p^{1/p}(E) = 0$. In the proof of the converse result one uses Theorem 3.24.

The following application concerns the analogue of the Fefferman-Stein theorem for the Besov classes $B_p^{1/p}$. This theorem affirms that a function $f$ on $\mathbb{T}$ is of bounded mean oscillation, that is,

$$\sup_I 1/m(I) \int_I |f - f_I|\, dm < \infty.$$

iff $f = u + \tilde{v}$ with $u, v \in L^\infty$ (cf. Fefferman-Stein [1]).

**Theorem 3.28.** *Every function* $f$ *in the space* $B_p^{1/p}$, $1 < p < \infty$, *admits the decomposition*

$$f = u + \tilde{v}$$

*with* $u, v \in B_p^{1/p} \cap L^\infty$.

This follows readily from Theorem 3.3.

# Chapter IV. Regularity Conditions for Discrete Time Stationary Gaussian Processes

In this Chapter the machinery developed in the previous Chapters of our survey will be used to obtain various regularity criteria for stationary Gaussian processes in terms of the spectral measure of the process. Recall that in Chap. II the problem of classification of processes by regularity conditions, stated in the usual probability language, was reduced to problems in the theory of Hankel operators. Now we are in a position to carry out the second, final step of our plan. For this we will on the one hand rely on the theory of Hankel and Toeplitz operators and on the other hand on the results on best approximation and on restoration of unimodular functions, as obtained in Chap. III.

We will consider only the case of discrete time. Moreover all stationary Gaussian processes are assumed to be regular. In other words, the spectral measure of the process has to be absolutely continuous with respect to Lebesgue measure and its density $w$ has to satisfy the condition $\log w \in L^1$. In this case $w = |h|^2$ where $h$ is an outer function in $H^2$. For the sake of uniformity we keep this notation throughout the Chapter. Let us remark that although the spectral density of a stationary gaussian process automatically satisfies the condition $w(\zeta) = w(\bar{\zeta})$, $\zeta \in \mathbb{T}$, for the problems of this Chapter, this turns out to be inessential. Therefore we may assume that $w$ is any non-negative summable function with $\log w \in L^1$.

Let $\mathscr{P}_0$ denote orthogonal projection in $L^2(w)$ onto the subspace $H^2_-(w)$ and $\mathscr{P}^n$ orthogonal projection onto $z^n H^2(w)$. Recall that the regularity properties of the process can be stated in terms of the operators $\mathscr{P}_0 \mathscr{P}^n$ (cf. Chap. II); in particular

$$\rho_n = |\mathscr{P}_0 \mathscr{P}^n|, \tag{1}$$

is the cosine of the angle between the subspaces $H^2_-(w)$ and $z^n H^2(w)$.

Let us remark that the important operators $\mathscr{P}_0 \mathscr{P}^n$ were for the first time considered in Gel'fand-Yaglom [1] and Yaglom [2]. In particular, in Yaglom [2] a description was given of the densities for those stationary Gaussian processes for which the operators $\mathscr{P}_0 \mathscr{P}^n$ have finite rank.

In Helson-Sarason [1] there was given a description of the spectral densities of completely regular processes, i.e. processes such that $\lim_{n\to\infty} \rho_n = 0$. Later Sarason [15] refined this description. In Sect. 4.1 we give a new proof of the theorem of Helson-Sarason. After the appearance of the papers Fefferman-Stein [1] and Sarason [16] describing the spaces $L^\infty + \widetilde{L^\infty}$ and $C(\mathbb{T}) + \widetilde{C(\mathbb{T})}$ in terms of mean oscillations it became possible to reformulate the Helson-Sarason theorem in local terms. Let us remark that before the appearance of Helson-Sarason [1] Ibragimov [3], [1], [2], [4] had obtained several sufficient conditions for the complete regularity of a process and one necessary one possessing a local character (cf. Ibragimov-Rozanov [1], Chap. V, Sect. 3–4). For a long time it was not clear whether the necessary condition of Ibragimov really was sufficient. Although Sarason [16] proved that this was not the case, it is nevertheless interesting to relate Ibragimov's condition to the condition $\log w \in VMO$.

The Helson-Sarason theorem will in Sect. 4.2 be applied to the study of processes satisfying a condition of rapid mixing. In particular, there we give a simpler proof of Ibragimov's theorem characterizing the spectral densities of processes such that the sequence $\{\rho_n\}_{n\geq 0}$ falls off as a power or exponentially. We likewise consider intermediate rates of decrease.

Another interesting approach to this theory, differing from ours, appeared not long ago in Arocena-Cotlar-Sadosky [51].

Two of the pioneering works on stationary Gaussian processes were the papers Kolmogorov-Rozanov [1] and Helson-Szegö [1]. In the first of these the estimating of the quantity

$$\sup_{A \in \mathscr{A}, B \in \mathscr{B}} |P(A \cap B) - P(A)P(B)|$$

was reduced to the estimating of the angles between the Gaussian subspaces corresponding to the $\sigma$-algebras $\mathscr{A}$ and $\mathscr{B}$ (cf. Theorem 2.14). In the second one the processes with a non-zero angle between past and future were characterized. Later Helson and Sarason [1], generalizing this result, gave a description of the weights for which the angle between the subspaces $H^2_-(w)$ and $z^n H^2(w)$, $n \geq 0$, is positive. In Sect. 4.3 we offer a new proof of the Helson-Sarason theorem.

In Ibragimov-Solev [2] the $\mathfrak{S}_2$-regular processes were described. They used orthogonal polynomials and the asymptotic formula for Toeplitz determinants. In Sect. 4.4 we give a new proof of this result, which also allows us to get a description of $\mathfrak{S}_p$-regular processes, $1 \leq p < \infty$. Moreover we obtain results concerning rapid decrease of the norms $\|\mathscr{P}_0 \mathscr{P}^n\|_{\mathfrak{S}_p}$, $n \to \infty$.

**4.1. Complete Regularity.** The proof of the following theorem, describing the spectral densities of completely regular processes will be based on the theorem of restoration of unimodular functions in $VMO$ (cf. Sect. 3.1). Recall that the numbers $\rho_n$ are defined by inequality (1).

**Theorem 4.1** (Helson-Sarason [1], Sarason [15]). *The following statements are equivalent.*

1. $\lim_n \rho_n = 0$.
2. *The function $w$ admits the representation*

$$w = |Q|^2 \exp \varphi,$$

*where $\varphi$ is a real function in $VMO$ and $Q$ a polynomial with all its roots on $\mathbb{T}$.*

The proof of the theorem will be preceded by two lemmata of a technical nature.

**Lemma 4.2.** *We have* $\lim_n \rho_n = 0$ *iff* $\mathscr{P}_0 \mathscr{P}^m \in \mathfrak{S}_\infty$ *for some number $m$, $m \in \mathbb{Z}_+$. If* $\mathscr{P}_0 \mathscr{P}^m \in \mathfrak{S}_\infty$ *for some $m$ then* $\mathscr{P}_0 \mathscr{P}^m \in \mathfrak{S}_\infty$ *for all $m$, $m \in \mathbb{Z}_+$.*

*Proof.* Lemma 2.15 and Nehari's Theorem (cf. Lecture VIII, Sect. 2) allow us to contend that

$$\rho_n = |H_{z^n \bar{h}/h}| = \operatorname{dist}_{L^\infty}(z^n \bar{h}/h, H^\infty) = \operatorname{dist}_{L^\infty}(\bar{h}/h, z^n H^\infty)$$

so that by Sarason's Theorem (cf. Lecture VIII, Sect. 3, Lemma on $H^\infty + C(\mathbb{T})$) the relation $\lim_n \rho_n = 0$ is equivalent to the inclusion $\bar{h}/h \in H^\infty + C(\mathbb{T})$. By Lemma 2.15 the inclusions $\mathscr{P}_0 \mathscr{P}^m \in \mathfrak{S}^\infty$ and $H_{z^m \bar{h}/h} \in \mathfrak{S}^\infty$ are equivalent. The proof is finished by invoking Hartman's Theorem (cf. Appendix 4, Sect. 22). □

**Lemma 4.3.** *Let $Q$ be a polynomial of degree $N$ all of whose roots sit on $\mathbb{T}$. Then on the unit circle $\mathbb{T}$ (outside the set of roots) holds the relation*

$$\bar{Q}/Q = c\bar{z}^N$$

*with $c \in \mathbb{T}$.*

*Proof.* It is sufficient to consider the case of the polynomial $Q(z) = z - \gamma, \gamma \in \mathbb{T}$. We have

$$\bar{Q}/Q = \frac{\bar{z} - \bar{\gamma}}{z - \gamma} = -\bar{\gamma} \cdot \bar{z}. \ \square$$

*Proof of the theorem.* $2 \Rightarrow 1$. Let us define an auxiliary outer function $h_1$ by the formula $h_1 = \exp\left(\frac{1}{2}\varphi + \frac{i}{2}\tilde{\varphi}\right)$. Then $h = c_0 Q h_1$ with $c_0 \in \mathbb{T}$ (the function $Q \cdot h_1$ is outer and $|Q|^2 \cdot |h_1|^2 = |h|^2$). By Lemma 4.3 we have

$$\bar{h}/h = \bar{c}_0^2 \cdot \frac{\bar{Q}}{Q} \cdot \frac{\bar{h}_1}{h_1} = c\bar{z}^N \exp(-i\varphi), \ c \in \mathbb{T}.$$

Since $\varphi$ is a real function in $VMO$, by Theorem 1.31 we find $\bar{h}/h \in QC$. Now the inclusion $H_{\bar{h}/h} \in \mathfrak{S}_\infty$ follows from Hartman's Theorem (Appendix 4, Sect. 22).

$1 \Rightarrow 2$. By Lemma 4.2 we have $H_{\bar{h} \cdot h^{-1}} \in \mathfrak{S}_\infty$, i.e. $P_- \bar{h} \cdot h^{-1} \in VMO$. As the Toeplitz operator $T_{\bar{h} \cdot h^{-1}}$ has a dense range in $H^2$ (cf. Lemma 3.4) we see from Lemma 3.5.1 that $\bar{h} \cdot h^{-1} \in VMO$.

In view of Appendix 4, Sect. 57 we can find an integer $N$ and a real function $\omega$ in $VMO$ such that

$$\bar{h}/h = \bar{z}^N \cdot e^{i\omega}.$$

As the operator $T_{\bar{h} \cdot h^{-1}}$ is Fredholm (cf. Appendix 4, Sect. 50, 53) and as further $\operatorname{clos} T_{\bar{h} \cdot h^{-1}} H^2 = H^2$ then $N = \operatorname{ind} T_{\bar{h} \cdot h^{-1}} \geq 0$. Defining an auxiliary outer function $h_1$ by the formula $h_1 = \exp\left\{-\frac{1}{2}\tilde{\omega} + \frac{i}{2}\omega\right\}$ we see that the Toeplitz operator $T_{\bar{h}_1 \cdot h_1^{-1}}$ is invertible (cf. Appendix 4, Sect. 165.35). The function $h_1$ cannot tend to zero "too strongly". Therefore, loosely speaking, the zeros of the polynomial $Q$ must be located at "strong" zeros of $h$. This reasoning can be given a rigorous meaning and this is precisely the route chosen by Helson and Sarason [1]. However, it turns out that one can find the desired polynomial also by purely algebraic considerations. For this we require the following lemma.

**Lemma 4.4.** *Consider a pair $(h, h_1)$ of outer functions in $H^2$ such that the operator $T_{\bar{h}_1 \cdot h_1^{-1}}$ is invertible and*

$$\bar{h}/h = z^N \bar{h}_1/h_1, \ N \in \mathbb{Z}_+. \tag{2}$$

*Then there exists a polynomial of degree $N$ with its root on $\mathbb{T}$ such that $h = Q \cdot h_1$.*

*Proof.* As $\operatorname{clos} T_{\bar{h}/h} H^2 = H^2$ and $\operatorname{ind} T_{\bar{h}/h} = N$ we must have $\dim \operatorname{Ker} T_{\bar{h}/h} = N$. The Toeplitz operator $T_{\bar{z}}$ coincides with the inverse shift operator $S^*$, $S^* f =$

$(f(z) - f(0))/z$, in $H^2$. With the aid of the preceding formula we can extend it to the class of analytic functions on the unit disk. Let us consider in $H^2$ the auxiliary operator $A$:

$$Af = hS^*\frac{1}{h}f,\ f \in H^2.$$

A simple computation shows that this operator is bounded:

$$Af = h\bar{z}\left(\frac{f}{h} - \frac{f(0)}{h(0)}\right) = \bar{z}f - \frac{f(0)}{h(0)}\bar{z}\cdot h.$$

The finite dimensional subspace $E = \operatorname{Ker} T_{\bar{h}/h}$ is invariant for $A$. In fact, if $f \in E$ then

$$T_{\bar{h}/h}Af = P_+\bar{z}\frac{\bar{h}}{h}f - \frac{h(0)}{h(0)}P_+\overline{zh} = S^*T_{\bar{h}/h}f = \mathbb{O}.$$

Let $F$ be the minimal annihilating polynomial for the operator $A|E$. Then $F(A)|E = \mathbb{O}$ and $\deg F \le N$. It is clear that

$$F(A)f = hF(S^*)\frac{1}{h}f,\ f \in H^2.$$

It follows from (2) that $z^{N-1}h_1 \in E$. Therefore

$$\mathbb{O} = F(A)z^{N-1}h_1 = hS^*z^N F(1/z)\frac{h_1}{h}. \tag{3}$$

Set $g(z) = z^N F(1/z)h_1/h$. Then it follows from (3) that $z^{-1}\cdot(g(z) - g(0)) = \mathbb{O}$. Hence $g \equiv$ const. Consequently $h = Qh_1$ with $Q = \text{const}\, z^N F(1/z)$. The polynomial $Q$ has no roots in $\mathbb{D}$, because $h$ is outer. Set $Q = Q_1Q_2$ where $Q_1$ is a polynomial with the roots on $\mathbb{T}$ and $Q_2$ one with the roots outside the closed unit disk. Set $n = \deg Q_1$. We have (cf. Lemma 4.3)

$$\frac{\bar{h}}{h} = \frac{\bar{Q}_1}{Q_1}\cdot\frac{\bar{Q}_2}{Q_2}\cdot\frac{\bar{h}_1}{h_1} = \text{const}\,\bar{z}^n\cdot\frac{\bar{Q}_2}{Q_2}\cdot\frac{\bar{h}_1}{h_1}$$

and, consequently,

$$\begin{aligned}\operatorname{ind} T_{\bar{h}/h} &= \operatorname{ind}(T_{\bar{z}^n}T_{\bar{Q}_2}T_{\bar{h}_1/h_1}T_{Q_2^{-1}}) \\ &= n + \operatorname{ind} T_{\bar{Q}_2} + \operatorname{ind} T_{\bar{h}_1/h_1} + \operatorname{ind} T_{Q_2^{-1}} \\ &= n,\end{aligned}$$

since the operators $T_{\bar{Q}_2}$ and $T_{Q_2^{-1}}$ apparently are invertible (cf. Appendix 4, Sect. 47). Hence $\deg Q_2 = N - n = 0$. □

*Remark.* Notice that the representation of the function $w$ in the form $w = |Q|^2e^{\varphi}$, $Q$ a polynomial with all roots on $\mathbb{T}$ and $\varphi \in VMO$, is unique up to a constant factor. In fact, if $|Q_1|^2e^{\varphi_1} = |Q_2|^2e^{\varphi_2}$ then

$$\int_{\mathbb{T}}\left|\frac{Q_1}{Q_2}\right|^2 = \int_{\mathbb{T}} \exp(\varphi_2 - \varphi_1)\, dm < +\infty$$

(cf. Appendix 4, Sect. 165.35). Therefore the polynomial $Q_2$ must divide the polynomial $Q_1$. Analogously $Q_1$ divides $Q_2$.

**4.2. Rapid Decrease of the Regularity Coefficients.** In this Section we consider the dependence of the spectral properties of the density of the process on the rate of decrease of the numbers $\rho_n$. Our main tool will be the theorems on the restoration of unimodular functions in the classes $\Lambda_\alpha$, $B^\alpha_{\infty,q}$, $G_\alpha$, $AC_\delta$, as established in Chap. III.

**Theorem 4.5** (Ibragimov [2]). *Let* $\alpha > 0$. *The following statements are equivalent.*

1. $\rho_n \le \mathrm{const} \cdot n^{-\alpha}$, $n > 0$.
2. *The function* $w$ *admits the factorization*

$$w = |Q|^2 e^\varphi$$

*where* $Q$ *is a polynomial with its roots on* $\mathbb{T}$ *and* $\varphi$ *a real function in* $\Lambda_\alpha$.

*The proof* of the theorem will be based on a description of the class $\Lambda_\alpha$ given in the Review of Function Classes. Let $f \in L^\infty$. Then

$$P_- f \in \Lambda_\alpha \Leftrightarrow \mathrm{dist}_{L^\infty}(z^n f, H^\infty) \le \mathrm{const}\, n^{-\alpha}. \tag{4}$$

$2 \Rightarrow 1$. Let $h$ and $h_1$ be outer functions in $H^2$ such that $|h|^2 = w$ and $h = Qh_1$. It follows from Lemma 4.3 that

$$\frac{\bar{h}}{h} = \frac{\bar{Q}}{Q}\cdot\frac{\bar{h}_1}{h_1} = c\bar{z}^N \exp(-i\tilde{\varphi}),\ c \in \mathbb{T}.$$

As the operator of passing to the harmonic conjugate is bounded in the space $\Lambda_\alpha$, we have $\exp(-i\tilde{\varphi}) \in \Lambda_\alpha$ and $\bar{h}/h \in \Lambda_\alpha$. As $P_-\bar{h}/h \in \Lambda_\alpha$ we then see from (4) that

$$\rho_n = |H_{\bar{h}\cdot h^{-1} z^n}| = \mathrm{dist}_{L^\infty}(z^n \cdot \bar{h}/h, H^\infty) \le \mathrm{const}\, n^{-\alpha}.$$

$1 \Rightarrow 2$. By the Helson-Sarason theorem (cf. Sect. 4.1) $w = |Q|^2 e^\varphi$ where $Q$ is a polynomial with its roots on $\mathbb{T}$ and $\varphi = \mathrm{Re}\,\varphi \in VMO$. Let us show that $\varphi \in \Lambda_\alpha$. Set $h_1 = \exp\left(\frac{1}{2}\varphi + \frac{i}{2}\tilde{\varphi}\right)$ and let $h$ be an outer function such that $|h|^2 = w$. Then $h = \mathrm{const}\, Qh_1$ so that (cf. Lemma 4.3)

$$\rho_n = |H_{z^n \bar{h}/h}| = \mathrm{dist}\left(\frac{\bar{h}}{h} z^n, H^\infty\right) = \mathrm{dist}\left(z^{n-N}\frac{\bar{h}_1}{h_1}, H^\infty\right)$$

with $N = \deg Q$. Consequently $\mathrm{dist}\left(z^n \frac{\bar{h}_1}{h_1}, H^\infty\right) \le \mathrm{const}\, n^{-\alpha}$ and (4) allows us to conclude that $P_-\bar{h}_1/h_1 \in \Lambda_\alpha$.

As $\mathrm{clos}(T_{\bar{h}_1/h_1} H^2) = H^2$ (cf. Lemma 3.4) then by Theorem 3.14 (cf. likewise

Example 1 in Sect. 3.4) $\overline{h_1}/h_1 = \exp(-i\tilde{\varphi}) \in \Lambda_\alpha$. Hence (cf., for example, Theorem 3.23) $\tilde{\varphi} \in \Lambda_\alpha$ and finally $\varphi \in \Lambda_\alpha$. □

*Remark.* The above reasoning apparently allows us to conclude that $\lim_{n\to\infty} \rho_n n^\alpha = 0$ iff $w = |Q|^2 \exp(\varphi)$ where $Q$ is a polynomial with its roots on $\mathbb{T}$ and $\varphi$ a real function in $\Lambda_\alpha$.

The proof of Theorem 4.6 below is copied on the proof of Theorem 4.5. The single difference is that in the appropriate places one has to utilize the theorem of restoration of unimodular function of class $P_- B^s_{\infty q}$ (cf. Sect. 3.4, Example 2) along with the description of that class in terms of the sequence $\{\operatorname{dist}_{L^\infty}(z^n f, H^\infty)\}$ (cf. Review of Function Classes).

**Theorem 4.6.** *Let $s > 1$, $1 \le q < \infty$. The following statements are equivalent.*
1. $\sum_{n\ge 1} n^s \rho_n^q < \infty$.
2. *The function $w$ admits the representation*

$$w = |Q|^2 \exp \varphi$$

*where $Q$ is a polynomial with the roots on $\mathbb{T}$ and $\varphi$ a real function in $B_q^{(s+1)q}$.*

In order to formulate the following theorem we need the definition of the Gevrey classes. Let $\alpha > 0$, $M_n = n^{n/\alpha}$. The Gevrey class of order $\alpha$ is the Carleman class $C(C(\mathbb{T}), \{M_n\})$ (cf. Sect. 3.4, Example 5). This class will be denoted by $G_\alpha$. It is well-known (and not hard to verify) that

$$f \in G_\alpha \Leftrightarrow \exists K, \delta > 0 : |\hat{f}(n)| \le K \exp(-\delta |n|^{\frac{\alpha}{1+\alpha}}),\ n \in \mathbb{Z}. \tag{5}$$

**Theorem 4.7.** *Let $0 < \beta < 1$. The following statements are equivalent.*
1. *There exist constants $K$ and $c$ such that*

$$\rho_n \le K \exp(-cn^\beta), \quad n \in \mathbb{Z}_+.$$

2. *The function $w$ admits the representations*

$$w = |Q|^2 \exp(\varphi)$$

*where $Q$ is a polynomial with all the roots on $\mathbb{T}$, $\varphi$ being a real function of Gevrey class $G_{\beta/(1-\beta)}$.*

We list some preliminaries needed for this theorem, leaving the proof itself as an excercise for the reader.

a) Let $f \in L^\infty$. Then

$$P_- f \in G_\alpha \Leftrightarrow \exists K, c > 0 : \operatorname{dist}_{L^\infty}(z^n f, H^\infty) \le K \exp(-cn^{\alpha/(1+\alpha)}),\ n \in \mathbb{Z}_+$$

(this is not hard to get from (5)).

b) Let $\varphi$ be a real function in $L^\infty$. Then

$$\varphi \in G_\alpha \Leftrightarrow \exp(i\varphi) \in G_\alpha$$

(cf. Lemma 3.19, Theorem 3.23 and Theorem 3.18).

c) Let $h$ be an outer function in $H^2$ such that $P_- \bar{h}/h \in G_\alpha$. Then $\bar{h}/h \in G_\alpha$ (cf. Theorem 3.18).

The following theorem shows that an exponential growth of the numbers $\rho_n$ corresponds to an analytic weight.

**Theorem 4.8** (Ibragimov [2]). *Let* $0 < \delta < 1$. *The following statements are equivalent.*

1. *For every number* $\delta_1, \delta < \delta_1$, *there exists a constant* $K$ *such that* $\rho_n \le K\delta_1^n$, $n \in \mathbb{Z}$.
2. $w \in AC_\delta$.

Before proving this theorem we state a characterization of the class $P_- AC_\delta$. Let $f \in L^\infty$. Then

$$P_- f \in AC_\delta \Leftrightarrow \forall \delta_1 > \delta \; \exists K > 0 : \operatorname{dist}_{L^\infty}(z^n f, H^\infty) \le K\delta_1^n, \quad n \in \mathbb{Z}_+. \tag{6}$$

It is easy to deduce this statement from the following characterization of $AC_\delta$ in terms of Fourier coefficients:

$$f \in AC_\delta \Leftrightarrow \forall \delta_1 > \delta \; \exists K > 0 : |\hat{f}(n)| \le K\delta_1^n, \quad n \in \mathbb{Z}.$$

*Proof of Theorem* 4.8. For every polynomial $Q$ holds $|Q|^2 \in AC_\delta$. In fact, it is sufficient to consider the case $Q = z - \lambda$, $\lambda \in \mathbb{C}$. But then

$$|Q(\zeta)|^2 = (\zeta - \lambda)(\bar{\zeta} - \bar{\lambda}) = (\zeta - \lambda)(1/\zeta - \bar{\lambda}), \quad \zeta \in \mathbb{T},$$

and therefore $|Q|^2 \in AC_\delta$.

$1 \Rightarrow 2$. We have $w = |Q|^2 |h|^2$ where $Q$ is a polynomial with the roots on $\mathbb{T}$ and $h$ an outer function such that $h \in H^2$ and $1/h \in H^2$. Condition 1 means that

$$\operatorname{dist}_{L^\infty}\left(z^n \frac{\bar{h}}{h}, H^\infty\right) \le K\delta_1^n, \quad n \in \mathbb{Z}_+.$$

Consequently (cf. (6)) $P_- \bar{h}/h \in AC_\delta$. By Theorem 3.20 $|h|^2 \in AC_\delta$ but then also $w = |Q|^2 |h|^2 \in AC_\delta$.

$2 \Rightarrow 1$. The function $w$ has only a finite number of zeros on $\mathbb{T}$. Let $Q$ be a polynomial with the roots on $\mathbb{T}$ such that $w = |Q|^2 |h|^2$ where $h$ is an outer function invertible in the algebra $H^\infty$. It is then clear that $|h|^2 \in AC_\delta$. In view of Theorem 3.20 $P_- \bar{h}/h \in AC_\delta$ and, consequently, for each number $\delta_1$, $\delta_1 > \delta$, we can find a number $K$ such that

$$\operatorname{dist}_{L^\infty}\left(\frac{\bar{h}}{h} z^n, H^\infty\right) \le K\delta_1^n, \quad n \in \mathbb{Z}_+$$

(cf. (6)). Therefore

$$\begin{aligned}\rho_n &= \operatorname{dist}_{L^\infty}\left(\frac{\bar{Q}}{Q} \cdot \frac{\bar{h}}{h} z^n, H^\infty\right) \\ &= \operatorname{dist}_{L^\infty}\left(\frac{\bar{h}}{h} z^{n - \deg Q}, H^\infty\right) \le K\delta_1^{n - \deg Q}, \quad n \ge \deg Q. \; \square\end{aligned}$$

In conclusion, let us consider a limiting case of rapid mixing, namely the case when the subspaces $H_-^2(w)$ and $z^n H^2(w)$ are orthogonal for all sufficiently large values of $n$.

**Theorem 4.9.** *Let* $m \in \mathbb{Z}_+$. *The following statements are equivalent.*
1. $\rho_n = 0$.
2. $w = |Q|^2$, *$Q$ a polynomial with* $\deg Q \le n$.

*Proof.* $2 \Rightarrow 1$. As $|\zeta - \lambda|^2 = |1 - \zeta\bar{\lambda}|^2$ for $\lambda \in \mathbb{D}$, $\zeta \in \mathbb{T}$, we may without loss of generality assume that the polynomial $Q$ has no root in $\mathbb{D}$. Therefore $Q$ is also an outer function and consequently $z^n \frac{\bar{Q}}{Q} \in H^\infty$. But then $\rho_n = \operatorname{dist}_{L^\infty}\left(z^n \frac{\bar{Q}}{Q}, H^\infty\right) = 0$.

$1 \Rightarrow 2$. Let $h$ be an outer function with $|h|^2 = w$. We show that $\widehat{|h^2|}(k)$ for $k > n$. It is clear that $\widehat{|h^2|}(k) = \widehat{|h^2|}(-k)$. We now get for $k > n$

$$\widehat{|h^2|}(-k) = \int_{\mathbb{T}} z^k |h^2|\, dm = \int_{\mathbb{T}} z^k \frac{\bar{h}}{h} h^2\, dm = 0$$

as by Statement 1 $z^n \bar{h}/h \in H^\infty$. □

This theorem is in the book Ibragimov-Rozanov [1].

**4.3. Positive Angle Between the Subspaces $H^2_-(w)$ and $z^n H^2(w)$.** In this Section we give conditions on the weight function $w$ which are necessary and sufficient for the angle between the subspace $H^2(w)$ and $z^n H^2(w)$ to be non-zero. Recall that this is equivalent to the inequality $\rho_n < 1$. The following theorem was for $n = 0$ established by Helson-Szegö [1] and for arbitrary $n$ in $\mathbb{Z}_+$ by Helson-Sarason [1]. We offer here a new proof of it.

**Theorem 4.10.** *Let* $n \in \mathbb{Z}_+$. *The following statements are equivalent.*
1. $\rho_n < 1$.
2. *The function $w$ admits the representation*

$$w = |Q|^2 \exp(\varphi + \tilde{\psi}),$$

*where $Q$ is a polynomial with all roots on $\mathbb{T}$, $\deg Q \le n$, $\varphi, \psi$ being real functions in $L^\infty$, $\|\psi\|_{L^\infty} \le \pi/2$.*

*Proof.* Let us first deduce Statement 1 from Statement 2. Consider the outer function $h = Q \exp((\varphi + \tilde{\psi} + i\tilde{\varphi} - i\psi)/2)$ with $w = |h|^2$. We have

$$\frac{\bar{h}}{h} = \frac{\bar{Q}}{Q} \exp(-i(\tilde{\varphi} + \psi)) = \lambda z^{-\deg Q} \exp(-i(\tilde{\varphi} + \psi)), |\lambda| = 1.$$

By the Devinatz-Widom Theorem (cf. Appendix 4, Sect. 38) the Toeplitz operator $T_{\exp(-i(\tilde{\varphi}+\psi))}$ is invertible. Consequently $|H_{\exp(-i(\tilde{\varphi}+\psi))}| < 1$ (cf. Appendix 4, Sect. 36). Consequently

$$\rho_n = \|H_{z^n \bar{h}/h}\| = \|H_{z^n - \deg Q_{\exp(-i(\tilde{\varphi}+\psi))}}\| < 1,$$

because $n \ge \deg Q$ and $\|H_{\exp(-i(\tilde{\varphi}+\psi))}\| < 1$.

Next let us show that 2 follows from 1. By assumption $|H_{z^n \bar{h}/h}| < 1$. Let $m$ be a negative number such that $|H_{z^m \bar{h}/h}| < 1$ but $|H_{z^{m-1} \bar{h}/h}| = 1$. There does exist such a number, because apparently $\|H_{\bar{z}\bar{h}/h} h\|_2 = \|h\|_2$ and, consequently, $|H_{\bar{z}\bar{h}/h}| = 1$.

The Toeplitz operator $T_{z^m\bar{h}/h}$ is left invertible while the operator $T_{z^{m-1}\bar{h}/h}$ is not (cf. Appendix 4, Sect. 36). In view of Lecture VIII, Sect. 6, d) we conclude from this that $T_{z^m\bar{h}/h}$ is invertible.

Let $h_1$ be an outer function such that $\bar{h}_1/h_1 = z^m\bar{h}/h$. Then by the Devinatz-Widom Theorem (cf. Appendix 4, Sect. 38) $\bar{h}_1/h_1 = \exp(i(\psi - \tilde{\varphi} - c)$ where $c \in \mathbb{R}$, $\varphi$, $\psi$ real in $L^\infty$, $\|\varphi\|_{L^\infty} < \pi/2$. Hence $|h_1|^2 = \exp(\varphi + \tilde{\psi})$. In order to finish the proof we have to invoke Lemma 4.4 which shows that there exists a polynomial $Q$ of degree $m$ with all roots on $\mathbb{T}$ such that $h = Qh_1$ so that finally $w = |h|^2 = |Q|^2 \exp(\varphi + \tilde{\psi})$. □

*Remark* 1. The representation of $w$ in the form $|Q|^2 \exp(\varphi + \tilde{\psi})$, $\varphi, \psi \in L^\infty$, $\|\psi\|_{L^\infty} < \pi/2$ is unique up to a multiplicative factor. This is seen exactly as in the remark following Theorem 4.1.

*Remark* 2. A non-negative function $F$ has the representation

$$F = \exp(\varphi + \tilde{\psi}),\ \varphi, \psi \in L^\infty,\ \|\psi\|_{L^\infty} < \pi/2$$

iff $F$ satisfies Muckenhoupt's $(A_2)$ condition, i.e.

$$\sup_I \left(\frac{1}{m(I)} \int_I F\,dm\right)\left(\frac{1}{m(I)} \int_I \frac{1}{F}\,dm\right) < \infty,$$

where the supremum extends over all arcs $I$ on $\mathbb{T}$. For the proof of this result see Hunt-Muckenhoupt-Wheeden [1], Coifman-Fefferman [1].

Consequently Theorem 4.10 may be reformulated as follows. The angle between $z^n H^2(w)$ and $H^2_-(w)$ is non-zero iff $w$ has the representation $w = |Q|^2 F$ where $Q$ is a polynomial with all its roots on $\mathbb{T}$, $\deg Q \le n$, and $F$ a function satisfying Muckenhoupt's $(A_2)$ condition.

**4.4. $\mathfrak{S}_p$-Regularity.** In this Section we obtain a characterization of the $\mathfrak{S}_p$-regular processes in terms of the spectral density. To this end we will use the description of the Hankel operators of class $\mathfrak{S}_p$ (cf. Theorem 1.1), the theorem on restoration of unimdular functions in the classes $B_p^{1/p}$ and, finally, the description of the arguments of unimodular functions of the class $B_p^{1/p}$. Recall that we obtained this description in Sect. 3.5 on the basis of the theorem of best approximation by analytic functions in $B_p^{1/p}$.

Moreover we get a description of those processes for which the norms $\|\mathscr{P}_0\mathscr{P}^n\|_{\mathfrak{S}_p}$ falls off with a given rate.

Recall that $\mathscr{P}_0$ is orthogonal projection in $L^2(w)$ onto $H^2_-(w)$ and $\mathscr{P}^n$ orthogonal projection in $L^2(w)$ onto $z^n H^2(w)$.

**Theorem 4.11.** *Let $1 \le p < \infty$. The following statements are equivalent.*

1. *$\mathscr{P}_0\mathscr{P}^m \in \mathfrak{S}_p$ for some $m$ in $\mathbb{Z}_+$.*
2. *$\mathscr{P}_0\mathscr{P}^m \in \mathfrak{S}_p$ for any $m$ in $\mathbb{Z}_+$.*
3. *The function $w$ admits the representation*

$$w = |Q|^2 \exp\varphi,$$

*where $Q$ is a polynomial with the roots on $\mathbb{T}$, $\varphi$ a real function in $B_p^{1/p}$.*

Recall that the $\mathfrak{S}_\infty$-regular processes were described in the Helson-Sarason theorem (Theorem 4.1).

*Proof of Theorem* 4.11. In view of Lemma 2.15 $P_0 P^m \in \mathfrak{S}_p$ iff $H_{z^m \bar{h}/h} \in \mathfrak{S}_p$. This is equivalent to $P_- \bar{h}/h \in B_p^{1/p}$ (cf. Theorem 1.1). Thus Statements 1 and 2 are equivalent.

Assume that 3 is fulfilled. Then

$$\frac{\bar{h}}{h} = \frac{\bar{Q}}{Q} \exp(-i\tilde{\varphi}) = \operatorname{const} \bar{z}^{\deg Q} \exp(-i\tilde{\varphi}).$$

By Corollary 3.24 $\exp(-i\varphi) \in B_p^{1/p}$ whence $P_- \bar{h}/h \in B_p^{1/p}$, as $P_- B_p^{1/p} \subset B_p^{1/p}$. But this again is equivalent to $H_{\bar{h}/h} \in \mathfrak{S}_p$. Thus $\mathscr{P}_0 \mathscr{P}^0 \in \mathfrak{S}_p$.

Finally assume that condition 2 is fulfilled. By Theorem 4.1 $w = |Q|^2 \exp \varphi$ where $\varphi$ is a real function in $VMO$. We have

$$\frac{\bar{h}}{h} = \frac{\bar{Q}}{Q} \exp(-i\tilde{\varphi}) = \operatorname{const} z^{-\deg Q} \exp(-i\tilde{\varphi}).$$

In view of condition 2 $P_- \bar{h}/h \in B_p^{1/p}$. Consequently $P_- \exp(-i\varphi) \in B_p^{1/p}$. But then in view of Corollary 3.5 $\exp(-i\tilde{\varphi}) \in B_p^{1/p}$, as $\exp(-i\tilde{\varphi}) \in VMO$.

The condition $\varphi \in VMO$ entails that ind $T_{\exp(-i\tilde{\varphi})} = 0$ and therefore in view of Corollary 3.24 $\tilde{\varphi} \in B_p^{1/p}$ and, the operator of passing to the harmonic conjugate being continuous in the space $B_p^{1/p}$, likewise $\varphi \in B_p^{1/p}$. □

The following theorem describes those processes for which the norms $\|\mathscr{P}_0 \mathscr{P}^n\|_{\mathfrak{S}_p}$ decrease as a power.

**Theorem 4.12.** *Let* $1 \le p < \infty$, $\alpha > 0$. *The following Statements are equivalent.*

1. $\|\mathscr{P}_0 \mathscr{P}^n\|_{\mathfrak{S}_p} \le \operatorname{const} n^{-\alpha}$, $n > 0$.
2. *The function w admits the representation*

$$w = |Q|^2 \exp \varphi$$

*where Q is a polynomial with all roots on* $\mathbb{T}$, $\varphi$ *a real function in* $B_{p,\infty}^{\alpha+1/p}$.

Theorem 4.12 is proved by the standard scheme using the restoration theorem for unimodular functions of class $B_{p,\infty}^{\alpha}$ along with the following lemma, where $S^*$ the inverse shift operator, $S^* f = \dfrac{f - f(0)}{z}$.

**Lemma 4.13.** *Let* $f \in P_+ B_p^{1/p}$. *The following statements are equivalent.*

a. $f \in B_{p,\infty}^{\alpha+1/p}$,

b. $\|(S^*)^m f\|_{B_p^{1/p}} \le \operatorname{const} m^{-\alpha}$, $m > 0$.

*Proof of the lemma.* Assume that Condition a is fulfilled. In order to establish Statement b it suffices to prove that $\|(S^*)^{2^n} f\|_{B_p^{1/p}} \le \operatorname{const} 2^{-n\alpha}$, $n > 0$. Condition a means that

$$\|f * W_n\|_{L^p} \le \operatorname{const} 2^{-n(\alpha+1/p)}.$$

It is then easy to see that

$$\|(S^*)^{2^n}f * W_m\|_{L^p} \le \text{const}\, 2^{-m(\alpha+1/p)}$$

for $m > n$. It is also obvious that

$$\|(S^*)^{2^n}f * W_m\|_{L^p} \le \text{const}\, 2^{-n(\alpha+1/p)}$$

for $m \le n$. We thus obtain

$$\begin{aligned}
\|(S^*)^{2^n}f\|^p_{B^{1/p}_p} &\le \text{const} \sum_{m\ge 0} 2^m \|(S^*)^{2^n}f * W_m\|^p_{L^p} \\
&= \text{const} \sum_{m=0}^{n} 2^m \|(S^*)^{2^n}f * W_m\|^p_p + \text{const} \sum_{m>n} 2^m \cdot \|(S^*)^{2^n}f * W_m\|^p_p \\
&\le \text{const} \sum_{m=0}^{n} 2^m \cdot 2^{-np(\alpha+1/p)} + \text{const} \sum_{m>n} 2^m \cdot 2^{-mp(\alpha+1/p)} \\
&\le \text{const}\, 2^{-np\alpha} + \text{const}\, 2^{-np\alpha} \\
&\le \text{const}\, 2^{-np\alpha}.
\end{aligned}$$

Consequently $\|(S^*)^{2^n}f\|^p_{B^{1/p}_p} \le \text{const}\, 2^{-n\alpha}$.

Conversely let Statement b be fulfilled. Then we get

$$\begin{aligned}
\|f * W_n\|^p_p &= \|(S^*)^{2^{n-2}}f * ((S^*)^{2^{n-2}}W_n)\|^p_p \\
&= \|(S^*)^{2^{n-2}}f * (W_{n-2} + W_{n-1} + W_n + W_{n+1}) * ((S^*)^{2^{n-2}}W_n)\|^p_p \\
&\le \text{const}\, \|(S^*)^{2^{n-2}}f * (W_{n-2} + W_{n-1} + W_n + W_{n+1})\|^p_p \\
&\le \text{const}\, 2^{-n} \|(S^*)^{2^{n-2}}f\|^p_{B^{1/p}_p} \\
&\le \text{const}\, 2^{-n} \cdot 2^{-np\alpha} \\
&= \text{const}\, 2^{-n(1+p\alpha)}.
\end{aligned}$$

Hence $f \in B^{\alpha+1/p}_{p\infty}$. □

In a similar way one may prove the following result.

**Theorem 4.14.** *Let* $1 \le p < \infty$, $1 \le q < \infty$, $s > -1$. *The following statements are equivalent.*

1. $\sum_{n\ge 1} n^s \|\mathscr{P}_0\mathscr{P}^n\|^q_{\mathfrak{S}_p} < \infty$.
2. *The function w admits the representation*

$$W = |Q|^2 \exp\varphi$$

*where Q is a polynomial with the roots on* $\mathbb{T}$, $\varphi$ *a real function of class* $B^{(s+1)/q+1/p}_{p,q}$. □

*Remark.* It is not hard to see that an exponential decrease of the norms $\|\mathscr{P}_0\mathscr{P}^n\|_{\mathfrak{S}_p}$ corresponds to analyticity of the function $w$.

## Review of Function Classes

Let us first define the scale of *Hölder spaces* $\Lambda_\alpha$, $\alpha > 0$. Let $0 < \alpha < 1$. Then

$$\Lambda_\alpha = \{f \in C(\mathbb{T}) : |f(\zeta_1) - f(\zeta_2)| \leq C_f |\zeta_1 - \zeta_2|, \zeta_1, \zeta_2 \in \mathbb{T}\}.$$

By $\Lambda_1$ we denote the *Zygmund class*

$$\Lambda_1 = \{f \in C(\mathbb{T}) : |f(e^{i(x+t)}) + f(e^{i(x-t)}) - 2f(e^{ix})| \leq C_f |t|, x, t \in \mathbb{R}\},$$

not the Lipschitz class. For $\alpha > 1$ we set

$$\Lambda_\alpha = \{f \in C^{(n)}(\mathbb{T}) : f^{(n)} \in \Lambda_{\alpha-n}\},$$

where $n$ is an integer such that $n < \alpha \leq n + 1$. The space $\Lambda_\alpha$ equipped with the natural norm is a non-separable Banach space. The closure of all trigonometrical polynomials in $\Lambda_\alpha$ will be denoted by $\lambda_\alpha$.

Let $s > 0$ and $1 \leq p, q \leq \infty$. We define the *Besov class* $B^s_{pq}$ by the formula

$$B^s_{pq} = \left\{ f \in L^p : \int_{-\pi}^{\pi} \| \Delta^n_t f \|^q_{L^p} / |t|^{1+sq}\, dt < \infty \right\}, \quad q < \infty,$$

$$B^s_{p\infty} = \left\{ f \in L^p : \sup_{t \neq 0} \| \Delta^n_t f \|_{L^p} / |t|^s < \infty \right\},$$

where $n$ is an integer such that $s < n$; $\Delta^n_t = \Delta_t \Delta^{n-1}_t$, $(\Delta_t f)(e^{ix}) = f(e^{i(x+t)}) - f(e^{ix})$. It is well-known that this definition is independent of the choice of $n$, $n > s$. In what follows we set $B^s_p \stackrel{\text{def}}{=} B^s_{pp}$.

Notice that $\Lambda_\alpha = B^\alpha_{\infty\infty}$.

Next we give another description of the Besov classes in terms of special kernels $W_n$, $n \in \mathbb{Z}$. If $n > 0$ then $W_n$ is a trigonometric polynomial such that $\hat{W}_n(2^n) = 1$, $\hat{W}_n \equiv 0$ outside $(2^{n-1}, 2^{n+1})$, $\hat{W}_n$ linear on $[2^{n-1}, 2^n]$ and $[2^n, 2^{n+1}]$. If $n < 0$ we set $W_n = \overline{W_{-n}}$. We define $W_0(\zeta) = \bar{\zeta} + 1 + \zeta$. It is plain that

$$f = \sum_{n \in \mathbb{Z}} f * W_n$$

for every trigonometric polynomial. If we represent $W_n$, $n > 0$, as a linear combination of two translates of the Fejér kernel we see that $\| W_n \|_{L^1} \leq 3/2$, $n \in \mathbb{Z}$.

We have now the following description of Besov classes:

$$f \in B^s_{pq} \Leftrightarrow \left( \left( \sum_{n \in \mathbb{Z}} 2^{|n|s} \| W_n * f \|_{L^p} \right)^q \right)^{1/q} < \infty, \quad q < \infty,$$

$$f \in B^s_{p\infty} \Leftrightarrow \sup_{n \in \mathbb{Z}} 2^{|n|s} \| W_n * f \|_{L^p} < \infty.$$

With the aid of these formulae we may define $B^s_{pq}$ for $s \leq 0$ too. In this case $B^s_{pq}$ consists of distributions of course.

With the aid of this representation it is easy to describe the dual of a Besov class $B^s_{pq}$, $q < \infty$. Indeed $(B^s_{pq})^* = B^{-s}_{p'q'}$, $p' = \dfrac{p}{p-1}$, $q' = \dfrac{q}{q-1}$. Here the duality between $B^s_{pq}$ and $B^{-s}_{p'q'}$ is given by the formula

$$(f, g) = \sum_{n \in \mathbb{Z}} \hat{f}(n)\hat{g}(n),$$

where $f$ is a trigonometric polynomial and $g \in B^{-s}_{p'q'}$.

On the set $\mathscr{D}'$ of all distributions on $\mathbb{T}$ we define now the *operator of fractional integration* $I_\alpha$, $\alpha \in \mathbb{R}$:

$$I_\alpha f = \sum_{n \in Z} (1 + |n|)^{-\alpha} \hat{f}(n) z^n.$$

The following relationship holds true: $I_\alpha B^s_{pq} = B^{s+\alpha}_{pq}$. Let us consider the subspace

$$(B^s_{pq})_A = P_+ B^s_{pq} = \left\{ f \in B^s_{pq} : f = \sum_{n \geq 0} \hat{f}(n) z^n \right\}$$

of the Besov space consisting of analytic functions. This subspace may be described as the set of all analytic functions $f$ in $\mathbb{D}$ such that

$$\int_0^1 (1 - r)^{nq - sq - 1} \| f_r^{(n)} \|^q_{L^p} \, dr < \infty, \quad q < \infty,$$

$$\sup_{0<r<1} (1 - r)^{n-s} \| f_r^{(n)} \|_{L^p} < \infty, \quad q = \infty,$$

where, as above, $n$ is an integer subject to the inequality $n > s$, further $f_r^{(n)}(\zeta) = f^{(n)}(r\zeta)$, $\zeta \in \mathbb{T}$.

The *Sobolev classes* $W^n_p$, $n \in \mathbb{Z}_+$, were defined by Sobolev as $\{f : f^{(n)} \in L^p\}$. With the aid of the operators $I_\alpha$ we now define the *Bessel potential spaces* $\mathscr{L}^s_p \stackrel{\text{def}}{=} I_s(L^p)$, $1 \leq p < \infty$, $s \in \mathbb{R}$, generalizing the Sobolev spaces.

It is well-known that $B^s_p \subset \mathscr{L}^s_p$ for $1 \leq p \leq 2$ and $\mathscr{L}^s_p \subset B^s_p$ for $p \geq 2$.

The proofs of the results described above can be located in the books (S.M.) Nikol'skiĭ [1], Bergh-Löfström [1], Stein [1], Triebel [1].

We require also the following characterization of the function class $P_- B^s_{pq}$, $s > 0$.

**Theorem.** *Let $f \in P_- L^p$, $1 \leq p \leq \infty$. Then $f \in P_- B^s_{pq}$ iff*

$$\sum_{n \geq 0} (2^{ns} \operatorname{dist}_{L^p}(z^{2^n} f, H^p))^q < \infty, \quad q < \infty,$$

$$\sup_{n \geq 0} 2^{ns} \operatorname{dist}_{L^\infty}(z^{2^n} f, H^p) < \infty, \quad q = \infty.$$

*The proof* will for simplicity be given in the case of the space $\Lambda_\alpha = B^\alpha_{\infty\infty}$, $\alpha > 0$ (the general case differs from this only by somewhat longer computations). Assume first that

$$\sup_{n \geq 1} 2^{n\alpha} \operatorname{dist}_{L^\infty}(z^{2^n} f, H^\infty) < \infty.$$

Let us observe that $f * W_{-n} = W_{-n} * (f - \bar{z}^{2^{n-1}} g)$ for any $n > 0$. Therefore $\| f * W_{-n} \|_{L^\infty} \leq \frac{3}{2} \cdot \operatorname{dist}_{L^\infty}(f, \bar{z}^{2^{n-1}} H^\infty)$.

Finally let $P_- f \in \Lambda_\alpha$. It is then clear that

$$\operatorname{dist}_{L^\infty}(f, \bar{z}^{2^{n-1}} H^\infty) \leq \sum_{k \geq n} \| f * W_{-k} \|_{L^\infty} \leq \text{const} \cdot 2^{-n\alpha}. \ \square$$

# Bibliography

Adamyan, V. M., Arov, D. Z., Kreĭn, M. G.
[1] Infinite Hankel block matrices and some related continuation problems. Izv. Akad. Nauk Armjan. SSR Ser. Mat. **6**, 87–112 (1971) [Russian]
[2] On infinite Hankel matrices and the generalized problems of Carathéodory-Fejér and F. Riesz. Funkcional. Anal. i Priložen. **2**, 1–19 (1968) [Russian]
[3] Infinite Hankel matrices and the generalized problems of Carathéodory-Fejér and I. Schur. Funkcional. Anal. i Priložen. **2**, 1–17 (1968) [Russian]
[4] On bounded operators commuting with contractions of class $C_{00}$ of unit non-unitarity class. Funkcional. Anal. i Priložen. **3**, 86–87 (1969) [Russian]
[5] Analytic properties of Schmidt pairs of an Hankel operator and the generalized Schur-Takagi problem. Mat. Sb. **85 (128)**, 33–73 (1971) [Russian]

Agmon, S.
[1] Sur un problème de translations. C. R. Acad. Sci. Paris **229**, 540–542 (1949)

Ahern, P. R., Clark, D. N.
[1] On functions orthogonal to invariant subspaces. Acta Math. **124**, 191–204 (1970)
[2] Radial limits and invariant subspaces. Amer. J. Math. **92**, 332–342 (1970)

Ahieser, N. I.
[1] Theory of approximation (2nd ed). Moscow: Nauka 1965 [Russian] (German translation (2nd ed): Berlin: Akademie-Verlag 1967; English translation (1st ed): New York: Frederick Ungar 1956)
[2] The classical moment problem. Moscow: GIFML 1961 [Russian] (English translation: Edinburgh-London: Oliver & Boyd 1965)

Ahieser, N. I., Glazman, I. M.
[1] Theory of linear operators in Hilbert space. Moscow: Nauka 1966 [Russian] (German translation (2nd ed): Berlin, Akademie-Verlag, 1960); English translation: New York: Frederick Ungar 1961)

Airapetyan, G. M.
[1] Multiple interpolation and basisness for some biorthogonal systems of rational functions in the Hardy classes $H_p$. Izv. Akad. Nauk SSSR Ser. Mat. **12**, 267–277 (1977) [Russian]

Alexander, H., Taylor, B. A., Williams, D. L.
[1] The interpolating sets for $A^\infty$. J. Math. Anal. Appl. **36**, 556–568 (1971)

Arocena, R., Cotlar, M., Sadosky, C.
[1] Weighted inequalities in $L^2$ spaces and lifting properties. Adv. in Math., Supplementary Studies **7A**, 95–128 (1981)

Arsene, Gr., Gheondea, A.
[1] Completing matrix contractions. J. Operator Theory **7**, 179–189 (1982)

Arveson, W. B.
[1] Interpolation problems in nest algebras. J. Functional Analysis **20**, 208–233 (1975)

Avdonin, S. A.
[1] On the question of Riesz bases of exponential functions in $L^2$. Vestnik Leningrad. Univ. Mat. Meh. Astronom. **13**, 5–12 (1974) [Russian]

Axler, S.
[1] Factorization of $L^\infty$ functions. Ann. Math. **106**, 567–572 (1977)

Axler, S., Berg, I. D., Jewell, N., Shields, A.
[1] Approximation by compact operators and the space $H^\infty + C$. Ann. Math. **109**, 601–612 (1979)

Axler, S., Chang. S.-Y. A., Sarason, D.
[1] Products of Toeplitz operators. Integral Equations Operator Theory **1**, 285–309 (1978)

Banach, S.
[1] Théorie des opérations linéaires. Warsaw, 1932
[2] Über einige Eigenschaften der lacunären trigonometrischen Reihen. Studia Math. **2**, 207–220 (1930)

Barbey, K., König, H.
[1] Abstract analytic function theory and Hardy algebras. Lecture Notes in Mathematics 593. Berlin-Heidelberg-New York: Springer 1977

Bari, N. K.
[1] Biorthogonal systems and bases in Hilbert space. Učen. Zap. Moskov. Univ. **4**, 69–107 (1951) [Russian]
[2] Trigonometric series. Moscow: GIFML 1961 [Russian] (English translation: Oxford etc.: Pergamon Press 1964)

Barria, J., Halmos, P. R.
[1] Asymptotic Toeplitz operators. Pre-print

Basor, E. L.
[1] A localization theorem for Toeplitz determinants. Indiana Univ. Math. J. **28**, 975–983 (1979)

Basor, E. L., Helton, J. W.
[1] A new proof of the Szegö limit theorem and new results for Toeplitz operators with discontinuous symbol. J. Operator Theory **3**, 23–39 (1980)

Bergh, J., Löfström, J.
[1] Interpolation spaces. An introduction. Grundlehren 226. Berlin-Heidelberg-New York: 1976

Berkovici, H., Foiaş, C., Sz.-Nagy, B.
[1] Compléments à l'étude des opérateurs de classe $C_0$. III. Acta Sci. Math. (Szeged) **37**, 313–322 (1975)

Berkovici, H., Voiculescu, D.
[1] Tensor operations on characteristic functions of $C_0$ contractions. Acta Sci. Math. (Szeged) **39**, 205–231 (1977)

Bernard, A.
[1] Une caractérisation de $C(X)$ parmis les algèbres de Banach. C. R. Acad. Sci. Paris **287**, A634–A635 (1968)

Beurling, A.
[1] On two problems concerning linear operators in Hilbert space. Acta Math. **81**, 239–255 (1949)

Beurling, A., Malliavin, P.
[1] On the closure of characters and the zeros of entire functions. Acta Math. **118**, 79–93 (1967)

Beurling, P., Carleson, L.
[1] Research on interpolation problems. Preprint, Uppsala, 1962

Birman, M. S., Solomyak, M. Z.
[1] Estimates for singular numbers of integral operators. Uspehi Mat. Nauk **32**, 17–84 (1977) [Russian]

Bishop, E.
[1] Spectral theory for operators in Banach space. Trans. Amer. Math. Soc. **86**, 414–445 (1957)

Bochner, S., Bohnenblust, H. F.
[1] Analytic functions with almost periodic coefficients. Ann. Math. **35**, 152–161 (1934)

Bonsall, F. F., Power, S. C.
[1] A proof of Hartman's theorem on compact operators. Math. Proc. Cambridge Philos. Soc. **78**, 447–450 (1975)

Böttcher, A., Silbermann, B.
[1] The asymptotic behavior of Toeplitz determinants for generating functions with zeros of integral orders. Math. Nachr. **102**, 79–105 (1981)
[2] Notes on the asymptotic behavior of block Toeplitz matrices and determinants. Math. Nachr. **98** (1980), 183–210
[3] Invertibility and asymptotics of Toeplitz matrices. Mathematische Forschung 17. Berlin: Akademie-Verlag 1983

Brodskiĭ, M. S.
[1] On Jordan blocks of infinite operators. Dokl. Akad. Nauk SSSR **111**, 926–929 (1956) [Russian]
[2] On a problem of I. M. Gel'fand. Uspehi Mat. Nauk **12 : 2**, 129–132 (1957) [Russian]
[3] On the unicellularity of the integration operator and a theorem by Titchmarsh. Uspehi Mat. Nauk **20 : 5**, 189–192 (1965) [Russian]
[4] On operators with nuclear imaginary components. Acta Sci. Math. (Szeged) **27**, 147–155 (1966) [Russian]
[5] On the triangular representation of completely continuous operators with a one point spectrum. Uspehi Mat. Nauk **16 : 1**, 135–141 (1961) [Russian]
[6] Triangular and Jordan representations of linear operators. Moscow: Nauka 1969 [Russian] (English translation: Providence: American Mathematical Society, 1971)

Brodskiĭ, M. S., Kisilevskiĭ, G. E.
[1] A criterion for unicellularity of Volterra operators with nuclear imaginary part. Izv. Akad. Nauk SSSR Ser. Mat. **30**, 1213–1228 (1966) [Russian]

Brodskiĭ, M. S., Livšic, M. S.
[1] Spectral analysis of non-selfadjoint operators and intermediate systems. Usp. Mat. Nauk **13 : 1**, 3–85 (1958) [Russian]

Brown, S.
[1] Some invariant subspaces for subnormal operators. Integral Equations Operator Theory **1**, 310–333 (1978)

Brown, A., Halmos, P. R.
[1] Algebraic properties of Toeplitz operators. J. Reine Angew. Math. **213**, 89–102 (1963/64)

Budyanu M. S., Gohberg, I. C.
[1] On the factorization problem in abstract Banach algebras. I, II. Mat. Issledov. (Kišinev) **2 : 2**, 25–51, **2 : 3**, 3–19 (1967) [Russian]

Bunce, J.
[1] The joint spectrum of commuting nonnormal operators. Proc. Amer. Math. Soc. **29**, 499–504 (1971)

Butz, J. R.
[1] Compact Hankel operators. Pre-print, Univ. of Oklahoma
[2] $s$-numbers of Hankel matrices. J. Functional Analysis **15**, 297–305 (1974)

Cambern, M.
[1] Invariant subspaces and extremum problems in spaces of vectorvalued functions. J. Math. Anal. Appl. **57**, 290–297 (1977)

Carathéodory, C.
[1] Funktionentheorie. Zweiter Band. Basel: Birkhäuser 1950

Carleman, T.
[1] L'intégrale de Fourier et questions qui s'y rattachent. Uppsala 1944
[2] Les fonctions quasi analytiques. Gauthier-Villars: Paris 1926

Carleson, L.
[1] An interpolation problem for bounded analytic functions. Amer. J. Math. **80**, 921–930 (1958)
[2] Interpolations by bounded analytic functions and the corona problem. Ann. Math. **76**, 547–559 (1962)
[3] The corona theorem. In: Lecture Notes in Mathematics 118, pp. 121–132. Berlin-Heidelberg-New York: Springer 1970

Carleson, L., Jacobs, S.
[1] Best uniform approximation by analytic functions. Ark. Mat. **10**, 219–229 (1972)

Carleson, L., Jones, P.
[1] Weighted norm inequalities and a theorem of Koosis. Preprint, Mittag-Leffler Institute, 1981

Chalmers, B. L.
[1] Some interpolation problems in Hilbert spaces. Michigan Math. J. **18**, 41–50 (1971)

Chang S.-Y. A.
[1] A characterization of Douglas subalgebras. Acta Math. **137**, 81–89 (1976)
[2] Structure of subalgebras between $L^\infty$ and $H^\infty$. Trans. Amer. Math. Soc. **227**, 319–332 (1977)
[3] A generalized area integral estimate and applications. Amer. J. Math. **99**, 530–578 (1977)

Chang, S.-Y. A., Marshall D. E.
[1] Some algebras of bounded analytic functions containing the disc algebra. In: Lecture Notes in Mathematics 604 (Banach spaces of analytic functions), pp. 12–20. Berlin-Heidelberg-New York: Springer 1977

Chow, K. N., Protas, D.
[1] The maximal ideal spaces of bounded analytic functions with finite Dirichlet integrals. Arch. Math. (Basel) **31**, 298–301 (1978)

Clancey, K.
[1] Seminormal operators. Lecture Notes in Mathematics 742. Berlin-Heidelberg-New York: Springer 1979

Clark, D. N.
[1] On interpolating sequences and the theory of Hankel and Toeplitz matrices. J. Functional Analysis **5**, 247–258 (1970)
[2] On the spectra of bounded Hermitian Hankel matrices. Amer. J. Math. **90**, 627–656 (1968)
[3] One dimensional perturbations of restricted shifts. J. Anal. Math. **25**, 169–191 (1972)
[4] On the point spectrum of a Toeplitz operator. Trans. Amer. Math. Soc. **126**, 251–266 (1967)
[5] On a similarity theory of rational Toeplitz operators. J. Reine Angew. Math. **320**, 6–31 (1980)
[6] On Toeplitz operators with loops. J. Operator Theory **4**, 37–54 (1980)
[7] On Toeplitz operators with loops, II. J. Operator Theory **7**, 109–123 (1982)
[8] On the structure of rational Toeplitz operators. In: Contributions to analysis and geometry (Baltimore, Md., 1980), pp. 63–72 Baltimore: John Hopkins Univ. Press 1981

Coburn. L. A.
[1] Weyl's theorem for nonnormal operators. Michigan Math. J. **13**, 285–288 (1966)
[2] The $C^*$-algebra generated by an isometry. I. Bull. Amer. Math. Soc. **73**, 722–726 (1967)
[3] The $C^*$-algebra generated by an isometry. II. Trans. Amer. Math. Soc. **137**, 211–217 (1969)

Coburn, L. A., Douglas, R. G.
[1] Translation operators on the half-line. Proc. Nat. Acad. Sci. U.S.A. **62**, 1010–1013 (1969)

Coburn, L. A., Douglas, R. G., Schaeffer, D. G., Singer, I. M.
[1] On $C^*$-algebras of operators on a half-space. Inst. Hautes Études Sci. Publ. Math. **40**, 69–79 (1971)

Coifman, R. R., Fefferman, C.
[1] Weighted norm inequalities for maximal functions and singular integrals. Studia Math. **51**, 241–250 (1974)

Collingwood, E. F., Lohwater, A. J.
[1] The theory of cluster sets. Cambridge Tracts in Math. and Math. Phys. 56. Cambridge: University Press 1966

Colojoara L., Foiaş, C.
[1] The theory of generalized spectral operators. New York: Gordon and Breach 1968

Davidson, K. R.
[1] On operators commuting with Toeplitz operators modulo the compact operators. J. Functional Analysis **24**, 291–302 (1977)

Davis, Ch.
[1] Some dilation and representation theorems. In: Proc. of the Second Internat. Symp. in West Africa on Functional Analysis and its applications. Kumasi, Ghana, 1979, pp. 159–182

Devinatz A.
[1] Toeplitz operators on $H^2$ spaces. Trans. Amer. Math. Soc. **112**, 304–317 (1964)
[2] On Wiener-Hopf operators. In: Functional Analysis. Proc. Conf., Irvine, Calif., 1966, pp. 81–118. London: Academic Press; Washington, D.C.: Thompson 1967

Devinatz, A., Shinbrot, M.
[1] General Wiener-Hopf operators. Trans. Amer. Math. Soc. **145**, 467–494 (1969)

Diestel, J., Uhl, J. J.
[1] Vector measures. Mathematical Surveys 15. Providence: Amer. Math. Soc. 1978

Dixmier, J.
[1] Les opérateurs permutables à l'opérateur intégral. Portugal. Math. **8**, 73–84 (1949)
[2] Les $C^*$ algèbres et leurs représentations. Deuxième éd. Cahiers scientifiques, Fasc. 29. Paris: Gauthier-Villars 1969

Domar, Y.
[1] On the analytic transform of bounded linear functionals on certain Banach algebras. Studia Math. **53**, 203–224 (1975)

Donoghue, W. F.
[1] The lattice of invariant subspaces of a completely continuous quasinilpotent transformation. Pacific J. Math. **7**, 1031–1035 (1957)

Douglas, R. G.
[1] Canonical models. In: Mathematical Surveys 13, pp. 161–218. Providence: Amer. Math. Soc. 1974
[2] Banach algebra techniques in operator theory. New York: Academic Press 1972
[3] Banach algebra techniques in the theory of Toeplitz operators. CBMS, n° 15. Providence: American Mathematical Society 1973
[4] Toeplitz and Wiener-Hopf operators in $H^\infty + C$. Bull. Amer. Math. Soc. **74**, 895–899 (1968)
[5] On the spectrum of Toeplitz and Wiener-Hopf operators. In: Abstract spaces and approximation. Proc. Conf., Oberwolfach, pp. 53–66. Basel: Birkhäuser 1969
[6] Local Toeplitz operators. Proc. London Math. Soc. **36**, 243–272 (1978)
[7] $C^*$-algebra extensions and $K$-homology. Princeton: Princeton Univ. Press 1980
[8] On the spectrum of a class of Toeplitz operators. J. Math. Mech. **18**, 433–435 (1968/69)

Douglas, R. G., Rudin, W.
[1] Approximation by inner functions. Pacific J. Math. **31**, 313–320 (1969)

Douglas, R. G., Sarason, D. E.
[1] Fredholm Toeplitz operators. Proc. Amer. Math. Soc. **26**, 117–120 (1970)
[2] A class of Toeplitz operators. Indiana Univ. Math. J. **20**, 891–895 (1971)

Douglas, R. G., Shapiro, H. S., Shields, A. L.
[1] Cyclic vectors and invariant subspaces of the backward shift operator. Ann. Inst. Fourier (Grenoble) **20**, 37–76 (1970)

Douglas, R. G., Taylor, J. L.
[1] Wiener-Hopf operators with measure kernels. In: Colloq. Math. Soc. Janos Bolyai, 5. Hilbert space operators (Proc. Internat. Conf., Tihany, Hungary, 1970), pp. 135–141. Amsterdam: North Holland 1972

Douglas, R. G., Widom, H.
[1] Toeplitz operators with locally sectorial symbols. Indiana Univ. Math. J. **20**, 385–388 (1970/71)

Dovbysh, L. N., Nikol'skiĭ, N. K., Sudakov, V. N.
[1] How good can a nonhereditarily complete family be? Zap. Naučn. Sem. Leningrad. Otdel. Mat. Inst. Steklov. (LOMI) **73**, 52–69 (1977) [Russian]

Dragiliev, M. M. Zaharyuta, V. P., Korobeĭnik, Yu. F.
[1] A duality connection between some questions of the theory of bases and interpolation. Dokl. Akad. Nauk SSSR **215**, 522–525 (1974) [Russian]

Dunford, N.
[1] Spectral operators. Pacific J. Math. **4**, 321–354 (1954)

Dunford, N., Schwartz, J.
[1] Linear operators. General theory. Part 1. New York: Wiley (Interscience) 1958
[2] Linear operators. Self-adjoint operators in Hilbert space. Part 2. New York: Wiley (Interscience) 1963
[3] Linear operators. Part 3. Spectral operators. New York: Wiley (Interscience) 1971

Duren, P. L.
[1] Theory of $H^p$ spaces. New York-London: Academic Press, 1970

Dym, H., McKean, H. P.
[1] Gaussian processes, function theory and the inverse spectral problem. New York: Academic Press 1976

Dym, H., Shapiro, I.
[1] Conditions for restricted translation operators to belong to $S_p$. Proc. Amer. Math. Soc. **63**, 251–258 (1977)

Dyn'kin, E. M.
[1] An operator calculus based on the Cauchy-Green formula. Zap. Naučn. Sem. Lenigrad. Otdel. Mat. Inst. Steklov. (LOMI) **30**, 33–39 (1972) [Russian]

Džrbašyan, M. M.
[1] Biorthogonal systems and solution of an interpolation problem with nodes of bounded multiplicity in the class $H^2$. Izv. Akad. Nauk Armjan. SSR Ser. Mat. **9**, 339–373 (1974) [Russian]
[2] On the problem of representability of analytic functions. Soobšč. Inst. Mat. Meh. Akad. Nauk Armjan. SSR **2**, 3–39 (1948) [Russian]
[3] Basisness of some biorthogonal systems and solution of the multiple interpolation problem in the class $H_+^p$. Dokl. Akad Nauk SSSR **234**, 517–520 (1977) [Russian]

Evgrafov, M. A.
[1] The interpolation problem of Abel-Gončarov. Moscow: GITTL, 1954 [Russian]

Fefferman, C., Stein E.
[1] $H^p$ spaces of several variables. Acta Math. **129**, 137–193 (1972)

Figiel, T., Johnson, W. B.
[1] The approximation property does not imply the bounded approximation property. Proc. Amer. Math. Soc. **41**, 197–200 (1973)

Fihtengol'c, G. M. (Fichtenholz, G. M.)
[1] Sur l'intégrale de Poisson et quelques questions qui s'y rattachent. Fund. Math. **13**, 1–33 (1929)

Fillmore, P. A. (ed.)
[1] Proc. Conf. Operator Theory. Lecture Notes in Mathematics 345. Berlin-Heidelberg-New York: Springer 1973

Foiaş, C.
[1] On the scalar parts of a decomposable operator. Rev. Roumaine Math. Pures Appl. **17**, 1182–1198 (1972)
[2] The class $C_0$ in the theory of decomposable operators. Rev. Roumaine Math. Pures Appl. **14**, 1433–1440 (1969)
[3] Spectral maximal spaces and decomposable operators in Banach space. Arch. Math. (Basel) **14**, 341–349 (1963)

Foiaş, C., Mlak, W.
[1] The extended spectrum of completely non-unitary contractions and the spectral mapping theorem. Studia Math. **26**, 239–245 (1966)

Foiaş, C., Williams, J. P.
[1] Some remarks on the Volterra operator. Proc. Amer. Math. Soc. **31**, 177–184 (1972)

Frolov V. D.
[1] On singular integral equations with measurable coefficients in spaces with weights. Mat. Issledov. Kišinev **5 : 1**, 141–151 (1970) [Russian]

Frostman, O.
[1] Potentiel d'équilibre et capacité des ensembles avec quelques applications à la théorie des fonctions. Med. Lunds Univ. Mat. Sem. **3**, 1–118 (1935)

Fuhrmann, P. A.
[1] On the corona theorem and its application to spectral problems in Hilbert space. Trans. Amer. Math. Soc. **132**, 55–66 (1968)
[2] On sums of Hankel operators. Proc. Amer. Math. Soc. **46**, 65–68 (1974)

Gabriel, R. M.
[1] Concerning integrals of moduli of regular functions along convex curves. Proc. London Math. Soc. **39**, 216–231 (1935)

Gahov, F. D.
[1] Boundary problems. Moscow: GIFML 1963 [Russian] (English translation: Oxford etc.: Pergamon Press 1966)

Gamelin, T. W.
[1] Uniform algebras. Englewood Cliffs: Prentice Hall 1969
[2] Wolff's proof of the Corona theorem. Israel J. Math. **37**, 113–120 (1980)

Gantmaher, F. R.
[1] The theory of matrices. (2nd edition.) Moscow: Nauka 1966 [Russian] (English translation: New York: Chelsea 1960)

Garnett, J. B.
[1] Two remarks on interpolation by bounded analytic functions. In: Banach spaces of analytic functions. Lecture Notes in Mathematics 604, pp. 32–40. Berlin-Heidelberg-New York: Springer 1977

[2] Bounded analytic functions. New York: Academic Press 1981

Gel'fand, I. M.
[1] Problem. Uspehi Mat. Nauk **5**, 133 (1938) [Russian]
[2] Remark on N. K. Bari's paper "Biorthogonal systems and bases in Hilbert space". Učen. Zap. Moskov. Univ. **4**, 224–225 (1951) [Russian]

Gel'fand, I. M., Yaglom, A. M.
[1] On the computation of the amount of information of a random function contained in another such function. Uspehi Mat. Nauk **12 : 2**, 3–52 (1957) [Russian]

Gellar, R., Page, L.
[1] Inner-outer factorizations of operators. J. Math. Anal. Appl. **61**, 151–158 (1977)

Ginzburg, Yu. P.
[1] On divisors and minorants of operator functions of bounded form. In: Mat. Issledovaniya, vol. 2, nr. 4, pp. 47–72. Kišinev: 1968 [Russian]
[2] On spectral subspaces of contractions with a slowly growing resolvent. In: Mat. Issledovaniya, vol. 5, nr. 4, pp. 45–62. Kišinev: 1970 [Russian]
[3] On almost invariant spectral properties of contractions and multiplicative properties of analytic operator functions. Funkcional. Anal. i Priložen. **5**, 32–41 (1971) [Russian]
[4] Multiplicative representations and minorants of bounded analytic operator functions. Funkcional. Anal. i Priložen. **1**, 9–23 (1967) [Russian]

Girsanov, I. V.
[1] On the spectra of dynamical systems generated by stationary Gaussian processes. Dokl. Akad. Nauk SSSR **119**, 851–853 (1958) [Russian]

Glazman, I. M.
[1] On the possibility of expansion with respect to the system of eigenelements of dissipative operators. Usp. Mat. Nauk **13 : 3**, 179–181 (1958) [Russian]

Glazman, I. M., Lyubič, Yu. I.
[1] Finite dimensional linear analysis. Moscow: Nauka 1969 [Russian] (English translation: Cambridge: MIT Press 1974)

Gohberg, I. C.
[1] On an application of the theory of normed rings to singular integral equations. Uspehi Mat. Nauk **7 : 2**, 149–156 (1952) [Russian]
[2] The problem of factorization of operator functions. Izv. Akad. Nauk SSSR **28**, 1055–1082 (1964) [Russian]
[3] On the theory of multivariate singular integral equations. Dokl. Akad. Nauk SSSR **133**, 1279–1282 (1960) [Russian]
[4] The problem of factorization in normed rings, functions of isometric and symmetric operators, and singular integral equations. Uspehi Mat. Nauk **19 : 1**, 71–124 (1964) [Russian]

Gohberg, I. C., Fel'dman, I. A.
[1] Convolution equations and projection methods for their solution. Moscow: Nauka 1971 [Russian] (English translation: Providence: American Mathematical Society 1974)
[2] On integral-difference Wiener-Hopf equations. Dokl. Akad. Nauk SSSR **183**, 25–28 (1968) [Russian]
[3] Integro-difference Wiener-Hopf equations. Acta Sci. Math. (Szeged) **30**, 199–224 (1969) [Russian]

Gohberg, I. C., Kreĭn, M. G.
[1] Introduction to the theory of linear non-selfadjoint operators in Hilbert space. Moscow: Nauka 1965 [Russian] (English translation: Providence: American Mathematical Society 1969)
[2] The theory of Volterra operators in Hilbert space and applications. Moscow: Nauka 1967 [Russian] (English translation: Providence: American Mathematical Society 1970)

[3] On a completeness criterion for the system of root vectors of a contraction. Usp. Mat. Nauk **16 : 1**, 78–82 (1963) [Russian]
[4] On a description of contractions similar to unitary ones. Funkcional. Anal. i Priložen. **1**, 38–60 (1967) [Russian]
[5] Systems of integral equations on the halfaxis with kernels depending on the difference of the arguments. Uspehi Mat. Nauk **13 : 2**, 3–72 (1958) [Russian]

Gohberg, I. C., Krupnik, N. Ya
[1] Introduction to the theory of one dimensional singular integral operators. Kišinev: Stiinca 1973 [Russian] (German translation: Basel: Birkhäuser 1979)
[2] On the algebra generated by Toeplitz matrices. Funkcional. Anal. i Priložen. **3 : 2**, 46–56 (1969) [Russian]
[3] On a local principle and algebras generated by Toeplitz matrices. An. Şti. Univ. "Al. I. Cuza" Iaşi Secţ. I a Mat. 1973. **19 : 1** [Russian]

Gohberg, I. C., Semencul, A. A.
[1] Toeplitz matrices composed by Fourier coefficients of functions with discontinuities of almost-periodic type. Mat. Issledov. Kišinev **5**, 63–83 (1970) [Russian]

Goluzin, G. M.
[1] Geometric theory of functions of a complex variable. Moscow: Nauka 1966 [Russian] (English translation: Providence: American Mathematical Society 1969)

Gončar, A. A.
[1] The rate of approximation with rational fractions and properties of functions. In: Trudy Mezdunarodnovo Kongressa Matematikov (Moscow 1966), pp. 329–356. Moscow: Mir 1968 [Russian]

Gorin, E. A.
[1] A function-algebra variant of the Bohr-van Kampen theorem. Mat. Sb. **82**, 260–272 (1970) [Russian]

Grenander, U., Szegö, G.
[1] Toeplitz forms and their applications. Calif. Monographs in Mathematical Sciences 2. Berkeley-Los Angeles 1958

Gribov, M. B., Nikol'skiĭ, N. K.
[1] Invariant subspaces and rational approximation. Zap. Naučn. Sem. Leningrad. Otdel. Mat. Inst. Steklov. (LOMI) **92**, 103–114 (1979) [Russian]

Grinblyum, M. M.
[1] Some basis theorems in spaces of type B. Dokl. Akad. Nauk. SSSR **31**, 428–432 (1941) [Russian]
[2] On the representation of spaces of type B in the form of direct sums of subspaces. Dokl. Akad. Nauk SSSR **70**, 749–752 (1950) [Russian]

Guillory, C. J., Sarason, D.
[1] Division in $H^\infty + C$. Mich. Math. J. **28**, 173–181 (1981)

Gurariĭ, V. I., Macaev, V. I.
[1] Lacunary subspaces of powers in the spaces $C$ and $L^p$. Izv. Akad. Nauk SSSR Ser. Mat. **30**, 3–14 (1966) [Russian]

Hadamard, J.
[1] Leçons sur la propagation des ondes. Paris: Hermann 1903

Halmos, P. R.
[1] Shifts on Hilbert spaces. J. Reine Angew. Math. **208**, 102–112 (1961)
[2] A Hilbert space problem book. Graduate Texts in Mathematics. Berlin-Heidelberg-New York: Springer 1967
[3] Normal dilations and extensions of operators. Summa Brasil. Math. **2**, 125–134 (1950)

[4] Quadratic interpolation. J. Operator Theory **7**, 303–306 (1982)

Hartman, P.
[1] On completely continuous Hankel matrices. Proc. Amer. Math. Soc. **9**, 862–866 (1958)

Hartman, P., Wintner, A.
[1] The spectra of Toeplitz's matrices. Amer. J. Math. **76**, 867–882 (1954)
[2] On the spectra of Toeplitz's matrices. Amer. J. Math. **72**, 359–366 (1950)

Havin, V. P.
[1] The spaces $H^\infty$ and $L^1/H_0^1$. Zap. Naučn. Sem. Leningrad. Otdel. Mat. Inst. Stekl. (LOMI) **39**, 120–148 (1974) [Russian]
[2] On factorization of analytic functions smooth up to the boundary. Zap. Naučn. Sem. Leningrad. Otdel. Mat. Inst. Steklov. (LOMI) **22**, 202–205 (1971) [Russian]

Havin, V. P., Hruščev, S. V., Nikol'skiĭ, N. K. (editors)
[1] Linear and complex analysis problem book. 199 research problems. Lecture Notes in Mathematics 1043. Berlin-Heidelberg-New York: Springer 1984

Havinson, S. Ya.
[1] On an extremal problem in the theory of analytic functions. Uspehi Mat. Nauk **4:4**, 158–159 (1949) [Russian]
[2] On some extremal problems in the theory of analytic functions. Učen. Zap. Moskov. Gos. Univ. Mat. **148**, 133–143 (1951) [Russian]

Helson, H.
[1] Méthodes complexes et méthodes de Hilbert en analyse de Fourier. Publications Orsay 1967
[2] Lectures on invariant subspaces. New York: Academic Press 1964
[3] Analyticity on compact Abelian groups. In: Lectures at Instr. Conf. on Algebras in Analysis, Birmingham, 1973, pp. 1–79. London Mathematical Society

Helson, H., Lowdenslager, D.
[1] Invariant subspaces. In: Proc. Intern. Symp. Linear Spaces, Jerusalem, 1960, pp. 251–262. Oxford: Pergamon Press 1961

Helson, H., Sarason, D.
[1] Past and future. Math. Scand. **21**, 5–16 (1967)

Helson, H., Szegö, G.
[1] A problem of prediction theory. Ann. Mat. Pura Appl. **51**, 107–138 (1960)

Hilden, H. M., Wallen, L. J.
[1] Some cyclic and non-cyclic vectors of certain operators. Indiana Univ. Math. J. **23**, 557–565 (1974)

Hirshman, I. I., Jr.
[1] The asymptotic behavior of finite section Wiener-Hopf equations. In: Abstract spaces and approximation (Proc. Conf., Oberwolfach, 1968), pp. 31–52. Basel: Birkhäuser 1969

Hoffman, K.
[1] Banach spaces of analytic functions. Englewood Cliffs: Prentice-Hall 1962

Hoffman, M. J.
[1] Operators commuting with Toeplitz and Hankel operators modulo the compact operators. J. Operator Theory **6**, 51–54 (1981)

Hruščev, S. V.
[1] Perturbation theorems for bases of exponentials and the Muckenhoupt condition. Dokl. Akad. Nauk SSSR **247**, 44-47 (1979) [Russian]

Hruščev, S. V., Nikol'skiĭ, N. K., Pavlov, B. S.
[1] Unconditional bases of exponentials and of reproducing kernels. In: Complex analysis and

spectral theory. Lecture Notes in Mathematics 864, pp. 214–335. Berlin-Heidelberg-New York: Springer 1981

Hruščev, S. V., Vol'berg, A. P.
[1] A generalization of Koosis-Lax interior compactness theorem. J. Operator Theory **8**, 197–208 (1982)

Hunt, R., Muckenhoupt, B., Wheeden, R.
[1] Weighted norm inequalities for conjugate functions and Hilbert transform. Trans. Amer. Math. Soc. **176**, 227–251 (1973)

Ibragimov, I. A.
[1] On stationary Gaussian sequences with the strong mixing property. Dokl. Akad. Nauk SSSR **147**, 1282–1284 (1962) [Russian]
[2] On the spectrum of stationary Gaussian sequences with the strong mixing condition. I. Necessary conditions. Teor. Verojatnost. i Primenen. **10**, 95–116 (1965) [Russian]
[3] On the spectral functions of some classes stationary Gaussian processes. Dokl. Akad. Nauk SSSR **137**, 1046–1048 (1961) [Russian]
[4] On the spectrum of stationary Gaussian processes with the strong mixing property. II. Sufficient conditions. Rate of mixing. Teor. Verojatnost. i Primenen. **15**, 24–37 (1970) [Russian]

Ibragimov, I. A., Linnik Yu. V.
[1] Independent and stationarily dependent variables. Moscow: Nauka 1965 [Russian] (English translation: Groningen: Walters-Noordhoff 1971)

Ibragimov, I. A., Rozanov, Yu. A.
[1] Gaussian stochastic processes. Moscow: Nauka 1970 [Russian] (English translation: New York-Heidelberg-Berlin: Springer 1978)

Ibragimov, I. A., Solev, V. N.
[1] On a regularity condition for stationary Gaussian sequences. Zap. Naučn. Sem. Leningrad. Otdel. Mat. Inst. Stekl. (LOMI) **12**, 113–125 (1969) [Russian]

Iohvidov, I. S.
[1] Hankel and Toeplitz matrices and forms. Moscow: Nauka 1974 [Russian] (English translation: Boston: Birkhäuser 1982)

Ismagilov, R. S.
[1] On the spectrum of Toeplitz matrices. Dokl. Akad. Nauk SSSR **149**, 769–772 (1963) [Russian]

Ivanov, S. L., Pavlov, B. S.
[1] Carleson series of resonances in Regge's problem. Izv. Akad. Nauk SSSR Ser. Mat. **42**, 26–55 (1978) [Russian]

Jerison, D.
[1] Sur un théorème d'interpolation de R. Nevanlinna. C. R. Acad. Sci. Paris Sér. A **282**, 1291–1293 (1976)

Jones, P.
[1] Geometric constructions with ratios of Blaschke products and functions of bounded mean oscillation. Preprint, 1979
[2] Estimates for the corona problem. J. Functional Analysis **39**, 162–181 (1980)
[3] $L^\infty$ estimates for $\bar\partial$-problem in a half-plane. Acta Math. **150**, 137–152 (1983)

Kabaĭla, V. P.
[1] On interpolation of functions of class $H_\delta$. Uspehi Mat. Nauk **13:1**, 181–188 (1958) [Russian]

Kac, M.
[1] Theory and application of Toeplitz forms. In: Summer institute of spectral theory and statistical mechanics. Brookhaven National Laboratory 1965
[2] Probability and related topics in physics. London: Interscience 1959

Kacnel'son, V. È.
[1] On basisness conditions for systems of root vectors of some classes of operators. Funkcional. Anal. i Prilož. **1**, 39–51 (1967) [Russian]
[2] On bases of exponential functions in $L^2$. Funkcional. Anal. i Prilož. **5**, 37–47 (1971) [Russian]

Kadec, M. I.
[1] The exact value of the Paley-Wiener constant. Dokl. Akad. Nauk SSSR **155**, 1253–1254 (1964) [Russian]

Kahane, J.-P.
[1] Séries de Fourier absolument convergentes. Erg. Math. 50. Berlin-Heidelberg-New York: Springer 1970
[2] Best approximation in $L^1(T)$. Bull. Amer. Math. Soc. **80**, 788–804 (1974)

Kalisch, G. K.
[1] A functional analysis proof of Titchmarsh's theorem on convolution. J. Math. Anal. Appl. **5**, 176–183 (1962)
[2] Théorème de Titchmarsh sur la convolution et opérateurs de Volterra. In: Sém. P. Lelong, Fac. Sci. Paris. 1962–1963, pp. 5/01–5/06

Kato, T.
[1] Perturbation theory for linear operators. Grundlehren der math. Wissenschaften, 144. Berlin-Heidelberg-New York: Springer 1967

Katznelson, Y.
[1] An introduction to harmonic analysis. New York: Wiley 1968

Kisilevskiĭ, G. E.
[1] Invariant subspaces of dissipative Volterra operators with nuclear imaginary components. Izv. Akad. Nauk SSSR Ser. Math. **32**, 3–23 (1968) [Russian]
[2] Volterra operators with nuclear imaginary components. Ukrain. Mat. Ž. **27**, 794–797 (1975) [Russian]

Kislyakov, S. V.
[1] Projections onto the set of Hankel matrices. Zap. Naučn. Sem. Leningrad. Otdel. Mat. Inst. Steklov. (LOMI) **126**, 109–116 (1983) [Russian]

Kober, H.
[1] Approximation of continuous functions by integral functions of finite order. Trans. Amer. Math. Soc. **61**, 293–306 (1947)

Kolmogorov, A. N., Rozanov, Yu. A.
[1] On conditions for strong mixing of a stationary Gaussian process. Teor. Verojatnost. i Primenen. **5**, 221–227 (1960) [Russian]

Koosis, P. J.
[1] Sur l'approximation ponderée par des polynômes et par des sommes d'exponentielles imaginaires. Ann. Sci. École Norm. Sup. **81**, 387–409 (1964)
[2] Interior compact spaces of functions on a half-line. Comm. Pure Appl. Math. **10**, 583–615 (1957)
[3] Introduction to $H_p$ spaces. With an appendix on Wolff's proof of the corona theorem. London Mathematical Society Lecture Note Series 40. London etc.: Cambridge University Press 1980
[4] Moyennes quadratiques pondérées de fonctions périodiques et de leurs conjuguées harmoniques. C. R. Acad. Sci. Paris Sér. A **291**, 255–257 (1980)

Köthe, G., Toeplitz, O.
[1] Lineare Räume mit unendlich vielen Koordinaten und Ringe unendlicher Matrizen. J. Reine Angew. Math. **171**, 251–270 (1934)

Kreĭn, M. G.
[1] Integral equations on the halfaxis with a kernel depending on the difference of the arguments. Uspehi Mat. Nauk **13 : 5**, 3–120 (1958) [Russian]

[2] The theory of self-adjoint extensions of semi-bounded Hermitian operators and applications. Mat. Sb. **20**, 431–495 (1947), **21**, 365–404 (1947) [Russian]
[3] On a special ring of functions. Dokl. Akad. Nauk SSSR **29**, 355–359 (1940) [Russian]
[4] On some new Banach algebras and Wiener-Lewy type theorems for Fourier series and Fourier integrals. Mat. Issled. Kišinev **1**, 82–109 (1966) [Russian]
[5] On an extrapolation problem of A. N. Kolmogorov. Dokl. Akad. Nauk SSSR **46**, 339–342 (1945) [Russian]

Kreĭn, M. G., Nudelman, A. A.
[1] The Markov moment problem and extremal problems. Moscow: Nauka 1973 [Russian] (English translation: Providence: American Mathematical Society 1977)

Kriete, T. L. III
[1] A generalized Paley-Wiener theorem. J. Math. Anal. Appl. **36**, 529–555 (1971)
[2] Fourier transforms and chains of inner functions. Duke Math. J. **40**, 131–143 (1973)

Kronecker, L.
[1] Zur Theorie der Elimination einer Variablen aus zwei algebraischen Gleichungen. Königl. Preuss. Akad. Wiss. (Berlin) 1881, 535–600 (= Werke II, pp. 113–192)

Krupnik, N. Ya.
[1] A Noether criterion for singular integral operators. Soobš. Akad. Nauk Gruzin. SSR **80**, 533–536 (1975) [Russian]

Krupnik, N. Ya., Fel'dman, I. A.
[1] On the impossibility of the introduction of a matrix symbol in some operator algebras. In: Linear operators and integral equations (Mat. Issled. 61), pp. 75–85. Kišinev: Stinca 1981 [Russian]

Krupnik, N. Ya., Spigel', E. M.
[1] On perturbation of singular integral operators. In: Non-selfadjoint operators (Mat. Issled. 42), pp. 113–123, Kišinev: Stinca 1976 [Russian]

Lamperti, J.
[1] Probability. New York-Amsterdam: Benjamin 1966

Langer, H.
[1] Ein Zerspaltungssatz für Operatoren im Hilbertraum. Acta Sci. Math. (Szeged) **12**, 441–445 (1961)

Lax, P. D.
[1] Translation invariant spaces. Acta Math. **101**, 163–178 (1959)

Lax, P. D., Phillips, R. S.
[1] Scattering theory. New York: Academic Press (1967)
[2] Scattering theory. Bull. Amer. Math. Soc. **70**, 130–142 (1964)

Lee, M., Sarason, D. E.
[1] The spectra of some Toeplitz operators. J. Math. Anal. Appl. **33**, 529–543 (1971)

Leggett, R.
[1] On the invariant subspace structure of compact dissipative operators. Indiana Univ. Math. J. **22**, 919–928 (1973)

Levin, B. Ya.
[1] Distribution of roots of entire functions. Moscow: GITTL 1956 [Russian] (German translation: Berlin: Akademie-Verlag 1962)
[2] On bases of exponentials in $L^2$. Zap. Mat. Otd. Fiz.-Mat. Fak. Har'k. Un-ta i Har'k. Mat. Obšč. **27**, 39–48 (1961) [Russian]

Levinson, N.
[1] Gap and density theorems. Amer. Math. Soc. Coll. Publ. 26. New York: Amer. Math. Soc. 1940

Levitan, B. M.
[1] Almost-periodic functions. Moscow: Nauka 1973 [Russian]

Lindenstrauss, J., Zippin, M.
[1] Banach spaces with a unique unconditional basis. J. Functional Anal. **3**, 115–125 (1969)

Lions, J.-L., Peetre, J.
[1] Sur une classe d'espaces d'interpolation. Inst. Hautes Études Sci. Publ. Math. **19**, 5–68 (1964)

Littlewood, J. E.
[1] A mathematician's miscellany. London: Methuen 1953

Litvinčuk, G. S.
[1] Boundary problems and singular integral equations with shift. Moscow: Nauka 1977 [Russian]

Livšic, M. S.
[1] On a class of linear operators in Hilbert space. Mat. Sb. **19**, 239–262 (1946) [Russian]
[2] On spectral expansions of linear non-selfadjoint operators. Mat. Sb. **34**, 145–198 (1954) [Russian]

Lorch, E. R.
[1] Bicontinuous linear transformations in certain vector spaces. Bull. Amer. Math. Soc. **45**, 564–569 (1939)

Lubin, A.
[1] Isometries of $*$-invariant subspaces. Trans. Amer. Math. Soc. **190**, 405–415 (1974)

Lyance, V. È.
[1] On a generalization of the notion of spectral measure. Mat. Sb. **61**, 80–120 (1963) [Russian]

Lyubič, Yu. I., Macaev, V. I.
[1] On operators with a separable spectrum. Mat. Sb. **56**, 433–468 (1962) [Russian]

Macaev, V. I.
[1] On a class of completely continuous operators. Dokl. Akad. Nauk SSSR **139**, 548–551 (1961) [Russian]

Macintyre, A. J., Rogosinski W.W.
[1] Extremum problems in the theory of analytic functions. Acta Math. **82**, 275–325 (1950) [Russian]

Malyšev, V. A.
[1] Wiener-Hopf equations and their application to probability theory. In: Itogi nauki i tehn., teor. ver., mat statist., Teoret. kibern. 13, pp. 5–35. 1976

Markus, A. S.
[1] The problem of spectral synthesis for operators with a point spectrum. Izv. Akad. Nauk SSSR Ser. Mat. **34**, 662–688 (1970) [Russian]
[2] On bases of root vectors of dissipative operators. Dokl. Akad. Nauk SSSR **132**, 524–527 (1960) [Russian]

Markuševič, A. I.
[1] Theory of analytic functions. 2 volumes. Moscow: Nauka 1968 [Russian] (English translation (3 volumes!): Englewood Cliffs: Prentice-Hall 1965–67)

Marshall, D.
[1] Subalgebras of $L^\infty$ containing $H^\infty$. Acta Math. **137**, 91–98 (1976)

Maruyama, G.
[1] The harmonic analysis of stationary stochastic processes. Mem. Fac. Sci. Kyushu Univ. **4**, 45–106 (1949)

Masani, P.

[1] Wiener's contributions to generalized harmonic analysis, prediction theory and filter theory. Bull. Amer. Math. Soc. **72**, 73–125 (1966)

Masani, P., Wiener, N.

[1] The prediction theory of multivariate stochastic processes. I. The regularity condition. Acta Math. **98**, 111–150 (1957)

[2] The prediction theory of multivariate stochastic processes. II. The linear predictor. Acta Math. **99**, 93–137 (1958)

Maslov, V. P.

[1] Operational methods. Moscow: Nauka 1973 [Russian] (English translation: Moscow: Mir Publishers 1976)

Matveev, R. F.

[1] On the regularity of discrete time multivariate stationary stochastic processes. Dokl. Akad. Nauk SSSR **126**, 713–715 (1959) [Russian]

McKenna, P. J.

[1] Discrete Carleson measures and some interpolation problems. Mich. Math. J. **24**, 311–319 (1977)

Mihlin, S. G.

[1] Numerical realization of variational methods. Moscow: Nauka 1966 [Russian] (English translation: Groningen: Wolters-Nordhoff 1971)

[2] Singular integral equations. Uspehi Mat. Nauk **3**, 29–112 (1948)

[3] Multivariate singular integrals and integral equations. Moscow: GIFML 1962 [Russian] (English translation: Oxford etc.: Pergamon Press 1965)

Mityagin, B. S.

[1] Equivalence of bases in Hilbert scales. Studia Math. **37**, 111–137 (1971) [Russian]

Moeller, J. W.

[1] Translation invariant spaces with zero-free spectra. Duke Math. J. **31**, 99–108 (1964)

[2] On the spectra of some translation invariant spaces. J. Math. Anal. Appl. **4**, 276–296 (1962)

Moore, B. III

[1] The Szegö infimum. Proc. Amer. Math. Soc. **29**, 55–62 (1971)

Moore, B. III. Nordgren, E.

[1] On quasi-equivalence and quasi-similarity. Acta. Sci. Math. (Szeged) **34**, 311–316 (1973)

Muhly, P. S.

[1] Compact operators in the commutant of a contraction. J. Functional Anal. **8**, 197–224 (1971)

Mukminov, B. R.

[1] On expansions in eigenfunctions of dissipative kernels. Dokl. Akad. Nauk SSSR **99**, 499–502 (1954) [Russian]

Müller, V.

[1] On Jordan models of $C_0$-contractions. Acta Sci. Math. (Szeged) **40**, 309–313 (1978)

Muschelišvili, N. I.

[1] Singular integral equations. Moscow: GIFML 1962 [Russian] (English translation (2nd edition): Groningen: Noordhoff 1953)

Naftalevič, A. G.

[1] On interpolation functions of bounded type. Učen. Zap. Vil'nyus. Univ. **5**, 5–27 (1956) [Russian]

Naĭmark, M. A.
[1] Normed rings. Moscow: Nauka, 1968 [Russian] (English translation: Groningen: Noordhoff 1964)
[2] Positive definite operator functions on a commutative group. Izv. Akad. Nauk SSSR Ser. Mat. **7**, 237–244 (1943) [Russian]

Nakazi, T.
[1] Invariant subspaces of weak *-Dirichlet algebras. Pacif. J. Math. **69**, 151–167 (1977)

Natanson, I. P.
[1] Theory of functions of a real variable. Moscow: Gostehizdat 1957 [Russian] (German translation: Berlin: Akademie-Verlag 1954)

Nehari, Z.
[1] On bounded bilinear forms. Ann. Math. **65**, 153–162 (1957)

Nelson, E.
[1] Probability theory and Euclidean field theory. Constructive quantum field theory. Lecture Notes in Physics 25. Berlin-Heidelberg-New York: Springer, 1973

Nevanlinna, R.
[1] Eindeutige analytische Funktionen. 2. Aufl. Berlin-Göttingen-Heidelberg: Springer 1953
[2] Über beschränkte analytische Funktionen. Ann. Acad. Sci. Fenn. **32 : 7** (1929)

Neville, C.
[1] Invariant subspaces of Hardy classes on infinitely connected open surfaces. Mem. Amer. Math. Soc. **160** (1975)

Newman, D. J.
[1] Interpolation in $H^\infty$. Trans. Amer. Math. Soc. *92*, 501–507 (1959)

Nieto, J. I.
[1] Sur un théorème d'interpolation de J. Lions and J. Peetre. Can. Math. Bull. **14**, 373–376 (1971)

Nikol'skiĭ, N. K.
[1] On invariant subspaces of unitary operators. Vestnik Leningrad. Univ. Mat. Meh. Astronom. **19**, 36–43 (1966) [Russian]
[2] Selected problems on weighted approximation and spectral analysis. Trudy Mat. Inst. Steklov. **120** (1974) [Russian]
[3] On perturbations of the spectrum of unitary operators. Mat. Sb. **5**, 341–349 (1969) [Russian]
[4] Invariant subspaces in the theory of operators and in the theory of functions. In: Advances in sciences and technology. Mathematical analysis 12, pp. 199–412. Moscow: Izd. VINITI 1974 [Russian]
[5] Two problems of spectral synthesis. Zap. Naučn. Sem. Leningrad. Otdel. Mat. Inst. Steklov. (LOMI) **81**, 139–141 (1978) [Russian]
[6] Operators and approximation. Zap. Naučn. Sem. Leningrad. Otdel. Mat. Inst. Steklov. (LOMI) **81**, 94–95 (1978) [Russian]
[7] The current status of the problem of spectral analysis and synthesis. In: Operator theory in function spaces. Proceedings of the first school-symposium on linear operators and functional spaces, pp. 240–282. Novosibirsk: Izdat. Nauka Sibirsk. Otdel. 1977
[8] Lectures on the shift operator. II. Zap. Naučn. Sem. Leningrad. Otdel. Mat. Inst. Steklov. (LOMI) **47**, 90–119 (1974) [Russian]
[9] Five problems on invariant subspaces. Zap. Naučn. Sem. Leningrad. Otdel. Mat. Inst. Steklov. (LOMI) **23**, 115–125 (1971) [Russian]
[10] Bases of invariant subspaces and operator interpolation. Trudy Mat. Inst. Steklov. **130**, 50–123 (1977) [Russian]
[11] Bases of exponentials and values of reproducing kernels. Dokl. Akad. Nauk SSSR **252**, 1316–1320 (1980) [Russian]

Nikol'skiĭ, N. K., Havin, V. P., Hruščev, S. V. (composers and editors)
[1] Studies in linear operators and function theory. 99 unsolved problems in linear and complex analysis. Zap. Naučn. Sem. Leningrad. Otdel. Mat. Inst. Steklov. (LOMI) **81** (1978) [Russian]

Nikol'skiĭ, N. K., Pavlov, B. S.
[1] Bases of eigenvectors of completely non-unitary contractions and the characteristic function. Izv. Akad. Nauk SSSR Ser. Mat. **34**, 90–133 (1970) [Russian]
[2] Expansions in eigenfunctions of non-unitary operators and the characteristic function. Zap. Naučn. Sem. Leningrad. Otdel. Mat. Inst. Steklov. (LOMI) **11**, 150–203 (1968) [Russian]

Nikol'skiĭ, N. K., Vasyunin, V. I.
[1] Classification of $H^2$-functions by their cyclicity degree. Izv. Akad. Nauk SSSR **47**, 942–960 (1983) [Russian]

Nikol'skiĭ, S. M.
[1] Approximation of functions of several variables and imbedding theorems. Moscow: Nauka 1977 [Russian] (English translation: Grundlehren 205. Berlin-Heidelberg-New York: Springer 1975

Nordgren, E. A.
[1] On quasi-equivalence of matrices over $H^\infty$. Acta Sci. Math. (Szeged) **34**, 301–310 (1973)
[2] The ring $N^+$ is not adequate. Acta Sci. Math. (Szeged) **36**, 203–204 (1974)

Nyman, B.
[1] On the one-dimensional translation group and semi-group in certain function spaces. Thesis. Uppsala 1950

Oleĭnik, V. L.
[1] Imbedding theorems for weighted classes of harmonic and analytic functions. Zap. Naučn. Sem. Leningrad. Otdel. Mat. Inst. Steklov. (LOMI) **47**, 120–137 (1974) [Russian]

Orlicz, W.
[1] Über unbedingte Konvergenz in Funktionenräumen. Studia Math. **2**, 41–47 (1933)

Øyma, K.
[1] Extremal interpolatory functions in $H^\infty$. Proc. Amer. Math. Soc. **64**, 272–276 (1977)

Page, L. B.
[1] Applications of the Sz.-Nagy and Foiaş lifting theorem. Indiana Univ. Math. J. **20**, 135–145 (1970)
[2] Bounded and compact vectorial Hankel operators. Trans. Amer. Math. Soc. **150**, 529–539 (1970)

Parfenov, O. G.
[1] The asymptotics of singular numbers of integral operators with Cauchy and Bergman kernels. Funkcional. Anal. i Priložen. **12 : 2**, 88 (1978) [Russian]

Parodi, M.
[1] La localisation de nombres charactéristiques de matrices et ses applications. Traité de Phys. Théor. et Phys. Mathématique 12. Paris: Gauthier-Villars 1959

Parrott, S.
[1] On a quotient norm and the Sz.-Nagy-Foias lifting theorem. J. Functional Analysis **30**, 311–328 (1978)

Pavlov, B. S.
[1] Basis conditions for systems of exponentials and Muckenhoupt's condition. Dokl. Akad. Nauk SSSR **247**, 37–40 (1979) [Russian]
[2] Scattering theory and the "non-physical sheet" for systems of ordinary differential equations. Dokl. Akad. Nauk SSSR **193**, 36–39 (1970) [Russian]
[3] Spectral analysis of a differential operator with a "smeared out" boundary condition. In: Problems of mathematical physics. Nr. 6, pp. 101–119. Izd. LGU, 1973 [Russian]

PDO

[1] Pseudo-differential operators. A collection of papers. Moscow: Mir 1967 [Russian]

Peetre, J., Svensson, E.

[1] On the generalized Hardy's inequality of McGehee, Pigno and Smith and the problem of interpolation between BMO and a Besov space. Math. Scand. **54**, 221–241 (1984)

Peller, V. V.

[1] An analogue of J. von Neuman's inequality, the isometric dilatation of contractions and approximation with isometries in spaces of bounded functions. Trudy Mat. Inst. Steklov. **155**, 103–150 (1980) [Russian]

[2] Nuclearity of Hankel operators. LOMI pre-prints E-1-79. Leningrad 1979

[3] Hankel operators of class $\mathfrak{S}_p$ and applications (rational approximation, Gaussian processes, the majorant problem for operators). Mat. Sb. **113**, 538–581 (1980) [Russian]

[4] Vectorial Hankel operators, commutators and related operators of the Schatten – von Neumann class $\mathfrak{S}_p$. Integral Equations Operator Theory **5**, 244–272 (1982)

[5] Smooth Hankel operators and applications (the ideals $\mathfrak{S}_p$, Besov classes, stochastic processes). Dokl. Akad. Nauk SSSR **252**, 43–48 (1980) [Russian]

[6] A description of Hankel operators of class $\mathfrak{S}_p$ for $p > 0$, investigation of the rate of rational approximation, and other applications. Mat. Sb. **122**, 481–510 (1983) [Russian]

[7] Continuity properties of the averaging projection onto the set of Hankel matrices. J. Functional Analysis **53**, 74–83 (1983)

[8] Metric properties of the averaging projection onto the set of Hankel matrices. Dokl. Akad. Nauk SSSR **278**, 275–281 [Russian]

Peller, V. V., Hruščev, S. V.

[1] Hankel operators, best approximation and stationary Gaussian processes. Uspehi Mat. Nauk **37 : 1**, 53–124 (1982) [Russian]

Plessner, A. I.

[1] On semi-unitary operators. Dokl. Akad. Nauk SSSR **25**, 710–712 (1939) [Russian]

Potapov, V. P.

[1] On the multiplicative structure of $J$-nonexpanding matrix functions. Trudy Moskov. Mat. Obšč. **4**, 125–236 (1955) [Russian]

[2] Collected papers of V. P. Potapov. Private translation and edition by T. Ando. Sapporo: 1982

Power, S. C.

[1] Hankel operators on Hilbert space. Bull. London Math. Soc. **12**, 422–442 (1980)

[2] Fredholm Toeplitz operators and slow oscillation. Can. J. Math. **32**, 1058–1071 (1980)

[3] $C^*$-algebras generated by Hankel operators and Toeplitz operators. J. Functional Anal. **31**, 52–68 (1979)

[4] Hankel operators with PQC symbols and singular integral operators. Proc. London Math. Soc. **40**, 45–65 (1980)

Privalov, I. I.

[1] Boundary behavior of analytic functions. Moscow-Leningrad: GITTL 1950 [Russian] (German translation: Berlin: Deutscher Verlag der Wissenschaften 1956)

Prössdorf, S.

[1] Einige Klassen singulärer Gleichungen. Mathematische Reihe, Band 46. Basel-Stuttgart: Birkhäuser 1974

Prössdorf, S., Silbermann, B.

[1] Projektionsverfahren und die näherungsweise Lösung singulärer Gleichungen. Leipzig: Teubner 1977

Rabindranathan, M.

[1] On the inversion of Toeplitz operators. J. Math. Mech. **19**, 195–206 (1969/70)

Rajchman, A.
[1] Une classe de séries trigonométriques qui convergent presque partout vers zéro. Math. Ann. **101**, 686–700 (1929)

Rapoport, I. M.
[1] On a class of singular integral equations. Dokl. Akad. Nauk SSSR **59**, 1403–1406 (1948) [Russian]
[2] On a class of infinite systems of linear algebraic equations. Dokl. Akad. Nauk USSR, Viddil Fiziko-Mat. ta Him. Nauk **3**, 6–10 (1948) [Ukrainian]

Reed, M. C.
[1] Functional analysis and probability theory. In: Constructive Quantum Theory. Lecture Notes in Physics 25. Berlin-Heidelberg-New York: Springer 1973

Reed, M., Simon, B.
[1] Methods of modern mathematical physics. I. Functional analysis. New York-London: Academic Press 1972
[2] Methods of modern mathematical physics, II. Fourier analysis, self-adjointness. New York: Academic Press 1977

Riesz, F., Sz.-Nagy, B.
[1] Leçons d'Analyse Fonctionnelle. Troisième édition. Szeged: Akadémiai Kiado 1955

Roch, S., Silbermann, B.
[1] Reduktionsverfahren für Potenzen von Toeplitzoperatoren mit unstetigem Symbol. Wiss. Z. Techn. Hochsch. Karl-Marx-Stadt **24**, 289–294 (1982)

Rochberg, R.
[1] Trace ideal criteria for Hankel operators and commutators. Indiana Univ. Math. J. **31**, 913–925 (1982)

Rogosinski, W. W., Shapiro, H.
[1] On certain extremum problems for analytic functions. Acta Math. **90**, 287–318 (1953)

Rosenbaum, J. T.
[1] Simultaneous interpolation in $H^2$. Michigan Math. J. **14**, 65–70 (1967)
[2] Simultaneous interpolation in $H^2$ II. Pacif. J. Math. **27**, 607–610 (1968)

Rosenblum, M.
[1] Self-adjoint Toeplitz operators. In: Summer Inst. of Spectral Theory and Statistical Mech., Brookhaven National Laboratory, 1965. Upton, New York 1965
[2] A corona theorem for countably many functions.
[3] Vectorial Toeplitz operator and the Fejér-Riesz theorem. J. Math. Anal. Appl. **23**, 139–147 (1968)
[4] A concrete spectral theory for self-adjoint Toeplitz operators. Amer. J. Math. **87**, 709–718 (1965)
[5] The absolute continuity of Toeplitz's matrices. Pacific J. Math. **10**, 987–996 (1960)
[6] A spectral theory for self-adjoint singular integral operators. Amer. J. Math. **88**, 314–328 (1966)

Rosenthal, P.
[1] A note on unicellular operators. Proc. Amer. Math. Soc. **19**, 505–506 (1968)

Rozanov, Yu. A.
[1] Linear extrapolation of discrete time multivariate stochastic processes of rank 1. Dokl. Akad. Nauk SSSR *125*, 277–280 [Russian]
[2] Stationary stochastic processes. Moscow: Fizmatgiz 1963 [Russian] (English translation: San Fransisco etc.: Golden-Day 1967)

Rudin, W.
[1] Function theory in polydiscs. New York: Benjamin 1969
[2] The closed ideals in an algebra on analytic functions. Canad. J. Math. **9**, 426–434 (1957)

[3] Spaces of type $H^\infty + C$. Ann. Inst. Fourier (Grenoble) **25**, 99–125 (1975)
[4] Functional Analysis. New York: McGraw-Hill 1973

Rukšin, S. E.
[1] On N. K. Nikol'skiĭ's theorem concerning interpolation of germs of analytic functions with respect to a family of inner functions. Mat. Zametki **33**, 83–88 (1983) [Russian]

Sahnovič, L. A.
[1] Spectral analysis for Volterra operators and inverse problems. Dokl. Akad. Nauk SSSR **115**, 666–669 (1957) [Russian]
[2] Spectral analysis of operators of type $Kf = \int_0^x f(t)K(x-t)\,dt$. Izv. Akad. Nauk SSR Ser. Mat. **22**, 299–308 (1958) [Russian]
[3] On the reduction of non-selfadjoint operators to triangular form. Izv. Vysš. Učebn. Zaved. Matematika nr. 1, 1959, 180–186 [Russian]
[4] Study of the "triangular model" of non-selfadjoint operators. Izv. Vysš. Učebn. Zaved. Matematika nr. 4, 1959, 141–149

Šamoyan, F. A.
[1] Imbedding theorems connected with the problem of multiple interpolation. Izv. Akad. Nauk Armjan. SSR Ser. Mat. **11**, 124–131 (1976) [Russian]
[2] Division by an inner function in some spaces of analytic functions in the disk. Zap. Naučn. Sem. Leningrad. Otdel. Mat. Inst. Steklov. LOMI **22**, 206–208 (1971) [Russian]

Sarason, D.
[1] The $H^p$ spaces on an annulus. Mem. Amer. Math. Soc. **56** (1965)
[2] Invariant subspaces. In: Mathematical surveys 13. Topics in operator theory pp. 1–47. Providence: American Mathematical Society 1974
[3] Weak-star generators of $H^\infty$. Pacific. J. Math. **17**, 519–528 (1966)
[4] Weak-star density of polynomials. J. Reine Angew. Math. **252**, 1–15 (1972)
[5] A remark on the Volterra operator. J. Math. Anal. Appl. **12**, 244–246 (1965)
[6] Generalized interpolation in $H^\infty$. Trans. Amer. Math. Soc. **127**, 179–203 (1967)
[7] Invariant subspaces and unstarred operator algebras. Pacific J. Math. **17**, 511–518 (1966)
[8] On spectral sets having connected complement. Acta Sci. Math. (Szeged) **24**, 289–299 (1965)
[9] Function theory on the unit circle. Notes for lectures at a conference at Virginia Polytechnic Institute and State University, Blacksburg, Virginia, June 19–23, 1978
[10] On products of Toeplitz operators. Acta Sci. Math. (Szeged) **35**, 7–12 (1973)
[11] Toeplitz operators with piecewise quasicontinuous symbols. Indiana Univ. Math. J. **26**, 817–838 (1977)
[12] Algebras of functions on the unit circle. Bull. Amer. Math. Soc. **79**, 286–299 (1973)
[13] Toeplitz operators with semi-almost periodic symbols. Duke Math. J. **44**, 357–364 (1977)
[14] Algebras between $L^\infty$ and $H^\infty$. In: Lecture Notes in Mathematics 512, pp. 117–129. Berlin-Heidelberg-New York: Springer 1976
[15] Addendum to "Past and Future". Math. Scand. **30**, 62–64 (1972)
[16] Functions of vanishing mean oscillation. Trans. Amer. Math. Soc. **207**, 391–405 (1975)

Sawyer, E. T.
[1] Two weight norm inequalities for certain maximal and integral operators. In: Lecture Notes in Mathematics 908, pp. 102–127. Berlin-Heidelberg-New York: Springer-Verlag 1982

Schreiber, M.
[1] Remark on a paper of Kalisch. J. Math. Anal. Appl. **7**, 62–63 (1963)

Schubert, J.
[1] The corona theorem as an operator theorem. Proc. Amer. Math. Soc. **69**, 73–76 (1978)

Schur, I.
[1] Über Potenzreihen, die im Innern des Einheitskreises beschränkt sind. J. Reine Angew. Math. **147**, 205–232 (1917); **148**, 122–145 (1918)

Schwartz, L.
[1] Étude des sommes d'exponentielles réelles. Actualités Scientifiques et Industrielles 959. Paris: Hermann 1943

Segal, I.
[1] Tensor algebras over Hilbert spaces I. Trans. Amer. Math. Soc. **81**, 106–134 (1956)

Semmes, S.
[1] Trace ideal criteria for Hankel operators and applications to Besov spaces. Integral Equations Operator Theory 7, 241–281 (1984)

Shapiro, H. S.
[1] Generalized analytic continuation. In: Symposia on Theoretical Physics and Mathematics, vol. 8, pp. 151–163. Plenum Press 1968
[2] Some function-theoretic problems motivated by the study of Banach algebras. In: Proc. NRL Conference on Classical Function Theory, ed. F. Gross, pp. 95–113. Washington, D.C.: NRL 1970
[3] Extremal problems for polynomials and power series. Dissertation. M.I.T. 1952
[4] Applications of normed linear spaces to function-theoretic extremal problems. In: Lectures on functions of complex variables, ed. by W. Kaplan, pp. 399–404: Univ. of Michigan Press 1955

Shapiro, H. S., Shields, A.
[1] On some interpolation problems for analytic functions. Amer. J. Math. **83**, 513–532 (1961)
[2] Interpolation in Hilbert spaces of analytic functions. In: Proc. Conf. Func. Anal. (1960), pp. 109–110. Studia Math., Special Series 1963

Sherman, M. J.
[1] Invariant subspaces containing all analytic directions. J. Functional Analysis **3**, 164–173 (1969)

Shields, A.
[1] Weighted shift operators and analytic function theory. In: Math. Surveys 13, pp. 49–128. Providence: American Mathematical Society 1974

Šilov, G. E.
[1] On theory of ideals in normed rings of functions. Dokl. Akad. Nauk SSSR **27**, 900–903 (1940) [Russian]
[2] On regular normed rings. Trudy Mat. Inst. Steklov. **21** (118 pp.) (1947) [Russian]

Silbermann, B.
[1] Lokale Theorie des Reduktionsverfahrens für Toeplitz-Operatoren. Math. Nachr. **104**, 137–146 (1981)
[2] The strong Szegö limit theorem for a class of singular generating functions. Demonstratio Math. **14**, 647–667 (1981)

Simonenko, I. B.
[1] The Riemann boundary problem with measurable coefficients. Dokl. Akad. Nauk SSSR **135**, 538–541 (1960) [Russian]
[2] A new general method for studying linear operator equations of the type of singular integral equations. Dokl. Akad. Nauk SSSR **158**, 790–793 (1964) [Russian]
[3] Singular integral equations with continuous and piece-wise continuous symbols. Dokl. Akad. Nauk SSSR **159**, 279–282 (1964) [Russian]
[4] A new general method for studying linear operator equations of the type of singular integral equations. I, II. Izv. Akad. Nauk SSSR Ser. Mat. **29 : 3–4**, 567–586, 757–782 (1965) [Russian]
[5] Some general questions of the theory of the Riemann boundary problem. Izv. Akad. Nauk SSSR Ser. Mat. **32**, 1138–1146 (1968) [Russian]
[6] The Riemann boundary problem for $n$ pairs of functions with measurable coefficients and application to the study of singular integral equations in $L_p$ spaces with weights. Izv. Akad. Nauk SSSR Ser. Mat. **28**, 277–306 (1964) [Russian]

Sjödin, T.
[1] Capacities of compact sets in linear subspaces of $R^n$. Pacific J. Math. **78**, 261–266 (1978)

Smirnov, V. I.
[1] Sur la théorie des polynômes orthogonaux à une variable complexe. J. Leningrad. Fiz. Mat. Obšč. **2**, 155–179 (1928)

Šnyder, A. K.
[1] Sequence spaces and interpolation problems for analytic functions. Studia Math. **39**, 137–153 (1971)

Solev, V. N.
[1] Absolutely regular trajectories in Hilbert space. Zap. Naučn. Sem. Leningrad. Otdel. Mat. Inst. Steklov. (LOMI) **22**, 139–160 (1971) [Russian]

Solomyak, B. M.
[1] Calculuses annihilators and hyperinvariant subspaces. J. Operator Theory **9**, 341–370 (1983)

Spitkovskiĭ, I. M.
[1] On partial indices on continuous matrix functions. Dokl. Akad. Nauk SSSR **229**, 1059–1062 (1976) [Russian]
[2] On multipliers not destroying factorizability. Mat. Zametki **27**, 291–299 (1980) [Russian]
[3] Some estimates for the partial indices of measurable matrix functions. Mat. Sb. **111**, 227–248 (1980) [Russian]

Stein, E. M.
[1] Singular integral and differentiability properties of functions. Princeton: Princeton University Press 1970

Stephenson, K.
[1] Two counterexamples involving inner functions. Proc. Amer. Math. Soc. **73**, 22–24 (1979)

Stray, A.
[1] A formula by V. M. Adamjan, D. Z. Arov and M. G. Kreĭn. Proc. Amer. Math. Soc. **83**, 337–340 (1981)

Strichartz, R. S.
[1] Multipliers on fractional Sobolev spaces. J. Math. Mech. **16**, 1031–1060 (1967)

Suciu, I., Valusescu, I.
[1] Factorization of semi-spectral measures. Rev. Roumaine Math. Pures Appl. **6**, 773–793 (1976)
[2] Factorization theorems and prediction theory. Rev. Roumaine Math. Pures Appl. **23**, 1393–1423 (1978)

Švedenko, S. V.
[1] Interpolation in some Hilbert spaces of analytic functions. Mat. Zametki **15**, 101–112 (1974) [Russian]

Szász, O.
[1] Über die Approximation stetiger Funktionen durch lineare Aggregate von Potenzen. Math. Ann. **77**, 482–496 (1916)

Sz.-Nagy, B.
[1] A problem on operator valued analytic functions. Zap. Naučn. Sem. Leningrad. Otdel. Mat. Inst. Steklov. (LOMI) **81**, 99 (1978)
[2] Models of Hilbert space operators. Proc. Roy. Irish Acad. Sect. A **74**, 263–281 (1974)
[3] Diagonalisation of matrices over $H^\infty$. Acta Sci. Mat. (Szeged) **38**, 223–238 (1976)

Sz.-Nagy, B., Foiaş, C.
[1] Analyse harmonique des opérateurs de l'espace de Hilbert. Budapest: Akadémiai Kiadó 1967
[2] Modèle de Jordan pour une classe d'opérateurs dans l'espace de Hilbert. Acta Sci. Math. (Szeged) **31**, 91–115 (1970)

[3] Sur les contractions de l'espace de Hilbert. VII. Triangulations canoniques. Fonction minimum. Acta Sci. Math. (Szeged) **25**, 12–37 (1964)
[4] Sur les contractions de l'espace de Hilbert. VIII. Fonctions caractéristiques. Modèles fonctionnels. Acta. Sci. Math. (Szeged) **25**, 38–71 (1964)
[5] Échelles continues de sous-espaces invariants. Acta. Sci. Math. (Szeged) **28**, 213–220 (1967)
[6] Commutants de certains opérateurs. Acta Sci. Math. (Szeged) **29**, 1–17 (1968)
[7] Compléments à l'étude des opérateurs de classe $C_0$. Acta Sci. Math. (Szeged) **31**, 287–296 (1970)
[8] Compléments à l'étude des opérateurs de class $C_0$ II. Acta Sci. Math. (Szeged) **33**, 113–116 (1972)
[9] Local characterization of operators of class $C_0$. J. Functional Analysis **8**, 76–81 (1971)
[10] On contractions similar to isometries and Toeplitz operators. Ann. Acad. Sci. Fenn. A I **2**, 553–564 (1976)

Szücz, J.
[1] Diagonalization theorems for matrices over certain domains. Acta Sci. Math. (Szeged) **36**, 193–201 (1974)

Taylor, B. A., Williams, D. L.
[1] Interpolation of $l^q$ sequences by $H^p$ functions. Proc. Amer. Math. Soc. **34**, 181–186 (1972)

Toeplitz, O.
[1] Zur Theorie der quadratischen Formen von unendlichvielen Veränderlichen. Math. Ann. **70**, 351–376 (1911)

Tolokonnikov, V. A.
[1] Estimates in Carleson's corona theorem and finitely generated ideals in the algebra $H^\infty$. Funkcional. Anal. i Prilož. **14**, 85–86 (1980) [Russian]
[2] Estimates in Carleson's corona theorem. Ideals of the algebra $H^\infty$, the problem of Szekefalvi-Nagy. Zap. Naučn. Sem. Leningrad. Otdel. Mat. Inst. Steklov. (LOMI) **113**, 178–198 (1981) [Russian]
[3] Corona theorems in algebras of bounded analytic functions. Manuscript desposed in VINITI, Moscow, No. 251–84 DEP, 1984 [Russian]

Treĭl', S. R.
[1] A geometric approach to weighted estimates for the Hilbert transform. Funkcional. Anal. i Priložen. [to appear] [Russian]

Triebel, H.
[1] Spaces of Besov-Hardy-Sobolev type. Leipzig: Teubner Verlag 1978

Tumarkin, G. Č.
[1] Expansion of analytic functions in series of rational functions with a given set of poles. Izv. Akad. Nauk Armjan. SSR Ser. Fiz. Mat. **4**, 9–31 (1961) [Russian]
[2] Description of the class of functions permitting approximation with fractions with fixed poles. Izv. Akad. Nauk Armjan. SSR Ser. Fiz. Mat. **1**, 89–105 (1961) [Russian]
[3] On the convergence of sequences of analytic and meromorphic functions. Dokl. Akad. Nauk SSSR **129**, 280–283 (1959) [Russian]
[4] Conditions for uniform convergence and the convergence of boundary values of analytic and meromorphic functions with uniformly bounded characteristics. Sibirsk. Mat. Ž. **5**, 387–417 (1964) [Russian]

Uchiyama, A.
[1] Corona theorems for countably many functions and estimates for their solutions. Pre-print, Univ. of California, Los Angeles

Urbanik, K.
[1] Szegö's theorem for Orlicz spaces. Bull. Acad. Polon. Sci. Ser. Math. Phys. Astron. **14**, 503–509 (1966)

Vasyunin, V. I.
[1] Bases of rational functions and multiple interpolation. Zap. Naučn. Sem. Leningrad. Otdel. Mat. Inst. Steklov. (LOMI) **56**, 174–176 (1976) [Russian]
[2] Unconditionally convergent spectral expansions and interpolation problems. Trudy Mat. Inst. Steklov. **130**, 5–49 (1977) [Russian]
[3] Bases of eigensubspaces and nonclassical interpolation problems. Funkcional. Anal. i Prilož. **9**, 65–66 (1975) [Russian]
[4] On the number of Carleson series. Zap. Naučn. Sem. Leningrad. Otdel. Mat. Inst. Steklov. (LOMI) **65**, 178–182 (1976) [Russian]

Veršik, A. M.
[1] On the theory of normal dynamical systems. Dokl. Akad. Nauk SSSR **144**, 9–12 (1962) [Russian]
[2] On spectral and metric isomorphism of some normal dynamical systems. Dokl. Akad. Nauk SSSR **144**, 255–257 (1962) [Russian]

Vinogradov, S. A.
[1] Bases of exponentials and free interpolation in a space with an $L^p$ norm. Zap. Naučn. Sem. Leningrad. Otdel. Mat. Inst. Steklov. (LOMI) **65**, 17–68 (1976) [Russian]
[2] Multipliers of power series with the sequence of coefficients in $l^p$. Zap. Naučn. Sem. Leningrad. Otdel. Mat. Inst. Steklov. (LOMI) **39**, 30–39 (1974) [Russian]
[3] Interpolation problems for analytic functions continuous in the closed disk and for functions with the sequence of coefficients in $l^p$. Dissertation, Leningrad, 1968 [Russian]
[4] On interpolation of power series which are absolutely convergent on the boundary of the circle of convergence. Vestnik Leningrad. Univ. Mat. Meh. Astronom. Nr. 7, 1965, 30–44. [Russian]
[5] Some remarks on free interpolation by bounded and slowly growing analytic functions. Zap. Naučn. Sem. Leningrad. Otdel. Mat. Inst. Steklov. (LOMI) **126**, 35–46 (1983) [Russian]

Vinogradov, S. A., Goluzina, M. G., Havin, V. P.
[1] Multipliers and divisors of integrals of Cauchy-Stieltjes type. Zap. Naučn. Sem. Leningrad. Otdel. Mat. Inst. Steklov. (LOMI) **19**, 55–79 (1970) [Russian]

Vinogradov, S. A., Gorin, E. A., Hruščev, S. V.
[1] Interpolation in $H^\infty$ along P. Jones' lines. Zap. Naučn. Sem. Leningrad. Otdel. Mat. Inst. Steklov. (LOMI) **113**, 212–214 (1981) [Russian]

Vinogradov, S. A., Havin, V. P.
[1] Free interpolation in $H^\infty$ and some other classes of functions. I. Zap. Naučn. Sem. Leningrad. Otdel. Mat. Inst. Steklov. (LOMI) **47**, 15–54 (1974) [Russian]
[2] Free interpolation in $H^\infty$ and some other classes of functions. II. Zap. Naučn. Sem. Leningrad. Otdel. Mat. Inst. Steklov. (LOMI) **56**, 12–58 (1976) [Russian]

Vinogradov, S. A., Rukšin, S. E.
[1] On the free interpolation of germs of analytic functions in Hardy spaces. Zap. Naučn. Sem. Leningrad. Otdel. Mat. Inst. Steklov. (LOMI) **107**, 36–45 (1982) [Russian]

Vol'berg, A. L.
[1] Two remarks concerning the theorem of S. Axler, S.-Y.A. Chang and D. Sarason. J. Operator Theory **7**, 209–218 (1982)
[2] The comparison of the strength of integral norms on subspaces of pseudoreproducing functions in the space $H^p$. Uspehi Mat. Nauk **36** : *6*, 205–206 (1981) [Russian]

Vol'berg, A. L., Tolokonnikov, V. A.
[1] Some remarks on multipliers of Cauchy type integrals and Sarason algebras. Funkcional. Anal. i Priložen. **18**, 61–62 (1984) [Russian]

Walsh, J. L.
[1] Interpolation and approximation by rational functions in the complex domain. Amer. Math. Soc. Coll. Publ. 20. New York: Amer. Math. Soc. 1935

Wermer, J.
[1] On invariant subspaces of normal operators. Proc. Amer. Math. Soc. **3**, 270–277 (1952)
[2] Commuting spectral measures on Hilbert space. Pacif. J. Math. **4**, 355–361 (1954)

Widom, H.
[1] Inversion of Toeplitz matrices. III. Notices Amer. Math. Soc. **7**, 63 (1960)
[2] Inversion of Toeplitz matrices. II. Illinois J. Math. **4**, 88–99 (1960)
[3] Toeplitz operators on $H_p$. Pacific J. Math. **19**, 573–582 (1966)
[4] Toeplitz determinants with singular generating functions. Amer. J. Math. **95**, 333–383 (1973)
[5] Asymptotic expansions of determinants for families of trace class operators. Indiana Univ. Math. J. **27**, 449–478 (1978)
[6] Asymptotic behavior of block Toeplitz matrices and determinants. II. Advances in Math. **21**, 1–29 (1976)
[7] Hankel matrices. Trans. Amer. Math. Soc. **127**, 179–203 (1966)

Wiener, N.
[1] The Fourier integral and some of its applications. New York: Cambridge University Press 1933
[2] The homogeneous chaos. Amer. J. Math. **60**, 897–936 (1938)

Wiener, N., Paley, R.E.A.C.
[1] Fourier transforms in the complex domain. Amer. Math. Soc. Colloquium Publications 19. New York: Amer. Math. Soc. 1934

Wintner, A.
[1] Zur Theorie der beschränkten Bilinearformen. Math. Z. **30**, 228–282 (1929)

Wogen, W. R.
[1] On some operators with cyclic vectors. Indiana Univ. Math. J. **27**, 163–171 1(978)

Wold, W. R.
[1] A study in the analysis of stationary time series. Uppsala: Almquist och Wiksell 1938, 2 ed. 1954

Wolff, Th. H.
[1] Two algebras of bounded functions. Duke Math. J. **49**, 321–328 (1982)
[2] Counterexamples to two variants of the Helson-Szegö theorem. Submitted to Ark. Mat.

Yaglom, A. M.
[1] Introduction to the theory of stationary random functions. Uspehi Mat. Nauk **7 : 5**, 3–168 (1952) [Russian]
[2] Stationary Gaussian processes satisfying the strong mixing condition and best predictable functionals. In: Proc. of International Research Seminar, University of Calif., Berkeley, 1963. Berlin-Heidelberg-New York: Springer 1965

Yamazaki, S.
[1] Normed rings and unconditional bases in Banach spaces. Sci. Papers College Gen. Ed. Univ. Tokyo **14**, 1–10 (1964), **15**, 1–3 (1965)

Yosida, K.
[1] Functional analysis. Sixth ed. Grundlehren 123. Berlin-Heidelberg-New York: Springer 1980

Zygmund, A.
[1] Trigonometric series, I–II. Cambridge: University Press 1959

# List of Symbols

# Author Index

Page numbers in *italics* refer to the bibliography.

# Subject Index

MIX
Papier aus verantwortungsvollen Quellen
Paper from responsible sources
FSC® C105338

FSC
www.fsc.org

If you have any concerns about our products,
you can contact us on
**ProductSafety@springernature.com**

In case Publisher is established outside the EU,
the EU authorized representative is:
**Springer Nature Customer Service Center GmbH**
**Europaplatz 3, 69115 Heidelberg, Germany**

Printed by Libri Plureos GmbH
in Hamburg, Germany